IMAGING OF THE HUMAN BRAIN IN HEALTH AND DISEASE

IMAGING OF THE HUMAN BRAIN IN HEALTH AND DISEASE

Edited by

PHILIP SEEMAN, BERTHA MADRAS

ELSEVIER

Amsterdam • Boston • Heidelberg • London
New York • Oxford • Paris • San Diego
San Francisco • Sydney • Tokyo
Academic Press is an imprint of Elsevier

Academic Press is an imprint of Elsevier
The Boulevard, Langford Lane, Kidlington, Oxford OX5 1GB, UK
Radarweg 29, PO Box 211, 1000 AE Amsterdam, The Netherlands
225 Wyman Street, Waltham, MA 02451, USA
525 B Street, Suite 1800, San Diego, CA 92101-4495, USA

First edition 2014

Library of Congress Cataloging-in-Publication Data

Imaging of the human brain in health and disease / edited by Philip Seeman, Bertha Madras. – 1st edition.
p. ; cm.
Includes bibliographical references and index.
Summary: "Modern imaging techniques have allowed researchers to non-invasively peer into the human brain and investigate, among many other things, the acute effects and long-term consequences of drug abuse. Here, we review the most commonly used and some emerging imaging techniques in addiction research, explain how the various techniques generate their characteristic images and describe the rational that researchers use to interpret them. In addition, examples of seminal imaging findings are highlighted that illustrate the contribution of each imaging modality to the expansion in our understanding of the neurobiological bases of drug abuse and addiction, and how they can be parlayed in the future into clinical and therapeutic applications"– Provided by publisher.
ISBN 978-0-12-418677-4 (alk. paper)
I. Seeman, Philip, editor of compilation. II. Madras, Bertha, editor of compilation.
[DNLM: 1. Neuroimaging--methods. 2. Brain Chemistry--physiology. 3. Brain Diseases--radionuclide imaging. 4. Mental Disorders--radionuclide imaging. 5. Substance-Related Disorders--radionuclide imaging. WL 141.5.N47]
RC386.6.T65
616.8'047575--dc23

2013039184

British Library Cataloguing in Publication Data

A catalogue record for this book is available from the British Library

For information on all Academic Press publications
visit our web site at store.elsevier.com

ISBN: 978-0-12-418677-4

CONTENTS

LIST OF CONTRIBUTORS

Ruben Baler
National Institute on Drug Abuse, Bethesda, MD, USA

James Robert Brašić
The Russell H. Morgan Department of Radiology and Radiological Science, Johns Hopkins University School of Medicine, Baltimore, MD, USA; Section of High Resolution Brain Positron Emission Tomography Imaging, Division of Nuclear Medicine, The Johns Hopkins University School of Medicine, Baltimore, Maryland, USA

Guy Bormans
MoSAIC, Molecular Small Animal Imaging Center, KU Leuven, Leuven, Belgium; Laboratory for Radiopharmacy, KU Leuven, Leuven, Belgium

Cindy Casteels
Division of Nuclear Medicine, University Hospitals and KU Leuven, Leuven, Belgium; MoSAIC, Molecular Small Animal Imaging Center, KU Leuven, Leuven, Belgium

Sofia N. Chatziioannou
Department of Radiology, Nuclear Medicine Section, National and Kapodistrian University of Athens Medical School, Attikon General Hospital, Athens, Greece

Thilo Deckersbach
Department of Psychiatry, Massachusetts General Hospital, Boston, MA, USA

Lora Deuitch
Departments of Radiology, University of Pittsburgh, Pittsburgh, PA, USA

Darin D. Dougherty
Department of Psychiatry, Massachusetts General Hospital, Charlestown, MA, USA

Andre C. Felicio
Pacific Parkinson's Research Centre, Vancouver Hospital and Health Sciences Centre, University of British Columbia, Vancouver, BC, Canada

Joanna S. Fowler
Brookhaven National Laboratory, Upton, NY, USA

Boris Frolov
The Russell H. Morgan Department of Radiology and Radiological Science, Johns Hopkins University School of Medicine, Baltimore, MD, USA

Hironobu Fujiwara
Molecular Imaging Center, Department of Molecular Neuroimaging, National Institute of Radiological Sciences, Inage, Chiba, Japan

Camille Garcia-Ramos
Department of Medical Physics, University of Wisconsin–Madison, WI, USA

Emily Gean
The Russell H. Morgan Department of Radiology and Radiological Science, Johns Hopkins University School of Medicine, Baltimore, MD, USA

Noble George
The Russell H. Morgan Department of Radiology and Radiological Science, Johns Hopkins University School of Medicine, Baltimore, MD, USA

Sharmin Ghaznavi
Department of Psychiatry, Massachusetts General Hospital, Boston, MA, USA

Udi E. Ghitza
Center for the Clinical Trials Network, National Institute on Drug Abuse, National Institutes of Health, Bethesda, MD, USA

Roger N. Gunn
Imanova Limited, London, UK; Department of Medicine, Imperial College, London, UK; Department of Engineering Science, University of Oxford, UK

Christer Halldin
Department of Clinical Neuroscience, Karolinska Institutet, Centre for Psychiatry Research, Stockholm, Sweden

Jarmo Hietala
Turku PET Centre, Turku University Hospital and University of Turku, Turku, Finland; Department of Psychiatry, University of Turku, Turku, Finland

Jussi Hirvonen
Department of Radiology, Turku University Hospital and University of Turku, Turku, Finland; Turku PET Centre, Turku University Hospital and University of Turku, Turku, Finland

Andrew Horti
The Russell H. Morgan Department of Radiology and Radiological Science, Johns Hopkins University School of Medicine, Baltimore, MD, USA

Kiichi Ishiwata
Positron Medical Center, Tokyo Metropolitan Institute of Gerontology, Tokyo, Japan

Karin B. Jensen
Department of Psychiatry, Massachusetts General Hospital, Harvard Medical School, Boston, MA, USA

Robert M. Kessler
Department of Radiology and Radiological Sciences, Vanderbilt University School of Medicine, Nashville, TN, USA

Yuichi Kimura
Positron Medical Center, Tokyo Metropolitan Institute of Gerontology, Tokyo, Japan; Molecular Imaging Center, National Institute of Radiological Sciences, Chiba, Japan

Christian La
Department of Radiology, University of Wisconsin–Madison, WI, USA; Department of Medical Physics, University of Wisconsin–Madison, WI, USA; Neuroscience Training Program, University of Wisconsin–Madison, WI, USA

Koen Van Laere
Division of Nuclear Medicine, University Hospitals and KU Leuven, Leuven, Belgium; MoSAIC, Molecular Small Animal Imaging Center, KU Leuven, Leuven, Belgium

Marco L. Loggia
Athinoula A. Martinos Center for Biomedical Imaging, Massachusetts General Hospital, Harvard Medical School, Boston, MA, USA; Department of Psychiatry, Massachusetts General Hospital, Harvard Medical School, Boston, MA, USA

Masahiro Mishina
Positron Medical Center, Tokyo Metropolitan Institute of Gerontology, Tokyo, Japan; The Second Department of Internal Medicine, Nippon Medical School, Tokyo, Japan

Mona Mohamed
Division of Neuroradiology, The Russell H. Morgan Department of Radiology and Radiological Science, The Johns Hopkins University School of Medicine, Baltimore, Maryland, USA

Ayon Nandi
The Russell H. Morgan Department of Radiology and Radiological Science, Johns Hopkins University School of Medicine, Baltimore, MD, USA

Veena A. Nair
Department of Radiology, University of Wisconsin–Madison, WI, USA

Rajesh Narendran
Departments of Radiology, University of Pittsburgh, Pittsburgh, PA, USA; Department of Psychiatry, University of Pittsburgh, Pittsburgh, PA, USA

Yoshiro Okubo
Department of Neuropsychiatry, Nippon Medical School, Bunkyo-ku, Tokyo, Japan

Vivek Prabhakaran
Department of Radiology, University of Wisconsin–Madison, WI, USA; Neuroscience Training Program, University of Wisconsin–Madison, WI, USA; Department of Neurology, University of Wisconsin–Madison, WI, USA; Department of Psychiatry, University of Wisconsin–Madison, WI, USA; Medical Scientist Training Program, University of Wisconsin–Madison, WI, USA

Eugenii A. Rabiner
Imanova Limited, London, UK; Institute of Psychiatry, Kings College, London, UK

Emmanouil N. Rizos
Department of Psychiatry, National and Kapodistrian University of Athens Medical School, Attikon General Hospital, Athens, Greece

Muneyuki Sakata
Positron Medical Center, Tokyo Metropolitan Institute of Gerontology, Tokyo, Japan

Hitoshi Shimada
Molecular Imaging Center, Department of Molecular Neuroimaging, National Institute of Radiological Sciences, Inage, Chiba, Japan

Mark Slifstein
Department of Psychiatry, Columbia University, New York, NY, USA; Division of Translational Imaging, New York State Psychiatric Institute, New York, NY, USA

A. Jon Stoessl
Pacific Parkinson's Research Centre, Vancouver Hospital and Health Sciences Centre, University of British Columbia, Vancouver, BC, Canada

Tetsuya Suhara
Molecular Imaging Center, Department of Molecular Neuroimaging, National Institute of Radiological Sciences, Inage, Chiba, Japan

Hidehiko Takahashi
Molecular Imaging Center, Department of Molecular Neuroimaging, National Institute of Radiological Sciences, Inage, Chiba, Japan; Department of Psychiatry, Kyoto University Graduate School of Medicine, Sakyo-ku, Kyoto, Japan

Dardo Tomasi
National Institute on Alcohol Abuse and Alcoholism, Bethesda, MD, USA

Jun Toyohara
Positron Medical Center, Tokyo Metropolitan Institute of Gerontology, Tokyo, Japan

Andrea Varrone
Department of Clinical Neuroscience, Karolinska Institutet, Centre for Psychiatry Research, Stockholm, Sweden

Nora D. Volkow
National Institute on Drug Abuse, Bethesda, MD, USA

Gene-Jack Wang
Brookhaven National Laboratory, Upton, NY, USA

Dean F. Wong
The Russell H. Morgan Department of Radiology and Radiological Science, Johns Hopkins University School of Medicine, Baltimore, MD, USA; Department of Psychiatry, Johns Hopkins University School of Medicine, Baltimore, MD, USA; Department of Neuroscience, Johns Hopkins University School of Medicine, Baltimore, MD, USA; Department of Environmental Health Sciences, Johns Hopkins University School of Medicine, Baltimore, MD, USA; Carey School of Business, Johns Hopkins University School of Medicine, Baltimore, MD, USA

Brittany M. Young
Department of Radiology, University of Wisconsin–Madison, WI, USA; Neuroscience Training Program, University of Wisconsin–Madison, WI, USA; Medical Scientist Training Program, University of Wisconsin–Madison, WI, USA

Eram Zaidi
The Russell H. Morgan Department of Radiology and Radiological Science, Johns Hopkins University School of Medicine, Baltimore, MD, USA

CHAPTER ONE

Neuroimaging of Addiction

Nora D. Volkow[1], Gene-Jack Wang[2], Joanna S. Fowler[2], Dardo Tomasi[3] and Ruben Baler[1]

[1]National Institute on Drug Abuse, Bethesda, MD, USA
[2]Brookhaven National Laboratory, Upton, NY, USA
[3]National Institute on Alcohol Abuse and Alcoholism, Bethesda, MD, USA

1. INTRODUCTION

Scientific advances over the past 20 to 30 years have established drug addiction as a chronic brain disease (Leshner, 1997). Key evidence supporting this concept was produced by brain imaging studies of drug abusers obtained during or following various periods of drug exposure. These studies have provided information on drugs' neurobiological effects, helped explain the causes and mechanisms of vulnerability to drug abuse, and yielded important insights into abusers' subjective experiences and behaviors, including their difficulty to attain a sustained, relapse-free recovery. Clinicians may be able, in the not too distant future, to use brain imaging to evaluate the level and pattern of brain dysfunction in their addicted patients, helping them to tailor their treatments and to monitor their response to therapy.

The seven primary brain imaging techniques - structural magnetic resonance imaging (MRI), functional MRI, resting functional MRI, Diffusion Tensor Imaging (DTI), magnetic resonance spectroscopy (MRS), positron emission tomography (PET), and single photon emission computed tomography (SPECT) - reveal different aspects of brain structure and/or function (Bandettini, 2009; Detre and Floyd, 2001; Duyn and Koretsky, 2011; Johansen-Berg and Rushworth, 2009; Sharma and Ebadi, 2008). Individually, the techniques yield highly complementary information about brain anatomy and tissue composition; biochemical, physiological, and functional processes; neurotransmitter levels; energy utilization and blood flow; and drug distribution and kinetics. Together, and in combination with other research techniques they contribute to continuously improve our understanding of drug abuse and addiction.

2.1. MAGNETIC RESONANCE-BASED IMAGING TECHNIQUES

2.1.1. Structural Magnetic Resonance Imaging

Structural magnetic resonance imaging (sMRI) translates the local differences in water content into different shades of gray that serve to outline the shapes and sizes of the

Imaging of the Human Brain in Health and Disease
http://dx.doi.org/10.1016/B978-0-12-418677-4.00001-4

brain's various subregions. An MRI scanner delivers a specific radiofrequency that excites hydrogen atoms in the water molecule, which return some of this energy in the form of a characteristic nuclear magnetic resonance signal. Not all protons "resonate" in that way, but enough do such that the resulting computer-generated image constitutes a highly detailed map of the brain's tissues and structures. Thus, this tool can be used to discover the presence of abnormal tissue through the changes in tissue density or composition. Scientists examining an sMRI can readily distinguish between gray and white matter and other types of tissue—both normal, such as blood vessels, and abnormal, such as tumors—by their different shading and contrast with surrounding areas.

Such measurements can help scientists and doctors to home in on the regions that are most heavily affected by drugs. Importantly, these initial observations often guide additional investigations, using other research tools and techniques, to determine the reasons for the structural changes as well as their experiential and behavioral consequences. As explained below, sMRI studies have provided detailed evidence that chronic drug exposure can lead to both increases and reductions in the volume of specific brain regions.

Drug Exposure can Trigger Abnormalities in Prefrontal Cortex and Other Brain Regions

Numerous sMRI studies have documented that addictive drugs can cause volume and tissue composition changes in the prefrontal cortex (PFC), a brain region that supports logical thinking, goal-directed behaviors, planning, and self-control. These changes in turn are likely to be associated with drug abusers' cognitive and decision-making deficiencies. Related to this finding, another sMRI study found that individuals with a history of abusing multiple substances have smaller prefrontal lobes than did matched controls (Liu et al., 1998).

These findings add to the growing evidence associating prefrontal abnormalities with the abuse of various substances (Goldstein and Volkow, 2002; Stapleton et al., 1995; Volkow et al., 1991). For example, using sMRI, Schlaepfer and colleagues found that chronic substance abusers' frontal lobe tissues contained a lower proportion of white matter than those of matched controls did (Schlaepfer et al, 2006). Interestingly, similar deficits in white matter content have been found in individuals with other psychiatric disorders that tend to cooccur with substance abuse.

Pertaining to the abuse of stimulants, Kim and colleagues (Kim et al., 2006) documented a reduction in the gray-matter density in the right middle frontal cortex of abstinent methamphetamine abusers (Figure 1). A lower density correlated with a worse performance on a test that measures a person's ability to switch mental gears (Wisconsin Card Sorting Task). Gray matter was closer to normal in individuals who had been abstinent for >6 months than in others with a shorter period of abstinence.

In another sMRI study, cocaine abusers who had been abstinent for 20 days exhibited a reduced gray-matter density in the regions of the frontal cortex. Interestingly, no

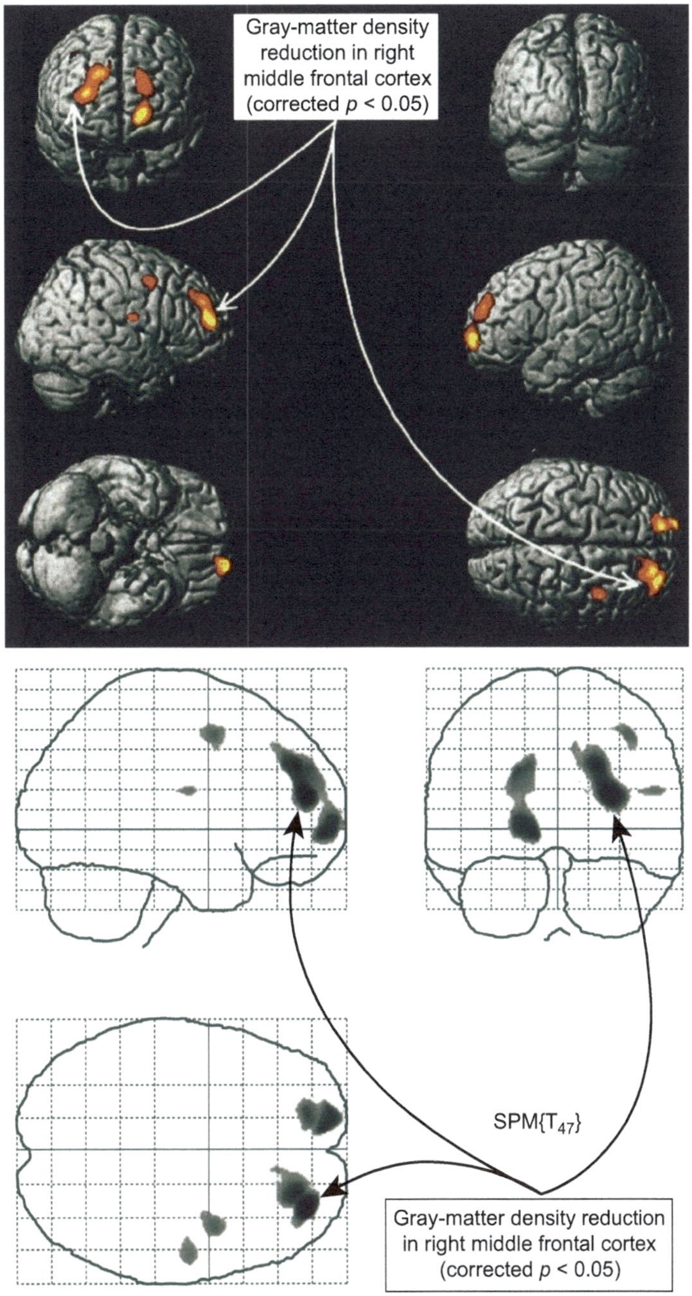

Figure 1 MRI: methamphetamine reduces gray matter. The yellow and red area in the central brain view indicates a reduced gray-matter density in the right middle frontal cortex. The same deficit is shown from other perspectives in the flanking views. *Reprinted with permission from Kim et al. (2006).*

differences were found with respect to white matter density (Matochik et al., 2003). With regards to other brain regions, several sMRI studies have shown an enlargement of the brain's basal ganglia in cocaine-dependent (Jacobsen et al., 2001) and methamphetamine-dependent (Chang et al., 2005; Jernigan et al., 2005) subjects compared with healthy subjects. This is similar to other observations made in schizophrenic subjects who were treated with typical antipsychotics (Gur et al., 1998).

The fact that stimulant drugs, such as cocaine or methamphetamine, and typical antipsychotics that occupy receptors for dopamine in the basal ganglia appear to cause an enlargement of the basal ganglia and are related to psychosis, support the hypothesis that the hyperstimulation of dopamine in basal ganglia structures is involved in psychosis. Finally, an automated morphometric analysis of MR images also showed that a group of chronic methamphetamine abusers had severe gray-matter deficits in cingulate, limbic, and paralimbic cortices. They also had smaller hippocampi than did nondrug abusers of drugs.

The hippocampus is a key site for memory storage, and the volume decrements correlated with a poorer performance on a word recall test (Thompson et al., 2004). Furthermore, sMRI studies have also reported amygdala volume reductions in cocaine addicts (Makris et al., 2004).

Alcohol abuse provides a case study on the utility of MRI to evaluate the structural damage that can result from the chronic use of a psychoactive substance. Investigators using sMRI have reported diminished cortical gray matter, most prominently in the PFC, in alcoholic patients in treatment (Pfefferbaum et al., 1998). In another study, researchers found that alcohol-dependent individuals had reduced whole brain, prefrontal cortical, and parietal cortical gray matter compared with controls (Fein et al., 2002). Two additional studies have shown alcoholics' frontal cortex and other structures beginning to recover their normal volumes within weeks of stopping drinking (Bendszus et al., 2001; O'Neill et al., 2001; Pfefferbaum et al., 1995) [see Mann et al. (Mann et al., 2001), for a comprehensive review on the brain imaging of alcoholism].

Another MRI study indicated that the amygdala, a brain structure that helps shape our emotional responses to experiences, is relatively smaller in children of alcoholics (Hill et al., 2001; Wrase et al., 2008), a finding that might be a clue to brain-dependent vulnerabilities to alcohol abuse disorders.

2.1.2. Functional MRI

Like sMRI, functional MRI (fMRI) produces images by applying a magnetic field and detecting the radiofrequency energy from the excited protons in water molecules. However, fMRI is an ultrafast technique that can image the whole brain in a second and has the ability to detect changes in the ratio of oxygenated to deoxygenated hemoglobin in the capillary bed of the brain by contrasting task and baseline conditions. Since neurons use oxygen as the main fuel source, this measure turns out to be a reliable proxy for brain activity. In an fMRI image, differences in oxygen content appear as variations in the signal intensity, which is referred to as blood oxygen level-dependent (BOLD) contrast.

In fMRI studies, researchers compare multiple images, which may be of single or different individuals. Images of a single individual taken under varying conditions—for example, at rest and then working on a cognitive task, such as a puzzle, or before and after taking a drug—enable researchers to map which brain regions were activated during the performance of that task or in response to experiences or chemical exposures.

Studies of individuals from different groups—for example, drug-addicted and nonaddicted—can reveal differences in the brain regions that the two groups tap into in order to perform identical tasks or respond to stimuli or exposures. In turn, the differences in brain activity patterns revealed by fMRI provide valuable information on a wide range of issues. For example, studies have correlated regional brain patterns in response to taking a drug with a vulnerability to drug abuse, addictive symptoms and behaviors, and long-term cognitive capacity.

Increasingly, fMRI is being used to investigate the pattern of interactions associated with a given task and how these differed as a function of performance and intersubject variability. This change in the emphasis, from the identification of specific brain region toward the identification of networks (regions working together) reflect the understanding that any given process in the brain results from the complex interactions of dynamic networks that are distributed between and within different brain regions.

Stimulant Effects Correlate with Brain Activity in Several Areas

Researchers have used fMRI to obtain detailed information about the roles of different brain areas in mediating cocaine-induced euphoria and subsequent craving and, more recently, about the involvement of functional networks in drug reward and addiction. In one investigation, volunteers given an infusion of cocaine reported a "high" during the brief period when a set of areas, including the caudate (an area of the basal ganglia), cingulate, and most of the lateral PFC showed higher levels of activity. The participants' reports of craving commenced when the euphoria subsided and persisted as long as a different set of brain areas—including the nucleus accumbens (NAc)—remained activated (Breiter et al., 1997; Breiter and Rosen, 1999). Two more recent studies also saw correlations between craving and NAc activity, although—possibly because of differences in study methods—the "high" was associated with a decreased rather than an increased brain activity in regions including the NAc, inferior frontal/orbitofrontal gyrus, and anterior cingulate. Craving correlated positively with the activity in these regions (Kufahl et al., 2005; Risinger et al., 2005).

Other fMRI studies demonstrated that a cocaine-addicted individual's vulnerability to cocaine-related cues has a neurological basis. For example, Wexler and colleagues

(Wexler et al., 2001) documented the activation of the anterior cingulate cortex (ACG), a region associated with emotional processing, while cocaine-addicted subjects watched videotapes containing cocaine-associated cues, even if they did not experience craving (Figure 2). The finding indicates that addicted individuals' emotional responses to cues have a subconscious component. The subjects also showed less activation in the frontal lobe relative to healthy subjects during the viewing of cocaine-cue tapes, suggesting that their ability to control their cue responses was inhibited.

Research with fMRI has linked chronic stimulant abusers' cognitive impairments to drug-related alterations in brain activation: In one study, methamphetamine dependence

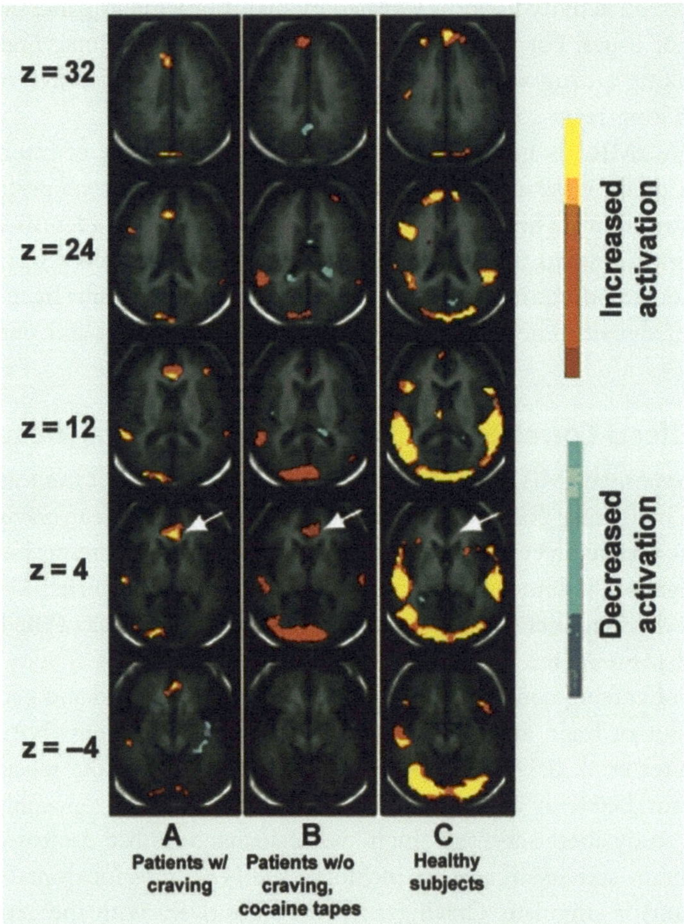

Figure 2 fMRI: The brain's response to cocaine cues. The arrows point to the anterior cingulate area, which activated (yellow) in cocaine-addicted patients (left) but not in healthy volunteers (right). *Reprinted with permission from Wexler et al. (2001).*

and poor decision making correlated with the reduced activation in the PFC (Paulus et al., 2002). In another study, investigators found that chronic cocaine abusers had abnormally low levels of activity in the midline areas of the anterior cingulate that are crucial for cognitive and behavioral control (Kaufman et al., 2003).

Genes Affect Responses to Drugs and Vulnerability to Abuse

More recently, innovative fMRI researchers have begun to explore the role of genes in drug abuse. In one such study, a gene variation that affects the metabolism of neurotransmitters, including dopamine and norepinephrine, appeared to influence the brain's response to amphetamine (Mattay et al., 2003). A similar fMRI study showed that individuals with a particular variation in the serotonin transporter gene experienced a greater activation of the amygdala, a region associated with fear and anxiety, in response to frightening stimuli (Hariri et al., 2002). This particular genetic variation is likely to increase sensitivity to stress and heighten vulnerability to drug abuse.

2.1.3. Magnetic Resonance Spectroscopy

Magnetic resonance spectroscopy (MRS) is a special application of fMRI that uses different scanner settings. In MRS, the magnetic pulses and radiofrequency energy are precisely calibrated so that they stimulate a specific nucleus of interest (e.g., proton, carbon, phosphorous, or fluorine) in the molecules located in the regions of interest of the brain. The sum of all the returning (resonance) signals is recorded and analyzed using sophisticated computer programs that separate the signals for each metabolite. The result of these signals can then be displayed as various metabolite peaks on a spectrum. Thus, in addition to creating structural (sMRI) and functional (fMRI) maps of the brain, magnetic resonance technology can also be used to detect and measure important chemical compounds in the brain.

To be detectable by MRS, however, a compound must respond in a unique way to magnetization and energy stimulation and it must be present at relatively high concentrations (typically in the millimolar range). If these two conditions are met, MRS scans can reveal the location and concentrations of target chemicals in the brain tissues (Ross et al., 1992). Among chemicals naturally present in the brain, two that can be studied with MRS are *N*-acetylaspartate (NAA), which researchers use as a gauge of neuronal cell health (De Stefano et al., 1995), and myoinositol, which is primarily present in support cells called glia (Brand et al., 1993), thus providing an index of glial health.

Other molecules that can be detected easily are choline compounds, which are involved in the synthesis of cell membranes, and creatine compounds, which are important for cells' energy maintenance. Among substances of abuse that penetrate the brain after being ingested or administered, alcohol is readily detectable with MRS (Hetherington et al., 1999). Researchers have used MRS to identify drug-related biochemical changes that indicate damage to the health and function of brain cells. Often, these studies focus on brain areas that preclinical models or neuropathology of drug users had shown to be affected. In some cases, biochemical changes have been directly correlated with cognitive and behavioral deficits (Cowan et al., 2009; Yang et al., 2009).

MRS-based Findings

One of the main contributions of MRS studies to the addiction field has been the discovery that drugs affect markers associated with inflammation, brain energy metabolism, and neuronal health. For example, Ernst and colleagues (2000) found that methamphetamine abusers exhibit reduced NAA concentrations in their basal ganglia and frontal white matter, compared to that nondrug abusers exhibit, a finding that could help explain the cognitive difficulties experienced by methamphetamine abusers, since the concentrations of NAA correlate with measures of cognitive function even in nondrug users (Rae et al., 1998).

Cocaine-dependent individuals also exhibit lower NAA levels, suggesting neuronal damage, as well as elevated creatine and myoinositol levels reflecting either increased glial cell activity or inflammation (Chang et al., 1999).

Other MRS research has evaluated possible interactions between human immunodeficiency virus (HIV) and drugs of abuse on brain metabolites. For example, a study found that methamphetamine abuse and HIV decreased brain NAA additively, especially in the striatum, while choline and myoinositol were further elevated in the frontal lobes (Chang et al., 2005). Chronic marijuana use and HIV infection are each separately associated with lower levels of glutamate, but together, they appear to moderate glutamate loss in the frontal white matter while exacerbating it in the basal ganglia (Chang et al., 2006).

Magnetic resonance techniques like MRS are especially useful for studying the effects of drugs in the pediatric population because they do not involve radiation. Smith and colleagues (Smith et al., 2001a, 2001b) conducted MRS studies of children who had been exposed to cocaine or methamphetamine prenatally and found that their total brain creatine levels were elevated, suggesting abnormalities in energy metabolism.

2.1.4. Resting State Functional MRI

Resting state fMRI is a relatively recent development that allows investigators to explore the modular nature of cortical function and to assess resting state functional connectivity (RFC). The images generated through this approach rely on the fact that the spontaneous low-frequency fluctuations in the BOLD signals emerging from some brain regions (at rest) are more highly correlated with one another than with fluctuations emanating from other regions (Raichle and Snyder, 2007). The resulting functional connectivity maps reveal the existence of temporally dynamic linkages within and among several brain regions that include inhibitory control, visual, auditory, default mode, dorsal attention, and sensorimotor networks (Damoiseaux et al., 2006). Moreover, the nonrandomness of such non-task-related, spontaneous brain activity suggests that its member functional networks are organized in a highly coherent fashion.

It is important to point out that the emergence of large RFC databases, across healthy populations will allow investigators to probe into the involvement of specific genes in the functional organization of the brain. In addition, it could also provide an extremely valuable, normative baseline against which to compare various disordered states of the brain that result from either biological, developmental and/or environmental perturbations as well as a means to monitor treatment effectiveness and predict clinical outcomes (Biswal et al., 2010). For example, early studies have found RFC abnormalities in the PFC (Kelly et al., 2011) and mesocorticolimbic circuitry in cocaine abusers (Gu et al., 2011). If corroborated by future studies, such results could add important information toward improving our understanding of the range of cognitive and behavioral disruptions seen in individuals addicted to cocaine.

2.1.5. Neurofeedback

Real-time fMRI is yet another technological development in the fMRI field that is poised to help expand treatment options. This technique emerged as a result of advances in imaging algorithms that enable very fast image processing, allowing researchers to feed that information back to the subjects while in the scanner. In this way, the real-time signal can be used for biofeedback-mediated retraining of neural circuits, such as the strengthening of frontal executive function (Berman et al., 2011). This approach, which has proven its utility for chronic pain patients, is being investigated as a possible treatment for addiction and other psychiatric disorders.

2.1.6. Diffusion Tensor Imaging

In addition to measuring the location of water molecules in the brain to generate a map of the gross brain structures, a recent MRI technique called diffusion tensor imaging (DTI) detects the directions of the water diffusion in the brain, which in turn reflects the placement of fine tissue structures and connecting tracts. DTI has been used to demonstrate in vivo changes in the brain development across different age groups (Gilmore et al., 2006; Huang et al., 2006) and in many different disease states.

Because this technique is relatively new, few DTI studies of the addicted brain have been performed. However, alcoholics have shown significant brain abnormalities on DTI (Daurignac et al., 2005; Pfefferbaum and Sullivan, 2005). Specifically, fractional anisotropy, a measure of the orientation of the water diffusion that reflects the coherence of the fiber tracks in the brain, was abnormally low in the white matter of alcoholics, while the apparent diffusion coefficient was higher than normal. These findings suggest increased intracellular or extracellular fluid and the possible disorganization of fiber structure (Pfefferbaum and Sullivan, 2005).

Similarly, Warner and colleagues (Warner et al., 2006) evaluated children (average age 10 years) who had been exposed to cocaine prenatally and found that, relative to unexposed children, they had a significantly higher average diffusion in the left frontal callosal and right frontal projection fibers, which suggests a disrupted or reduced maturation along these frontal white matter pathways. This highly sensitive technique to assess the changes in the microscopic environment of the brain will likely yield new, more informative data regarding drugs' effects on the brain. Because MRI techniques are noninvasive and nonradioactive, they are particularly valuable for monitoring the growth, development, and the effects of treatment over time and for studying children and adolescents since they do not rely on the use of radioactivity (as positron emission tomography (PET) and single photon emission computerized tomography (SPECT) technologies do, see below).

2.2. NUCLEAR MEDICINE IMAGING TECHNIQUES

2.2.1. PET and SPECT

PET and SPECT are referred to as "nuclear medicine techniques" because they involve the injection of molecules labeled with radioactive isotopes (o radiotracers) into the bloodstream of the person being studied. A PET or SPECT image displays the distribution of such radiotracers after they reach their targets in the brain and other organs. The energy emitted by the radiotracer interacts with detectors in the PET or SPECT instrument. The instrument's computers register the location of the radioisotope and use this information to construct a 3D map of the radiotracer's distribution in the brain or body.

Because the half-lives of the PET or SPECT isotopes are short, the net radiation dose is small, in the order of other medical diagnostic procedures; thus, studies can be carried out in healthy volunteers as well as in drug-addicted patients. However, PET and SPECT are not normally used in healthy children. PET and SPECT are actually similar technologies, differing mainly in their use of different types of isotopes in their radiotracers. PET radiotracers incorporate isotopes that emit beta positron ($\beta+$) radiation. One especially important set of PET radiotracers incorporates positron-emitting isotopes of the chemical elements of life, that is, carbon, oxygen, and nitrogen, into organic compounds in place of the naturally occurring nonradioactive elements. Substituting radioactive ^{11}carbon for nonradioactive ^{12}carbon in a drug molecule, for example, does not alter the drug's biochemical properties in the brain, but renders it detectable through PET imaging. ^{11}Carbon has a very short half-life (20.4 min), making it ideal for use in humans.

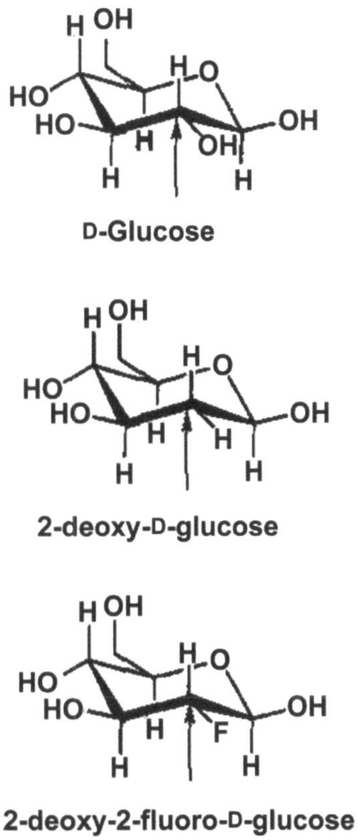

Figure 3 Chemists replace a hydroxyl (–OH) group on the glucose molecule with ^{18}fluorine to make ^{18}FDG, a radiotracer that is used to measure brain glucose metabolism. 18F decays by positron emission resulting in two energetic photons that are detected by a PET scanner to produce an image of glucose metabolism in the brain. *Reprinted with permission from Fowler and Ido (2002).*

One uniquely valuable PET tool is 2-deoxy-2-[18F]fluoro-D-glucose (^{18}FDG), a radiotracer used to measure brain glucose metabolism. ^{18}FDG consists of a glucose molecule in which the radioactive isotope fluorine-18 has been substituted for the naturally occurring hydroxyl group (Figure 3) (Fowler and Ido, 2002). A PET's ability to produce an image of glucose metabolism in the brain using 18FDG is a major advantage, as glucose, along with oxygen, is a major source of the organ's energy. SPECT radiotracers are labeled with single photon emitting radioisotopes. The most commonly used are ^{123}iodine and ^{99m}technetium.

Nuclear medicine techniques are particularly valuable in drug abuse research because they can measure the concentrations of molecules that are extremely low—in the nanomolar and picomolar range, one millionth to one billionth of the minimum amounts necessary for visibility in MRS (Fowler et al., 2003a; Kung et al., 2003). This level of sensitivity enables researchers to study drugs' effects on key components of cell-to-cell communication, including cell receptors, transporters, and enzymes involved in the synthesis or metabolism of neurotransmitters (Volkow et al., 2003b). In particular, many PET studies have explored the role of the neurotransmitter dopamine in drug abuse and addiction. The radiotracers for these studies piggyback on compounds that bind to various dopamine-interacting proteins on brain cells, including dopamine receptors, dopamine transporters, and dopamine-degrading and synthetic enzymes (Halldin et al., 2001; Rinne et al., 1995; Volkow et al., 1995, 1996a; Wong et al., 1993).

Researchers also use PET to study drug pharmacokinetics: A series of images taken at appropriate intervals provides a stop-action record of a drug's movement into and out of the brain, showing how much of the drug enters the brain, where it binds in the brain, and for how long it lingers (Fowler et al., 1999). This information is crucial because the rate at which a drug enters the brain largely determines its euphorigenic effects and addictiveness. PET can also be used to assess the rates of glucose metabolism, providing an alternative to functional MRI measurement of blood oxygen levels for determining cellular activity. A common use for SPECT is to measure brain blood flow. PET and SPECT brain imaging have perhaps shown their greatest value to date in helping researchers to analyze how drugs affect the neurotransmitter systems that link and coordinate brain cells. Much of this work has focused on the dopamine system, but researchers are also exploring the roles of other neurotransmitters in drug abuse and drugs' effects on cells' energy consumption and health.

2.2.2. PET and SPECT-based Findings

Dopamine Plays Key Roles in Drug Abuse Euphoria and Addiction

The neurotransmitter dopamine is highly concentrated in the striatum, which forms part of the brain's reward circuit. The ebb and flow of dopamine into these areas is the main determinant of how much pleasure we derive from our experiences; it also helps us focus our attention on what is important. PET studies have linked the drugs' presence and action here with their euphorigenic properties and their ability to attract the complete attention of the addicted individual, at the expense of most other natural reinforcers (Di Chiara, 1999; Di Chiara and Imperato, 1988; Leshner, 1997; Volkow et al., 2003a).

In one study, researchers used ^{11}carbon-labeled cocaine and PET imaging to track the movement of cocaine into and out of the cocaine abuser's brain while also recording the intensity of their highs. The results showed that the "high" spiked and subsided in close temporal correlation with cocaine's movement in and out of the striatum (Volkow et al., 1997) (Figure 4). Using the same experimental design with different radiotracers, PET and SPECT investigators established that cocaine, amphetamine, and methylphenidate, when given intravenously, produce their highs by massively increasing the amount of dopamine in the striatum (Drevets et al., 2001; Laruelle et al., 1995;

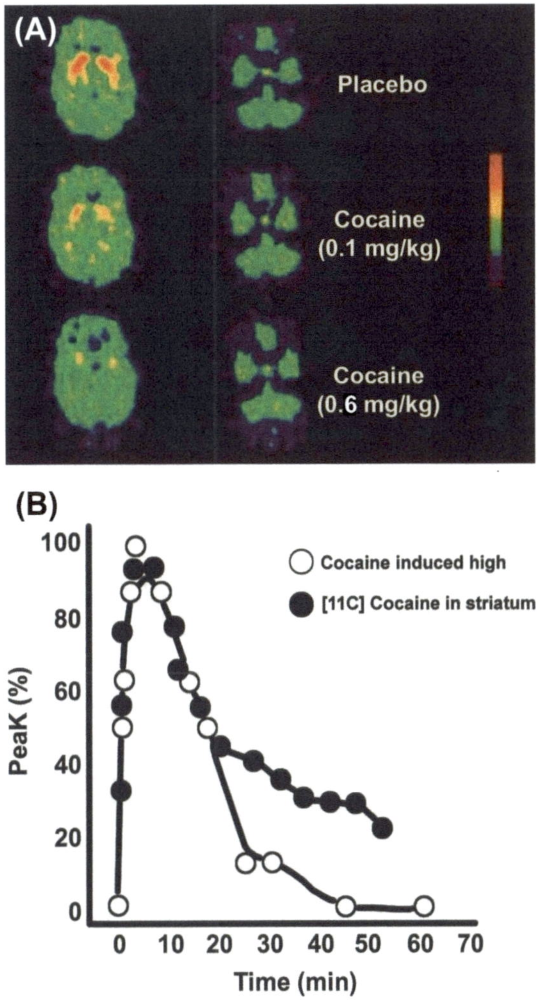

Figure 4 PET: Cocaine activity in the striatum underlies the cocaine high. (A) Study subjects were given a cocaine radiotracer that homes in toward the striatum, where dopamine D2/3 receptors are expressed at high concentrations. (B) PET showed that their reports of how high they felt rose and fell in very close parallel with the passage of the drug in and out of the brain's striatum. *Reprinted with permission from Volkow et al. (1997).*

Volkow et al., 1999). Previous studies with animals had suggested that this was probably the case, but nuclear medicine imaging enabled researchers to noninvasively document the correlation as it actually occurred in living human beings. Still, other PET studies have shown that the abuse liability of any drug actually depends on both the magnitude of the dopamine spike it produces and the speed with which dopamine rises and falls back to normal levels.

PET experiments have identified how one drug, cocaine, causes dopamine to surge: The drug interferes with the normal activity of a molecule called dopamine transporter that is expressed on the cell surface of the presynaptic neuron (Volkow et al., 1997). By doing so, it disrupts the equilibrium of dopamine release and reuptake that maintains the levels of this neurotransmitter within normal limits.

Chronic Methamphetamine Abuse Depletes Dopamine Function

PET studies have also shown that while acute methamphetamine temporarily hyperactivates the dopamine system, chronic exposure to the drug reduces the availability of dopamine transporters, which may indicate a loss of dopamine cells (Volkow et al., 2001b). Study participants with fewer dopamine transporters had a poorer memory and a slower motor function and another study reported that the loss of dopamine transporters was associated with psychoses (Sekine et al., 2003).

Stimulants Reduce the Cellular Activity in Brain Areas that Affect Judgment

PET studies have also been used to explore cocaine's impact on brain structures and activity, and their relationship to the addicted individuals' ability to regain function during and after treatment. Among the most significant results in this line of study are those that showed that cocaine (Volkow et al., 1993) and methamphetamine (Bolla et al., 2003; Volkow et al., 2001a) reduce the cellular activity in the orbitofrontal cortex (OFC), a brain area we rely on to make strategic, rather than impulsive, decisions. Patients with traumatic injuries in this area of the brain exhibit diverse cognitive deficits—aggressiveness, poor judgment of future consequences, inability to inhibit inappropriate responses—that are reminiscent of those seen among substance abusers (Bechara et al., 1994, 2001; Eslinger et al., 1992). The radiotracers used in these studies were 18FDG and oxygen-15 water, which measure the brain's consumption of its two main fuels, glucose and oxygen (Raichle et al., 1983).

In fact, a link between a lower metabolism in the OFC and poor judgment has been found in cocaine abusers (Bolla et al., 2003). The researchers took serial PET images, using oxygen-15 (^{15}O) water as the radiotracer, while cocaine abusers who had been abstinent for 25 days played a card game on a computer. Players who had used more cocaine prior to being abstinent exhibited a lower OFC activity and a poorer performance during the game.

Dopamine Receptor Levels May Determine Vulnerability to Abuse and Addiction

PET studies have demonstrated that abusers of alcohol (Volkow et al., 1996b), cocaine (Volkow et al., 1993, 1990), heroin (Wang et al., 1997), and methamphetamine (Volkow et al., 2001a) all have reduced levels of brain dopamine D2 receptors—one of the cell surface proteins activated by dopamine. These and other findings have given rise to the hypothesis that people with low levels of dopamine D2 receptors, as a result of genetic variability, life experience, or both, may present a higher risk for drug abuse and addiction. Scientists speculate that such individuals derive suboptimal levels of dopamine-mediated pleasure from ordinary activities and accomplishments and are therefore more susceptible to wanting to repeat the euphoria that follows the massive, drug-induced increase in striatal dopamine.

The μ-Opioid System Plays a Role in Cocaine Craving

PET studies have suggested that the symptoms of cocaine dependence and craving may be caused at least in part by the drug's effects on another neurotransmitter system, the one driven by the μ-opioid peptide. In one study (Zubieta et al., 1996), cocaine-addicted individuals who entered a clinic to quit the drug and remained there for a month of monitored abstinence filled out assessments of their mood and craving symptoms and underwent PET scans, once during their first 4 days in the clinic and again toward the end of the month. Using a potent opioid agonist as the radiotracer (^{11}carbon carfentanil), researchers found that the participants' symptom severity correlated with μ-opioid receptor levels in several brain areas. In interpreting their findings, the researchers suggested that cocaine may have depleted the body's natural opioids, stimulating either a compensatory production of more opioid receptors or increasing the avidity of existing receptors toward opioid molecules.

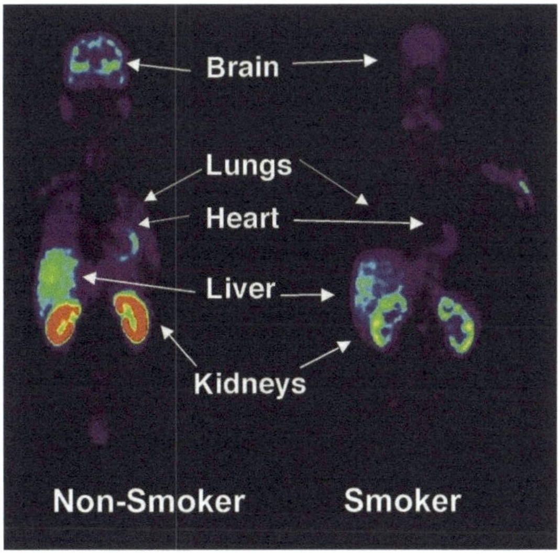

Figure 5 PET: Smoking reduces an important enzyme. In these composite PET images of smokers versus nonsmokers, the arrows demonstrate lower concentrations of the enzyme MAO in many of the smokers' organs. *Reprinted with permission from Fowler et al. (2003c).*

Nicotine is Not the Sole Culprit in Tobacco Addiction

PET imaging studies have confirmed the importance of dopamine signaling in nicotine abuse and addiction (Brody et al., 2004) while also highlighting the need to investigate the role of other chemicals present in tobacco smoke. For example, studies have found that one or more components of tobacco smoke can reduce the levels of monoamine oxidase A and B (MAO A and B), an enzyme that breaks down dopamine in the brain and throughout the body (Fowler et al., 1996a, 1996b, 2003b, 2003c, 2005) (Figure 5). This activity may lead to the exacerbation of the nicotine-induced dopamine dysregulation that reinforces the desire to smoke as well as to abuse other substances. In fact, recent preclinical studies show that the inhibition of MAO A enhances nicotine self-administration in animals (Guillem et al., 2005). While MAO A, rather than MAO B, inhibition increases nicotine reinforcement in rats (Guillem et al., 2006), a recent trial using the selective MAO B inhibitor selegiline (at a dose of 10 mg/day) safely enhanced smoking cessation rates compared with placebo in nicotine-dependent cigarette smokers (George et al., 2003).

The PET finding that smokers have relatively lower levels of MAO may help explain why smokers have a reduced risk of Parkinson's disease (Morens et al., 1995). When MAO metabolizes dopamine, a byproduct is hydrogen peroxide, a potential source of free radicals that can damage nerve cells. MAO-inhibiting compounds have been isolated from tobacco (Khalil et al., 2000) and have shown to be protective in a rodent model of Parkinson's disease (Castagnoli et al., 2002).

2.3. CLINICAL APPLICATIONS OF IMAGING

2.3.1. Medication Development and Imaging Studies

The information that magnetic resonance and nuclear imaging studies have yielded on the brain dynamics of addiction has become a primary source of medication development strategies. Direct clinical applications are still few, but recent studies suggest that the techniques may in the future enhance patient assessment and monitoring. Such studies, together with other research, overwhelmingly indicate that drug addiction is a disease of the brain, and thus, it must be viewed as a bona fide behavioral disorder. To be effective in the long term, treatments should focus on enhancing and restoring the disrupted dopamine function and processing among a widely distributed network of brain circuits (Figure 6), and take advantage of effective pharmacologic and/or behavioral approaches.

In the area of pharmacologic interventions in particular, imaging findings have suggested many possible new approaches. One strategy under active investigation takes its cue from the PET finding that stimulant drugs produce euphoria by causing a rapid dopamine spike and, in doing so, reduce abusers' ability to feel pleasure when their other, non-drug-related

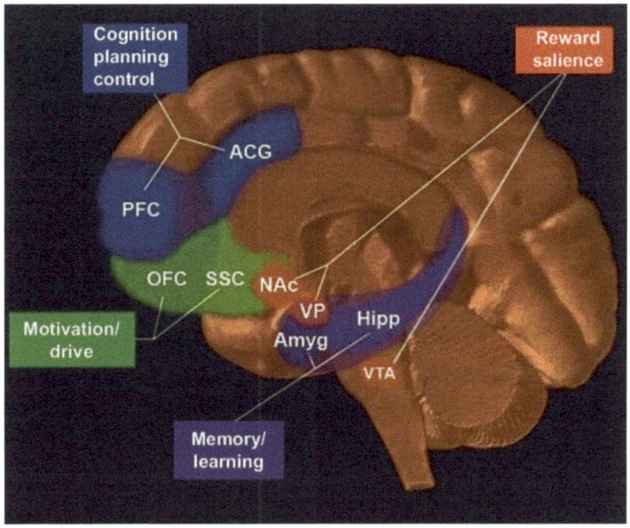

Figure 6 Major brain regions with roles in addiction. The prefrontal cortex (PFC) and the anterior cingulate cortex (ACG) are focal areas for cognition, planning, and inhibitory control. The orbitofrontal (OFC) and somatosensory cortex play key roles in supporting the motivation for goal-directed behaviors. The ventral tegmental area (VTA) and NAc are key components of the brain's reward system. The VTA, NA, amygdala, and hippocampus are the major components of the limbic system, which coordinates drives, emotions, and memories.

activities cause more modest, natural neurotransmitter elevations (Volkow et al., 2007). Researchers are identifying and testing medications that slightly increase the amount of dopamine that cells release when a person engages in normally rewarding activities in the hope that the boost will enable addicted individuals to once again begin to feel pleasure from them. For example, MAO B inhibitors and other medications fitting this criterion have been used successfully to treat smoking addiction (George et al., 2003).

Another medication strategy that stems from imaging evidence that dopamine spikes underlie drug euphoria seeks to reduce the stimulant high and the desire to repeat it by inhibiting the initial dopamine response to these drugs. In one form of this strategy, researchers are testing compounds that enhance the neurotransmitter gamma-aminobutyric acid, which has been shown to inhibit dopamine-releasing cells' response to drug-related cues (Di Ciano and Everitt, 2003). Preliminary clinical trials of this approach have yielded promising results (Brodie et al., 2003, Brodie et al., 2009, 2005). Still other medications interfere with the responses of dopamine-receiving cells and thereby attenuate the reinforcing effects of abused drugs. For example, selective cannabinoid receptor (CB1) antagonists have been shown to modulate both dopamine-releasing and dopamine-receiving cell responses in preclinical studies (De Vries et al., 2001; Julian et al., 2003).

A third strategy to counter drug-induced euphoria and its hold over individuals is based on a medication that activates the same neurotransmitter system coopted by an abused drug, but produces no sharp dopamine spike. Treatment of heroin addiction with methadone and buprenorphine exemplifies this approach (Kreek et al., 2002). Similar attempts to treat stimulant addiction have yet to produce positive results (Shearer, 2008). For example, replacement of cocaine with oral methylphenidate or oral amphetamine did not decrease cocaine consumption when compared with placebo in most drug-addicted individuals. However, treatment with oral methylphenidate did decrease drug consumption by patients suffering from comorbid addiction and attention deficit hyperactivity disorder (Grabowski et al., 1997).

2.3.2. Patient Assessment and Monitoring

Recent studies suggest that imaging has the potential to help clinicians determine the most appropriate level of treatment for individual patients and monitor their progress toward recovery. Results of a recent fMRI study performed on a group of men entering treatment for methamphetamine addiction revealed two contrasting patterns of brain activity (during a psychological task that required decision making) that predicted with a 90% accuracy which of the men would relapse within 1–3 years after completing treatment (Paulus et al., 2005). Those who relapsed exhibited less activity in the prefrontal lobe and also in regions not previously thought to play a role in addiction. Another study found that a more rapid response of the posterior cingulate to cocaine cues distinguished relapsers from nonrelapsers, even though both groups reported similarly intense cravings (Kosten et al., 2006).

Imaging researchers have also been documenting changes that appear to represent brain healing in response to treatment. One group has applied MRS to evaluate the effects of methadone maintenance therapy on heroin-addicted individuals (Silveri et al., 2004). The subjects' levels of certain metabolites involved in cellular energy production, which were abnormal at the beginning of the treatment, began to change over the first month. The researchers interpreted the metabolite changes as evidence that the switch from heroin to methadone might have improved the neurons' oxygen supply. This explanation may account for the findings from another study by the same research group that individuals' cognitive abilities improve during their first 2 months of methadone therapy (Gruber et al., 2006).

Similarly, studies have shown that, while detoxified methamphetamine abusers have fewer dopamine transporters than do drug-naive, age-matched individuals, those who remained abstinent for 9 months recovered a significant fraction of transporters (Volkow et al., 2001b; Wang et al., 2004). Unfortunately, they did not exhibit a concomitant recovery from the cognitive and motor deficits associated with low transporter levels. PET studies with 18FDG also showed a significant recovery in brain glucose metabolism in methamphetamine abusers after protracted abstinence (Wang et al., 2004) (Figure 7).

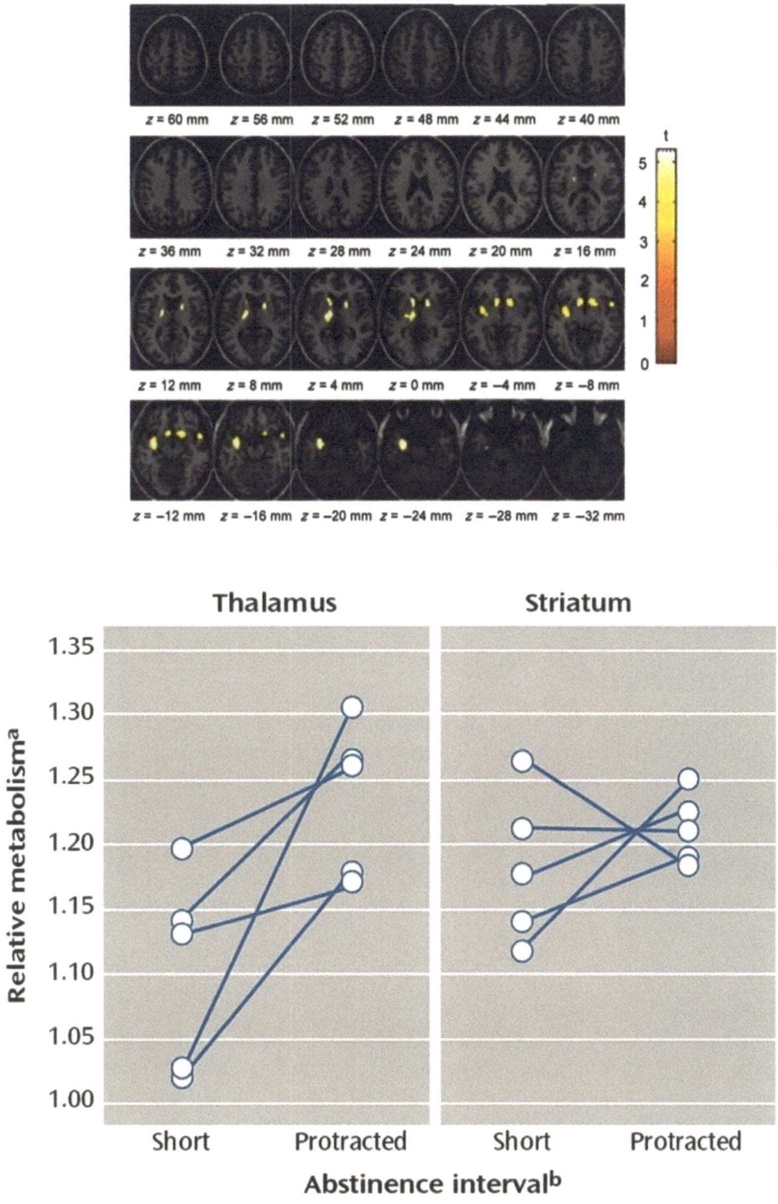

Figure 7 PET: Brain recovery occurs with methamphetamine abstinence. A. PET studies in methamphetamine abusers show that brain metabolism is depressed in the thalamus and striatum shortly after quitting methamphetamine. Protracted abstinence can lead to partial recovery, particularly in the thalamus. *Reprinted with permission from Wang et al. (2004).*

3. CONCLUSIONS

Brain imaging techniques enable researchers to observe drug effects while they are occurring in the brain and compare brain structure, function, and metabolism in drug-abusing and nonabusing individuals. The results to date have firmly established that drug addiction is a disease of the brain, causing measurable perturbations in many areas, including pathways affecting reward, inhibitory control, motivation, interoception, memory/learning, and emotion/stress. Ongoing studies continue to broaden our understanding of the dynamics underlying the development, symptoms, and consequences of addiction, as well as recovery.

While there is yet no clinically approved application for the imaging tools described here in ways that will improve outcomes for substance-abusing individuals, the translational opportunities created, offered, and presented by this suite of imaging technologies are incredibly exciting and represent a very active and potentially transformative area of research. Indeed, the impact of most of these advances on addiction research has been remarkable, and there is little doubt that, at the current rate of technological progress, the translation of these and emerging new imaging applications into the clinic is only a matter of time.

ACKNOWLEDGMENTS

Some of the work described in this article was performed at the Brookhaven National Laboratory under contract DEAC02-98CH10886 with the US Department of Energy and was supported by its Office of Biological and Environmental Research and by the National Institute on Drug Abuse (K24-DA16170 and Ko5-DA020001) and NIH GCRC (MO1RR10710).

REFERENCES

Bechara, A., Damasio, A.R., Damasio, H., Anderson, S.W., 1994. Insensitivity to future consequences following damage to human prefrontal cortex. Cognition 50 (1–3), 7–15.

Bechara, A., Dolan, S., Denburg, N., Hindes, A., Anderson, S.W., Nathan, P.E., 2001. Decision-making deficits, linked to a dysfunctional ventromedial prefrontal cortex, revealed in alcohol and stimulant abusers. Neuropsychologia 39 (4), 376–389.

Bendszus, M., Weijers, H.G., Wiesbeck, G., Warmuth-Metz, M., Bartsch, A.J., Engels, S., Boning, J., Solymosi, L., 2001. Sequential MR imaging and proton MR spectroscopy in patients who underwent recent detoxification for chronic alcoholism: correlation with clinical and neuropsychological data. AJNR Am. J. Neuroradiol. 22 (10), 1926–1932.

Berman, B.D., Horovitz, S.G., Venkataraman, G., Hallett, M., 2011. Self-modulation of primary motor cortex activity with motor and motor imagery tasks using real-time fMRI-based neurofeedback. Neuroimage.

Biswal, B.B., Mennes, M., Zuo, X.N., Gohel, S., Kelly, C., Smith, S.M., Beckmann, C.F., Adelstein, J.S., Buckner, R.L., Colcombe, S., Dogonowski, A.M., Ernst, M., Fair, D., Hampson, M., Hoptman, M.J., Hyde, J.S., Kiviniemi, V.J., Kotter, R., Li, S.J., Lin, C.P., Lowe, M.J., Mackay, C., Madden, D.J., Madsen, K.H., Margulies, D.S., Mayberg, H.S., McMahon, K., Monk, C.S., Mostofsky, S.H., Nagel, B.J., Pekar, J.J., Peltier, S.J., Petersen, S.E., Riedl, V., Rombouts, S.A., Rypma, B., Schlaggar, B.L., Schmidt, S., Seidler, R.D., Siegle, G.J., Sorg, C., Teng, G.J., Veijola, J., Villringer, A., Walter, M., Wang, L., Weng, X.C., Whitfield-Gabrieli, S., Williamson, P., Windischberger, C., Zang, Y.F., Zhang, H.Y., Castellanos, F.X., Milham, M.P., 2010. Toward discovery science of human brain function. Proc. Natl. Acad. Sci. USA 107 (10), 4734–4739.

Bolla, K.I., Eldreth, D.A., London, E.D., Kiehl, K.A., Mouratidis, M., Contoreggi, C., Matochik, J.A., Kurian, V., Cadet, J.L., Kimes, A.S., Funderburk, F.R., Ernst, M., 2003. Orbitofrontal cortex dysfunction in abstinent cocaine abusers performing a decision-making task. Neuroimage 19 (3), 1085–1094.

Brand, A., Richter-Landsberg, C., Leibfritz, D., 1993. Multinuclear NMR studies on the energy metabolism of glial and neuronal cells. Dev. Neurosci. 15 (3–5), 289–298.

Breiter, H.C., Gollub, R.L., Weisskoff, R.M., Kennedy, D.N., Makris, N., Berke, J.D., Goodman, J.M., Kantor, H.L., Gastfriend, D.R., Riorden, J.P., Mathew, R.T., Rosen, B.R., Hyman, S.E., 1997. Acute effects of cocaine on human brain activity and emotion. Neuron 19 (3), 591–611.

Breiter, H.C., Rosen, B.R., 1999. Functional magnetic resonance imaging of brain reward circuitry in the human. Ann. N.Y. Acad. Sci. 877, 523–547.

Brodie, J.D., Case, B.G., Figueroa, E., Dewey, S.L., Robinson, J.A., Wanderling, J.A., Laska, E.M., 2009. Randomized, double-blind, placebo-controlled trial of vigabatrin for the treatment of cocaine dependence in Mexican parolees. Am. J. Psychiatry 166 (11), 1269–1277.

Brodie, J.D., Figueroa, E., Dewey, S.L., 2003. Treating cocaine addiction: from preclinical to clinical trial experience with gamma-vinyl GABA. Synapse 50 (3), 261–265.

Brodie, J.D., Figueroa, E., Laska, E.M., Dewey, S.L., 2005. Safety and efficacy of gamma-vinyl GABA (GVG) for the treatment of methamphetamine and/or cocaine addiction. Synapse 55 (2), 122–125.

Brody, A.L., Olmstead, R.E., London, E.D., Farahi, J., Meyer, J.H., Grossman, P., Lee, G.S., Huang, J., Hahn, E.L., Mandelkern, M.A., 2004. Smoking-induced ventral striatum dopamine release. Am. J. Psychiatry 161 (7), 1211–1218.

Castagnoli, K., Steyn, S.J., Magnin, G., Van Der Schyf, C.J., Fourie, I., Khalil, A., Castagnoli Jr, N., 2002. Studies on the interactions of tobacco leaf and tobacco smoke constituents and monoamine oxidase. Neurotox. Res. 4 (2), 151–160.

Chang, L., Cloak, C., Yakupov, R., Ernst, T., 2006. Combined and independent effects of chronic marijuana use and HIV on brain metabolites. J. Neuroimmune Pharmacol. 1 (1), 65–76.

Chang, L., Ernst, T., Speck, O., Grob, C.S., 2005. Additive effects of HIV and chronic methamphetamine use on brain metabolite abnormalities. Am. J. Psychiatry 162 (2), 361–369.

Chang, L., Ernst, T., Strickland, T., Mehringer, C.M., 1999. Gender effects on persistent cerebral metabolite changes in the frontal lobes of abstinent cocaine users. Am. J. Psychiatry 156 (5), 716–722.

Cowan, R.L., Joers, J.M., Dietrich, M.S., 2009. N-acetylaspartate (NAA) correlates inversely with cannabis use in a frontal language processing region of neocortex in MDMA (Ecstasy) polydrug users: a 3 T magnetic resonance spectroscopy study. Pharmacol. Biochem. Behav. 92 (1), 105–110.

Damoiseaux, J.S., Rombouts, S.A., Barkhof, F., Scheltens, P., Stam, C.J., Smith, S.M., Beckmann, C.F., 2006. Consistent resting-state networks across healthy subjects. Proc. Natl. Acad. Sci. USA 103 (37), 13848–13853.

Daurignac, E., Toga, A., Jones, D., Aronen, H., Hommer, D., Jemigan, T., Krystal, J., Mathalon, D., 2005. Applications of morphometric and diffusion tensor magnetic resonance imaging to the study of brain abnormalities in the alcoholism spectrum. Alcohol. Clin. Exp. Res. 29 (1), 159–166.

De Stefano, N., Matthews, P.M., Arnold, D.L., 1995. Reversible decreases in N-acetylaspartate after acute brain injury. Magn. Reson. Med. 34 (5), 721–727.

De Vries, T.J., Shaham, Y., Homberg, J.R., Crombag, H., Schuurman, K., Dieben, J., Vanderschuren, L.J., Schoffelmeer, A.N., 2001. A cannabinoid mechanism in relapse to cocaine seeking. Nat. Med. 7 (10), 1151–1154.

Di Chiara, G., 1999. Drug addiction as dopamine-dependent associative learning disorder. Eur. J. Pharmacol. 375 (1–3), 13–30.

Di Chiara, G., Imperato, A., 1988. Drugs abused by humans preferentially increase synaptic dopamine concentrations in the mesolimbic system of freely moving rats. Proc. Natl. Acad. Sci. USA 85 (14), 5274–5278.

Di Ciano, P., Everitt, B.J., 2003. The GABA(B) receptor agonist baclofen attenuates cocaine- and heroin-seeking behavior by rats. Neuropsychopharmacology 28 (3), 510–518.

Drevets, W.C., Gautier, C., Price, J.C., Kupfer, D.J., Kinahan, P.E., Grace, A.A., Price, J.L., Mathis, C.A., 2001. Amphetamine-induced dopamine release in human ventral striatum correlates with euphoria. Biol. Psychiatry 49 (2), 81–96.

Eslinger, P.J., Grattan, L.M., Damasio, H., Damasio, A.R., 1992. Developmental consequences of childhood frontal lobe damage. Arch. Neurol. 49 (7), 764–769.

Fein, G., Di Sclafani, V., Cardenas, V.A., Goldmann, H., Tolou-Shams, M., Meyerhoff, D.J., 2002. Cortical gray matter loss in treatment-naive alcohol dependent individuals. Alcohol. Clin. Exp. Res. 26 (4), 558–564.

Fowler, J.S., Ding, Y.S., Volkow, N.D., 2003a. Radiotracers for positron emission tomography imaging. Semin. Nucl. Med. 33 (1), 14–27.

Fowler, J.S., Ido, T., 2002. Initial and subsequent approach for the synthesis of 18FDG. Semin. Nucl. Med. 32 (1), 6–12.

Fowler, J.S., Logan, J., Wang, G.J., Volkow, N.D., 2003b. Monoamine oxidase and cigarette smoking. Neurotoxicology 24 (1), 75–82.

Fowler, J.S., Logan, J., Wang, G.J., Volkow, N.D., Telang, F., Zhu, W., Franceschi, D., Pappas, N., Ferrieri, R., Shea, C., Garza, V., Xu, Y., Schlyer, D., Gatley, S.J., Ding, Y.S., Alexoff, D., Warner, D., Netusil, N., Carter, P., Jayne, M., King, P., Vaska, P., 2003c. Low monoamine oxidase B in peripheral organs in smokers. Proc. Natl. Acad. Sci. USA 100 (20), 11600–11605.

Fowler, J.S., Logan, J., Wang, G.J., Volkow, N.D., Telang, F., Zhu, W., Franceschi, D., Shea, C., Garza, V., Xu, Y., Ding, Y.S., Alexoff, D., Warner, D., Netusil, N., Carter, P., Jayne, M., King, P., Vaska, P., 2005. Comparison of monoamine oxidase a in peripheral organs in nonsmokers and smokers. J. Nucl. Med. 46 (9), 1414–1420.

Fowler, J.S., Volkow, N.D., Wang, G.J., Ding, Y.S., Dewey, S.L., 1999. PET and drug research and development. J. Nucl. Med. 40 (7), 1154–1163.

Fowler, J.S., Volkow, N.D., Wang, G.J., Pappas, N., Logan, J., MacGregor, R., Alexoff, D., Shea, C., Schlyer, D., Wolf, A.P., Warner, D., Zezulkova, I., Cilento, R., 1996a. Inhibition of monoamine oxidase B in the brains of smokers. Nature 379 (6567), 733–736.

Fowler, J.S., Volkow, N.D., Wang, G.J., Pappas, N., Logan, J., Shea, C., Alexoff, D., MacGregor, R.R., Schlyer, D.J., Zezulkova, I., Wolf, A.P., 1996b. Brain monoamine oxidase A inhibition in cigarette smokers. Proc. Natl. Acad. Sci. USA 93 (24), 14065–14069.

George, T.P., Vessicchio, J.C., Termine, A., Jatlow, P.I., Kosten, T.R., O'Malley, S.S., 2003. A preliminary placebo-controlled trial of selegiline hydrochloride for smoking cessation. Biol. Psychiatry 53 (2), 136–143.

Gilmore, J.H., Lin, W., Gerig, G., 2006. Fetal and neonatal brain development. Am. J. Psychiatry 163 (12), 2046.

Goldstein, R.Z., Volkow, N.D., 2002. Drug addiction and its underlying neurobiological basis: neuroimaging evidence for the involvement of the frontal cortex. Am. J. Psychiatry 159 (10), 1642–1652.

Grabowski, J., Roache, J.D., Schmitz, J.M., Rhoades, H., Creson, D., Korszun, A., 1997. Replacement medication for cocaine dependence: methylphenidate. J. Clin. Psychopharmacol. 17 (6), 485–488.

Gruber, S.A., Tzilos, G.K., Silveri, M.M., Pollack, M., Renshaw, P.F., Kaufman, M.J., Yurgelun-Todd, D.A., 2006. Methadone maintenance improves cognitive performance after two months of treatment. Exp. Clin. Psychopharmacol. 14 (2), 157–164.

Gu, H., Salmeron, B.J., Ross, T.J., Geng, X., Zhan, W., Stein, E.A., Yang, Y., 2011. Mesocorticolimbic circuits are impaired in chronic cocaine users as demonstrated by resting-state functional connectivity. Neuroimage 53 (2), 593–601.

Guillem, K., Vouillac, C., Azar, M.R., Parsons, L.H., Koob, G.F., Cador, M., Stinus, L., 2005. Monoamine oxidase inhibition dramatically increases the motivation to self-administer nicotine in rats. J. Neurosci. 25 (38), 8593–8600.

Guillem, K., Vouillac, C., Azar, M.R., Parsons, L.H., Koob, G.F., Cador, M., Stinus, L., 2006. Monoamine oxidase A rather than monoamine oxidase B inhibition increases nicotine reinforcement in rats. Eur. J. Neurosci. 24 (12), 3532–3540.

Gur, R.E., Maany, V., Mozley, P.D., Swanson, C., Bilker, W., Gur, R.C., 1998. Subcortical MRI volumes in neuroleptic-naive and treated patients with schizophrenia. Am. J. Psychiatry 155 (12), 1711–1717.

Halldin, C., Gulyas, B., Langer, O., Farde, L., 2001. Brain radioligands–state of the art and new trends. Q. J. Nucl. Med. 45 (2), 139–152.

Hariri, A.R., Mattay, V.S., Tessitore, A., Kolachana, B., Fera, F., Goldman, D., Egan, M.F., Weinberger, D.R., 2002. Serotonin transporter genetic variation and the response of the human amygdala. Science 297 (5580), 400–403.

Hetherington, H.P., Telang, F., Pan, J.W., Sammi, M., Schuhlein, D., Molina, P., Volkow, N.D., 1999. Spectroscopic imaging of the uptake kinetics of human brain ethanol. Magn. Reson. Med. 42 (6), 1019–1026.

Hill, S.Y., De Bellis, M.D., Keshavan, M.S., Lowers, L., Shen, S., Hall, J., Pitts, T., 2001. Right amygdala volume in adolescent and young adult offspring from families at high risk for developing alcoholism. Biol. Psychiatry 49 (11), 894–905.

Huang, H., Zhang, J., Wakana, S., Zhang, W., Ren, T., Richards, L.J., Yarowsky, P., Donohue, P., Graham, E., van Zijl, P.C., Mori, S., 2006. White and gray matter development in human fetal, newborn and pediatric brains. Neuroimage 33 (1), 27–38.

Jacobsen, L.K., Giedd, J.N., Gottschalk, C., Kosten, T.R., Krystal, J.H., 2001. Quantitative morphology of the caudate and putamen in patients with cocaine dependence. Am. J. Psychiatry 158 (3), 486–489.

Jernigan, T.L., Gamst, A.C., Archibald, S.L., Fennema-Notestine, C., Mindt, M.R., Marcotte, T.D., Heaton, R.K., Ellis, R.J., Grant, I., 2005. Effects of methamphetamine dependence and HIV infection on cerebral morphology. Am. J. Psychiatry 162 (8), 1461–1472.

Julian, M.D., Martin, A.B., Cuellar, B., Rodriguez De Fonseca, F., Navarro, M., Moratalla, R., Garcia-Segura, L.M., 2003. Neuroanatomical relationship between type 1 cannabinoid receptors and dopaminergic systems in the rat basal ganglia. Neuroscience 119 (1), 309–318.

Kaufman, J.N., Ross, T.J., Stein, E.A., Garavan, H., 2003. Cingulate hypoactivity in cocaine users during a GO-NOGO task as revealed by event-related functional magnetic resonance imaging. J. Neurosci. 23 (21), 7839–7843.

Kelly, C., Zuo, X.N., Gotimer, K., Cox, C.L., Lynch, L., Brock, D., Imperati, D., Garavan, H., Rotrosen, J., Castellanos, F.X., Milham, M.P., 2011. Reduced interhemispheric resting state functional connectivity in cocaine addiction. Biol. Psychiatry 69 (7), 684–692.

Khalil, A.A., Steyn, S., Castagnoli Jr, N., 2000. Isolation and characterization of a monoamine oxidase inhibitor from tobacco leaves. Chem. Res. Toxicol. 13 (1), 31–35.

Kim, S.J., Lyoo, I.K., Hwang, J., Chung, A., Hoon Sung, Y., Kim, J., Kwon, D.H., Chang, K.H., Renshaw, P.F., 2006. Prefrontal grey-matter changes in short-term and long-term abstinent methamphetamine abusers. Int. J. Neuropsychopharmacol. 9 (2), 221–228.

Kosten, T.R., Scanley, B.E., Tucker, K.A., Oliveto, A., Prince, C., Sinha, R., Potenza, M.N., Skudlarski, P., Wexler, B.E., 2006. Cue-induced brain activity changes and relapse in cocaine-dependent patients. Neuropsychopharmacology 31 (3), 644–650.

Kreek, M.J., LaForge, K.S., Butelman, E., 2002. Pharmacotherapy of addictions. Nat. Rev. Drug Discov. 1 (9), 710–726.

Kufahl, P.R., Li, Z., Risinger, R.C., Rainey, C.J., Wu, G., Bloom, A.S., Li, S.J., 2005. Neural responses to acute cocaine administration in the human brain detected by fMRI. Neuroimage 28 (4), 904–914.

Kung, H.F., Kung, M.P., Choi, S.R., 2003. Radiopharmaceuticals for single-photon emission computed tomography brain imaging. Semin. Nucl. Med. 33 (1), 2–13.

Laruelle, M., Abi-Dargham, A., van Dyck, C.H., Rosenblatt, W., Zea-Ponce, Y., Zoghbi, S.S., Baldwin, R.M., Charney, D.S., Hoffer, P.B., Kung, H.F., et al., 1995. SPECT imaging of striatal dopamine release after amphetamine challenge. J. Nucl. Med. 36 (7), 1182–1190.

Leshner, A.I., 1997. Addiction is a brain disease, and it matters. Science 278 (5335), 45–47.

Liu, X., Matochik, J.A., Cadet, J.L., London, E.D., 1998. Smaller volume of prefrontal lobe in polysubstance abusers: a magnetic resonance imaging study. Neuropsychopharmacology 18 (4), 243–252.

Makris, N., Gasic, G.P., Seidman, L.J., Goldstein, J.M., Gastfriend, D.R., Elman, I., Albaugh, M.D., Hodge, S.M., Ziegler, D.A., Sheahan, F.S., Caviness Jr., V.S., Tsuang, M.T., Kennedy, D.N., Hyman, S.E., Rosen, B.R., Breiter, H.C., 2004. Decreased absolute amygdala volume in cocaine addicts. Neuron 44 (4), 729–740.

Mann, K., Agartz, I., Harper, C., Shoaf, S., Rawlings, R.R., Momenan, R., Hommer, D.W., Pfefferbaum, A., Sullivan, E.V., Anton, R.F., Drobes, D.J., George, M.S., Bares, R., Machulla, H.J., Mundle, G., Reimold, M., Heinz, A., 2001. Neuroimaging in alcoholism: ethanol and brain damage. Alcohol. Clin. Exp. Res. 25 (5 Suppl. ISBRA), 104S–109S.

Matochik, J.A., London, E.D., Eldreth, D.A., Cadet, J.L., Bolla, K.I., 2003. Frontal cortical tissue composition in abstinent cocaine abusers: a magnetic resonance imaging study. Neuroimage 19 (3), 1095–1102.

Mattay, V.S., Goldberg, T.E., Fera, F., Hariri, A.R., Tessitore, A., Egan, M.F., Kolachana, B., Callicott, J.H., Weinberger, D.R., 2003. Catechol O-methyltransferase val158-met genotype and individual variation in the brain response to amphetamine. Proc. Natl. Acad. Sci. USA 100 (10), 6186–6191.

Morens, D.M., Grandinetti, A., Reed, D., White, L.R., Ross, G.W., 1995. Cigarette smoking and protection from Parkinson's disease: false association or etiologic clue? Neurology 45 (6), 1041–1051.

O'Neill, J., Cardenas, V.A., Meyerhoff, D.J., 2001. Effects of abstinence on the brain: quantitative magnetic resonance imaging and magnetic resonance spectroscopic imaging in chronic alcohol abuse. Alcohol. Clin. Exp. Res. 25 (11), 1673–1682.

Paulus, M.P., Hozack, N.E., Zauscher, B.E., Frank, L., Brown, G.G., Braff, D.L., Schuckit, M.A., 2002. Behavioral and functional neuroimaging evidence for prefrontal dysfunction in methamphetamine-dependent subjects. Neuropsychopharmacology 26 (1), 53–63.

Paulus, M.P., Tapert, S.F., Schuckit, M.A., 2005. Neural activation patterns of methamphetamine-dependent subjects during decision making predict relapse. Arch. Gen. Psychiatry 62 (7), 761–768.

Pfefferbaum, A., Sullivan, E.V., 2005. Disruption of brain white matter microstructure by excessive intracellular and extracellular fluid in alcoholism: evidence from diffusion tensor imaging. Neuropsychopharmacology 30 (2), 423–432.

Pfefferbaum, A., Sullivan, E.V., Mathalon, D.H., Shear, P.K., Rosenbloom, M.J., Lim, K.O., 1995. Longitudinal changes in magnetic resonance imaging brain volumes in abstinent and relapsed alcoholics. Alcohol. Clin. Exp. Res. 19 (5), 1177–1191.

Pfefferbaum, A., Sullivan, E.V., Rosenbloom, M.J., Mathalon, D.H., Lim, K.O., 1998. A controlled study of cortical gray matter and ventricular changes in alcoholic men over a 5-year interval. Arch. Gen. Psychiatry 55 (10), 905–912.

Rae, C., Lee, M.A., Dixon, R.M., Blamire, A.M., Thompson, C.H., Styles, P., Talcott, J., Richardson, A.J., Stein, J.F., 1998. Metabolic abnormalities in developmental dyslexia detected by 1H magnetic resonance spectroscopy. Lancet 351 (9119), 1849–1852.

Raichle, M.E., Martin, W.R., Herscovitch, P., Mintun, M.A., Markham, J., 1983. Brain blood flow measured with intravenous H2(15)O. II. Implementation and validation. J. Nucl. Med. 24 (9), 790–798.

Raichle, M.E., Snyder, A.Z., 2007. A default mode of brain function: a brief history of an evolving idea. Neuroimage 37 (4), 1083–1090 discussion 1097–1089.

Rinne, J.O., Laihinen, A., Nagren, K., Ruottinen, H., Ruotsalainen, U., Rinne, U.K., 1995. PET examination of the monoamine transporter with [11C]beta-CIT and [11C]beta-CFT in early Parkinson's disease. Synapse 21 (2), 97–103.

Risinger, R.C., Salmeron, B.J., Ross, T.J., Amen, S.L., Sanfilipo, M., Hoffmann, R.G., Bloom, A.S., Garavan, H., Stein, E.A., 2005. Neural correlates of high and craving during cocaine self-administration using BOLD fMRI. Neuroimage 26 (4), 1097–1108.

Ross, B., Kreis, R., Ernst, T., 1992. Clinical tools for the 90s: magnetic resonance spectroscopy and metabolite imaging. Eur. J. Radiol. 14 (2), 128–140.

Schlaepfer, T.E., Lancaster, E., Heidbreder, R., Strain, E.C., Kosel, M., Fisch, H.U., Pearlson, G.D., 2006. Decreased frontal white-matter volume in chronic substance abuse. Int. J. Neuropsychopharmacol. 9 (2), 147–153.

Sekine, Y., Minabe, Y., Ouchi, Y., Takei, N., Iyo, M., Nakamura, K., Suzuki, K., Tsukada, H., Okada, H., Yoshikawa, E., Futatsubashi, M., Mori, N., 2003. Association of dopamine transporter loss in the orbitofrontal and dorsolateral prefrontal cortices with methamphetamine-related psychiatric symptoms. Am. J. Psychiatry 160 (9), 1699–1701.

Shearer, J., 2008. The principles of agonist pharmacotherapy for psychostimulant dependence. Drug Alcohol Rev. 27 (3), 301–308.

Silveri, M.M., Pollack, M.H., Diaz, C.I., Nassar, L.E., Mendelson, J.H., Yurgelun-Todd, D.A., Renshaw, P.F., Kaufman, M.J., 2004. Cerebral phosphorus metabolite and transverse relaxation time abnormalities in heroin-dependent subjects at onset of methadone maintenance treatment. Psychiatry Res. 131 (3), 217–226.

Smith, L.M., Chang, L., Yonekura, M.L., Gilbride, K., Kuo, J., Poland, R.E., Walot, I., Ernst, T., 2001a. Brain proton magnetic resonance spectroscopy and imaging in children exposed to cocaine in utero. Pediatrics 107 (2), 227–231.

Smith, L.M., Chang, L., Yonekura, M.L., Grob, C., Osborn, D., Ernst, T., 2001b. Brain proton magnetic resonance spectroscopy in children exposed to methamphetamine in utero. Neurology 57 (2), 255–260.

Stapleton, J.M., Morgan, M.J., Phillips, R.L., Wong, D.F., Yung, B.C., Shaya, E.K., Dannals, R.F., Liu, X., Grayson, R.L., London, E.D., 1995. Cerebral glucose utilization in polysubstance abuse. Neuropsychopharmacology 13 (1), 21–31.

Thompson, P.M., Hayashi, K.M., Simon, S.L., Geaga, J.A., Hong, M.S., Sui, Y., Lee, J.Y., Toga, A.W., Ling, W., London, E.D., 2004. Structural abnormalities in the brains of human subjects who use methamphetamine. J. Neurosci. 24 (26), 6028–6036.

Volkow, N.D., Chang, L., Wang, G.J., Fowler, J.S., Ding, Y.S., Sedler, M., Logan, J., Franceschi, D., Gatley, J., Hitzemann, R., Gifford, A., Wong, C., Pappas, N., 2001a. Low level of brain dopamine D2 receptors in methamphetamine abusers: association with metabolism in the orbitofrontal cortex. Am. J. Psychiatry 158 (12), 2015–2021.

Volkow, N.D., Chang, L., Wang, G.J., Fowler, J.S., Leonido-Yee, M., Franceschi, D., Sedler, M.J., Gatley, S.J., Hitzemann, R., Ding, Y.S., Logan, J., Wong, C., Miller, E.N., 2001b. Association of dopamine transporter reduction with psychomotor impairment in methamphetamine abusers. Am. J. Psychiatry 158 (3), 377–382.

Volkow, N.D., Ding, Y.S., Fowler, J.S., Wang, G.J., Logan, J., Gatley, S.J., Schlyer, D.J., Pappas, N., 1995. A new PET ligand for the dopamine transporter: studies in the human brain. J. Nucl. Med. 36 (12), 2162–2168.

Volkow, N.D., Fowler, J.S., Gatley, S.J., Logan, J., Wang, G.J., Ding, Y.S., Dewey, S., 1996a. PET evaluation of the dopamine system of the human brain. J. Nucl. Med. 37 (7), 1242–1256.

Volkow, N.D., Fowler, J.S., Wang, G.J., 2003a. The addicted human brain: insights from imaging studies. J. Clin. Invest. 111 (10), 1444–1451.

Volkow, N.D., Fowler, J.S., Wang, G.J., 2003b. Positron emission tomography and single-photon emission computed tomography in substance abuse research. Semin. Nucl. Med. 33 (2), 114–128.

Volkow, N.D., Fowler, J.S., Wang, G.J., Hitzemann, R., Logan, J., Schlyer, D.J., Dewey, S.L., Wolf, A.P., 1993. Decreased dopamine D2 receptor availability is associated with reduced frontal metabolism in cocaine abusers. Synapse 14 (2), 169–177.

Volkow, N.D., Fowler, J.S., Wang, G.J., Swanson, J.M., Telang, F., 2007. Dopamine in drug abuse and addiction: results of imaging studies and treatment implications. Arch. Neurol. 64 (11), 1575–1579.

Volkow, N.D., Fowler, J.S., Wolf, A.P., Hitzemann, R., Dewey, S., Bendriem, B., Alpert, R., Hoff, A., 1991. Changes in brain glucose metabolism in cocaine dependence and withdrawal. Am. J. Psychiatry 148 (5), 621–626.

Volkow, N.D., Fowler, J.S., Wolf, A.P., Schlyer, D., Shiue, C.Y., Alpert, R., Dewey, S.L., Logan, J., Bendriem, B., Christman, D., et al., 1990. Effects of chronic cocaine abuse on postsynaptic dopamine receptors. Am. J. Psychiatry 147 (6), 719–724.

Volkow, N.D., Wang, G.J., Fischman, M.W., Foltin, R.W., Fowler, J.S., Abumrad, N.N., Vitkun, S., Logan, J., Gatley, S.J., Pappas, N., Hitzemann, R., Shea, C.E., 1997. Relationship between subjective effects of cocaine and dopamine transporter occupancy. Nature 386 (6627), 827–830.

Volkow, N.D., Wang, G.J., Fowler, J.S., Logan, J., Gatley, S.J., Wong, C., Hitzemann, R., Pappas, N.R., 1999. Reinforcing effects of psychostimulants in humans are associated with increases in brain dopamine and occupancy of D(2) receptors. J. Pharmacol. Exp. Ther. 291 (1), 409–415.

Volkow, N.D., Wang, G.J., Fowler, J.S., Logan, J., Hitzemann, R., Ding, Y.S., Pappas, N., Shea, C., Piscani, K., 1996b. Decreases in dopamine receptors but not in dopamine transporters in alcoholics. Alcohol. Clin. Exp. Res. 20 (9), 1594–1598.

Wang, G.J., Volkow, N.D., Chang, L., Miller, E., Sedler, M., Hitzemann, R., Zhu, W., Logan, J., Ma, Y., Fowler, J.S., 2004. Partial recovery of brain metabolism in methamphetamine abusers after protracted abstinence. Am. J. Psychiatry 161 (2), 242–248.

Wang, G.J., Volkow, N.D., Fowler, J.S., Logan, J., Abumrad, N.N., Hitzemann, R.J., Pappas, N.S., Pascani, K., 1997. Dopamine D2 receptor availability in opiate-dependent subjects before and after naloxone-precipitated withdrawal. Neuropsychopharmacology 16 (2), 174–182.

Warner, T.D., Behnke, M., Eyler, F.D., Padgett, K., Leonard, C., Hou, W., Garvan, C.W., Schmalfuss, I.M., Blackband, S.J., 2006. Diffusion tensor imaging of frontal white matter and executive functioning in cocaine-exposed children. Pediatrics 118 (5), 2014–2024.

Wexler, B.E., Gottschalk, C.H., Fulbright, R.K., Prohovnik, I., Lacadie, C.M., Rounsaville, B.J., Gore, J.C., 2001. Functional magnetic resonance imaging of cocaine craving. Am. J. Psychiatry 158 (1), 86–95.

Wong, D.F., Yung, B., Dannals, R.F., Shaya, E.K., Ravert, H.T., Chen, C.A., Chan, B., Folio, T., Scheffel, U., Ricaurte, G.A., et al., 1993. In vivo imaging of baboon and human dopamine transporters by positron emission tomography using [11C]WIN 35,428. Synapse 15 (2), 130–142.

Wrase, J., Makris, N., Braus, D.F., Mann, K., Smolka, M.N., Kennedy, D.N., Caviness, V.S., Hodge, S.M., Tang, L., Albaugh, M., Ziegler, D.A., Davis, O.C., Kissling, C., Schumann, G., Breiter, H.C., Heinz, A., 2008. Amygdala volume associated with alcohol abuse relapse and craving. Am. J. Psychiatry 165 (9), 1179–1184.

Yang, S., Salmeron, B.J., Ross, T.J., Xi, Z.X., Stein, E.A., Yang, Y., 2009. Lower glutamate levels in rostral anterior cingulate of chronic cocaine users – a (1)H-MRS study using TE-averaged PRESS at 3 T with an optimized quantification strategy. Psychiatry Res. 174 (3), 171–176.

Zubieta, J.K., Gorelick, D.A., Stauffer, R., Ravert, H.T., Dannals, R.F., Frost, J.J., 1996. Increased mu opioid receptor binding detected by PET in cocaine-dependent men is associated with cocaine craving. Nat. Med. 2 (11), 1225–1229.

CHAPTER TWO

Brain PET Imaging in the Cannabinoid System

Dean F. Wong[1,2,3,4,5], Ayon Nandi[1], Eram Zaidi[1], Emily Gean[1], Andrew Horti[1], Boris Frolov[1], Noble George[1] and James R. Brasic[1]

[1]The Russell H. Morgan Department of Radiology and Radiological Science, Johns Hopkins University School of Medicine, Baltimore, MD, USA
[2]Department of Psychiatry, Johns Hopkins University School of Medicine, Baltimore, MD, USA
[3]Department of Neuroscience, Johns Hopkins University School of Medicine, Baltimore, MD, USA
[4]Department of Environmental Health Sciences, Johns Hopkins University School of Medicine, Baltimore, MD, USA
[5]Carey School of Business, Johns Hopkins University School of Medicine, Baltimore, MD, USA

1. INTRODUCTION

The cannabinoid system consists of two receptors that have been characterized and cloned: cannabinoid type 1 receptor (CB_1R), predominantly located in the neural tissue and, to a lesser extent, in the peripheral tissues; and CB_2, which is found primarily in the peripheral tissues (Howlett et al., 2002). CB_1Rs have been implicated in reward and neuropsychiatric illnesses. Antagonists or inverse agonists of cannabinoid receptors have been shown to reduce eating in obese individuals and to decrease drug taking in individuals with substance abuse problems (Janero, 2012; Rodriguez et al., 2011; Gorelick et al., 2011).

Evidence suggests that the reward system, specifically the dopaminergic system, interacts significantly with cannabinoid systems via the basal ganglia (El Khoury et al., 2012). CB_1R is involved in the pathways of nondrug rewards such as food, but is also involved in the rewarding properties of drugs. CB_1R antagonists and inverse agonists reduce the rewarding effects of food (Cota et al., 2003) and drugs of abuse (Le Foll and Goldberg, 2004; Caillé et al., 2007). Further, there is evidence to suggest that the cannabinoid system regulates emotional homeostasis and, hence, psychiatric disorders such as schizophrenia and mood disorders, via noradrenergic circuits (Carvalho and Van Bockstaele, 2012; Häring et al., 2012).

Positron emission tomography (PET) imaging allows one to track the binding of neurotransmitters to neuroreceptors, and is thus used to study cannabinoid receptors. This chapter will review PET tracers used in cannabinoid research and the applications of such findings to the treatment of obesity and other eating disorders, substance abuse, schizophrenia, and mood disorders.

Imaging of the Human Brain in Health and Disease
http://dx.doi.org/10.1016/B978-0-12-418677-4.00002-6

2.1. FIRST ATTEMPTS AT IMAGING CB_1R WITH PET: A HISTORICAL PERSPECTIVE

Creating an effective method of imaging the central CB_1R using PET has been attempted since the initial studies utilizing ^{18}F-labeled D9-tetrahydrocannabinol (D9-THC) in 1991 (Charalambous et al., 1991). However, the quantitative PET imaging of CB_1R in the mammalian brain of both animal and human models is yet to be accomplished despite ample CB_1R quantity and density (Howlett et al., 2002). To date, CB_1R imaging attempts have been met with significant difficulties due to poor blood–brain barrier (BBB) permeability, nonspecificity of binding, fast washout, and poor target-to-background ratios.

The quantification of CB_1R is particularly complicated due to the poor permeability of the BBB. While the factors contributing to BBB permeation are not fully understood, it is certain to be affected by lipophilicity, hydrogen bonding capacity, molecular charge, size, shape, and flexibility. It is well established that an optimal lipophilicity is a requirement for sufficient BBB permeability (Banks and Kastin, 1985). Due to their hydrophobic interaction with the receptor-binding site, cannabinoids exhibit high lipophilicities, which greatly exceed the optimal range (Clark, 2003).

Lipophilicity is also dependent on the binding site's location within the cell membrane (Shire et al., 1999). A correlation between affinity and lipophilicity has been shown for inverse agonists of CB_1R (e.g. rimonabant) (Katoch-Rouse et al., 2003). This discovery further implicates the difficulties in developing a high-affinity CB_1R radioligand with relatively low lipophilicity, resulting in a lower nonspecific binding and increased BBB permeation.

Lipophilicity is a major determinant of the effectiveness of a developed radiotracer. Increased lipophilicity is directly associated with greater nonspecific binding, and thus, low binding potential (BP), despite adequate ligand binding affinity (Waterhouse, 2003). Many PET/Single Positron Emission Computed Tomography (SPECT) CB_1R radioligands, which are predominantly analogs of rimonabant and agonist WIN 55, 212-2, possess greater lipophilicities than do radiotracers for imaging other central nervous system receptors, and therefore still cannot be successfully imaged (Mathews et al., 2002). As of 2000, the most efficacious CB_1R radiotracer ratio of target-to-nontarget region was in the mouse brain using ^{18}F-SR144385 (Mathews et al., 2000), and in 2002 in the rhesus monkey brain using ^{11}C-NIDA41020 (Katoch-Rouse et al., 2002). Additionally, the first attempt at quantifying CB_1R in the living human brain yielded a low BP using the radioligand ^{123}I-AM281 (Berding et al., 2004).

A recent study synthesized ^{11}C-JHU75528 (4-cyano-1-(2,4-dichlorophenyl)-5-(4-$^{11}CH_3$-methoxyphenyl)-*N*-(piperidin-1-yl)-1*H*-pyrazole-3-carboxamide) ([^{11}C]

OMAR), a novel analog of rimonabant with a unique combination of a higher binding affinity and membranes from HEK-293 cells expressing the human recombinant CB_1R ($K_i = 11$ nM) and a lower lipophilicity than those of rimonabant (Horti et al., 2006).

2.2. THE CURRENTLY AVAILABLE PET RADIOLIGANDS

^{11}C-JHU75528 ([^{11}C]OMAR)

The radiotracer [^{11}C]OMAR, an analog of rimonabant, is a selective CB_1R antagonist developed at Johns Hopkins University. Compared to other CB_1R tracers, OMAR has a higher affinity and reduced lipophilicity in the human brain (Horti et al., 2006).

[^{11}C]OMAR demonstrated excellent passage through the BBB and readily enters the human brain. The radiotracer showed good uptake in the regional brain distribution that is consistent with that of CB_1R and correlated well with postmortem studies and CB_1R have been linked to neuropsychiatric illnesses, such as schizophrenia. [^{11}C]OMAR has been shown to be a viable tracer for CB_1R in subjects with schizophrenia (Wong et al., 2010) (Figures 1 and 2).

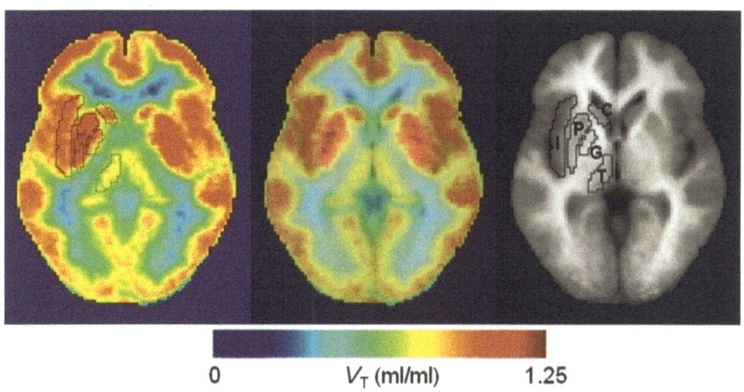

Figure 1 Transaxial images of the distribution volume (V_T) of [^{11}C]OMAR, mean of 10 healthy subjects in a standard space (left panel), a standard magnetic resonance imaging (MRI; right), and merged image (Middle). Volumes of interest of the selected structures in the standard space are shown on the V_T image and MRI. Regions are insula (I), putamen (P), globus pallidus (G), thalamus (T), and caudate nucleus (C). The highest uptake was in the globus pallidus, with a mean of 1.47 ± 0.25 (SD), with a range of 0.95–1.82 in individual subjects. The cingulate cortex (mean 1.23 ± 0.16 (SD), range of 1.01–1.49) and the putamen (mean 1.32 ± 0.20 (SD), range of 1.05–1.63) also showed high binding. The intermediate uptake was in the hippocampus, cerebellum, and cortex. The lowest accumulation of radioactivity was seen in the white matter, the pons, and the thalamus (Fan et al., 2006). *Image courtesy of Wong et al. (2010).*

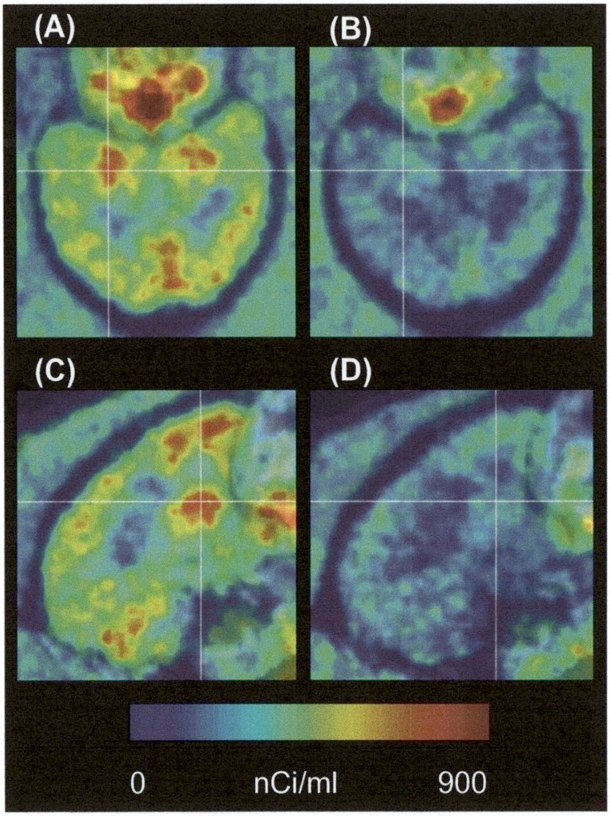

Figure 2 [^{11}C]OMAR baboon PET images on aligned MRI images. (A) Transaxial image, baseline. (B) Transaxial image, blocking with rimonabant (1 mg/kg, intravenously). (C) Sagittal image, baseline. (D) Sagittal image, blocking with rimonabant. *Courtesy of Horti et al. (2006).*

MK-9470

[^{18}F]MK-9470 (*N*-[2-(3-cyano-phenyl)-3-(4-(2-[^{18}F]fluorethoxy)phenyl)-1-methylpropyl]-2-(5-methyl-2-pyridyloxy)-2-methyl propanamide) is an inverse agonist at the CB_1R receptor and has a high specificity and high affinity for the receptor (Lin et al., 2008).

PET imaging studies in rhesus monkeys showed high brain uptake and a distribution pattern generally consistent with that seen in autoradiographic studies, which showed high specific binding in the cerebral cortex, cerebellum, caudate/putamen, globus pallidus, substantia nigra, and hippocampus (Burns, 2007).

In the human brain, [^{18}F]MK-9470 exhibited relatively slow brain kinetics reaching a plateau at approximately 120 min after bolus injection. Tracer uptake was observed in all gray matter regions and remained relatively constant from 120 to 360 min after tracer injection (Van Laere, 2008a). The greatest levels of uptake were observed in the striatum, frontal cortex, and posterior cingulate, whereas intermediate uptake was seen in the cerebellum, and the lowest uptake was observed in the thalamus and hippocampus.

To characterize the kinetics and reproducibility of the CB_1R tracer [^{18}F]MK-9470 in the human brain, [^{18}F]MK-9470 data were analyzed using reversible models and the distribution volume, tracer binding, and fractional uptake rate (FUR) were estimated. The effect of blood flow on these parameters was also evaluated. Results showed that [^{18}F]MK-9470-specific binding can be accurately determined using FUR values and required a short scan 90–120 min after tracer administration. Using a few venous samples, [^{18}F]MK-9470 plasma kinetics can also be assessed (Sanabria-Bohórquez, 2010).

[^{18}F]MK-9470 PET has been shown to bind to CB_1R in patient populations, specifically in subjects with eating disorders. In a study of women diagnosed with anorexia nervosa (AN) and bulimia compared with healthy controls, [^{18}F]MK-9470 PET imaging showed that global CB_1R availability was significantly increased in cortical and subcortical brain areas in AN patients compared with healthy control subjects and regional CB_1R availability increased in the insula in both AN and bulimia nervosa patients and the inferior frontal and temporal cortex in AN patients only (Gérard, 2011).

[^{18}F]MK-9470 has also been used to test the relationship between human temperament traits and regional cerebral CB_1R availability. Low baseline cerebral CB_1R availability was related to a high novelty-seeking personality, in particular to extravagance. This was the most pronounced in the amygdala (Van Laere, 2010). [^{18}F]MK-9470 has also been used to assess the binding to CB_1R in normal aging. CB_1R binding increased with aging but only in women. This was the most pronounced in the basal ganglia, lateral temporal cortex, and limbic system, especially in the hippocampus. Men showed a higher [^{18}F]MK-9470 binding than did women in the clusters of the limbic system and cortico–striato–thalamic–cortical circuit (Van Laere, 2008b). In rat brain, [^{18}F]MK-9470 PET showed no effect of chronic nicotine administration on CB_1R binding (Gérard, 2010).

[^{11}C]MePPEP

[^{11}C]MePPEP ((3*R*,5*R*)-5-(3-methoxyphenyl)-3-((*R*)-1-phenyl-ethylamino)-1-(4-trifluoromethyl-phenyl)-pyrrolidin-2-one) ([^{11}C]MePPEP) is a radiotracer that binds to CB_1R. In nonhuman primates, it has high CB_1R affinity and moderately high lipophilicity. The regional uptake is dependent on the distribution of CB_1R, with the highest uptake in the striatum and the cerebellum but a low uptake in the thalamus and the pons. Preblockade/displacement with two CB_1R selective agents, benzamide and rimonabant, demonstrates the majority of brain uptake in regions with high receptor densities. It was specific and reversibly bound to CB_1R in the high-binding regions. [^{11}C]MePPEP demonstrates fast clearance from plasma. Regional brain uptake was quantified as distribution volume relative to the concentration in plasma (Yasuno et al., 2008).

In the human brain, [^{11}C]MePPEP brain uptake of radioactivity was high and washed out very slowly. A two-tissue compartment model produced numbers of distribution volume that were well identifiable and stable between 60 and 210 min (Terry et al., 2009).

2.3. APPLICATIONS OF CANNABINOID IMAGING

2.3.1. Overview

Endocannabinoids are involved in reward systems, and with emotional processing. CB_1R, in particular, plays a role in conditions including schizophrenia, obesity, drug and alcohol abuse, and depression. Imaging, then, can inform pharmacologic treatment development.

2.3.2. Schizophrenia

A current theory posits that cannabis use may cause or exacerbate psychosis. In situ radioligand binding and autoradiography revealed an increased density of cannabinoid-1 receptor binding in the dorsolateral prefrontal cortex in schizophrenia, regardless of cannabis ingestion, suggesting that schizophrenia patients have altered endocannabinoid systems (Dean et al., 2001). In support of this is the finding that the polymorphism of the CNR1 gene that encodes the human CB_1R is associated with schizophrenia, and, particularly, the hebephrenic subtype (Ujike et al., 2002).

Additionally, CB_1R, together with dopamine (DA), are involved in the ventral tegmental area (VTA)–basolateral amygdala (BLA)–medial prefrontal cortex (mPFC) pathway critical in emotionally salient associative learning and memory processing. Disturbances in this pathway are associated with schizophrenia (Laviolette and Grace, 2006). Abnormalities in the CB_1R of patients with schizophrenia are also found in the posterior cingulate cortex (PCC), with an increase in receptor density in schizophrenia patients, compared with that in controls. CB_1R located presynaptically on gamma-aminobutyric acid (GABA) interneurons in the PCC and modulate GABA neurotransmission. PCC in schizophrenic patients showed weaker activation during semantic memory tasks, than in controls. Memory is also compromised in both schizophrenia and cannabis use (Newell et al., 2006). The anterior cingulate cortex (ACC) is involved in normal motivation and attention. Motivation and attention impairment are primary negative symptoms in schizophrenia, and is also an effect of long-term cannabis use. In the ACC of postmortem schizophrenia patients, there is increased CB_1R binding compared to that in controls (Zavitsanou et al., 2004).

2.3.3. Eating Disorders

CB_1R are also involved in eating behavior. CB_1R knockout mice eat less than do wild types, and CB_1R antagonists reduce the food intake in wild-type mice, but not in knockout mice (Di Marzo et al., 2001). CB_1R antagonists and inverse agonists also suppress operant responding for food rewards (Cooper, 2004). In a 12-week study in mice with CB_1R inverse agonist, taranabant, the animals lost weight over the entire range of tested doses (Addy et al., 2008). CB_1R antagonists to aid weight loss are being developed in humans (Cota et al., 2003). The most advanced of such compounds is the inverse agonist, rimonabant, with which the reduction in body weight, waist circumference, and improved lipid and glucose metabolism in obese individuals has been observed (Vickers and Kennett, 2005). CB_1R research may also help treat forms of anorexia. Restricting type AN and binging/purging type are associated with distinct alleles of the CNR1 gene (Siegfried et al., 2004).

2.3.4. Drug and Alcohol Abuse

There is also evidence that CB_1Rs are involved in drug abuse. In alcohol-dependent men, CB_1R density (V_T), measured with [^{11}C]OMAR, was elevated by 20% in several regions, compared to those of controls (Neumeister et al., 2012). In another study, CB_1R antagonists decreased the drug seeking of THC, nicotine, cocaine (Le Foll and Goldberg, 2004), ethanol, and heroin (Caillé et al., 2007). CB_1R antagonists such as rimonabant have potential for clinical application.

2.3.5. Depression

There are alterations in endocannabinoid levels and in the density of CB_1R in the prefrontal cortex of depressed and alcoholic suicide victims (Vinod and Hungund, 2006).

2.3.6. Posttraumatic Stress Disorder

Posttraumatic stress disorder patients administered [^{11}C]OMAR PET scans had elevated CB_1 binding compared to controls in brain areas associated with the condition (Bailey et al., 2012).

3. CONCLUSIONS

There is a dearth of knowledge about the endocannabinoid system and its role in neuropsychiatric disorders. Further investigation of cannabinoids in the brain requires PET imaging with effective radioligands. Three relatively new second-generation radiotracers currently employed for brain cannabinoid receptor imaging are an improvement over their predecessors, but still lack adequate binding affinity, lipophilicity, and specificity. Currently, analogs of [^{11}C]OMAR with improved characteristics are being developed and tested (Caillé et al., 2007). Refinement of such CB_1 ligands will ultimately aid in the investigation and treatment of disorders including schizophrenia, eating disorders, substance abuse, depression, and posttraumatic stress disorder.

REFERENCES

Addy, C., Wright, H., Laere, K.V., et al., 2008. The acyclic CB_1R inverse agonist taranabant mediates weight loss by increasing energy expenditure and decreasing caloric intake. Cell Metab. 7, 69–78.

Bailey, C., Normandin, M., Henry, S., Greene, A., Potenza, M., Carson, R., Huang, Y., Neumeister, A. The Use of [^{11}C]OMAR and positron emission tomography to investigate the pathophysiology of posttraumatic stress disorder. Presented at the Neuroreceptor Mapping Meeting in Baltimore, MD, August 2012.

Banks, W.A., Kastin, A.J., 1985. Peptides and the blood–brain barrier: lipophilicity as a predictor of permeability. Brain Res. Bull. 15 (3), 287–292.

Berding, G., Muller-Vahl, K., Schneider, U., et al., 2004. [^{123}I]AM281 single-photon emission computed tomography imaging of central cannabinoid CB_1 receptors before and after Δ^9-tetrahydrocannabinol therapy and whole-body scanning for assessment of radiation dose in tourette patients. Biol. Psychiatry 55, 904–915.

Burns, H.D., 2007. [^{18}F]MK-9470, a positron emission tomography (PET) tracer for in vivo human PET brain imaging of the cannabinoid-1 receptor. Proc. Natl. Acad. Sci. USA 104, 9800–9805.

Caillé, S., Alvarez-Jaimes, L., Polis, I., et al., 2007. Specific alterations of extracellular endocannabinoid levels in the nucleus accumbens by ethanol, heroin, and cocaine self-administration. J. Neurosci. 27 (14), 3695–3702.

Carvalho, A.F., Van Bockstaele, E.J., 2012. Cannabinoid modulation of noradrenergic circuits: implications for psychiatric disorders. Prog. Neuropsychopharmacol. Biol. Psychiatry (Epub ahead of print).

Charalambous, A., Marciniak, G., Shiue, C.Y., et al., 1991. PET studies in the primate brain and biodistribution in mice using (-)-5′-^{18}F-Δ^9-THC. Pharmacol. Biochem. Behav. 40, 503–507.

Clark, D.E., 2003. In silico prediction of blood–brain barrier permeation. Drug Discov. Today 8, 927–933.

Cooper, S.J., 2004. Endocannabinoids and food consumption: comparisons with benzodiazepine and opioid palatability-dependent appetite. Eur. J. Pharmacol. 500, 37–49.

Cota, D., Marsicano, G., Lutz, B., et al., 2003. Endogenous cannabinoid system as a modulator of food intake. Int. J. Obes. Relat. Metab. Disord. 27 (3), 289–301.

Dean, B., Sundram, S., Bradbury, R., et al., 2001. Studies on [^{3}H]CP-55940 binding in the human central nervous system: regional specific changes in density of cannabinoid-1 receptors associated with schizophrenia and cannabis use. Neuroscience 103, 9–15.

Di Marzo, V., Goparaju, S.K., Wang, L., et al., 2001. Leptin-regulated endocannabinoids are involved in maintaining food intake. Nature 410, 822–825.

El Khoury, M., Gorgievski, V., Moutsimilli, L., et al., 2012. Interaction between the cannabinoid and dopaminergic system: evidence from animal studies. Prog. Neuropsychopharmacol. Biol. Psychiatry 38 (1), 36–50 (Epub ahead of print).

Fan, H., Ravert, H.T., Holt, D., et al., 2006. Synthesis of 1-(2,4-dichlorophenyl)-4-cyano-5-(4-[^{11}C]methoxyphenyl)-*N*-(piperidin-1-yl)-1*H*-pyrazole-3-carboxamide ([^{11}C]JHU75528) and 1-(2bromophenyl)-4-cyano-5-(4-[^{11}C]methoxyphenyl)-*N*-(piperidin-1-yl)-1*H*-pyrazole-3-carboxamide ([^{11}C]JHU75575) as potential radioligands for PET imaging of cerebral cannabinoid receptor. J. Labelled Compd. Radiopharm. 49, 1021–1036.

Gérard, N., 2010. Influence of chronic nicotine administration on cerebral type 1 cannabinoid receptor binding: an in vivo micro-PET study in the rat using [^{18}F]MK-9470. J. Mol. Neurosci. 42, 162–167.

Gérard, N., 2011. Brain type 1 cannabinoid receptor availability in patients with anorexia and bulimia nervosa. Biol. Psychiatry 70, 777–784.

Gorelick, D.A., Goodwin, R.S., Schwilke, E., et al., 2011. Antagonist-elicited cannabis withdrawal in humans. J. Clin. Psychopharmacol. 31 (5), 603–612.

Häring, M., Guggenhuber, S., Lutz, B., 2012. Neuronal populations mediating the effects of endocannabinoids on stress and emotionality. Neuroscience 204, 145–158.

Horti, A.G., Fan, H., Kuwabara, H., Hilton, J., Ravert, H.T., Holt, D.P., et al., 2006. ^{11}C-JHU75528 a radiotracer for PET imaging of CB_1 cannabinoid receptors. J. Nucl. Med. 47, 1689–1696.

Howlett, A.C., Barth, F., Bonner, T.I., et al., 2002. International Union of Pharmacology. XXVII. Classification of cannabinoid receptors. Pharmacol. Rev. 54, 161–202.

Janero, D.R., 2012. Cannabinoid-1 receptor (CB_1R) blockers as medicines: beyond obesity and cardiometabolic disorders to substance abuse/drug addiction with CB_1R neutral antagonists. Expert Opin. Emerging Drugs, 1–13.

Katoch-Rouse, R., Chefer, S.I., Pavlova, O.A., et al., 2002. Development of C-11-NIDA-41020: a potential radiotracer for studying cerebral cannabinoid receptors (CB_1) by PET. Presented at: IX Symposium on the Medical Applications of Cyclotrons; May 25–28; Turku, Finland.

Katoch-Rouse, R., Pavlova, O.A., Caulder, T., Hoffman, A.F., Mukhin, A.G., Horti, A.G., 2003. Synthesis, structure–activity relationship, and evaluation of SR141716 analogues: development of central cannabinoid receptor ligands with lower lipophilicity. J. Med. Chem. 46, 642–645.

Laviolette, S.R., Grace, A.A., 2006. The roles of cannabinoid and dopamine receptor systems in neural emotional learning circuits: implications for schizophrenia and addiction. Cell. Mol. Life Sci. 63, 1597–1613.

Le Foll, B., Goldberg, S.R., 2004. Rimonabant, a CB_1 antagonist, blocks nicotine-conditioned place preferences. Neuroreport 15 (13), 2139–2143.

Lin, L.S., Ha, S., Ball, R.G., et al., 2008. Conformational analysis and receptor docking of *N*-[(1*S*,2*S*)-3-(4-chlorophenyl)-2-(3-cyanophenyl)-1-methylpropyl]-2-methyl-2-{[5-(trifluoromethyl)pyridin-2-yl]oxy}propanamide (Taranabant, MK-0364), a novel, acyclic cannabinoid-1 receptor inverse agonist. J. Med. Chem. 51, 2108–2114.

Mathews, W.B., Scheffel, U., Finley, P., et al., 2000. Biodistribution of [^{18}F] SR144385 and [^{18}F] SR147963: selective radioligands for positron emission tomographic studies of brain cannabinoid receptors. Nucl. Med. Biol. 27, 757–762.

Mathews, W.B., Scheffel, U., Rauseo, P.A., et al., 2002. Carbon-11 labeled radioligands for imaging brain cannabinoid receptors. Nucl. Med. Biol. 29, 671–677.

Newell, K.A., Deng, C., Huang, X., 2006. Increased cannabinoid receptor density in the posterior cingulate cortex in schizophrenia. Exp. Brain Res. 172, 556–560.

Neumeister, A., Normandin, M.D., Murrough, J.W., et al., May 2, 2012. Positron emission tomography shows elevated cannabinoid CB_1 receptor binding in men with alcohol dependence. Alcohol. Clin. Exp. Res. 36 (12), 2104–2109 (Epub ahead of print).

Rodriguez, J.S., Boctor, S.Y., Flores, L.C., et al., 2011. Local pretreatment with the cannabinoid CB_1 receptor antagonist AM251 attenuates methamphetamine intra-accumbens self-administration. Neurosci. Lett. 489 (3), 187–191.

Sanabria-Bohórquez, S.M., 2010. Kinetic analysis of the cannabinoid-1 receptor PET tracer [^{18}F]MK-9470 in human brain. Eur. J. Nucl. Med. Mol. Imaging 37, 920–933.

Shire, D., Calandra, B., Bouaboula, M., et al., 1999. Cannabinoid receptor interactions with the antagonists SR141716A and SR144528. Life Sci. 65, 627–635.

Siegfried, Z., Kanyas, K., Latzer, Y., et al., 2004. Association study of cannabinoid receptor gene (CNR1) alleles and anorexia nervosa: differences between restricting and binging/purging subtypes. Am. J. Med. Genet. B (Neuropsychiatr. Genet.) 125B, 126–130.

Terry, G.E., Liow, J.S., Zoghbi, S.S., et al., 2009. Quantitation of cannabinoid CB_1 receptors in healthy human brain using positron emission tomography and an inverse agonist radioligand. NeuroImage 48, 362–370.

Ujike, H., Takaki, M., Nakata, K., et al., 2002. CNR1, central cannabinoid receptor gene, associated with susceptibility to hebephrenic schizophrenia. Mol. Psychiatry 7, 515–518.

Van Laere, K., 2008a. Whole-body biodistribution and radiation dosimetry of the human cannabinoid type-1 receptor ligand ^{18}F-MK-9470 in healthy subjects. Nucl. Med. 49, 439–445.

Van Laere, K., 2008b. Gender-dependent increases with healthy aging of the human cerebral cannabinoid-type 1 receptor binding using [^{18}F]MK-9470 PET. NeuroImage 39, 1533–1541.

Van Laere, K., 2010. Relationship of type 1 cannabinoid receptor availability in the human brain to novelty-seeking temperament. Eur. J. Nucl. Med. Mol. Imaging 37, 917–919.

Vickers, S.P., Kennett, G.A., 2005. Cannabinoids and the regulation of ingestive behavior. Curr. Drug Targets 6 (2), 215–223.

Vinod, K.Y., Hungund, B.L., 2006. Role of the endocannabinoid system in depression and suicide. Trends Pharmacol. Sci. 27 (10), 539–545.

Waterhouse, R.N., 2003. Determination of lipophilicity and its use as a predictor of blood–brain barrier penetration of molecular imaging agents. Mol. Imaging Biol. 5, 376–389.

Wong, D.F., Kuwabara, H., Horti, A.G., et al., 2010. Quantification of cerebral cannabinoid receptors subtype 1 (CB_1) in healthy subjects and schizophrenia by the novel PET radioligand [^{11}C]OMAR. NeuroImage 52 (4), 1505–1513.

Yasuno, F., Brown, A.K., Zoghbi, S.S., et al., 2008. The PET radioligand [^{11}C]MePPEP binds reversibly and with high specific signal to cannabinoid CB_1 receptors in nonhuman primate brain. Neuropsychopharmacology 33, 259–269.

Zavitsanou, K., Garrick, T., Huang, X., 2004. Selective antagonist [^3H]SR141716A binding to cannabinoid CB_1 receptors in increased in the anterior cingulate cortex in schizophrenia. Prog. Neuropsychopharmacol. Biol. Psychiatry 28, 355–360.

CHAPTER THREE

Brain Imaging of Cannabinoid Receptors

Cindy Casteels[1,2], Guy Bormans[2,3] and Koen Van Laere[1,2]

[1]Division of Nuclear Medicine, University Hospitals and KU Leuven, Leuven, Belgium
[2]MoSAIC, Molecular Small Animal Imaging Center, KU Leuven, Leuven, Belgium
[3]Laboratory for Radiopharmacy, KU Leuven, Leuven, Belgium

1. INTRODUCTION

1.1. The Endocannabinoid System

1.1.1. Components, Working Mechanism, and Distribution

Marijuana has been used in numerous cultures throughout recorded history. While the folkloric use of marijuana as medicine has been around since ancient times, it was largely during the nineteenth century that this substance was assimilated into the standard armamentarium of official medical practice. From about 1850 to 1900, several published testimonials and case histories indicated that marijuana could ameliorate neurological symptoms. In 1964, Gaoni and Mechoulam (Mechoulam and Hanus, 2000) identified (−)-Δ^9-tetrahydrocannabinol (Δ^9-THC) as the major psychoactive constituent of marijuana. Since that time, about 60 other cannabinoids have been identified (Turner et al., 1980), and many hundreds of analogs of Δ^9-THC have been synthesized in the laboratory (Razdan, 1986). However, it was not until the 1990s that the receptors responsible for many of the actions of Δ^9-THC were identified (Devane et al., 1988) and cloned (Matsuda et al., 1990). Since then, the knowledge on the endogenous cannabinoid system and its physiology, pharmacology, and therapeutic potential has expanded enormously.

The so-called endocannabinoid system (ECS) consists of cannabinoid receptors, endogenous ligands, as well as the proteins for their biosynthesis, degradation, and transport.

To date, two cannabinoid receptors, type 1 (CB_1) and type 2 (CB_2), have been identified by molecular cloning and are unambiguously established as mediators of the biological effects induced by cannabinoids, either synthetic or endogenously produced (Matsuda et al., 1990; Munro et al., 1993). CB_1 and CB_2 receptors are heptahelical transmembrane $G_{i/o}$-coupled receptors that share 44% protein identity and display different pharmacological profiles and patterns of expression. The majority of cannabinoid effects on the central nervous system (CNS) are mediated by the CB_1 receptor. In the brain, the CB_1 receptor is found in areas controlling motor, cognition, emotional, and

Imaging of the Human Brain in Health and Disease
http://dx.doi.org/10.1016/B978-0-12-418677-4.00003-8

sensory functions, i.e. the hippocampus, basal ganglia, cerebellum, cortex, and olfactory bulb (Herkenham et al., 1990; Tsou et al., 1998a). In these regions, the CB_1 receptor is abundantly expressed presynaptically. Evidence of the presence of CB_1 receptors on the dendrites and soma of neurons is not convincing (Freund et al., 2003). CB_1 receptors are additionally found at low levels on various astrocytes, oligodendrocytes, and neural stem cells (Aguado et al., 2005), while in peripheral tissues, CB_1 receptors are expressed in the heart, uterus, testis, liver, adipose tissue, and small intestine (Maccarrone et al., 2001; Nong et al., 2001; Klein et al., 2003; Massa et al., 2005). CB_2 receptor expression seems to be confined to cells of the immune system. Under non-pathological conditions, CB_2 receptors are primarily expressed by leukocytes (Munro et al., 1993). Recently, they were also localized in the endocrine pancreas (Juan-Pico et al., 2006), bone (Ofek et al., 2006), and brainstem (Van Sickle et al., 2005), with the brainstem localized at much lower levels than the CB_1 receptor.

By coupling to $G_{i/o}$ proteins, cannabinoid receptors regulate the activity of several membrane proteins and signal transduction pathways. Both CB_1 and CB_2 receptors inhibit cyclic adenosine 5′-monophosphate formation and activate mitogen-activated protein kinase (Pertwee, 1997). In addition, CB_1 receptors activate ion channels, such as A-type and inwardly rectifying potassium channels, and inhibit voltage-sensitive N-type and P/Q-type calcium (Ca^{2+}) channels (Deadwyler et al., 1995; Hampson et al., 1995). An important functional consequence of the regulation of ionic currents is the inhibition of neurotransmitter release. Studies have indicated that CB_1 receptor activation decreases the release of glutamate (Grundy et al., 2001) as well as modulates γ-aminobutyric acid (GABA)-ergic transmission in several brain areas by either effect on GABA release or actions on the GABA transporter (Szabo et al., 1998).

The family of endogenous ligands, termed endocannabinoids, is expanding. There are at least five different arachidonoyl derivatives, which can activate the cannabinoid receptors. Anandamide (AEA) and 2-arachidonoylglycerol (2-AG) are the two best studied members. 2-Arachidonoylglyceryl ether (Hanus et al., 2001), *O*-arachidonoylethanolamide (Porter et al., 2002), and *N*-arachidonoyl dopamine (Huang et al., 2002) have been identified more recently as endogenous ligands and their classification as true endocannabinoids awaits further biochemical and pharmacological characterization. The affinity of AEA and 2-AG for the human CB_1 receptor is fairly low: between 26 and 209 nM for AEA and even above 10 μM for 2-AG (Steffens et al., 2005). Also for the human CB_2 receptor their affinity is not high, within the 0.6–3.5 μM range (Gonsiorek et al., 2000). Besides the activation of these receptors, AEA and 2-AG are also able to activate other molecular targets still under characterization, such as not-well-defined cannabinoid receptors (Breivogel et al., 2001; Begg et al., 2005). Furthermore AEA, but not 2-AG, behaves as a weak ligand to the type 1 vanilloid receptor (now called transient receptor potential vanilloid 1, TRPV1) (Jung et al., 1999). TRPV1 is a ligand-gated and nonselective cationic channel, activated by molecules derived from plants, such as

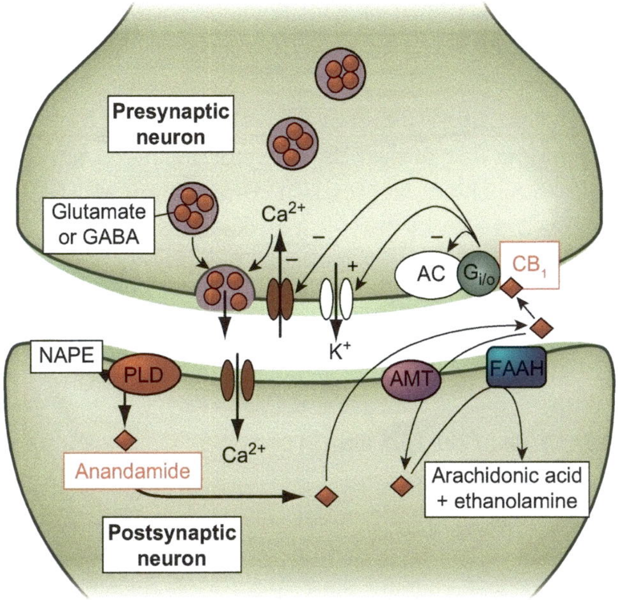

Figure 1 Schematic representation of endocannabinoid signaling, as illustrated for AEA influx of Ca^{2+} in the postsynaptic cells, activates phospholipase D (PLD), which acts on *N*-arachidonoyl phosphatidylethanolamine (NAPE) to produce AEA. AEA leaves the postsynaptic cell and activates presynaptic cannabinoid CB_1 receptors. $G_{i/o}$ protein activation leads to inhibition of adenylate cyclase (AC), opening of presynaptic potassium (K^+) channel, and inhibition of presynaptic Ca^{2+} influx, decreasing the probability of neurotransmitter release. AEA is transported into the postsynaptic cell via a membrane transporter (AMT) and is degraded by FAAH to arachidonic acid and ethanolamine. *Adapted from Benarroch (2007).*

the pungent component of "hot" red peppers capsaicin (Jordt and Julius, 2002). Since TRPV1 is expressed in peripheral sensory fibers and also in several nuclei of the CNS (Marinelli et al., 2003), the endovanilloid activity of AEA may also play a role in physiological control of brain function.

Endocannabinoids act as retrograde signals at CNS synapses as shown in Figure 1. In contrast to classical neurotransmitters, endocannabinoids are not stored in vesicles but are produced in dendrites on "on demand". This is the result of a biosynthetic mechanism relying on the existence of phospholipid precursors of these compounds and of Ca^{2+}-sensitive lipases for the conversion of the precursors in the endocannabinoid products. The biosynthesis of endocannabinoids is immediately followed by their release and the activation of presynaptically located CB_1 receptors. The life span of endocannabinoids in the extracellular space is limited by a rapid elimination process consisting of selective uptake into the postsynaptic cell and subsequent degradation. AEA is inactivated by reuptake via the AEA membrane transporter (AMT) and by intracellular enzymatic degradation by fatty acid amide hydrolase (FAAH)-mediated hydrolysis (Giuffrida et al., 2001). 2-AG undergoes similar FAAH-mediated hydrolysis (Ueda et al., 2000) and carrier-mediated transmembranal transport (Beltramo and Piomelli, 2000), but recent evidence has demonstrated the existence of another enzyme involved in the degradation of 2-AG, monoacylglycerol lipase (Dinh et al., 2002). FAAH is distributed in brain areas in a pattern corresponding that of the CB_1 receptor, i.e. high concentrations in hippocampus, basal ganglia, cerebellum, and cerebral cortex (Tsou et al., 1998b).

Interestingly, COX-2 is also identified in endocannabinoid transmission and forms prostacannabinoids that in turn participate in cell signaling (Kozak et al., 2002). Indirect evidence additionally supports the presence of an endocannabinoid transporter on postsynaptic membranes; however, no protein structure has been elucidated to confirm its existence (Fowler, 2012).

1.1.2. Physiological Role of the ECS within the CNS

The ECS plays an important role in the CNS by regulating brain networks and synaptic transmission, leading to several central actions such as the control of cognition, pain, perception, movement, drug addiction, and memory consolidation (for review see Di Marzo et al. (1998)). Generally, the current view on the ECS is that it acts as a broad-spectrum modulator and that it is likely the most important and ubiquitous feedback mechanism for neurotransmission (Katona and Freund, 2012). This broad-spectrum modulatory effect is largely caused by the interference of endo- and exocannabinoids with classical neurotransmitter signaling. As mentioned above, the ECS interferes with release or reuptake of neurotransmitters produced by presynaptic terminals, and provides a physiological feedback mechanism able to reduce synaptic inputs onto the stimulated neuron in a highly selective and restricted manner. This "retrograde signaling" of AEA and 2-AG may result in depolarization-induced suppression of inhibition at GABA-ergic synapses, and in depolarization-induced suppression of excitation at glutamatergic synapses. The specific role of endocannabinoids in different cell types and brain regions may vary depending on temporal and spatial patterns of neuronal activity, kinetics of endocannabinoid biosynthesis, transport, and degradation. Next to its role in synaptic transmission, CB_1 receptors also have antiinflammatory and neuroprotective properties (Drysdale and Platt, 2003).

Less is known about the physiological role of CB_2 receptors. CB_2 receptor expression is related to the immune system and is not widespread in other organs. In immune-mediated pathologies, its expression is even 10 times that of basal levels, primarily in activated microglia (Maresz et al., 2005). The activation of glial cells, in addition to excitotoxicity, mitochondrial failure, and oxidative stress, is a basic event in neuroinflammatory conditions. CB_2 receptor activation has been shown to induce immune-modulatory and neuroprotective effects (Cabral and Marciano-Cabral, 2005). CB_2 receptors are also implicated in mediating pain transmission, in immune cell function, and communication (Hsieh et al., 2011; Cencioni et al., 2010). Their expression is also correlated with the stage of tumor malignancy (Ellert-Miklaszewska et al., 2007; Calatozzolo et al., 2007), but their exact role in these noninflammatory conditions remains unclear to date.

1.2. ECS IN BRAIN PATHOLOGY

1.2.1. Eating Disorders

The ECS has been recognized as an important component in many different homeostatic processes in living organisms, including food intake, energy metabolism, and reward processing. Animal studies showed that CB_1 receptor-knockout mice are leaner than control animals due to reduced food intake, and the same mice are resistant to diet-induced obesity when subjected to a highly caloric diet without their energy uptake being reduced (Cota et al., 2003; Ravinet et al., 2004). The CB_1 receptor is highly expressed in the hypothalamic nuclei, which are important in the control and regulation of energy balance and body weight. Besides this homeostatic role, CB_1 receptors are also abundant on the neurons of the mesolimbic system modulating reward or the hedonic aspects of food intake (Di Marzo and Matias, 2005). Administration of endocannabinoids systemically or in hypothalamic nuclei induces food intake and endocannabinoid concentrations vary depending on the feeding status (Fride et al., 2005; Benarroch, 2010). In addition, different alleles of CNR1 and FAAH have been associated with anorexia nervosa (AN) (Siegfried et al., 2004; Monteleone et al., 2009) and obesity (Benzinou et al., 2008).

Compiled evidence has led to the likeliness of a hyperactive ECS in obesity, while in anorectic condition the ECS may be underactive (Di Marzo and Matias, 2005; Arias, 2007; Sipe et al., 2010). In this perspective, some therapeutic application of ECS modulation for regulation of feeding behavior in eating disorders and obesity have already been applied or are under investigation. Since about 20 years, the synthetic Δ^9-THC analog and CB_1 receptor agonist nabilone is available as Dronabinol for the treatment of chemotherapy-induced nausea in cancer patients and later also for anorexia in cancer and acquired immunodeficiency syndrome patients (Di Marzo and Matias, 2005). CB_1 receptor antagonists have been developed for treatment of obesity and metabolic syndrome. Rimonabant (Acomplia, Sanofi-Aventis) was put forward as supporting therapy for weight reduction in obese patients or for reduction of associated risks. A 1-year treatment led not only to significant weight loss but also to improvement in cardiovascular and metabolic risk factors such as high-density lipoprotein cholesterol or triglyceride concentrations. However, clinical trials also reported an elevated risk of neurological and psychiatric side effects such as depression, anxiety, suicidality, and convulsions (Pi-Sunyer et al., 2006; Van Gaal et al., 2008). Even if the side effects of administered doses of another CB_1 receptor antagonist, taranabant (Merck), seemed manageable, similar conclusions were drawn for studies with this compound (Addy et al., 2008). As a consequence, the European Medicine Agency withdrew rimonabant as antiobesity medication and development programs for other potential antiobesity CB_1 receptor antagonists were discontinued (Le Foll et al., 2009).

1.2.2. Neurodegenerative Diseases: HD, PD, and AD

The ECS plays a central role in several neurodegenerative diseases.

In Huntington disease (HD), one of the earliest neurochemical alterations is the loss of CB_1 receptors in the basal ganglia, an alteration that significantly precedes the development of identifiable striatal neuropathology (Glass et al., 2000). CB_1 receptor mRNA levels are also decreased in the absence of neuronal loss in the lateral striatum, cortex, and hippocampus of transgenic mouse models of HD (Lastres-Becker et al., 2002a; Naver et al., 2003; McCaw et al., 2004; Ovan-Wright and Robertson, 2000; Dowie et al., 2009). In the HD94 transgenic mice also decreases in the number of basal ganglia-specific binding sites and the activation of GTP-binding proteins by CB_1 receptor agonists were noticed (Lastres-Becker et al., 2002a). Loss of CB_1 receptors in the basal ganglia not only occurred in transgenic mice HD models but also in rats after local intrastriatal application of 3-nitropropionic acid (NP), a toxin that reproduces the mitochondrial complex II deficiency, characteristic of HD patients (Lastres-Becker et al., 2002b). Delaying the onset of HD symptoms by enriched environments has been shown to selectively slow down the loss of CB_1 receptors in experimental HD (Glass et al., 2004). The mechanism by which HD promotes this loss of CB_1 receptors is suggested to be through interactions between mutant htt and nuclear transcription factors. Blázquez and coworkers showed in striatal cells that mutant htt downregulates CB_1 receptors by controlling gene promoter activity via repressor element 1-silencing transcription factor and sensitizes cells to excitotoxic damage (Blázquez et al., 2011).

In addition, the number of activated microglia in the striatum and cortex correlates with the extent and progression of neuronal loss (Sapp et al., 2001). Lesion models of HD showed CB_2 receptor upregulation at the lesioned side and CB_2 receptor agonists showed neuroprotective effects in this model (Sagredo et al., 2009). CB_2 receptor-knockout mice are more susceptible to neuroinflammation, with enhanced microglia activation, aggravated disease symptomatology, and reduced life span (Palazuelos et al., 2009).

Numerous ex vivo studies have also suggested a profound disturbance of the ECS in Parkinson disease (PD) (Benarroch, 2007; Brotchie, 2003; Van der Stelt and Di Marzo, 2003). Postmortem CB_1 levels are increased in striatal tissue from 1-methyl-4-phenyl-1,2,3,6-tetrahydropyridine (MPTP)-lesioned parkinsonian monkeys and PD patients (Lastres-Becker et al., 2001). Endocannabinoid levels are increased in the cerebrospinal fluid of untreated PD patients compared with non-PD controls (Pisani et al., 2010). An ECS-dependent form of synaptic plasticity was found to be absent in PD animal models (Kreitzer and Malenka, 2007). Particular attention has also focused on the possible role of the ECS in levodopa-induced dyskinesias (LID). The CB_1 receptor agonist nabilone reduced LID in MPTP-lesioned monkeys and PD patients (Sieradzan et al., 2001).

In addition, an overexpression of CB_2 receptors in the 6-hydroxydopamine PD model has been reported. Agonism of CB_2 resulted in neuroprotection against behavioral and neurochemical alterations (Ternianov et al., 2012).

In Alzheimer's disease (AD), both CB_1 and CB_2 receptors have been found localized on/near amyloid plaques (Benito et al., 2003). Selective agonists for the CB_2 receptor blocked the activation of cultured microglia by amyloid (Ramirez et al., 2005). In postmortem studies, [^{125}I]SD7015, a selective CB_1 receptor ligand, showed an inverse correlation between CB_1 receptor density and Braak tau staging (Farkas et al., 2012). Δ^9-THC inhibited acetylcholinesterase and prevented the aggregation of amyloid-β peptide (Eubanks et al., 2006).

1.2.3. Drug Addiction

Several preclinical and clinical studies have suggested potential therapeutic effects for CB_1 receptor antagonists in the treatment of drug dependence (Le Foll and Goldberg, 2005). A postmortem study showed that the density of CB_1 receptors is significantly lower in long-term cannabis users as compared to healthy controls in the striatum, nucleus accumbens, and hippocampal regions (Villares, 2007). Δ^9-THC has been found to increase smoking cessation in humans (Soyka et al., 2008; Huestis et al., 2001; Cahill and Ussher, 2011). The effects from chronic alcohol exposure to rodents on the ECS are less consistent, with some studies reporting increased endocannabinoid levels in reward-related brain areas (Gonzalez et al., 2002a; Vinod et al., 2006; Caille et al., 2007), whereas others observed either a decrease (Gonzalez et al., 2004; Gonzalez et al., 2002a) or no change (Gonzalez et al., 2004; Caille et al., 2007). In line with these, CB_1 receptor expression, density, and functionality were shown to be either downregulated and reduced in the hippocampus and caudate putamen of rodents chronically exposed to ethanol (Vinod et al., 2006; Ortiz et al., 2004) or unchanged (Gonzalez et al., 2002b). Genetic polymorphism in the CB_1 receptor encoding gene, *CNR1*, has been associated with risk of alcohol or drug use (Zhang et al., 2004; Hutchison et al., 2008).

Recently, a potential role for CB_2 receptors in drug addiction has been suggested (Morales and Bonci, 2012; Racil-Fernandez et al., 2012). Systemic administration of JWH133 and GW405833, two selective CB_2 receptor agonists inhibited the self-administration of cocaine in wild-type and CB_1-deficient mice, but not in CB_2-deficient mice. Pretreatment with the CB_2 receptor antagonist AM630 prevented this effect.

1.2.4. Depression

Recent reviews have dealt with the potential involvement of the ECS and CB_1 receptors in various psychiatric disorders, including major depression (Ashton and Moore, 2011; Parolaro et al., 2010). A deficiency has been proposed in the neurobiology of major depression and suicidal behavior, and consequently, activation of cannabinoid signaling may result in antidepressant effects (Gorzalka and Hill, 2011). In the context of major depression, the CB_1 receptor-deficient mouse has recently been proposed as a useful animal model for some forms of depression mainly associated with alterations of the serotonergic and other systems (Valverde and Torrens, 2012). In line with this preclinical model, the CB_1 receptor antagonist rimonabant (SR141716A) has been reported to induce depressive symptoms in humans (Nissen et al., 2008). Paradoxically, CB_1 receptor activity appears to be increased in a number of animal models of depression (Gorzalka and Hill, 2011). Also conflicting is the fact that activation and blockade of the CB_1 receptor result in antidepressant responses in the same behavioral tasks (Bambico et al., 2010; Griebel et al., 2005). Moreover, conventional antidepressant drugs have been shown to facilitate (e.g. fluoxetine) or inhibit (citalopram) CB_1 receptor signaling in laboratory animals (Gorzalka and Hill, 2011). Interestingly, chronic fluoxetine treatment in rats has been reported to increase cortical CB_1 receptor signaling through the activation of 5-hydroxytryptamine (HT) 1A receptors (Mato et al., 2010), which indicates the existence of a functional interaction between these two inhibitory receptors in the action of the antidepressant. At present, further studies are still needed to clarify the putative role of the ECS and CB_1 receptors in relevant animal models of depression as well as in the mechanisms of action and efficacy of antidepressant drugs, including their interactions with the classic monoaminergic systems in the brain.

1.2.5. Schizophrenia

The use of cannabis has also since long been associated with schizophrenia and a genetic susceptibility for it has been found in patients who developed psychotic symptoms after cannabis use (Caspi et al., 2005; Ho et al., 2011). Genetic studies reported an association between the CB_1 receptor encoding gene, *CNR1*, and schizophrenia, especially the hebephrenic type (Ujike et al., 2002). Large epidemiological studies showed an increased risk for developing schizophrenia on the use of cannabis (Fernandez-Espejo et al., 2009; Malone et al., 2010). Signaling through the CB_1 receptor has been found to be involved in the pathogenesis of schizophrenia and its symptoms (Vinod and Hungund, 2006; Laviolette and Grace, 2006; Ujike and Morita, 2004; Dean et al., 2001). Some studies showed increased CB_1 receptor binding in the frontal and cingulate cortex of patients with schizophrenia (Dean et al., 2001), while others demonstrated reductions in the mRNA and protein levels of this receptor (Newell et al., 2006). Antipsychotics increased the CB_1 receptor expression in rat brain (Secher et al., 2010).

1.2.6. Brain Tumors

Cannabinoid drugs have also recently become candidates in the field of oncology (Herrera et al., 2005). While nontransformed astrocytes express only the CB_1 receptor, both types of functional cannabinoid receptors have been found in several established human glioblastoma cell lines, as well as in primary cultures derived from the most malignant brain tumor, glioblastoma multiforme (Galve-Roperh et al., 2000; Sanchez et al., 2001). Immunohistochemical analysis of low- and high-grade human glioma surgical specimens revealed increased CB_2 receptor expression in tumor cells, invading microglia/macrophages, and endothelial cells of the tumor blood vessels, as compared to nontumor brain samples (Ellert-Miklaszewska et al., 2007; Sanchez et al., 2001; Schley et al., 2009). The proportion of malignant tumors expressing high levels of CB_2 receptor was over twofold higher than that seen in the tumors of lower grade (Ellert-Miklaszewska et al., 2007), indicating that the extent of CB_2 expression correlates with the tumor malignancy grade. Interestingly, some benign pediatric astrocytic tumors, such as subependymal giant cell astrocytoma, which may occasionally cause mortality owing to progressive growth in some patients, also displayed high CB_2 immunoreactivity. Moreover, as observed by Sanchez et al., CB_2 receptor immunoreactivity markedly prevailed over detected CB_1 receptor levels in grade IV astrocytomas (Sanchez et al., 2001). An enhancement of cannabinoid receptor expression in malignant vs healthy tissues might suggest a possible role of the ECS in the tonic suppression of cell divisions and cancer growth. This hypothesis is partly supported by the finding of increased levels of anandamide and decreased levels of endocannabinoid metabolizing enzymes (i.e. FAAH) in human glioblastoma compared to human nontumor brain tissue (Petersen et al., 2005).

1.2.7. Future Perspectives

In summary, although advances in defining the role of endocannabinoids in both normal and pathological condition have provided rationale for endocannabinoid target therapy among others in HD, PD, AD, drug addiction, and schizophrenia, experimental evidence has also indicated that the effects of CB_1/CB_2 receptor agonists or antagonists on the symptoms are complex, among others due to regional expression differences and target specificity (Papa, 2008). For instance, an agonist may facilitate or oppose the action of dopamine in PD depending on the interacting mechanism elicited, whether in the striatum or at an extrastriatal site. Experimental evidence has also indicated that the effects of drugs on symptoms depend on the disease severity and/or probably on the dose and gender (Sundram, 2006). Given these multiple variable, the results of currently available experiments/trials still need to be interpreted with caution, since, so far, they have not provided consistent data to draw definite conclusions. In this perspective, in vivo human studies using molecular imaging may provide new insights to better define their use in the above pathologies.

2.1. MOLECULAR IMAGING OF THE ECS

2.1.1. Overview

Today, in vivo molecular imaging is playing an important role in biomedical research. Molecular imaging allows visualization, three-dimensional (3D) localization, and quantification of molecular processes at the cellular and subcellular level within intact living organisms that can be repeated over time in the same subjects (Massoud and Gambhir, 2003). Many molecular processes can be targeted, including receptors (density and drug occupancy (Burns et al., 2007)), transporter and enzyme activity, gene expression (Willmann et al., 2009), metabolite concentration, protein–protein interaction (Lake et al., 2012), transcriptional activity (Pouliot et al., 2011), signal transduction, and apoptosis (Wang et al., 2013). It provides us with new tools to objectively evaluate both in humans and animals the neurobiological and potential clinical impact of the ECS. Functional imaging of brain cannabinoid receptors using positron emission tomography (PET) and single-photon emission computed tomography (SPECT) may lead to the identification of novel diagnostic biomarkers and guide dose occupancy studies in drug development research. These techniques can also be used to examine the interaction between ECS and other neurotransmitter systems both in control and pathological condition.

2.1.2. Positron Emission Tomography

PET is a quantitative imaging modality which uses small amounts of pharmaceuticals labeled with positron-emitting isotopes. Because of their similarity to naturally occurring atoms in the human body, positron emitters such as carbon-11 (^{11}C), nitrogen-13 (^{13}N), and fluorine-18 (^{18}F) can be incorporated in biological molecules, without significantly influencing their physiological and biochemical interaction. The biological properties of these molecules determine the (patho)physiological significance of the images.

Positron-emitting isotopes such as ^{18}F have an excess of protons in their nucleus and are therefore unstable. The excess proton transforms to a neutron under the emission of a positron, which is the positively charged antiparticle of an electron (b^+ decay). Upon decay of the isotope, the emitted positron scatters in the tissue for a distance of 2–3 mm, depending on its energy, before it collides with an electron and both annihilate, thereby creating a pair of 511-keV photons, emitted in opposite directions, which can be detected using a PET camera (Figure 2). The coincidence detection of two photons by two opposite detectors within the PET camera indicates that somewhere along the imaginary line between the detectors, a positronium has decayed, allowing 3D reconstruction of the distribution pattern of the isotope in the body. The spatial

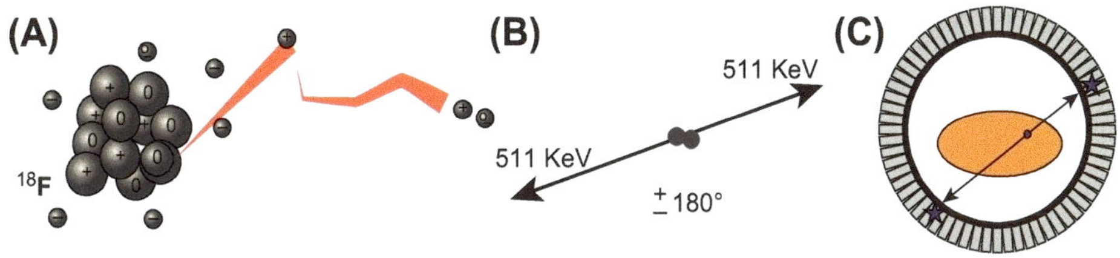

Figure 2 The unstable nucleus of ^{18}F decays by emitting a positron (A). This positron combines with an electron resulting in a positronium that annihilates hereby generating a pair of photons of 511 keV each traveling in nearly opposite directions (B). These photons can be detected by detector pairs installed in a PET camera (C).

resolution of state-of-the-art clinical scanners is 4–6 mm, whereas preclinical scanners achieve 1–2 mm. The sensitivity of PET is in the picomolar range (10^{-11} to 10^{-12} mol/l) (Massoud and Gambhir, 2003), allowing the detection of subpharmacological amounts of radioligand. The major disadvantage of PET is the necessity for most of the isotopes to be generated on site by a dedicated cyclotron because of their short half-life (^{11}C, ~20 min; ^{18}F, ~110 min) and the need for a subsequent elaborate radiopharmaceutical synthesis of the radioligands.

2.1.3. Single-Photon Emission Tomography

SPECT is based on the detection of a single photon. By acquiring planar projection images from many different angles the 3D distribution of the isotope can be reconstructed. The most common isotopes in SPECT, technetium-99m (^{99m}Tc) and iodine-123 (^{123}I), have half-lives of 6 hours and 13 hours respectively, allowing to study processes taking place on a moderately long time scale. As SPECT requires physical collimation of the incident photon, its detection sensitivity is lower than in PET by typically an order of magnitude. Another disadvantage of SPECT is the limited temporal resolution as it requires multiple planar images from different angles, typically acquired in ~30 minutes. Modern small animal devices have acquisition times ranging from 1 to 10 minutes. A major advantage of SPECT is the possibility to image two different isotopes at the same time provided they have different photon energies, allowing multiplexing of two different molecular signals. The spatial resolution of state-of-the-art clinical scanners is between 7 and 15 mm depending on camera and collimator characteristics. The spatial resolution of preclinical scanners can be as low as 0.6 mm, although there is a trade-off between sensitivity and resolution (Vastenhouw and Beekman, 2007).

2.1.4. PET and SPECT Requirements for CNS Application

Radioligands for imaging brain receptors must fulfill a variety of characteristics. These required properties are listed in Table 1 (Pike, 2009).

Table 1 Properties of the Ideal Receptor Radioligand

Properties	**Criteria**
Radiochemistry	Reliable/reproducible labeling
	High yield
Pharmacology/pharmacodynamics	High affinity and selectivity for its target
	Low nonspecific binding
	Reversible target binding
	Fast plasma clearance
Specifically for CNS	No radiometabolites in the brain
	Low or fast-reversible protein binding in plasma
	Efficient BBB penetration
	• Ideal lipophilicity
	• No substrate for BBB efflux carriers
	Suitable brain pharmacokinetics in relation to radiolabel half-life
	Safe to administer in low tracer dose

Radioligands should exhibit a high affinity for their targets, preferably having a dissociation rate constant (K_d) in the subnanomolar to low nanomolar range. The ideal affinity for a radioligand depends on the expression level of the target receptor and should be at least 5- to 10-fold higher than the receptor expression (B_{max}). High selectivity for its target is also required, preferably 100-fold less affinity for any other binding site which is expressed in the same level.

The blood–brain barrier (BBB) prevents hydrophilic and large molecules from entering the brain and is thus a major hurdle in developing drugs that target the CNS. Favorable molecule characteristics for entering the brain are a low molecular mass (<600 Da), a polar surface area (PSA) less than 90 Å^2 and lipophilicity, $\log D_{7.4}$, between 1 and 3. However, molecules having a higher $\log D$ and good BBB penetration have also been described. Nevertheless, too lipophilic molecules often are substrates of efflux pumps (such as P-glycoproteins (P-gp) or multidrug resistance-associated proteins) at the BBB, thereby hampering their uptake in the CNS. Moreover, when the tracer is highly lipophilic, the formulation of an intravenous (IV) injection will be more challenging due to absorption of molecules on the surface of vials and syringes, and in most cases, high nonspecific binding to lipophilic membranes and proteins in vivo is observed.

High plasma protein binding is disadvantageous for the brain uptake of the radioligand as only the unbound fraction diffuses across the BBB. There are, however, many examples of radioligands that despite extensive protein binding still show sufficient brain uptake probably due to rapid dissociation of tracer from plasma proteins continuously replenishing the unbound fraction that diffuses over the BBB.

Furthermore, after crossing the BBB, the tracer should show reversible (allowing an equilibrium binding and subsequently kinetic analysis) and specific receptor binding, low nonspecific binding in the brain, fast brain washout from nonspecific regions, and no toxicity. Neither should there be formation of radiometabolites that cross the BBB nor should there be metabolism in the brain, as the presence of radiometabolites in the brain will contaminate the specificity of the signal in the CNS and hamper accurate quantification.

In addition to the above-mentioned imaging properties, radioligands must be safe for human use. Important safety requirements are low radiation burden and low toxicity of the radioligand.

The manufacturing process of a good PET/SPECT radioligand must reliably yield a sufficient dose of the final radiolabeled drug product, with high radiochemical and chemical purity, specific radioactivity, and formulation as a sterile and apyrogenic solution in saline.

2.2. IMAGING OF CANNABINOID RECEPTORS

2.2.1. CB_1 Receptor

Radioligands that have been developed for imaging brain CB_1 receptors to date may be broadly categorized as first- and second-generation compounds. First-generation compounds were limited by their low specific binding and poor brain uptake. More recently developed compounds have succeeded in combining a high binding affinity with more moderate lipophilicity and have been used in preclinical and clinical investigations since 2006. The physicochemical and in vivo imaging properties of the first- and second-generation compounds are listed in Table 2.

2.2.1.1. First-Generation CB_1 Receptor Radioligands

2.2.1.1.1. (−)-5′-[^{18}F]-Δ^8-THC

The first success in developing CB_1 receptor PET radioligands was established with a fluor-18 analog of Δ^8-THC (Charalambous et al., 1991). The uptake of (−)-5′-[^{18}F]-Δ^8-THC in the mouse brain reached the value of 2.2%ID/g at 5 min postinjection, but radioactivity rapidly declined. Clearance of radioactivity from the baboon brain was also rapid and the regional brain distribution of (−)-5′-[^{18}F]-Δ^8-THC was similar for CB_1 receptor-rich and CB_1 receptor-poor regions, indicating nonspecific binding. This could be due to its low binding affinity ($K_i = 57$ nM) and its high lipophilicity ($\log D = 6.9$) (Compton et al., 1993; Bonhomme-Faivre et al., 2008).

2.2.1.1.2. Analogs of Rimonabant (SR141716)

Most reported efforts at developing ligands for in vivo imaging of CB_1 receptors have focused on radioligands with structural similarity to the CB_1 inverse agonist SR141716. Of these, [^{123}I]AM251, the deschloro-iodo derivative of SR141716 manifested high in vitro binding affinity ($K_i = \sim 1$ nM) (Figure 3) (Lan et al., 1999b). In comparison to

Table 2 In vitro and In vivo Imaging Properties of the First- and Second-Generation Radioligands for Imaging of the CB_1 Receptor

First Generation	Molecular Weight	CB_1 Binding Affinity, K_i (nM)	Relative CB_1 Binding Affinity, $^1K_{rel}$	CB_2 Binding Affinity, K_i (nM)	Lipophilicity, $\log D_{7.4}$	Polar Surface Area (PSA)	Target-to-Nontarget Ratio: Mouse	Target-to-Nontarget Ratio: Monkey	Target-to-Nontarget Ratio: Human
[^{123}I]AM251	555	0.6–1	0.3–2	2290	5.3	50	1.5; 2.2; 1.7[2]	Low brain uptake	–
[^{123}I]AM281	557	4.5	2.5	4200	3.7	59	~1.9[2]	~1.4[3]	0.21[4]
[^{124}I]AM281							–	–	0.37[4]
[^{18}F]NIDA54	400	0.7	–	–	3.8[5]	25	–	0.7	–
[^{18}F]PipISB	492	1.5[6]	–	–	5.0	80	–	0.9	–

Second Generation	Molecular Weight	Radiochemical Yield	K_i (nM)/ K_{rel}	Selectivity CB_1/CB_2	Lipophilicity, $\log D_{7.4}$	PSA	Target-to-Nontarget Ratio	Putamen Uptake, %SUV	Time to Reach Steady State
[^{11}C]JHU75528 In baboon and human	470	16 ± 5%	11/0.3	250–480[7]	3.3; 3.6[8]	83	Putamen/Pons = 2.5 Putamen/Thalamus = 2.1	140–240	60–80 min
[^{11}C]MePPEP In rhesus monkey	489	2.5 ± 1.1%	0.6[6]	726	4.8	84	Putamen/Pons = 2.4 Putamen/Thalamus = 1.9	400–600	>120 min

[18F]FMEP-d_2 In rhesus monkey and human	454	–	0.2	669	6.0	42	Putamen/Pons = 1.8	500–600	~90 min
[18F]MK-9470 In rhesus monkey and human	474	4.6 ± 0.3%	0.7	44–60	4.7	42	Putamen/Pons = 1.8 Putamen/Thalamus = 1.9	120–160	120–180 min

[1]Precise inhibition assay conditions differ between laboratories. Therefore, K_{rel}, a ratio of the K_i for the test compound to that of SR141716 from the same laboratory and by the same method; provides a sense of comparative affinity.
[2]Cerebellum/brainstem ratio.
[3]Cerebellum baseline/cerebellum block with SR141716 ratio.
[4]Binding potential.
[5]Experimental data.
[6]In vitro functional binding activity (K_b).
[7]$K_i^{CB_2}$ = 2700 and 5250 nM.
[8]Two different methods. Note: the lowest CB_1 receptor density (nontarget) is found in the pons and thalamus. The lipophilicity is expressed by the log D value at pH 7.4. The maximum permeability of CNS drugs is often seen if $\log D_{7.4} = 0–4$.

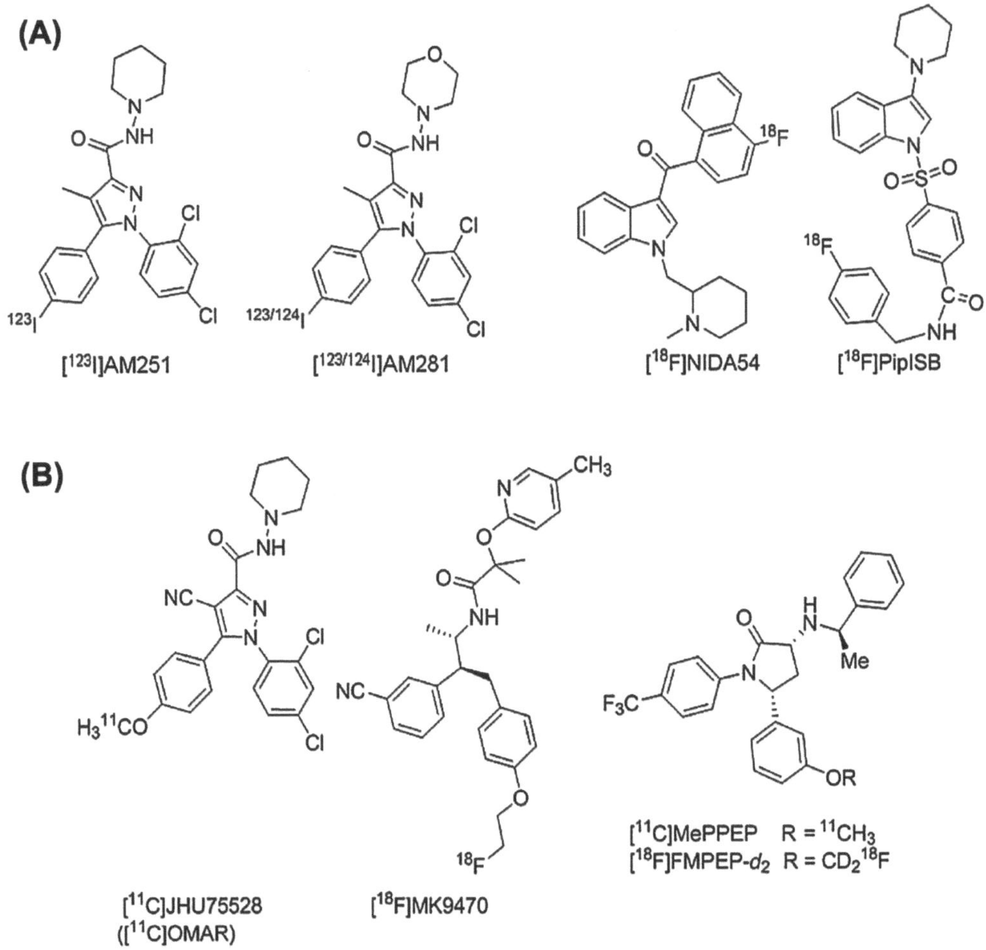

Figure 3 Structures of first-generation (A) and second-generation (B) radioligands for CB_1 receptor imaging.

[^{3}H]CP55,940, a nonclassical CB_1 receptor radioligand with equal affinity for CB_2, [^{123}I]AM251, displayed a better signal-to-noise ratio, high specific radioactivity, and shorter film exposure times that are characteristics for ^{123}I vs ^{3}H. Mouse studies with [^{123}I]AM251 demonstrated that the radioligand modestly enters the brain and specifically labels cerebral CB_1 receptors (Gatley et al., 1996). Unfortunately, [^{123}I]AM251 failed to show significant uptake in the baboon brain (Gatley et al., 1996).

In contrast, the structurally related compound, [^{123}I]AM281, is less lipophilic and showed good accumulation of radioactivity in the cerebellum, cortex, and striatum (Figure 3) (Lan et al., 1999a; Gatley et al., 1998). Pretreatment with SR141716 demonstrated that [^{123}I]AM281 specifically bound to CB_1 receptors in baboon cerebellum, but the signal-to-noise ratio was moderate. Substantial nonspecific binding was also observed in most brain regions. When [^{123}I]AM281 was used in a clinical study of Δ^9-THC-therapy in Tourette patients low specific signal and no significant effect on Δ^9-THC-therapy was observed (Berding et al., 2004). Low brain uptake and poor contrast was also seen using the PET analog [^{124}I]AM281 in one schizophrenic patient (Berding et al., 2006). The low specific signal, high nonspecific binding, and radiosafety concerns of ^{124}I were the main reasons why [^{124}I]AM281 was not pursued in clinical settings.

2.2.1.1.3. Radiolabeled Aminoalkylindole Derivatives and Structurally Related Compounds

The synthetic cannabinoid agonist WIN55,212-2 is a widely used neuropharmacological probe (Kuster et al., 1993). Analogs of this aminoalkylindole compound have been synthesized (Eissenstat et al., 1995; Huffman and Padgett, 2005). There were several attempts to develop PET and SPECT aminoalkylindole-based radioligands, but the imaging properties of the radioligands were not sufficient for further quantitative imaging studies.

The most notable PET radioligands among aminoalkylindole derivatives were [^{18}F]NIDA54 (Willis et al., 2005) and [^{18}F]PipISB (Finnema et al., 2009). NIDA54, the ligand with the best binding affinity of a series of 3-naphthoyl-substituted aminoalkylindoles, was radiolabeled with fluor-18 via the corresponding bromo precursor. The maximal uptake of [^{18}F]NIDA54 in the mouse brain was 1.2%ID/g and the hippocampus-to-brainstem ratio reached the value of 1.7 at 70 min postinjection. [^{18}F]NIDA54 manifests greater binding affinity, a lower molecular weight (MW), PSA, and lipophilicity than [^{18}F]PipISB, but both radioligands exhibited comparable in vivo specific binding in monkey brain (Finnema et al., 2009; Willis et al., 2005). Neither [^{18}F]NIDA54 nor [^{18}F]PipISB was a sufficiently good CB_1 receptor radioligands in animal experiments to be pursued into clinical investigations.

2.2.1.2. Currently Available CB_1 Receptor Radioligands for Human CB_1 Receptor Imaging: "Second-Generation Compounds"

The second-generation ligands exhibit a better combination of high binding affinity and reduced lipophilicity than the first-generation ligand. As a result, the radioligands manifest a greater percentage of specific binding vs nonspecific binding.

2.2.1.2.1. [^{11}C]JHU75528 or [^{11}C]OMAR

In 2006, the John Hopkins PET group reported the synthesis of 4-cyano-1-(2,4-dichlorophenyl)-5-(4-[^{11}C]methoxyphenyl)-*N*-(pirrolidin-1-yl)-1*H*-pyrazole-3-carboxamide ([^{11}C]JHU75528 or [^{11}C]OMAR), the first CB_1 receptor radioligand for quantitative PET studies (Fan et al., 2009; Horti et al., 2006) (Figure 3). [^{11}C]OMAR is a structural analog of SR141716 with comparable binding affinity for the CB_1 receptor and good CB_1/CB_2 selectivity. The lipophilicity of [^{11}C]OMAR is lower than that of SR141716 as the hydrophobic methyl and chlorine substituents of SR141716 are replaced with cyano and methoxy groups. The PSA value of [^{11}C]OMAR is 83.

Extracellular electrophysiological recordings of rodent brain slices revealed that OMAR reverses the effects of the CB_1 receptor agonist WIN55,212-2 on glutamate release in the striatal brain slices, pointing to functional antagonist properties (Fan et al., 2009). Radiosynthesis of [^{11}C]OMAR is performed by [^{11}C]methylation of the corresponding phenol precursor (Horti et al., 2006). In mice and baboon studies, [^{11}C] OMAR showed promising results. When compared to the previously reported CB_1 receptor radioligands of the first generation, the target-to-nontarget ratio of [^{11}C] OMAR was substantially higher (Horti et al., 2006; Fan et al., 2009). [^{11}C]OMAR also readily entered the brain to specifically and selectively label cerebral CB_1 receptors. The specific binding of [^{11}C]OMAR in vivo can be saturated by pretreatment of nonlabeled OMAR of SR141716 in a dose-dependent manner (Horti et al., 2006). Various central noncannabinoid drugs did not reduce regional CB_1 receptor binding, indicating that [^{11}C]OMAR does not bind to other central receptors such as D_1-, D_2-, D_3-, $5HT_{2A}$- ,$5HT_{1C/2C}$-, opioid, and $\alpha_4\beta_2$-nACh.

Small animal PET studies with [^{11}C]OMAR showed 50% higher brain uptake in wild-type mice vs CB_1-knockout animals (Herance et al., 2011). [^{11}C]OMAR is converted to several hydrophilic radiometabolites in mice and baboon blood, but only a fraction of these radiometabolites penetrate the BBB, i.e. ~6% in mice, and likely correspond to [^{11}C]CO_2 (Horti et al., 2006). Kinetic analysis of baboon PET data demonstrated binding potential (BP) values of ~1.4 in CB_1 receptor-rich regions at baseline, and a substantially lower BP value of 0.4 upon blocking. Steady state was reached in baboon brain before the end of the 90 min scan, which is a substantial advantage of this radioligand (Horti et al., 2006).

Human studies with [^{11}C]OMAR have been reported. [^{11}C]OMAR was used under an Investigational New Drug Application, approved by US Food and Drug Administration, to quantify brain CB_1 receptors in healthy subjects and patients with schizophrenia (Wong et al., 2010a). Brain activity in baseline dynamic PET scans peaked at approximately 20 min postinjection and decreased gradually thereafter, which confirmed reversibility of the radioligand binding. PET modeling using arterial sampling resulted in a total distribution volume (V_T) of 1.47 ± 0.25 for the globus pallidus. Cingulate cortex and putamen also showed high binding, i.e. V_T of 1.23 ± 0.16 and 1.32 ± 0.20, respectively. In contrast, white matter, pons, and thalamus were the regions with the lowest accumulation of [^{11}C]OMAR. V_T values of [^{11}C]OMAR were found to be higher in patients with schizophrenia as compared to healthy controls, but the difference did not reach significance. Within the schizophrenia patient group, V_T values did not correlate with the Brief Psychotic Rating Score; however, when scores were taken as ratio of positive over negative symptoms, a positive correlation was found with V_T in several areas of the neocortex. Additionally, [^{11}C]OMAR binding in healthy subjects tended to decline with increasing age in nearly all regions but only attained statistical significance in the globus pallidus, the region with the highest V_T.

Human PET studies using the CB_1 receptor antagonist AVE1625 showed the feasibility of $[^{11}C]$OMAR for the evaluation of drug occupancy studies (Wong et al., 2010b). Because of the short half-life of $[^{11}C]$OMAR, baseline and occupancy scans can be performed on the same day, an advantage of $[^{11}C]$OMAR over fluorine-18-labeled radioligands for drug development research.

Several analogs of OMAR were synthesized by the same group with the aim to obtain CB_1 receptor radioligands with a higher binding affinity and a lower lipophilicity than $[^{11}C]$OMAR (Fan et al., 2009). $[^{11}C]$JHU76609 was studied in mice along with $[^{11}C]$OMAR. Its regional distribution was similar and its target-to-nontarget ratio was significantly higher (+20%). A blocking study with SR141716 confirmed the in vivo specific CB_1 binding of $[^{11}C]$JHU76609. To date, no baboon studies with this radioligand showing whether the imaging properties of $[^{11}C]$JHU76609 are superior to those of $[^{11}C]$OMAR have been reported.

2.2.1.2.2. $[^{11}C]$MePPEP and $[^{18}F]$FMPEP-d_2

The Eli Lilly, NIH, and Karolinska University collaborative group developed two structurally related non-SR141716-based CB_1 receptor PET radioligands for human studies, $[^{11}C]$MePPEP and $[^{18}F]$FMPEP-d_2 (Figure 3) (Terry et al., 2010b; Terry et al., 2009). Both compounds behave as CB_1 receptor inverse agonists.

The functional binding activity of MePPEP is impressively high ($K_i = 0.66\,nM$) (Yasuno et al., 2008). Due to a high lipophilicity, $[^{11}C]$MePPEP is not very soluble in saline and requires formulation with Tween-80. The high lipophilicity is likely responsible for its low free fraction in monkey plasma (Yasuno et al., 2008). Its PSA value of 84 is in the acceptable range for passive BBB diffusion.

Small animal PET studies in wild-type and CB_1 receptor-knockout mice demonstrated that ~65% of total brain uptake represents specific binding (Terry et al., 2008). Blocking studies in the rodent brain also demonstrated that the CB_1 receptor inverse agonist SR141716 has a much higher in vivo potency to displace $[^{11}C]$MePPEP as compared to various other CB_1 receptor agonists. This is consistent with the existence of different binding sites for agonist and inverse agonists on the CB_1 receptor. $[^{11}C]$MePPEP is not a substrate for P-gp and it can be used to quantify CB_1 expression in rat brain within 70 min of scan time using arterial sampling and total distribution volume as outcome measure (Terry et al., 2008). About 13% of the total brain radioactivity in rats corresponded to radiometabolites, likely overestimating values of V_T (Terry et al., 2008).

In Rhesus monkey studies, $[^{11}C]$MePPEP showed reasonable target-to-nontarget ratios (1.9–2.4). Brain uptake of $[^{11}C]$MePPEP is high, but clearance is relatively slow, requiring 120–150 min of scanning (Yasuno et al., 2008).

When performing PET studies with [^{11}C]MePPEP in healthy subjects, a high brain uptake, a slow washout from brain, and a distribution pattern consistent with that of CB_1 receptors ex vivo was also observed (Terry et al., 2009). Quantification of [^{11}C]MePPEP in humans was also feasible using V_T as outcome measure. However, its precision and accuracy was highly influenced by the slow brain kinetics of [^{11}C]MePPEP and by the very low fraction of free radioligand in plasma. Further studies determined that CB_1 quantification was not limited by the measurements from brain, but rather by the measurements from plasma (Terry et al., 2009). Thus, the relatively poor precision on plasma measurements contributed to the higher inter- and intrasubject variability the authors reported on V_T values in comparison to the use of brain uptake alone, as an alternative quantification method. Nevertheless, measurements of V_T were reported to be stable over time and the interclass correlation coefficient of V_T, which allows one to distinguish variability between subjects from within subjects, was also good to excellent (0.97) and better than brain uptake alone (0.77). Based on the findings, the authors stated that quantification of [^{11}C]MePPEP using V_T is still superior to the use of brain uptake, despite its poor precision on plasma measurements.

The above-mentioned difficulties with the quantification of [^{11}C]MePPEP were the driving force behind the development of its derivate, [^{18}F]FMPEP-d_2. The binding affinity of [^{18}F]FMPEP-d_2 is comparable to that of [^{11}C]MePPEP, i.e. 0.2 nM (Terry et al., 2010b). To reduce the de-[^{18}F]fluorination commonly seen with [^{18}F]fluoromethoxy-labeled compounds and consequently the high uptake of [^{18}F]fluoride in the skull bone, two deuterium atoms were introduced into the molecule's structure. [^{18}F]FMPEP-d_2 had high uptake in the monkey brain, with greater than 80% specific binding (Terry et al., 2010b). High brain uptake with [^{18}F]FMPEP-d_2 was also observed in humans, in whom V_T was well identified within approximately 60 min. Retest variability of plasma measurements was good (16%); consequently, V_T had a good retest variability (14%), intersubject variability (26%), and interclass correlation coefficient (0.89). When changes of receptor density were simulated, V_T was shown to be more sensitive than brain uptake alone for detecting both increases and decreases. Acquisition times longer than 120 min were vulnerable to brain influx of radiometabolites. Radioactivity also accumulated in the skull throughout the entire scan but was thought to be an insignificant source of data contamination (<2% of total radioactivity) (Terry et al., 2010b).

One limitation of [^{18}F]FMPEP-d_2 is that its plasma free fraction in humans is also low (~0.63%). Due to this nature, small differences in free fraction would be disproportionately large in percentage. Correcting V_T for free fraction, which is suggested by the authors to be a more correct quantification, added too much noise to the final outcome measurements and was therefore excluded. Nevertheless, the authors still recommended accounting in future studies for potential changes in free fraction, particularly with pharmacological challenges, as a proportional amount could be displaced from plasma proteins and may enter the brain.

Whole-body imaging studies using [^{11}C]MePPEP and [^{18}F]FMPEP-d_2 demonstrated the radiation burden to be acceptable for multiple brain studies per year, i.e. 4.6 ± 0.3 μSv/MBq and 19.7 ± 2.1 μSv/MBq, respectively. Brain uptake of both radioligands was ~7–8% of the total dose. Receptor occupancy is estimated to be well below 1%. [^{11}C]MePPEP undergoes exclusive hepatobiliary excretion, while [^{18}F]FMPEP-d_2 is also excreted through the urinary tract (Terry et al., 2010a).

2.2.1.2.3. [^{18}F]MK-9470

Mid-past decade, Merck researchers disclosed a series of acyclic, non-SR141716-based CB_1 receptor inverse agonists (Lin et al., 2006). Subsequent conformational analysis and receptor docking studies showed that the same binding area of CB_1 was targeted as for SR141716 (Lin et al., 2008). The pharmacological lead compound of the series, taranabant, demonstrated a substantially higher affinity (K_i = 0.4 nM) than SR141716. The optimization of the taranabant structure for PET application led to a fluoroalkyl analog MK9470 with high CB_1 receptor affinity (K_i = 0.7 nM) (Figure 3) (Lin et al., 2006). The CB_1 receptor subtype selectivity of the compound MK9470 was about 60-fold greater than CB_2 receptors. MK9470 is not a substrate for P-gp, but PSA and log D values were relatively high. In vitro affinity testing with a battery of over 100 known targets yielded no off-target activity below the micromolar level (Burns et al., 2007). [^{18}F]MK-9470 is prepared by [^{18}F]fluoroethylation of the corresponding phenol precursor (Burns et al., 2007). The radiochemical yield is about 4% with a specific activity of >10 GBq/μmol.

[^{18}F]MK-9470 binds with relatively slow kinetics to the CB_1 receptor and is readily displaced by unlabeled MK9470 in rats (Casteels et al., 2012), by CB_1 receptor antagonists such as the SR141716-analog AM251, and by MK0364, an acyclic inverse agonist structurally analogous to MK9470 in monkey brain (Burns et al., 2007). Blocking studies in rats showed that ~56% of the radioligand binding in brain was CB_1 receptor specific. A polar radiometabolite was detected in plasma and brain. The brain radiometabolite concentration was uniform across the whole brain and remained relatively constant from 10 to 240 min, ~13% of the total activity concentration. The kinetics of [^{18}F]MK-9470 in rat brain can be modeled using a one-tissue compartment model with and without constrained radiometabolite input (Casteels et al., 2012).

In the rhesus monkey, the time–activity curve of [^{18}F]MK-9470 showed no to very slow washout, indicating that slow dissociation kinetics may hamper easy quantification (Burns et al., 2007). After 120 min, the target-to-nontarget ratio was about 1.9.

The human biodistribution and radiation dosimetry study showed that brain uptake was about 5% of the injected activity and that [^{18}F]MK-9470 showed predominantly hepatobiliary excretion and an average effective dose of 22.9 μSv/MBq (Van Laere et al., 2008b).

After IV injection in humans, [^{18}F]MK-9470 radioactivity in brain increased throughout the scanning period (120 min), even though radioligand concentration in arterial plasma decreased throughout the length of the scan (Burns et al., 2007). The highest uptake was observed in the striatum, frontal cortex, and posterior cingulate; intermediate uptake was seen in the cerebellum; and the lowest uptake is observed in the thalamus, pons, and hippocampus. Despite the favorable properties of [^{18}F]MK-9470 for imaging brain CB_1 receptors, including good brain uptake and low nonspecific binding, its slow kinetics was a challenge for modeling acceptable outcome measures such as V_T within clinically applicable measurement times (Sanabria-Bohorquez et al., 2010). A reversible two-tissue compartment model using a global k_4 value was necessary to describe brain kinetics. Both V_T and $V_{ND}k_3$ were estimated satisfactorily and their test–retest variability was between 10% and 30%. The irreversible macroparameter K_i modeled the data well. The linear relationship between K_i and $V_{ND}k_3$ demonstrated that K_i also provides a reliable index of receptor binding. Fractional uptake ratios (FUR) which can be measured using a limited set of venous samples, were shown to be equivalent to K_i. More simplified brain uptake measurements (standardized uptake value (SUV) and modified SUV (mSUV)) were reasonably well correlated to FUR. The authors concluded that in cases when plasma measurements are not statistically different, SUV and mSUV are sufficient outcome measures. Using this analytical method, we found that [^{18}F]MK-9470 had good precision (test–retest variability <7%) and intersubject variability (16–35%).

Also CB-119, a [^{11}C]methoxy analog of MK-9470, was developed by Merck as a carbon-11 alternative for PET imaging (Hamill et al., 2009). Although its binding characteristics were nearly identical to [^{18}F]MK-9470, it did not achieve an equilibrium within scanning time and was therefore abandoned for further clinical development.

2.2.1.3. (Pre)Clinical Applications of CB_1 PET Imaging

As the second-generation of CB_1 radioligands exhibited a better combination of good binding affinity and lipophilicity, they were subsequently used in pathophysiological settings to investigate CB_1 receptor expression alterations. To date, most data on CB_1 receptor imaging in preclinical and clinical conditions, apart from the work described on tracer validation, make use of [^{18}F]FMPEP-d_2, [^{11}C]-OMAR, and [^{18}F]MK-9470. We have listed below the main in vivo findings of these studies.

[^{18}F]FMPEP-d_2 has been used to demonstrate a reversible downregulation of CB_1 receptors in chronic cannabis users (Figure 4) (Hirvonen et al., 2012). V_T of [^{18}F]FMPEP-d_2 was ~20% lower in cannabis smokers than in control subjects in all neocortical but not in subcortical regions. Among the cannabis smokers, years of smoking negatively correlated with V_T in cortical regions: subjects who had smoked cannabis for longer time had smaller V_T than subjects who had smoked cannabis for a shorter time. This correlation was not confounded by effects of age. Following about 1 month of cannabis abstinence, V_T increased in all the affected regions, with the exception of the hippocampus, pointing to reversibility of the downregulation. Body mass index (BMI) was used as an adjustment variable in all findings as BMI negatively correlated to V_T. In addition, BMI or cannabis use did not correlate with the free tracer fraction in plasma, which is a potential confounding variable (see Section 3.1.2.2 for more detail). Neither values of V_T

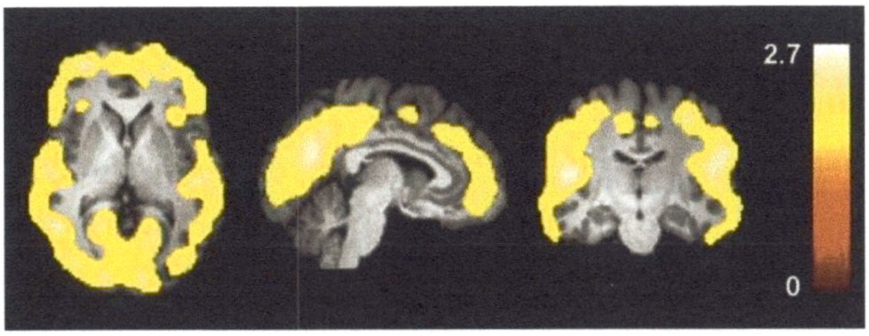

Figure 4 Statistical parametric mapping (SPM) analysis shows lower V_T in chronic daily cannabis smokers ($n = 30$) than in control subjects ($n = 28$) at baseline as a large single cluster that includes only cortical regions. This cluster comprised 67,513 voxels, had a maximum *t*-value of 2.8 at (−34, −78, 16), and had a cluster-level corrected *P*-value of 0.043. Color bar represents *t*-value in each voxel within the significant cluster. *Published with permission from Hirvonen et al. (2012).*

nor the amount of recovery of V_T correlated with age of first use, amount of daily use, or clinical scales rating withdrawal or craving. Tobacco smoking was more prevalent among cannabis smokers than among control subjects, but did not show effect on V_T.

The same group used [^{18}F]FMPEP-d_2 PET in alcohol-dependent patients in early and protracted abstinence and in comparison to control subjects without alcohol use to characterize brain CB_1 receptor alterations in these conditions (Hirvonen et al., in press). Overall, V_T of [^{18}F]FMPEP-d_2 was lower in patients with alcohol dependence than in control subjects. The magnitude of the decreases in BMI-adjusted V_T ranged from 21% in the hippocampus to 35% in the white matter. Among patients with alcohol dependence, years of alcohol use correlated negatively with V_T in the midbrain, cingulate cortex, and putamen: patients who had been drinking longer had lower V_T than those who had been drinking for a shorter period. This correlation was not confounded by effects of age. Whole-brain V_T did not correlate with alcohol dependence severity, measures of craving, drinks per day, withdrawal scores, benzodiazepine dose received, or ratings of depression and anxiety. In contrast to their findings in cannabis users, protracted abstinence did not change V_T in any brain region and thus remained decreased at least 2–4 weeks into abstinence. Irrespective of the diagnostic status, C allele carriers at rs2023239 had higher CB_1 receptor binding compared with noncarriers.

First PET studies with [^{18}F]MK-9470 in humans have been performed in drug occupancy trials with new antiobesity medication indicating that the doses of taranabant evaluated in a 12-week weight-loss efficacy study were associated with brain CB_1 receptor occupancies of approximately 10–40% (Addy et al., 2008).

Later on, the Leuven group used [^{18}F]MK-9470 in (pre)clinical settings to investigate CB_1 receptor variability with gender and normal aging (Van Laere et al., 2008a; Van Laere et al., 2009) and in studying pathophysiological conditions (Gerard et al., 2011; Van Laere et al., 2010; Goffin et al., 2011; Van der Schueren et al., 2012). We demonstrated gender-dependent increases of [^{18}F]MK-9470 uptake in women but not in men, particularly in the basal ganglia, lateral temporal cortex, and limbic system (Van Laere et al., 2008a).

In an attempt to explain the large natural intersubject variability, we investigated CB_1 receptor availability in relation to temperament traits (Van Laere et al., 2009). Voxel-based analysis demonstrated that novelty-seeking traits were inversely correlated with global cerebral CB_1 receptor availability, most pronounced in the amygdala. Since novelty seeking is associated with thrill seeking, addictive behavior, and impulsive aggression (Hiroi and Agatsuma, 2005), their link with $[^{18}F]$MK-9470 PET, particularly in the amygdala, is intriguing.

In HD, parametric images of partial-volume–corrected receptor availability showed a widespread decrease of CB_1 in HD patients, compared with controls (Figure 5). The reductions of CB_1 availability ranged from ~15% in the cerebellum to ~25% in the frontal cortex. Interestingly, these reductions were observed irrespective of the disease stage, duration, CAG repeat, or patient age, but related to disease burden in $[(CAG)_n$xage] in prefrontal areas. The mechanism by which HD promotes this loss of CB_1 receptors is suggested to be caused by interactions between mutant htt and nuclear transcription factors of CB_1 (Blázquez et al., 2011). Blázquez and coworkers showed in striatal cells that mutant htt downregulates CB_1 receptors by controlling gene promoter activity via repressor element 1-silencing transcription factor and sensitizes cells to excitotoxic damage. In line with this, $[^{18}F]$MK-9470 PET may thus be a useful tool to assess early symptomatic, potentially presymptomatic HD. Our group found similar reductions in the brain uptake of $[^{18}F]$MK-9470 in a transgenic rat model of HD in comparison to wild-type littermates (Casteels et al., 2011). However, in the latter study, reductions were only detected in the basal ganglia. Rats unilaterally lesioned

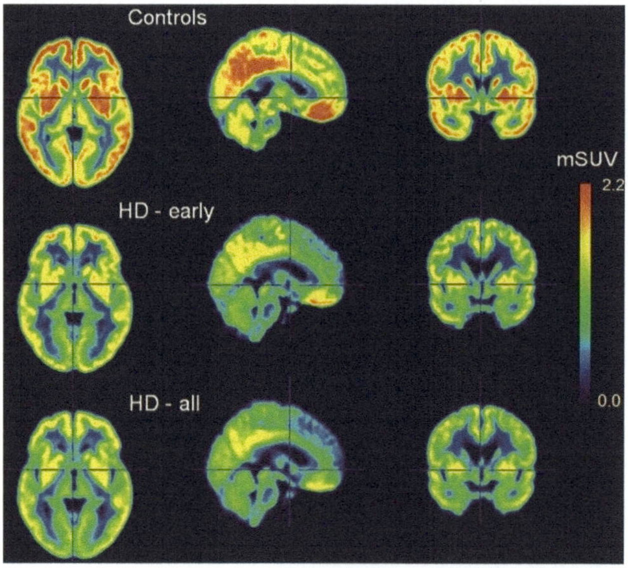

Figure 5 Cross-sectional, partial-volume corrected, modified standardized uptake value (mSUV) parametric images, averaged for all controls ($n=14$), early HD patients ($n=7$), and total group of HD patients ($n=19$).

with quinolinic acid in the left caudate putamen to resemble the cell death characteristic of HD demonstrated a disproportional modest decrease in [^{18}F]MK-9470 on the ipsilateral side as compared to the lesion severity (Casteels et al., 2010c). In contrast, an increase in [^{18}F]MK-9470 binding was observed on the contralateral side and in the cerebellum, of which the latter corresponded to improved functional outcome.

In patients with PD, CB_1 receptor availability using [^{18}F]MK-9470 PET was altered in a region-specific pattern (Van Laere et al., 2012). Voxel-based analysis revealed a decrease in CB_1 receptor availability in the bilateral ventral mesencephalic region, comprising the substantia nigra. This reduction of CB_1 availability was not correlated with disease duration and equally present in early disease. When early PD patients were compared with controls, increased CB_1 receptor availability, relative to total brain uptake, was observed in the putamen, anterior insula, prefrontal cortex, hippocampal area, and midcingulate. Interestingly, the putaminal changes were only significant on the side contralateral to the clinically affected body side. Similar results were found when comparing the advanced PD group with controls, with additional clusters in the lower prefrontal and midcingulate cortex. Putaminal changes were now bilateral, although still most pronounced contralaterally to the clinically most affected body side. Regionally increased CB_1 receptor availability in PD could be explained by compensatory processes due to imbalances in the glutamatergic–dopaminergic pathways. Comparison of PD patients with and without LID did not reveal any differences in [^{18}F]MK-9470 binding.

In a rodent model of PD using 6-hydroxydopamine, regional alterations in CB_1 receptor binding were detected as well, pointing to decreased [^{18}F]MK-9470 uptake in the bilateral caudate putamen and in the contralateral cerebellum relative to the lesioned side. [^{18}F]MK-9470 uptake in the cerebellum inversely correlated to remaining dopaminergic neurons (Casteels et al., 2010b).

In patients with AN, a whole-brain increase of CB_1 receptor availability was observed compared to healthy volunteers and to patients with bulimia nervosa (BN) (Figure 6) (Gerard et al., 2011). The average gray matter increases were 24.5% in comparison to control subjects and 14.7% compared with the BN group.

[^{18}F]MK-9470 PET uptake was relatively increased in the insula for both AN and BN and also in areas of frontotemporal cortex for AN. The increase in CB_1 availability was interpreted as reactive to a hypoactive endocannabinoid transmission (Monteleone et al., 2009).

[^{18}F]MK-9470 PET in patients with temporal lobe epilepsy due to hippocampal sclerosis provided a first in vivo perspective on the role of CB_1 receptors in a locally hyperexcitable and excitotoxic environment (Goffin et al., 2011). An increased CB_1 receptor availability in the ipsilateral temporal lobe was observed in patients, which correlated negatively with the latency since last seizure before scanning and positively to the number of seizures in the month before scanning. These changes in brain regions that are involved during complex partial seizures in refractory mesial temporal lobe epilepsy might be a protective mechanism of neurons against hyperexcitability and seizure

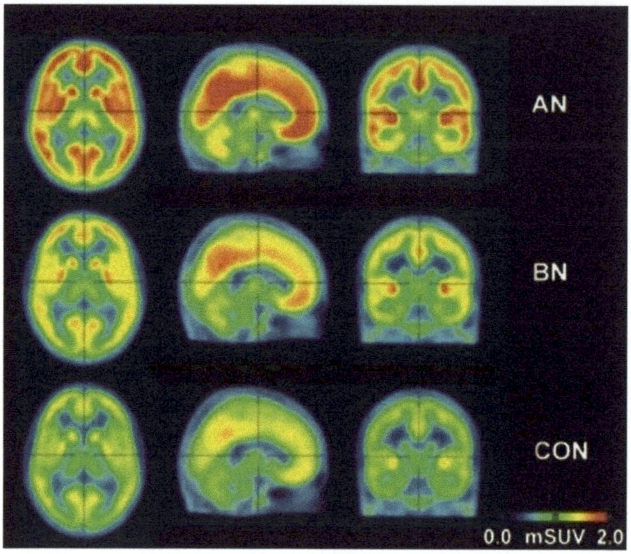

Figure 6 Mean parametric images of global [^{18}F]MK-9470 binding in the anorexia nervosa (AN), bulimia nervosa (BN), and control group (CON). Modified standard uptake values (mSUV) are indicated by the color bar.

activity, contribute to the process of epileptogenesis, or play a role in the surround inhibition and prevention of seizure propagation.

Finally, in a photothrombotic stroke model, [^{18}F]MK-9470 PET showed a strong increase in CB_1 binding 24 and 72 h after stroke in the cortex surrounding the lesion, pointing to a relatively early upregulation. Nevertheless, no increase in CB_1 binding was observed with the early 3-h intervention window (Vandeputte et al., 2012).

Preclinical imaging also allowed performing in vivo pharmacological interference studies. The effects on CB_1 binding were evaluated for nicotine (Gerard et al., 2010), valproic acid and levetiracetam (Goffin et al., 2008), for L-DOPA and bromocriptine (Casteels et al., 2010d), as well as for isoflurane and pentobarbital anesthesia (Casteels et al., 2010a). These studies reported no change in CB_1 for nicotine, levetiracetam, L-DOPA, and bromocriptine, while valproic acid caused a whole-brain increase. [^{18}F]MK-9470 uptake values obtained applying global scaling were decreased in the cortex under isoflurane and pentobarbital anesthesia, while an increase was seen in the cerebellum for both of them (Casteels et al., 2010a).

2.2.1.4. Future Perspectives

The three CB_1 receptor radioligands that have been used clinically (as presented above) have their own pros and cons. [^{11}C]OMAR with its relatively fast kinetics has the advantage to enable performing multiple CB_1 PET scans on the same day in the same subject. CB_1 receptor density can be quantified using V_T and its fast kinetics is attributed to a modest binding affinity, likely also contributing to lower brain uptake as compared to the others. [^{18}F]MK-9470 has high brain uptake, good image contrast, and has been studied in several pathophysiological conditions, despite its slow kinetics limiting quantification at clinically feasible times. [^{18}F]FMPEP-d_2 has high brain uptake that can be quantified by V_T, without large methodological variability as was seen for [^{11}C]MPEPP.

As demonstrated for [^{11}C]MePPEP, current CB_1 receptor radioligands are not displaceable in vivo by endogenous or synthetic agonists. Therefore, development of agonist radioligands that demonstrate preference for the high-affinity state of CB_1 and are more susceptible to competition by endocannabinoids would thus be of great value for future research. Their development would provide tremendous new insights onto the ECS, particularly on the regional selectivity of the endocannabinoid (EC) synthesis following inhibition of its degradation; on the interaction of EC with GABA, glutamate, and dopamine; and on the effect of acute stress on EC levels.

So far, the status of the CB_1 receptor has been explored in neurological diseases, substance dependence, eating disorders, schizophrenia, epilepsy, and personality traits, whereas studies involving mood or anxiety disorders, dementia, neuroinflammation, or non-CNS diseases have yet to be reported. Non-CNS applications may be liver fibrosis, metabolic diseases such as diabetes, and cancer. At present, there is not sufficient data to determine whether currently available radioligands would be suitable for extracerebral application.

2.2.2. CB_2 Receptor

Compared to CB_1 receptor radioligands, the development of radioligands for brain CB_2 receptor imaging is still in its pioneering phase. Only a limited number of publications are available at the moment and are focused on the development of PET ligands. Present CB_2 receptor radioligands can be categorized into four groups, depending on structure characteristics, i.e. derivates of triaryl bis-sulfones, of indoles, of quinolines, and of thiazoles. The physicochemical and in vivo imaging properties of currently available CB_2 receptor radioligands are listed in Table 3.

2.2.2.1. Triaryl Bis-Sulfones

In 2005, a new class of CB_2 receptor inverse agonists, namely, the triaryl bis-sulfones, was reported (Lavey et al., 2005). The compound with the best combination of CB_2 receptor affinity (K_i hCB_2 = 0.4 nM) and selectivity for CB_2 receptor over CB_1 receptor (K_i hCB & K_i hCB_2 = 2262) was Sch225336 (Shankar et al., 2005). Sch225336 showed immunomodulatory characteristics in both in vitro and in vivo experiments (Lunn et al., 2006). Schering-Plough Research Institute managed to label Sch225336 with sulfur-35. Sulfur-35 is a long-lived ($T_{1/2}$ = 87 days) β^--emitting isotope, thus not suitable for in vivo imaging, but useful for in vitro studies. Labeling with sulfur-35 has the advantage of achieving higher specific activities compared to tritium labeling, and thereby providing more sensitive detection of CB_2 receptors (Gonsiorek et al., 2006). Autoradiography studies incubating [^{35}S]Sch225336 with human spleen resulted in labeling areas representing splenic white pulp (Gonsiorek et al., 2006).

Table 3 Physicochemical and Imaging Properties of CB_2 Receptor Radioligands

	[^{11}C]Sch225336	[^{11}C]X1	[^{11}C]GW405833	[^{18}F]FE-GW405833	[^{11}C]NE40	[^{11}C]A836339	[^{18}F]X2
Molecular weight (MW)	540	395	447	356	372	310	398
log $D_{7.4}$	2.2	4.6	2.5	2.8	3.9	3.4	4.4
Polar suface area (PSA)	132	39	43.7	43.7	80.4	42	77
K_i hCB_2 R (nM)	4.5	0.3	35	27.2	9.6	0.7	3.4
K_i hCB_1 R (nM)	78.5	3333	6450	10,000	>1000	>600	–
Brain uptake[1]	0.1	–	1.4	1.3	1.7	–	–
Brain washout[2]	1	–	14	1.9	4.3	–	–
Spleen uptake[1]	0.5	–	1.0	0.9	0.9	–	–
Spleen clearance[2]	5	–	2	3	1.3	–	–

[1]%ID 2 min postinjection in mice.
[2]%ID 2 min postinjection/60 min postinjection ratio in mice.

In cooperation with Schering-Plough, Evens et al. reported labeling of Sch225336 with carbon-11 (Figure 7) (Evens et al., 2008). Demethylation of [^{11}C]Sch225336 with boron tribromide provided two isomeric monomethoxy derivates, one of which was used for methylation with [^{11}C]methyl iodide providing [^{11}C]Sch225336. Despite its favorable log *D* value of 2.15, [^{11}C]Sch225336 exhibited low brain uptake in mice. This is in accord with its high PSA value and high MW of 540 that are above the conventional limits for passive BBB diffusion. Moreover, its brain uptake increased following the administration of cyclosporine A, which inhibits several BBB efflux transporters such as P-gp, indicating that [^{11}C]Sch225336 is an efflux transporter substrate. No specific binding of [^{11}C]Sch225336 to spleen tissue or blood cells was observed in vivo in mice, in contrast to the results with [^{35}S]Sch225336 (Gonsiorek et al., 2006). No further studies on [^{11}C]Sch225336 have been published.

In addition, the Chiba University PET Center selected from a series of triaryl CB_2 radioligands [^{11}C]-X1 (Fujinaga et al., 2010). [^{11}C]-X1 exhibited a high binding affinity of 0.3 nM. [^{11}C]-X1 showed good BBB penetration in mice, but its slow brain washout suggests substantial nonspecific binding that can be explained by its high lipophilicity.

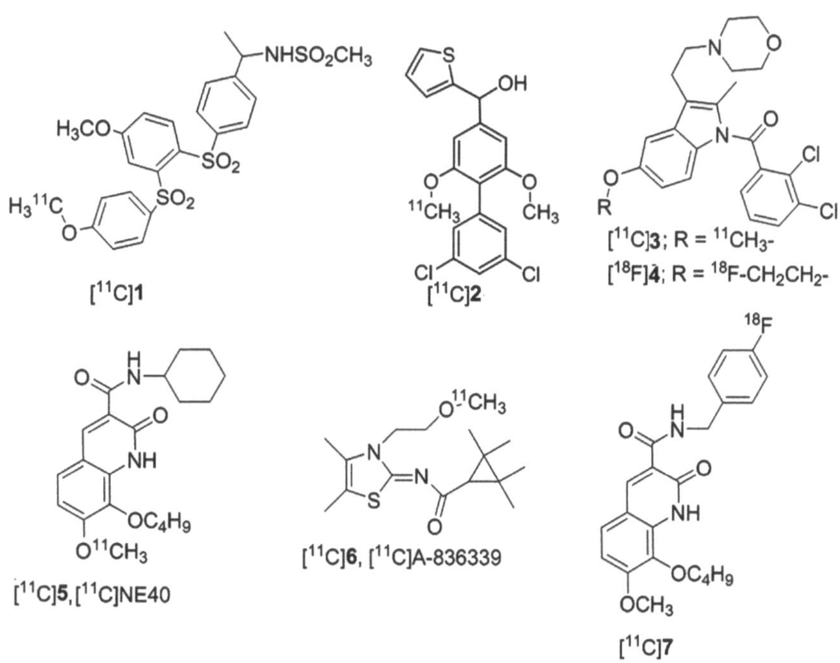

Figure 7 Structures of currently available PET radioligands for CB_2 receptor imaging.

2.2.2.2. Indole Derivates

GW405833 is a well-known CB_2 receptor agonist with high affinity and selectivity (Valenzano et al., 2005). It has been used in several preclinical studies to investigate the antiinflammatory properties and antinociceptive characteristics of CB_2 receptor agonists (LaBuda et al., 2005; Whiteside et al., 2005; Hu et al., 2009). The Leuven group labeled GW405833 with carbon-11 and also synthesized the [^{18}F]fluoroethyl derivate ([^{18}F]FE-GW405833) (Figure 7) (Evens et al., 2011). Both [^{11}C]GW405833 and [^{18}F]FE-GW405833 showed moderate affinity for the CB_2 receptor and good BBB permeability. Unfortunately, [^{18}F]FE-GW405833 had much slower washout of radioactivity from the mouse brain than [^{11}C]GW405833, which was due to a large fraction of brain radiometabolites (50% at 30 min after injection), limiting its further development. None of these tracers showed spleen retention in mice or rats.

Taking advantage of the low expression of CB_2 receptors in healthy brain, this target can be used for developing a brain reporter gene system to further validate the abovementioned radioligands in vivo (Vandeputte et al., 2011). Both [^{11}C]GW405833 and [^{18}F]FE-GW405833 showed higher binding in the striatum where CB_2 receptor expression was induced by stereotactic injection of the vector in comparison to the contralateral sham-injected striatum. Binding was displaceable by IV injection of unlabeled GW405833, 20 min after tracer injection, thereby confirming reversibility of tracer binding. However, due to their relatively low binding affinity, both [^{11}C]GW405833 and [^{18}F]FE-GW405833 are unlikely candidates for further imaging in neuroinflammatory conditions. Also, [^{11}C]GW405833 showed slow washout and high nonspecific binding in healthy monkey brain (Vandeputte et al., 2011).

GW842166X is a CB_2 receptor agonist that showed promising antihyperalgesia properties in animal pain models and that has entered human trials for the treatment of inflammatory pain (Giblin et al., 2009). It is thought that GW842166X cross the BBB to perform its analgesic actions. Therefore, a PET study using carbon-11 labeled GW842166X was conducted (http://clinicaltrials.gov). However, no further data on the radiosynthesis or outcome of the study have been published to date.

2.2.2.3. Quinoline Derivates

Quinoline derivates have been extensively studied as CB_2 receptor radioligands. Of them, JT3-907 is a well-characterized CB_2 receptor inverse agonist (Iwamura et al., 2001). The Leuven group synthesized 2-oxo-7-[^{11}C]methoxy-8-butyloxy-1,2-dihydroquinoline-3-carboxylic acid cyclohexylamide (NE40) and 2-oxo-7-[^{18}F]fluoroethoxy-8-butyloxy-1,2-dihydroquinoline-3-carboxylic acid cyclohexylamide (Figure 7) (Evens et al., 2009). By shortening the lipophilic carbon tail from a pentoxy to a butoxy group, a decrease of nonspecific binding related to the lipophilicity of the tracer was envisaged. In competition binding studies, both compounds showed low nanomolar affinity for the CB_2 receptor. This was in line with the results of in vivo biodistribution studies in normal mice, where both tracers showed high spleen uptake and spleen retention. This spleen uptake was inhibited by pretreatment of mice with 1-(2,4-dichlorophenyl)-6-fluoro-*N*-piperidin-1-yl-1,4-dihydroindeno[1,2-*c*]pyrazole-3-carboxamide, a potent rodent CB_2 receptor inverse agonist (Mussinu et al., 2003), thereby strongly suggesting that retention in spleen is specific for CB_2 receptor binding.

In the rat model with local human (h)-CB_2 receptor expression, [^{11}C]NE40 demonstrated specific and reversible binding to hCB_2 receptors. [^{11}C]NE40 was successfully evaluated in a substantial number of preclinical safety studies (Evens et al., 2012). [^{11}C]NE40 is also the first CB_2 receptor PET radioligand that was selected for human PET studies. In healthy human brain, [^{11}C]NE40 exhibited rapid uptake and washout. Preliminary findings demonstrated predominantly hepatobiliary excretion and an effective dose of 4.4 μSv/MBq in three healthy males (Ahmad, personal communication). Apart from liver and intestines, [^{11}C]NE40 retention was observed in lymph nodes and spleen, indicating CB_2 receptor binding. Clinical PET studies using [^{11}C]NE40 in pathological conditions are ongoing.

In addition, Turkman and coworkers from the MD Anderson Cancer Center have also published a series of 2-oxoquinoline derivates (Turkman et al., 2011) and reported one as a suitable candidate for PET studies (Turkman et al., 2012). Designed to be a metabolically stable fluorine-18 compound, [^{18}F]-X2 (7-methoxy-8-butoxy-2-oxo-1,2-dihydroquinoline-3-carboxylic acid-(4-fluorobenzyl)amide-[^{18}F]) had poor uptake in the spleen and only about 50% specific binding on CB_2 receptor-positive tumor cells transfected in mouse. Preclinical and clinical applications of this compound are hampered by its low solubility.

2.2.2.4. Thiozole Derivates

The John Hopkins University group recently synthesized [^{11}C]-A836339 [2,2,3,3 tetramethyl-cyclopropanecarboxylic acid [3-(2-methoxy-ethyl)-4,5-dimethyl-3*H*-thiazol-(2*Z*)ylidene]-amide], a selective CB_2 receptor agonist with high binding affinity, moderate lipophilicity, and an adequate PSA value for CNS application (Table 2) (Horti et al., 2010). [^{11}C]-A836339 shows in healthy CD1 mice-specific binding in the spleen. It also exhibits good BBB permeability and subtle amount of specific binding in healthy mice brain. This is in agreement with the low expression of CB_2 receptors in this condition (Munro et al., 1993). Specific binding of [^{11}C]-A836339 was further studied in two animal models of neuroinflammation, a lipopolysaccharide (LPS)-induced mouse model and a transgenic amyloid mouse model of AD (APPswe/PS1dE9 mice). Pretreatment studies showed that ~78–84% of the brain radioactivity in LPS-treated mice was specific for the CB_2 receptor. Similar values of CB_2 receptor expression levels have been reported previously in this model (Mukhopadhyay et al., 2006). This high cerebral uptake of [^{11}C]-A836339 may, however, in part be associated with dysfunction of the BBB.

Brain distribution studies of [^{11}C]-A836339 in the AD mouse model showed that [^{11}C]-A836339 display in vivo ~29–33% of specific binding in various brain regions, which is consistent with the distribution of Aβ plaques in this model (Benito et al., 2003) (Table 3).

2.2.2.5. *Future Perspectives*

Of the above radioligands, quinoline derivate [^{11}C]NE40 seems to exhibit the most promising tracer characteristics, due to a combination of specific CB_2 receptor binding to mouse and rat spleen, high brain uptake, and in vivo binding to human CB_2 receptors. However, the specificity may be insufficient, as ideally the affinity of the radioligand for its target should be 5–10 times smaller compared to the B_{max} ($B_{max}/K_d > 5$–10). Consequently, to be able to visualize the rat and mouse spleen CB_2 receptor density, radioligands with subnanomolar affinity are likely required, as a B_{max} value of 0.71 pmol/mg for rat spleen and a B_{max} of 0.31 pmol/mg for mouse spleen have been described using [^{3}H]CP55,940 (Govaerts et al., 2004). Absolute measures of CB_2 receptor expression levels have also not yet been reported in neuroinflammatory conditions. It remains difficult to estimate the required affinity for this purpose and further study of this aspect is needed. Ongoing (pre)clinical studies will further clarify the usefulness of currently available radioligands.

3. CONCLUSIONS

In this chapter, we have described the potential of the available PET/SPECT radioligands for imaging the brain cannabinoid receptors, encompassing their successful development, some of their pitfalls, and the receptor quantification specifics. Akin to its description as a nonclassical neurotransmitter system, the development of suitable radioligands has challenged pharmaceutical development toward the boundaries. Direct in vivo imaging of the ECS will, however, provide an essential window onto basic operation of the normal brain, neurotransmitter feedback loops, and its role in a wide variety of neuropsychiatric diseases either by the modulatory aspects of the CB_1 receptor or by contribution of the CB_2 receptor in neuroinflammatory and neuroprotective response.

REFERENCES

Addy, C., et al., 2008. The acyclic CB_1R inverse agonist taranabant mediates weight loss by increasing energy expenditure and decreasing caloric intake. Cell Metab. 7, 68–78.

Aguado, T., Monory, K., Palazuelos, J., Stella, N., Cravatt, B., Lutz, B., Marsicano, G., Kokaia, Z., Guzman, M., Galve-Roperh, I., 2005. The endocannabinoid system drives neural progenitor proliferation. FASEB J. 19, 1704–1706.

Arias, H.F., 2007. Cannabinoids in eating disorders and obesity. Mol. Neurobiol. 36, 113–128.

Ashton, C.H., Moore, P.B., 2011. Endocannabinoid system dysfunction in mood and related disorders. Acta Psychiatr. Scand. 124, 250–261.

Bambico, F.R., Nguyen, N.T., Katz, N., Gobbi, G., 2010. Chronic exposure to cannabinoids during adolescence but not during adulthood impairs emotional behaviour and monoaminergic neurotransmission. Neurobiol. Dis. 37, 641–655.

Begg, M., Pacher, P., Batkai, S., Osei-Hyiaman, D., Offertaler, L., Mo, F.M., Liu, J., Kunos, G., 2005. Evidence for novel cannabinoid receptors. Pharmacol. Ther. 106, 133–145.

Beltramo, M., Piomelli, D., 2000. Carrier-mediated transport and enzymatic hydrolysis of the endogenous cannabinoid 2-arachidonylglycerol. Neuroreport 11, 1231–1235.

Benarroch, E., 2007. Endocannabinoids in basal ganglia circuits: implications for Parkinson disease. Neurology 69, 306–309.

Benarroch, E.E., 2010. Neural control of feeding behavior: overview and clinical correlations. Neurology 74, 1643–1650.

Benito, C., Nunez, E., Tolon, R.M., Carrier, E.J., Rabano, A., Hillard, C.J., Romero, J., 2003. Cannabinoid CB_2 receptors and fatty acid amide hydrolase are selectively overexpressed in neuritic plaque-associated glia in Alzheimer's disease brains. J. Neurosci. 23, 11136–11141.

Benzinou, M., Chevre, J.C., Ward, K.J., Lecoeur, C., Dina, C., Lobbens, S., Durand, E., Delplanque, J., Horber, F.F., Heude, B., Balkau, B., Borch-Johnsen, K., Jorgensen, T., Hansen, T., Pedersen, O., Meyre, D., Froguel, P., 2008. Endocannabinoid receptor 1 gene variations increase risk for obesity and modulate body mass index in European populations. Hum. Mol. Genet. 17, 1916–1921.

Berding, G., Muller-Vahl, K., Schneider, U., Gielow, P., Fitschen, J., Stuhrmann, M., Harke, H., Buchert, R., Donnerstag, F., Hofmann, M., Knoop, B.O., Brooks, D.J., Emrich, H.M., Knapp, W.H., 2004. [^{123}I]AM281 single-photon emission computed tomography imaging of central cannabinoid CB_1 receptors before and after Δ^9-tetrahydrocannabinol therapy and whole-body scanning for assessment of radiation dose in tourette patients. Biol. Psychiatry 55, 904–915.

Berding, G., Schneider, U., Gielow, P., Buchert, R., Donnerstag, F., Brandau, W., Knapp, W.H., Emrich, H.M., Muller-Vahl, K., 2006. Feasibility of central cannabinoid CB_1 receptor imaging with [^{124}I]AM281 PET demonstrated in a schizophrenic patient. Psychiatry Res. 147, 249–256.

Blázquez, C., et al., 2011. Loss of striatal type 1 cannabinoid receptors is a key pathogenic factor in Huntington's disease. Brain 134, 119–136.

Bonhomme-Faivre, L., Benyamina, A., Reynaud, M., Farinotti, R., Abbara, C., 2008. Disposition of Delta tetrahydrocannabinol in CF1 mice deficient in mdr1a P-glycoprotein. Addict. Biol. 13, 295–300.

Breivogel, C.S., Griffin, G., Di Marzo, V., Martin, B.R., 2001. Evidence for a new G protein-coupled cannabinoid receptor in mouse brain. Mol. Pharmacol. 60, 155–163.

Brotchie, J.M., 2003. CB_1 cannabinoid receptor signalling in Parkinson's disease. Curr. Opin. Pharmacol. 3, 54–61.

Burns, H.D., et al., 2007. [^{18}F]MK-9470, a positron emission tomography (PET) tracer for in vivo human PET brain imaging of the cannabinoid-1 receptor. Proc. Natl. Acad. Sci. USA 104, 9800–9805.

Cabral, G.A., Marciano-Cabral, F., 2005. Cannabinoid receptors in microglia of the central nervous system: immune functional relevance. J. Leukoc. Biol. 78, 1192–1197.

Cahill, K., Ussher, M.H., 2011. Cannabinoid type 1 receptor antagonists for smoking cessation. Cochrane Database Syst. Rev., CD005353.

Caille, S., varez-Jaimes, L., Polis, I., Stouffer, D.G., Parsons, L.H., 2007. Specific alterations of extracellular endocannabinoid levels in the nucleus accumbens by ethanol, heroin, and cocaine self-administration. J. Neurosci. 27, 3695–3702.

Calatozzolo, C., Salmaggi, A., Pollo, B., Sciacca, F.L., Lorenzetti, M., Franzini, A., Boiardi, A., Broggi, G., Marras, C., 2007. Expression of cannabinoid receptors and neurotrophins in human gliomas. Neurol. Sci. 28, 304–310.

Caspi, A., Moffitt, T.E., Cannon, M., McClay, J., Murray, R., Harrington, H., Taylor, A., Arseneault, L., Williams, B., Braithwaite, A., Poulton, R., Craig, I.W., 2005. Moderation of the effect of adolescent-onset cannabis use on adult psychosis by a functional polymorphism in the catechol-O-methyltransferase gene: longitudinal evidence of a gene X environment interaction. Biol. Psychiatry 57, 1117–1127.

Casteels, C., Bormans, G., Van Laere, K., 2010a. The effect of anaesthesia on [^{18}F]MK-9470 binding to the type 1 cannabinoid receptor in the rat brain. Eur. J. Nucl. Med. Mol. Imaging 37, 1164–1173.

Casteels, C., Lauwers, E., Baitar, A., Bormans, G., Baekelandt, V., Van Laere, K., 2010b. In vivo type 1 cannabinoid receptor mapping in the 6-hydroxydopamine lesion rat model of Parkinson's disease. Brain Res. 1316, 153–162.

Casteels, C., Martinez, E., Bormans, G., Camon, L., de Vera, N., Baekelandt, V., Planas, A.M., Van Laere, K., 2010c. Type 1 cannabinoid receptor mapping with [^{18}F]MK-9470 PET in the rat brain after quinolinic acid lesion: a comparison to dopamine receptors and glucose metabolism. Eur. J. Nucl. Med. Mol. Imaging 37, 2354–2363.

Casteels, C., Vanbilloen, B., Vercammen, D., Bosier, B., Lambert, D.M., Bormans, G., Van, L.K., 2010d. Influence of chronic bromocriptine and levodopa administration on cerebral type 1 cannabinoid receptor binding. Synapse 64, 617–623.

Casteels, C., Koole, M., Celen, S., Bormans, G., Van Laere, K., 2012. Preclinical evaluation and quantification of [^{18}F]MK-9470 as a radioligand for PET imaging of the type 1 cannabinoid receptor in rat brain. Eur. J. Nucl. Med. Mol. Imaging 39, 1467–1477.

Casteels, C., Vandeputte, C., Rangarajan, J.R., Dresselaers, T., Riess, O., Bormans, G., Maes, F., Himmelreich, U., Nguyen, H., Van Laere, K., 2011. Metabolic and type 1 cannabinoid receptor imaging of a transgenic rat model in the early phase of Huntington disease. Exp. Neurol. 229, 440–449.

Cencioni, M.T., Chiurchiu, V., Catanzaro, G., Borsellino, G., Bernardi, G., Battistini, L., Maccarrone, M., 2010. Anandamide suppresses proliferation and cytokine release from primary human T-lymphocytes mainly via CB_2 receptors. PLoS One 5, e8688.

Charalambous, A., Marciniak, G., Shiue, C.Y., Dewey, S.L., Schlyer, D.J., Wolf, A.P., Makriyannis, A., 1991. PET studies in the primate brain and biodistribution in mice using (-)-5′-^{18}F-delta 8-THC. Pharmacol. Biochem. Behav. 40, 503–507.

Compton, D.R., Rice, K.C., De Costa, B.R., Razdan, R.K., Melvin, L.S., Johnson, M.R., Martin, B.R., 1993. Cannabinoid structure–activity relationships: correlation of receptor binding and in vivo activities. J. Pharmacol. Exp. Ther. 265, 218–226.

Cota, D., Marsicano, G., Tschop, M., Grubler, Y., Flachskamm, C., Schubert, M., Auer, D., Yassouridis, A., Thone-Reineke, C., Ortmann, S., Tomassoni, F., Cervino, C., Nisoli, E., Linthorst, A.C., Pasquali, R., Lutz, B., Stalla, G.K., Pagotto, U., 2003. The endogenous cannabinoid system affects energy balance via central orexigenic drive and peripheral lipogenesis. J. Clin. Invest. 112, 423–431.

Deadwyler, S.A., Hampson, R.E., Mu, J., Whyte, A., Childers, S., 1995. Cannabinoids modulate voltage sensitive potassium A-current in hippocampal neurons via a cAMP-dependent process. J. Pharmacol. Exp. Ther. 273, 734–743.

Dean, B., Sundram, S., Bradbury, R., Scarr, E., Copolov, D., 2001. Studies on [^{3}H]CP-55940 binding in the human central nervous system: regional specific changes in density of cannabinoid-1 receptors associated with schizophrenia and cannabis use. Neuroscience 103, 9–15.

Devane, W.A., Dysarz, F.A.I., Johnson, M.R., Melvin, L.S., Howlett, A.C., 1988. Determination and characterization of a cannabinoid receptor in rat brain. Mol. Pharmacol. 34, 605–613.

Di Marzo, V., Matias, I., 2005. Endocannabinoid control of food intake and energy balance. Nat. Neurosci. 8, 585–589.

Di Marzo, V., Melck, D., Bisogno, T., De Petrocellis, L., 1998. Endocannabinoids: endogenous cannabinoid receptor ligands with neuromodulatory action. Trends Neurosci. 21, 521–528.

Dinh, T.P., Carpenter, D., Leslie, F.M., Freund, T.F., Katona, I., Sensi, S.L., Kathuria, S., Piomelli, D., 2002. Brain monoglyceride lipase participating in endocannabinoid inactivation. Proc. Natl. Acad. Sci. USA 99, 10819–10824.

Dowie, M.J., Bradshaw, H.B., Howard, M.L., Nicholson, L.F., Faull, R.L., Hannan, A.J., Glass, M., 2009. Altered CB_1 receptor and endocannabinoid levels precede motor symptom onset in a transgenic mouse model of Huntington's disease. Neuroscience 163, 456–465.

Drysdale, A.J., Platt, B., 2003. Cannabinoids: mechanisms and therapeutic applications in the CNS. Curr. Med. Chem. 10, 2719–2732.

Eissenstat, M.A., Bell, M.R., D'Ambra, T.E., Alexander, E.J., Daum, S.J., Ackerman, J.H., Gruett, M.D., Kumar, V., Estep, K.G., Olefirowicz, E.M., 1995. Aminoalkylindoles: structure–activity relationships of novel cannabinoid mimetics. J. Med. Chem. 38, 3094–3105.

Ellert-Miklaszewska, A., Grajkowska, W., Gabrusiewicz, K., Kaminska, B., Konarska, L., 2007. Distinctive pattern of cannabinoid receptor type II (CB_2) expression in adult and pediatric brain tumors. Brain Res. 1137, 161–169.

Eubanks, L.M., Rogers, C.J., Beuscher, A.E., Koob, G.F., Olson, A.J., Dickerson, T.J., Janda, K.D., 2006. A molecular link between the active component of marijuana and Alzheimer's disease pathology. Mol. Pharm. 3, 773–777.

Evens, N., Bosier, B., Lavey, B.J., Kozlowski, J.A., Vermaelen, P., Baudemprez, L., Busson, R., Lambert, D.M., Van, L.K., Verbruggen, A.M., Bormans, G.M., 2008. Labelling and biological evaluation of [^{11}C] methoxy-Sch225336: a radioligand for the cannabinoid-type 2 receptor. Nucl. Med. Biol. 35, 793–800.

Evens, N., Muccioli, G.G., Houbrechts, N., Lambert, D.M., Verbruggen, A.M., Van, L.K., Bormans, G.M., 2009. Synthesis and biological evaluation of carbon-11- and fluorine-18-labeled 2-oxoquinoline derivatives for type 2 cannabinoid receptor positron emission tomography imaging. Nucl. Med. Biol. 36, 455–465.

Evens, N., Vandeputte, C., Coolen, C., Janssen, P., Sciot, R., Baekelandt, V., Verbruggen, A.M., Debyser, Z., Van, L.K., Bormans, G.M., 2012. Preclinical evaluation of [^{11}C]NE40, a type 2 cannabinoid receptor PET tracer. Nucl. Med. Biol. 39, 389–399.

Evens, N., Vandeputte, C., Muccioli, G.G., Lambert, D.M., Baekelandt, V., Verbruggen, A.M., Debyser, Z., Van Laere, K., Bormans, G.M., 2011. Synthesis, in vitro and in vivo evaluation of fluorine-18 labelled FE-GW405833 as a PET tracer for type 2 cannabinoid receptor imaging. Bioorg. Med. Chem. 19, 4499–4505.

Fan, H., Kotsikorou, E., Hoffman, A.F., Ravert, H.T., Holt, D., Hurst, D.P., Lupica, C.R., Reggio, P.H., Dannals, R.F., Horti, A.G., 2009. Analogs of JHU75528, a PET ligand for imaging of cerebral cannabinoid receptors (CB_1): development of ligands with optimized lipophilicity and binding affinity. Eur. J. Med. Chem. 44, 593–608.

Farkas, S., Nagy, K., Palkovits, M., Kovacs, G.G., Jia, Z., Donohue, S., Pike, V., Halldin, C., Mathe, D., Harkany, T., Gulyas, B., Csiba, L., 2012. [^{125}I]SD-7015 reveals fine modalities of CB_1 cannabinoid receptor density in the prefrontal cortex during progression of Alzheimer's disease. Neurochem. Int. 60, 286–291.

Fernandez-Espejo, E., Viveros, M.P., Nunez, L., Ellenbroek, B.A., Rodriguez De, F.F., 2009. Role of cannabis and endocannabinoids in the genesis of schizophrenia. Psychopharmacology (Berlin) 206, 531–549.

Finnema, S.J., Donohue, S.R., Zoghbi, S.S., Brown, A.K., Gulyas, B., Innis, R.B., Halldin, C., Pike, V.W., 2009. Evaluation of [^{11}C]PipISB and [^{18}F]PipISB in monkey as candidate radioligands for imaging brain cannabinoid type-1 receptors in vivo. Synapse 63, 22–30.

Fowler, C.J., 2012. Anandamide uptake explained? Trends Pharmacol. Sci. 33, 181–185.

Freund, T.F., Katona, I., Piomelli, D., 2003. Role of endogenous cannabinoids in synaptic signaling. Physiol. Rev. 83, 1017–1066.

Fride, E., Bregman, T., Kirkham, T.C., 2005. Endocannabinoids and food intake: newborn suckling and appetite regulation in adulthood. Exp. Biol. Med. (Maywood) 230, 225–234.

Fujinaga, M., Kumata, K., Yanamoto, K., Kawamura, K., Yamasaki, T., Yui, J., Hatori, A., Ogawa, M., Yoshida, Y., Nengaki, N., Maeda, J., Zhang, M.R., 2010. Radiosynthesis of novel carbon-11-labeled triaryl ligands for cannabinoid-type 2 receptor. Bioorg. Med. Chem. Lett. 20, 1565–1568.

Galve-Roperh, I., Sanchez, C., Cortes, M.L., Gomez del, P.T., Izquierdo, M., Guzman, M., 2000. Anti-tumoral action of cannabinoids: involvement of sustained ceramide accumulation and extracellular signal-regulated kinase activation. Nat. Med. 6, 313–319.

Gatley, S.J., Gifford, A.N., Volkow, N.D., Lan, R., Makriyannis, A., 1996. ^{123}I-labeled AM251: a radioiodinated ligand which binds in vivo to mouse brain cannabinoid CB_1 receptors. Eur. J. Pharmacol. 307, 331–338.

Gatley, S.J., Lan, R., Volkow, N.D., Pappas, N., King, P., Wong, C.T., Gifford, A.N., Pyatt, B., Dewey, S.L., Makriyannis, A., 1998. Imaging the brain marijuana receptor: development of a radioligand that binds to cannabinoid CB_1 receptors in vivo. J. Neurochem. 70, 417–423.

Gerard, N., Ceccarini, J., Bormans, G., Vanbilloen, B., Casteels, C., Goffin, K., Bosier, B., Lambert, D.M., Van, L.K., 2010. Influence of chronic nicotine administration on cerebral type 1 cannabinoid receptor binding: an in vivo micro-PET study in the rat using [^{18}F]MK-9470. J. Mol. Neurosci. 42, 162–167.

Gerard, N., Pieters, G., Goffin, K., Bormans, G., Van Laere, K., 2011. Brain type 1 cannabinoid receptor availability in patients with anorexia and bulimia nervosa. Biol. Psychiatry 70, 777–784.

Giblin, G.M., Billinton, A., Briggs, M., Brown, A.J., Chessell, I.P., Clayton, N.M., Eatherton, A.J., Goldsmith, P., Haslam, C., Johnson, M.R., Mitchell, W.L., Naylor, A., Perboni, A., Slingsby, B.P., Wilson, A.W., 2009. Discovery of 1-[4-(3-chlorophenylamino)-1-methyl-1*H*-pyrrolo[3,2-c]pyridin-7-yl]-1-morpholin-4-ylmethanone (GSK554418A), a brain penetrant 5-azaindole CB_2 agonist for the treatment of chronic pain. J. Med. Chem. 52, 5785–5788.

Giuffrida, A., Beltramo, M., Piomelli, D., 2001. Mechanisms of endocannabinoid inactivation: biochemistry and pharmacology. J. Pharmacol. Exp. Ther. 298, 7–14.

Glass, M., Dragunow, M., Faull, R.L., 2000. The pattern of neurodegeneration in Huntington's disease: a comparative study of cannabinoid, dopamine, adenosine and $GABA_A$ receptor alterations in the human basal ganglia in Huntington's disease. Neuroscience 97, 505–519.

Glass, M., van Dellen, A., Blakemore, C., Hannan, A.J., Faull, R.L., 2004. Delayed onset of Huntington's disease in mice in an enriched environment correlates with delayed loss of cannabinoid CB_1 receptors. Neuroscience 123, 207–212.

Goffin, K., Bormans, G., Casteels, C., Bosier, B., Lambert, D.M., Grachev, I.D., Van Paesschen, W., Van Laere, K., 2008. An in vivo [^{18}F]MK-9470 microPET study of type 1 cannabinoid receptor binding in Wistar rats after chronic administration of valproate and levetiracetam. Neuropharmacology 54, 1103–1106.

Goffin, K., Van Paesschen, W., Van Laere, K., 2011. In vivo activation of endocannabinoid system in temporal lobe epilepsy with hippocampal sclerosis. Brain 134, 1033–1040.

Gonsiorek, W., Hesk, D., Chen, S.C., Kinsley, D., Fine, J.S., Jackson, J.V., Bober, L.A., Deno, G., Bian, H., Fossetta, J., Lunn, C.A., Kozlowski, J.A., Lavey, B., Piwinski, J., Narula, S.K., Lundell, D.J., Hipkin, R.W., 2006. Characterization of peripheral human cannabinoid receptor (hCB$_2$) expression and pharmacology using a novel radioligand, [^{35}S]Sch225336. J. Biol. Chem. 281, 28143–28151.

Gonsiorek, W., Lunn, C., Fan, X., Narula, S., Lundell, D., Hipkin, R.W., 2000. Endocannabinoid 2-arachidonyl glycerol is a full agonist through human type 2 cannabinoid receptor: antagonism by anandamide. Mol. Pharmacol. 57, 1045–1050.

Gonzalez, S., Cascio, M.G., Fernandez-Ruiz, J., Fezza, F., Di Marzo, V., Ramos, J.A., 2002a. Changes in endocannabinoid contents in the brain of rats chronically exposed to nicotine, ethanol or cocaine. Brain Res. 954, 73–81.

Gonzalez, S., Fernandez-Ruiz, J., Sparpaglione, V., Parolaro, D., Ramos, J.A., 2002b. Chronic exposure to morphine, cocaine or ethanol in rats produced different effects in brain cannabinoid CB$_1$ receptor binding and mRNA levels. Drug Alcohol Depend. 66, 77–84.

Gonzalez, S., Valenti, M., de Miguel, R., Fezza, F., Fernandez-Ruiz, J., Di Marzo, V., Ramos, J.A., 2004. Changes in endocannabinoid contents in reward-related brain regions of alcohol-exposed rats, and their possible relevance to alcohol relapse. Br. J. Pharmacol. 143, 455–464.

Gorzalka, B.B., Hill, M.N., 2011. Putative role of endocannabinoid signaling in the etiology of depression and actions of antidepressants. Prog. Neuropsychopharmacol. Biol. Psychiatry 35, 1575–1585.

Govaerts, S.J., Hermans, E., Lambert, D.M., 2004. Comparison of cannabinoid ligands affinities and efficacies in murine tissues and in transfected cells expressing human recombinant cannabinoid receptors. Eur. J. Pharm. Sci. 23, 233–243.

Griebel, G., Stemmelin, J., Scatton, B., 2005. Effects of the cannabinoid CB$_1$ receptor antagonist rimonabant in models of emotional reactivity in rodents. Biol. Psychiatry 57, 261–267.

Grundy, R.I., Rabuffetti, M., Beltramo, M., 2001. Cannabinoids and neuroprotection. Mol. Neurobiol. 24, 29–51.

Hamill, T.G., Lin, L.S., Hagmann, W., Liu, P., Jewell, J., Sanabria, S., Eng, W., Ryan, C., Fong, T.M., Connolly, B., Vanko, A., Hargreaves, R., Goulet, M.T., Burns, H.D., 2009. PET imaging studies in rhesus monkey with the cannabinoid-1 (CB$_1$) receptor ligand [^{11}C]CB-119. Mol. Imaging Biol. 11, 246–252.

Hampson, R.E., Evans, G.J., Mu, J., Zhuang, S.Y., King, V.C., Childers, S.R., Deadwyler, S.A., 1995. Role of cyclic AMP dependent protein kinase in cannabinoid receptor modulation of potassium "A-current" in cultured rat hippocampal neurons. Life Sci. 56, 2081–2088.

Hanus, L., bu-Lafi, S., Fride, E., Breuer, A., Vogel, Z., Shalev, D.E., Kustanovich, I., Mechoulam, R., 2001. 2-arachidonyl glyceryl ether, an endogenous agonist of the cannabinoid CB$_1$ receptor. Proc. Natl. Acad. Sci. USA 98, 3662–3665.

Herance, R., Rojas, S., Abad, S., Jimenez, X., Gispert, J.D., Millan, O., Martin-Garcia, E., Burokas, A., Serra, M.A., Maldonado, R., Pareto, D., 2011. Positron emission tomographic imaging of the cannabinoid type 1 receptor system with [^{11}C]OMAR ([^{11}C]JHU75528): improvements in image quantification using wild-type and knockout mice. Mol. Imaging.

Herkenham, M., Lynn, A.B., Little, M.D., Johnson, M.R., Melvin, L.S., de Costa, B.R., Rice, K.C., 1990. Cannabinoid receptor localization in brain. Proc. Natl. Acad. Sci. USA 87, 1932–1936.

Herrera, B., Carracedo, A., ez-Zaera, M., Guzman, M., Velasco, G., 2005. p38 MAPK is involved in CB$_2$ receptor-induced apoptosis of human leukaemia cells. FEBS Lett. 579, 5084–5088.

Hiroi, N., Agatsuma, S., 2005. Genetic susceptibility to substance dependence. Mol. Psychiatry 10, 336–344.

Hirvonen, J., Goodwin, R.S., Li, C.T., Terry, G.E., Zoghbi, S.S., Morse, C., Pike, V.W., Volkow, N.D., Huestis, M.A., Innis, R.B., 2012. Reversible and regionally selective downregulation of brain cannabinoid CB$_1$ receptors in chronic daily cannabis smokers. Mol. Psychiatry 17, 642–649.

Hirvonen, J., Zanotti-Fregonara, P., Umhau, J.C., George, D.T., Rallis-Frutos, D., Lyoo, C.H., Li, C.T., Hines, C.S., Sun, H., Terry, G.E., Morse, C., Zoghbi, S.S., Pike, V.W., Innis, R.B., Heilig, M., Aug, 2013. Reduced cannabinoid CB$_1$ receptor binding in alcohol dependence measured with positron emission tomography. Mol Psychiatry. 18(8), 916–921. doi: 10.1038/mp.2012.100.

Ho, B.C., Wassink, T.H., Ziebell, S., Andreasen, N.C., 2011. Cannabinoid receptor 1 gene polymorphisms and marijuana misuse interactions on white matter and cognitive deficits in schizophrenia. Schizophr. Res. 128, 66–75.

Horti, A.G., Fan, H., Kuwabara, H., Hilton, J., Ravert, H.T., Holt, D.P., Alexander, M., Kumar, A., Rahmim, A., Scheffel, U., Wong, D.F., Dannals, R.F., 2006. ^{11}C-JHU75528: a radiotracer for PET imaging of CB_1 cannabinoid receptors. J. Nucl. Med. 47, 1689–1696.

Horti, A.G., Gao, Y., Ravert, H.T., Finley, P., Valentine, H., Wong, D.F., Endres, C.J., Savonenko, A.V., Dannals, R.F., 2010. Synthesis and biodistribution of [^{11}C]A-836339, a new potential radioligand for PET imaging of cannabinoid type 2 receptors (CB_2). Bioorg. Med. Chem. 18, 5202–5207.

Hsieh, G.C., Pai, M., Chandran, P., Hooker, B.A., Zhu, C.Z., Salyers, A.K., Wensink, E.J., Zhan, C., Carroll, W.A., Dart, M.J., Yao, B.B., Honore, P., Meyer, M.D., 2011. Central and peripheral sites of action for CB_2 receptor mediated analgesic activity in chronic inflammatory and neuropathic pain models in rats. Br. J. Pharmacol. 162, 428–440.

Hu, B., Doods, H., Treede, R.D., Ceci, A., 2009. Depression-like behaviour in rats with mononeuropathy is reduced by the CB_2-selective agonist GW405833. Pain 143, 206–212.

Huang, S.M., Bisogno, T., Trevisani, M., Al-Hayani, A., De, P.L., Fezza, F., Tognetto, M., Petros, T.J., Krey, J.F., Chu, C.J., Miller, J.D., Davies, S.N., Geppetti, P., Walker, J.M., Di, M.V., 2002. An endogenous capsaicin-like substance with high potency at recombinant and native vanilloid VR_1 receptors. Proc. Natl. Acad. Sci. USA 99, 8400–8405.

Huestis, M.A., Gorelick, D.A., Heishman, S.J., Preston, K.L., Nelson, R.A., Moolchan, E.T., Frank, R.A., 2001. Blockade of effects of smoked marijuana by the CB_1-selective cannabinoid receptor antagonist SR141716. Arch. Gen. Psychiatry 58, 322–328.

Huffman, J.W., Padgett, L.W., 2005. Recent developments in the medicinal chemistry of cannabimimetic indoles, pyrroles and indenes. Curr. Med. Chem. 12, 1395–1411.

Hutchison, K.E., Haughey, H., Niculescu, M., Schacht, J., Kaiser, A., Stitzel, J., Horton, W.J., Filbey, F., 2008. The incentive salience of alcohol: translating the effects of genetic variant in CNR1. Arch. Gen. Psychiatry 65, 841–850.

Iwamura, H., Suzuki, H., Ueda, Y., Kaya, T., Inaba, T., 2001. In vitro and in vivo pharmacological characterization of JTE-907, a novel selective ligand for cannabinoid CB_2 receptor. J. Pharmacol. Exp. Ther. 296, 420–425.

Jordt, S.E., Julius, D., 2002. Molecular basis for species-specific sensitivity to "hot" chili peppers. Cell 108, 421–430.

Juan-Pico, P., Fuentes, E., Bermudez-Silva, F.J., Javier Diaz-Molina, F., Ripoll, C., Rodriguez De, F.F., Nadal, A., 2006. Cannabinoid receptors regulate Ca^{2+} signals and insulin secretion in pancreatic beta-cell. Cell Calcium 39, 155–162.

Jung, J., Hwang, S.W., Kwak, J., Lee, S.Y., Kang, C.J., Kim, W.B., Kim, D., Oh, U., 1999. Capsaicin binds to the intracellular domain of the capsaicin-activated ion channel. J. Neurosci. 19, 529–538.

Katona, I., Freund, T.F., 2012. Multiple functions of endocannabinoid signaling in the brain. Annu. Rev. Neurosci. 35, 529–558.

Klein, T.W., Newton, C., Larsen, K., Lu, L., Perkins, I., Nong, L., Friedman, H., 2003. The cannabinoid system and immune modulation. J. Leukoc. Biol. 74, 486–496.

Kozak, K.R., Crews, B.C., Morrow, J.D., Wang, L.H., Ma, Y.H., Weinander, R., Jakobsson, P.J., Marnett, L.J., 2002. Metabolism of the endocannabinoids, 2-arachidonylglycerol and anandamide, into prostaglandin, thromboxane, and prostacyclin glycerol esters and ethanolamides. J. Biol. Chem. 277, 44877–44885.

Kreitzer, A.C., Malenka, R.C., 2007. Endocannabinoid-mediated rescue of striatal LTD and motor deficits in Parkinson's disease models. Nature 445, 643–647.

Kuster, J.E., Stevenson, J.I., Ward, S.J., D'Ambra, T.E., Haycock, D.A., 1993. Aminoalkylindole binding in rat cerebellum: selective displacement by natural and synthetic cannabinoids. J. Pharmacol. Exp. Ther. 264, 1352–1363.

LaBuda, C.J., Koblish, M., Little, P.J., 2005. Cannabinoid CB_2 receptor agonist activity in the hindpaw incision model of postoperative pain. Eur. J. Pharmacol. 527, 172–174.

Lake, M.C., Nguyen, Q.D., Ali, S., Aboagye, E.O., 2012. Development of a novel molecular sensor for imaging estrogen receptor-coactivator protein–protein interactions. PLoS One 7, e44160.

Lan, R., Gatley, J., Lu, Q., Fan, P., Fernando, S.R., Volkow, N.D., Pertwee, R., Makriyannis, A., 1999a. Design and synthesis of the CB_1 selective cannabinoid antagonist AM281: a potential human SPECT ligand. AAPS PharmSci 1, E4.

Lan, R., Liu, Q., Fan, P., Lin, S., Fernando, S.R., McCallion, D., Pertwee, R., Makriyannis, A., 1999b. Structure–activity relationships of pyrazole derivatives as cannabinoid receptor antagonists. J. Med. Chem. 42, 769–776.

Lastres-Becker, I., Berrendero, F., Lucas, J.J., Martin-Aparicio, E., Yamamoto, A., Ramos, J.A., Fernandez-Ruiz, J.J., 2002a. Loss of mRNA levels, binding and activation of GTP-binding proteins for cannabinoid CB_1 receptors in the basal ganglia of a transgenic model of Huntington's disease. Brain Res. 929, 236–242.

Lastres-Becker, I., Hansen, H.H., Berrendero, F., de Miguel, R., Perez-Rosado, A., Manzanares, J., Ramos, J.A., Fernandez-Ruiz, J., 2002b. Alleviation of motor hyperactivity and neurochemical deficits by endocannabinoid uptake inhibition in a rat model of Huntington's disease. Synapse 44, 23–35.

Lastres-Becker, I., Cebeira, M., de Ceballos, M.L., Zeng, B.Y., Jenner, P., Ramos, J.A., Fernandez-Ruiz, J.J., 2001. Increased cannabinoid CB_1 receptor binding and activation of GTP-binding proteins in the basal ganglia of patients with Parkinson's syndrome and of MPTP-treated marmosets. Eur. J. Neurosci. 14, 1827–1832.

Lavey, B.J., Kozlowski, J.A., Hipkin, R.W., Gonsiorek, W., Lundell, D.J., Piwinski, J.J., Narula, S., Lunn, C.A., 2005. Triaryl bis-sulfones as a new class of cannabinoid CB_2 receptor inhibitors: identification of a lead and initial SAR studies. Bioorg. Med. Chem. Lett. 15, 783–786.

Laviolette, S.R., Grace, A.A., 2006. The roles of cannabinoid and dopamine receptor systems in neural emotional learning circuits: implications for schizophrenia and addiction. Cell. Mol. Life Sci. 63, 1597–1613.

Le Foll, B., Goldberg, S.R., 2005. Cannabinoid CB_1 receptor antagonists as promising new medications for drug dependence. J. Pharmacol. Exp. Ther. 312, 875–883.

Le Foll, B., Gorelick, D.A., Goldberg, S.R., 2009. The future of endocannabinoid-oriented clinical research after CB_1 antagonists. Psychopharmacology (Berlin) 205, 171–174.

Lin, L.S., Ha, S., Ball, R.G., Tsou, N.N., Castonguay, L.A., Doss, G.A., Fong, T.M., Shen, C.P., Xiao, J.C., Goulet, M.T., Hagmann, W.K., 2008. Conformational analysis and receptor docking of *N*-[(1*S*,2*S*)-3-(4-chlorophenyl)-2-(3-cyanophenyl)-1-methylpropyl]-2-methyl-2-{[5-(trifluoromethyl)pyridin-2-yl]oxy}propanamide (taranabant, MK-0364), a novel, acyclic cannabinoid-1 receptor inverse agonist. J. Med. Chem. 51, 2108–2114.

Lin, L.S., et al., 2006. Discovery of *N*-[(1*S*,2*S*)-3-(4-chlorophenyl)-2-(3-cyanophenyl)-1-methylpropyl]-2-methyl-2-{[5-(trifluoromethyl)pyridin-2-yl]oxy}propanamide (MK-0364), a novel, acyclic cannabinoid-1 receptor inverse agonist for the treatment of obesity. J. Med. Chem. 49, 7584–7587.

Lunn, C.A., Fine, J.S., Rojas-Triana, A., Jackson, J.V., Fan, X., Kung, T.T., Gonsiorek, W., Schwarz, M.A., Lavey, B., Kozlowski, J.A., Narula, S.K., Lundell, D.J., Hipkin, R.W., Bober, L.A., 2006. A novel cannabinoid peripheral cannabinoid receptor-selective inverse agonist blocks leukocyte recruitment in vivo. J. Pharmacol. Exp. Ther. 316, 780–788.

Maccarrone, M., Valensise, H., Bari, M., Lazzarin, N., Romanini, C., Finazzi-Agro, A., 2001. Progesterone up-regulates anandamide hydrolase in human lymphocytes: role of cytokines and implications for fertility. J. Immunol. 166, 7183–7189.

Malone, D.T., Hill, M.N., Rubino, T., 2010. Adolescent cannabis use and psychosis: epidemiology and neurodevelopmental models. Br. J. Pharmacol. 160, 511–522.

Maresz, K., Carrier, E.J., Ponomarev, E.D., Hillard, C.J., Dittel, B.N., 2005. Modulation of the cannabinoid CB_2 receptor in microglial cells in response to inflammatory stimuli. J. Neurochem. 95, 437–445.

Marinelli, S., Di Marzo, V., Berretta, N., Matias, I., Maccarrone, M., Bernardi, G., Mercuri, N.B., 2003. Presynaptic facilitation of glutamatergic synapses to dopaminergic neurons of the rat substantia nigra by endogenous stimulation of vanilloid receptors. J. Neurosci. 23, 3136–3144.

Massa, F., Storr, M., Lutz, B., 2005. The endocannabinoid system in the physiology and pathophysiology of the gastrointestinal tract. J. Mol. Med. (Berl.) 83, 944–954.

Massoud, T.F., Gambhir, S.S., 2003. Molecular imaging in living subjects: seeing fundamental biological processes in a new light. Genes Dev. 17, 545–580.

Mato, S., Vidal, R., Castro, E., Diaz, A., Pazos, A., Valdizan, E.M., 2010. Long-term fluoxetine treatment modulates cannabinoid type 1 receptor-mediated inhibition of adenylyl cyclase in the rat prefrontal cortex through 5-hydroxytryptamine 1A receptor-dependent mechanisms. Mol. Pharmacol. 77, 424–434.

Matsuda, L.A., Lolait, S.J., Brownstein, M.J., Young, A.C., Bonner, T.I., 1990. Structure of a cannabinoid receptor and functional expression of the cloned cDNA. Nature 346, 561–564.

McCaw, E.A., Hu, H., Gomez, G.T., Hebb, A.L., Kelly, M.E., Ovan-Wright, E.M., 2004. Structure, expression and regulation of the cannabinoid receptor gene (CB_1) in Huntington's disease transgenic mice. Eur. J. Biochem. 271, 4909–4920.

Mechoulam, R., Hanus, L., 2000. A historical overview of chemical research on cannabinoids. Chem. Phys. Lipids 108, 1–13.

Monteleone, P., Bifulco, M., Di, F.C., Gazzerro, P., Canestrelli, B., Monteleone, F., Proto, M.C., Di, G.M., Grimaldi, C., Maj, M., 2009. Association of CNR1 and FAAH endocannabinoid gene polymorphisms with anorexia nervosa and bulimia nervosa: evidence for synergistic effects. Genes Brain Behav. 8, 728–732.

Morales, M., Bonci, A., 2012. Getting to the core of addiction: Hooking CB_2 receptor into drug abuse? Nat. Med. 18, 504–505.

Mukhopadhyay, S., Das, S., Williams, E.A., Moore, D., Jones, J.D., Zahm, D.S., Ndengele, M.M., Lechner, A.J., Howlett, A.C., 2006. Lipopolysaccharide and cyclic AMP regulation of CB_2 cannabinoid receptor levels in rat brain and mouse RAW 264.7 macrophages. J. Neuroimmunol. 181, 82–92.

Munro, S., Thomas, K.L., bu-Shaar, M., 1993. Molecular characterization of a peripheral receptor for cannabinoids. Nature 365, 61–65.

Mussinu, J.M., Ruiu, S., Mule, A.C., Pau, A., Carai, M.A., Loriga, G., Murineddu, G., Pinna, G.A., 2003. Tricyclic pyrazoles. Part 1: synthesis and biological evaluation of novel 1,4-dihydroindeno[1,2-*c*]pyrazol-based ligands for CB_1 and CB_2 cannabinoid receptors. Bioorg. Med. Chem. 11, 251–263.

Naver, B., Stub, C., Moller, M., Fenger, K., Hansen, A.K., Hasholt, L., Sorensen, S.A., 2003. Molecular and behavioral analysis of the R6/1 Huntington's disease transgenic mouse. Neuroscience 122, 1049–1057.

Newell, K.A., Deng, C., Huang, X.F., 2006. Increased cannabinoid receptor density in the posterior cingulate cortex in schizophrenia. Exp. Brain Res. 172, 556–560.

Nissen, S.E., Nicholls, S.J., Wolski, K., Rodes-Cabau, J., Cannon, C.P., Deanfield, J.E., Despres, J.P., Kastelein, J.J., Steinhubl, S.R., Kapadia, S., Yasin, M., Ruzyllo, W., Gaudin, C., Job, B., Hu, B., Bhatt, D.L., Lincoff, A.M., Tuzcu, E.M., 2008. Effect of rimonabant on progression of atherosclerosis in patients with abdominal obesity and coronary artery disease: the STRADIVARIUS randomized controlled trial. JAMA 299, 1547–1560.

Nong, L., Newton, C., Friedman, H., Klein, T.W., 2001. CB_1 and CB_2 receptor mRNA expression in human peripheral blood mononuclear cells (PBMC) from various donor types. Adv. Exp. Med. Biol. 493, 229–233.

Ofek, O., Karsak, M., Leclerc, N., Fogel, M., Frenkel, B., Wright, K., Tam, J., ttar-Namdar, M., Kram, V., Shohami, E., Mechoulam, R., Zimmer, A., Bab, I., 2006. Peripheral cannabinoid receptor, CB_2, regulates bone mass. Proc. Natl. Acad. Sci. USA 103, 696–701.

Ortiz, S., Oliva, J.M., Perez-Rial, S., Palomo, T., Manzanares, J., 2004. Chronic ethanol consumption regulates cannabinoid CB_1 receptor gene expression in selected regions of rat brain. Alcohol Alcohol. 39, 88–92.

Ovan-Wright, E.M., Robertson, H.A., 2000. Cannabinoid receptor messenger RNA levels decrease in a subset of neurons of the lateral striatum, cortex and hippocampus of transgenic Huntington's disease mice. Neuroscience 98, 705–713.

Palazuelos, J., Aguado, T., Pazos, M.R., Julien, B., Carrasco, C., Resel, E., Sagredo, O., Benito, C., Romero, J., Azcoitia, I., Fernandez-Ruiz, J., Guzman, M., Galve-Roperh, I., 2009. Microglial CB_2 cannabinoid receptors are neuroprotective in Huntington's disease excitotoxicity. Brain 132, 3152–3164.

Papa, S.M., 2008. The cannabinoid system in Parkinson's disease: multiple targets to motor effects. Exp. Neurol. 211, 334–338.

Parolaro, D., Realini, N., Vigano, D., Guidali, C., Rubino, T., 2010. The endocannabinoid system and psychiatric disorders. Exp. Neurol. 224, 3–14.

Pertwee, R.G., 1997. Pharmacology of cannabinoid CB_1 and CB_2 receptors. Pharmacol. Ther. 74, 129–180.

Petersen, G., Moesgaard, B., Schmid, P.C., Schmid, H.H., Broholm, H., Kosteljanetz, M., Hansen, H.S., 2005. Endocannabinoid metabolism in human glioblastomas and meningiomas compared to human non-tumour brain tissue. J. Neurochem. 93, 299–309.

Pi-Sunyer, F.X., Aronne, L.J., Heshmati, H.M., Devin, J., Rosenstock, J., 2006. Effect of rimonabant, a cannabinoid-1 receptor blocker, on weight and cardiometabolic risk factors in overweight or obese patients: RIO-North America: a randomized controlled trial. JAMA 295, 761–775.

Pike, V.W., 2009. PET radiotracers: crossing the blood–brain barrier and surviving metabolism. Trends Pharmacol. Sci. 30, 431–440.

Pisani, V., Moschella, V., Bari, M., Fezza, F., Galati, S., Bernardi, G., Stanzione, P., Pisani, A., Maccarrone, M., 2010. Dynamic changes of anandamide in the cerebrospinal fluid of Parkinson's disease patients. Mov. Disord. 25, 920–924.

Porter, A.C., Sauer, J.M., Knierman, M.D., Becker, G.W., Berna, M.J., Bao, J., Nomikos, G.G., Carter, P., Bymaster, F.P., Leese, A.B., Felder, C.C., 2002. Characterization of a novel endocannabinoid, virodhamine, with antagonist activity at the CB_1 receptor. J. Pharmacol. Exp. Ther. 301, 1020–1024.

Pouliot, F., Karanikolas, B.D., Johnson, M., Sato, M., Priceman, S.J., Stout, D., Sohn, J., Satyamurthy, N., deKernion, J.B., Wu, L., 2011. In vivo imaging of intraprostatic-specific gene transcription by PET. J. Nucl. Med. 52, 784–791.

Racil-Fernandez, A., Trigo, J.M., Garcia-Gutierrez, M.S., Ortega-Alvaro, A., Ternianov, A., Navarro, D., Robledo, P., Berbel, P., Maldonado, R., Manzanares, J., 2012. Decreased cocaine motor sensitization and self-administration in mice overexpressing cannabinoid CB_2 receptors. Neuropsychopharmacology 37, 1749–1763.

Ramirez, B.G., Blázquez, C., Gomez del, P.T., Guzman, M., de Ceballos, M.L., 2005. Prevention of Alzheimer's disease pathology by cannabinoids: neuroprotection mediated by blockade of microglial activation. J. Neurosci. 25, 1904–1913.

Ravinet, T.C., Delgorge, C., Menet, C., Arnone, M., Soubrie, P., 2004. CB_1 cannabinoid receptor knockout in mice leads to leanness, resistance to diet-induced obesity and enhanced leptin sensitivity. Int. J. Obes. Relat. Metab. Disord. 28, 640–648.

Razdan, R.K., 1986. Structure–activity relationships in cannabinoids. Pharmacol. Rev. 38, 75–149.

Sagredo, O., Gonzalez, S., Aroyo, I., Pazos, M.R., Benito, C., Lastres-Becker, I., Romero, J.P., Tolon, R.M., Mechoulam, R., Brouillet, E., Romero, J., Fernandez-Ruiz, J., 2009. Cannabinoid CB_2 receptor agonists protect the striatum against malonate toxicity: relevance for Huntington's disease. Glia 57, 1154–1167.

Sanabria-Bohorquez, S.M., Hamill, T.G., Goffin, K., De Lepeleire, I., Bormans, G., Burns, H.D., Van Laere, K., 2010. Kinetic analysis of the cannabinoid-1 receptor PET tracer [^{18}F]MK-9470 in human brain. Eur. J. Nucl. Med. Mol. Imaging 37, 920–933.

Sanchez, C., de Ceballos, M.L., Gomez del, P.T., Rueda, D., Corbacho, C., Velasco, G., Galve-Roperh, I., Huffman, J.W., Cajal, S., Guzman, M., 2001. Inhibition of glioma growth in vivo by selective activation of the CB_2 cannabinoid receptor. Cancer Res. 61, 5784–5789.

Sapp, E., Kegel, K.B., Aronin, N., Hashikawa, T., Uchiyama, Y., Tohyama, K., Bhide, P.G., Vonsattel, J.P., DiFiglia, M., 2001. Early and progressive accumulation of reactive microglia in the Huntington disease brain. J. Neuropathol. Exp. Neurol. 60, 161–172.

Schley, M., Stander, S., Kerner, J., Vajkoczy, P., Schupfer, G., Dusch, M., Schmelz, M., Konrad, C., 2009. Predominant CB_2 receptor expression in endothelial cells of glioblastoma in humans. Brain Res. Bull. 79, 333–337.

Secher, A., Husum, H., Holst, B., Egerod, K.L., Mellerup, E., 2010. Risperidone treatment increases CB_1 receptor binding in rat brain. Neuroendocrinology 91, 155–168.

Shankar, B.B., Lavey, B.J., Zhou, G., Spitler, J.A., Tong, L., Rizvi, R., Yang, D.Y., Wolin, R., Kozlowski, J.A., Shih, N.Y., Wu, J., Hipkin, R.W., Gonsiorek, W., Lunn, C.A., 2005. Triaryl bis-sulfones as cannabinoid-2 receptor ligands: SAR studies. Bioorg. Med. Chem. Lett. 15, 4417–4420.

Siegfried, Z., Kanyas, K., Latzer, Y., Karni, O., Bloch, M., Lerer, B., Berry, E.M., 2004. Association study of cannabinoid receptor gene (CNR1) alleles and anorexia nervosa: differences between restricting and binging/purging subtypes. Am. J. Med. Genet. B Neuropsychiatr. Genet. 125B, 126–130.

Sieradzan, K.A., Fox, S.H., Hill, M., Dick, J.P., Crossman, A.R., Brotchie, J.M., 2001. Cannabinoids reduce levodopa-induced dyskinesia in Parkinson's disease: a pilot study. Neurology 57, 2108–2111.

Sipe, J.C., Scott, T.M., Murray, S., Harismendy, O., Simon, G.M., Cravatt, B.F., Waalen, J., 2010. Biomarkers of endocannabinoid system activation in severe obesity. PLoS One 5, e8792.

Soyka, M., Koller, G., Schmidt, P., Lesch, O.M., Leweke, M., Fehr, C., Gann, H., Mann, K.F., 2008. Cannabinoid receptor 1 blocker rimonabant (SR 141716) for treatment of alcohol dependence: results from a placebo-controlled, double-blind trial. J. Clin. Psychopharmacol. 28, 317–324.

Steffens, M., Zentner, J., Honegger, J., Feuerstein, T.J., 2005. Binding affinity and agonist activity of putative endogenous cannabinoids at the human neocortical CB_1 receptor. Biochem. Pharmacol. 69, 169–178.

Sundram, S., 2006. Cannabis and neurodevelopment: implications for psychiatric disorders. Hum. Psychopharmacol. 21, 245–254.

Szabo, B., Dorner, L., Pfreundtner, C., Norenberg, W., Starke, K., 1998. Inhibition of GABAergic inhibitory postsynaptic currents by cannabinoids in rat corpus striatum. Neuroscience 85, 395–403.

Ternianov, A., Perez-Ortiz, J.M., Solesio, M.E., Garcia-Gutierrez, M.S., Ortega-Alvaro, A., Navarrete, F., Leiva, C., Galindo, M.F., Manzanares, J., 2012. Overexpression of CB_2 cannabinoid receptors results in neuroprotection against behavioral and neurochemical alterations induced by intracaudate administration of 6-hydroxydopamine. Neurobiol. Aging 33, 421 e1–e16.

Terry, G., Liow, J.S., Chernet, E., Zoghbi, S.S., Phebus, L., Felder, C.C., Tauscher, J., Schaus, J.M., Pike, V.W., Halldin, C., Innis, R.B., 2008. Positron emission tomography imaging using an inverse agonist radioligand to assess cannabinoid CB_1 receptors in rodents. NeuroImage 41, 690–698.

Terry, G.E., Hirvonen, J., Liow, J.S., Seneca, N., Tauscher, J.T., Schaus, J.M., Phebus, L., Felder, C.C., Morse, C.L., Pike, V.W., Halldin, C., Innis, R.B., 2010a. Biodistribution and dosimetry in humans of two inverse agonists to image cannabinoid CB_1 receptors using positron emission tomography. Eur. J. Nucl. Med. Mol. Imaging 37, 1499–1506.

Terry, G.E., Hirvonen, J., Liow, J.S., Zoghbi, S.S., Gladding, R., Tauscher, J.T., Schaus, J.M., Phebus, L., Felder, C.C., Morse, C.L., Donohue, S.R., Pike, V.W., Halldin, C., Innis, R.B., 2010b. Imaging and quantitation of cannabinoid CB_1 receptors in human and monkey brains using ^{18}F-labeled inverse agonist radioligands. J. Nucl. Med. 51, 112–120.

Terry, G.E., Liow, J.S., Zoghbi, S.S., Hirvonen, J., Farris, A.G., Lerner, A., Tauscher, J.T., Schaus, J.M., Phebus, L., Felder, C.C., Morse, C.L., Hong, J.S., Pike, V.W., Halldin, C., Innis, R.B., 2009. Quantitation of cannabinoid CB_1 receptors in healthy human brain using positron emission tomography and an inverse agonist radioligand. NeuroImage 48, 362–370.

Tsou, K., Brown, S., Sanudo-Pena, M.C., Mackie, K., Walker, J.M., 1998a. Immunohistochemical distribution of cannabinoid CB_1 receptors in the rat central nervous system. Neuroscience 83, 393–411.

Tsou, K., Nogueron, M.I., Muthian, S., Sanudo-Pena, M.C., Hillard, C.J., Deutsch, D.G., Walker, J.M., 1998b. Fatty acid amide hydrolase is located preferentially in large neurons in the rat central nervous system as revealed by immunohistochemistry. Neurosci. Lett. 254, 137–140.

Turkman, N., Shavrin, A., Ivanov, R.A., Rabinovich, B., Volgin, A., Gelovani, J.G., Alauddin, M.M., 2011. Fluorinated cannabinoid CB_2 receptor ligands: synthesis and in vitro binding characteristics of 2-oxoquinoline derivatives. Bioorg. Med. Chem. 19, 5698–5707.

Turkman, N., Shavrin, A., Paolillo, V., Yeh, H.H., Flores, L., Soghomonian, S., Rabinovich, B., Volgin, A., Gelovani, J., Alauddin, M., 2012. Synthesis and preliminary evaluation of [^{18}F]-labeled 2-oxoquinoline derivatives for PET imaging of cannabinoid CB_2 receptor. Nucl. Med. Biol. 39, 593–600.

Turner, C.E., ElSohly, M.A., Boeren, E.G., 1980. Constituents of *Cannabis sativa* L. XVII. A review of the natural constituents. J. Nat. Prod. 43, 169–234.

Ueda, N., Puffenbarger, R.A., Yamamoto, S., Deutsch, D.G., 2000. The fatty acid amide hydrolase (FAAH). Chem. Phys. Lipids 108, 107–121.

Ujike, H., Morita, Y., 2004. New perspectives in the studies on endocannabinoid and cannabis: cannabinoid receptors and schizophrenia. J. Pharmacol. Sci. 96, 376–381.

Ujike, H., Takaki, M., Nakata, K., Tanaka, Y., Takeda, T., Kodama, M., Fujiwara, Y., Sakai, A., Kuroda, S., 2002. CNR1, central cannabinoid receptor gene, associated with susceptibility to hebephrenic schizophrenia. Mol. Psychiatry 7, 515–518.

Valenzano, K.J., Tafesse, L., Lee, G., Harrison, J.E., Boulet, J.M., Gottshall, S.L., Mark, L., Pearson, M.S., Miller, W., Shan, S., Rabadi, L., Rotshteyn, Y., Chaffer, S.M., Turchin, P.I., Elsemore, D.A., Toth, M., Koetzner, L., Whiteside, G.T., 2005. Pharmacological and pharmacokinetic characterization of the cannabinoid receptor 2 agonist, GW405833, utilizing rodent models of acute and chronic pain, anxiety, ataxia and catalepsy. Neuropharmacology 48, 658–672.

Valverde, O., Torrens, M., 2012. CB_1 receptor-deficient mice as a model for depression. Neuroscience 204, 193–206.

Van der Schueren, B.J., Van Laere, K., Gerard, N., Bormans, G., De Hoon, J.N., 2012. Interictal type 1 cannabinoid receptor binding is increased in female migraine patients. Headache 52, 433–440.

Van der Stelt, M., Di Marzo, V., 2003. The endocannabinoid system in the basal ganglia and in the mesolimbic reward system: implications for neurological and psychiatric disorders. Eur. J. Pharmacol. 480, 133–150.

Van Gaal, L.F., Scheen, A.J., Rissanen, A.M., Rossner, S., Hanotin, C., Ziegler, O., 2008. Long-term effect of CB_1 blockade with rimonabant on cardiometabolic risk factors: two year results from the RIO-Europe Study. Eur. Heart J. 29, 1761–1771.

Van Laere, K., Casteels, C., Dhollander, I., Goffin, K., Grachev, I., Bormans, G., Vandenberghe, W., 2010. Widespread decrease of type 1 cannabinoid receptor availability in Huntington disease in vivo. J. Nucl. Med. 51, 1413–1417.

Van Laere, K., Casteels, C., Lunskens, S., Goffin, K., Grachev, I.D., Bormans, G., Vandenberghe, W., 2012. Regional changes in type 1 cannabinoid receptor availability in Parkinson's disease in vivo. Neurobiol. Aging 33, 620–628.

Van Laere, K., Goffin, K., Bormans, G., Casteels, C., Mortelmans, L., de Hoon, J., Grachev, I., Vandenbulcke, M., Pieters, G., 2009. Relationship of type 1 cannabinoid receptor availability in the human brain to novelty-seeking temperament. Arch. Gen. Psychiatry 66, 196–204.

Van Laere, K., Goffin, K., Casteels, C., Dupont, P., Mortelmans, L., de Hoon, J., Bormans, G., 2008a. Gender-dependent increases with healthy aging of the human cerebral cannabinoid-type 1 receptor binding using [^{18}F]MK-9470 PET. NeuroImage 39, 1533–1541.

Van Laere, K., Koole, M., Sanabria Bohorquez, S.M., Goffin, K., Guenther, I., Belanger, M.J., Cote, J., Rothenberg, P., De Lepeleire, I., Grachev, I.D., Hargreaves, R.J., Bormans, G., Burns, H.D., 2008b. Whole-body biodistribution and radiation dosimetry of the human cannabinoid type-1 receptor ligand ^{18}F-MK-9470 in healthy subjects. J. Nucl. Med. 49, 439–445.

Van Sickle, M.D., Duncan, M., Kingsley, P.J., Mouihate, A., Urbani, P., Mackie, K., Stella, N., Makriyannis, A., Piomelli, D., Davison, J.S., Marnett, L.J., Di Marzo, V., Pittman, Q.J., Patel, K.D., Sharkey, K.A., 2005. Identification and functional characterization of brainstem cannabinoid CB_2 receptors. Science 310, 329–332.

Vandeputte, C., Casteels, C., Struys, T., Koole, M., van Veghel, D., Evens, N., Gerits, A., Dresselaers, T., Lambrichts, I., Himmelreich, U., Bormans, G., Van Laere, K., 2012. Small-animal PET imaging of the type 1 and type 2 cannabinoid receptors in a photothrombotic stroke model. Eur. J. Nucl. Med. Mol. Imaging 39, 1796–1806.

Vandeputte, C., Evens, N., Toelen, J., Deroose, C.M., Bosier, B., Ibrahimi, A., Van der Perren, A., Gijsbers, R., Janssen, P., Lambert, D.M., Verbruggen, A., Debyser, Z., Bormans, G., Baekelandt, V., Van Laere, K., 2011. A PET brain reporter gene system based on type 2 cannabinoid receptors. J. Nucl. Med. 52, 1102–1109.

Vastenhouw, B., Beekman, F., 2007. Submillimeter total-body murine imaging with U-SPECT-I. J. Nucl. Med. 48, 487–493.

Villares, J., 2007. Chronic use of marijuana decreases cannabinoid receptor binding and mRNA expression in the human brain. Neuroscience 145, 323–334.

Vinod, K.Y., Hungund, B.L., 2006. Cannabinoid-1 receptor: a novel target for the treatment of neuropsychiatric disorders. Expert Opin. Ther. Targets 10, 203–210.

Vinod, K.Y., Yalamanchili, R., Xie, S., Cooper, T.B., Hungund, B.L., 2006. Effect of chronic ethanol exposure and its withdrawal on the endocannabinoid system. Neurochem. Int. 49, 619–625.

Wang, M.W., Wang, F., Zheng, Y.J., Zhang, Y.J., Zhang, Y.P., Zhao, Q., Shen, C.K., Wang, Y., Sun, S.H., 2013. An in vivo molecular imaging probe ^{18}F-Annexin B1 for apoptosis detection by PET/CT: preparation and preliminary evaluation. Apoptosis 18, 238–247.

Whiteside, G.T., Gottshall, S.L., Boulet, J.M., Chaffer, S.M., Harrison, J.E., Pearson, M.S., Turchin, P.I., Mark, L., Garrison, A.E., Valenzano, K.J., 2005. A role for cannabinoid receptors, but not endogenous opioids, in the antinociceptive activity of the CB_2-selective agonist, GW405833. Eur. J. Pharmacol. 528, 65–72.

Willis, P.G., Pavlova, O.A., Chefer, S.I., Vaupel, D.B., Mukhin, A.G., Horti, A.G., 2005. Synthesis and structure–activity relationship of a novel series of aminoalkylindoles with potential for imaging the neuronal cannabinoid receptor by positron emission tomography. J. Med. Chem. 48, 5813–5822.

Willmann, J.K., Paulmurugan, R., Rodriguez-Porcel, M., Stein, W., Brinton, T.J., Connolly, A.J., Nielsen, C.H., Lutz, A.M., Lyons, J., Ikeno, F., Suzuki, Y., Rosenberg, J., Chen, I.Y., Wu, J.C., Yeung, A.C., Yock, P., Robbins, R.C., Gambhir, S.S., 2009. Imaging gene expression in human mesenchymal stem cells: from small to large animals. Radiology 252, 117–127.

Wong, D.F., Kuwabara, H., Horti, A.G., Raymont, V., Brasic, J., Guevara, M., Ye, W., Dannals, R.F., Ravert, H.T., Nandi, A., Rahmim, A., Ming, J.E., Grachev, I., Roy, C., Cascella, N., 2010a. Quantification of cerebral cannabinoid receptors subtype 1 (CB_1) in healthy subjects and schizophrenia by the novel PET radioligand [^{11}C]OMAR. NeuroImage 52, 1505–1513.

Wong, D.F., Kuwabara, H., Hussain, B., Horti, A.G., Brasic, J., Raymont, V., Ye, W., Guevara, M., Cascella, N., 2010b. Evaluation of occupancy of cannabinoid CB_1 receptors by a novel antagonist drug with [^{11}C] OMAR and PET. NeuroImage 52, S64.

Yasuno, F., Brown, A.K., Zoghbi, S.S., Krushinski, J.H., Chernet, E., Tauscher, J., Schaus, J.M., Phebus, L.A., Chesterfield, A.K., Felder, C.C., Gladding, R.L., Hong, J., Halldin, C., Pike, V.W., Innis, R.B., 2008. The PET radioligand [^{11}C]MePPEP binds reversibly and with high specific signal to cannabinoid CB_1 receptors in nonhuman primate brain. Neuropsychopharmacology 33, 259–269.

Zhang, P.W., Ishiguro, H., Ohtsuki, T., Hess, J., Carillo, F., Walther, D., Onaivi, E.S., Arinami, T., Uhl, G.R., 2004. Human cannabinoid receptor 1: 5′ exons, candidate regulatory regions, polymorphisms, haplotypes and association with polysubstance abuse. Mol. Psychiatry 9, 916–931.

CHAPTER FOUR

Human Brain Imaging of Opioid Receptors: Application to CNS Biomarker and Drug Development

Udi E. Ghitza
Center for the Clinical Trials Network, National Institute on Drug Abuse, National Institutes of Health, Bethesda, MD, USA

1. INTRODUCTION

This chapter reviews the emerging opportunities enabled by modern human brain imaging technologies and associated research to advance the understanding of the role of central nervous system (CNS) opioid receptor (OR)-mediated molecular processes in the areas of pain and substance use disorder research and treatment.

The CNS OR system has been shown to be involved in analgesia, substance use disorders, cancer, and other conditions, and this role has heightened pharmaceutical companies' interest in the discovery of OR-acting ligands for research and development (R&D) purposes (Smith et al., 2004; Madar et al., 2007; Stohler and Zubieta, 2001). The pharmaceutical industry is interested in identifying compounds with improved selective molecular therapeutic targeting of ORs and abuse-resistant formulations, given the large market and various uses for them. The importance of developing abuse-resistant formulations of analgesics for the treatment of pain is evidenced by the high rates of diversion and abuse of prescription opioid analgesics in the United States (Volkow et al., 2011; Volkow and McLellan, 2011). Positron emission tomography (PET)-radiolabeled opioid ligands are currently being used in drug development studies to identify new therapeutic targets and agents with optimal pharmacokinetic and pharmacodynamic properties consistent with abuse-resistant, controlled-release formulations in order to lower the risk of drug development failure and to select best lead-optimization candidates with necessary safety profiles.

The substantive risk pharmaceutical companies face for the medication development for CNS disorders is evident from the high rate of failure and stagnation in the development and seeking regulatory approval of novel molecular entities as putative therapeutics, in which pharmacological candidate compounds that are efficacious in preclinical studies and proceed to large-scale clinical trials often do not show clinical efficacy and limited abuse potential (Paul et al., 2010). The pharmaceutical industry and US Food and Drug Administration's (FDA's) Critical Path Initiative (http://www.fda.gov/oc/

Imaging of the Human Brain in Health and Disease
http://dx.doi.org/10.1016/B978-0-12-418677-4.00004-X

initiatives/criticalpath/) have come to the realization that molecular brain imaging with nuclear radiolabeling techniques such as PET, particularly radiolabeling of CNS drugs for receptor and neurotransmitter system imaging, will facilitate lead optimization and CNS drug development (Bergstrom et al., 2003; Hammond et al., 2003; Perkins and Frier, 2004; Wang and Maurer, 2005; Wong et al., 2009).

Recent reviews have detailed various ways that human PET imaging of ORs, using radiolabeling of molecular imaging agents with positron-emitting radionuclides (such as ^{11}C ($t^{1/2}$ = 20.3 min) or ^{11}F ($t^{1/2}$ = 109.7 min)) can provide unique valuable insight into receptor occupancy mechanisms of action impacting the course of chronic diseases observed in humans, such as pain and substance use disorders, with respect to assessing the pertinence of preclinical model systems utilized in basic research and their predictive validity to these clinical phenomena (Cosgrove, 2010; Stohler and Zubieta, 2010). A lack of fundamental knowledge concerning the pathogenesis of disease and the absence of validated biomarkers for these complex diseases—which often include heterogeneous phenotypes and patient subpopulations each manifesting distinct prognosis and response to treatment—impedes the timely identification and development of effective pharmacotherapeutics.

Application of these brain imaging technologies may also shorten time to market and cut costs by providing needed validated pharmacodynamic biomarkers that could enable an objective evaluation of drug action and its impact on CNS OR systems. In so doing, the use of these brain imaging tools may aid in bridging the gap between basic science findings and the mechanistic knowledge of clinical phenomena regarding complex disease (Bergstrom et al., 2003; Perkins and Frier, 2004; Wang and Maurer, 2005; Brooks, 2005; Wong et al., 2009). In these ways, the brain receptor imaging of ORs serves as an example of a translational research tool, which may facilitate the comprehension of molecular mechanisms involved in brain function and the identification and pursuit of drug development targets.

Implementing molecular imaging of ORs early in the drug development process has value in Phase I and II clinical trials to determine optimal dosing schedules and to demonstrate proof of concept by testing exploratory hypotheses, which assess a relationship between target receptor occupancy and drug-induced biological changes postulated to confer a clinical benefit (Bergstrom et al., 2003; Perkins and Frier, 2004; Wang and Maurer, 2005; Brooks, 2005). Tracking the fate of a radiolabeled drug candidate directly by well-established radioligand tracer-binding assays of CNS ORs and determining the relationship between drug exposure (dose or plasma concentration) and pharmacokinetics and pharmacodynamics of drug delivery to target receptor occupancy in target and nontarget tissues may provide an insight into optimal dosing regimens and metabolism in early-phase clinical trials and exploratory studies.

Also, a series of drug candidates may be screened in preclinical studies against known well-established radiotracers for the biological sites or processes of interest. This approach allows for a systematic empirical evaluation of selectivity, delivery, and determination of fractional occupancy of ORs as predictors of clinical efficacy (van Waarde, 2000; Passchier et al, 2002; Gatley et al., 2003). In so doing, such brain imaging tools may accelerate informed implementation of phase II clinical drug development studies. Molecular and functional brain imaging technologies of OR systems may also be utilized to link mechanisms of complex disease related to OR binding and signaling to those mechanisms discovered by the efforts of molecular biologists, using preclinical experimental models (Stohler and Zubieta, 2010).

The present review describes the status of radiotracers for opioid PET human brain imaging. It then reviews selected studies as examples demonstrating the power of such brain imaging tools in several areas of clinical trials research and their application toward establishing biomarkers for clinical end points of efficacy in these areas.

2.1. CNS OPIOID RECEPTORS

Almost 40 years ago, specific CNS receptors for endogenous opioid peptide ligands were discovered by Snyder and colleagues using radioligand-binding assays on isolated brain tissue (Pert and Snyder, 1973). The ORs of three major types (mu (μ), kappa (κ), and delta (δ)) are distributed throughout the CNS, and they contain ligand-binding sites for brain imaging studies of opioid neurotransmission. Each OR type belongs to the guanine nucleotide-binding protein-coupled receptor heterotrimeric family, which has seven α-helical transmembrane domains, has a highly conserved amino acid sequence, and has been cloned (Waldhoer et al., 2004; Pasternak, 2005). Effector processes of these receptors include ion channels, adenylate cyclase, mitogen-activated protein kinase, and phospholipase C (Law et al., 2000; Waldhoer et al., 2004).

Three classes of endogenous opioid-receptor ligands are (1) β-endorphin (derived from proopiomelanocortin), which has an affinity for the μ-ORs and δ-ORs, (2) methionine–enkephalin and leucine–enkephalin, which also have a pronounced affinity for μ-ORs and δ-ORs and a very low affinity for the κ-ORs, and (3) the dynorphins, which preferentially bind to κ-ORs. The OR agonists selective for μ-ORs and δ-ORs tend to produce analgesic and rewarding effects in humans. In contrast, OR agonists that are selective for k-ORs tend to produce dysphoric effects. The m-ORs are distributed widely throughout the CNS and are the main receptors through which morphine and other opioid-mediated analgesics act to induce their antinociceptive effects. At higher doses of morphine and its derivatives, other OR-mediated neurotransmission of enkephalins at these receptors also mediates the antinociceptive effects of these analgesics. Additional information regarding CNS ORs, their ligands, OR-mediated signaling mechanisms, and cellular adaptations of these receptors in response to the administration of OR agonists has been the subject of recent reviews and is detailed elsewhere (Lever, 2007; Henriksen and Willoch, 2008).

2.2. OR-RADIOLABELED TRACERS

In 1975, Pert and Snyder showed that ORs may be labeled in vivo following an intravenous injection of an opiate antagonist, 3H-naloxone (Pert and Snyder, 1975). This important finding was a precursor to the development of autoradiography methods for the visualization of ORs. This study was the first report that demonstrated the feasibility of localizing, quantifying, and labeling the CNS OR in vivo under conditions that would allow its imaging with PET and single-photon emission computed tomography (SPECT) in the human brain. The results of this study also helped to advance principles that are currently utilized in developing new receptor-binding radiotracers: the use of brain regions with a negligible content of receptors as reference regions for nonspecific binding which could be subtracted from binding in other regions to assess regional specific binding, different in vivo kinetics for specific and nonspecific binding, the demonstration of various receptor-binding criteria such as appropriate regional distribution, pharmacology, and saturability under in vivo conditions, and others. However, a problem with $[^{3}H]$naloxone as an in vivo ligand is its low ratio of specific to nonspecific binding due, in part, to its modest binding affinity for the OR.

Since then, the PET imaging of human ORs has been made possible with the development of higher affinity radiolabeled tracers, such as $[^{11}C]$ Carfentanil (μ), $[^{18}F]$ cyclofoxy (μ, k), $[^{11}C]$ methylnaltrindole (d), $[^{11}C]$ diprenorphine (μ, d, k), and $[^{18}F]$ fluoroethyldiprenorphine (μ, d, k). Consequently, since 1984, PET imaging studies of ORs in normal human volunteers; drug-dependent individuals; and persons with pain, eating disorders, seizures, or affective disorders have been active areas of research.

A central principle to the development of new probes is the use of high binding affinity-radiolabeled tracers, which have a slow receptor dissociation rate and result in the prolonged retention of these ligands at CNS ORs. These tracers must exhibit a high selectivity for their target receptor compared to other low nonspecific binding in brain regions devoid of the target receptor. These tracers also must show rapid crossing through the blood–brain barrier and need to be metabolized in a manner that does not interfere with ongoing measurements (Frost, 2003).

$[^{11}C]$ diprenorphine has been successfully utilized as a radiolabeled tracer of CNS ORs and has high specific to nonspecific binding ratios without the need to wash tissue sections to remove nonspecific binding. However, the use of $[^{11}C]$ diprenorphine as a radioligand for in vivo labeling poses a problem. Diprenorphine has nonselective properties in vivo to bind to different OR subtypes (μ, d, and k) because it has a similar binding affinity for these subtypes. This nonselectivity of $[^{11}C]$ diprenorphine limits its ability to discriminate the changes in binding due to the release of endogenous opioids, since some of which are subtype specific (e.g., dynorphin) while others are not (e.g., Met-enkephalin). Also $[^{11}C]$ diprenorphine exhibits low sensitivity to full opiate agonists (Hume et al., 2007).

Therefore, the development of selective and sensitive positron-emitting d-OR and k-OR radiolabeled tracers for in vivo labeling continues to be a desirable priority.

2.3. PET IMAGING OF μ-OR

[^{11}C] Carfentanil is a potent and selective mu-opioid-receptor (μ-OR) agonist (Stahl et al., 1977), a fentanyl derivative with a high affinity (K_I = 0.051 nM, 37 °C) for the CNS μ-OR. It has been used since the mid-1980s in PET studies as a radiolabeled tracer to assess in vivo changes in the μ-OR availability in the CNS (Frost et al., 1985). The Logan graphical reference region analysis method with occipital input as a reference region has been used as a validated method to obtain noninvasive estimates of [^{11}C] Carfentanil percent receptor occupancy (Bmax/K_D) and has been found to be relatively insensitive to the alterations in regional cerebral blood flow.

The occipital cortex is used as a reference region because it has negligible specific binding to [^{11}C] Carfentanil and has low expression of μ-ORs (Hiller and Fan, 1996). Percent receptor occupancy (Bmax/K_D) or binding potential (BP_{ND}) is utilized for the quantification of PET scans as a measure of μ-OR availability and reflects the ratio of receptor concentration to the receptor's affinity for the radiotracer. Activation of the CNS endogenous μ-OR system has been shown to result in a reduction of the binding potential of [^{11}C] Carfentanil to μ-OR, due to the displacement of this radiolabeled tracer from μ-OR during endogenous opioid release. This displacement may be due to (1) a competition between the radiolabeled tracer and the endogenous opioid neurotransmitter, (2) receptor internalization and recycling, (3) alterations in the conformational state of the receptors (from high to low affinity states to bind to the radioligand agonist) associated with high concentrations in the synaptic cleft of the endogenous ligand (Stohler and Zubieta, 2010).

[^{11}C] Carfentanil has been used in healthy control subjects to determine normative values and to evaluate the effect of gender and age on brain μ-OR availability. It has also been instrumental as a radiolabeled tracer in human PET studies that advance our understanding, on a brain-system level, the impact of alterations in brain μ-OR-mediated neurotransmission on psychiatric illnesses, such as unipolar major depression and bulimia nervosa, chronic cocaine or alcohol use disorders, and chronic pain, where studies have shown pathophysiological changes in the CNS opioid system playing a key role in these diseases and the treatment of these disorders. For instance, in a series of studies utilizing validated psychophysical tools together with PET imaging of μ-OR binding potential using [^{11}C] Carfentanil, Zubieta and colleagues and others evaluated the antinociceptive effects mediated by the CNS μ-opioid system (Stohler and Zubieta, 2010).

Converging data from those studies have uncovered on a regional level a distributed neural network of brain regions regulated by the μ-OR system mediating the affective experience of sustained pain as well as the expectancies of pain relief by placebo. These regions include the anterior cingulate and prefrontal cortex, the anterior and posterior insular cortex, orbitofrontal cortex, ventral striatum, ventral pallidum, thalamus, hypothalamus, amygdala, periaqueductal gray (PAG), and others. This brain circuitry plays an important role in reward processing and motivation, as well as the integration of emotional and cognitive information (Rainville et al., 1997; Zubieta et al., 2001, 2005; Scott et al., 2008; reviewed in Stohler and Zubieta, 2010).

Another radiolabeled tracer that has been used in the PET-binding studies of μ-OR is [^{18}F] cyclofoxy, which was first used by Cohen et al. in human studies. [^{18}F] cyclofoxy is a μ-OR and k-OR opioid antagonist (Pert et al., 1984; Cohen et al., 1988). It has been used in PET imaging studies in substance use treatment research to explore the OR binding in methadone-maintained patients (Kling et al., 2000). However, a limitation of the use of [^{18}F] cyclofoxy as a radiolabeled tracer to assess μ-OR binding potential is that it displays nonselective receptor pharmacology to this receptor, in contrast to [^{11}C] Carfentanil.

2.4. PET IMAGING OF δ-OR AND κ-OR

Neurons with a high expression of the δ-OR are found in many brain regions that form parts of CNS pain circuitry, including the PAG, hypothalamus, insular cortex, amygdala, and rostroventral medulla. Microinjection of δ-OR agonists into these regions induces analgesia, which may be reversed by OR antagonists. Quantifying δ-OR distribution in the CNS may be accomplished through the use of receptor-specific positron-emitting radiolabeled tracers with a high affinity and selectivity to these receptors. [^{11}C] methylnaltrindole is a potent highly selective δ-OR antagonist derived from naltrexone (Portoghese et al., 1990; Yamamura et al., 1992; Contreras et al., 1993). [^{11}C] methylnaltrindole has a high binding affinity to the δ-OR (K_I = 0.02 nM), and is 700-fold selective for δ-OR over μ-OR sites. It also exhibits a very high (>3000-fold) selectivity for d-OR over k-OR sites (Portoghese et al., 1990). PET studies of healthy human volunteers showed the selectivity of [^{11}C] methylnaltrindole for δ-OR. In brain sites rich in δ-OR, [^{11}C] methylnaltrindole exhibits kinetics typical to high binding affinity, and in receptor-poor sites, [^{11}C] methylnaltrindole exhibits kinetics that reflect a rapid washout of nonspecific binding. The extent of δ-OR antagonist's blockade of [^{11}C] methylnaltrindole's effects is enhanced with increasing δ-OR receptor density. [^{11}C] methylnaltrindole regional binding distribution correlates highly with a known human brain distribution of δ-OR. The utility of δ-OR human PET imaging with [^{11}C] methylnaltrindole has been shown in epilepsy and pain research.

The κ-OR system has been implicated in suppressing the antinociceptive actions of opioids acting at the CNS μ-OR. The κ-OR is the most prevalent OR in the human brain. The κ-OR dynorphin blocks the effects of μ-OR agonists (e.g., morphine and [D-Ala2, N-MePhe4, Gly-ol]-enkephalin (DAMGO)) in the CNS. Furthermore, multiple studies have reported that μ-OR and κ-OR systems have antagonistic interactions in a number of brain regions. [11C]GR-103,545 is a promising radiolabeled tracer for the PET imaging of κ-OR and may provide opportunities for PET studies into changes in κ-OR binding (Ravert et al., 1999; Talbot et al., 2005).

2.5. PET IMAGING STUDIES OF THE μ-OR AND DRUG USE

The brain opioid system (e.g., the μ-OR system) has been linked to the positive reinforcing effects of most abused drugs including cocaine and in mediating other reward-related behaviors. [^{11}C] Carfentanil PET studies have been conducted to assess CNS μ-OR binding and availability. In cocaine abusers studied 1–4 days after their last use of cocaine, increased brain μ-OR binding in several regions of the mesocorticolimbic dopamine circuitry has been shown to be associated with more severe craving for cocaine, with upregulation in these receptors persisting after 4 weeks of monitored abstinence (Zubieta et al., 1996). Regional μ-OR-binding upregulation has also been assessed over 12 weeks of monitored cocaine abstinence on a closed research ward. As seen in Figure 1, μ-OR binding in the anterior cingulate cortex of cocaine abusers increased and remained elevated over 12 weeks of cocaine abstinence.

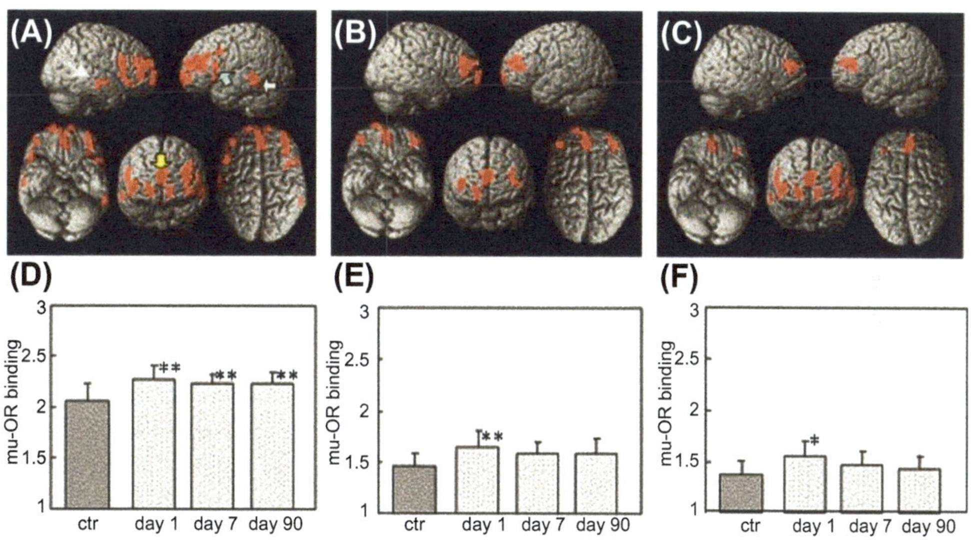

Figure 1 Increased regional brain normalized mu-opioid-receptor (mOR, "μ-OR" in the figure) binding in cocaine-dependent subjects over 12 weeks of enforced abstinence. Three-dimensional brain renderings of regions (shown in red) with significant increases in the mOR binding in cocaine-dependent subjects versus 16 healthy, nondrug-using subjects, after 1 day ($n = 17$) (A), 7 days ($n = 16$) (B), or 90 days ($n = 16$) (C) of enforced cocaine abstinence on a closed research ward. (D–F) Means and standard deviations of mOR-binding values for regions in the bilateral anterior cingulate cortex (yellow arrow in A), left dorsolateral prefrontal cortex (blue arrow in A), and right lateral temporal cortex (white arrow in A), respectively. Display threshold for mOR binding is $p < 0.005$ (uncorrected); cluster size $k > 50$. **$p < 0.001$, *$p < 0.005$ versus control subjects by t test (from the region of interest analysis). *Adapted from Gorelick et al. (2005); Permission to reproduce the figure obtained from the publisher, Elsevier.*

This increased binding was positively correlated with the abusers' self-reported cocaine craving, suggesting a role for the brain μ-OR in this craving (Gorelick et al., 2005). In a subsequent study, cocaine abusers with increased μ-OR binding in frontal and temporal cortices at 1 and 12 weeks of monitored cocaine abstinence relapsed sooner, compared to those abusers who did not exhibit elevated binding (Gorelick et al., 2008). Cocaine abusers exhibiting a smaller reduction in μ-OR binding in these regions between 1 and 12 weeks also relapsed sooner compared to those abusers whose regional μ-OR binding dropped to normal levels over that period. Relapse was defined by two consecutive days of cocaine use after leaving the supervised clinical research facility. These correlations remained significant when taking into account standard clinical variables, such as drug and alcohol use and employment status. Furthermore, there was a significant positive correlation between brain μ-OR binding of cocaine abusers at the time of leaving the facility and the severity of their relapse (expressed as more days of cocaine use in the first month following a relapse). Moreover, elevated brain regional μ-OR binding at 7 days cocaine abstinence was correlated to relapse even when the relapse episode occurred over a year after the day 7 PET scan. Lastly, Ghitza et al. (2010) extended these results to cocaine abusers engaged in 12 weeks of medication-free psychosocial outpatient treatment. As seen in Figures 2 and 3, cocaine abusers who exhibited a greater μ-OR binding in various regions of the mesocorticolimbic dopamine system prior to treatment entry exhibited greater cocaine use and shorter periods of abstinence during treatment.

Collectively, these results indicate a fundamentally important role for the brain endogenous opioid system in cocaine addiction, and that brain regional μ-OR levels may influence drug relapse by mediating cocaine craving. These findings provide a proof-of-concept rationale for targeting the brain opioid system in the development of pharmacotherapeutics to treat cocaine use disorders. Furthermore, they suggest that brain μ-OR binding could serve as a predictor for patient response to treatment of cocaine addiction. They suggest that it could be useful as a diagnostic tool to predict the treatment outcome and identify patients who may be at a greater risk for cocaine relapse and thus in need of more intensive intervention. These results are pertinent to randomized controlled trials demonstrating a key role for brain μ-OR activation in psychostimulant use disorders. For instance, buprenorphine, a partial μ-OR agonist, has been shown to reduce cocaine use in outpatients concurrently dependent on cocaine and opiates (Montoya et al., 2004) and naltrexone, a μ-OR antagonist, has been demonstrated to reduce amphetamine use in amphetamine-dependent outpatients (Jayaram-Lindstrom et al., 2008).

[^{11}C] Carfentanil has also been used in the area of drug targeting, in which brain imaging with this tracer has been utilized to quantify and demonstrate drug bioavailability. For instance, Weerts et al. (2008) used [^{11}C] Carfentanil to assess in recently abstinent alcohol-dependent subjects whether clinical response variability to 50 mg po naltrexone treatment is directly related to the individual variability in naltrexone's inhibition of the

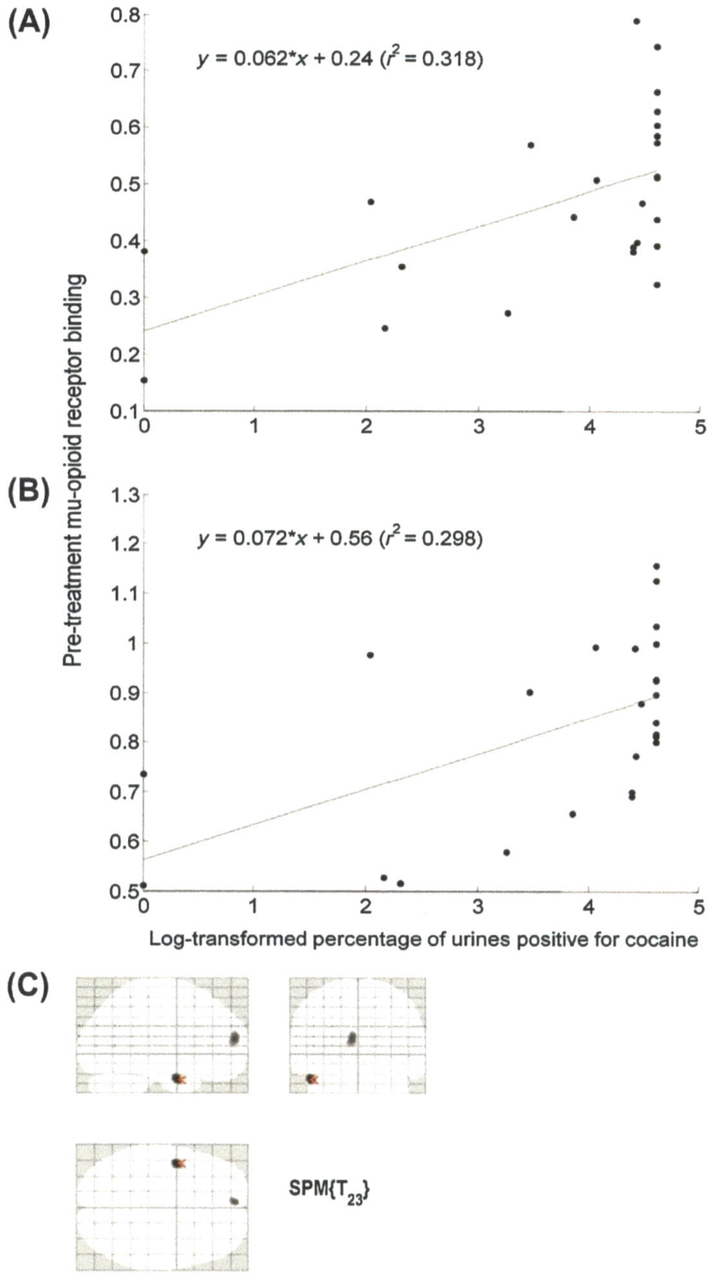

Figure 2 Association of regional μ-OR binding before treatment and the overall percentage of urine specimens positive for cocaine during the first month of treatment. Pretreatment μ-OR binding in the (A) right middle frontal gyrus and (B) left medial frontal gyrus shows a significant positive correlation with the percentage of urine specimens positive for cocaine. The lines are least squares best fit to data points. Display threshold for μ-OR binding is $p<0.005$ (uncorrected); cluster size $k>50$. (C) Location of peak regional associations. SPM, statistical parametric mapping. *Adapted from Ghitza et al. (2010); Permission to reproduce the figure obtained from the publisher, Elsevier.*

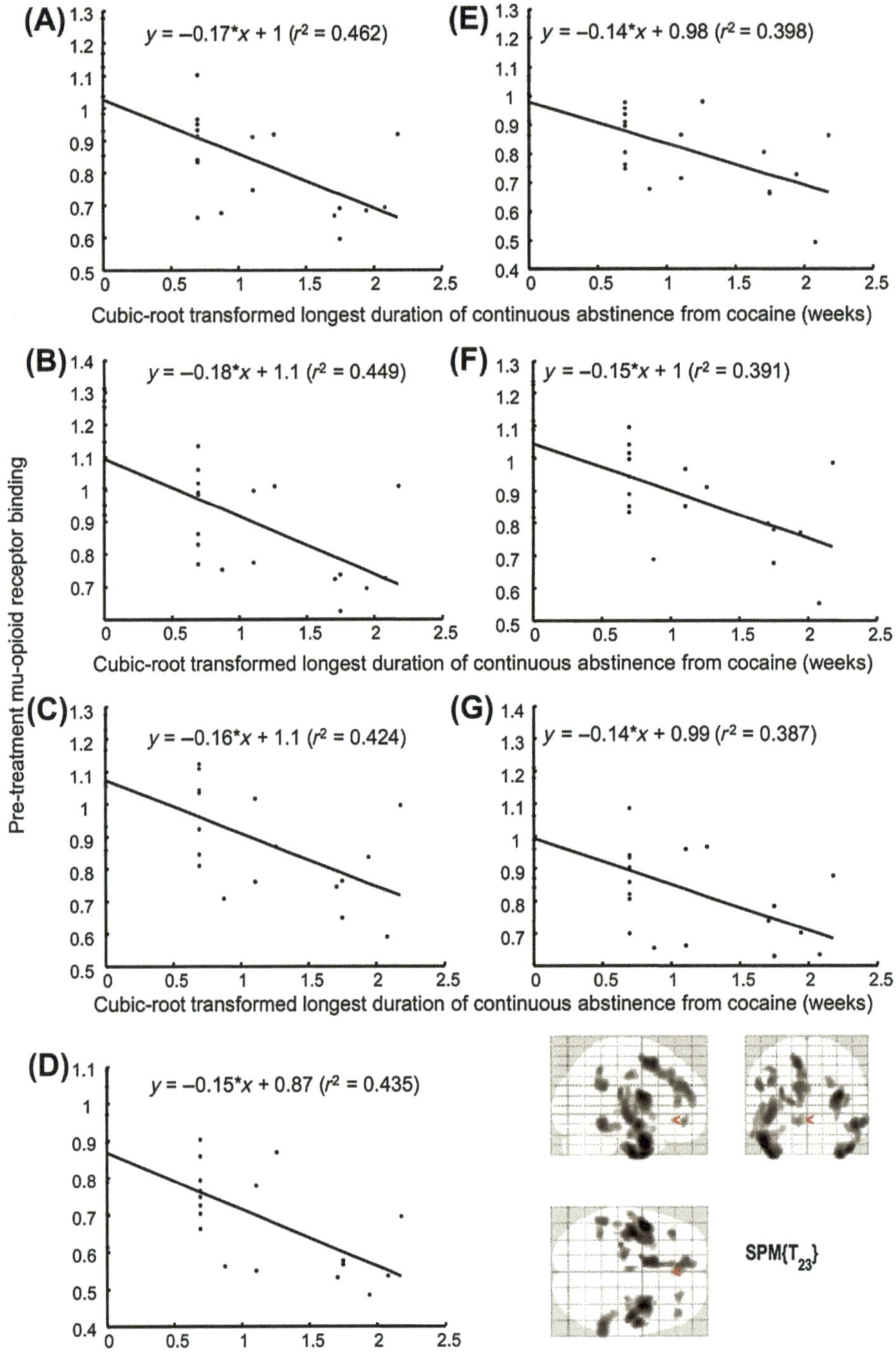
(A)
y = −0.17*x + 1 (r² = 0.462)
(E)
y = −0.14*x + 0.98 (r² = 0.398)
Cubic-root transformed longest duration of continuous abstinence from cocaine (weeks)
(B)
y = −0.18*x + 1.1 (r² = 0.449)
(F)
y = −0.15*x + 1 (r² = 0.391)
Cubic-root transformed longest duration of continuous abstinence from cocaine (weeks)
(C)
y = −0.16*x + 1.1 (r² = 0.424)
(G)
y = −0.14*x + 0.99 (r² = 0.387)
Cubic-root transformed longest duration of continuous abstinence from cocaine (weeks)
Pre-treatment mu-opioid receptor binding
(D)
y = −0.15*x + 0.87 (r² = 0.435)
SPM{T23}

μ-OR and the d-OR. Naltrexone is an FDA-approved medication for the treatment of alcohol dependence, but there is considerable variability in its clinical efficacy among alcohol-dependent subjects. In the study by Weerts et al. (2008), the dose of 50 mg po naltrexone produced nearly complete inhibition of μ-OR but only partial inhibition of δ-OR. Higher doses of naltrexone are necessary to produce more complete inhibition of the δ-OR, and those higher doses have been shown to attenuate the subjective effects of alcohol to a greater degree than with lower doses (McCaul et al., 2000). The authors postulate that lower percent inhibition of δ-OR by 50 mg naltrexone (the FDA-approved dose for treatment of alcohol dependence) across subjects could be responsible for individual differences in treatment outcomes to naltrexone and the observed variability in its efficacy for this indication. The subjects in this study will be further assessed during follow-up visits and continued naltrexone treatment. Future studies will, through PET brain imaging using [^{11}C] Carfentanil, examine the specific relationship between μ-OR and δ-OR blockade by naltrexone and treatment outcomes to naltrexone in alcohol-dependent subjects.

Using PET imaging with [^{11}C] Carfentanil, a recent study in cigarette smokers has shown that individual variations in brain regional μ-OR binding potential and associated differences in the expression of single nucleotide polymorphisms (SNPs) of the human MOR gene (OPRM1 A118G) mediate individual differences in the reinforcing effects of nicotine. Ray et al. (2011) demonstrated that a genetic variation in a human μ-OR gene (*OPRM1* A118G) is linked to differences in the in vivo μ-OR binding potential in brain regions within the mesocorticolimbic dopamine system, a brain system necessary for the reinforcing effects of abused drugs. Notably, this study showed that certain SNPs of this human gene are associated with alterations in cigarette smoking behaviors in these cigarette smokers. In human smokers prescreened for genotype (12 A/A, 10 */G) of this gene, cigarette smokers homozygous for the wild-type *OPRM1 A allele* exhibited significantly higher levels of μ-OR binding potential (relative to smokers carrying the G allele) in left anterior cingulate cortex, left thalamus, and bilateral amygdala. Among G allele carriers of this gene, the degree of subjective reward difference (for denicotinized versus nicotine cigarettes) was significantly associated with the μ-OR

Figure 3 Association of regional μ-OR binding before treatment and duration (weeks) of longest abstinence from cocaine achieved during treatment. Pretreatment μ-OR binding in the (A) left middle temporal gyrus extending to the left sublobar insular gyrus, (B) right middle temporal gyrus, (C) left medial frontal gyrus, (D) right middle temporal gyrus, (E) left middle frontal gyrus, (F) left anterior cingulate gyrus, and (G) right middle frontal gyrus shows a significant negative correlation with the duration of longest abstinence achieved during treatment. The lines are the least squares best fit to data points. Display threshold for μ-OR binding is $p < 0.001$ (uncorrected); cluster size $k > 50$. SPM, statistical parametric mapping. *Adapted from Ghitza et al. (2010); Permission to reproduce the figure obtained from the publisher, Elsevier.*

binding potential difference in the right amygdala, caudate, anterior cingulate cortex, and thalamus. Subjective nicotine reward in *OPRM1 G* allele carriers was associated with a reduced μ-OR binding potential in these brain regions. The specific mechanism underlying this link between reduced μ-OR binding potential, *OPRM1 G* allele SNP, and altered cigarette smoking behavior in humans remains to be determined and will be the subject of future studies. A useful advancement to the field would be future studies that test whether variations in μ-OR binding potential predict the severity of nicotine dependence, degree of response to smoking cessation treatment regimens, and likelihood of relapse to smoking in treatment seekers.

2.6. CHARACTERIZING ACTIONS OF PUTATIVE OR-ACTING ANALGESICS: COMBINED MOLECULAR IMAGING OF OCCUPANCY OF μ-OR IN TANDEM WITH REWARD-RELATED OR PAIN-RELATED FUNCTIONAL BRAIN ACTIVATION IN HUMANS

Combined functional and molecular brain imaging is a useful tool that could be used in the early evaluation of therapeutic potential of novel compounds to assess target OR-binding potential and the functional role of novel OR antagonists and inverse agonists in modulating reward-related behaviors and in pharmacologically characterizing their mechanism of action relative to active comparators. As an example, Rabiner et al. (2011) have used combined PET and functional magnetic resonance imaging (fMRI) in humans to compare the pharmacological properties of a μ-OR antagonist, naltrexone, and an experimental inverse μ-OR agonist GSK1521498. They employed [^{11}C] Carfentanil PET imaging to assess μ-OR binding potential over a range of doses and measured the plasma concentrations of both drugs. In the same scanning session, they also used fMRI to measure food reward-related activation of brain regions implicated in appetitive reward-related behaviors, such as the amygdala and ventral striatum. They compared GSK1521498 to naltrexone in terms of their target receptor occupancy of μ-OR, pharmacokinetics, and functional efficacy in response to presentation of reward-related food stimuli. They showed that relative to naltrexone, this novel compound has a direct relationship between plasma concentration and μ-OR binding potential, and is more efficacious in terms of reducing activation of the amygdala and ventral striatum evoked by food-related-reward stimuli. Strengths of this study design and methodology include the fact that the PET and fMRI data were acquired from the same subjects in the same scanning session. Notably, the effect of GSK1521498 on amygdala activation was still significant even after controlling for individual differences in baseline activation. Future studies assessing the relationship between target receptor occupancy, functional activation of target areas, and behavioral and therapeutic effects of candidate compounds, compared to FDA-approved drugs, would be helpful in accelerating the development and assessment of novel pharmacological treatments.

Combined brain imaging tools may also have utility in establishing pharmacodynamic biomarkers, which allow for distinguishing drugs with versus drugs without clinical analgesic efficacy. For instance, Upadhyay et al. (2011) recently utilized combined pharmacological/functional magnetic resonance imaging (phMRI/fMRI) methodology in human volunteers to establish the changes in functional connectivity of CNS pain circuitry distinguishing the actions of an effective, FDA-approved analgesic, buprenorphine, versus a failed analgesic, aprepritant. Buprenorphine is a partial μ-OR agonist while aprepitant is a neurokinin-1 (NK1) antagonist. The phMRI is a brain imaging tool used to assess the effects of pharmacological agents on the resting blood oxygenation level-dependent (BOLD) signal, whereas fMRI is used to measure the BOLD signal in response to evoked stimulation subsequent to administration of, in this case, drug or placebo. Using phMRI, the investigators found that activation induced solely by buprenorphine was present in regions expressing μ-OR within the CNS pain circuitry. In contrast, aprepitant-induced activation was observed in regions expressing NK1 receptors within that CNS circuitry. Buprenorphine versus aprepitant were also differentiated by distinct fMRI signals in regions of the CNS pain circuitry following evoked pain. After buprenorphine administration, there was a significant correlation between functional connectivity changes in CNS pain-processing pathways, evoked pain fMRI measures, subjective pain ratings, and peak drug concentration. The pharmacological action of buprenorphine, but not aprepitant, altered the functional connectivity (functional interaction) of the CNS pain-processing pathway between the anterior cingulate cortex and thalamus, which was concomitant to a reduction in perceived pain collected after employing an evoked pain fMRI paradigm. Combined phMRI/fMRI tools may be utilized to identify the neuroanatomical location of an analgesic effect for a specific pharmacological compound, when used in conjunction with standard clinical methods such as validated subjective measures of pain/pain ratings. By aiding in the identification of the unique pharmacology of clinical effective versus ineffective medications, this combined phMRI/fMRI approach may aid drug development and lead to optimization by further understanding of functional connectivity changes in known target CNS pathways, which may serve as an objective validated biomarker for the clinical efficacy of putative analgesics. An insight into pharmacodynamic biomarkers, once validated, which are distinctly associated with the action of effective medications on the target brain pathway implicated in chronic pain may provide useful surrogate clinical end points for the treatment of this condition in patients.

3. CONCLUSIONS

Most CNS disorders lack effective pharmacological treatments. Pharmaceutical companies are often hesitant to invest in medication development for CNS disorders, since there are few clinically validated targets or molecular biomarkers. There is also a long record of failure due to the lack of sufficient compounds with optimal potency, pharmacokinetics, selectivity of drug-like effects at the target receptors, bioavailability, and necessary safety profiles. In fact, on average, only 8% of new molecular compounds proceed from the preclinical candidate selection stage through FDA approval to launch, costing an average of $1.8 billion per compound (Paul et al., 2010). Among the reasons cited for failure are poor targeting, lack of clinical efficacy, and/or adverse side effects. The major goals are to improve R&D productivity and medication development success rate, and to reduce cost. While validated preclinical animal models are essential for the medication development process, extrapolation to human patients rarely is direct nor is it straightforward, particularly regarding CNS disorders where basic mechanistic knowledge is incomplete.

A critical need for improvement is the optimal transitioning of novel drug candidates from early preclinical stages to clinical testing and swift reinvestment of resources from failed drug candidates to more promising preclinical leads. In CNS medication development, the lessons learned from early drug discovery phase successes and failures need to be applied rapidly and in an informed manner in order to efficiently establish the drug dose for optimal target receptor occupancy needed to achieve clinical efficacy without adverse side effects. Applying these lessons is essential for safety testing to meet the standards of the FDA for clinical testing (Paul et al., 2010).

Molecular human brain imaging has become a key method for the development of molecular biomarkers to facilitate this chemical optimization process of identification of compounds that have desirable selectivity, safety, bioavailability, and pharmacokinetic and pharmacodynamic properties targeting specific CNS receptor systems in an acceptable drug-like manner. Molecular brain imaging also plays a critical role in defining the spectrum of relevant CNS targets for drug development across therapeutic areas. In so doing, it offers a methodology for derisking drug development at each stage of the process and reducing phase II and III attrition by focusing the drug development pipeline on leads with sufficient affinity and specificity to appropriate CNS targets relevant to the disease in question.

Brain imaging validated biomarkers that provide reliable predictors of treatment-outcome clinical end points are needed to reduce associated R&D costs by serving as objective endpoint surrogates to purely subjective clinical end points of efficacy, which may take considerably longer to collect and assess. In this manner, they may enable the timely estimation of the likelihood of the efficacy of lead candidate compounds. In so doing, biomarker imaging may reduce the cycle time within each drug development stage through the acceleration of clinical trials as objective imaging data are obtained early on in this process in small subject groups, thus overcoming the time and cost limitations of large, longer, and costlier trials with purely subjective clinical end points of efficacy (Frost, 2008).

Human brain imaging of ORs through PET, SPECT, and integrated methods, which combine functional with quantitative molecular imaging tools, serve as an indispensable bridge between basic laboratory research and clinical development of pharmacotherapeutics and validated biomarkers for complex disease processes. Clinical testing and validation of such biomarkers and new drug candidates may be accelerated by incorporating brain imaging early into exploratory proof-of-concept studies in human volunteers with or without the disease being evaluated. Establishing validated molecular biomarkers requires collaborative R&D among biotech companies, pharmaceutical companies, universities, contract research organizations, radiopharmaceutical diagnostic companies (including producers and distributors), and others.

Examples of such collaborative research efforts to establish biomarkers for efficacy are proof-of-concept studies led by the Quantitative Imaging Biomarker Alliance (QIBA) (http://www.rsna.org/Research/QIBA/) and the Biomarkers Consortium (http://www.biomarkersconsortium.org/) both of which endeavor, partly through human brain imaging tools and methods, to discover, develop, quantify, and validate biomarkers to support their use in clinical trials, new drug development, preventative medicine, and medical diagnostics. Also, appropriate utilization of these proof-of-concept approaches may facilitate the selection and validation of drug candidates, targets, and hypotheses about receptor occupancy mechanisms of action thereby enhancing the likelihood of success and helping to derisk CNS drug development ventures.

When combined with tools measuring the impact of effective therapeutics on alterations in resting brain functional connectivity of a known CNS circuitry implicated in disease conditions, human brain imaging tools may accelerate the discovery of novel psychodynamic biomarkers of the mechanisms of actions distinguishing effective from ineffective therapeutics. In turn, a knowledge of target-specific pharmacodynamic biomarkers, once validated, distinctly associated with the action of effective medications on the target brain pathway involved in the disease phenomenology may provide useful surrogate clinical end points to treat a condition in patients.

By such means, the brain imaging of OR methods described in this review and highlighted by examples from recent studies may be used to link preclinical research demonstrating a novel disease mechanism to viable molecular targets or processes of interest related to the clinical outcome for human disease, against which promising drug candidates can be screened in exploratory studies. Such features of brain imaging approaches in proof-of-concept studies are in line with the goals outlined in the FDA's Critical Path Initiative (http://www.fda.gov/oc/initiatives/criticalpath/) to promote national efforts to accelerate the translation of knowledge from early discovery to lead optimization to informed drug development based upon the validation of results in such studies (Wong et al., 2009).

Future directions for human PET and SPECT brain imaging research include finding suitable ligands that could serve as radiolabeled tracers, which provide appropriate CNS receptor binding. This is challenging because of the scarcity of compounds that have the capacity to be administered intravenously, cross the blood–brain barrier, and bind with appropriate affinity and specificity to the target receptors in the CNS, relative to other receptors. Useful ligands also exhibit suitable pharmacokinetic properties and a sufficient signal-to-noise ratio to enable quantitative mathematical modeling using established brain imaging analysis methodologies that would permit their utility in proof-of-concept studies in the early discovery precompetitive stage of the drug development process. Joint development and sharing of these radiotracers is emerging as an important research area in recent years (an example is the Biomarker Consortium) and is considered by both pharmaceutical companies and academia as important endeavors to facilitate the establishment, recognition, and regulatory acceptance of such tracers as imaging biomarkers to aid in CNS drug development and the treatment of CNS diseases.

ACKNOWLEDGMENTS

The author is an employee of the Center for the Clinical Trials Network of the National Institute on Drug Abuse, National Institutes of Health, the funding agency for the National Drug Abuse Treatment Clinical Trials Network. The opinions in this chapter are those of the author and do not represent the official position of the US government.

REFERENCES

Bergstrom, M., Grahnen, A., Langstrom, B., 2003. Positron emission tomography microdosing: a new concept with application in tracer and early clinical drug development. Eur. J. Clinic. Pharmacol. 59, 357–366.

Brooks, D.J., 2005. Positron emission tomography and single-photon emission computed tomography in central nervous system drug development. Neuro. Rx. 2, 226–236.

Cohen, R.M., Carson, R.E., Channing, M., Nordahl, T., Gross, M., Hauck-Newman, A., et al., 1988. F-18-cyclofoxy PET imaging in man. J. Nucl. Med. 29, 796.

Contreras, P.C., Tam, L., Drower, E., Rafferty, M.F., 1993. [^{3}H] naltrindole: a potent and selective ligand for labeling delta-opioid receptors. Brain Res. 604, 160–164.

Cosgrove, K.P., 2010. Imaging receptor changes in human drug abusers. Curr. Top. Behav. Neurosci. 3, 199–217.

Frost, J.J., Wagner Jr., H.N., Dannals, R.F., Ravert, H.T., Links, J.M., Wilson, A.A., Burns, H.D., Wong, D.F., McPherson, R.W., Rosenbaum, A.E., et al., 1985. Imaging opiate receptors in the human brain by positron tomography. J. Comput. Assist. Tomogr. 9, 231–236.

Frost, J.J., 2003. Molecular imaging of the brain: a historical perspective. Neuroimag. Clin. N. Am. 13, 653–658.

Frost, J.J., 2008. Molecular imaging to biomarker development in neuroscience. Ann. N.Y. Acad. Sci. 1144, 251–255.

Gatley, S.J., Volkow, N.D., Fowler, J.S., Ding, Y.S., Logan, J., Wang, G.J., Gifford, A.N., 2003. Positron emission tomography and its use to image the occupancy of drug binding sites. Drug Dev. Res. 59, 194–207.

Ghitza, U.E., Preston, K.L., Epstein, D.H., Kuwabara, H., Endres, C.J., Bencherif, B., Boyd, S.J., Copersino, M.L., Frost, J.J., Gorelick, D.A., 2010. Brain mu-opioid receptor binding predicts treatment outcome in cocaine-abusing outpatients. Biol. Psychiatry 68, 697–703.

Gorelick, D.A., Kim, Y.K., Bencherif, B., Boyd, S.J., Nelson, R., Copersino, M., Endres, C.J., Dannals, R.F., Frost, J.J., 2005. Imaging brain mu-opioid receptors in abstinent cocaine users: time course and relation to cocaine craving. Biol. Psychiatry 57, 1573–1582.

Gorelick, D.A., Kim, Y.K., Bencherif, B., Boyd, S.J., Nelson, R., Copersino, M.L., Dannals, R.F., Frost, J.J., 2008. Brain mu-opioid receptor binding: relationship to relapse to cocaine use after monitored abstinence. Psychopharmacology (Berl). 200, 475–486.

Hammond, L.A., Denis, L., Salman, U., Jerabek, P., Thomas Jr., C.R., Kuhn, J.G., 2003. Positron emission tomography (PET): expanding the horizons of oncology drug development. Invest. New Drugs 21, 309–340.

Henriksen, G., Willoch, F., 2008. Imaging of opioid receptors in the central nervous system. Brain 131, 1171–1196.

Hiller, J.M., Fan, L.Q., 1996. Laminar distribution of the multiple opioid receptors in the human cerebral cortex. Neurochem. Res. 21, 1333–1345.

Hume, S.P., Lingford-Hughes, A.R., Nataf, V., Ahmad, R., Davies, A.N., Nutt, D.J., 2007. Low sensitivity of the positron emission tomography ligand [11C] diprenorphine to agonist opiates. J. Pharmacol. Exp. Ther. 322, 661–667.

Jayaram-Lindstrom, N., Hammarberg, A., Beck, O., Franck, J., 2008. Naltrexone for the treatment of amphetamine dependence: a randomized, placebo-controlled trial. Am. J. Psychiatry 165, 1442–1448.

Kling, M.A., Carson, R.E., Borg, L., Zametkin, A., Matochik, J.A., Schluger, J., Herscovitch, P., Rice, K.C., Ho, A., Eckelman, W.C., Kreek, M.J., 2000. Opioid receptor imaging with positron emission tomography and [(18)F]cyclofoxy in long-term, methadone-treated former heroin addicts. J. Pharmacol. Exp. Ther. 295, 1070–1076.

Law, P.Y., Wong, Y.H., Loh, H.H., 2000. Molecular mechanisms and regulation of opioid receptor signaling. Annu. Rev. Pharmacol. Toxicol. 40, 389–430.

Lever, J.R., 2007. PET and SPECT imaging of the opioid receptor system: receptors, radioligands and avenues for drug discovery and development. Curr. Pharm. Des. 13, 33–49.

Madar, I., Bencherif, B., Lever, J., Heitmiller, R.F., Yang, S.C., Brock, M., Brahmer, J., Ravert, H., Dannals, R., Frost, J.J., 2007. Imaging d- and m-opioid receptors by PET in lung carcinoma patients. J. Nucl. Med. 48, 207–213.

McCaul, M.E., Wand, G.S., Rohde, C., Lee, S.M., 2000. Serum 6-betanaltrexol levels are related to alcohol responses in heavy drinkers. Alcohol. Clin. Exp. Res. 24, 1385–1391.

Montoya, I.D., Gorelick, D.A., Preston, K.L., Schroeder, J.R., Umbricht, A., Cheskin, L.J., et al., 2004. Randomized trial of buprenorphine for treatment of concurrent opiate and cocaine dependence. Clin. Pharmacol. Ther. 75, 34–48.

Passchier, J., Gee, A., Willemsen, A., Vaalburg, W., van Waarde, A., 2002. Measuring drug-related receptor occupancy with positron emission tomography. Methods 27, 278–286.

Pasternak, G.W., 2005. Molecular biology of opioid analgesia. J. Pain Symptom Manage. 29, S2–9.

Paul, S.M., Mytelka, D.S., Dunwiddie, C.T., Persinger, C.C., Munos, B.H., Lindborg, S.R., Schacht, A.L., 2010. How to improve R&D productivity: the pharmaceutical industry's grand challenge. Nat. Rev. Drug Discov. 9, 203–214.

Perkins, A.C., Frier, M., 2004. Radionuclide imaging in drug development. Curr. Pharm. Des. 10, 2907–2921.

Pert, C.B., Snyder, S.H., 1973. Opiate receptor: demonstration in nervous tissue. Science 179, 1011–1014.

Pert, C.B., Snyder, S.H., 1975. Identification of opiate receptor binding in intact animals. Life Sci. 16, 1623–1634.

Pert, C.B., Danks, J.A., Channing, M.A., Eckelman, W.C., Larson, S.M., Bennett, J.M., Burke Jr., T.R., Rice, K.C., 1984. 3-[18F]Acetylcyclofoxy: a useful probe for the visualization of opiate receptors in living animals. FEBS Lett. 177, 281–286.

Portoghese, P.S., Sultana, M., Takemori, A.E., 1990. Naltrindole 5'-isothiocyanate: a nonequilibrium, highly selective delta opioid receptor antagonist. J. Med. Chem. 33, 1547–1548.

Rabiner, E.A., Beaver, J., Makwana, A., Searle, G., Long, C., Nathan, P.J., Newbould, R.D., Howard, J., Miller, S.R., Bush, M.A., Hill, S., Reiley, R., Passchier, J., Gunn, R.N., Matthews, P.M., Bullmore, E.T., 2011. Pharmacological differentiation of opioid receptor antagonists by molecular and functional imaging of target occupancy and food reward-related brain activation in humans. Mol. Psychiatry 16, 826–835 785.

Rainville, P., Duncan, G.H., Price, D.D., Carrier, B., Bushnell, M.C., 1997. Pain affect encoded in human anterior cingulate but not somatosensory cortex. Science 277, 968–971.

Ravert, H.T., Mathews, W.B., Musachio, J.L., Scheffel, U., Finley, P., Dannals, R.F., 1999. [11C]-methyl 4-[(3,4-dichlorophenyl)acetyl]-3-[(1-pyrrolidinyl)-methyl]-1-piperazinecarboxylate ([11C]GR89696): synthesis and in vivo binding to kappa opiate receptors. Nucl. Med. Biol. 26, 737–741.

Ray, R., Ruparel, K., Newberg, A., Wileyto, E.P., Loughead, J.W., Divgi, C., Blendy, J.A., Logan, J., Zubieta, J.K., Lerman, C., 2011. Human Mu Opioid Receptor (OPRM1 A11 G) polymorphism is associated with brain mu-opioid receptor binding potential in smokers. Proc. Natl. Acad. Sci. U. S. A. 31 (108), 9268–9673.

Scott, D.J., Stohler, C.S., Egnatuk, C.M., Wang, H., Koeppe, R.A., Zubieta, J.K., 2008. Placebo and nocebo effects are defined by opposite opioid and dopaminergic responses. Arch. Gen. Psych. 65, 220–231.

Smith, J.P., Conter, R.L., Bingaman, S.I., Harvey, H.A., Mauger, D.T., Ahmad, M., Demers, L.M., Stanley, W.B., McLaughlin, P.J., Zagon, I.S., 2004. Treatment of advanced pancreatic cancer with opioid growth factor: phase I. Anticancer Drugs 15, 203–209.

Stahl, K., van Bever, W., Janssen, P., Simon, E.J., 1977. Receptor affinity and pharmacological potency of a series of narcotic analgesic, anti-diarrheal and neuroleptic drugs. Eur. J. Pharmacol. 46, 199–205.

Stohler, C.S., Zubieta, J.K., 2010. Pain imaging in the emerging era of molecular medicine. Methods Mol. Biol. 617, 517–537.

Talbot, P.S., Narendran, R., Butelman, E.R., Huang, Y., Ngo, K., Slifstein, M., Martinez, D., Laruelle, M., Hwang, D.R., 2005. 11C-GR103545, a radiotracer for imaging kappa-opioid receptors in vivo with PET: synthesis and evaluation in baboons. J. Nucl. Med. 46, 484–494.

Upadhyay, J., Anderson, J., Schwarz, A.J., Coimbra, A., Baumgartner, R., Pendse, G., et al., 2011. Imaging drugs with and without clinical analgesic efficacy. Neuropsychopharmacology 36, 2659–2673 Epub 2011 Aug 17.

Van Waarde, A., 2000. Measuring receptor occupancy with PET. Curr. Pharm. Des. 6, 1593–1610.

Volkow, N.D., McLellan, T.A., 2011. Curtailing diversion and abuse of opioid analgesics without jeopardizing pain treatment. JAMA 305, 1346–1347.

Volkow, N.D., McLellan, T.A., Cotto, J.H., Karithanom, M., Weiss, S.R., 2011. Characteristics of opioid prescriptions in 2009. JAMA 305, 1299–1301.

Wang, J., Maurer, L., 2005. Positron emission tomography: applications in drug discovery and drug development. Curr. Top. Med. Chem. 5, 1053–1075.

Waldhoer, M., Bartlett, S.E., Whistler, J.L., 2004. Opioid receptors. Ann Rev. Biochem. 73, 953–990.

Weerts, E.M., Kim, Y.K., Wand, G.S., Dannals, R.F., Lee, J.S., Frost, J.J., McCaul, M.E., 2008. Differences in d- and µ-opioid receptor blockade measured by positron emission tomography in naltrexone-treated recently abstinent alcohol-dependent subjects. Neuropsychopharmacology 33, 653–665.

Wong, D.F., Tauscher, J., Grunder, G., 2009. The role of imaging in proof of concept for CNS drug discovery and development. Neuropsychopharmacol. Rev. 34, 187–203.

Yamamura, M.S., Horvath, R., Toth, G., Otvos, F., Malatynska, E., Knapp, R.J., Porreca, F., Hruby, V.J., Yamamura, H.I., 1992. Characterization of [3H] naltrindole binding to delta opioid receptors in rat brain. Life Sci. 50, PL119–124.

Zubieta, J.K., Gorelick, D.A., Stauffer, R., Ravert, H.T., Dannals, R.F., Frost, J.J., 1996. Increased mu opioid receptor binding detected by PET in cocaine-dependent men is associated with cocaine craving. Nat. Med. 2, 1225–1229.

Zubieta, J.K., Smith, Y.R., Bueller, J.A., Xu, Y., Kilbourn, M.R., Jewett, D.M., et al., 2001. Regional mu opioid receptor regulation of sensory and affective dimensions of pain. Science 293, 311–315.

Zubieta, J.K., Bueller, J.A., Jackson, L.R., Scott, D.J., Xu, Y., Koeppe, R.A., et al., 2005. Placebo effects mediated by endogenous opioid activity on mu-opioid receptors. J. Neurosci. 25, 7754–7762.

CHAPTER FIVE

Brain Imaging of Sigma Receptors

Muneyuki Sakata, Jun Toyohara, and Kiichi Ishiwata
Positron Medical Center, Tokyo Metropolitan Institute of Gerontology, Tokyo, Japan

1. INTRODUCTION

Sigma receptors were once categorized as a subtype of opioid receptors, because the effects of a prototypical opioid agonist, SKF 10,047 (*N*-allyl-normetazocine), were independent of those of other opioid receptor agonists (Martin et al., 1976). Later, it was determined that sigma receptors have unique binding sites that are different from those of other opioid receptors, and at least two subtypes ($sigma_1$ and $sigma_2$) were classified pharmacologically (Bowen et al., 1989).

The $sigma_1$ receptor was subsequently cloned, and its molecular conformation was revealed (Hanner et al., 1996). In contrast, the $sigma_2$ receptor is yet to be cloned. $Sigma_1$ receptors mainly reside on the endoplasmic reticulum (ER), and dynamically translocate inside cells (Hayashi and Su, 2003a, 2003b). Furthermore, Hayashi and Su (2007) found that $sigma_1$ receptors are enriched at the mitochondria-associated ER membrane. From these findings, Hayashi and Su (2007) proposed that $sigma_1$ receptors function as receptor chaperones.

Very recently, the putative $sigma_2$ binding site was identified as progesterone receptor membrane component 1 (PGRMC1) (Xu et al., 2011). Like the $sigma_1$ receptors, PGMRC1s were localized at the ER and mitochondria.

Endogenous ligands for the $sigma_1$ and $sigma_2$ receptors have not been fully identified, and those for the $sigma_1$ receptor were first proposed to be neurosteroids, including pregnenolone sulfate and progesterone (Su et al., 1988). Recently, the endogenous hallucinogenic trace amine *N*,*N*-dimethyltryptamine was found to be a potent endogenous ligand for the $sigma_1$ receptor (Fontanila et al., 2009).

$Sigma_1$ receptors are widely distributed in the central nervous system (CNS) and at the periphery (Hayashi and Su, 2004). Peripherally, both subtypes are overexpressed in rapidly proliferating cells, such as those in various types of cancer, and the functional roles of the receptors, especially $sigma_2$, have also been the focus of tumor imaging studies (van Waarde et al., 2010). In the CNS, the $sigma_1$ receptor may function as a modulator of signal transduction in neurotransmitter systems such as *N*-methyl-D-aspartate receptors. $Sigma_1$ agonists show antiamnesic and neuroprotective effects (van Waarde et al., 2011); therefore, they are novel targets for the treatment of neuropsychiatric diseases (schizophrenia, depression, and cognition) and brain ischemia.

Imaging of the Human Brain in Health and Disease
http://dx.doi.org/10.1016/B978-0-12-418677-4.00005-1

In this chapter, in vitro and in vivo investigations related to the brain imaging of sigma receptors are reviewed.

2.1. POSTMORTEM STUDIES

Several postmortem studies on the distribution and densities of sigma receptors in human brains have been reported. Weissman et al. (1988) reported that the densities of the sigma sites in the brain were the highest in the cerebellum, nucleus accumbens, and cerebral cortex, using a membrane-binding assay with a nonselective radioligand [^{3}H]haloperidol in the presence of spiperone to block binding to dopamine D_2 receptors. The binding is moderate in the caudate nucleus and substantia nigra and low in the thalamic pontine nuclei.

Autoradiographic distribution determination using nonselective [^{3}H]DTG gave slightly different results: high prevalence in the substantia nigra pars compacta and cerebellum and moderate prevalence in the hippocampal dentate granular cells, striatum, and pineal and pituitary glands (Jansen et al., 1991). In the neocortex, binding was high in the laminae II–IVA and low in the midzone. These two radioligands bind to both sigma$_1$ and sigma$_2$ receptor subtypes. Subtype-selective regional distribution in the brain has been reported in nonhuman primates but not in humans. Mash and Zabetian (1992) quantified the binding of [^{3}H]3-PPP (sigma$_1$ subtype) within 36 brain regions in monkeys (*Macaca mulatta*) and demonstrated a striking enrichment of sigma binding sites over the paralimbic belt cortices, medial and central nuclei of the amygdala, and hippocampal formation. Moderate sigma receptor densities were observed in the caudate and putamen.

The effects of age on the sigma receptor have been noted. Kornhuber et al., (1996) reported an 8.5% decrease per decade in [^{3}H](+)-pentazocine binding (sigma$_1$ subtype) in the frontal cortex. Altered densities of sigma receptors were observed in patients with schizophrenia and Alzheimer's disease (AD). Using nonselective [^{3}H]haloperidol in the presence of spiperone, Weissman et al. (1991) first reported reductions in the density of sigma binding sites in schizophrenic patients on antipsychotic medication; these reductions were the most prominent in the temporal cerebral cortex followed by the parietal cortex. Shibuya et al. (1992) measured sigma receptor binding with [^{3}H]DTG in 17 areas of the cerebral cortex and found no significant differences between schizophrenic patients and controls, with the exception of the superior parietal cortex in which the binding significantly increased in the schizophrenic group. In addition, no significant differences were observed between off-drug and on-drug schizophrenic patients in any brain areas. Helmeste et al. (1996) reported no differences in [^{3}H]nemonapride binding in the presence of spiperone in the frontal cortex and cerebellum; however, they observed a decrease in the caudate. Discrepancy in these results may be due to differences in schizophrenia types, medication period, differences in quantification methods (receptor-binding assay or autoradiography), and differences in radioligands.

In the diseased brain of patients with AD, [^{3}H]DTG binding was measured in seven hippocampal regions by autoradiography; an average reduction of 26% was found, which correlated with an average 29% pyramidal cell loss in the same region (Jansen et al., 1993).

It is noted that all postmortem examinations of human brains have measured both $sigma_1$- and $sigma_2$-binding sites using nonselective radioligands. An exception would be a study on aging effects, which used $sigma_1$-selective [^{3}H](+)-pentazocine (Kornhuber et al., 1996). Subtype selectivity of the tritiated sigma receptor ligands described above is summarized in Table 1.

Table 1 PET and SPECT Ligands for Imaging of Sigma Receptors

Radioligands	Affinity (K_i, nM)[1] Sigma$_1$	Sigma$_2$	Selectivity Sigma$_2$/Sigma$_1$	Reference
Tritiated ligands used for the in vitro binding assay (Walker et al., 1990)[2]				
(+)-SKF 10,047	44.8	4263		Walker et al. (1990)
(+)-3-PPP	5.1	442		Walker et al. (1990)
Haloperidol	0.30	36.1		Walker et al. (1990)
Nemonapride	8.4	9.6		Ujike et al. (1996)
(+)-Pentazocine	2.0	456		Walker et al. (1990)
DTG	11.9	37.6		Walker et al. (1990)
Radioligands used for human and related compounds				
[^{11}C]SA4503	17.4 (IC_{50}) (*p*, *g*)	1784 (IC_{50}) (*dp*, *g*)	106	Matsuno et al. (1996)
	10.1[3]	1520[3]		
	4.4 (*p*, *r*)	242 (*dp*, *rl*)	55.0	Shiba et al. (2006)
	4.6 (*p*, *g*)	61.2 (*dp*, *g*)	13.3	Lever et al. (2006)
	4.0 (*p*, *r*)	55.7 (*dc*, *r*)	13.9	Hirata et al. (2006)
[^{18}F]FM-SA4503	6.4 (*p*, *r*)	250 (*dp*, *rl*)	39.0	Kawamura et al. (2007)
[^{18}F]FE-SA4503	3.8[1] (*p*, *g*)	1.8[1] (*dp*, *g*)	0.5	Kawamura et al. (2003a)
	8.0 (*p*, *g*)	113.2 (*dp*, *g*)	14.2	Lever et al. (2006)
[^{18}F]*o*I-SA4503	6.4 (*p*, *r*)	10.2 (*dc*, *r*)	1.6	Hirata et al. (2006)
[^{18}F]*m*I-SA4503	8.9 (*p*, *r*)	54.5 (*dc*, *r*)	6.1	Hirata et al. (2006)
[^{18}F]FPS	4.3	144	33	Collier et al. (1996)
	0.5 (K_d)			Zhao et al. (2005)
[^{18}F]SFE	24.3 (*p*, *g*)	361 (*dp*, *g*)	14.9	Waterhouse et al. (1997)
	5.0 (K_d)			Zhao et al. (2005)
[^{123}I]TPCNE	0.67(*p*, *g*)	38.8 (*g*, *dp*)	57.9	Waterhouse et al. (1997)

[1]The affinity was expressed as K_i except when indicating IC_{50} or K_d in parenthese.
Radioligands used for the binding assay: *p*, [^{3}H](+)-pentazocine; *dp*, [^{3}H]1,3-di(2-tolyl)guanidine ([^{3}H]DTG) in the presence of (+)-pentazocine to block the $sigma_1$-binding site; *dc*,[^{3}H]DTG in the presence of carbetapentane to block the $sigma_1$-binding site.
Receptor membrane source used for the binding assay: *g*, guinea pig brain; *r*, rat brain; *rl*, rat liver.
[2]Inhibitory dissociation constants for [^{3}H]DTG binding site 1 and site 2 in guinea pig brain membrane as quoted by Walker et al. (1990), except for the nemonapride (Ujike et al., 1996). Haloperidol and nemonapride have a high affinity for the dopamine D_2 receptor, and the tritiated analogs were used for the sigma receptor-binding assay in the presence of a dopamine D_2 receptor blocker such as spiperone.
[3]K_i values derived from IC_{50} values according to the equation: $K_i = IC_{50}/(1 + C/K_i)$, where C is the concentration of the radioligand and each K_d is the dissociation constant of the corresponding radioligand. K_d = 6.9 nM of [^{3}H](+)-pentazocine to $sigma_1$ (Matsuno et al., 1996), K_d = 29.2 nM [^{3}H]DTG to $sigma_2$ (Senda et al., 1995).

2.2. RADIOLIGANDS FOR IMAGING SIGMA RECEPTORS

Since the early 1990s, different classes of compounds, such as benzomorphans, phenyl piperidines, *N,N*-disubstituted pyrrolidinylethylamines, and *N,N'*-disubstituted piperazines, have been evaluated as radioligands for use in the imaging of sigma receptors by positron emission tomography (PET) and single photon emission computed tomography (SPECT). A comprehensive overview of the development of radioligands for the different sigma receptor subtypes has been presented in some recent reviews (Hashimoto and Ishiwata, 2006; Berardi et al., 2009; Mach and Wheeler, 2009). The binding properties of radioligands used clinically and those of related ligands are summarized in Table 1.

The first radioligand used for imaging $sigma_1$ receptors in the CNS was $[^{11}C]$SA4503 (Figure 1), which was developed by Ishii et al. (2001). SA4503 has a high and selective affinity for the $sigma_1$ receptor (IC_{50}, 17.4 nM; selectivity for $sigma_1$ vs $sigma_2$ receptors, 103) but a low affinity for 36 neuroreceptors, ion channels, and second messenger systems (Matsuno et al., 1996; Matsuno and Mita, 1998).

Later, three groups reinvestigated the binding properties of SA4503 and reported slightly different affinities and selectivities: K_i, 4.0–4.6 nM; selectivity for $sigma_1$ vs $sigma_2$ receptors, 13.3–55.0 (Lever et al., 2006; Shiba et al., 2006; Hirata et al., 2006). A large variation was observed in the K_i values of SA4503 and related compounds for the $sigma_2$ receptor subtype. Density (B_{max}) of the sigma receptor was estimated to be 30–600 fmol/mg of protein (~3–60 nM) in the human brain (Weissman et al., 1988; Kornhuber et al., 1996; Jansen et al., 1993). Theoretically, radioligands with nanomolar affinity (0.3–6 nM)

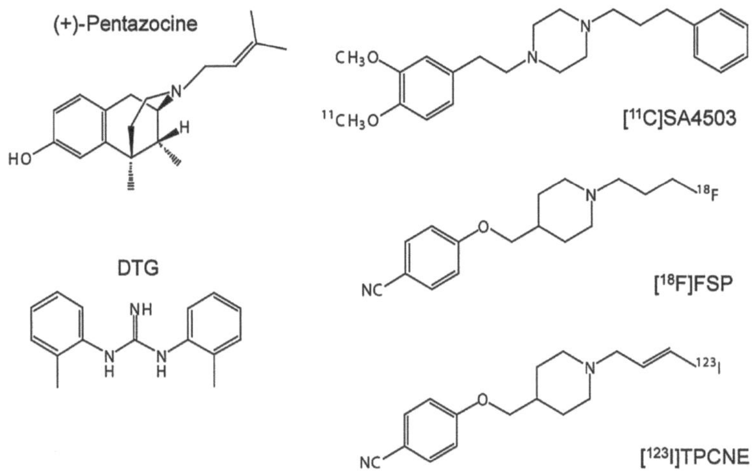

Figure 1 Chemical structure of sigma receptor ligands.

will be suitable for sigma receptor imaging. From this viewpoint, SA4503 has a suitable range of affinity for quantitative in vivo imaging.

In human studies, the use of [^{11}C]SA4503-PET allowed the successful visualization of the sigma$_1$ receptor in the brain as described below, but the affinity of [^{11}C]SA4503 was slightly high for measurement of the sigma$_1$ receptor on the time scale of the PET scan using [^{11}C]-labeled tracer with a short half-life (20.4 min; Sakata et al., 2008).

The second PET ligand that was applied in human studies was [^{18}F]FPS, which was developed by Waterhouse and coworkers. [^{18}F]FPS was not cleared fast enough from the CNS to reach transient equilibrium by 4 h after administration in healthy volunteers. This is because of its high affinity for the sigma$_1$ receptor (K_d = 0.5 nM) (Waterhouse et al., 2004; Zhao et al., 2005). In an effort to improve the CNS clearance for in vivo PET studies, Zhao et al. (2005) synthesized and evaluated [^{18}F]SFE (K_d = 5 nM), an analog of [^{18}F]FPS that exhibits a lower affinity for sigma$_1$.

As an SPECT ligand, the sigma$_1$ receptor-selective [^{123}I]TPCNE (K_i = 0.67 nM) was used in a human study (Stone et al., 2006). High brain uptake was reduced by haloperidol pretreatment, suggesting the specific binding of [^{123}I]TPCNE to sigma$_1$ receptors. However, the time–activity data were best described by an irreversible model. Thus, no further studies for this ligand have been conducted.

Radioiodinated analogs of SA4503 were also developed for SPECT ligands, namely, *o*I-SA4503 and *m*I-SA4503 (designated as *o*-BON and *m*-BON, respectively, Hirata et al., 2006). The affinity and selectivity for the sigma$_1$ receptor were lower for *m*I-SA4503 than for SA4503 (K_i, 8.9 vs 4.0 nM; selectivity for sigma$_1$ vs sigma$_2$ receptors, 6.1 vs 13.9). Although Hirata et al. (2008) have used these compounds in a tumor study on rodents, no clinical studies have been conducted to date.

A review of the clinical trials for a few sigma$_1$ receptor radioligands shows that the ligands with affinities of K_i = 5–10 nM for the sigma$_1$ receptor may be preferable for human studies, and that [^{18}F]FM-SA4503 may be more suitable than SA4503 (Kawamura et al., 2007). However, it is noted that many sigma receptor ligands have an affinity for the vesicular acetylcholine transporter (Efange, 2000) and the sterol delta8–delta7 isomerase, also known as emopamil-binding protein (Berardi et al., 2001). The affinity of [^{11}C]SA4503 for VAChT (K_i = 50 nM, Shiba et al., 2006) and emopamil-binding protein (K_i = 1.7 nM, Berardi et al., 2001) was reported; however, [^{11}C]SA4503 did not seem to bind to VAChT in the rat brain in vivo (Ishiwata et al., 2006a).

Recently, we confirmed that the brain uptake of [^{11}C]SA4503 in mice was not blocked by trifluoperazine with a high affinity for emopamil-binding protein (unpublished data). On the other hand, for other PET and SPECT ligands presented in Table 5.1, no data regarding these two binding sites are available. Selectivity toward these binding sites should be considered in designing new selective ligands, although the relationship between the sigma receptor and sterol delta8–delta7 isomerase has not clearly elucidated.

2.3. HUMAN PET IMAGING OF THE SIGMA$_1$ RECEPTOR

2.3.1. Imaging of the Sigma$_1$ Receptor in the Human Brain

Three radioligands, [^{11}C]SA4503 (Ishii et al., 2001; Sakata et al., 2007), [^{18}F]FPS (Waterhouse et al., 2004), and [^{123}I]TPCNE (Stone et al., 2006), were used for the in vivo investigations of the density of the sigma$_1$ receptor in the human brain by PET or SPECT; however, [^{18}F]FPS and [^{123}I]TPCNE showed irreversible binding profiles during the scan periods because of their high affinities. No further clinical studies were conducted thereafter. At present, clinical studies on sigma$_1$ receptor imaging in living human brains are limited to [^{11}C]SA4503 (Toyohara et al., 2009). The uptake of [^{11}C]SA4503 in the brain was widely but discretely distributed with different concentrations. The regional tissue time–activity curves of [^{11}C]SA4503 in the brain showed a gradual increase in radioactivity over 30 min and then reached a plateau, whereas the radioactivity in plasma decreased very rapidly (Figure 2).

In the traditional PET pharmacokinetics analysis for neuroreceptor radioligands, the ratio of k_3 (association constant rate between the free plus nonspecifically and specifically bound compartments) over k_4 (dissociation constant rate between the free plus nonspecifically and specifically bound compartments) in a two-tissue compartmental model (Mintun et al., 1984) is often used as the binding potential, which is currently defined as the specific binding potential relative to nondisplaceable binding (BP_{ND}) (Innis et al., 2007). Because sigma$_1$

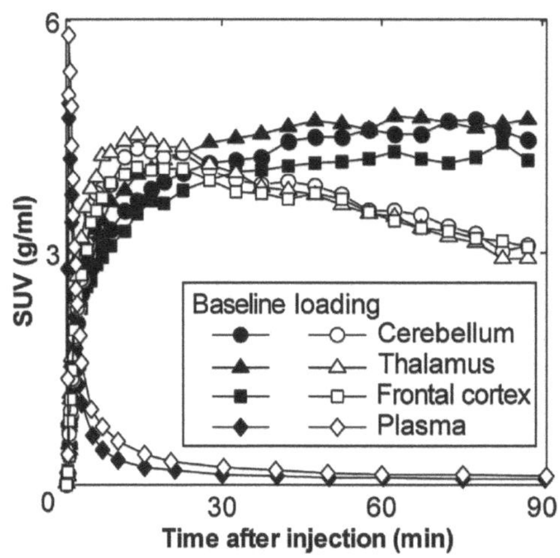

Figure 2 Time–activity curves in the healthy human brain and plasma after intravenous injection of [^{11}C]SA4503 under baseline and haloperidol-loading (3 mg) conditions. Decay-corrected radioactivity was expressed as the standardized uptake value (SUV: regional activity divided by administered dose per body weight). Haloperidol was administered orally 18 h before the injection of [^{11}C]SA4503. The distribution of the radioligands could depend on regional blood flow because of high permeability and slow dissociation.

receptors are distributed in whole-brain regions, no suitable reference region with negligible specific binding is available. Therefore, in the kinetic analysis of [^{11}C]SA4503, arterial blood sampling measurements corrected with the metabolites analysis are essential for obtaining the input function. In general, however, the variability of BP_{ND} estimated from direct compartmental analysis without any constraints is clearly larger than that with indirect methods, such as a reference tissue model analysis (Vilkman et al., 2000). In the case of [^{11}C]SA4503, slow kinetics due to its slightly higher affinity (Figure 2) enlarged the variability of BP_{ND} in the time scale of the PET scan using a ^{11}C-labeled tracer with a short half-life (20.4 min).

An alternative parameter for the evaluation of [^{11}C]SA4503 binding is the total volume of distribution (V_T) estimated by the linear graphical method (Logan et al., 1990), although V_T includes both specific and nonspecific binding (Kimura et al., 2007). The estimation of V_T ($K_1/k_2 \times (1 + k_3/k_4)$) in compartmental analysis tends to be more stable than that of BP_{ND}. Figure 5.3 shows representative V_T images of [^{11}C]SA4503 in a healthy brain. The binding of [^{11}C]SA4503 was high in the cerebellar cortex, moderate in the temporal and parietal cortices, and low in the caudate and putamen (Sakata et al., 2007). These distribution patterns resemble the distribution patterns of radioligand binding in primate brains, where sigma receptors are widely distributed with different densities, using an in vitro binding assay and autoradiography (Weissman et al., 1988; Shibuya et al., 1992; Mash and Zabetian, 1992).

An interaction between sigma$_1$ receptors and steroids has been established (Collier et al., 2007; Su et al., 1988). Currently, we do not have any information about the relationship between plasma hormone levels and [^{11}C]SA4503 binding in a living human brain. If the binding of [^{11}C]SA4503 is sensitive to competition by endogenous steroids,

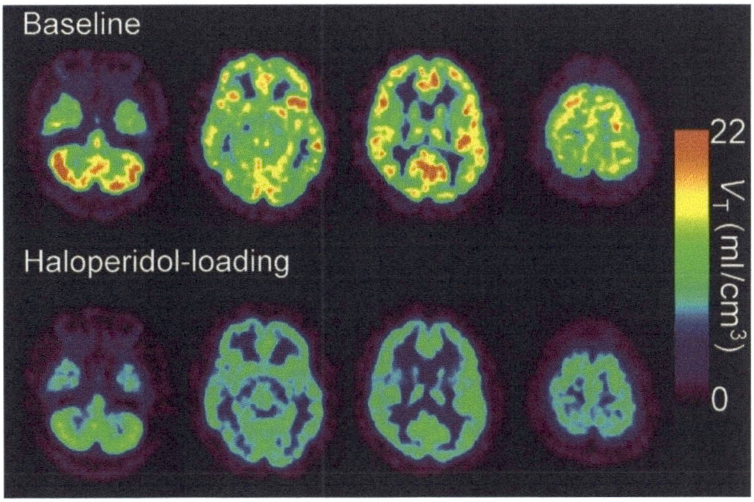

Figure 3 Image of the total distribution volume (V_T) of [^{11}C]SA4503 in the human brain under baseline (upper) and haloperidol-loading conditions (lower). The binding of [^{11}C]SA4503 was considerably reduced after the oral administration of haloperidol (3 mg, 18 h prior to the PET scan).

the intrasubject variability in BP_{ND} of [^{11}C]SA4503 in women during the menstrual cycle and intersubject variability in aged subjects, including postmenopausal subjects, may be found. Therefore, plasma steroid levels should be taken into account in PET studies of patients and healthy volunteers.

Preclinical studies confirmed an age-dependent increase in the $sigma_1$ receptors in the brains of rats and nonhuman primates (Ishiwata et al., 2003; Kawamura et al., 2003b). These results contrast strikingly with the age-dependent decreases in several other receptors in the brain. For investigation of the neuroprotective functions of $sigma_1$ receptors, a PET study with [^{11}C]SA4503 for the evaluation of the aging process in the human brain would be of great interest.

2.3.2. Imaging of the $Sigma_1$ Receptor in Disease

AD, which is a progressive, degenerative disorder of the CNS, is the most common form of dementia. Mishina et al. (2008) performed [^{11}C]SA4503-PET in AD patients in the early stages of the disease. Figure 4 shows representative [^{11}C]SA4503 images compared with those for glucose metabolism evaluated with [^{18}F]FDG. Although the results may have been influenced by atrophy or changes in regional blood flow, a widespread decrease in [^{11}C]SA4503 binding in the brain was found; BP_{ND} for [^{11}C]SA4503 was significantly lower in the cerebellum, thalamus, and frontal, temporal, and occipital cortices of the AD patients than in those of the normal subjects. In other regions of the AD brain, BP_{ND} showed a tendency to decline.

Parkinson's disease (PD) is also a progressive degenerative disorder of the CNS, characterized clinically by resting tremors, sluggish movements, cogwheel rigidity, and postural instability. These symptoms are mainly caused by insufficient dopamine synthesis

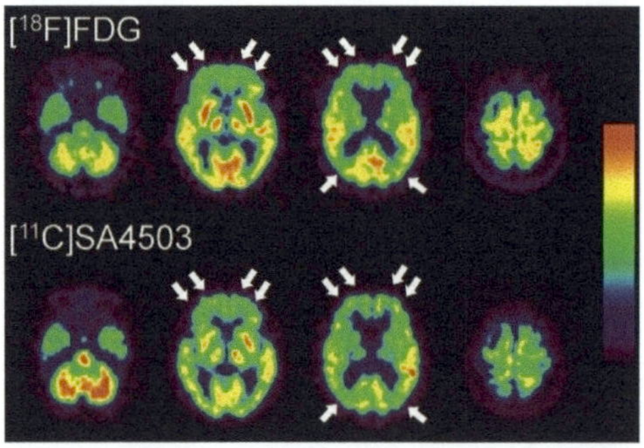

Figure 4 Glucose metabolism and densities of $sigma_1$ receptors in the brain of a patient with alzheimer's disease (AD). Glucose metabolism (upper): static image acquired 45–51 min after injection of [^{18}F]FDG. $Sigma_1$ receptors (lower): volume of total distribution (V_T) derived from Logan graphical analysis of a 90-min dynamic scan of [^{11}C]SA4503. In this patient, the uptake of [^{18}F]FDG was primarily decreased in the frontal and parietal cortex, and the distribution of [^{11}C]SA4503 was consistent with the reduction.

in the dopaminergic neurons. Some studies have suggested that $sigma_1$ receptors are involved in modulating the synthesis and release of dopamine (Chaki et al., 1998). Using [^{11}C]SA4503-PET, Mishina et al. (2005) investigated whether $sigma_1$ receptors were involved in the damage to the dopaminergic system in patients with PD, who showed low densities of dopamine transporters and normal or high densities of dopamine D_2 receptors in the putamen using [^{11}C]CFT and [^{11}C]raclopride, respectively. Although the differences in the BP_{ND} of [^{11}C]SA4503 between normal and PD patients were not clear because of the large intersubject variability, BP_{ND} in PD patients was significantly lower on the more affected side of the anterior putamen than on the less affected side.

2.3.3. Measurement of $Sigma_1$ Receptor Occupancy rates in the Human Brain

While the physiological and pathophysiological roles of sigma receptors remain under investigation, a number of neuropsychiatric drugs are known to have moderate to high affinities for sigma receptors. Moreover, $sigma_1$ receptor ligands represent a new class of therapeutic agents for neuropsychiatric disorders (Hashimoto and Ishiwata, 2006; Ishikawa and Hashimoto, 2010). Therefore, sigma receptor occupancy rates of therapeutic drugs in the living human brain are of great interest. In a feasibility study, Ishiwata et al. (2006b) measured $sigma_1$ receptor and dopamine D_2 receptor occupancy rates by haloperidol in the human brain by PET using [^{11}C]SA4503 and [^{11}C]raclopride, respectively. $Sigma_1$ receptor occupancy rates were approximately 80% after the oral administration of 3 mg haloperidol, whereas dopamine D_2 receptor occupancy rates were approximately 60%.

With respect to drug occupancy studies for drugs with a good potential for high receptor occupancy, graphical analysis using a Lassen plot (Lassen et al., 1995; Cunningham et al., 2010) may be applicable for the evaluation of specific binding and occupancy rate of the drugs, thus providing more stable estimates. The Lassen plot is based on the regional changes in V_T between baseline and drug-loading conditions. The assumptions for the analysis are that nonspecific binding is homogeneous, occupancy is the same in all regions of a regression line, and there is a steady state of occupancy for the duration of the scan. A typical Lassen plot for the haloperidol-loading study is shown in Figure 5, and the parametric V_T images of [^{11}C]SA4503 for baseline and loading conditions (Figure 3) show that the binding of [^{11}C]SA4503 was globally decreased by haloperidol-loading.

Ishikawa et al. measured the $sigma_1$ receptor occupancy rates of two therapeutic drugs using [^{11}C]SA4503-PET. Selective serotonin reuptake inhibitors (SSRIs) are the treatment of choice for many disorders, including major depressive disorder, dysthymia, obsessive–compulsive disorder, and obsessive–compulsive spectrum disorders. Fluvoxamine has moderate affinity ($K_i = 36$ nM) for $sigma_1$ receptors in addition to the main affinity for serotonin reuptake sites (Narita et al., 1996). A single administration of therapeutic doses of fluvoxamine (50–200 mg) decreased [^{11}C]SA4503 binding in the human

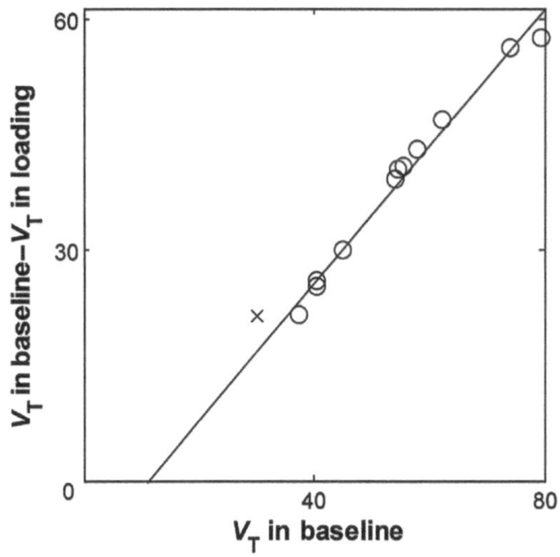

Figure 5 Typical lassen plot analysis applied to a haloperidol-loading study. In this plot, the *x*-axis is the volume of the total distribution (V_T) under baseline conditions and the *y*-axis is the difference between V_T under loading and baseline conditions. The circles represent data from the gray matter regions (cerebral cortices, basal ganglia, and cerebellar cortex) and the cross represents data from the white matter region (centrum semiovale). The clear linear relationship suggests homogeneous nonspecific bindings across the gray matter regions investigated, and the *x*-intercept and gradient of the regression line ($y=0.89(x-11.3)$) represent the volume of free plus nonspecifically bound radioligand (the volume of distribution of the non-displaceable uptake (VND) (or the non-displaceable compartment in 2-tissue compartment model)) and the occupancy rate, respectively. Moreover, the absence of data near the *x*-axis indicates that no regions, including the white matter, are available as a true reference region devoid of specific binding, and that most [^{11}C]SA4503 binding under baseline conditions was specific.

brain in a dose-dependent manner, whereas that of paroxetine (20 mg), another SSRI with a very low affinity for sigma$_1$ receptors ($K_i=1893$), did not (Ishikawa et al., 2007). A similar occupancy study of an antipsychotic drug was also preliminarily described (van Waarde et al., 2011). The second example evaluated was donepezil, an acetylcholinesterase inhibitor that is the most widely prescribed drug for AD. This drug has a high affinity for sigma receptors ($IC_{50}=14$ nM measured with nonselective [^{3}H]DTG) (Kato et al., 1999). A single administration of donepezil (5 or 10 mg) decreased [^{11}C]SA4503 binding in the human brain in a dose-dependent manner (Ishikawa et al., 2009). The sigma$_1$ receptor occupancy rates by fluvoxamine and donepezil were approximately 60% and 75%, respectively.

3. CONCLUSIONS

Sigma receptors may play important roles in the pathophysiology of some neurological and neuropsychiatric disorders. A number of radioligands have been developed and evaluated for use in the imaging of sigma receptors in the CNS by

PET and SPECT. [^{11}C]SA4503 can be used to generate new findings regarding brain functions in patients with AD and PD. Moreover, [^{11}C]SA4503-PET demonstrated that some therapeutic drugs for neurological and neuropsychiatric disorders have an affinity for $sigma_1$ receptors in the human brain. Sigma receptor imaging could be useful for pathophysiological studies of the human brain and help in the discovery and the verification of potential therapeutic agents for several CNS disorders.

ABBREVIATIONS

AD Alzheimer's disease
BP_{ND} specific binding potential relative to nondisplaceable binding
CNS central nervous system
CFT *2β*-carbomethoxy-*3β*-(4-fluorophenyl)tropane
DTG 1,3-di(2-toly1)guanidine
ER endoplasmic reticulum
FDG 2-deoxy-2-fluoro-D-glucose
FM-SA4503 1-[2(4-fluoromethoxy-3-methoxyphenyl)ethyl]-4-(3-phenylpropyl)piperazine
FPS 1-(3-fluoropropyl)-4-(4-cyanophenoxymethyl)piperidine
***m*I-SA4503** 1-[2-(3,4-dimethoxyphenyl)ethyl]-4-[3-(3-iodophenyl)propy1]piperazine
***o*I-SA4503** 1-[2-(3,4-dimethoxyphenyl)ethyl]-4-[3-(2-iodophenyl)propy1]piperazine
PD Parkinson's disease
PET positron emission tomography
PGRMC1 progesterone receptor membrane component 1
3-PPP (+)-3-(3-hydroxyphenyl)-*N*-(1-propy1)piperidine
SA4503 1-[2-(3,4-dimethoxyphenyl)ethyl]-4-(3-phenylpropyl)piperazine
SFE 1-(2-fluoroethyl)-4-[4-cyanophenoxy]methyl]piperidine
SKF 10,047 (2*R*,6*R*,11*R*)-3-ally-1,2,3,4,5,6-hexahydro-6,11-dimethyl-2,6-methano-3-benzazocine-8-ol
SPECT single photon emission computed tomography
SSRI selective serotonin reuptake inhibitor
SUV standardized uptake value
TPCNE 1-(*trans*-[^{123}I]iodopropen-2-yl)-4-(4-cyanophenoxymethyl)piperidine
V_T total volume of distribution

REFERENCES

Berardi, F., Ferorelli, S., Colabufo, N.A., Leopoldo, M., Perrone, R., Tortorella, V., 2001. A multireceptorial binding reinvestigation on an extended class of s ligands: *N*-[ω-(indan-1-yl and tetralin-1-yl)alkyl] derivatives of 3,3-dimethylpiperidine reveal high affinities towards σ_1 and EBP sites. Bioorg. Med. Chem. 9, 1325–1335.
Berardi, F., Abate, C., Ferorelli, S., Colabufo, N.A., Perrone, R., 2009. 1-Cyclohexylpiperazine and 3,3-dimethylpiperidine derivatives as sigma-1 (σ_1) and sigma-2 (σ_2) receptor ligands: a review. Cent. Nerv. Syst. Agents Med. Chem. 9, 205–219.
Bowen, W.D., Hellewell, S.B., McGarry, K.A., 1989. Evidence for a multi-site model of the rat brain sigma receptor. Eur. J. Pharmacol. 163, 309–318.
Chaki, S., Okuyama, S., Ogawa, S., Tomisawa, K., 1998. Regulation of NMDA-induced [^{3}H]dopamine release from rat hippocampal slices through sigma-1 binding sites. Neurochem. Int. 33, 29–34.

Collier, T.L., O'Brien, J.C., Waterhouse, R.N., 1996. Synthesis of [^{18}F]-1-(3-fluoropropyl)-4-(4-cyanophenoxymethyl)piperidine: a potential sigma-1 receptor radioligands for PET. J. Label. Compd. Radiopharm. 38, 785–794.

Collier, T.L., Waterhouse, R.N., Kassiou, M., 2007. Imaging sigma receptors: applications in drug development. Curr. Pharm. Des. 13, 51–72.

Cunningham, V.J., Rabiner, E.A., Slifstein, M., Laruelle, M., Gunn, R.N., 2010. Measuring drug occupancy in the absence of a reference region: the Lassen plot re-visited. J. Cereb. Blood Flow Metab. 30, 46–50.

Efange, S.M.N., 2000. *In vivo* imaging of the vesicular acetylcholine transporter and the vesicular monoamine transporter. FASEB J. 14, 2401–2413.

Fontanila, D., Johannessen, M., Hajipour, A.R., Cozzi, N.V., Jackson, M.B., Ruoho, A.E., 2009. The hallucinogen *N, N*-dimethyltryptamine (DMT) is an endogenous sigma$_1$ receptor regulator. Science 323, 934–937.

Hanner, M., Moebius, F.F., Flandorfer, A., Knaus, H.G., Striessnig, J., Kempner, E., Glossmann, H., 1996. Purification, molecular cloning, and expression of the mammalian sigma$_1$-binding site. Proc. Natl. Acad. Sci. USA 93, 8072–8077.

Hashimoto, K., Ishiwata, K., 2006. Sigma receptor ligands: possible application as therapeutic drugs and as radiopharmaceuticals. Curr. Pharm. Des. 12, 3857–3876.

Hayashi, T., Su, T.P., 2003a. Intracellular dynamics of sigma-1 receptors (sigma-1 binding sites) in NG108-15 cells. J. Pharmacol. Exp. Ther. 306, 726–733.

Hayashi, T., Su, T.P., 2003b. Sigma-1 receptors (sigma-1 binding sites) from raft-like microdomains and target lipid droplets on the endoplasmic reticulum: roles in endoplasmic reticulum lipid compartmentalization and export. J. Pharmacol. Exp. Ther. 306, 718–725.

Hayashi, T., Su, T.P., 2004. Sigma-1 receptor ligands: potential in the treatment of neuropsychiatric disorders. CNS Drugs 18, 269–284.

Hayashi, T., Su, T.P., 2007. Sigma-1 receptor chaperones at the ER-mitochondrion interface regulate Ca^{2+} signaling and cell survival. Cell 131, 596–610.

Helmeste, D.M., Tang, S.W., Bunney Jr., W.E., Potkin, S.G., Jones, E.G., 1996. Decrease in sigma but no increase in striatal dopamine D_4 sites in schizophrenic brains. Eur. J. Pharmacol. 314 (Suppl.), R3–R5.

Hirata, M., Mori, T., Soga, S., Umeda, T., Ohmomo, Y., 2006. Synthesis and in vitro evaluation of iodinated derivatives of piperazine as a new ligand for sigma receptor imaging by single photon emission computed tomography. Chem. Pharm. Bull. 54, 407–475.

Hirata, M., Mori, T., Umeda, T., Abe, T., Yamamoto, T., Ohmomo, Y., 2008. Evaluation of radioiodinated 1-[2-(3,4-Dimethoxyphenyl)ethyl]-4-(2-iodophenylpropyl)piperazine as a tumor diagnostic agent with functional sigma receptor imaging by single photon emission computed tomography. Biol. Pharm. Bull. 31, 879–883.

Innis, R.B., et al., 2007. Consensus nomenclature for in vivo imaging of reversibly binding radioligands. J. Cereb. Blood Flow Metab. 27(9), 1533–1539.

Ishii, K., Ishiwata, K., Kimura, Y., Kawamura, K., Oda, K., Senda, M., 2001. Mapping of sigma$_1$ receptors in living human brain. NeuroImage 13 (Suppl.), S984.

Ishikawa, M., Ishiwata, K., Ishii, K., Kimura, Y., Sakata, M., Naganawa, M., Oda, K., Miyatake, R., Fujisaki, M., Shimizu, E., Shirayama, Y., Iyo, M., Hashimoto, K., 2007. High occupancy of sigma-1 receptors in the human brain after single oral administration of fluvoxamine: a positron emission tomography study using [^{11}C]SA4503. Biol. Psychiatry 62, 878–883.

Ishikawa, M., Sakata, M., Ishii, K., Kimura, Y., Oda, K., Toyohara, J., Wu, J., Ishiwata, K., Iyo, M., Hashimoto, K., 2009. High occupancy of sigma$_1$ receptors in the human brain after single oral administration of donepezil: a positron emission tomography study using [^{11}C]SA4503. Int. J. Neuropsychopharmacol. 12, 1127–1131.

Ishikawa, M., Hashimoto, K., 2010. The role of sigma-1 receptors in the pathophysiology of neuropsychiatric diseases. J. Receptor Ligand Channel Res. 3, 25–36.

Ishiwata, K., Kobayashi, T., Kawamura, K., Matsuno, K., 2003. Age-related changes of the binding of [^{3}H] SA4503 to sigma$_1$ receptors in the rat brain. Ann. Nucl. Med. 17, 73–77.

Ishiwata, K., Kawamura, K., Yajima, K., QingGeLeTu, Mori, H., Shiba, K., 2006a. Evaluation of (+)-*p*-[^{11}C]methylvesamicol for mapping sigma$_1$ receptors: a comparison with [^{11}C]SA4503. Nucl. Med. Biol. 33, 543–548.

Ishiwata, K., Oda, K., Sakata, M., Kimura, Y., Kawamura, K., Sasaki, T., Naganawa, M., Chihara, K., Okubo, Y., Ishii, K., 2006b. A feasibility study of [^{11}C]SA4503-PET for evaluating sigma$_1$ receptor occupancy by neuroleptics: the binding of haloperidol to sigma$_1$ and dopamine D$_2$-like receptors. Ann. Nucl. Med. 20, 569–573.

Jansen, K.L., Faull, R.L., Dragunow, M., Leslie, R.A., 1991. Autoradiographic distribution of sigma receptors in human neocortex, hippocampus, basal ganglia, cerebellum, pineal and pituitary glands. Brain Res. 559, 17–27.

Jansen, K.L., Faull, R.L., Storey, P., Leslie, R.A., 1993. Loss of sigma binding sites in the CA1 area of the anterior hippocampus in Alzheimer's disease correlates with CA1 pyramidal cell loss. Brain Res. 623, 299–302.

Kato, K., Hayako, H., Ishihara, Y., Marui, S., Iwane, M., Miyamoto, M., 1999. TAK-147, an acetylcholinesterase inhibitor, increases choline acetyltransferase activity in cultured rat septal cholinergic neurons. Neurosci. Lett. 260, 5–8.

Kawamura, K., Elsinga, P.H., Kobayashi, T., Ishii, S., Wang, W.F., Matsuno, K., Vaalburg, W., Ishiwata, K., 2003a. Synthesis and evaluation of ^{11}C- and ^{18}F-labeled 1-[2-(4-alkoxy-3-methoxyphenyl)ethyl]-4-(3-phenylpropyl)piperazines as sigma receptor ligands for positron emission tomography studies. Nucl. Med. Biol. 30, 273–284.

Kawamura, K., Kimura, Y., Tsukada, H., Kobayashi, T., Nishiyama, S., Kakiuchi, T., Ohba, H., Harada, N., Matsuno, K., Ishii, K., Ishiwata, K., 2003b. An increase of sigma receptors in the aged monkey brain. Neurobiol. Aging 24, 745–752.

Kawamura, K., Tsukada, H., Shiba, K., Tsuji, C., Harada, N., Kimura, Y., Ishiwata, K., 2007. Synthesis and evaluation of fluorine-18-labeled SA4503 as a selective sigma$_1$ receptor ligand for positron emission tomography. Nucl. Med. Biol. 34, 571–577.

Kimura, Y., Naganawa, M., Sakata, M., Ishikawa, M., Mishina, M., Oda, K., Ishii, K., Ishiwata, K., 2007. Distribution volume as an alternative to the binding potential for sigma$_1$ receptor imaging. Ann. Nucl. Med. 21, 533–535.

Kornhuber, J., Schoppmeyer, K., Bendig, C., Riederer, P., 1996. Characterization of [^{3}H]pentazocine binding sites in post-mortem human frontal cortex. J. Neural Transm. 103, 45–53.

Lassen, N.A., Bartenstein, P.A., Lammertsma, A.A., Prevett, M.C., Turton, D.R., Luthra, S.K., Osman, S., Bloomfield, P.M., Jones, T., Patsalos, P.N., O'Connell, M.T., Duncan, J.S., Andersen, J.V., 1995. Benzodiazepine receptor quantification in vivo in humans using [^{11}C]flumazenil and PET: application of the steady-state principle. J. Cereb. Blood Flow Metab. 10, 740–747.

Lever, J.R., Gustafson, J.L., Xu, R., Allmon, R.L., Lever, S.Z., 2006. Sigma$_1$ and sigma$_2$ receptor binding affinity and selectivity of SA4503 and fluoroethyl SA4503. Synapse 59, 350–358.

Logan, J., Fowler, J.S., Volkow, N.D., Wolf, A.P., Dewey, S.L., Schlyer, D.J., MacGregor, R.R., Hitzemann, R., Bendriem, B., Gatley, S.J., Christman, D.R., 1990. Graphical analysis of reversible radioligand binding from time-activity measurements applied to [*N*-^{11}C-methyl]-(–)-cocaine PET studies in human subjects. J. Cereb. Blood Flow Metab. 10, 740–747.

Mach, R.H., Wheeler, K.T., 2009. Development of molecular probes for imaging sigma-2 receptors in vitro and in vivo. Cent. Nerv. Syst. Agents Med. Chem. 9, 230–245.

Martin, W.R., Eades, C.G., Thompson, J.A., Huppler, R.E., Gilbert, P.E., 1976. The effects of morphine- and nalorphine- like drugs in the nondependent and morphine-dependent chronic spinal dog. J. Pharmacol. Exp. Ther. 197, 517–532.

Mash, D.C., Zabetian, C.P., 1992. Sigma receptors are associated with cortical limbic areas in the primate brain. Synapse 12, 195–205.

Matsuno, K., Nakazawa, M., Okamoto, K., Kawashima, Y., Mita, S., 1996. Binding properties of SA4503, a novel and selective sigma$_1$ receptor agonist. Eur. J. Pharmacol. 306, 271–279.

Matsuno, K., Mita, S., 1998. SA4503: a novel sigma$_1$ receptor agonist. CNS Drug Rev. 4, 1–24.

Mintun, M.A., Raichle, M.E., Kilbourn, M.R., Wooten, G.F., Welch, M.J., 1984. A quantitative model for the in vivo assessment of drug binding sites with positron emission tomography. Ann. Neurol. 15, 217–227.

Mishina, M., Ishiwata, K., Ishii, K., Kitamura, S., Kimura, Y., Kawamura, K., Oda, K., Sasaki, T., Sakayori, O., Hamamoto, M., Kobayashi, S., Katayama, Y., 2005. Function of sigma$_1$ receptors in Parkinson's disease. Acta Neurol. Scand. 112, 103–107.

Mishina, M., Ohyama, M., Ishii, K., Kitamura, S., Kimura, Y., Oda, K., Kawamura, K., Sasaki, T., Kobayashi, S., Katayama, Y., Ishiwata, K., 2008. Low density of $sigma_1$ receptors in early Alzheimer's disease. Ann. Nucl. Med. 22, 151–156.

Narita, N., Hashimoto, K., Tomitaka, S., Minabe, Y., 1996. Interactions of selective serotonin reuptake inhibitors with subtypes of σ receptors in rat brain. Eur. J. Pharmacol. 307, 117–119.

Sakata, M., Kimura, Y., Naganawa, M., Oda, K., Ishii, K., Chihara, K., Ishiwata, K., 2007. Mapping of human cerebral $sigma_1$ receptors using positron emission tomography and [^{11}C]SA4503. NeuroImage 35, 1–8.

Sakata, M., Kimura, Y., Naganawa, M., Ishikawa, M., Oda, K., Ishii, K., Hashimoto, K., Chihara, K., Ishiwata, K., 2008. Shortened protocol in practical [^{11}C]SA4503-PET studies for $sigma_1$ receptor quantification. Ann. Nucl. Med. 22, 143–146.

Senda, T., Matsuno, K., Mita, S., 1995. Differences in the high affinity sites of sigma receptors between guinea pig and rat brain. Neurosci. Res. Commun. 17, 97–105.

Shiba, K., Ogawa, K., Ishiwata, K., Yajima, K., Mori, H., 2006. Synthesis and binding affinities of methyl vesamicol analogs for the acetylcholine transporter and sigma receptor. Bioorg. Med. Chem. 14, 2620–2626.

Shibuya, H., Mori, H., Toru, M., 1992. Sigma receptors in schizophrenic cerebral cortices. Neurachem. Res. 17, 983–990.

Stone, J.M., Arstad, E., Erlandsson, K., Waterhouse, R.N., Ell, P.J., Pilowsky, L.S., 2006. [^{123}I]TPCNE – a novel SPET tracer for the sigma-1 receptor: first human studies and in vivo haloperidol challenge. Synapse 60, 109–117.

Su, T.P., London, E.D., Jaffe, J.H., 1988. Steroid binding at sigma receptors suggests a link between endocrine, nervous, and immune systems. Science 240, 219–221.

Toyohara, J., Sakata, M., Ishiwata, K., 2009. Imaging of $sigma_1$ receptors in the human brain using PET and [^{11}C]SA4503. Cent. Nerv. Syst. Agents Med. Chem. 9, 190–196.

Ujike, H., Akiyama, K., Kuroda, S., 1996. [^{3}H]YM-09151-2 (nemonapride), a potent radioligand for both $sigma_1$ and $sigma_2$ receptor subtypes. Neuroreport 7, 1057–1061.

Vilkman, H., Kajander, J., Nagren, K., Oikonen, V., Syvalahti, E., Hietala, J., 2000. Measurement of extrastriatal D_2-like receptor binding with [^{11}C]FLB 457 – a test–retest analysis. Eur. J. Nucl. Med. 27, 1666–1673.

van Waarde, A., Rybczynska, A.A., Ramakrishnan, N., Ishiwata, K., Elsinga, P.H., Dierckx, R.A., 2010. Sigma receptors in oncology: therapeutic and diagnostic applications of sigma ligands. Curr. Pharm. Des. 16, 3519–3537.

van Waarde, A., Ramakrishnan, N.K., Rybczynska, A.A., Elsinga, P.H., Ishiwata, K., Nijholt, I.M., Luiten, P.G., Dierckx, R.A., 2011. The cholinergic system, sigma-1 receptors and cognition. Behav. Brain Res. 221, 543–554.

Walker, J.M., Bowen, W.D., Walker, F.O., Matsumoto, R.R., De Costa, B., Rice, K.C., 1990. Sigma receptors: biology and function. Pharmacol. Rev. 42, 355–402.

Waterhouse, R.N., Mardon, K., Giles, K.M., Collier, T.L., O'Brien, J.C., 1997. Halogenated 4-(phenoxymethyl) piperidines as potential radiolabeled probes for sigma-1 receptors: in vivo evaluation of [^{123}I]-1-(iodopropen-2-yl)-4-[(4-cyanophenoxy)methyl]piperidine. J. Med. Chem. 40, 1657–1667.

Waterhouse, R.N., Nobler, M.S., Zhou, Y., Chang, R.C., Morales, O., Kuwabawa, H., Kumar, A., van Heertum, R.V., Wong, D.F., Sackeim, H.A., 2004. First evaluation of the sigma-1 receptor radioligand [^{18}F]1-3-fluoropropyl-4-((4-cyanophenoxy)-methyl)piperidine ([^{18}F]FPS) in Humans. NeuroImage 22 (Suppl. 2), T29–T30.

Weissman, A.D., Su, T.P., Hedreen, J.C., London, E.D., 1988. Sigma receptors in post-mortem human brains. J. Pharmacol. Exp. Ther. 247, 29–33.

Weissman, A.D., Casanova, M.F., Kleinman, J.E., London, E.D., de Souza, E.B., 1991. Selective loss of cerebral cortical sigma, but not PCP binding sites in schizophrenia. Biol. Psychiatry 29, 41–54.

Xu, J., Zeng, C., Chu, W., Pan, F., Rothfuss, J.M., Zhang, F., Tu, Z., Zhou, D., Zeng, D., Vangveravong, S., Johnston, F., Spitzer, D., Chang, K.C., Hotchkiss, R.S., Hawkins, W.G., Wheeler, K.T., Mach, R.H., 2011. Identification of the PGRMC1 protein complex as the putative sigma-2 receptor binding. Nat. Commun. 2, 380.

Zhao, J., Chang, R., Carambot, P., Waterhouse, R.N., 2005. Radiosynthesis and in vivo study of [^{18}F]1-(2-fluoroethyl)-4-[(cyanophenoxy)me thyl]piperidine: a promising new sigma-1 receptor ligand. J. Label. Compd. Radiopharm. 48, 547–555.

CHAPTER SIX

Human Brain Imaging of Acetylcholine Receptors

Jun Toyohara, Muneyuki Sakata and Kiichi Ishiwata
Positron Medical Center, Tokyo Metropolitan Institute of Gerontology, Tokyo, Japan

1. INTRODUCTION

1.1. Acetylcholine

Acetylcholine (ACh) is the founding member of the class of biochemicals known as neurotransmitters (Cooper et al., 2003). ACh is widely distributed within the central nervous system (CNS) and is involved in a number of important processes, including roles in sensory and motor processing, sleep, nociception, mood, stress response, attention, arousal, memory, motivation, and reward (Lucas-Meunier et al., 2003; Perry et al., 1999b; Smythies, 2005).

The neurotransmitter ACh is synthesized from acetyl-coenzyme A and choline by choline acetyltransferase (ChAT), an enzyme primarily expressed by cholinergic neurons (Smythies, 2005). ChAT is a specific marker for identifying cholinergic neurons (Oda, 1999). Following its release into the synaptic cleft, ACh is hydrolyzed by acetylcholinesterase (AChE). This enzyme is the target of a group of medications called AChE inhibitors, which increase the levels of ACh in the synaptic cleft (Lane et al., 2006).

In the synaptic cleft, ACh interacts with muscarinic ACh receptor (mAChR) and nicotinic ACh receptor (nAChR), named based on their differential sensitivity to the exogenous ligands muscarine and nicotine, respectively. These receptors differ in both their structure and function.

In the CNS, the cholinergic system includes cholinergic projection neurons and cholinergic interneurons (Mesulam, 2004; Perry et al., 1999b). There are major cholinergic projections from the basal forebrain to the entire cerebral cortex, hippocampus, and amygdala. Basal forebrain cholinergic neurons are located in the medial septal nucleus, vertical and horizontal limb nuclei of Broca, and nucleus basalis of Meynert. The thalamus receives projections from the brain stem cholinergic neurons, which are located in the pedunculopontine tegmental and laterodorsal tegmental nuclei. The striatum contains intrinsic cholinergic neurons. The cerebellum, however, has neither major cholinergic projections nor intrinsic cholinergic systems; therefore, this region can be used as a reference for single photon emission computed tomography (SPECT) and positron emission tomography (PET) imaging of cholinergic systems.

Imaging of the Human Brain in Health and Disease
http://dx.doi.org/10.1016/B978-0-12-418677-4.00006-3

1.2. Muscarinic Acetylcholine Receptors

The mAChRs are a family of G protein-coupled receptors (Clader and Wang, 2005; Gu, 2002). The mAChRs represent approximately 90% of the cholinergic receptors. There are five different mAChRs that are divided in two groups based on their G protein coupling mechanisms: M1, M3, and M5 vs. M2 and M4 (Caufield and Birdsall, 1998; Eglen, 2006). The M1, M4, and M5 receptors are predominantly expressed in the CNS, whereas M2 and M3 receptors are found in both the peripheral nervous system and CNS (Wess, 2004). The M1, M3, and M5 receptors are coupled with the Gq/G11 family, whereas the M2 and M4 receptors prefer Gi/Go-type G proteins (Caufield and Birdsall, 1998). M1, M3, and M5 receptors are found postsynaptically and they activate several effectors, including phospholipase C, intracellular calcium, inositol triphosphatase, and mitogen-activated protein kinase (MAPK) (Wess et al., 2007). M2 and M4 receptors are found both pre- and postsynaptically, and inhibit adenylyl cyclase and voltage-operated calcium channels and activate MAPK and G protein-activated inwardly rectifying potassium channels (Wess et al., 2007). These signaling cascades are essential for many neuronal functions, including neuronal differentiation, survival, and synaptic plasticity.

1.2.1. M_1 Receptors

While a wide range of behavioral, physiological, and biochemical effects mediated by mAChR have been identified in the brain, the identities of the molecular subtypes, together with the lack of drugs showing molecular specificity, have made it impossible to determine the individual roles of M_1–M_5 in the brain. Pirenzepine is used to operationally define pharmacological M_1 receptors, although this antagonist shows a <10-fold difference in the affinity between M_1 and M_4 receptors (Dorje et al., 1991). The M_1 receptor is widely expressed in the brain including the cerebral cortex, hippocampus, and striatum (Oki et al., 2005). The M_1 receptor, the predominant mAChR in the CNS, has been implicated in learning and memory processes. Antagonism of central M_1 receptors with intrahippocampal pirenzepine impaired the spatial memory in a rat model (Messer et al., 1990). In addition, mice lacking the M_1 receptor exhibit defects in a number of cognitive processes (Anagostaras et al., 2003), and M_1 receptor agonists ameliorate learning and memory impairment in animal models of Alzheimer's disease (AD) (Fisher et al., 2003). Therefore, M_1 agonists are considered to have the potential for treatment of AD and other cognitive disorders (Langmead et al., 2008).

1.2.2. M_2 Receptors

M_2 receptors are located presynaptically and regulate the release of ACh and other neurotransmitters, including dopamine (Felder et al., 2000; Smythies, 2005). M_2 receptors are widely expressed in the CNS (Oki et al., 2005), including the brain stem, thalamus, cortex, hippocampus, and striatum, where they control ACh release. The M_2 pharmacological class is usually defined by AF-DX 384 or related antagonists, compounds that have virtually identical affinities for M_2 and M_4 receptors (Dorje et al., 1991). Mice devoid of M_2 receptor activity show cognitive deficits (Tazavara et al., 2003). Moreover, M_2 knockout (KO) mice showed deficits in behavioral flexibility, working memory, and hippocampal plasticity (Seeger et al., 2004). Most importantly, M_2 receptors are lost in the cerebral cortex in AD and Parkinson's disease (PD) (Aubert et al., 1992; Quirion, 1993). The M_2 receptor is a potential treatment target for AD and pain (Langmead et al., 2008).

1.2.3. M_3 Receptors

The M_3 receptor is less densely distributed in the CNS, and its function has not been characterized in detail (Wess, 2004). M_3 KO mice show normal cognition and behavior (Yamada et al., 2001).

1.2.4. M_4 Receptors

The M_4 receptor is found in the midbrain, cortex, hippocampus, and striatum (Oki et al., 2005). M_4 receptors are the most abundant in the striatum (Hersch et al., 1994; Levy et al., 1991; Yasuda et al., 1993) and preferentially coexpressed with dopamine D_1 receptors in striatonigral projection neurons (Ince et al., 1997; Santiago and Potter, 2001). M_4 receptors are concentrated in postsynaptic sites and can be found in the postsynaptic density micro-domain, and therefore, this receptor is sensitive to synaptic input changes and plays a pivotal role in regulating synaptic strength and efficacy (Yasuda et al., 1993). Indeed, M_4 receptors vigorously regulate normal cellular and synaptic activities in the striatum. Stimulation of M_4 receptors is thought to reduce the activity of neurons in the dopaminergic ventral tegmental area (VTA), leading to reduced dopamine release in the nucleus accumbens (Tzavara et al., 2004). The M_4 receptor is a potential treatment target for schizophrenia, PD, and possibly also drug addiction (Langmead et al., 2008; Scarr and Dean, 2008).

1.2.5. M_5 Receptors

The M_5 receptor is the last of the human mAChR family to be cloned (Bonner et al., 1988). M_5 receptors are expressed on dopaminergic neurons in the substantia nigra and the VTA, suggesting that they may play a role in the modulation of dopaminergic transmission (Reever et al., 1997). Therefore, the M_5 receptor is a potential treatment target for stimulant addiction, PD, and schizophrenia (Langmead et al., 2008). The levels of M_5 receptor expression in the rat brain are consistently low at all ages (Wall et al., 1992). In addition, M_5 expression was shown to be unchanged in contrast to the marked reduction in M_2 and upregulation of M_4 in AD patients (Flynn et al., 1995).

1.3. Nicotinic Acetylcholine Receptors

The nAChR is a ligand-gated ion channel, which is permeable to sodium, potassium, and calcium ions. This receptor shows relatively fast responses (within milliseconds), and is excitatory (Clader and Wang, 2005; Dani and Bertrand, 2007; Mihailescu and Drucker-Colin, 2000). Most nAChRs in the CNS are located presynaptically, and they modulate the release of ACh, dopamine, serotonin, glutamate, gamma-amino butyric acid, and norepinephrine (Dani and Bertrand, 2007). The nAChR can also be located postsynaptically, such as those on the dopaminergic neurons in the VTA. The nAChR is a pentamer made up of combinations of 12 subunits (α2–α10 and β2–β4). Most nAChRs contain heteromeric α4β2 or homomeric α7 subunits (Dani and Bertrand, 2007). These subtypes are best characterized in terms of ligand selectivity, as they can be studied by the receptor-binding technique: [^{3}H]cytisine or [^{3}H]nicotine can label α4β2 nAChR, while [^{125}I]α-bungarotoxin or [^{3}H]methyllycaconitine ([^{3}H]MLA) can be used to label α7 nAChR (Romaneli et al., 2007). Other subtypes (e.g., α3 and α6) have been reported in the brain (Guan et al., 2000; Perez et al., 2010), and the α6 subtype may be involved in the pathophysiology of PD (Perez et al., 2010). Despite its lower number, loss of the α3 subtype, consistent with the observations regarding α4 and α7 nAChR subtypes, was also observed in the brains of AD patients (Guan et al., 2000).

1.3.1. α4β2 nAChR

The α4β2 subtype is the most prominent nAChR subtype in the mammalian brain with the highest density of receptors in the thalamus, intermediate density in cortical regions and caudate, and the lowest density in the cerebellum (Lukas et al., 1999; Pimlott et al., 2004). The α4β2 nAChR is involved in various brain functions, including cognition, memory, and behavior. The α4β2 nAChR accounts for >90% of the high-affinity binding sites for nicotine in the rat brain (Flores et al., 1992). Nicotine withdrawal was shown to reduce brain reward function in rats (Epping-Jordan et al., 1998). β2-Containing receptors are critical for the addictive as well as cognitive performance-enhancing properties of nicotine (Picciotto and Corrigall, 2002; Picciotto et al., 1995). Mice lacking the β2 subunit show deficits in executive function, including hyperactivity and impairment of behavioral flexibility (Granon and Changeux, 2006; Granon et al., 2003). On the other hand, the role of the α4 subunit in cognitive processes remains to be elucidated. There is strong evidence for a role of α4β2 nAChR and alteration of the receptor density in CNS disorders, such as AD, PD, schizophrenia, depression, attention deficit hyperactivity disorder, and tobacco dependence (Holladay et al., 1997; Paterson and Nordberg, 2000; Pimlott et al., 2004).

1.3.2. α7 nAChR

α7 is the second most abundant nAChR subtype in the brain after the α4β2 subtype (Paterson and Nordberg, 2000). High concentrations of α7 nAChR are present in the pons, hippocampus, and colliculi of the mouse brain (Whiteaker et al., 1999). In the rat brain, high densities of α7 nAChR are found in the hippocampus and cortical areas, whereas they are expressed at lower levels in the striatum and cerebellum (Davies et al., 1999). Although the distribution of α7 nAChR in primates has still not been completely determined, the available data suggest that it does not differ markedly from that in rodents. The α7 nAChRs show the highest density in the thalamic nuclei with a moderate density in the hippocampus, prefrontal cortex, caudate, putamen, and substantia nigra in the monkey brain (Kulak and Schneider, 2004; Kulak et al., 2006).

α7 nAChR KO mice did not show any morphological CNS abnormalities (Orr-Urtreger et al., 1997; Paylor et al., 1998). However, behavioral tests indicated some cognitive deficits in sustained attention (Hoyle et al., 2006; Young et al., 2007), working memory (Fernandes et al., 2006), and performance under high attentional demand (Keller et al., 2005). Several lines of evidence suggest that α7 nAChRs play a role in the pathophysiology of neuropsychiatric diseases, such as schizophrenia, AD, anxiety, depression, and drug addiction (Freedman et al., 1994; Levin et al., 2006; Nomikos et al., 2000; Nordberg, 2001). Therefore, α7 nAChR is an attractive therapeutic target for these diseases (Hashimoto et al., 2005; Toyohara and Hashimoto, 2010).

2.1. POSTMORTEM STUDIES

2.1.1. Postmortem Studies of mAChR

In the normal adult brain, the levels of total mAChRs were relatively high in the M_1-predominant striatal and cortical areas, compared to that in the thalamic, nigral, and cerebellar regions and the spinal cord, where M_2 are predominant (Perry et al., 1989; Piggott et al., 2002). Significant decreases in the density of M_1 and M_2 receptors were found in the cerebral cortex, while the M_1 and M_2 receptor densities increased with age in the thalamus (Nordberg et al., 1992). In alcoholics, the density of mAChR is decreased (up to 40%) in the hippocampus and temporal cortex (Freund and Ballinger, 1989; Nordberg et al., 1982). M_1 receptor level is elevated in the temporal and parietal cortex of subjects with dementia with Lewy bodies (DLB) (Perry et al., 1990a, b), especially in those with persistent delusions (Ballard et al., 2000). In contrast, M_1 receptor levels are significantly reduced in the striatum in DLB (Piggott et al., 2003). In schizophrenia, M_1 receptor levels are significantly decreased in the dorsolateral prefrontal cortex (Dean et al., 2007). In AD, the M_2 receptor level is moderately reduced in the cortex (Roberson and Harrell, 1997), particularly in the hippocampus, and elevated in the striatum (Rodriguez-Puertas et al., 1997). The M_2 receptor level is reduced in the anteroprincipal nucleus in progressive supranuclear palsy (PSP) (Warren et al., 2007). Warren et al. reported that the M_4 receptor level is reduced in the mediodorsal nucleus in PSP, DLB, and AD (Warren et al., 2007). M_4 receptor levels are also reduced in the subthalamic nucleus in patients with PSP (Warren et al., 2007).

2.1.2. Postmortem Studies of nAChR

nAChR-binding sites, predominantly the α4β2 subtype, are concentrated in the thalamic, neocortical, and striatal regions (Perry et al., 1989). Postmortem studies suggested reductions in α4β2 nAChR with aging (Flynn and Mash, 1986; Nordberg et al., 1992).

Postmortem autoradiographic studies using nicotinic probes [^{3}H]-(−)-nicotine and [^{3}H]-epibatidine revealed significant reductions in α4β2 nAChR in AD (Flynn and Mash, 1986; Kellar et al., 1987; London et al., 1989; Nordberg and Winblad, 1986; Nordberg et al., 1992; Warpman and Nordberg, 1995; Whitehouse et al., 1988) PD, (Martin-Ruiz et al., 2000; Whitehouse et al., 1988), and autism (Lee et al., 2002) compared to that in healthy controls. Interestingly, the reduction of α4β2 nAChR may occur very early in the course of AD (Marutle et al., 1999). The addictive nature of cigarette smoking can be attributed to the reinforcing properties of the α4β2 nAChR agonist, nicotine. Indeed, increases in α4β2 nAChR levels have been reported in the brains of human smokers (Benwell et al., 1988; Bress et al., 1997b; Perry et al., 1999a). Postmortem studies suggested that there are no significant alterations in α4β2 nAChR in schizophrenia (Court et al., 1999; Freedman et al., 1995).

Postmortem studies showed that α7 nAChR is present at a high density in the thalamic nuclei and at a moderate density in the hippocampus, cortex, and cerebellum (Breese et al., 1997a; Falk et al., 2003). In the frontal cortex, the α7 subunit mRNA level decreases significantly with age (Utsugisawa et al., 1999). Postmortem studies also demonstrated altered levels of α7 nAChR in the brain in schizophrenia (Freedman et al., 1995; Marutle et al., 2001) and AD (Burghaus et al., 2000; Court et al., 2001; Guan et al., 2000; Hellström-Lindhal et al., 1999; Wevers et al., 2000).

2.2. RADIOLIGANDS FOR IMAGING AChRs

Cholinergic neurotransmission and function in the living human brain can be evaluated noninvasively by SPECT and PET using suitable radioligands. Over the past three decades, numerous radiotracers have been developed for the evaluation of the cholinergic system in the human brain by SPECT and PET (Eckelman, 2006; Giboureau et al., 2010; Horti and Villemagne, 2006; Kikuchi et al., 2007; Toyohara et al., 2010b). These include several radioligands suitable for mapping mAChR and nAChR, presynaptic vesicular ACh transporter, and AChE. ChAT, the enzyme that synthesizes ACh, is a selective and specific marker for cholinergic perikarya and axons. At present, however, no radiotracers are available for the measurement of ChAT activity by SPECT and PET. This chapter focuses mainly on mAChR and nAChR imaging by SPECT and PET, and discusses the potential usefulness of these techniques in the diagnosis of CNS disease.

2.2.1. mAChR Ligands

Several radiotracers have been developed for mapping mAChR (Dewey et al., 1990a; Eckelman et al., 1984; Weinberger et al., 1990; Zubieta et al., 1998). These radiotracers are generally limited by the lack of selectivity for mAChR subtypes (M_1–M_4), with the exception of 3-((3-[^{18}F]fluoropropyl)thio)-4-(1-methyl-1,2,5,6-tetrahydropyridin-3-yl)-1,2,5-thiadiazole ([^{18}F]FP-TZTP), which appears to bind predominantly to the

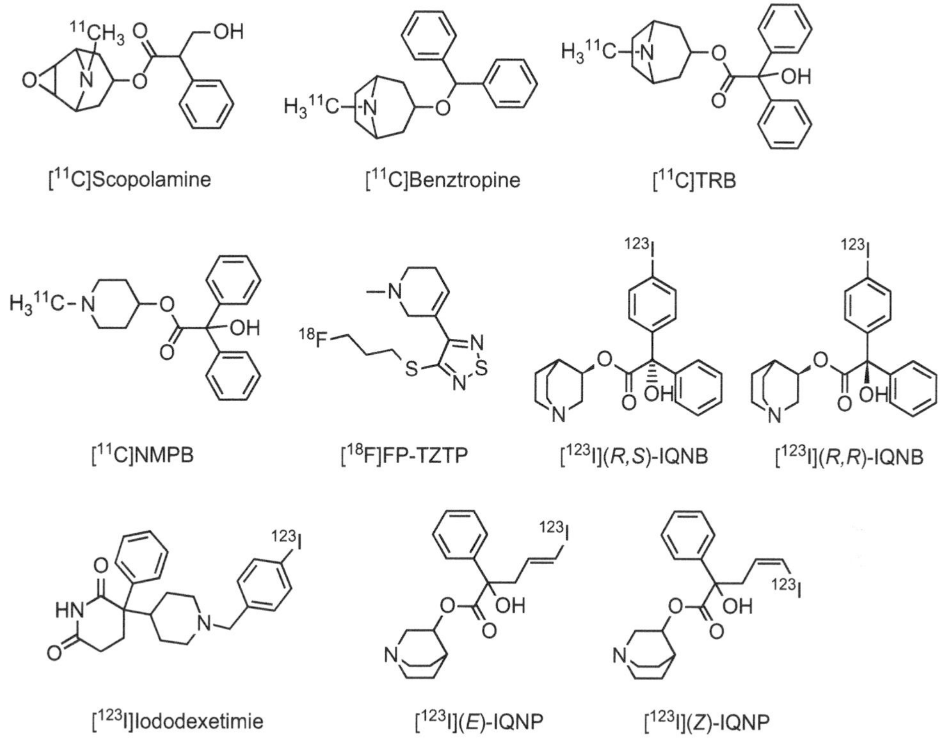

Figure 1 Structures of clinically available radioligands for in vivo mAChR imaging.

M_2 subtype (Carson et al., 1998). The structures, in vitro receptor affinities, and some in vivo characteristics of these radioligands are summarized in Figure 1 and Table 1.

The densities (B_{max}) of M_1 and M_2 receptors are estimated to be 60–500 and 30–70 fmol/mg of protein in the human brain, respectively (Quirion, 1993). Theoretically, radioligands with nanomolar affinity (3–50 nM) will be suitable as mAChR-specific radioligands for in vivo imaging (Eckelman et al., 2006). The in vitro affinities of the first developed non-subtype-selective radioligands (e.g., radiolabeled QNB, scopolamine, benztropine, and dexetimide) were too high—1 nM or less depending on the assay system. Although a high-affinity ligand creates a high target-to-nontarget ratio, the dependence on the blood flow and transport across the blood–brain barrier (BBB) was evident compared to receptor density. In addition, the distribution kinetics of these radioligands in the brain was slow, and several hours were required after injection to reach a steady state. These drawbacks complicate and hinder simple and reliable quantification approaches. Therefore, appropriate pharmacokinetic modeling techniques or a long wait after injection, mainly for ^{123}I-labeled tracers, to separate the parameters for the flow and transport from receptor density were needed. The suboptimal kinetics of these radioligands is thought to be associated with the slow permeability rate through the BBB and low off-rate of the receptor–ligand complexes.

Table 1 Clinically Evaluated Radioligands for mAChR Imaging with SPECT and PET

Radioligand	Affinity (nM) M$_1$	M$_2$	M$_3$	M$_4$	Peak Equilibrium Time in the Cerebral Cortex	BP of the Cerebral Cortex or Uptake Ratio (Ctx CB)[1]	Reference
[^{123}I](*R*,*S*)-IQNB	0.5[2]		1.3[2]		>24 h	No information	Zeeberg et al. (1991)
[^{123}I](*R*,*R*)-IQNB	0.01[2]	0.05[2]	0.05[2] 5.3[3]		>7 h	3.6–4.5	Hiramastu et al. (1994); Norbury et al. (2004); Zeeberg et al. (1991)
[^{123}I]IDEX	14[2], Total				12 h	4 (8 h)	Matsumura et al. (1991)
[^{123}I]*E*-IQNP	0.4[3]	29.6[3]		0.4[3]	34 h	9 (BP of Ctx) 4.4 (24 h)	McPherson et al. (1995); Nobuhara et al. (2001)
[^{123}I]Z-IQNP	0.1[3]	0.4[3]			>48 h 90–150[6] min	1.9[5] (60–80 min)	Bergström et al. (1999); Nobuhara et al. (2001)
[^{11}C]Scopolamine	1.3[4], Total				>120 min	2.3 (70–90 min)	Frey et al. (1992); Otto et al. (1989)
[^{11}C]Benztropine	2.1[3]				>70 min	1.5 (60 min)	Dewey et al. (1990a); Katz et al. (1999)
[^{11}C]TRB	0.7[4], Total				>100 min 10[6] min	9.7[6] 3[5] (50–80 min)	Koeppe et al. (1994); Mulholland et al. (1992)
[^{11}C]NMPB	1.6[4]				>100 min	5.5 (90–110 min)	Otto et al. (1989); Zubieta et al. (1998)
[^{18}F]FP-TZTP	7.4[3]	2.2[3]			20[6] min	Unity 1.6[5,6]	Carson et al. (1998); Kiesewetter et al. (1995)

[1]Ctx = cerebral cortex, CB = cerebellum.
[2]K_d.
[3]K_i.
[4]IC_{50}.
[5]Monkey data.
[6]V_T ratio.

Radioligands with a lower binding affinity generally show faster in vivo brain kinetics, but they also exhibit insufficient specific binding. Therefore, optimal affinity is important for ideal mAChR radioligands with fast kinetics. Eckelman reviewed the details of radioligand development for mAChR imaging covering the literature published before 2004 (Eckelman, 2006). However, there have been no clinical reports of new mAChR radioligands since 2004.

2.2.1.1. α4β2 nAChR Ligands

The density (B_{max}) of α4β2 nAChR is the highest in the thalamus (10–60 fmol/mg of protein) and lower in other brain regions, such as the cerebral cortex (4–21 fmol/mg of protein) (Hall et al., 1993; Marutle et al., 1998; Sihver et al., 1998). Therefore, successful in vivo imaging of α4β2 nAChR has only been accomplished with radioligands based on the epibatidine and A-85380 structures, which have in vitro K_d values in the 0.05–0.15 nM range (Eckelman et al., 2006; Shiver et al., 2000). Although low-affinity [^{11}C]nicotine (K_i = 1 nM) has been used to study α4β2 nAChR in the human brain (Kadir et al., 2006, 2007; Nordberg et al., 1990, 1995), the image quality of [^{11}C]nicotine was poor due to high nonspecific binding and rapid metabolism (Muzik et al., 1998; Nybäck et al., 1994). Thus, [^{11}C]nicotine is not a suitable tracer for PET studies of α4β2 nAChR in the human brain. Studies with the very high-affinity epibatidine analogs yielded the first high-quality images of α4β2 nAChR in an animal brain (Musachio et al., 1997; Villemagen et al., 1997). However, these compounds showed high toxicity (Ding et al., 1999) due to their high affinity for ganglionic α3β4 nAChR (Gao et al., 2008). Less toxic radiolabeled analogs of A-85380, [^{18}F]2-FA, [^{18}F]6-FA, and [^{123}I]5-IA (Figure 2) are now available for the quantitative imaging of α4β2 nAChR in the human brain (Dolci et al., 1999; Horti and Villemange, 2006).

However, there has been only one report of experiments with [^{18}F]6-FA in humans (Ding et al., 2004). The paucity of human PET studies with [^{18}F]6-FA may be due to the relatively high toxicity of this compound (Sheffel et al, 2000). The drawbacks of these ligands are their slow brain distribution kinetics and relatively low binding potential (BP) in the human brain (Table 2).

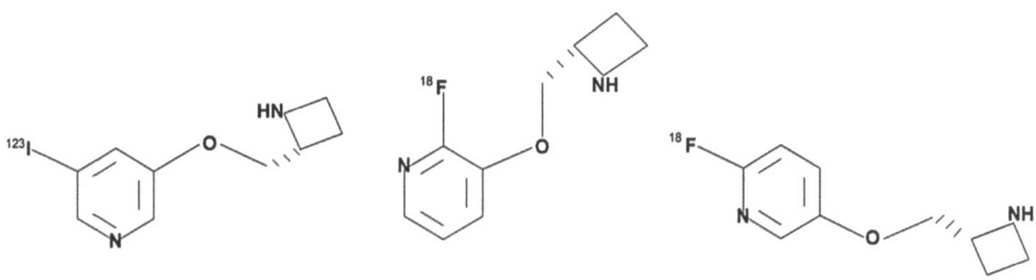

Figure 2 Structures of clinically available radioligands for in vivo α4β2 nAChR imaging.

Table 2 Clinically Evaluated Radioligands for Imaging α4β2 nAChR with SPECT and PET

Radioligand	Affinity (nM)			Species, Time of the Highest Level of Radioactivity Accumulation in the Thalamus	BP of the Thalamus or V_T ratio (Th/CB)[1]	Reference
	α4β2	α7	α3β4			
5-[^{123}I]IA	0.011[2]			Monkey, 1.5 h Human, 100 min	3 2.1–2.2[4]	Chefer et al. (1998); Fujita et al. (2002); Horti et al. (1999)
2-[^{18}F]FA	0.05[2] 1.33[3]		3680[3]	Monkey, 2 h Human, 2 h	2.0[4] (BP)	Bottlaender et al. (2003); Chefer et al. (1999, 2003); Kimes et al. (2003)
6-[^{18}F]FA	0.026[3] 0.028[2]	170[3]	14[3]	Monkey, 60 min Human, 2 h	3.5[4] 3.1 (BP)	Ding et al. (2000, 2004); Gundisch et al. (2005); Horti et al. (2000)

[1]Th = thalamus, CB = cerebellum.
[2]K_d.
[3]K_i.
[4]DVR in baboons.

Quantitative studies require many hours for these tracers to reach a steady state in the brain. In addition, the BP of these compounds is low in extrathalamic regions (BP < 0.6–0.8), including the cerebral cortex, which has lower densities of the receptor. The importance of imaging of extrathalamic α4β2 nAChR has emerged from the demonstration of altered densities of cortical and striatal α4β2 nAChR in neurodegenerative diseases (Pimlott et al., 2004) and schizophrenia (Ochoa and Lasalde-Dominicci, 2007). These drawbacks of the currently available α4β2 nAChR radioligands have initiated the development of radioligands with faster brain kinetics or a higher BP by several research groups. Recent progress in the development of α4β2 nAChR radioligands has been reviewed extensively by Horti et al. (2010) and Horti and Wong (2009).

2.2.1.2. α7 nAChR Ligands

The development of a radioligand that specifically depicts α7 nAChR has been problematic due to its low density in the brain. The density (B_{max}) of α7 nAChR is estimated to be 5–70 fmol/mg of protein in rodents (Davies et al., 1999; Whiteaker et al., 1999), 4–50 fmol/mg of protein in nonhuman primates (Kulak and Schneider, 2004; Kulak et al., 2006), and 2–30 fmol/mg of protein in the human brain (Court et al., 1995, 2001; Falk et al., 2003; Marutle et al., 2001). Due to this low density of α7 nAChR, radioligands require very high binding affinities to produce images with sufficient signal-to-noise ratios (i.e., specific-to-nonspecific binding ratios). Theoretically, radioligands with subnanomolar affinity (<0.02–0.3 nM) will be required to develop α7 nAChR-specific radioligands for in vivo imaging (Eckelman et al., 2006).

Both α7 nAChR and 5-hydroxytryptamine-3 ($5\text{-}HT_3$) receptors are members of the ligand-gated ion channel superfamily (Le Novère and Changeux, 1995; Ortells and Lunt, 1995). These two receptors share the greatest similarity within the family, with an approximately 30% sequence identity (Maricq et al., 1991). In fact, α7 nAChR agonists exhibit crossreactivity with $5\text{-}HT_3$ receptors more often than they do with other nAChRs. Therefore, the low (or ideally negligible) affinity for $5\text{-}HT_3$ receptors is of major concern for developing α7 nAChR-selective radioligands. Furthermore, pharmacological doses of agonists themselves may cause agonist-induced conformational changes in the receptor, resulting in receptor desensitization. Therefore, the contribution of the accompanying pharmacological effects of the agonist radioligand to quantify α7 nAChR may exist. Hence, an ideal in vivo radiotracer for α7 nAChR should be an antagonist. The α-bungarotoxin and MLA are well known as specific α7 nAChR antagonists. However, due to their large molecular weights, they have difficulty in passing through the BBB, which makes them poor radioligands (Davies et al., 1999; James et al., 1980; Navarro et al., 2000). A radiolabeled antagonist for α7 nAChR with a sufficiently high affinity for in vivo imaging is yet to be identified. Consequently, the development of α7 nAChR imaging agents with better properties, including substantially high signal-to-noise ratio, has been challenging. Although several compounds exhibited potential as candidates for α7 nAChR imaging (Figure 3 and Table 3), no compounds other than

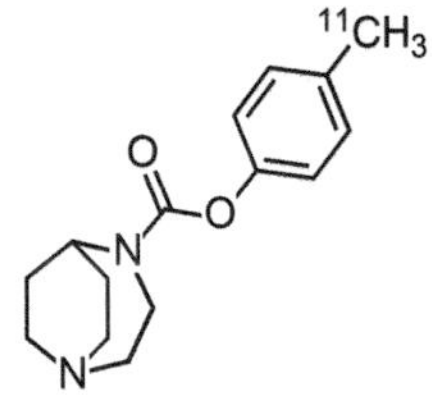

[^{11}C]CHIBA-1001

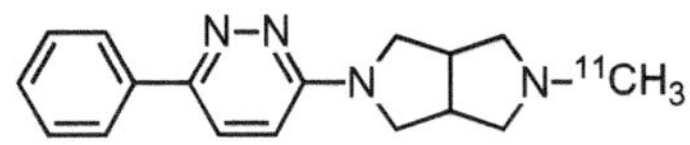

[^{11}C]A-582941

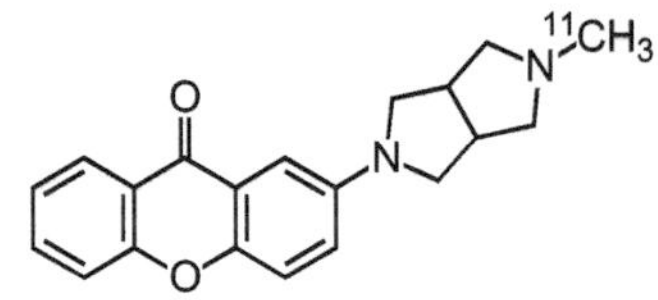

[^{11}C]A-844606

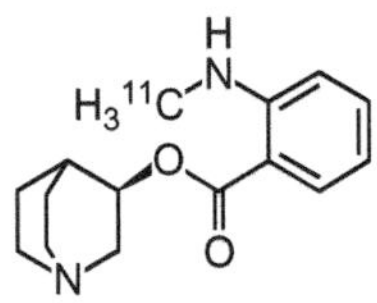

[^{11}C](*R*)-MeQAA

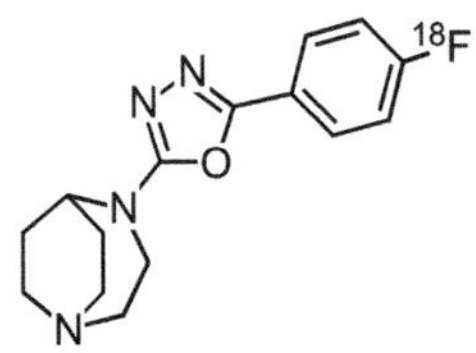

[^{18}F]NS10743

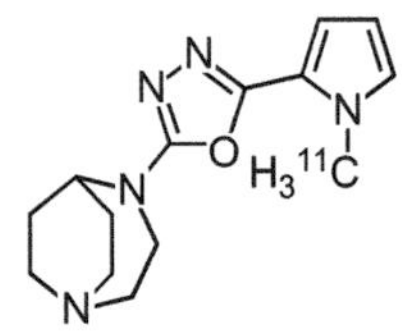

[^{11}C]NS14492

Figure 3 Structures of radioligands for in vivo α7 nAChR imaging.

Table 3 Preclinically Evaluated Radioligands for Imaging α7 nAChR with PET

Radioligand	Affinity (nM) α7	α4β2	5-HT$_3$	Species, Timing of the Highest Accumulation of Radioactivity in the Thalamus	BP of the Thalamus or V_T Ratio (Th/CB)[1]	Reference
[^{11}C]CHIBA-1001	45.8[2] 193.4[3]	>10,000[2]	>1000[2]	Monkey, 10 min Human, 15 min	1.3[5]	Hashimoto et al. (2008); Toyohara et al. (2009)
[^{11}C]A-582941	10.8[4]	>100,000[4]	154[4]	Monkey, 10 min	1.3	Toyohara et al. (2010a)
[^{11}C]A-844606	11[2]	>30,000[2]	–	Monkey, >90 min	1.5	Toyohara et al. (2010a)
[^{11}C](*R*)-MeQAA	40.6[4]	4800[4]	125[4]	Monkey, 12.5 min	1.3	Ogawa et al. (2010, 2011)
[^{11}C]NS10743	11.6[4] 9.3[3]	>10000[4]	855[4]	Pig, 8–10 min	0.8 (BP) 1.3	Deuther-Conard et al. (2009, 2011)
[^{11}C]NS14492	2.2[4]	2800[4]	170[4]	Pig, 10 min	1.4	Ettrup et al. (2011)

[1]Th = thalamus, CB = cerebellum.
[2]IC_{50}.
[3]K_d.
[4]K_i.
[5]Human data.

4-[^{11}C]methylphenyl 1,4-diazabicyclo[3.2.2]nonane-4-carboxylate (CHIBA-1001) have progressed to the clinical trial stage (Toyohara et al., 2010b). The details of radioligand development for imaging α7 nAChR covering the literature published before 2009 were reviewed by Toyohara et al. (2010b).

2.3. HUMAN SPECT/PET IMAGING OF mAChR

2.3.1. Healthy Subjects

Approximately 10 different types of radioligands have been tested in humans. Most were nonselective for mAChR subtypes and showed a distribution corresponding to the pattern of total mAChR density in the brain (Figure 4 and Table 4).

The first radioligand for mapping mAChR in a human brain was the SPECT ligand [^{123}I](*R*,*S*)-IQNB (Eckelman et al., 1984). Most human studies performed to date have used this ligand. In healthy subjects, the regional distributions of [^{123}I](*R*,*S*)-IQNB binding were consistent with postmortem studies of the relative distribution of mAChR in the normal human brain, showing a high accumulation in the basal ganglia, occipital cortex, and insular cortex, low uptake in the thalamus, and a very low uptake in the cerebellum (Weinberger et al., 1991). The diastereomer [^{123}I](*R*,*R*)-IQNB with a higher affinity was first injected into humans in 1995 (Hiramatsu et al., 1995). The estimated BP of [^{123}I](*R*,*R*)-IQNB was the highest in two cortical regions, intermediate in the parotid gland, and lowest in the cerebellum. Quantitative mAChR binding can be estimated by a static scan approximately 7 h postinjection (Norbury et al., 2004).

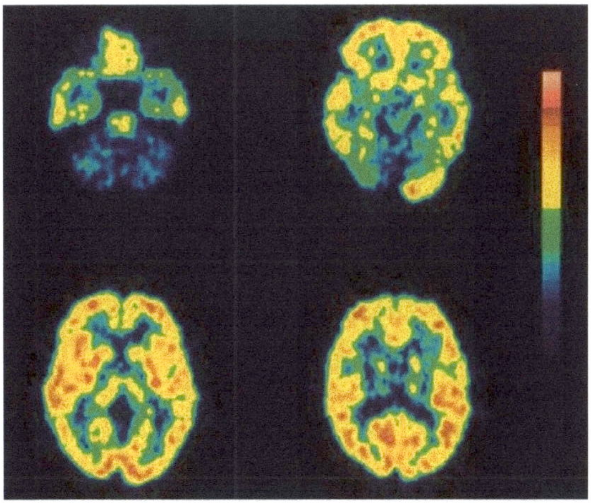

Figure 4 Representative PET images of 1-[^{11}C]Methylpiperidin-4-yl 2-Hydroxy-2,2-Diphenylacetate (NMPB) in a healthy subject. Static images acquired from 40 to 60 min after injection. *Courtesy of Dr Shinotoh, Asahi Hospital for Neurological Disorders and Rehabilitation, Chiba, Japan.*

Table 4 Concentrations (Nanomoles) of AChR Subtypes in Various Brain Regions

	Striatum	Frontal Cortex	Hippocampus	Thalamus	Medulla	Pons	Cerebellum
Total mAChR[1]	232	180	150.7	72.2	44.5	44.5	17.8
M_1 receptor	72.8	70.4	76.1	13.8	2.6	2.6	0.42
M_2 receptor	30.1	39.3	27.5	36.1	36.4	36.4	15.8
M_3 receptor	16.3	20.7	16.2	5.16	2.9	2.86	1.05
M_4 receptor	113	46.7	30.8	17.2	2.6	2.6	0.35
α4β2 nAChR[2]	3.0	3.5	–	9.0	–	–	1.0
α7 nAChR[3]	0.9	4.5	7.0	4.8	–	–	0.5

[1]Data from Zeeberg (1999).
[2]Data from Perry et al. (1995).
[3]Data from Davies et al. (1999).

The first human trial of 1′-(4-[^{123}I]iodobenzyl)-3-phenyl-[3,4′-bipiperidine]-2,6-dione (IDEX) was reported in 1992 (Müller-Gärther et al., 1992). The level of radioactivity increased over 7–12 h in the neocortex, striatum, and thalamus, but decreased immediately in the cerebellum. The radioactivity distribution was correlated with the mAChR concentrations in the corresponding brain regions, with the highest level in the neostriatum, followed by the neocortex, thalamus, and cerebellum. They confirmed the specific binding of pixel-by-pixel [^{123}I]IDEX images by subtracting images showing the nonspecific binding of its pharmacologically inactive enantiomer, [^{123}I]iodolevetimide.

There has been only one report of human experiments with IQNP (Nobuhara et al., 2001). [^{123}I]*E*-(*R*,*R*)-IQNP is a radioligand with modest selectivity for the M_1 and M_4 subtypes, whereas [^{123}I]*Z*-(*R*,*R*)-IQNP is nonsubtype selective. In healthy subjects, [^{123}I]*E*-(*R*,*R*)-IQNP SPECT showed a high uptake in the neocortex and striatum, an intermediate uptake in the thalamus, and a low uptake in the pons and cerebellum, consistent with the rank order for the regional densities of M_1 and M_4 subtypes in vitro. In contrast, the uptake of [^{123}I]*Z*-(*R*,*R*)-IQNP was high in the neocortex and striatum, intermediate in the thalamus and pons, and low in the cerebellum, in agreement with the densities of M_1, M_2, and M_4 subtypes determined in vitro. The total distribution volume (V_T) of [^{123}I]*Z*-(*R*,*R*)-IQNP in the cerebellum was five times higher than that of [^{123}I]*E*-(*R*,*R*)-IQNP. This relatively high binding of [^{123}I]*Z*-(*R*,*R*)-IQNP may be explained by M_2 receptor binding in the cerebellum.

With regard to PET ligands, [^{11}C]benztropine uptake was observed in the frontal, parietal, occipital, and temporal cortices as well as in subcortical structures, including the corpus striatum and thalamus, in normal human subjects. The uptake increased in all brain regions over an 80-min experimental period, except in the cerebellum where the uptake peaked at 25 min postinjection (Dewey et al., 1990a).

A group at the University of Michigan extensively studied the kinetic modeling of [^{11}C]scopolamine and [^{11}C]TRB. Frey et al. (1992) studied the pharmacokinetic modeling of [^{11}C]scopolamine in healthy subjects. They confirmed the significant contribution of ligand delivery and extraction in the late distribution of [^{11}C]scopolamine, reconciling the discrepancy between receptor levels and tracer concentrations. They further analyzed the next compound [^{11}C]TRB (Koeppe et al., 1994). Using a variety of approaches, including the single rate scan technique, they concluded that reliable receptor density information can be obtained from kinetic [^{11}C]TRB PET studies, with complicated compartmental methods. [^{11}C]Scopolamine and [^{11}C]TRB have not been used in studies in human patients.

The first report of a human trial of [^{11}C]NMPB was in 1993 (Suhara et al., 1993). The uptake of [^{11}C]NMPB increased continuously over the 60-min experimental period in all brain areas with the exception of the cerebellum. Zubieta et al. (1998) evaluated the quantification of mAChR with [^{11}C]NMPB. They found that the initial cerebral distribution corresponded to the pattern of perfusion. However, at progressively longer postinjection intervals, the regional distribution of [^{11}C]NMPB was consistent with that of mAChR, highest in the striatum and cerebral cortex, intermediate in the thalamus and pons, and lowest in the cerebellum. The use of nonlinear least-squares fitting to two- and three-compartment models provided an accurate distinction of receptor binding estimates from tracer delivery.

[^{18}F]FP-TZTP is the only radioligand available for clinical studies of M_2 receptor-selective imaging. In the first human trials, V_T of [^{18}F]FP-TZTP was estimated by a one-tissue compartment model (Podruchny et al., 2003). The fairly uniform V_T values across cortical, subcortical, and cerebellar areas were consistent with in vitro studies of M_2 receptor distribution in humans (Flynn and Mash, 1993; Cortes et al., 1986, 1987).

2.3.2. Effects of Aging

In accordance with postmortem findings, an age-dependent decline of mAChR binding throughout the cerebrum was observed in human studies of [^{11}C]benztropine (Dewey et al., 1990b), [^{11}C]NMPB (Suhara et al., 1993; Yoshida et al., 2000; Zubieta et al., 2001), [^{123}I](*R,S*)-IQNB (Weinberger et al., 1991), and [^{123}I](*R,R*)-IQNB (Norbury et al., 2004, 2005). In contrast, Lee et al. (1996) reported no substantial age-related changes in the cerebrum in a human PET study with [^{11}C]TRB. These contrasting results may be due not only to differences in the radioligands, that is, a paucity of subtype-selective ligands, but also due to the differences in the analysis methods employed. Lee et al. (1996) used kinetic analysis data normalized by whole-brain receptor density. They found only a minor (8%) loss of receptor density in the frontal cortex. Norbury et al. (2004) reported that kinetic analysis and normalization relative to the cerebellum are effective for detecting the age-related reduction of [^{123}I](*R,R*)-IQNB binding in frontal, orbitofrontal, and parietal regions. However, normalizing to the whole brain failed to detect age-related changes in any region. Hence, for nonselective mAChR radioligands, normalizing to a region of negligible specific binding (cerebellum) significantly improves sensitivity when compared to global normalization.

Interestingly, the V_T of M_2-selective [^{18}F]FP-TZTP was increased throughout the brain in aged subjects (Podruchny et al., 2003). This indicated that there is either an age-related increase in the density of unoccupied M_2 receptors and/or an increase in the affinity constant of M_2 receptors and/or an increase in nonspecific binding in elderly subjects. The greater number of unoccupied receptors may result from an increase in the total number of M_2 receptors and/or from a reduction in the number of unoccupied receptors. A lower synaptic concentration of ACh would be expected to lead to fewer occupied receptors as well as an adaptive increase in the density. Postmortem studies showed that ChAT was decreased in older subjects (Nordberg and Winbland, 1981; Perry et al., 1993), which results in a lower concentration of ACh in the synapses. Furthermore, [^{18}F]FP-TZTP may compete with ACh as an agonist radioligand. Indeed, treatment with the AChE inhibitor, physostigmine, which probably raises synaptic ACh concentration, lowered the [^{18}F]FP-TZTP V_T in nonhuman primates (Carson et al., 1998). Therefore, a lower concentration of ACh in the synapses of some older subjects is one possible explanation for the increased binding of [^{18}F]FP-TZTP. Subsequently, Cohen et al. (2003) reported that gray matter V_T for [^{18}F]FP-TZTP was significantly higher in apolipoprotein E-e4 (APOE-e4) allele-positive aged normal subjects than in those negative for this allele, whereas there were no differences in global cerebral blood flow (CBF). A lower concentration of ACh in the synapses of APOE-e4-positive older individuals is a likely explanation for the greater [^{18}F]FP-TZTP V_T. The APOE-e4 allele confers an increased susceptibility to age-related memory problems and AD. Cohen et al. (2006) further examined the effects of aging and the APOE-e4 allele on the response to the AChE inhibitor, physostigmine, using [^{18}F]FP-TZTP. They reported that both age and APOE-e4 genotype were significant contributors to the response to the physostigmine-induced decrease in [^{18}F]FP-TZTP V_T. No interaction between age and APOE-e4 genotype was observed. These findings suggested that the APOE-e4 genotype is likely to contribute to cholinergic function prior to aging.

2.3.3. Sex-Related Changes

Little is known about the sex-related differences in cerebral mAChR. Changes in the affinity and number of mAChR during the estrous cycle have been reported in rats (van Huizen et al., 1994). The loss of ovarian function has a negative impact on the basal forebrain cholinergic neurons (Gibbs, 1998). In addition, females are at a greater risk of AD than are males (Henderson, 1997). In a human study, the regional [^{11}C]NMPB uptake ratio (vs cerebellum uptake) was higher in females than in males throughout all cerebral regions (Yoshida et al., 2000). The slope of the age-dependent decline in the [^{11}C]NMPB uptake ratio was steeper in females than in males, but the difference was not significant.

Norbury et al. (2007) examined the effects of long-term estrogen therapy and aging on the brain with M_1/M_4-selective [^{123}I](*R*,*R*)-IQNB SPECT in healthy women. Compared to young women, postmenopausal women (estrogen users and never-users combined) had a significantly lower mAChR density in all brain regions. Long-term estrogen users also had higher mAChR densities than did never-users in all brain regions, and the differences were statistically significant in the left striatum and hippocampus, lateral frontal cortex, and thalamus. Moreover, in the estrogen users, [^{123}I](*R*,*R*)-IQNB binding in the left hippocampus and temporal cortex showed a significant positive correlation with the plasma estradiol level. However, there was no significant relationship between receptor binding and cognitive function in any of the groups.

2.3.4. Pharmacology

Receptor occupancy rates by antipsychotic drugs in the human brain have been investigated extensively by SPECT and PET. These in vivo techniques are very useful for evaluating the therapeutic effects of drugs, for determining appropriate dosages, and for the development of new drugs.

A group at the National Institute of Radiological Sciences studied the interactions of anticholinergic agents with mAChR using [^{11}C]NMPB. Shinotoh et al. (1994) evaluated the feasibility of measuring mAChR occupancy by anticholinergic medication with [^{11}C]NMPB PET. Trihexyphenidyl is used for the symptomatic treatment of PD and binds to the M_1 receptors. There was a mean 28% inhibition of [^{11}C]NMPB binding in the brain of PD patients after a single dose of 4 mg of trihexyphenidyl, which was assumed to reflect mAChR occupancy. Further, Sudo et al. (1999) measured mAChR occupancy by oral administration of biperiden in eight subjects. Biperiden is an anticholinergic-type antiparkinsonian agent, which shows a prominent central blocking effect on M_1 receptors. The mAChR occupancy was 10%–45% in the frontal cortex 3 h after the administration of 4 mg of biperiden; the occupancy rate was correlated with the plasma biperiden concentration in a curvilinear manner.

Atypical antipsychotics, such as clozapine, olanzapine, and risperidone, are now considered first-line treatments for schizophrenia. These drugs antagonize a broad range of receptors, including dopamine, serotonin, and mAChR. Substantial occupancy of mAChR by atypical antipsychotics may explain the lower incidence rates of extrapyramidal side effects (EPS) associated with these drugs.

A group at the National Institute of Mental Health studied the interactions of atypical antipsychotics with mAChR using [^{123}I](*R*,*R*)-IQNB. Olanzapine has potent antimuscarinic properties in vitro ($K_i = 2.5$ nM for M_1) (Bymaster et al., 1996). Raedler et al. (2000) reported that both low-dose (5 mg/day) and high-dose (20 mg/day) olanzapine treatment showed a significantly lower [^{123}I](*R*,*S*)-IQNB binding in the cortex, thalamus, and pons than observed in drug-free patients. Receptor occupancy by olanzapine was shown to be dose dependent.

Clozapine has a higher affinity ($K_i = 1.4$ nM for M_1) than olanzapine (Bymaster et al., 1996). Raedler et al. (2003a) reported that [^{123}I](*R,S*)-IQNB binding was lower in the basal ganglia, cortex, thalamus, and pons in subjects treated with clozapine compared to that in unmedicated patients. Receptor occupancy by clozapine was related to the daily dose of clozapine.

Furthermore, this group performed a direct comparison of mAChR availability in vivo under treatment with olanzapine and clozapine (Raedler, 2007). A significantly lower [^{123}I](*R,S*)-IQNB binding was found with a moderate dose (mean 275 mg, range 200–450 mg) of clozapine than with a high dose (20 mg) of olanzapine in the basal ganglia, cortex, thalamus, and pons. These results suggest that clozapine shows a stronger blockade of mAChR than olanzapine.

Risperidone has negligible affinity ($K_i > 10,000$ nM) for mAChR (Bymaster et al., 1996). It showed a higher incidence of EPS than did olanzapine (Duggan et al., 2000). In addition, anticholinergic symptoms, such as dizziness and dry mouth, were found to be more common in association with olanzapine treatment than with risperidone treatment (Duggan et al., 2000). Lavalaye et al. (2001) compared the mAChR occupancy by olanzapine and risperidone in schizophrenia patients using [^{123}I]IDEX SPECT. Patients stabilized on olanzapine showed significantly lower striatal and cortical [^{123}I]IDEX binding than did controls. In contrast, risperidone treatment induced a slightly lower [^{123}I]IDEX binding only in the striatum region. The results of these three receptor occupancy studies were compatible with the rates of anticholinergic side effects seen with clozapine, olanzapine, and risperidone in clinical practice.

ACh deficit remains the most consistent neurotransmitter abnormality found in AD, and various therapeutic agents targeting this deficit have been developed. The effect of treatment with the AChE inhibitor tacrine (80 mg/day) on mAChR was assessed in one AD patient using [^{11}C]benztropine (Nordberg et al., 1997). After 3 months of treatment, there was a decrease in the [^{11}C]benztropine binding, which was interpreted as a temporary downregulation of mAChR because the reduction was no longer evident at 10 months. Two groups have performed similar studies using [^{123}I](*R,R*)-IQNB to monitor the changes in mAChR density during AChE treatment with donepezil. Kemp et al. (2003) reported a greater decrease in binding in the placebo group than in the donepezil-treated group for 4 months. As [^{123}I](*R,R*)-IQNB has a high affinity, its competition with endogenous ACh would not confound the results. Therefore, this result suggested a neuroprotective role of donepezil treatment. Brown et al. (2003) also studied patients treated with donepezil, but compared them with normal controls rather than patients given only placebo. They found that [^{123}I](*R,R*)-IQNB binding was not correlated with response, as measured by the standard clinical test. However, there was an inverse relationship between [^{123}I](*R,R*)-IQNB binding in the insular cortex and the response.

2.4. HUMAN SPECT/PET IMAGING OF mAChR IN THE DISEASED BRAIN

There have been numerous attempts to evaluate mAChR density using SPECT or PET imaging in various neurological, neurodegenerative, and dementing disorders (Table 5).

Most such studies have been carried out in AD or dementia, as cholinergic dysfunctions are thought to be among the factors contributing to cognitive deterioration.

The first clinical trials of AD patients were conducted by Weinberger's group using [^{123}I](*R,S*)-IQNB. They found clear differences in the images of normal and AD patients at 21 h after injection of [^{123}I](*R,S*)-IQNB (Weinberger et al., 1991, 1992). AD patients showed focal cortical defects in either the frontal or posterior temporal cortex. In accordance with the postmortem data, most SPECT/PET studies support the reduction of the mAChR density in AD patients; however, others have reported no changes (Wyper et al., 1993; Zubieta et al., 2001). The discrepancies in these results may have been due to the heterogeneity of the populations studied, differences in the quantification methods employed, and paucity of radioligands with different subtype selectivities. For example, Wyper et al. (1993) did not find [^{123}I](*R,S*)-IQNB binding exceeding the corresponding total functional regional perfusion deficit in six AD patients, but they observed increased deficits in two more severely affected patients. Therefore, a major reduction in the postsynaptic mAChR density may occur only at a very late stage of AD.

Despite these positive findings, mAChR ligands have not been used extensively in clinical settings. One major reason for this is the inability to impact clinical care. Yoshida et al. (1998) reported that a more marked decrease in the glucose metabolism in the cortical region was seen in mild and moderate AD compared to [^{11}C]NMPB binding. This study suggested that 2-deoxy-2-[^{18}F]fluoro-D-glucose (FDG) PET was the most sensitive method for detecting degenerative lesions in patients with AD. As [^{18}F]FDG PET with standard acquisition protocols for dementia is now widely available, the change in cerebral glucose metabolism is recognized as a feasible outcome variable for clinical drug trials (Herholz, 2010; Weiner et al., 2010). Furthermore, the development of radiopharmaceuticals for β-amyloid aggregation, a hallmark of AD, has facilitated in vivo amyloid imaging studies in the living human brain (Rabinovici and Jaguest, 2009).

PD, epilepsy, narcolepsy, and schizophrenia have also been suggested to involve the cholinergic systems. A group at the National Institute of Radiological Sciences studied the mAChR status in nondemented PD using [^{11}C]NMPB. Asahina et al. (1995) reported that [^{11}C]NMPB binding was 20% higher in the frontal cortex of PD patients than in controls (Figure 5).

They concluded that the hypersensitivity of mAChR in the frontal cortex of PD patients may be a response to the loss of ascending cholinergic input to this region, and

Table 5 SPECT and PET Studies Using Radioligands for mAChRs in Human Neurological Disease

Radiotracer	Disease	Receptor Change	Conclusions	References
$[^{123}I](R,S)$-IQNB	AD	Decrease	The $[^{123}I](R,S)$-IQNB activity ratio was reduced in AD patients compared with that in a normal age-matched subject.	Holman et al. (1985)
$[^{123}I](R,S)$-IQNB	Dementia	Decrease	AD patients had obvious focal cortical defects of $[^{123}I]$ (R,S)-IQNB binding in either the frontal or posterior temporal cortex.	Weinberger et al. (1991)
$[^{123}I](R,S)$-IQNB	Dementia	Decrease	The $[^{123}I](R,S)$-IQNB scans tended to show larger defects than did $[^{18}F]$FDG scans in demented patients.	Weinberger et al. (1992)
$[^{123}I](R,S)$-IQNB	AD	Decrease	Deficits in $[^{123}I](R,S)$-IQNB binding exceeding the corresponding perfusion deficit were not found in AD patients. Only the most severely affected patients showed significant deficits in $[^{123}I](R,S)$-IQNB binding.	Wyper et al. (1993)
$[^{123}I](R,R)$-IQNB	AD	Decrease	Significant reductions in $[^{123}I](R,R)$-IQNB uptake were observed in AD compared to that in the controls. No significant correlations were identified between imaging data and clinical variables.	Pakrasi et al. (2007)
$[^{123}I]$IDEX	AD	Decrease	AD patients showed a significantly less $[^{123}I]$IDEX binding in left temporal and right temporoparietal cortexes compared with controls, independent of age, sex, and cortical atrophy.	Claus et al. (1997)
$[^{11}C]$NMPB	AD	Decrease	A significant $[^{11}C]$NMPB binding decrease and more severe regional glucose metabolism decrease in the cortical region were seen in mild and moderate AD.	Yoshida et al. (1998)
$[^{11}C]$NMPB	AD	No change	No regional changes in the mAChR concentrations were observed in AD.	Zubieta et al. (2001)
$[^{123}I]$IDEX	AD	Decrease	SPM analysis found a $[^{123}I]$IDEX binding deficit in the posterior cingulated cortex in mild AD patents.	Boundy et al. (2005)
$[^{123}I]$IDEX	Epilepsy	Decrease	In the anterior hippocampus ipsilateral to the electrical focus, $[^{123}I]$IDEX binding was decreased by 40% compared with the contralateral hippocampus.	Müller-Gärtner et al. (1993)

[^{123}I]IDEX	Epilepsy	No change	The reduction in the hippocampal [^{123}I]IDEX binding in patients with epilepsy was due to a decrease in the hippocampal volume rather than to a decrease in the receptor concentration.	Weckesser et al. (1997)
[^{123}I]IDEX	Epilepsy	Decrease	[^{123}I]IDEX binding was altered in medial temporal lobe structures in TLE. [^{123}I]IDEX SPECT appears to be superior to interictal CBF SPECT and complementary to MRI for seizure focus localization.	Boundy et al. (1996)
[^{123}I]IDEX	Epilepsy	Decrease	The 6 h [^{123}I]IDEX SPECT scan is a viable alternative to [^{18}F]FDG PET imaging in seizure onset localization in TLE.	Mohamed et al. (2005)
[^{11}C]NMPB	Narcolepsy	No change	There were no differences in the [^{11}C]NMPB binding between control and drug-naive/free patients.	Sudo et al. (1998)
[^{123}I](*R*,*S*)-IQNB	Schizophrenia	Decrease	[^{123}I](*R*,*S*)-IQNB binding was significantly lower in schizophrenia than in normal subjects. Positive symptoms were correlated negatively with [^{123}I](*R*,*S*)-IQNB binding in the striatum and frontal cortex.	Raedler et al. (2003b)
[^{11}C]NMPB	PD	Increase	[^{11}C]NMPB binding was 20% higher in the frontal cortex of PD patients than in controls.	Asahina et al. (1995)
[^{11}C]NMPB	PD PSP	Increase (PD) No change (PSP)	[^{11}C]NMPB binding was significantly higher in the frontal cortex of PD patients than in controls. PSP patients had no significant changes in any cortical region.	Asahina et al. (1998)
[^{123}I](*R*,*R*)-IQNB	DLB PDD	Increase	Significant elevation of [^{123}I](*R*,*R*)-IQNB binding in the occipital lobe was associated with DLB and PDD.	Colloby et al. (2006)
[^{18}F]FP-TZTP	BD MDD	Decrease (BD) No change (MDD)	The [^{18}F]FP-TZTP binding in subjects with BD was reduced relative to both healthy controls and subjects with MDD, to an extent that correlated with depressive symptoms.	Cannon et al. (2006)
[^{18}F]FP-TZTP	BD MDD	Decrease	Reduced [^{18}F]FP-TZTP V_T in BD is associated with genetic variation within $CHRM_2$.	Cannon et al. (2011)

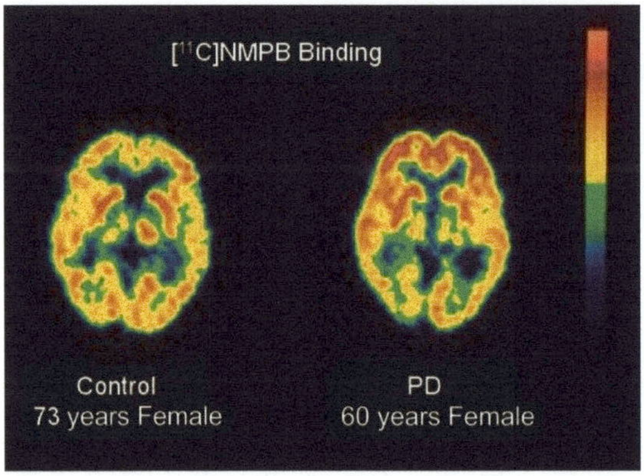

Figure 5 PET images of [^{11}C]NMPB in a control subject and a patient with PD. Static images acquired from 40 to 60 min after injection. High-level accumulation of radioactivity was noted in the frontal cortex of the PD patient, while the distribution of radioactivity was homogeneous in the control. *Courtesy of Dr Shinotoh, Asahi Hospital for Neurological Disorders and Rehabilitation, Chiba, Japan.*

may be related to frontal lobe dysfunction in PD. They further compared the mAChR status in PSP and PD (Asahina et al., 1998). In accordance with the results of previous studies, [^{11}C]NMPB binding was significantly higher for the frontal cortex in the patients with PD than in the controls. In contrast, the patients with PSP showed no significant changes in [^{11}C]NMPB binding in any cerebral cortical region. This observation suggested that the cerebral cortical cholinergic system may not play a major role in the cognitive dysfunction in PSP. Colloby et al. (2006) investigated the differences in the distribution of mAChR using [^{123}I](*R*,*R*)-IQNB in DLB, PD with dementia (PDD), and age-matched controls. They found a significant elevation of [^{123}I](*R*,*R*)-IQNB binding in the right occipital lobe in DLB and in the right and left occipital lobes in PDD. These observations may be related to the visual disturbances that are prevalent in these disorders. In contrast to the elevated mAChR binding in nondemented PD, PDD showed a significant loss of uptake in frontal regions and temporal lobes bilaterally that was not present in DLB.

There is persuasive evidence that mAChRs play a role in the generation of seizures. Tissues obtained from the epileptic human cortex show a loss of cholinergic fibers and elevated ChAT and AChE activity. Müller-Gärtner et al. (1993) reported the first images of the mAChR status in the temporal lobes of four patients with complex partial seizures. In the anterior hippocampus ipsilateral to the electrical focus, [^{123}I]IDEX binding was decreased by 40% compared to that in the contralateral hippocampus. These preliminary data indicated a regionally specific change in mAChR in the anterior hippocampus in complex partial seizures of temporal lobe origin. In a larger study, Boundy et al. (1996) reported that 78% of patients with temporal lobe epilepsy (TLE) had an asymmetrical [^{123}I]IDEX binding in the temporal lobe of seizure origin. Mohamed et al. (2005) further validated the optimal scan time for interictal [^{123}I]IDEX SPECT imaging. They reported that a 6-h [^{123}I]IDEX scan correctly predicted the temporal lobe of origin of seizures. Furthermore, [^{123}I]IDEX SPECT was superior to interictal [^{18}F]FDG PET in seizure onset localization. Patients with TLE often present a reduction in hippocampal volume. Weckesser et al. (1997) investigated the influence of hippocampal atrophy on the quantification of mAChR by SPECT in patients with TLE. They concluded that the reduction in hippocampal [^{123}I]IDEX binding in patients with epilepsy is due to a decrease in the hippocampal volume rather than to a decrease in receptor concentration.

Muscarinic neural transmission serves as the main executive system in rapid eye movement (REM) sleep. To investigate the function of the mAChR in narcolepsy, 11 drug-naive/free patients with narcolepsy and 21 normal controls were investigated by PET with [^{11}C]NMPB (Sudo et al., 1998). There were no differences in the [^{11}C]NMPB binding between the control and drug-naive/free patients. [^{11}C]NMPB binding in the thalamus was decreased after medication, but only to a small degree compared to the effects of anticholinergic drugs. This result does not support the notion that the mAChR is the main site of action of pharmacotherapy in the marked clinical improvement of human cataplexy.

Postmortem studies have implicated the central muscarinic ACh system in schizophrenia. Raedler et al. (2003b) reported that [^{123}I](*R,S*)-IQNB binding was significantly reduced in the cortex, basal ganglia, and thalamus in patients with unmedicated schizophrenia in comparison with normal subjects. The reductions ranged from 33% in the caudate to 20% in the occipital cortex. Positive symptoms of schizophrenia were shown to be negatively correlated with [^{123}I](*R,S*)-IQNB binding in the striatum and frontal cortex. These results indicate a reduction in mAChR in vivo in unmedicated patients with schizophrenia, confirming the results of postmortem studies and adding further evidence that the muscarinic system is involved in the pathophysiology of schizophrenia.

A variety of indirect evidence has implicated the M_2 receptor in the pathogenesis of depressive symptoms arising in major depressive disorder (MDD) and bipolar disorder (BD). Cannon et al. (2006) found a significantly lower [^{18}F]FP-TZTP binding in BD compared with that in both MDD and control groups. The extent of [^{18}F]FP-TZTP binding changes was correlated with depressive symptoms. This reduction of [^{18}F]FP-TZTP binding could be accounted for either by a reduction in M_2 receptor density or affinity or by elevation of endogenous ACh levels. Furthermore, Cannon et al. (2011) investigated the effects of six single-nucleotide polymorphisms (SNPs) in the cholinergic muscarinic M_2 receptor gene (CHRM$_2$) on M_2 receptor binding. Genetic variation in CHRM$_2$ is associated with the risk of developing depression. In healthy controls, one SNP (rs324650) showed a significant allelic effect on V_T in the pregenual and subgenual anterior cingulate cortices in the order AA < AT < TT. In contrast, V_T was significantly smaller in BD subjects with the TT genotype than in those with the AT genotype in these regions. The BD subjects homozygous for the T-allele also showed a markedly smaller V_T (by 27%–37% across regions) than did healthy controls of the same genotype. Post hoc analyses suggested that T homozygosity was associated with a more severe clinical course, as manifested by a lower socioeconomic function, poorer spatial recognition memory, and a greater likelihood of having attempted suicide. These data represent novel preliminary evidence that reduced M_2 receptor V_T in BD is associated with genetic variation within $CHRM_2$. The differential impact of M_2 receptor polymorphism at rs324650 in the BD and healthy control samples suggested interactive effects with an as-yet unidentified vulnerability factor for BD.

2.5. HUMAN SPECT/PET IMAGING OF α4β2 nAChR

2.5.1. Imaging in Healthy Subjects

Fujita et al. (2002) reported the first [^{123}I]5-IA SPECT imaging study in 10 healthy subjects. Whole-brain activity peaked at 15 min, and reached 5% of the injected dose. The radioactivity distribution was consistent with the known regional distribution of α4β2 nAChR (Figure 6).

The accuracy of [^{123}I]5-IA SPECT for quantification of α4β2 nAChR was validated by the good agreement between α4β2 nAChR density in the postmortem human brain and regional V_T obtained by the compartmental modeling of [^{123}I]5-IA brain uptake and corresponding arterial input function (Mamede et al., 2004). Furthermore, the feasibility and reproducibility of the bolus-plus-constant-infusion paradigm for the equilibrium modeling of [^{123}I]5-IA using SPECT in healthy nonsmokers was also confirmed (Staley et al., 2005). In addition, there were no significant differences in the [^{123}I]5-IA binding between male and female nonsmokers or in female nonsmokers across the menstrual cycle (Cosgrove et al., 2007). However, sex-related differences in [^{123}I]5-IA metabolism and plasma protein binding were also evident and significantly affected the regional brain activity (kilobecquerels per milliliter) and brain/plasma ratio. Therefore, measuring plasma radiotracer levels and free fraction are necessary in studies that include both sexes.

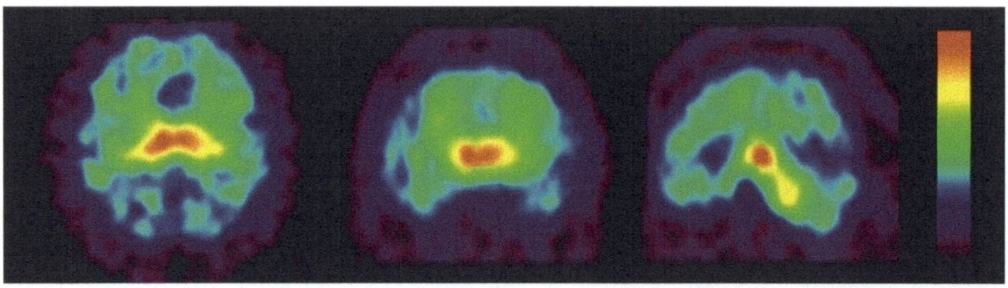

Figure 6 Representative SPECT images of [^{123}I]5-IA in a healthy subject. Axial, coronal, and sagittal static images acquired 4 h after injection. The distribution of [^{123}I]5-IA was associated with the known distribution of α4β2 nAChR in the human brain: thalamus > brain stem > cerebellum > cortex. *Courtesy of Dr Ishizu, Kyoto University, Kyoto, Japan.*

Kimes et al. (2003) and Bottlaender et al. (2003) reported the first human PET studies of [^{18}F]2-FA in healthy nonsmokers. In these studies, the radioactivity in the brain peaked at 25 min after injection. Consistent with the results of postmortem studies in the human brain, the levels of [^{18}F]2-FA in the brain were the highest in the thalamus, intermediate in the midbrain, pons, cerebellum, and cortex, and lowest in the white matter. Further studies validated the use of both graphical and compartmental analysis approaches for the quantification of α4β2 nAChR with [^{18}F]2-FA (Gallezot et al., 2005; Mitkovski et al., 2005).

Ding et al. (2004) reported the first [^{18}F]6-FA PET study in five healthy nonsmokers. In this study, the radioactivity distribution was consistent with the known regional distribution of α4β2 nAChRs. The distribution volume ratio (DVR) and BP were calculated using a one-compartment model with the corpus callosum as the reference region. The BP value of [^{18}F]6-FA in the thalamus (3.1) was slightly higher than that of [^{18}F]2-FA. A total 4-h study with an actual scan time of 2 h is adequate for sufficient and reproducible kinetic analysis.

2.5.2. Effects of Aging

Postmortem studies have suggested that there are reductions in neuronal α4β2 nAChR with aging. Consistent with the results of postmortem studies, Mitsis et al. (2007; 2009a) reported an inverse correlation between age and regional [^{123}I]5-IA V_T of healthy nonsmokers (15–85 years old), with the decline ranging from 18% (occipital cortex) to 36% (thalamus), or by up to 5% per decade of life. In contrast, Ellis et al. (2009a) found no significant age-related differences in regional [^{18}F]2-FA V_T. Furthermore, no significant correlations were found between cognitive measures and [^{18}F]2-FA V_T.

2.5.3. Effects of Smoking

Nicotine, the addictive component of tobacco smoke, initiates its actions in the brain through the nAChR. In particular, receptors of the most prevalent subtype, α4β2, mediate the reinforcing properties of nicotine. Postmortem studies of the human brain have suggested that the number of nAChRs is higher in smokers than in nonsmokers or in former smokers (Breese et al., 1997b; Perry et al., 1999a). Consistent with these findings, greater α4β2 nAChR densities were observed in the living human brains of smokers using [^{18}F]2-FA PET (Mukhin et al., 2008; Wüllner et al., 2008).

Adaptations of α4β2 nAChR in response to repeated and protracted nicotine exposure likely contribute to the addictive effect of cigarette smoking. Therefore, an understanding of the adaptive changes in the α4β2 nAChR density during acute abstinence from smoking and their relationship to tobacco smoking behavior may advance the development of improved pharmacotherapies to aid in smoking cessation.

Staley et al. (2006) investigated the changes in the α4β2 nAChR availability after 6.8 days of abstinence in smokers using [^{123}I]5-IA SPECT. In smokers, [^{123}I]5-IA uptake was significantly elevated throughout the cerebral cortex (26%–36%) and in the striatum (27%) compared to that in nonsmokers, suggesting higher levels of α4β2 nAChRs in recently abstinent smokers. The α4β2 nAChR availability in recently abstinent smokers was correlated with the number of days since the last cigarette and the urge to smoke to relieve withdrawal symptoms but not the severity of nicotine dependence, severity of nicotine withdrawal, or the desire to smoke. The higher levels of brain α4β2 nAChRs during early abstinence indicated that smokers face a significant increase in receptor availability normally activated by nicotine when they quit smoking. Greater α4β2 nAChR availability during early abstinence may impact the ability of smokers to continue to refrain form smoking. Mamede et al. (2007) investigated the dynamic changes in α4β2 nAChRs in the living human brain after abstinence. They found that V_T of [^{123}I]5-IA in the brains of smokers was decreased by 34% after 4 h of abstinence, increased by 26% at 10 days, and decreased to the level of nonsmokers at 21 days. These data suggested that the upregulation of α4β2 nAChR in smokers after chronic exposure to nicotine recovered to the level in nonsmokers by around 21 days of abstinence. Similar results were also reported by Cosgrove et al. (2009), who concluded that a higher α4β2 nAChR availability persists for up to one month of abstinence and normalizes to the levels in nonsmokers by 6–12 weeks of abstinence from tobacco smoking. The longer period of the upregulation of α4β2 nAChR may be due to the high degree of heterogeneity in the subject population compared to that in the previous study in male subjects (Mamede et al., 2007).

The α4β2 nAChR occupancy rates after the administration of nicotine cigarettes were measured satisfactorily by [^{123}I]5-IA SPECT (Esterlis et al., 2010a) or [^{18}F]2-FA PET (Brody et al., 2006, 2009). These studies consistently demonstrated that most of the receptors remained occupied throughout the day in a typical daily smoker. In addition, this technique can detect substantial changes in the receptor occupancy associated with denicotinized (0.05 mg of nicotine) cigarette smoking (26%) (Brody et al., 2009) and exposure to secondhand smoke (19%) (Brody et al., 2011).

The nicotine inhaler reduces cravings by mimicking the behavioral component of cigarettes and delivering controlled doses of nicotine, which binds to α4β2 nAChRs. Esterlis et al. (2011) applied [^{123}I]5-IA SPECT to measuring α4β2 nAChR occupancy after the administration of nicotine via an inhaler, and analyzed the relationship between receptor occupancy and changes in cravings for tobacco smoking and withdrawal symptoms. They found that the use of a nicotine inhaler produced 56% occupancy of α4β2 nAChRs at 2–5 h postchallenge. On the other hand, the use of a cigarette produced a significantly higher receptor occupancy of 68% at 1.5–5 h postchallenge. In addition, there was a significant decrease in the withdrawal symptoms in postnicotine inhaler users. These results indicated significant differences in receptor occupancy rates by nicotine between inhaler and cigarette smoking, and confirmed the ability of the nicotine inhaler to relieve withdrawal symptoms.

The high rate of comorbidity of tobacco smoking with alcohol drinking suggests common neural substrate-mediated addictive disorders between these two substances. Esterlis et al. (2010b) hypothesized that α4β2 nAChR availability would be altered by alcohol in heavy drinking nonsmokers; however, they found no significant difference in α4β2 nAChR availability between heavy drinking and control nonsmokers.

Nicotine and tobacco smoking administration studies have demonstrated antinociceptive effects mediated by the α4β2 nAChR. Cosgrove et al. (2010) examined the relationship between α4β2 nAChR availability and nociception during acute withdrawal in human tobacco smokers. They reported that increased pain sensitivity was negatively associated with a higher α4β2 nAChR availability after 7–13 days of abstinence from smoking. Similar associations were not observed with pain tolerance.

2.5.4. Imaging in the Diseased Brain

Due to the important relationships between nicotine and memory functions, and the known α4β2 nAChR abnormalities seen in postmortem studies of AD patients, [^{123}I]5-IA and [^{18}F]2-FA studies have been explored in dementia (Table 6).

Many studies have indicated a decrease in the α4β2 nAChR binding in demented patients; however, others have reported no such changes. The variability in these results is most likely due to variations in the sensitivity of quantification methods employed and in the severity of illness among the subjects.

Most of the positive results were obtained using [^{123}I]5-IA. O'Brien et al. (2007) used region-of-interest analysis with cerebellar normalization. They found significant bilateral reductions of [^{123}I]5-IA binding in the cerebral cortical and striatal regions in AD compared to that in controls. Statistical parametric mapping (SPM) is a statistical technique for examining the differences in brain activity and is a very powerful and sensitive tool for identifying small changes. In DLB patients, significant reductions in relative [^{123}I]5-IA uptake were identified in the frontal, striatal, temporal, and cingulate regions compared to controls (O'Brien et al., 2008). In contrast, elevation of [^{123}I]5-IA uptake in the occipital cortex was observed in DLB relative to that in controls, and was associated with a recent history of visual hallucinations in DLB subjects. Using SPM analysis, Terrière et al. (2010) found that amnestic mild cognitive impairment (MCI) patients showed discrete reductions in the [^{123}I]5-IA uptake in the medial temporal cortex. A correlation with cognitive impairment was found in the left temporoparietal areas, bilateral temporolimbic areas, and right parahippocampal gyrus within the patient group. Furthermore, SPM analysis revealed that a reduced [^{123}I]5-IA uptake in the left superior, middle, and inferior frontal gyri and prepost central and anterior cingulate regions were significantly correlated with the decline in the executive function in a pooled group that comprised AD and DLB patients (Colloby et al., 2010). These findings suggest that α4β2 nAChR imaging may be useful for monitoring disease progression and response to treatment in cases of dementia. One study was unable to demonstrate the loss of α4β2 nAChRs using [^{123}I]5-IA in vivo in mild AD and amnestic MCI (Mitsis et al., 2009b). The authors concluded that α4β2 nAChR loss and other cholinergic reductions in AD are late-stage phenomena.

Ellis et al. (2008) examined the loss of α4β2 nAChRs in the early AD brain using [^{18}F]2-FA PET. Brain regional [^{18}F]2-FA binding was assessed by V_T. In contrast to postmortem

Table 6 SPECT and PET Studies Using Radioligands for the $\alpha4\beta2$ nAChRs in Human Neurological Disease

Radiotracer	Disease	Receptor Change	Conclusions	References
[^{123}I]5-IA	AD MCI	No change	No group differences of $\alpha4\beta2$ nAChR binding for any brain region.	Mitsis et al. (2009b)
[^{123}I]5-IA	AD	Decrease	$\alpha4\beta2$ nAChR binding is reduced in AD	O'Brien et al. (2007)
[^{123}I]5-IA	AD DLB	Decrease	$\alpha4\beta2$ nAChR binding was decreased in the left superior, middle, and frontal gyri, and prepost central and anterior cingulated regions.	Colloby et al. (2010)
[^{123}I]5-IA	DLB	Decrease	$\alpha4\beta2$ nAChR binding was decreased in the frontal, striatal, temporal, and cingulate regions, but increased in the occipital cortex of patients with visual hallucinations.	O'Brien et al. (2008)
[^{123}I]5-IA	MCI	Decrease	[^{123}I]5-IA binding in the left temporoparietal and bilateral temporolimbic areas, and right parahippocampal gyrus was correlated with cognitive impairment.	Terrière et al. (2010)
[^{18}F]2-FA	AD MCI	Decrease	Both patients with AD and MCI showed a significant reduction in the [^{18}F]2-FA binding in typical AD-affected brain regions. [^{18}F]2-FA binding was correlated with the severity of cognitive impairment. Only MCI patients who converted to AD later in the clinical course showed reductions in [^{18}F]2-FA binding.	Kendziorra et al. (2011)
[^{18}F]2-FA	Early AD	No change	No significant differences were found in $\alpha4\beta2$ nAChR V_T before and after galantamine treatment. The treatment-induced improvement in cognition was not correlated with regional or global $\alpha4\beta2$ nAChR V_T.	Ellis et al. (2009b)

[18F]2-FA	Early AD	No change	There was no evidence of in vivo α4β2 nAChR loss in early AD despite significant cognitive impairment. No correlation was found between α4β2 nAChR and cognitive performance.	Ellis et al. (2008)
[123I]5-IA	PD	Decrease	Human subjects with early to moderate PD without dementia demonstrated widespread decrements in the density of nAChR.	Fujita et al. (2006)
[18F]2-FA	PD	Decrease	Nondemented PD showed a reduction of [18F]2-FA V_T in the striatum and substantial nigra.	Kas et al. (2009)
[18F]2-FA	PD	Decrease	There was a broad reduction of α4β2 nAChR availability in patients with PD. Reduced α4β2 nAChR binding within the subcortical and cortical regions was associated with the severity of mild cognitive or depressive symptoms.	Meyer et al. (2009)
[123I]5-IA	Vascular dementia	Decrease	α4β2 nAChR binding is decreased in the dorsal thalamus, and right caudate, but increased in the cuneus.	Colloby et al. (2011)
[123I]5-IA	PTSD	Increase	Never-smoking PTSD patients compared to never-smoking healthy controls showed a significantly higher [123I]5-IA binding in the mesotemporal cortex. Among all PTSD patients, there was a significant correlation between the reexperiencing symptom cluster and thalamic [123I]5-IA binding.	Czermak et al. (2008)
[18F]2-FA	ADNFLE	Increase	α4β2 nAChRs binding was increased in the epithalamus, ventral mesencephalon, and cerebellum, but decreased in the right dorsolateral prefrontal region in ADNFLE.	Picard et al. (2006)

data, they found no evidence of in vivo α4β2 nAChR loss in early AD despite significant cognitive impairment. Ellis et al. (2009b) further examined whether treatment with AChE inhibitors induced α4β2 nAChR changes using [^{18}F]2-FA PET. They found no significant differences in [^{18}F]2-FA V_T before and after treatment. The treatment-induced improvement in cognition was not correlated with regional or global [^{18}F]2-FA V_T. From these two results, they concluded that the preservation of α4β2 nAChR in early AD and changes in α4β2 nAChR may not be responsible for the improvements in cognition following the AChE inhibitor treatment in patients with mild AD. However, these conclusions may have been misleading due to the low sensitivity of the [^{18}F]2-FA V_T analysis to determine small changes in receptor density. Kendziorra et al. (2011) reported that patients with mild AD and those with amnestic MCI showed significant reductions in [^{18}F]2-FA BP in typically AD-affected brain regions. They used the corpus callosum as the most suitable reference region. Furthermore, [^{18}F]2-FA BP was correlated with the severity of cognitive impairment. Interestingly, only MCI patients who converted to AD later in the clinical course showed reductions in [^{18}F]2-FA BP. Therefore, Kendziorra et al. (2011) concluded that [^{18}F]2-FA PET is a sensitive and useful tool for detecting reductions in α4β2 nAChR levels.

The nAChRs have close interactions with the dopaminergic system and play critical roles in cognitive function. Fujita et al. (2006) found a widespread significant decrease (~10%) of [^{123}I]5-IA V_T in both the cortical and subcortical regions in cases of nondemented PD. Kas et al. (2009) also found significant decreases in [^{18}F]2-FA V_T in the striatum (~10%) and substantia nigra (~15%) in nondemented PD. They also reported that there was no correlation between [^{18}F]2-FA V_T and severity of PD evaluated by 6-[^{18}F]fluoro-L-DOPA. Cognitive and depressive disorders are frequently noted in patients with PD. Meyer et al. (2009) conducted a [^{18}F]2-FA PET study to assess the availability of α4β2 nAChRs and their relationship to mild cognitive and mild depressive symptoms. Widespread reductions in [^{18}F]2-FA BP were observed in PD patients compared to that in healthy controls. Reductions in [^{18}F]2-FA BP within the subcortical and cortical regions of PD patients were associated with the severity of mild cognitive or depressive symptoms. These results provided evidence for a role of cholinergic neurotransmission in psychiatric comorbidity of PD.

Imaging of α4β2 nAChR has also been reported in patients with vascular dementia, posttraumatic stress disorder (PTSD), and epilepsy. In vascular dementia, the reduction of [^{123}I]5-IA uptake was confined to subcortical regions, such as the dorsal thalamus and caudate, unlike the cortical reductions in AD or widespread decrease in PD (Colloby et al., 2011). An increase in the [^{123}I]5-IA uptake in the cuneus was also demonstrated in vascular dementia. This may represent a compensatory response to reduced cholinergic activity in the dorsal thalamus.

Memory deficits and hyperarousal are defining symptoms of PTSD. The nAChRs have been strongly implicated in memory dysfunctions and regulation of arousal. Czermak et al. (2008) reported that never-smoking PTSD patients showed a significantly higher [^{123}I]5-IA binding in the mesotemporal cortex compared to never-smoking healthy controls. Among all PTSD patients, there was a significant correlation between reexperiencing symptom clusters and thalamic [^{123}I]5-IA binding. These findings suggest the involvement of α4β2 nAChR in the pathophysiology of PTSD.

[^{18}F]2-FA PET imaging in cases of autosomal dominant nocturnal frontal lobe epilepsy (ADNFLE) demonstrated a significantly elevated V_T in the mesencephalon, pons, and cerebellum compared with that of controls (Picard et al., 2006). These observations, together with the results of [^{18}F]FDG PET, suggested that the mesencephalon may be related to arousal.

2.6. HUMAN SPECT/PET IMAGING OF α7 nAChR

2.6.1. Imaging in Healthy Subjects

With the exception of a previous study using the PET radioligand [^{11}C]CHIBA-1001, there have been no reports of α7 nAChR imaging in the human brain (Toyohara et al., 2009). Figure 7 shows a representative [^{11}C]CHIBA-1001 image of the brain of a healthy subject.

A 90-min dynamic scan showed the rapid accumulation and gradual washout of radioactivity in the brain. The radioactivity peaked about 15 min after administration. The metabolism of [^{11}C]CHIBA-1001 was relatively slow, with the percentage of unchanged form remaining in the plasma >80% at 60 min after administration. [^{11}C]CHIBA-1001 PET data were described well with a one-tissue two-compartment model, and the direct derivation of the BP in a two-tissue three-compartment model was unstable. As the radioactivity after the administration of [^{11}C]CHIBA-1001 was widely distributed throughout the brain, there was no reference region for the indirect derivation of BP.

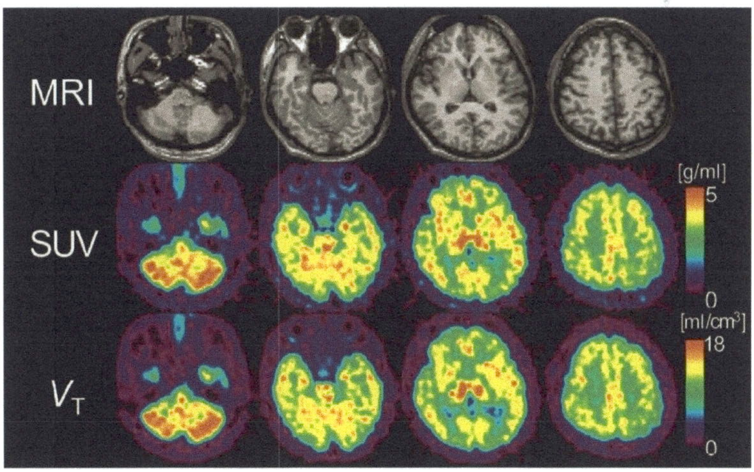

Figure 7 PET images of the human brain with [^{11}C]CHIBA-1001. Upper: magnetic resonance imaging (MRI) of the corresponding slices. Middle: Static images acquired from 40 to 60 min after injection of [^{11}C]CHIBA-1001 expressed as the Standardized Uptake Value (SUV). Lower: A parametric V_T image of [^{11}C]CHIBA-1001 generated using Logan graphical analysis. The data from 30 to 90 min were subjected to Logan plot analysis.

Therefore, V_T was calculated for the quantitative analysis of [^{11}C]CHIBA-1001. The highest V_T of [^{11}C]CHIBA-1001 was found in the thalamus; however, regional differences in brain radioactivity were small. Although CHIBA-1001 has been reported to show no affinity (inhibition < 50%) for 28 standard receptor binding profile (Hashimoto et al., 2008), the nonspecific binding of [^{3}H]CHIBA-1001 is relatively high in the membranes from postmortem human brain samples (Tanibuchi et al., 2010).

2.6.2. Pharmacology

Very recently, Ishikawa et al. (2011) examined the receptor occupancy in the human brain after a single oral administration of the 5-HT_3 antagonists tropisetron and ondansetron used as treatment in patients with chemotherapy-induced or postoperative nausea and vomiting. Tropisetron is a partial agonist of α7 nAChRs with high affinity (K_i = 6.9 nM) and is effective in the treatment of auditory P50 deficits in patients with schizophrenia (Koike et al., 2005; Shiina et al., 2010), whereas ondansetron has a low affinity for α7 nAChR (K_i > 10,000 nM) (Macro et al., 2001; Papke et al., 2004). A single oral administration of tropisetron, but not ondansetron, was shown to decrease the V_T of [^{11}C] CHIBA-1001 in the human brain (Figure 8).

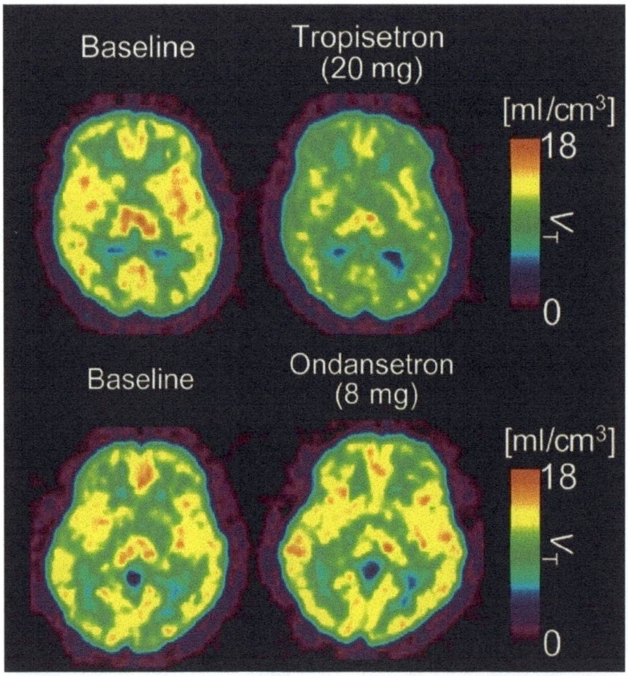

Figure 8 V_T images of [^{11}C]CHIBA-1001 PET before and after the single oral administration of tropisetron or ondansetron. Upper pair: V_T images at baseline (left) and at tropisetron (20 mg) loading (right) in the same subject. Lower pair: V_T images at baseline (left) and ondansetron (8 mg) loading (right) in another subject.

These data suggest that tropisetron, but not ondansetron, can bind to α7 nAChRs in the human brain.

3. CONCLUSIONS

Since the early 1980s, several high-affinity radioligands for mapping of mAChRs using SPECT and PET have been developed. Although these high-affinity tracers have high target/nontarget ratios, they show a dependence on blood flow and transport across the BBB. Therefore, appropriate pharmacokinetic modeling or long wait times after injection are necessary to separate the parameters for the flow and transport from receptor density. Most of the available mAChR radiotracers lack selectivity for the mAChR subtypes (M_1–M_4), with the exception being [^{18}F]FP-TZTP, which appears to bind predominantly to the M_2 subtype. In accordance with the results of postmortem studies, mAChR imaging identified changes in mAChR availability associated with aging, neurodegenerative diseases, and other neurological disorders. In addition, a number of clinical studies have elucidated the mechanisms of action of pharmaceuticals, such as anticholinergic drugs, antipsychotic agents, and AChE inhibitors.

The extremely high-affinity and reduced-toxicity A-85380 derivatives, [^{18}F]2-FA, [^{18}F]6-FA, and [^{123}I]5-IA, are now being employed for the evaluation of α4β2 nAChR subtypes in the human brain. The drawbacks of these ligands are their slow brain distribution kinetics and relatively low BP in the human brain. A number of clinical studies have identified changes of α4β2 nAChR associated with aging and neurodegenerative diseases. Imaging of α4β2 nAChR is a powerful tool for understanding the adaptive changes in α4β2 nAChR density during abstinence from tobacco smoking. Furthermore, α4β2 nAChR occupancy after the administration of nicotine cigarettes was measured satisfactorily by in vivo imaging. This technique was sufficiently sensitive to detect substantial changes in receptor occupancy associated with exposure to secondhand smoke.

With the exception of a previous study using the PET radioligand [^{11}C]CHIBA-1001, there have been no reports of α7 nAChR imaging in the human brain. Further studies are required to identify and develop radioligands for α7 nAChR with better properties, including a substantially higher binding affinity and lower nonspecific binding.

ABBREVIATIONS

A-582941 2-methyl-5-(6-phenylpyridazin-3-yl)octahydropyrrolo[3,4-*c*]pyrrole
A-844606 2-(5-methylhexahydropyrrolo[3,4-*c*]pyrrol-2(1*H*)-yl)-9*H*-xanthen-9-one
A-85380 3-(2(S)-azetidinylmethoxy)pyridine
ACh acetylcholine
AChE acetylcholinesterase
AChR acetylcholine receptor
AD Alzheimer's disease
ADNFLE autosomal dominant nocturnal frontal lobe epilepsy
APOE-ε4 apolipoprotein E-ε4

BBB blood–brain barrier
BD bipolar disorder
Benztropine 3-(benzhydryloxy)-8-methyl-8-azabicyclo[3.2.1]octane
BP binding potential
CBF cerebral blood flow
ChAT choline acetyltransferase
CHIBA-1001 4-methylphenyl 1,4-diazabicyclo[3.2.2]nonane-4-carboxylate
$CHRM_2$ cholinergic muscarinic M_2 receptor gene
CNS central nervous system
DLB dementia with Lewy bodies
DVR distribution volume ratio
EPS extrapyramidal side effects
2-FA (*S*)-3-(azetidin-2-ylmethoxy)-2-fluoropyridine
6-FA (*R*)-5-(azetidin-2-ylmethoxy)-2-fluoropyridine
FDG 2-deoxy-2-fluoro-D-glucose
FP-TZTP 3-((3-fluoropropyl)thio)-4-(1-methyl-1,2,5,6-tetrahydropyridin-3-yl)-1,2,5-thiadiazole
5-HT_3 5-hydroxytryptamine-3
5-IA (*R*)-3-(azetidin-2-ylmethoxy)-5-iodopyridine
IDEX 1′-(4-iodobenzyl)-3-phenyl-[3,4′-bipiperidine]-2,6-dione
(*R*,*R*)-IQNB (*R*)-(*R*)-quinuclidin-3-yl 2-hydroxy-2-(4-iodophenyl)-2-phenylacetate
(*R*,*S*)-IQNB (*S*)-(*R*)-quinuclidin-3-yl 2-hydroxy-2-(4-iodophenyl)-2-phenylacetate
(*E*)-IQNP (*E*)-quinuclidin-3-yl 2-hydroxy-5-iodo-2-phenylpent-4-enoate
(*Z*)-IQNP (*Z*)-quinuclidin-3-yl 2-hydroxy-5-iodo-2-phenylpent-4-enoate
KO knockout
mAChR muscarinic acetylcholine receptor
MCI mild cognitive impairment
MDD major depressive disorder
(*R*)-MeQAA (*R*)-quinuclidin-3-yl 2-(methylamino)benzoate
MLA methyllycaconitine
MRI magnet resonance imaging
nAChR nicotinic acetylcholine receptor
NMPB 1-methylpiperidin-4-yl 2-hydroxy-2,2-diphenylacetate
NS10743 2-(1,4-diazabicydo[3.2.2]nonan-4-yl)-5-(4-fluorophenyl)-1,3,4-oxadiazole
NS14492 2-(1,4-diazabicyclo[3.2.2]nonan-4-yl)-5-(1-methyl-1*H*-pyrrol-2-yl)-1,3,4-oxadiazole
PD Parkinson's disease
PDD Parkinson's disease with dementia
PET positron emission tomography
PSP progressive supranuclear palsy
PTSD posttraumatic stress disorder
Scopolamine 9-methyl-3-oxa-9-azatricyclo[3.3.1.$0^{2,4}$]nonan-7-yl 3-hydroxy-2-phenylpropanoate
SNP single-nucleotide polymorphism (SNP)
SPECT single photon emission computed tomography
SPM statistical parametric mapping
TLE temporal lobe epilepsy
TRB 8-methyl-8-azabicyclo[3.2.1]octan-3-yl 2-hydroxy-2,2-diphenylacetate
V_T total distribution volume
VTA ventral tegmental area

REFERENCES

Anagostaras, S.G., Murphy, G.G., Hamilton, S.E., Mitchell, S.L., Rahnama, N.P., Nathanson, N.M., Silvia, A.K., 2003. Selective cognitive dysfunction in acetylcholine M_1 muscarinic receptor mutant mice. Nat. Neurosci. 6, 51–58.

Asahina, M., Shinotoh, H., Hirayama, K., Suhara, T., Shishido, F., Inoue, O., Tateno, Y., 1995. Hypersensitivity of cortical muscarinic receptors in Parkinson's disease demonstrated by PET. Acta Neurol. Scand. 91, 437–443.

Asahina, M., Suhara, T., Shinotoh, H., Inoue, O., Suzuki, K., Hattori, T., 1998. Brain muscarinic receptors in progressive supranuclear palsy and Parkinson's disease: a positron emission tomographic study. J. Neurol. Neurosurg. Psychiatry 65, 155–163.

Aubert, I., Araujo, D.M., Cecyre, D., Robitaille, Y., Gauthier, S., Quirion, R., 1992. Comparative alterations of nicotinic and muscarinic binding sites in Alzheimer's disease and Parkinson's disease. J. Neurochem. 58, 529–541.

Ballard, C., Piggott, M., Johnson, M., Cairns, N., Perry, R., McKeith, I., Jaros, E., O'Brien, J., Holmes, C., Perry, E., 2000. Delusions associated with elevated muscarinic M_1 receptor binding in dementia with Lewy bodies. Ann. Neurol. 48, 868–876.

Benwell, M.E.M., Balfour, D.J.K., Anderson, J.M., 1988. Evidence that tobacco smoking increases the density of (-)-[^{3}H]nicotine binding in human brain. J. Neurochem. 50, 1243–1247.

Bergström, K.A., Halldin, C., Savonen, A., Okubo, Y., Hiltunen, J., Nobuhara, K., Swahn, C.G., Karlsson, P., McPherson, D., Knapp Jr., F.F., Larsson, S., Schnell, P.O., Farde, L., 1999. Iodine-123 labelled *Z*-(*R*, *R*)-IQNP: a potent radioligand for visualization of M_1 and M_2 muscarinic acetylcholine receptors in Alzheimer's disease. Eur. J. Nucl. Med. 26, 1482–1485.

Bonner, T.I., Young, A.C., Bran, M.R., Buckley, N.J., 1988. Cloning and expression of the human and rat M_5 muscarinic acetylcholine receptor gene. Neuron 1, 403–410.

Bottlaender, M., Valette, H., Roumenov, D., Dollé, F., Coulon, C., Ottaviani, M., Hinnen, F., Ricard, M., 2003. Biodistribution and radiation dosimetry of ^{18}F-fluoro-A-85380 in healthy volunteers. J. Nucl. Med. 44, 596–601.

Boundy, K.L., Barnden, L.R., Katsifis, A.G., Rowe, C.C., 2005. Reduced posterior cingulated binding of I-123 iodo-dexetimide to muscarinic receptors in mild Alzheimer's disease. J. Clin. Neurosci. 12, 421–425.

Boundy, K.L., Rowe, C.C., Black, A.B., Kitchener, M.I., Barnden, L.R., Sebben, R., Kneebone, A., Kassiou, M., Katsifis, A., Lambrecht, R., 1996. Localization of temporal lobe epileptic foci with iodine-123 iododexetimide cholinergic neuroreceptor single-photon emission computed tomography. Neurology 47, 1015–1020.

Breese, C.R., Adams, C., Logel, J., Drebing, C., Rollins, Y., Barnhart, M., Sullivan, B., Demasters, B.K., Freedman, R., Leonard, S., 1997a. Comparison of the regional expression of nicotinic acetylcholine receptor α7 mRNA and [^{125}I]-a-bungarotoxin binding in human postmortem brain. J. Comp. Neurol. 387, 385–398.

Breese, C.R., Marks, M.J., Logel, J., Adams, C.E., Sullivan, B., Collins, A.C., Leonard, S., 1997b. Effect of smoking history on [^{3}H]nicotine binding in human postmortem brain. J. Pharmacol. Exp. Ther. 282, 7–13.

Brody, A.L., Mandelkern, M.A., Costello, M.R., Abrams, A.L., Scheibal, D., Farahi, J., London, E.D., Olmstead, R.E., Rose, J.E., Mukhin, A.G., 2009. Brain nicotinic acetylcholine receptor occupancy: effect of smoking a denicotinized cigarette. Int. J. Neuropsychopharmacol. 12, 305–316.

Brody, A.L., Mandelkern, M.A., London, E.D., Khan, A., Kozman, D., Costello, M.R., Vellios, E.E., Archie, M.M., Bascom, R., Mukhin, A.G., 2011. Effect of secondhand smoke on occupancy of nicotinic acetylcholine receptors in brain. Arch. Gen. Psychiatry 68, 953–960.

Brody, A.L., Mandelkern, M.A., London, E.D., Olmstead, R.E., Farahi, J., Scheibal, D., Jou, J., Allen, V., Tiongson, E., Chefer, S.I., Koren, A.O., Mukhin, A.G., 2006. Cigarette smoking saturates brain α4β2 nicotinic acetylcholine receptors. Arch. Gen. Psychiatry 63, 907–915.

Brown, D., Chisholm, J.A., Owens, J., Pimlott, S., Patterson, J., Wyper, D., 2003. Acetylcholine muscarinic receptors and response to anti-cholinesterase therapy in patients with Alzheimer's disease. Eur. J. Nucl. Med. 30, 296–300.

Burghaus, L., Schutz, U., Krempel, U., de Vos, R.A., Jansen-Steur, E.N., Wevers, A., Lindstrom, J., Schroder, H., 2000. Quantitative assessment of nicotinic acetylcholine receptor proteins in the cerebral cortex of Alzheimer's disease. Brain Res. Mol. Brain Res. 76, 385–388.

Bymaster, F.P., Calligaro, D.O., Falcone, J.F., Marsh, R.D., Moore, N.A., Tye, N.C., Seeman, P., Wong, D.T., 1996. Radioreceptor binding profile of the atypical antipsychotic olanzapine. Neuropsychopharmacology 14, 87–96.

Cannon, D.M., Carson, R.E., Nugent, A.C., Eckelman, W.C., Kiesewetter, D.O., Williams, J., Rollis, D., Drevets, M., Gandhi, S., Solorio, G., Drevets, W.C., 2006. Reduced muscarinic type 2 receptor binding in subjects with bipolar disorder. Arch. Gen. Psychiatry 63, 741–747.

Cannon, D.M., Klaver, J.K., Gandhi, S.K., Solorio, G., Peck, S.A., Erickson, K., Akula, N., Savitz, J., Eckelman, W.C., Furey, M.L., Sahakian, B.J., McMahon, F.J., Drevets, W.C., 2011. Genetic variation in cholinergic muscarinic-2 receptor gene modulates M_2 receptor binding in vivo and accounts for reduced binding in bipolar disorder. Mol. Psychiatry 16, 407–418.

Carson, R.E., Kiesewetter, D.O., Jagoda, E., Der, M.G., Herscovitch, P., Eckelman, W.C., 1998. Muscarinic cholinergic receptor measurements with [^{18}F]FP-TZTP: control and competition studies. J. Cereb. Blood Flow Metab. 18, 1130–1142.

Caulfield, M.P., Birdsall, N.J., 1998. International union of pharmacology. XVII. Classification of muscarinic acetylcholine receptors. Pharmacol. Rev. 50, 279–290.

Chefer, S.I., Horti, A.G., Koren, A.O., Gundisch, D., Liks, J.M., Kurian, V., Dannals, R.F., Mukhin, A.G., London, E.D., 1999. 2-[^{18}F]F-A-85380: a PET radioligand for α4β2 nicotinic acetylcholine receptors. Neuroreport 10, 2715–2721.

Chefer, S.I., Horti, A.G., Lee, K.S., Kore, A.O., Jones, D.W., Gorey, J.G., Links, J.M., Mukhin, A.G., Weinberger, D.R., London, E.D., 1998. In vivo imaging of brain nicotinic acetylcholine receptors with 5-[^{123}I]iodo-A85380 using single photon emission computed tomography. Life Sci. 63, L355–L360.

Chefer, S.I., London, E.D., Koren, A.O., Pavlova, O.A., Kurian, V., Kimes, A.S., Horti, A.G., Mukhin, A.G., 2003. Graphical analysis of 2-[^{18}F]FA binding to nicotinic acetylcholine receptors in rhesus monkey brain. Synapse 48, 25–34.

Clader, J.W., Wang, Y., 2005. Muscarinic receptor agonists and antagonists in the treatment of Alzheimer's disease. Curr. Pharm. Des. 11, 3353–3361.

Claus, J.J., Dubois, E.A., Booij, J., Habraken, J., de Munck, J.C., van Herk, M., Verbeeten Jr., B., van Royen, E.A., 1997. Demonstration of a reduction in muscarinic receptor binding in early Alzheimer's disease using iodine-123 dexetimide single-photon emission tomography. Eur. J. Nucl. Med. 24, 602–608.

Cohen, R.M., Carson, R.E., Filbey, F., Szczepanik, J., Sunderland, T., 2006. Age and APOE-e4 genotype influence the effect of physostigmine infusion on the in-vivo distribution volume of the muscarinic-2-receptor dependent tracer [^{18}F]FP-TZTP. Synapse 60, 86–92.

Cohen, R.M., Podruchny, T.A., Bokde, A.L., Carson, R.E., Herscovitch, P., Kiesewetter, D.O., Eckelman, W.C., Sunderland, T., 2003. Higher in vivo muscarinic-2 receptor distribution volumes in aging subjects with an apolipoprotein E-e4 allele. Synapse 49, 150–156.

Colloby, S.J., Pakrasi, S., Firbank, M.J., Perry, E.K., Piggott, M.A., Owens, J., Wyper, D.J., McKeith, I.G., Burn, D.J., Williams, E.D., O'Brien, J.T., 2006. In vivo SPECT imaging of muscarinic acetylcholine receptors using (*R, R*) ^{123}I-QNB in dementia with Lewy bodies and Parkinson's disease dementia. Neuroimage 33, 423–429.

Colloby, S.J., Perry, E.K., Pakrasi, S., Pimlott, S.L., Wyper, D.J., McKeith, I.G., Williams, E.D., O'Brien, J.T., 2010. Nicotinic ^{123}I-5IA-85380 single photon emission computed tomography as a predictor of cognitive progression in Alzheimer's disease and dementia with Lewy bodies. Am. J. Geriatr. Psychiatry 18, 86–90.

Colloby, S.J., Firbank, M.J., Pakrasi, S., Perry, E.K., Pimlott, S.L., Wyper, D.J., McKeith, I.G., Williams, E.D., O'Brien, J.T., 2011. Alterations in nicotinic α4β2 receptor binding in vascular dementia using ^{123}I-5IA-85380 SPECT: comparison with regional cerebral blood flow. Neurobiol. Aging 32, 293–301.

Cooper, J.R., Bloom, F.E., Roth, R.H., 2003. The Biochemical Basis of Neuropharmacology, eighth ed. Oxford University Press, Oxford, New York.

Cortés, R., Probst, A., Palacios, J.M., 1987. Quantitative light microscopic autoradiographic localization of cholinergic muscarinic receptors in the human brain: forebrain. Neuroscience 20, 65–107.

Cortés, R., Probst, A., Tobler, H.J., Palacios, J.M., 1986. Muscarinic cholinergic receptor subtypes in the human brain. II. Quantitative autoradiographic studies. Brain Res. 362, 239–253.

Cosgrove, K.P., Batis, J., Bois, F., Maciejewski, P.K., Esterlis, I., Kloczynski, T., Stiklus, S., Krishnan-Sarin, S., O'Malley, S., Perry, E., Tamagnan, G., Seibyl, J.P., Staley, J.K., 2009. β2-Nicotinic acetylcholine receptor availability during acute and prolonged abstinence from tobacco smoking. Arch. Gen. Psychiatry 66, 666–676.

Cosgrove, K.P., Esterlis, I., McKee, S., Bois, F., Alagille, D., Tamagnan, G.D., Seibyl, J.P., Krishnan-Sarin, S., Staley, J.K., 2010. Beta2* nicotinic acetylcholine receptors modulate pain sensitivity in acutely abstinent tobacco smokers. Nicotine Tob. Res. 12, 535–539.

Cosgrove, K.P., Mitsis, E.M., Bois, F., Frohlich, E., Tamagnan, G.D., Krantzler, E., Perry, E., Maciejewski, P.K., Epperson, C.N., Allen, S., O'malley, S., Mazure, C.M., Seibyl, J.P., van Dyck, C.H., Staley, J.K., 2007. ^{123}I-5-IA-85380 SPECT imaging of nicotinic acetylcholine receptor availability in nonsmokers: effects of sex and menstrual phase. J. Nucl. Med. 48, 1633–1640.

Court, J., Martin-Ruiz, C., Piggott, M., Spurden, D., Griffiths, M., Perry, E., 2001. Nicotinic receptor abnormalities in Alzheimer's disease. Biol. Psychiatry 49, 175–184.

Court, J.A., Perry, E.K., Spurden, D., Griffiths, M., Kerwin, J.M., Morris, C.M., Johnson, M., Oakley, A.E., Birdsall, N.J.M., Clementin, F., Perry, R.H., 1995. The role of the cholinergic system in the development of the human cerebellum. Brain Res. Dev. Brain Res. 90, 159–167.

Court, J., Spurden, D., Lloyd, S., McKeith, I., Ballard, C., Cairns, N., Kerwin, R., Perry, R., Perry, E., 1999. Neuronal nicotinic receptors in dementia with Lewy bodies and schizophrenia: a-bungarotoxin and nicotine binding in the thalamus. J. Neurochem. 73, 1590–1597.

Czermak, C., Staley, J.K., Kasserman, S., Bois, F., Young, T., Henry, S., Tamagnan, G.D., Seibyl, J.P., Krystal, J.H., Neumeister, A., 2008. β2 nicotinic acetylcholine receptor availability in post-traumatic stress disorder. Int. J. Neuropsychopharmacol. 11, 419–424.

Dani, J.A., Bertrand, D., 2007. Nicotinic acetylcholine receptors and nicotinic cholinergic mechanisms of the central nervous system. Annu. Rev. Pharmacol. Toxicol. 47, 699–729.

Davies, A.R., Hardick, D.J., Blagbrough, I.S., Potter, B.V., Wolstenholme, A.J., Wonnacott, S., 1999. Characterization of the binding of [^{3}H]methyllycaconitine: a new radioligand for labeling α7-type neuronal nicotinic acetylcholine receptors. Neuropsychopharmacology 38, 679–690.

Dean, B., McLeod, M., Keriakous, D., McKenzie, J., Scarr, E., 2007. Decreased muscarinic1 receptors in the dorsolateral prefrontal cortex of subjects with schizophrenia. Mol. Psychiatry 7, 1083–1091.

Deuther-Conrad, W., Fischer, S., Hiller, A., Becker, G., Cumming, P., Xiong, G., Funke, U., Sabri, O., Peters, D., Brust, P., 2011. Assessment of α7 nicotinic acetylcholine receptor availability in juvenile pig brain with [^{18}F]NS10743. Eur. J. Nucl. Med. Mol. Imaging 38, 1541–1549.

Deuther-Conrad, W., Fischer, S., Hiller, A., Nielse, E.Ø., Timmermann, D.B., Steinbach, J., Sabri, O., Peters, D., Brust, P., 2009. Molecular imaging of α7 nicotinic acetylcholine receptors: design and evaluation of the potent radioligand [^{18}F]NS10743. Eur. J. Nucl. Med. Mol. Imaging 36, 791–800.

Dewey, S.L., MacGregor, R.R., Brodie, J.D., Bendriem, B., King, P.T., Volkow, N.D., Schlyer, D.J., Fowler, J.S., Wolf, A.P., Gatley, S.J., Hitzemann, R., 1990a. Mapping muscarinic receptors in human and baboon brain using [*N*-^{11}C-methyl]-benztropine. Synapse 5, 213–223.

Dewey, S.L., Volkow, N.D., Logan, J., MacGregor, R.R., Fowler, J.S., Schlyer, D.J., Bendriem, B., 1990b. Age-related decreases in muscarinic cholinergic receptor binding in the human brain measured with positron emission tomography (PET). J. Neurosci. Res. 27, 569–575.

Ding, Y.-S., Fowler, J.S., Logan, J., Wang, G.J., Telang, F., Garza, V., Biegon, A., Pareto, D., Rooney, W., Shea, C., Alexoff, D., Volkow, N.D., 2004. 6-[^{18}F]Fluoro-A-85380, a new PET tracer for the nicotinic acetylcholine receptor: studies in the human brain and in vivo demonstration of specific binding in white matter. Synapse 53, 184–189.

Ding, Y., Liu, N., Wang, T., Marecek, J., Garza, V., Ojima, I., Fowler, J.S., 2000. Synthesis and evaluation of 6-[^{18}F]fluoro-3-(2(*S*)-azetidinylmethoxy)pyridine as a PET tracer for nicotinic acetylcholine receptors. Nucl. Med. Biol. 27, 381–389.

Ding, Y.S., Molina, P.E., Fowler, J.S., Logan, J., Volkow, N.D., Kuhar, M.J., Carroll, F.I., 1999. Comparative studies of epibatidine derivatives [^{18}F]NFEP and [^{18}F]N-methyl-NFEP: kinetics, nicotine effect, and toxicity. Nucl. Med. Biol. 26, 139–148.

Dolci, L., Dolle, F., Valette, H., Vaufrey, F., Fuseau, C., Bottlaender, M., Crouzel, C., 1999. Synthesis of a fluorine-18 labeled derivative of epibatidine for in vivo nicotinic acetylcholine receptor PET imaging. Bioorg. Med. Chem. 7, 467–479.

Dörje, F., Wess, J., Lambrecht, G., Tacke, R., Mutschler, E., Brann, M.R., 1991. Antagonist binding profiles of five cloned human muscarinic receptor subtypes. J. Pharmacol. Exp. Ther. 256, 727–733.

Duggan, L., Fenton, M., Dardennes, R.M., El-Dosoky, A., Indran, S., 2000. Olanzapine for schizophrenia. Cochrane Database Syst. Rev. 2, CD001359.

Eckelman, W.C., 2006. Imaging of muscarinic receptors in the central nervous system. Curr. Pharm. Des. 12, 3901–3913.

Eckelman, W.C., Kilbourn, M.R., Mathis, C.A., 2006. Discussion of targeting proteins in vivo: in vitro guidelines. Nucl. Med. Biol. 33, 449–451.

Eckelman, W.C., Reda, R.C., Rzeszotarski, W.J., Gibson, R.R., Hill, T., Holman, B.L., Budinger, B.L., Conklin, J.J., Eng, R., Grissom, M.P., 1984. External imaging of cerebral muscarinic acetylcholine receptors. Science 20, 291–293.

Eglen, R.M., 2006. Muscarinic receptor subtypes in neuronal and non-neuronal cholinergic function. Auton. Autacoid Pharmacol. 26, 219–233.

Ellis, J.R., Nathan, P.J., Villemagne, V.L., Mulligan, R.S., Ellis, K.A., Tochon-Danguy, H.J., Chan, J.G., O'keefe, G.J., Bradley, J., Savage, G., Rowe, C.C., 2009a. The relationship between nicotinic receptors and cognitive functioning in healthy aging: an in vivo positron emission tomography (PET) study with 2-[^{18}F]fluoro-A-85380. Synapse 63, 752–763.

Ellis, J.R., Nathan, P.J., Villemagne, V.L., Mulligan, R.S., Saunder, T., Young, K., Smith, C.L., Welch, J., Woodward, M., Wesnes, K.A., Savage, G., Rowe, C.C., 2009b. Galantamine-induced improvements in cognitive function are not related to alterations in α4β2 nicotinic receptors in early Alzheimer's disease as measured in vivo by 2-[^{18}F]fluoro-A-85380 PET. Psychopharmacology (Berl.) 202, 79–91.

Ellis, J.R., Villemagne, V.L., Nathan, P.J., Mulligan, R.S., Gong, S.J., Chan, J.G., Sachinidis, J., O'Keefe, G.J., Pathmaraj, K., Wenses, K.A., Savage, G., Rowe, C.C., 2008. Relationship between nicotinic receptors and cognitive function in early Alzheimer's disease: a 2-[^{18}F]fluoro-A-85380 PET study. Neurobiol. Learn. Mem. 90, 404–412.

Epping-Jordan, M.P., Watkins, S.S., Koob, G.F., Markou, A., 1998. Dramatic decreases in brain reward function during nicotine withdrawal. Nature 393, 76–79.

Esterlis, I., Cosgrove, K.P., Batis, J.C., Bois, F., Stiklus, S.M., Perkins, E., Seibyl, J.P., Carson, R.E., Staley, J.K., 2010a. Quantification of smoking-induced occupancy of β2-nicotinic acetylcholine receptors: estimation of nondisplaceable binding. J. Nucl. Med. 51, 1226–1233.

Esterlis, I., Cosgrove, K.P., Petrakis, I.L., McKee, S.A., Bois, F., Krantzler, E., Stiklus, S.M., Perry, E.B., Tamagnan, G.D., Seibyl, J.P., Krystal, J.H., Staley, J.K., 2010b. SPECT imaging of nicotinic acetylcholine receptors in nonsmoking heavy alcohol drinking individuals. Drug Alcohol Depend. 108, 146–150.

Esterlis, I., Mitsis, E.M., Batis, J.C., Bois, F., Picciotto, M.R., Stiklus, S.M., Kloczynski, T., Perry, E., Seibyl, J.P., McKee, S., Staaley, J.K., Cosgrove, K.P., 2011. Brain β2*-nicotinic acetylcholine receptor occupancy after use of a nicotine inhaler. Int. J. Neuropsychopharmacol. 14, 389–398.

Ettrup, A., Mikkelsen, J.D., Lehel, S., Madsen, J., Nielsen, E.Ø., Palner, M., Timmermann, D.B., Peters, D., Knudsen, G.M., 2011. ^{11}C-NS14492 as a novel PET radioligand for imaging cerebral α7 nicotinic acetylcholine receptors: in vivo evaluation and drug occupancy measurements. J. Nucl. Med. 52, 1–8.

Falk, L., Nordberg, A., Seiger, A., Kjaeldgaard, A., Hellström-Lindahl, E., 2003. Higher expression of α7 nicotinic acetylcholine receptors in human fetal compared to adult brain. Brain Res. Dev. Brain Res. 142, 151–160.

Felder, C.C., Bymaster, F.P., Ward, J., DeLapp, N., 2000. Therapeutic opportunities for muscarinic receptors in the central nervous system. J. Med. Chem. 43, 4333–4353.

Fernandes, C., Hoyle, E., Dempster, E., Schalkwyk, L.C., Collier, D.A., 2006. Performance deficit of α7 nicotinic receptor knockout mice in a delayed matching-to-place task suggests a mild impairment of working/episodic-like memory. Genes Brain Behav. 5, 433–440.

Fisher, A., Pittel, Z., Haring, R., Bar-Ner, N., Kliger-Spatz, M., Natan, N., Egozi, I., Sonego, H., Marcovitch, I., Brandeis, R., 2003. M_1 muscarinic agonists can modulate some of the hallmarks in Alzheimer's disease. J. Mol. Neurosci. 20, 349–356.

Flores, C.M., Rogers, S.W., Pabreza, L.A., Wolfe, B.B., Kellar, K.J., 1992. A subtype of nicotinic cholinergic receptor in rat brain is composed of α4 and β2 subunits and is up-regulated by chronic nicotine treatment. Mol. Pharmacol. 41, 31–37.

Flynn, D.D., Ferali-Dileo, G., Mash, D.C., Levey, A.I., 1995. Differential regulation of molecular subtypes of muscarinic receptors in Alzheimer's disease. J. Neurochem. 64, 1888–1891.

Flynn, D.D., Mash, D.C., 1986. Characterization of L-[^{3}H]nicotine binding in human cerebral cortex: comparison between Alzheimer's disease and the normal. J. Neurochem. 47, 1948–1954.

Flynn, D.D., Mash, D.C., 1993. Distinct kinetic binding properties of *N*-[^{3}H]-methylscopolamine afford differential labeling and localization of M_1, M_2, and M_3 muscarinic receptor subtypes in primate brain. Synapse 14, 283–296.

Freedman, R., Adler, L.E., Bickford, P., Byerley, W., Coon, H., Cullum, C.M., Griffith, J.M., Harris, J.G., Leonard, S., Miller, C., Myles-Worsley, M., Nagamoto, H.T., Rose, G., Waldo, M., 1994. Schizophrenia and nicotinic receptors. Harv. Rev. Psychiatry 2, 179–192.

Freedman, R., Hall, M., Adler, L.E., Leonard, S., 1995. Evidence in postmortem brain tissue for decreased numbers of hippocampal nicotinic receptors in schizophrenia. Biol. Psychiatry 38, 22–33.

Freund, G., Ballinger Jr, W.E., 1989. Loss of muscarinic cholinergic receptors from the temporal cortex of alcohol abusers. Metab. Brain Dis. 4, 121–141.

Frey, K.A., Koeppe, R.A., Mulholland, G.K., Jewett, D., Hichwa, R., Ehrenkaufer, R.L., Carey, J.E., Wieland, D.M., Kuhl, D.E., Agranoff, B.W., 1992. In vivo muscarinic receptor imaging in human brain with [^{11}C] scopolamine and positron emission tomography. J. Cereb. Blood Flow Metab. 12, 147–154.

Fujita, M., Ichise, M., Zoghbi, S.S., Liow, J.-S., Ghose, S., Vines, D.C., Sangare, J., Lu, J.-Q., Cropley, V.L., Iida, H., Kim, K.M., Cohen, R.M., Bara-Jimenez, W., Ravina, B., Innis, R.B., 2006. Widespread decrease of nicotinic acetylcholine receptors in Parkinson's disease. Ann. Neurol. 59, 174–177.

Fujita, M., Seibyl, J.P., Vauoel, B., Tamagnan, G., Early, M., Zoghbi, S.S., Baldwin, R.M., Horti, A.G., Koren, A.O., Mukhin, A.G., Khan, S., Bozkurt, A., Kimes, A., London, E.D., Innis, R.B., 2002. Whole-body biodistribution, radiation absorbed dose, and brain SPET imaging with [^{123}I]5-I-A-85380 in healthy human subjects. Eur. J. Nucl. Med. 29, 183–190.

Gallezot, J.D., Bottlaender, M., Grégoire, M.C., Roumenov, D., Deverre, J.R., Coulon, C., Ottaviani, M., Dollé, F., Syrota, A., Valette, H., 2005. In vivo imaging of human cerebral nicotinic acetylcholine receptors with 2-^{18}F-fluoro-A-85380 and PET. J. Nucl. Med. 46, 240–247.

Gao, Y., Kuwabara, H., Spivak, C.E., Xiao, Y., Kellar, K., Ravert, H.T., Kumar, A., Alexander, M., Hilton, J., Wong, D.F., Dannals, R.F., Horti, A.G., 2008. Discovery of (-)-7-methyl-2-exo-[3¢(6-[18F] fluoropyridin-2-yl)-5¢-pyridinyl]-7-azabicyclo[2.2.2]heptanes, a radiolabeled antagonist for cerebral nicotinic acetylcholine receptor (α4β2-nAChR) with optimal positron emission tomography imaging properties. J. Med. Chem. 51, 4751–4764.

Gibbs, R.B., 1998. Impairment of basal forebrain cholinergic neurons associated with aging and long-term loss of ovarian function. Exp. Neurol. 151, 289–302.

Giboureau, N., Som, I.M., Boucher-Arnold, A., 2010. PET radioligands foe vesicular acetylcholine transporter (VAChT). Curr. Top. Med. Chem. 10, 1569–1583.

Granon, S., Changeux, J.P., 2006. Attention-deficit/hyperactivity disorder: a plausible mouse model? Acta Paediatr. 95, 645–649.

Granon, S., Faure, P., Changeux, J.P., 2003. Executive and social behaviors under nicotinic receptor regulation. Proc. Natl. Acad. Sci. U. S. A. 100, 9596–9601.

Gu, Q., 2002. Neuromodulatory transmitter system in the cortex and their role in cortical plasticity. Neuroscience 111, 815–835.

Guan, Z.Z., Zhang, X., Ravid, R., Nordberg, A., 2000. Decreased protein levels of nicotinic receptor subunits in the hippocampus and temporal cortex of patients with Alzheimer's disease. J. Neurochem. 74, 237–243.

Gundisch, D., Koren, A.O., Horti, A.G., Pavlova, O.A., Kimes, A.S., Mukhin, A.G., London, E.D., 2005. In vitro characterization of 6-[^{18}F]fluoro-A-85380, a high-affinity ligand for α4β2* nicotinic acetylcholine receptors. Synapse 55, 89–97.

Hall, M., Zerbe, L., Leonard, S., Freedman, R., 1993. Characterization of [^{3}H]cytosine binding to human brain membrane preparations. Brain Res. 600, 127–133.

Hashimoto, K., Koike, K., Shimizu, E., Iyo, M., 2005. α7 Nicotinic receptor agonists as potential therapeutic drugs for schizophrenia. Curr. Med. Chem. CNS Agents 5, 171–184.

Hashimoto, K., Nishiyama, S., Ohba, H., Matsuo, M., Kobayashi, T., Takahagi, M., Iyo, M., Kitashoji, T., Tsukada, H., 2008. [^{11}C]CHIBA-1001 as a novel PET ligand for α7 nicotinic receptors in the brain: a PET study in conscious monkeys. PLos One 3, e3231.

Hellström-Lindhal, E., Mousavi, M., Zhang, X., Ravid, R., Nordberg, A., 1999. Regional distribution of nicotinic receptor subunit mRNAs in human brain: comparison between Alzheimer and normal brain. Bain Res. Mol. Brain Res. 66, 94–103.

Henderson, V.W., 1997. The epidemiology of estrogen replacement therapy and Alzheimer's disease. Neurology 48, S27–S35.

Herholz, K., 2010. Cerebral glucose metabolism in preclinical and prodromal Alzheimer's disease. Expert Rev. Neurother. 10, 1667–1673.

Hersch, S.M., Gutekunst, C.A., Rees, H.D., Heilman, C.J., Levey, A.I., 1994. Distribution of m1–m4 muscarinic receptor proteins in the rat striatum: light and electron microscopic immunocytochemistry using subtype-specific antibodies. J. Neurosci. 14, 3351–3363.

Hiramatsu, Y., Eckelman, W.C., Baum, B.J., 1994. Interaction of iodinated quinuclidinyl benzilate enantiomers with M_3 muscarinic receptors. Life Sci. 54, 1777–1783.

Hiramatsu, Y., Eckelman, W.C., Carrasquillo, J.A., Miletich, R.S., Valdez, I.H., Kurrasch, R.H., Macynski, A.A., Paik, C.H., Neumann, R.D., Baum, B.J., 1995. Kinetic analysis of muscarinic receptors in human brain and salivary gland in vivo. Am. J. Physiol. 286, R1491–R1499.

Holman, B.L., Gibson, R.E., Hill, T.C., Eckelman, W.C., Albert, M., Reba, R.C., 1985. Muscarinic acetylcholine receptors in Alzheimer's disease. in vivo imaging with iodine 123-labeled 3-quinuclidinyl-4-iodobenzilate and emission tomography. JAMA 254, 3063–3066.

Holladay, W.M., Dart, M.J., Lynch, J.K., 1997. Neuronal nicotinic acetylcholine receptors as targets for drug discovery. J. Med. Chem. 40, 4169–4194.

Horti, A.G., Chefer, S.I., Mukhin, A.G., Koren, A.O., Gundisch, D., Links, J.M., Kurian, V., Dannals, R.F., London, E.D., 2000. 6-[^{18}F]Fluoro-A-85380, a novel radioligand for in vivo imaging of central acetylcholine receptors. Life Sci. 67, 463–469.

Horti, A.G., Gao, Y., Kuwabara, H., Dannals, R.F., 2010. Development of radioligands with optimized imaging properties for quantification of nicotinic acetylcholine receptors by positron emission tomography. Life Sci. 86, 575–584.

Horti, A.G., Koren, A.O., Lee, K.S., Mukhin, A.G., Vaupel, D.B., Kimes, A.S., Stratton, M., London, E.D., 1999. Radiosynthesis and preliminary evaluation of 5-[$^{123/125}$I]iodo-2-(2(*S*)-azetidinylmethoxy)pyridine: a radioligand for nicotinic acetylcholine receptors. Nucl. Med. Biol. 26, 175–182.

Horti, A.G., Villemagne, V.L., 2006. The quest for Eldorado: development of radioligands for in vivo imaging of nicotinic acetylcholine receptors in human brain. Curr. Pharm. Des. 12, 3877–3900.

Horti, A.G., Wong, D.F., 2009. Clinical perspective and recent development of PET radioligands for imaging cerebral nicotinic acetylcholine receptors. PET Clin. 4, 89–100.

Hoyle, E., Genn, R.F., Fernandes, C., Stolerman, I.P., 2006. Impaired performance of α7 nicotinic receptor knockout mice in five-choice serial reaction time task. Psychopharmacology (Berl) 189, 211–223.

Van Huizen, F., March, D., Cynader, M.S., Shaw, C., 1994. Muscarinic receptor characteristics and regulation in rat cerebral cortex: changes during development, aging and the oestrous cycle. Eur. J. Neurosci. 6, 237–243.

Ince, E., Ciliax, B.J., Levey, A.I., 1997. Differential expression of D_1 and D_2 dopamine and m4 muscarinic acetylcholine receptor proteins in identified striatonigral neurons. Synapse 27, 357–366.

Ishikawa, M., Sakata, M., Toyohara, J., Oda, K., Ishii, K., Wu, J., Yoshida, T., Iyo, M., Ishiwata, K., and Hashimoto, K., 2011. Occupancy of α7 nicotinic acetylcholine receptors in the brain by tropisetron: a positron emission tomography study using [^{11}C]CHIBA-1001 in healthy human subjects. Clin. Psychopharmacol. Neurosci. 9, 111–116.

James, R.W., Bersinger, N.A., Schwendimann, B., Fulpius, W., 1980. Characterization of iodinated derivatives of a-bungarotoxin. Hoppe. Seylers Z. Physiol. Chem. 361, 1517–1524.

Kadir, A., Almkvist, O., Wall, A., Långström, B., Nordberg, A., 2006. PET imaging of cortical ^{11}C-nicotine binding correlates with the cognitive function of attention in Alzheimer's disease. Psychopharmacology (Berl) 188, 509–520.

Kadir, A., Darreh-Shori, T., Almkvist, O., Wall, A., Långström, B., Nordberg, A., 2007. Changes in brain ^{11}C-nicotine binding sites in patients with mild Alzheimer's disease following rivastigmine treatment as assessed by PET. Psychopharmacology (Berl.) 191, 1005–1014.

Kas, A., Bottlaender, M., Gallezot, J.D., Vidailhet, M., Villafane, G., Grégorie, C., Coulon, D., Valette, H., Dollé, F., Ribeiro, M.-J., Hantraye, H., Remy, P., 2009. Decrease of nicotinic receptors in the nigrostriatal system in Parkinson's disease. J. Cereb. Blood Flow Metab. 29, 1601–1608.

Katz, J.L., Izenwasser, S., Kline, R.H., Allen, A.C., Newman, A.H., 1999. Novel 3a-diphenylmethoxytropane analogs: selective dopamine uptake inhibitors with behavioral effects distinct from those of cocaine. J. Pharmacol. Exp. Ther. 288, 302–315.

Keller, J.J., Keller, A.B., Bowers, B.J., Wehner, J.M., 2005. Performance of α7 nicotinic receptor null mutants is impaired in appetitive learning measured in a signaled nose poke task. Behav. Brain Res. 162, 143–152.

Kellar, K.J., Whitehouse, P.J., Martino-Barrows, A.M., Marcus, K., Price, D.L., 1987. Muscarinic and nicotinic cholinergic binding sites in Alzheimer's disease and cerebral cortex. Brain Res. 436, 62–68.

Kemp, P.M., Holmes, C., Hoffmann, S., Wilkinson, S., Zivanovic, M., Thom, J., Bolt, L., Fleming, J., Wilkinson, D.G., 2003. A randomized placebo controlled study to assess the effects of cholinergic treatment on muscarinic receptors in Alzheimer's disease. J. Neurol. Neurosurg. Psychiatry 74, 1567–1570.

Kendziorra, K., Wolf, H., Meyer, P.M., Barthel, H., Hesse, S., Becker, G.A., Luthardt, J., Schildan, A., Patt, M., Sorger, D., Seese, A., Gerts, H.-J., Sabri, O., 2011. Decreased cerebral α4β2* nicotinic acetylcholine receptor availability in patients with mild cognitive impairment and Alzheimer's disease assessed with positron emission tomography. Eur. J. Nucl. Med. Mol. Imaging 38, 515–525.

Kiesewetter, D.O., Lee, J., Lang, L., Park, S.G., Paik, C.H., Eckelman, W.C., 1995. Preparation of ^{18}F-labeled muscarinic agonist with M_2 selectivity. J. Med. Chem. 38, 5–8.

Kikuchi, T., Okamura, T., Fukushi, K., Takahashi, K., Toyohara, J., Okada, M., Zhang, M.-R., Irie, T., 2007. Cerebral acetylcholinesterase imaging: development of radioprobes. Curr. Top. Med. Chem. 7, 1790–1799.

Kimes, A.S., Horti, A.G., London, E.D., Chefer, S.I., Contoreggi, C., Emst, M., Friello, P., Koren, A.O., Kurian, V., Matochik, J.A., Pavlova, O., Vaupel, D.B., Mukhin, A.G., 2003. 2-[^{18}F]F-A-85380: PET imaging of brain nicotinic acetylcholine receptors and whole body distribution in humans. FASEB. J. 17, 1331–1333.

Koeppe, R.A., Frey, K.A., Mulholland, G.K., Kilnourn, M.R., Buck, A., Lee, K.S., Kuhl, D.E., 1994. [^{11}C] Tropanyl benzilate-binding to muscarinic cholinergic receptors: methodology and kinetic modeling alternatives. J. Cereb. Blood Flow Metab. 14, 85–99.

Koike, K., Hashimoto, K., Takai, N., Shimizu, E., Komatsu, N., Watanabe, H., Nakazato, M., Okamura, N., Stevens, K.F., Freedman, R., Iyo, M., 2005. Tropisetron improves deficits in auditory P50 suppression in schizophrenia. Schizophr. Res. 76, 67–72.

Kulak, J.M., Carroll, F.I., Schneider, J.S., 2006. [^{125}I]iodomethyllycaconitine binds to α7 nicotinic acetylcholine receptors in monkey brain. Eur. J. Neurosci. 23, 2604–2610.

Kulak, J.M., Schneider, J.S., 2004. Differences in α7 nicotinic acetylcholine receptor binding in motor symptomatic and asymptomatic MPTP-treated monkeys. Brain Res. 999, 193–202.

Lane, R.M., Potkin, S.G., Enz, A., 2006. Targeting acetylcholinesterase and butylcholinesterase in dementia. Int. J. Neuropsychopharmacol. 9, 101–124.

Langmead, C.J., Watson, J., Reavill, C., 2008. Muscarinic acetylcholine receptors as CNS drug targets. Pharmacol. Ther. 117, 232–243.

Lavalaye, J., Booji, J., Linszen, D.H., Reneman, L., van Royen, E.A., 2001. Higher occupancy of muscarinic receptors by olanzapine than resperidone in patients with schizophrenia. A [^{123}I]-IDX SPECT study. Psychopharmacology 156, 53–57.

Lee, K.S., Frey, K.A., Koeppe, R.A., Buck, A., Mulholland, K., Kuhl, D.E., 1996. In vivo quantification of cerebral muscarinic receptors in normal human aging using positron emission tomography and [^{11}C] tropanyl benzilate. J. Cereb. Blood Flow Metab. 16, 303–310.

Lee, M., Martin-Ruiz, C., Graham, A., Court, J., Jaros, E., Perry, R., Iversen, P., Bauman, M., Perry, E., 2002. Nicotinic receptor abnormalities in the cerebellar cortex in autism. Brain 125, 1483–1495.

Levin, E.D., McClernon, F.J., Rezvani, A.H., 2006. Nicotinic effects on cognitive function: behavioral characterization, pharmacological specification, and anatomic localization. Psychopharmacology (Berl.) 184, 523–539.

Levy, A.I., Kitt, C.A., Simonda, W.F., Price, D.L., Brann, M.R., 1991. Identification and localization of muscarinic receptor proteins in brain with subtype-specific antibodies. J. Neurosci. 11, 3218–3226.

London, E.D., Ball, M.J., Waller, S.B., 1989. Nicotinic binding sites in cerebral cortex and hippocampus in Alzheimer's dementia. Neurochem. Res. 14, 745–750.

Lucas-Meunier, E., Fossier, P., Baux, G., Amar, M., 2003. Cholinergic modulation of the cortical neuronal network. Pflugers Arch. 446, 17–29.

Lukas, R.J., Changeux, J.P., Le Novere, N., Albuquerque, E.X., Balfour, D.J., Berg, D.K., Bertrand, D., Chiappinelli, V.A., Clarke, P.B., Collins, A.C., Dani, J.A., Grady, S.R., Kellar, K.J., Lindstrom, J.M., Marks, M.J., Quik, M., Taylor, P.W., Wonnacott, S., 1999. International union of pharmacology. XX. Current status of the nomenclature for nicotinic acetylcholine receptors and their subunits. Pharmacol. Rev. 51, 397–401.

Macor, J.E., Gurley, D., Lanthorn, T., Loch, J., Mack, R.A., Mullen, G., Tran, O., Wright, N., Gordon, J.C., 2001. The 5-HT_3 antagonist tropisetron (ICS 205-930) is a potent and selective α7 nicotinic receptor partial agonist. Bioorg. Med. Chem. Lett. 11, 319–321.

Mamede, M., Ishizu, K., Ueda, M., Mukai, T., Iida, Y., Fukuyama, H., Saga, T., Saji, H., 2004. Quantification of human nicotinic acetylcholine receptors with ^{123}I-5IA SPECT. J. Nucl. Med. 45, 1458–1470.

Mamede, M., Ishizu, K., Ueda, M., Mukai, T., Iida, Y., Kawashima, H., Fukuyama, H., Togashi, K., Saji, H., 2007. Temporal changes in human nicotinic acetylcholine receptor after smoking cessation: 5IA SPECT study. J. Nucl. Med. 48, 1829–1835.

Maricq, A.V., Peterson, A.S., Brake, A.J., Myers, R.M., Julius, D., 1991. Primary structure and functional expression of the 5-HT_3 receptor, a serotonin-gated ion channel. Science 254, 432–437.

Martin-Ruiz, C.M., Piggott, M., Gotti, C., Lindstrom, J., Mendelow, A.D., Siddique, M.S., Perry, R.H., Perry, E.K., Court, J.A., 2000. Alpha and beta nicotinic acetylcholine receptors subunits and synaptophysin in putamen from Parkinson's disease. Neuropharmacology 39, 2830–2839.

Marutle, A., Warpman, U., Bogdanovic, N., Lannfelt, L., Nordberg, A., 1999. Neuronal nicotinic receptor deficits in Alzheimer patients with Swedish amyloid precursor protein 670/671 mutation. J. Neurochem. 72, 1161–1169.

Marutle, A., Warpman, U., Bogdanovic, N., Nordberg, A., 1998. Regional distribution of subtypes of nicotinic receptors in human brain and effect of aging studied by (+)-[^{3}H]epibatidine. Brain Res. 801, 143–149.

Marutle, A., Zhang, X., Court, J., Piggot, M., Johnson, M., Perry, R., Perry, E., Nordberg, A., 2001. Laminar distribution of nicotinic receptor subtypes in cortical regions in schizophrenia. J. Chem. Neuroanat. 22, 115–126.

Matsumura, K., Uno, Y., Scheffel, U., Wilson, A.A., Dannals, R.F., Wagner Jr, H.N., 1991. In vitro and in vivo characterization of 4-[^{125}I]iododexetimide binding to muscarinic cholinergic receptors in the rat heart. J. Nucl. Med. 32, 76–80.

McPherson, D.W., Lambert, C.R., Jahn, K., Sood, V., McRee, R.C., Zeeberg, B., Reba, R.C., Knapp Jr, F.F., 1995. Resolution and in vitro and initial in vivo evaluation of isomers of iodine-125-labeled 1-azabicyclo [2.2.2]oct-3-yl a-hydroxyl-a-(1-iodo-1-propen-3-yl)-a-phenylacetate: a high-affinity ligand for the muscarinic receptor. J. Med. Chem. 38, 3908–3917.

Mesulam, M.M., 2004. The cholinergic innervations of the human cerebral cortex. Prog. Brain Res. 145, 67–78.

Messer Jr, W.S., Bohnett, M., Stibbe, J., 1990. Evidence for a preferential involvement of M_1 muscarinic receptors in representational memory. Neurosci. Lett. 116, 184–189.

Meyer, P.M., Strecker, K., Kendziorra, K., Becker, G., Hesse, S., Woelpl, D., Hensel, A., Patt, M., Sorger, D., Wegner, F., Lobsien, D., Barthel, H., Burst, P., Gertz, H.J., Sabri, O., Schwarz, J., 2009. Reduced $\alpha4\beta2^*$-nicotinic acetylcholine receptor binding and its relationship to mild cognitive and depressive symptoms in Parkinson disease. Arch. Gen. Psychiatry 66, 866–877.

Mihailescu, S., Drucker-Colin, R., 2000. Nicotine, brain nicotinic receptors, and neuropsychiatric disorders. Arch. Med. Res. 31, 131–144.

Mitkovski, S., Villemagne, V.L., Novakovic, K.E., O'Keefe, G., Tochon-Danguy, H., Mulligan, R.S., Dickinson, K.L., Saunder, T., Gregoire, M.C., Bottlaender, M., Dolle, F., Rowe, C.C., 2005. Simplified quantification of nicotinic receptors with 2[^{18}F]F-A-85380 PET. Nucl. Med. Biol. 32, 585–591.

Mitsis, E.M., Cosgrove, K.P., Staley, J.K., Bois, F., Frohlich, E.B., Tamagnan, G.D., Estok, K.M., Seibyl, J.P., van Dyck, C.H., 2009a. Age-related decline in nicotinic receptor availability with [^{123}I]5-IA-85380 SPECT. Neurobiol. Aging 30, 1490–1497.

Mitsis, E.M., Cosgrove, K.P., Staley, J.K., Frohlich, E.B., Bois, F., Tamagnan, G.D., Estok, K.M., Seibyl, J.P., van Dyck, C.H., 2007. [^{123}I]5-IA-85380 SPECT imaging of β2-nicotinic acetylcholine receptor availability in the aging human brain. Ann. N.Y. Acad. Sci. 1097, 168–170.

Mitsis, E.M., Reech, K.M., Bois, F., Tamagnan, G.D., MacAvoy, M.G., Seibyl, J.P., Staley, J.K., van Dyck, C.H., 2009b. ^{123}I-5-IA-85380 SPECT imaging of nicotinic receptors in Alzheimer disease and mild cognitive impairment. J. Nucl. Med. 50, 1455–1463.

Mohamed, A., Eberl, S., Fulham, M.J., Kassiou, M., Zaman, A., Henderson, D., Beveridge, S., Constable, C., Lo, S.K., 2005. Sequential ^{123}I-iododexetimide scans in temporal lobe epilepsy: comparison with neuroimaging scans (MR imaging and ^{18}F-FDG PET imaging). Eur. J. Nucl. Med. Mol. Imaging 32, 180–185.

Mukhin, A.G., Kimes, A.S., Chefer, S.I., Matochik, J.A., Contoreggi, C.S., Horti, A.G., Vaupel, D.B., Pavlova, O., Stein, E.A., 2008. Greater nicotinic acetylcholine receptor density in smokers than in nonsmokers: a PET study with 2-^{18}F-FA-85380. J. Nucl. Med. 49, 1628–1635.

Mulholland, G.K., Otto, C.A., Jewett, D.M., Kilbourn, M.R., Koeppe, R.A., Sherman, P.S., Petry, N.A., Carey, J.E., Atkinson, E.R., Archer, S., Frey, K.A., Kuhl, D.E., 1992. Synthesis, rodent biodistribution, dosimetry, metabolism, and monkey images of carbon-11-labeled (+)-2a-tropanyl benzilate: a central muscarinic receptor imaging agent. J. Nucl. Med. 33, 423–430.

Müller-Gärtner, H.W., Mayberg, H.S., Fisher, R.S., Lesser, R.P., Wilson, A.A., Ravert, H.T., Dannals, R.F., Wagner Jr., H.N., Uematsu, S., Frost, J.J., 1993. Decreased hippocampal muscarinic cholinergic receptor binding measured by ^{123}I-iododexetimide and single-photon emission computed tomography in epilepsy. Ann. Neurol. 34, 235–238.

Müller-Gärtner, H.W., Wilson, A.A., Dannals, R.F., Wagner Jr., H.N., Frost, J.J., 1992. Imaging muscarinic cholinergic receptors in human brain in vivo with SPECT, [^{123}I]4-iododexetimide, and [^{123}I]4-iodolevetimide. J. Cereb. Blood Flow Metab. 12, 562–570.

Musachio, J.L., Villemagne, V.L., Scheffel, U., Stathis, M., Finley, P., Horti, A., London, E.D., Dannals, R.F., 1997. [$^{125/123}$I]IPH: a radioiodinated analog of epibatidine for in vivo studies of nicotinic acetylcholine receptors. Synapse 26, 392–399.

Muzic Jr., R.F., Berridge, M.S., Friedland, R.P., Zhu, N., Nelson, A.D., 1998. PET quantification of specific binding of carbon-11-nicotine in human brain. J. Nucl. Med. 39, 2048–2054.

Navarro, H.A., Zhong, D., Abraham, P., Xu, H., Carroll, F.I., 2000. Synthesis and pharmacological characterization of [^{125}I]iodomethyllycaconitine ([^{125}I]iodo-MLA). A new ligand for the α7 nicotinic acetylcholine receptor. J. Med. Chem. 43, 142–145.

Nobuhara, K., Farde, L., Halldin, C., Karlsson, P., Swahn, C.G., Olsson, H., Bergström, K.A., Larsson, S.A., Schnell, P.O., McPherson, D.W., Savonen, A., Hiltunen, J., Sedvall, G., 2001. SPET imaging of central muscarinic acetylcholine receptors with iosine-123 labelled *E*-IQNP and *Z*-IQNP. Eur. J. Nucl. Med. 28, 13–24.

Nomikos, G.G., Schilström, B., Hildebrand, B.E., Panagis, G., Grenhoff, J., Svensson, T.H., 2000. Role of α7 nicotinic receptors in nicotine dependence and implications for psychiatric illness. Behav. Brain Res. 113, 97–103.

Norbury, R., Travis, M.J., Erlandsson, K., Waddington, W., Ell, P.J., Murphy, D.G., 2007. Estrogen therapy and brain muscarinic receptor density in healthy females: a SPET study. Horm. Behav. 51, 249–257.

Norbury, R., Travis, M.J., Erlandsson, K., Waddington, W., Owens, J., Ell, P.J., Murphy, D.G., 2004. SPET imaging of central muscarinic receptors with (*R, R*)[^{123}I]-I-QNB: methodological consideration. Nucl. Med. Biol. 31, 583–590.

Norbury, R., Travis, M.J., Erlandsson, K., Waddington, W., Owens, J., Pimlott, S., Ell, P.J., Murphy, D.G., 2005. In vivo imaging of muscarinic receptors in the aging female brain with (*R, R*)[^{123}I]-I-QNB and single photon emission tomography. Exp. Gerontol. 40, 137–145.

Nordberg, A., 2001. Nicotinic receptor abnormalities of Alzheimer's disease: therapeutic implications. Biol. Psychiatry 49, 200–210.

Nordberg, A., Alafuzoff, I., Winblad, B., 1992. Nicotinic and muscarinic subtypes in the human brain: changes with aging and dementia. J. Neurosci. Res. 31, 103–111.

Nordberg, A., Hartvig, P., Lilja, A., Viitanen, M., Amberla, K., Lundqvist, H., Andersson, Y., Ulin, J., Winblad, B., Långström, B., 1990. Decreased uptake and binding of ^{11}C-nicotine in brain of Alzheimer patients as visualized by positron emission tomography. J. Neural Transm. Park. Dis. Dement. Sect. 2, 215–224.

Nordberg, A., Larsson, C., Perdahl, E., Winbland, B., 1982. Cholinergic activity in hippocampus in chronic alcoholism. Drug Alcohol Depend. 10, 333–344.

Nordberg, A., Lundqvist, H., Harting, P., Andersson, A., Johansson, M., Hellström-Lindhai, E., Långström, B., 1997. Imaging of nicotinic and muscarinic receptors in Alzheimer's disease: effect of tacrine treatment. Dement. Geriatr. Cogn. Disord. 8, 78–84.

Nordberg, A., Lundqvist, H., Hartvig, P., Lilja, A., Långström, B., 1995. Kinetic analysis of regional (*S*) (-)^{11}C-nicotine binding in normal and Alzheimer brains-in vivo assessment using positron emission tomography. Alzheimer Dis. Assoc. Disord. 9, 21–27.

Nordberg, A., Winbland, B., 1981. Cholinergic receptors in human hippocampus—regional distribution and variance with age. Life Sci. 29, 1937–1944.

Nordberg, A., Winblad, B., 1986. Reduced number of [^{3}H]nicotine and [^{3}H]acetylcholine binding sites in the frontal cortex of Alzheimer's brains. Neurosci. Lett. 72, 115–119.

Le Novère, N., Changeux, J.P., 1995. Molecular evolution of the nicotinic acetylcholine receptor: an example of multigene family in excitable cells. J. Mol. Evol. 40, 155–172.

Nybäck, H., Halldin, C., Ahlin, A., Curvall, M., Eriksson, L., 1994. PET studies of the uptake of (*S*)- and (*R*)-[^{11}C]nicotine in the human brain: difficulties in visualizing specific receptor binding in vivo. Psychopharmacology (Berl.) 115, 31–36.

O'Brien, J.T., Colloby, S.J., Pakrasi, S., Perry, E.K., Pimlott, S.L., Wyper, D.J., McKeith, I.G., Williams, E.D., 2007. $\alpha4\beta2$ nicotinic receptor status in Alzheimer's disease using ^{123}I-5IA-85380 single-photon-emission computed tomography. J. Neurol. Neurosurg. Psychiatry 78, 356–362.

O'Brien, J.T., Colloby, S.J., Pakrasi, S., Perry, E.K., Pimlott, S.L., Wyper, D.J., McKeith, I.G., Williams, E.D., 2008. Nicotinic $\alpha4\beta2$ receptor binding in dementia with Lewy bodies using ^{123}I-5IA-85380 SPECT demonstrates a link between occipital changes and visual hallucinations. Neuroimage 15, 1056–1063.

Ochoa, E.L., Lasalde-Dominicci, J., 2007. Cognitive deficits in schizophrenia: focus on neuronal nicotinic acetylcholine receptors and smoking. Cell Mol. Neurobiol. 27, 609–639.

Oda, Y., 1999. Choline acetyltransferase: the structure, distribution and pathologic changes in the central nervous system. Pathol. Int. 49, 921–937.

Ogawa, M., Nishiyama, S., Tsukada, H., Hatano, K., Fuchigami, T., Yamaguchi, H., Matsushima, Y., Ito, K., Magata, Y., 2010. Synthesis and evaluation of new imaging agent for central nicotinic acetylcholine receptor $\alpha7$ subtype. Nucl. Med. Biol. 37, 347–355.

Ogawa, M., Tsukada, H., Nishiyama, S., Oba, H., Kakiuchi, T., Harada, N., Magata, T., 2011. Quantification of central nicotinic acetylcholine receptor $\alpha7$ subtype with [^{11}C](*R*)MeQAA in the monkey brain. J. Label. Compd. Radiopharm. 54, S253.

Oki, T., Takagi, Y., Inagaki, S., Taketo, M.M., Manabe, T., Matsui, M., Yamada, S., 2005. Quantitative analysis of binding parameters of [^{3}H]N-methylscopolamine in central nervous system of muscarinic acetylcholine receptor knockout mice. Brain Res. Mol. Brain Res. 133, 6–11.

Orr-Urtreger, A., Goldner, F.M., Saeki, M., Lorenzo, I., Goldberg, L., De Biasi, M., Dani, J.A., Patrick, J.W., Beaudet, A.L., 1997. Mice deficient in the $\alpha7$ neuronal nicotinic acetylcholine receptor lack a-bungarotoxin binding sites and hippocampal fast nicotinic currents. J. Neurosci. 17, 9165–9171.

Ortells, M.O., Lunt, G.G., 1995. Evolutionary history of the ligand-gated ion-channel superfamily of receptors. Trends Neurosci. 18, 121–127.

Otto, C.A., Multholland, G.K., Perry, S.E., Combs, R., Sherman, P.S., Fisher, S.J., 1989. In vitro and ex vivo evaluation of cyclic aminoalkyl benzilates as potential emission tomography ligands for the muscarinic receptor. Int. J. Radiat. Appl. Instrum. B. 16, 51–55.

Pakrasi, S., Colloby, S.J., Firbank, M.J., Perry, E.K., Wyper, D.J., Owens, J., McKeith, I.G., Williams, E.D., O'Brien, J.T., 2007. Muscarinic acetylcholine receptor status in Alzheimer's disease assessed using (*R, R*) ^{123}I-QNB SPECT. J. Neurol. 254, 907–913.

Papke, R.L., Porter Papke, J.K., Rose, G.M., 2004. Activity of $\alpha7$-selective agonists at nicotinic and serotonin 5-HT_3 receptors expressed in *Xenopus* oocytes. Bioorg. Med. Chem. Lett. 14, 1849–1853.

Paterson, D., Nordberg, A., 2000. Neuronal nicotinic receptors in the human brain. Prog. Neurobiol. 61, 75–111.

Paylor, R., Nquyen, M., Crawley, J.N., Patrick, J., Beaudet, A., Orr-Urtreger, A., 1998. $\alpha7$ Nicotinic receptor subunits are not necessary for hippocampal-dependent learning or sensorimotor gating: a behavioral characterization of *Acra7*-deficient mice. Learn. Mem. 5, 302–316.

Perez, X.A., Bordia, T., McIntosh, J.M., Quik, M., 2010. α6β2* and α4β2* nicotinic receptors both regulate dopamine signaling with increased nigrostriatal damage: relevance to Parkinson's disease. Mol. Pharmacol. 78, 971–980.

Perry, D.C., Dávila-García, M.I., Stockmeier, C.A., Kellar, K.J., 1999a. Increased nicotinic receptors in brains from smokers: membrane binding and autoradiography studies. J. Pharmacol. Exp. Ther. 289, 1545–1552.

Perry, E.K., Kerwin, J., Perry, R.H., Irving, D., Blesses, G., Fairbairn, A., 1990a. Cerebral cholinergic activity is related to the incidence of visual hallucinations in senile dementia of Lewy body type. Dementia 1, 2–4.

Perry, E.K., Marshall, E., Perry, R.H., Irving, D., Smith, C.J., Blesses, G., Fairbairn, A.F., 1990b. Cholinergic and dopaminergic activities in senile dementia of Lewy body type. Alzheimer Dis. Accos. Disord. 4, 87–95.

Perry, E.K., Morris, C.M., Court, J.A., Chang, A., Fairbairn, A.F., McKeith, I.G., Irving, D., Brown, A., Perry, R.H., 1995. Alteration in nicotinic binding sites in Parkinson's disease, Lewy body dementia and Alzheimer's disease: possible index of early neuropathology. Neuroscience 64, 385–395.

Perry, E.K., Piggott, M.A., Court, J.A., Johnson, M., Perry, R.H., 1993. Transmitters in the developing and senescent human brain. Ann. N.Y. Acad. Sci. 695, 69–72.

Perry, E.K., Smith, C.J., Perry, R.H., Whitford, C., Johnston, M., Birdsall, N.J., 1989. Regional distribution of muscarinic and nicotinic cholinergic receptor binding activities in the human brain. J. Chem. Neuroanat. 2, 189–199.

Perry, E., Walker, M., Grace, J., Perry, R., 1999b. Acetylcholine in mind: a neurotransmitter correlate of consciousness? Trends Neurosci. 22, 273–280.

Picard, F., Bruel, D., Servent, D., Saba, W., Fruchart-Gaillard, C., Schöihorn-Peyronneau, M.-A., Roumenov, D., Brodtkorb, E., Zuberi, S., Gambardella, A., Steinborn, B., Hufnagel, A., Valette, H., Bottlaender, M., 2006. Alteration on the in vivo nicotinic receptor density in ADNFLE patients: a ET study. Brain 129, 2047–2060.

Picciotto, M.R., Corrigall, W.A., 2002. Neuronal systems underlying behaviors related to nicotine addiction: neural circuits and molecular genetics. J. Neurosci. 22, 3338–3341.

Picciotto, M.R., Zoli, M., Lena, C., Bessis, A., Lallemand, Y., Le Novere, N., Vincent, P., Pich, E.M., Brulet, P., Changeux, J.P., 1995. Abnormal avoidance learning in mice lacking functional high-affinity nicotine receptor in the brain. Nature 374, 65–67.

Piggott, M.A., Owens, J., O'Brien, J., Colloby, S., Fenwick, J., Wyper, D., Jaros, E., Johnson, M., Perry, R.H., Perry, E.K., 2003. Muscarinic receptors in basal ganglia in dementia with Lewy bodies, Parkinson's disease and Alzheimer's disease. J. Chem. Neuroanat. 25, 161–173.

Piggott, M., Owens, J., O'Brien, J., Paling, S., Wyper, D., Fenwick, J., Johnson, M., Perry, R., Perry, E., 2002. Comparative distribution of binding of the muscarinic receptor ligands pirenzepine, AF-DX 384, (R, R)-I-QNB and (R, S)-I-QNB to human brain. J. Chem. Neuroanat. 24, 211–223.

Pimlott, S.L., Piggott, M., Owens, J., Greally, E., Court, J.A., Jaros, E., Perry, R.H., Perry, E.K., Wyper, D., 2004. Nicotinic acetylcholine receptor distribution in Alzheimer's disease, dementia with Lewy bodies, Parkinson's disease, and vascular dementia: in vitro binding study using 5-[^{125}I]-A-85380. Neuropsychopharmacology (Berl.) 29, 108–116.

Podruchny, T.A., Connolly, C., Bokde, A., Herscovitch, P., Eckelman, W.C., Kiesewetter, D.O., Sunderland, T., Carson, R.E., Cohen, R.M., 2003. In vivo muscarinic 2 receptor imaging in cognitively normal young and older volunteers. Synapse 48, 39–44.

Quirion, R., 1993. Cholinergic markers in Alzheimer's disease and the autoregulation of acetylcholine release. J. Psychiatr. Neurosci. 18, 226–234.

Rabinovici, G.D., Jagust, W.J., 2009. Amyloid imaging in aging and dementia: testing the amyloid hypothesis in vivo. Behav. Neurol. 21, 117–128.

Raedler, T.J., 2007. Comparison of the in-vivo muscarinic cholinergic receptor availability in patients treated with clozapine and olanzapine. Int. J. Neuropsychopharmacol. 10, 275–280.

Raedler, T.J., Knable, M.B., Jones, D.W., Lafargue, T., Urbina, R.A., Egan, M.F., Pickar, D., Weinberger, D.R., 2000. In vivo olanzapine occupancy of muscarinic acetylcholine receptors in patients with schizophrenia. Neuropsychopharmacology 23, 56–68.

Raedler, T.J., Knable, M.B., Jones, D.W., Urbina, R.A., Egan, M.F., Weinberger, D.R., 2003a. Central muscarinic acetylcholine receptor availability in patients treated with clozapine. Neuropsychopharmacology 28, 1531–1537.

Raedler, T.J., Knable, M.B., Jones, D.W., Urbina, R.A., Gorey, J.G., Lee, K.S., Egan, M.F., Coppola, R., Weinberger, D.R., 2003b. In vivo determination of muscarinic acetylcholine receptor availability in schizophrenia. Am. J. Psychiatry 160, 118–127.

Reever, C.M., Ferrari-DiLeo, G., Flynn, D.D., 1997. The M_5 receptor subtype: fact or fiction? Life Sci. 60, 1105–1112.

Roberson, M.R., Harrell, L.E., 1997. Cholinergic activity and amyloid precursor protein metabolism. Brain Res. Brain Res. Rev. 25, 50–69.

Rodriguez-Puertas, R., Pascul, J., Vilaro, T., Pazos, A., 1997. Autoradiographic distribution of M_1, M_2, M_3, and M_4 muscarinic receptor subtypes in Alzheimer's disease. Synapse 26, 341–350.

Romanelli, M.N., Gratteri, P., Guandalini, L., Martini, E., Bonaccini, C., Gualtieri, F., 2007. Central nicotinic receptor: structure, function, ligands, and therapeutic potential. Chem. Med. Chem. 2, 746–767.

Santiago, M.P., Potter, L.T., 2001. Biotinylated m4-toxin demonstrates more M_4 muscarinic receptor protein on direct than indirect striatal projection neurons. Brain Res. 894, 12–20.

Scarr, E., Dean, B., 2008. Muscarinic receptor: do they have a role on the pathology and treatment of schizophrenia? J. Neurochem. 107, 1188–1195.

Scheffel, U., Horti, A.G., Koren, A.O., Ravert, H.T., Banta, J.P., Finley, P.A., London, E.D., Dannals, R.F., 2000. 6-[^{18}F]Fluoro-A85380: an in vivo tracer for the nicotinic acetylcholine receptor. Nucl. Med. Biol. 27, 51–56.

Seeger, T., Fedorova, I., Zheng, F., Miyakawa, T., Koustova, E., Gomeza, J., Basile, A.S., Alzheimer, C., Wess, J., 2004. M_2 muscarinic acetylcholine receptor knock-out mice show deficits in behavioral flexibility, working memory, and hippocampal plasticity. J. Neurosci. 24, 10117–10127.

Shiina, A., Shirayama, Y., Niitsu, T., Hashimoto, T., Yoshida, T., Hasegawa, T., Haraguchi, T., Kanahara, N., Shiraishi, T., Fujisaki, M., Fukami, G., Nakazato, M., Iyo, M., Hashimoto, K., 2010. A randomised, double-blind, placebo-controlled trial of tropisetron in patients with schizophrenia. Ann. Gen. Psychiatry 9, 27.

Shinotoh, H., Asahina, M., Inoue, O., Suhara, T., Hirayama, K., Tateno, Y., 1994. Effects of trihexyphenidyl and L-dopa on brain muscarinic cholinergic receptor binding measured by positron emission tomography. J. Neural Transm. Park. Dis. Dement. Sect. 7, 35–46.

Sihver, E., Gillberg, P.G., Nordberg, A., 1998. Laminar distribution of nicotinic receptor subtypes in human cerebral cortex as determined by [^{3}H](-)nicotine, [^{3}H]cyticine and [^{3}H]epibatidine in vitro autoradiography. Neuroscience 85, 1121–1133.

Sihver, W., Nordberg, A., Langstrom, B., Mukhin, A.G., Koren, A.O., Kimes, A.S., London, E.D., 2000. Development of ligands for in vivo imaging of cerebral nicotinic receptors. Behav. Brain Res. 113, 143–157.

Smythies, J., 2005. Section I. The cholinergic system. Int. Rev. Neurobiol. 64, 1–122.

Staley, J.K., Krishnan-Sarin, S., Cosgrove, K.P., Krantzler, E., Frohlich, E., Perry, E., Dubin, J.A., Estok, K., Brenner, E., Baldwin, R.M., Tamagnan, G.D., Seibyl, J.P., Jatlow, P., Picciotto, M.R., London, E.D., O'Malley, S., van Dyck, C.H., 2006. Human tobacco smokers in early abstinence have higher levels of β2*nicotinic acetylcholine receptors than nonsmokers. J. Neurosci. 26, 8707–8714.

Staley, J.K., van Dyck, C.H., Weinzimmer, D., Brenner, E., Baldwin, R.M., Tamagnan, G.D., Riccardi, P., Mitsis, E., Seibyl, J.P., 2005. ^{123}I-5IA-85380 SPECT measurement of nicotinic acetylcholine receptors in human brain by the constant infusion paradigm: feasibility and reproducibility. J. Nucl. Med. 46, 1466–1472.

Sudo, Y., Suhara, T., Honda, Y., Nakajima, T., Okubo, Y., Suzuki, K., Nakashima, Y., Yoshikawa, K., Okauchi, T., Sasaki, Y., Matsushita, M., 1998. Muscarinic cholinergic receptors in human narcolepsy: a PET study. Neurology 51, 1297–1302.

Sudo, Y., Suhara, T., Suzuki, K., Okubo, Y., Yoshikawa, K., Uchida, S., Sassa, T., Okauchi, T., Sasaki, Y., Matsushita, M., 1999. Muscarinic receptor occupancy by biperiden in living human brain. Life Sci. 64, L99–L104.

Suhara, T., Inoue, O., Kobayashi, K., Suzuki, K., Tateno, Y., 1993. Age-related changes in human muscarinic acetylcholine receptors measured by positron emission tomography. Neurosci. Lett. 149, 225–228.

Tanibuchi, Y., Wu, J., Toyohara, J., Fujita, Y., Iyo, M., Hashimoto, K., 2010. Characterization of [^{3}H]CHIBA-1001 binding to α7 nicotinic acetylcholine receptors in the brain from rat, monkey, and human. Brain Res. 1348, 200–208.

Terrière, E., Dempsey, M.F., Hermann, L.L., Tierney, K.M., Lonie, J.A., O'Carroll, R.E., Pimlott, S., Wyper, D.J., Herholz, K., Ebmeier, K.P., 2010. 5-123I-A-85380 binding to the $\alpha4\beta2$-nicotinic receptor in mild cognitive impairment. Neurobiol. Aging 31, 1885–1893.

Toyohara, J., Hashimoto, K., 2010. $\alpha7$ Nicotinic receptor agonists: potential therapeutic drugs for treatment of cognitive impairments in schizophrenia and Alzheimer's disease. Open Med. Chem. J. 4, 37–56.

Toyohara, J., Ishiwata, K., Sakata, M., Wu, J., Nishiyama, S., Tsukada, H., Hashimoto, K., 2010a. In vivo evaluation of $\alpha7$ nicotinic acetylcholine receptor agonists [^{11}C]A-582941 and [^{11}C]A-844606 in mice and conscious monkeys. PLos One 5, e8961.

Toyohara, J., Sakata, M., Wu, J., Ishikawa, M., Oda, K., Ishii, K., Iyo, M., Hashimoto, K., Ishiwata, K., 2009. Preclinical and the first clinical studies on [^{11}C]CHIBA-1001 for mapping $\alpha7$ nicotinic receptors by positron emission tomography. Ann. Nucl. Med. 23, 301–309.

Toyohara, J., Wu, J., Hashimoto, K., 2010b. Recent development of radioligands for imaging $\alpha7$ nicotinic acetylcholine receptors in the brain. Curr. Top. Med. Chem. 10, 1544–1557.

Tzavara, E.T., Bymaster, F.P., Davis, R.J., Wade, M.R., Perry, K.M., Wess, J., McKinzie, D.L., Felder, C., Nomikos, G.G., 2004. M_4 muscarinic receptors regulate the dynamics of cholinergic and dopaminergic neurotransmission: relevance to the pathophysiology and treatment of related CNS pathologies. FASEB. J. 18, 1410–1412.

Tazavara, E.T., Bymaster, F.P., Felder, C.C., Wade, M., Gomeza, J., Wess, J., Mckinzie, D.L., Nomikos, G.G., 2003. Dysregulated hippocampal acetylcholine neurotransmission and impaired cognition in M_2, M_4, and M_2/M_4 muscarinic receptor knockout mice. Mol. Psychiatry 8, 673–679.

Utsugisawa, K., Nagane, Y., Tohgi, H., Yoshimura, M., Ohba, H., Genda, Y., 1999. Changes with aging and ischemia in nicotinic acetylcholine receptor subunit $\alpha7$ mRNA expression in postmortem human frontal cortex and putamen. Neurosci. Lett. 270, 145–148.

Villemagne, V.L., Horti, A., Scheffel, U., Ravert, H.T., Finley, P., Clough, D.J., London, E.D., Wagner Jr., H.N., Dannals, R.F., 1997. Imaging nicotinic acetylcholine receptors with fluorine-18-FPH, an epibatidine analog. J. Nucl. Med. 38, 1737–1741.

Wall, S.J., Yasuda, R.P., Li, M., Ciesla, A., Wolf, B.B., 1992. The ontogeny of M_1–M_5 muscarinic receptor subtypes in rat forebrain. Dev. Brain Res. 66, 181–185.

Warpman, U., Nordberg, A., 1995. Epibatidine and ABT 418 reveal selective loss of $\alpha4\beta2$ nicotinic receptors in Alzheimer brains. Neuroreport 6, 2419–2423.

Warren, N.M., Piggott, M.A., Lees, A.J., Burn, D.J., 2007. Muscarinic receptors in the thalamus in progressive supranuclear palsy and other neurodegenerative disorders. J. Neuropathol. Exp. Neurol. 66, 399–404.

Weckesser, M., Hufnagel, A., Ziemons, K., Griessmeier, M., Sonnenberg, F., Hackländer, T., Langen, K.J., Holschbach, M., Elger, C.E., Müller-Gärtner, H., 1997. Effect of partial volume correction on muscarinic cholinergic receptor imaging with single-photon emission tomography in patients with temporal lobe epilepsy. Eur. J. Nucl. Med. 24, 1156–1161.

Weinberger, D.R., Gibson, R., Coppola, R., Jones, D.W., Molchan, S., Sunderland, T., Berman, K.F., Reba, R.C., 1991. The distribution of cerebral muscarinic acetylcholine receptors in vivo in patients with dementia. A controlled study with ^{123}IQNB and single photon emission tomography. Arch. Neurol. 48, 169–176.

Weinberger, D.R., Jones, D.W., Sunderland, T., Lee, K.S., Sexton, R., Gorey, J., Reba, R., 1992. In vivo imaging of cerebral muscarinic receptors with I-123 QNB and SPECT: studies in normal subjects and patients with dementia. Clin. Neuropharmacol. 15 (Suppl. 1), 194A–195A.

Weinberger, D.R., Mann, U., Gibson, R.E., Coppola, R., Jones, D.W., Braun, A.R., Berman, K.F., Sunderland, T., Reba, R.C., Chase, T.N., 1990. Cerebral muscarinic receptors in primary degenerative dementia as evaluated by SPECT with iodine-123-labeled QNB. Adv. Neurol. 51, 147–150.

Weiner, M.W., Aisen, P.S., Jack Jr., C.R., Jagust, W.J., Trojanowski, J.Q., Shaw, L., Saykin, A.J., Morris, J.C., Cairns, N., Beckett, L.A., Toga, A., Green, R., Walter, S., Soares, H., Snyder, P., Siemers, E., Potter, W., Cole, P.E., Schmidt, M., Alzheimer's Disease Neuroimaging Initiative, 2010. The Alzheimer's disease neuroimaging initiative: progress report and future plans. Alzheimers Dement. 6, 202–211.

Wess, J., 2004. Muscarinic acetylcholine receptor knockout mice: novel phenotypes and clinical implications. Annu. Rev. Pharmacol. Toxicol. 44, 423–450.

Wess, J., Eglen, R.M., Gautam, D., 2007. Muscarinic acetylcholine receptors: mutant mice provide new insights for drug development. Nat. Rev. Drug Discov. 6, 721–733.

Wevers, A., Burghaus, L., Moser, N., Witter, B., Steinlein, O.K., Schütz, U., Achnitz, B., Krempel, U., Nowacki, S., Pilz, K., Stoodt, J., Lindstrom, J., De Vos, R.A., Jansen Steur, E.N., Schröder, H., 2000. Expression of nicotinic acetylcholine receptors in Alzheimer's disease: postmortem investigations and experimental approaches. Behav. Brain Res. 113, 207–215.

Whiteaker, P., Davis, A.R., Marks, M.J., Blagbrough, I.S., Potter, B.V., Wolstenholme, A.J., Collins, A.C., Wonnacott, S., 1999. An autoradiographic study of the distribution of binding sites for the novel α7-selective nicotinic radioligand [^{3}H]-methyllycaconitine in the mouse brain. Eur. J. Neurosci. 11, 2689–2696.

Whitehouse, P.J., Martino, A.M., Wagster, M.V., Price, D.L., Mayeux, R., Atack, J.R., Kellar, K.J., 1988. Reductions in [^{3}H]nicotinic acetylcholine binding in Alzheimer's disease and Parkinson's disease: an autoradiographic study. Neurology 38, 720–723.

Wüllner, U., Gündisch, D., Herzog, H., Minnerop, M., Joe, A., Warnecke, M., Jessen, F., Schütz, C., Reinhardt, M., Eschner, W., Klockgether, T., Schmaljohann, J., 2008. Smoking upregulates α4β2* nicotinic acetylcholine receptors in the human brain. Neurosci. Lett. 430, 34–37.

Wyper, D.J., Brown, D., Patterson, J., Owens, J., Hunter, R., Teasdale, E., McCulloch, J., 1993. Deficits in iodine-labelled 3-quinuclidinyl benzilate binding in relation to cerebral blood flow in patients with Alzheimer's disease. Eur. J. Nucl. Med. 20, 379–386.

Yamada, M., Miyakawa, T., Duttaroy, A., Yamanaka, A., Moriguchi, T., Makita, R., Mckinzie, D.L., Felder, C.C., Deng, C.X., Faraci, F.M., Wess, J., 2001. Mice lacking the M_3 muscarinic acetylcholine receptor are hypophagic and lean. Nature 410, 207–212.

Yasuda, R.P., Ciesla, W., Flores, L.R., Wall, S.J., Li, M., Satkus, S.A., Weisstein, J.S., Spagnola, B.V., Wolfe, B.B., 1993. Development of antisera selective for m4 and m5 muscarinic cholinergic receptors: distribution of m4 and m5 receptors in rat brain. Mol. Pharmacol. 43, 149–157.

Young, J.W., Crawford, N., Kelly, J.S., Kerr, L.E., Marston, H.M., Spratt, C., Finlayson, K., Sharkey, J., 2007. Impaired attention is central to the cognitive deficits observed in alpha 7 deficient mice. Eur. Neuropsychopharmacol. 17, 145–155.

Yoshida, T., Kuwabara, Y., Ichiya, Y., Sasaki, M., Fukumura, T., Ichimiya, A., Takita, M., Ogomori, K., Masuda, K., 1998. Cerebral muscarinic acetylcholinergic receptor measurement in Alzheimer's disease patients on ^{11}C-*N*-methyl-4-piperidyl benzilate-comparison with cerebral blood flow and cerebral glucose metabolism. Ann. Nucl. Med. 12, 35–42.

Yoshida, T., Kuwabara, Y., Sasaki, M., Fukumura, T., Ichimiya, A., Takita, M., Ogomori, K., Ichiya, Y., Masuda, K., 2000. Sex-related differences in the muscarinic acetylcholinergic receptor in the healthy human brain – a positron emission tomography study. Ann. Nucl. Med. 14, 97–101.

Zeeberg, B.R., 1999. Pharmacokinetic computer simulation of the relationship between in vivo and in vitro neuroreceptor subtype selectivity of radioligands. Nucl. Med. Biol. 26, 803–809.

Zeeberg, B.R., Gitler, M.S., Baumgold, J., De la Cruz, R.A., Reba, R.C., 1991. Binding of radioiodinated SPECT ligand to transfected cell membranes expressing single muscarinic receptor subtypes. Biochem. Biophys. Res. Commun. 179, 768–775.

Zubieta, J.K., Koeppe, R.A., Frey, K.A., Kilbourn, M.R., Mangner, T.J., Foster, N.L., Kuhl, D.E., 2001. Assessment of muscarinic receptor concentrations in aging and Alzheimer disease with [^{11}C]NMPB and PET. Synapse 39, 275–287.

Zubieta, J.K., Koeppe, R.A., Mulholland, G.K., Kuhl, D.E., Frey, K.A., 1998. Quantification of muscarinic cholinergic receptors with [^{11}C]NMPB and positron emission tomography: method development and differentiation of tracer delivery from receptor binding. J. Cereb. Blood Flow Metab. 18, 619–631.

CHAPTER SEVEN

Human Brain Imaging of Adenosine Receptors

Masahiro Mishina[1,2], Yuichi Kimura[1,3] and Kiich Ishiwata[1]
[1]Positron Medical Center, Tokyo Metropolitan Institute of Gerontology, Tokyo, Japan
[2]The Second Department of Internal Medicine, Nippon Medical School, Tokyo, Japan
[3]Molecular Imaging Center, National Institute of Radiological Sciences, Chiba, Japan

1. INTRODUCTION

1.1. Adenosine

Adenosine is a purine nucleoside and is produced by conversion of intracellular and extracellular adenine nucleotides (Figure 1). The adenosine is rapidly converted to inosine and adenosine monophosphate, and the half-life of adenosine is reported to be 0.6–1.5 s in human plasma (Agteresch et al., 1999). It is involved in cellular energy metabolism via adenosine triphosphate and adenosine diphosphate, and in signal transduction as cyclic adenosine monophosphate (cAMP). Extracellular adenosine plays a role as an endogenous modulator of synaptic functions in the central nervous system (Dunwiddie and Masino, 2001; Latini and Pedata, 2001; Ishiwata et al., 2002b; Jacobson and Gao, 2006; Ishiwata et al., 2007). The effects are mediated by at least four adenosine receptor (AR) subtypes: A_1, A_{2A}, A_{2B}, and A_3 (Palmer and Stiles, 1995; Haas and Selbach, 2000; Dunwiddie and Masino, 2001; Fredholm et al., 2001). These four subtypes are members of the superfamily of G-protein-coupled receptors (Jacobson and Gao, 2006).

1.2. Adenosine A_1 Receptors

Adenosine A_1 receptors (A_1Rs) are widely distributed throughout the entire brain and inhibit adenylyl cyclase (Ferre et al., 1997; Svenningsson et al., 1997; Fukumitsu et al., 2005). The A_1Rs interact negatively with dopamine D_1 receptors in direct pathway neurons (Ferre et al., 1994; Yabuuchi et al., 2006). The A_1Rs are also presynaptic receptors that regulate the dopaminergic system (Yabuuchi et al., 2006). Adenosine can have neuroprotective effects on dopaminergic neurons via A_1Rs (Delle Donne and Sonsalla, 1994). Endogenous adenosine in the tuberomammillary nucleus suppresses the histaminergic system via A_1R to promote non-rapid eye movement sleep (Oishi et al., 2008). At peripheral nerve terminals in rodents, the A_1R activation produces antinociception by decreasing cAMP levels in the sensory nerve terminal (Sawynok, 1998). A_1R agonists may be effective for therapy of migraine and cluster headache because A_1R activation leads to neuronal

Imaging of the Human Brain in Health and Disease
http://dx.doi.org/10.1016/B978-0-12-418677-4.00007-5

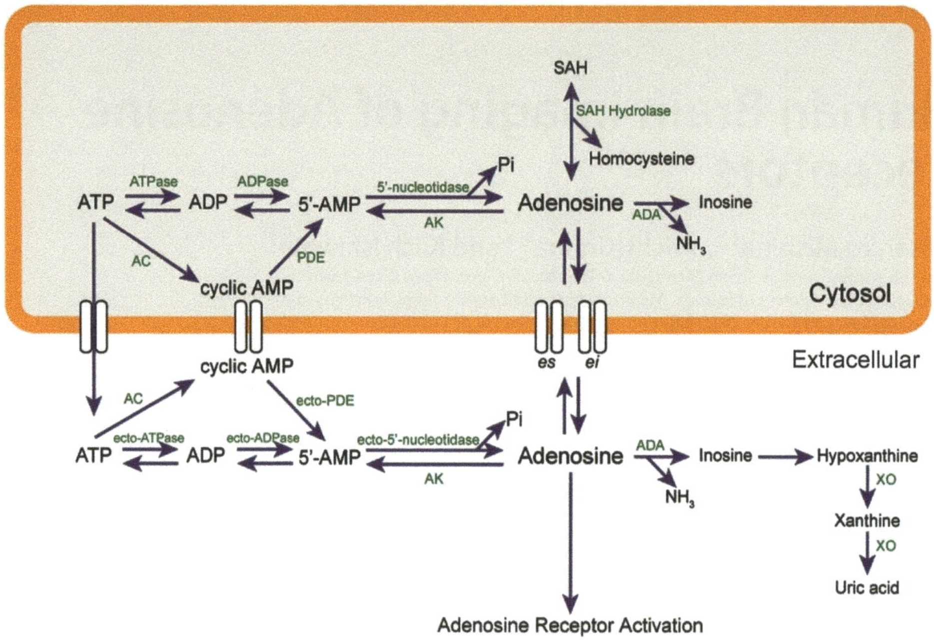

Figure 1 ***Pathways of adenosine metabolism.*** Uric acid is the final oxidation product of purine metabolism. AC, adenylyl cyclase; ADA, adenosine deaminase; ADP, adenosine diphosphate; AK, adenosine kinase; AMP, adenosine monophosphate; ATP, adenosine triphosphate; es, equilibrative-sensitive nucleoside transporters; ei, equilibrative-insensitive nucleoside transporters; PDE, phosphodiesterase; SAH, *S*-adenosyl homocysteine; XO, xanthine oxidase.

inhibition without concomitant vasoconstriction (Giffin et al., 2003; Zambrowicz et al., 2003). The A_1R inhibitor is expected to be useful to treat patients with congestive heart failure and hypotension in dialysis (Shah and Frishman, 2009). Efficacy of A_1R agonists in patients with Parkinson disease (PD) remains unclear (Kelsey et al., 2009).

1.3. Adenosine A_{2A} Receptors

Adenosine A_{2A} receptors ($A_{2A}Rs$) are abundant in dopamine-rich areas of the brain, such as the basal ganglia (Fredholm and Svenningsson, 2003). These receptors are known to stimulate adenylyl cyclase and interact negatively with dopamine D_2 receptors at the level of second messengers and beyond (Fredholm and Svenningsson, 2003). The $A_{2A}R$ has a key role in adenosine-mediated sleep-promoting effects (Urade et al., 2003). Recently, $A_{2A}R$ antagonists have attracted attention for nondopaminergic treatment of PD (Jenner et al., 2009).

1.4. Adenosine A_{2B} Receptors

Adenosine A_{2B} receptors ($A_{2B}Rs$) are positively coupled to both adenylyl cyclase and phospholipase C (Jacobson and Gao, 2006). The arachidonic acid pathway was recently demonstrated to be involved in $A_{2B}R$ activation (Donoso et al., 2005). Activation of the $A_{2B}R$ induces vasodilation in some vascular beds, such as the main pulmonary artery of guinea pigs (Eltzschig et al., 2003).

1.5. Adenosine A_3 Receptors

Adenosine A_3 receptors (A_3Rs) inhibit adenylyl cyclase and stimulate phospholipase C and calcium mobilization (Jacobson and Gao, 2006). Overexpression of A_3Rs decreases heart rate, preserves energetics, and protects ischemic hearts, and low-level expression of A_3Rs in the heart provides effective protection against ischemic injury without detectable adverse effects (Cross et al., 2002). Higher levels of A_3R expression lead to the development of dilated cardiomyopathy (Guo et al., 2001).

2.1. POSTMORTEM STUDIES

Postmortem studies of ARs in human brains are limited to A_1R and $A_{2A}R$, but not $A_{2B}R$ and A_3R (Ishiwata et al., 2007).

The first postmortem study of A_1Rs showed that the receptors were widely distributed throughout the brain, with the highest densities in the hippocampus, cerebral cortex, and striatum (Fastbom et al., 1987). Svenningsson et al. (1997) reported the distribution of A_1Rs and $A_{2A}Rs$ in whole-hemisphere sections and showed that $A_{2A}Rs$ were abundant in the putamen, nucleus caudatus, nucleus accumbens, and globus pallidus pars lateralis. Jennings et al. (2001) found colocalization of human equilibrative nucleoside transporter 1 and A_1Rs.

Postmortem studies in patients with Alzheimer disease (AD) reported a reduced density of A_1Rs in the hippocampus (Jansen et al., 1990; Kalaria et al., 1990; Jaarsma et al., 1991; Ulas et al., 1993), although the reduction was also observed in vascular dementia (Deckert et al., 1998). The density of A_1Rs in the striatum was also decreased in patients with AD (Ikeda et al., 1993). However, upregulation of A_1R was detected in the frontal cortex of patients with AD, using quantitative autoradiography and the radioligand [^{3}H] 8-cyclopentyl-1,3-dipropylxanthine (DPCPX) (Albasanz et al., 2008). Similar increases of A_1R density in the frontal cortex were also reported in patients with Pick disease (Albasanz et al., 2007) and Creutzfeldt-Jakob disease (Rodriguez et al., 2006).

A postmortem study found that the A_1Rs were reduced in the epileptic temporal cortex in patients with temporal lobe epilepsy (Glass et al., 1996), although upregulation of A_1R was found in the specimens of epileptogenic neocortex by surgical resection (Angelatou et al., 1993).

The density of $A_{2A}Rs$ in the basal ganglia was lower in patients with Huntington disease than in normal subjects (Martinez-Mir et al., 1991). The loss of $A_{2A}Rs$ in the caudate nucleus, putamen, and external globus pallidus was more dominant than that of dopamine D_2 receptor binding (Glass et al., 2000).

In PD, postmortem study reported that the density of $A_{2A}R$ binding sites was comparable to that found in normal subjects (Martinez-Mir et al., 1991). Using reverse transcription polymerase chain reaction and postmortem brain tissue, Hurley et al. (2000) reported that mRNA levels for $A_{2A}R$ of patients with PD were decreased in the caudate nucleus and anterior dorsal putamen and were increased in the substantia nigra pars reticulata. Another study using in situ hybridization and autoradiography suggested that increases of $A_{2A}Rs$ in the putamen were involved in the development of dyskinesia following long-term levodopa therapy in PD (Calon et al., 2004).

In patients with schizophrenia, postmortem studies using [^{3}H] 2-*p*-(2-carboxyethyl)-phenethylamino-5′-*N*-ethylcarboxamidoadenosine (CGS21680) reported that the $A_{2A}Rs$ were increased in the striatum (Kurumaji and Toru, 1998; Deckert et al., 2003) and that the increase of the receptor density correlated with the dose of antipsychotic medication (Deckert et al., 2003).

2.2. IMAGING OF A_1 AND A_{2A} RECEPTORS

2.2.1. A_1 Receptors in Normal Subjects

Although ARs have attracted much attention, until quite recently there has been little information regarding their presence in the living human brain. However, we developed a PET ligand for mapping A_1Rs, and we successfully visualized A_1Rs in living humans with PET and [^{11}C] 8-dicyclopropylmethyl-1-methyl-3-propylxanthine (MPDX, Figure 2A) (Ishiwata et al., 2002a; Fukumitsu et al., 2003; Kimura et al., 2004; Fukumitsu et al., 2005). The distribution volume of [^{11}C]MPDX was large in the striatum

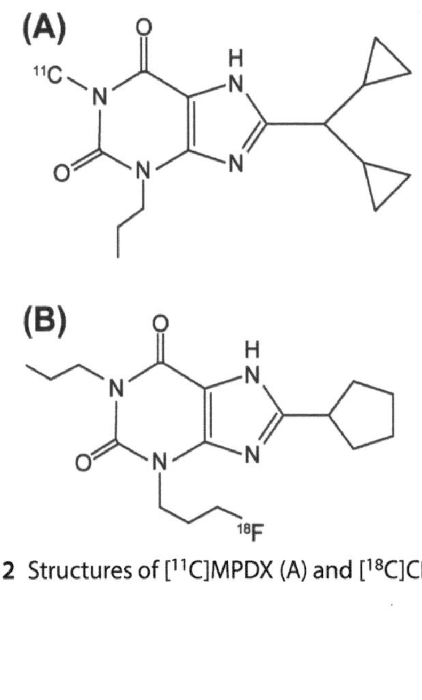

Figure 2 Structures of [^{11}C]MPDX (A) and [^{18}C]CPFPX (B).

and thalamus, moderate in the cerebral cortices and pons, and small in the cerebellum (Figure 4) (Fukumitsu et al., 2005).

Bauer et al. (2003) also successfully performed imaging of A_1Rs in the human brain by PET using a similar radioligand, [^{18}F] 8-cyclopentyl-3-(3-fluoropropyl)-1-propylxanthine (CPFPX, Figure 2B). They also confirmed the specific binding of [^{18}F]CPFPX to A_1Rs using unlabeled CPFPX (Meyer et al., 2006). [^{11}C]MPDX (inhibition constant [K_i] 4.2 nM for the rat forebrain membrane, Table 1) has a slightly lower affinity for A_1Rs than [^{18}F]CPFPX (K_i 1.26 nM for the cloned human A_1Rs) (Noguchi et al., 1997; Holschbach et al., 2002). Considering peripheral metabolism, the unchanged form of [^{18}F]CPFPX in the plasma decreased rapidly by 25% by 10 min after injection, whereas the percentages of unchanged [^{11}C]MPDX in the plasma remained high during the 60-min PET scan: 89% at 10 min and 75% at 60 min after injection. Thus, the in vivo stability of [^{11}C]MPDX is an advantage for the continuous input function for kinetic analysis.

A study using autoradiography reported that the binding ability of A_1R in the striatum was reduced in aged rats (Meerlo et al., 2004). A human PET study reported that the binding ability of [^{18}F]CPFPX was negatively correlated with age (Meyer et al., 2007).

A human PET study with [^{18}F]CPFPX showed that the distribution of A_1Rs was increased in subjects who were deprived of sleep for 24 h compared with that in controls with regular sleep and suggested that changes of A_1R expression contribute to homeostatic sleep regulation (Elmenhorst et al., 2007).

2.2.2. A_1 Receptors in Brain Diseases

Alzheimer's Disease (AD) is the most common form of dementia, and its pathological features include neuritic plaques composed of amyloid-β peptide fibrils, neurofibrillary tangles of hyperphosphorylated tau, and neurotransmitter deficits. Several studies have been reported for ARs in AD (Rahman, 2009). As mentioned above, postmortem studies have yielded conflicting results about variations of A_1Rs in AD. Interestingly, the distribution of A_1R has been reported to be associated with the pathology of AD (Angulo et al., 2003). A [^{11}C]MPDX PET study showed that the density of A_1Rs was decreased in the temporal cortex and thalamus of AD patients, while hypometabolism of glucose was observed in the parietotemporal cortex and posterior cingulate gyrus in 2-[18F] fluoro-2-deoxy-D-glucose ([^{18}F]FDG) PET (Fukumitsu et al., 2008). Further studies are expected to compare this with [^{11}C]MPDX PET and amyloid imaging.

An animal study of occlusion and reperfusion has found that decreased [^{11}C]MPDX binding to A_1Rs after reperfusion was a sensitive predictor of severe ischemic damage (Nariai et al., 2003), although human PET data were not available for A_1Rs in patients with stroke.

In patients with temporal lobe epilepsy, binding of [^{11}C]MPDX was significantly decreased in the mesial temporal lobe of the focus side outside the hippocampus (mainly the fusiform gyrus), whereas it was significantly increased in the frontal cortex of the focus side.

Table 1 In Vitro Affinity of PET Ligands for Adenosine Receptors

	Affinity (K_i nM)		Selectivity	Affinity (Ki nM)	Selectivity		Reference
	A_1	A_{2A}	A_{2A}/A_1 or A_1/A_{2A}	A_3	A_{2B}/A_{2A}	A_3/A_{2A}	
A_1 Receptor Ligand							
DPCPX	6.4 (***g, 1***)	590 (***g, 4***)	92				Shimada et al. (1991)
	0.17 (***b***, **K_d**)						Holschbach et al. (1998)
	2.58 (***h***, **K_d**)						Maemoto et al. (2004)
[11C]KF15372	3.0 (***g, 1***)	430 (***g, 4***)	140				Shimada et al. (1991)
[11C]MPDX	4.2 (***r, 1***)	>100 (***r, 5***)	>24				Noguchi et al. (1997)
[18F]CPFPX	0.18 (***b, 2***)	812 (***r, 5***)	4,500				Holschbach et al. (1998)
	0.63 (***r***, **K_d**)						Holschbach et al. (2002)
	1.26 (***h***, **K_d**)	940 (***h***, **K_d**)	>700				Holschbach et al. (2002)
[11C]FR194921	4.96 (***r, 2***)	>100 (***r, 5***)					Maemoto et al. (2004)
	2.91 (***h, 2***)	>100 (***h, 5***)	>34	>100 (***h, 4***)			Maemoto et al. (2004)
A_{2A} Receptor Ligand							
[11C]KF17837	62 (***r, 1***)	1.0 (***r, 5***)	62				Nonaka et al. (1994)
[11C]TMSX ([11C]KF 18446)	1,600 (***r, 1***)	5.9 (***r, 5***)	270				Ishiwata et al. (2000a)
[11C]KF21213	>10,000 (***r, 1***)	3.0 (***r, 5***)	>3,300				Wang et al. (2000)
[11C]KW-6002	150 (***r, 1***)	2.2 (***r, 5***)	68				Hirani et al. (2001)
SCH 58261	121 (***r, 1***)	2.3 (***r, 5***)	53	>1,000 (***r, 8***)			Zocchi et al. (1996a)
[11C]SCH442416	1,800 (***r, 2***)	0.50 (***r, 6***)	3,600	>10,000 (***r, 8***)		>20,000	Todde et al. (2000)
	1.111 (***h, 2***)	0.048 (***h, 6***)	23,000	>10,000 (***h, 8***)	>200,000	>200,000	Todde et al. (2000)
A_3 Receptor Ligand							
[18F]FE@SUPPY*	11,500 (***r, 3***)	7,330 (***r, 7***)		600 (***r, 8***)	19	12	Li et al. (1999)
				4.22 (***h, 8***)	2,700		Li et al. (1999)

Radioligands used for binding assay: ***1***, [3H]*N*6-cyclohexyladenosine; ***2***, [3H]DPCPX; ***3***, [3H]*R*-PIA; ***4***, [3H]NECA; ***5***, [3H]CGS 21680; ***6***, [3H]SCH 58261; ***8***, [125I]AB-MECA.
Receptor membrane source used for binding assay: ***b***, bovine; ***g***, guinea pig; ***h***, human; ***r***, rat.
*Recently developed [18F]FE@SUPPY maybe a candidate radioligand for A_3 subtype (Wadsak *et al.*, 2008), but has not been applied to humans and non-human primates.

The regional abnormality observed by [^{11}C]MPDX PET was different from that observed by already established PET methods measuring central benzodiazepine receptor density by [^{11}C]flumazenil PET and glucose metabolism by [^{18}F]FDG PET (in preparation).

The A_1Rs interact negatively with dopamine D_1 receptors in direct pathway neurons (Ferre et al., 1994; Yabuuchi et al., 2006), and are also presynaptic receptors that regulate the dopaminergic system (Yabuuchi et al., 2006). Although increased attention has been paid to the A_{2A}Rs as mentioned below, the roles of A_1Rs in patients with PD remain unclear (Kelsey et al., 2009). This issue is now under investigation. A recent study suggested that dyskinesia might involve not only A_{2A}Rs but also A_1Rs (Xiao et al., 2011).

Boy et al. (2008) reported that the density of A_1Rs was decreased in the cerebral cortex of patients with liver cirrhosis and hepatic encephalopathy.

2.2.3. A_{2A} Receptors in Normal Subjects

We also successfully visualized A_{2A}Rs in the human brain with PET and [^{11}C] (*E*)-8-(3,4,5-trimethoxystyryl)-1,3,7-trimethylxanthine (TMSX, Figure 3A) (Ishiwata et al., 2007; Mishina et al., 2007; Bauer and Ishiwata, 2009; Ishiwata et al., 2010). The specific binding of [^{11}C]TMSX to A_{2A}Rs was confirmed with a theophylline challenge (Ishiwata et al., 2005). We performed test–retest studies and optimized the kinetics for [^{11}C]TMSX PET in normal subjects, thus confirming good reproducibility of [^{11}C]TMSX PET in the putamen. Our studies have demonstrated that the binding of A_{2A}Rs is greatest in the anterior and posterior putamen, followed by the head of the caudate nucleus and thalamus, but is low in the cerebral cortex, especially the frontal lobe (Figure 5) (Ishiwata et al., 2005; Mishina et al., 2007). [^{11}C]TMSX PET has also shown a large binding potential in the striatum where A_{2A}Rs are abundant, as found in postmortem and nonhuman studies, but the binding potential of [^{11}C]TMSX is greater in the human thalamus than in other mammals.

A non-xanthine-type PET ligand, [^{11}C]SCH442416 (Figure 3C), was also available (Todde et al., 2000; Moresco et al., 2005), although human data are scarce (Ramlackhansingh et al., 2011).

As mentioned above, age-related decline was reported in A_1R (Meerlo et al., 2004; Meyer et al., 2007). Compared with those in young rats, however, the bindings of the A_{2A}R agonist and antagonist in the cortical membranes were increased in aged rats, although the bindings in the striatal membranes of aged rats were comparable to those in young rats (Lopes et al., 1999). Therefore, these studies suggested that the effects of aging on A_1R and A_{2A}R might be different.

2.2.4. A_{2A} Receptors in Brain Diseases

Parkinson's Disease (PD) is a progressive degenerative neurological disorder characterized clinically by resting tremor, bradykinesia, cogwheel rigidity, and postural instability (Lees et al., 2009). Recently, A_{2A}R antagonists have attracted attention as potential non-dopaminergic therapies for PD. Caffeine is a nonselective A_1R and A_{2A}R antagonist and

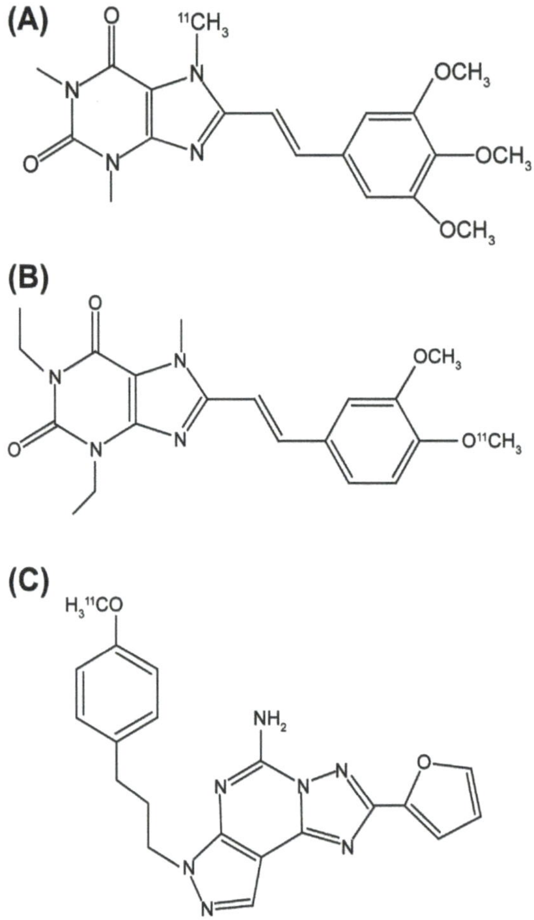

Figure 3 Structures of [^{11}C]TMSX ([^{11}C]KF18446, (A)), [^{11}C][4-O-methyl-^{11}C]KW-6002 (B), and [^{11}C]SCH442416 (C).

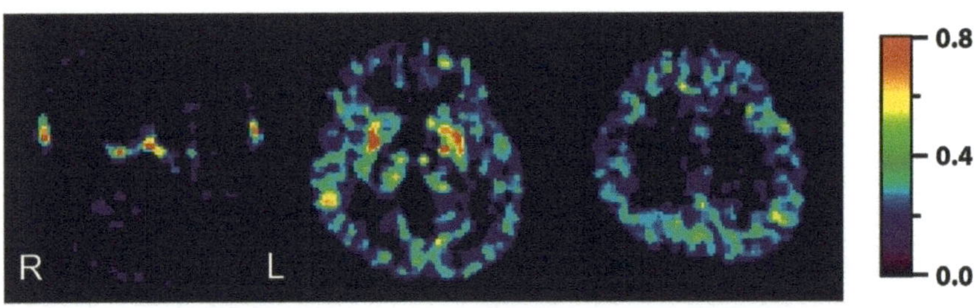

Figure 4 ***Brain images for 20-year-old man of MRI and BP_{ND} of MPDX.*** A_1Rs, as well as A_{2A}Rs, are enriched in the striatum and thalamus. Unlike the A_{2A}Rs, however, the A_1Rs are widely distributed in the cerebral cortex.

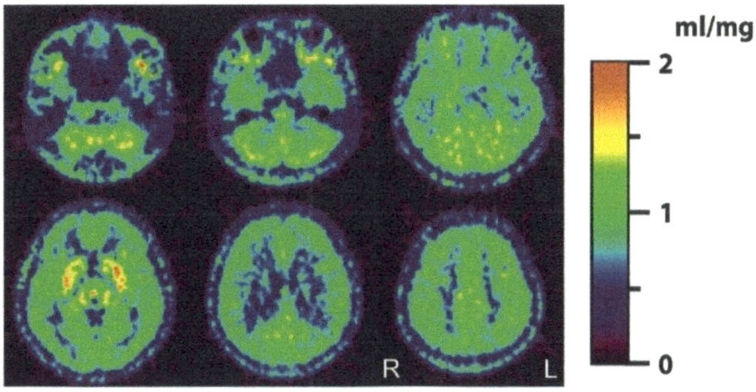

Figure 5 ***A parametric image for the total distribution volume of [^{11}C]TMSX PET image for a 21-year-old man.*** The parametric image was generated using graphical analysis. This image includes the non-specific binding of [^{11}C]TMSX. The image demonstrates that the distribution of [^{11}C]TMSX was high in the putamen and low in the cerebral cortex.

is known to reduce the risk of developing PD (Ross et al., 2000; Ascherio et al., 2001). In addition, theophylline, which is also a nonselective A_1R and $A_{2A}R$ antagonist, was expected to be a promising agent for the treatment of PD (Kostic et al., 1999). However, findings from clinical trials of both caffeine and theophylline have been unimpressive (Kulisevsky et al., 2002; Schwarzschild et al., 2002). The selective $A_{2A}R$ antagonists were developed as novel nondopaminergic agents for PD (Kase, 2001) and provide antiparkinsonian benefit without causing or worsening dyskinesia, which is one of the most inconvenient side effects of dopaminergic therapy (Bara-Jimenez et al., 2003). A recent study suggested that $A_{2A}R$ may give a-synuclein neurotoxicity and that the findings support the neuroprotective potential of $A_{2A}R$ antagonists (Kachroo and Schwarzschild, 2012). A postmortem study reported that the density of $A_{2A}R$ binding sites in PD was comparable to that found in normal subjects, while the density in the basal ganglia was lower in patients with Huntington chorea than in normal subjects (Martinez-Mir et al., 1991). However, another study using in situ hybridization and autoradiography suggested that $A_{2A}Rs$ were involved in the development of dyskinesia following long-term levodopa therapy in PD (Calon et al., 2004). Therefore, $A_{2A}Rs$ may be involved in the appearance of the side effects of antiparkinsonian agents. Uric acid, the final oxidation product of purine metabolism (Figure 1), may be involved in PD (Weisskopf et al., 2007; Schwarzschild et al., 2008; Ascherio et al., 2009; O'Reilly et al., 2010; Schwarzschild et al., 2011).

We investigated $A_{2A}Rs$ in the striata of drug-naïve patients with PD, and patients with mild drug-induced dyskinesia and PD, as well as alteration of $A_{2A}Rs$ after antiparkinsonian therapy using [^{11}C]TMSX PET (Mishina et al., 2011). Our study demonstrates that the putaminal binding ability of $A_{2A}Rs$ was increased in the patients with dyskinesia and that there was no significant difference in the striatal binding ability of $A_{2A}Rs$ between drug-naïve patients and normal controls (Figure 6). However, in drawing attention to the asymmetrical symptoms in drug-naïve patients, our study suggests

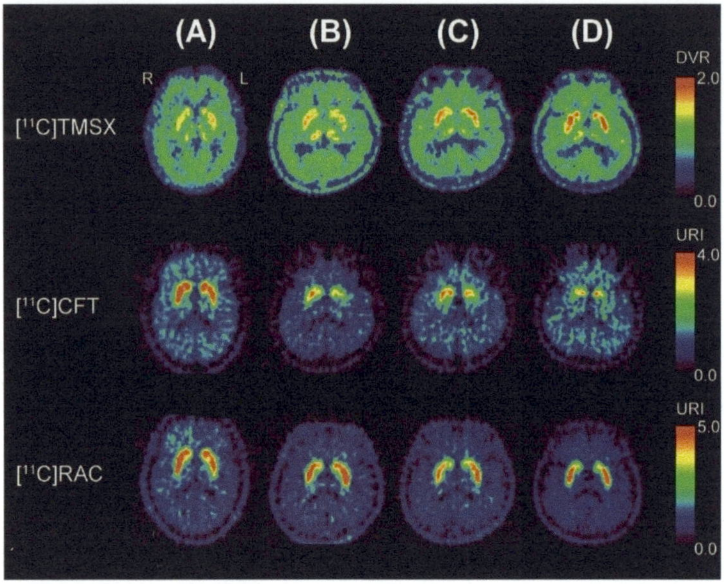

Figure 6 ***PET images for normal subjects (A), a drug-naïve patient with PD before (B) and after antiparkinsonian therapy (C), and a patient with drug-induced dyskinesia (D).*** The normal subjects are a 63-year-old female for [^{11}C]TMSX, and a 63-year-old male for [^{11}C]2β-carbomethoxy-3β-(4-fluorophenyl) tropane ([^{11}C]CFT) and [^{11}C]raclopride ([^{11}C]RAC, A). The drug-naïve patient is a 60-year-old male with right-dominant parkinsonism, who underwent two series of PET scans before (B) and after antiparkinsonian therapy (C). The patients did not develop dyskinesia at a second series of PET scans in the posttherapeutic state. The patient with dyskinesia is a 65-year-old male with right-dominant parkinsonism (D).

that A_{2A}Rs were asymmetrically downregulated in the putamen but not in the head of the caudate nucleus. The asymmetrical regulation of A_{2A}Rs seems to be involved in compensation for the decrease in dopamine. Our study also showed that the A_{2A}Rs were increased in the putamen after antiparkinsonian therapy.

Another PET study with [^{11}C]SCH442416 also reported that binding potentials for A_{2A}Rs were increased in patients with dyskinesia compared with those in patients without dyskinesia (Ramlackhansingh et al., 2011). Previous postmortem studies also showed that the putaminal binding ability of A_{2A}Rs was increased in PD patients with dyskinesia (Martinez-Mir et al., 1991; Calon et al., 2004; Morissette et al., 2006). Our study suggested that the increase in putaminal A_{2A}Rs after antiparkinsonian therapy preceded the development of dyskinesia in patients with PD (Mishina et al., 2011).

2.3. QUANTIFICATION OF ADENOSINE RECEPTORS

2.3.1. Kinetic Analysis for Quantitative Receptor Imaging

PET has a unique capability of visualizing the spatial distribution of administered radiopharmaceutical in tissues. Although PET gives us radioactivity concentration in

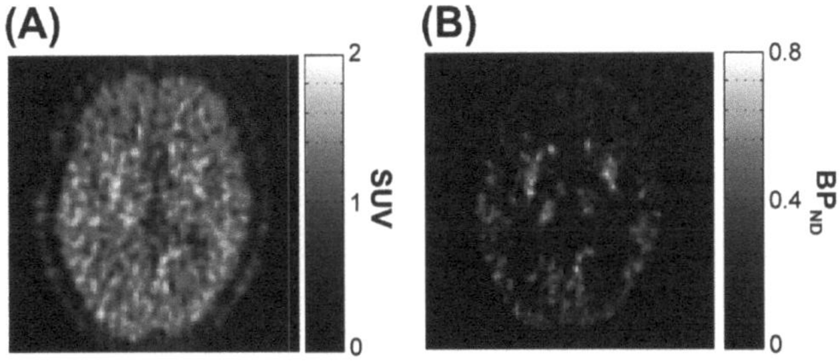

Figure 7 ***A static image in (A) and the parametric image derived from kinetic analysis in (B).*** (A) is an standardized uptake value (SUV) image derived from the administration of [^{11}C]MPDX. No physiologically meaningful distribution is observed. (B) is an image of a binding potential that describes the density of A_1Rs that is a result of kinetic analysis. A physiologically reasonable distribution is observed: high in the putamen and caudate, intermediate in major cortices, and low in the white matter.

Bq/ml, this is insufficient to quantify receptor density. Some general overviews for neuroreceptor quantification are given in this subsection, including for A_1Rs and A_{2A}Rs.

Molecules of ligands take two forms in tissues: those specifically bound to neuroreceptors and nondisplaceable forms composed of molecules free from receptors and molecules captured by nonspecific binding sites in tissues (Innis et al., 2007). Thus, the radioactivity measured by PET can originate from both forms, and the radioactivity emitted from the molecules bound to receptors has to be separated for quantification of receptor density. For this purpose, kinetic analysis is applied to a set of sequentially measured PET data (Watabe et al., 2006). For example, Figure 7(A) shows a static PET image with the administration of [^{11}C]MPDX developed for mapping A_1R. [^{11}C]MPDX accumulated broadly in the brain, and no physiologically acceptable distribution can be obtained. On the other hand, kinetic analysis provides a parametric image as shown in Figure 7(B), where the image represents the regional receptor densities.

In the kinetic analysis, a compartment model is utilized as shown in Figure 8. The compartment model for receptor quantification was originally proposed by Mintun et al. (1984), and it is widely used for receptor quantification with PET. In addition, various algorithms have been proposed to estimate the model. Molecules of administered radiopharmaceutical are assumed to take three forms: molecules in capillaries, nondisplaceable forms in tissues, and those specifically bound to receptors. These are pictured as compartments in Figure 8. The rate of transportation between the next compartment is represented as rate constants of $K_1 \sim k_4$ in the model.

For the kinetic analysis, sequential PET scans are conducted after the administration of radiopharmaceutical to acquire a time history of concentration in tissues (tissue time–activity curve; tTAC). Usually, sequential arterial blood samplings are also performed concurrently to derive the time history in arterial plasma (plasma time–activity curve; pTAC). A mathematical relationship between tTAC and pTAC is introduced from the compartment model analytically. The rate constants are therefore estimated using the measured tTAC and pTAC with the help of mathematical model estimation algorithms.

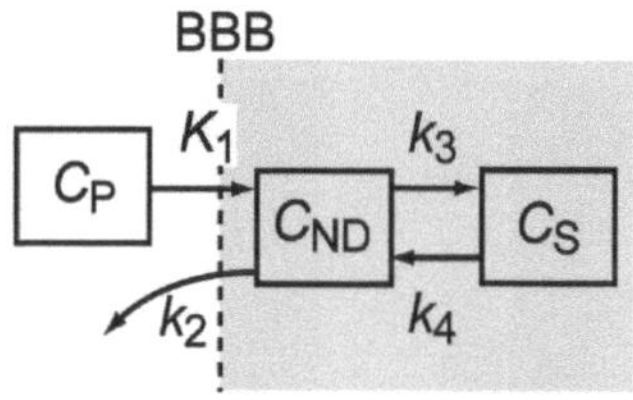

Figure 8 ***A compartment model for neuroreceptor quantification.*** The three compartments of C_P, C_{ND}, and C_S denote administered radioligands in arterial plasma, free or nonspecifically bound forms in tissues, and those specifically bound to a target neuroreceptor, respectively. C_{ND} and C_S are located in tissues, and therefore, they are shaded. The rate constants K_1 to k_4 represent the rate of transportation to the next compartment. BBB, blood–brain barrier.

The aimed receptor density can then be computed from the rate constants as a binding potential (Innis et al., 2007). In addition, selection of an estimation algorithm is performed on the basis of the kinetics of the investigated receptor and radioligand.

2.3.2. Algorithms for Receptor Quantification

We have several choices for the estimation algorithms. Nonlinear estimation algorithms (Hendrix and Tóth, 2010) require only minimal assumptions as to the behavior of the administered radiopharmaceutical; therefore, they can be applied widely in neuroreceptor quantification. However, the algorithms have some problems: mathematical instability and computational burden.

Some graphical plot approaches have been proposed. The Logan graphical analysis (Logan et al., 1990; Kimura et al., 2007) is a popular algorithm for quantitative neuroreceptor imaging. The algorithm can be implemented as a linear regression, and thus it is stable for noise in tTAC. It is also free from the problems that arise in nonlinear estimation algorithms. On the other hand, the initiation time when the line estimation is applied should be given a priori. Note that the Logan graphical analysis has another drawback; it underestimates neuroreceptor density owing to the noise in tTAC (Slifstein and Laruelle, 2000; Kimura et al., 2007). Hence, the algorithm is not suitable to visualize neuroreceptor density because the noise in voxel-based tTAC is worse than that in an averaged tTAC in a region of interest (ROI). However, some algorithms have been proposed to overcome this issue (Ichise et al., 2002; Ogden, 2003; Joshi et al., 2008; Shidahara et al., 2009).

Arterial blood sampling is another issue for neuroreceptor quantification with PET. In order to acquire pTAC, a set of sequential arterial blood samplings should be conducted concurrent with a PET scan. There are two problems in pTAC measurement: technical and clinical problems. The technical problem is related to metabolite analysis. pTAC is the radioactivity concentration of administered radiopharmaceutical in arterial plasma. The molecules of the radiopharmaceutical are usually metabolized, and then its biochemical properties are altered. Consequently, the fraction of the intact form of the radiopharmaceutical in arterial plasma should also be measured. High-performance liquid chromatography is usually utilized for the metabolite analysis, but it is not easy to carry out frequent and reliable measurements. From a clinical point of view, pTAC measurements are uncomfortable or painful for patients because a catheter is inserted into their femoral artery and is kept there during a PET scan, which lasts more than 90 min. Moreover, the radioactivity measurement in the sampled blood causes radiation exposure to medical staff. Consequently, it is favorable to omit the pTAC measurement.

The nonlinear estimation approaches require pTAC because it is an input to the compartment model and pTAC is necessary to compute tTAC on the basis of the compartment model. Meanwhile, some graphical plot approaches do not require pTAC (Logan et al., 1996). The algorithms use a reference region that is devoid of the investigated neuroreceptors instead of pTAC, the details of which are described in the next subsection. Another approach that can omit arterial blood sampling in neuroreceptor imaging is simplified reference tissue model (SRTM) (Lammertsma and Hume, 1996). SRTM assumes that the behaviors of radioligand in both a target and a reference tissue can be described using the compartment model that contains one tissue compartment and another compartment for blood pool. pTAC is common for both the tissues; thus, pTAC can be canceled using a mathematical formulation. SRTM is implemented using a nonlinear estimation algorithm, and its linearized version is available (Gunn et al., 1997).

2.3.3. Reference Region

The reference region is important for neuroreceptor quantification. It is defined as the region where the density of target neuroreceptors is zero or negligible and the distribution volume of a nondisplaceable compartment is the same among the reference region and target region (Innis et al., 2007). Here, distribution volume means the ratio of the density of radioligand between a target compartment, C_{ND} or C_S in Figure 8, and arterial plasma, C_P in Figure 8, in an equilibrium state. For example, the cerebellum is utilized for the reference region to measure dopamine D_2 receptors.

If a reference region can be defined in the brain, some algorithms that do not require pTAC are applicable, such as the Logan graphical approaches without pTAC and SRTM. It is thus possible to skip the arterial blood sampling. However, the approaches require some additional assumptions other than the existence of a reference region, and their applicability should be considered on the basis of the physiological properties of target receptors and the kinetics of the radiopharmaceutical.

Concerning ARs, A_1Rs are distributed globally in the whole brain (Svenningsson et al., 1997). It is difficult to find an ideal reference region for receptor quantification. However, the density in the cerebellum is lower than that in other brain regions, especially in humans compared to rodents. On the other hand, it is considered that A_{2A}Rs are rich in the striatum and very scarce in the cerebral cortex and cerebellum, but in the human brain, they seem to be present in the extrastriatal regions to a certain degree (Svenningsson et al., 1997). Again, an ideal reference region is not available for $A_{2A}R$ quantification. It is noted that extremely high affinity is required for the binding of radioligands to the neuroreceptors with very low densities. Therefore, depending on the affinities of the radioligands used, the regions with few neuroreceptors become candidates for reference regions. The negative influence of an imperfect reference region on the computed receptor density was investigated by Meyer et al. (2006), in which the estimated receptor density had a linear relationship with the true density.

2.3.4. Summary of Algorithms

Equation (1) is the Logan graphical analysis using pTAC,

$$\frac{\int_0^t C(u)\,du}{C(t)} = V_T \frac{\int_0^t C_P(u)\,du}{C(t)} + b_1 \left(t > t_1^*\right) \tag{1}$$

where C and C_P denote tTAC and pTAC, respectively. The two quantities appearing on both sides have a linear relationship, and its gradient of V_T is the total distribution volume, which is correlated with the neuroreceptor density (Innis et al., 2007). Note that the linear relationship holds after t_1^*, which should be specified when Logan graphical analysis is applied.

Two modifications that do not need pTAC are shown in Eqn (2) and Eqn (3).

In Eqn (2), BP_{ND} is the binding potential that is a function of the receptor density and affinity of a radioligand to a target receptor; therefore, it is utilized as a representative quantity of a receptor density. k_2^R and C^R denote the tTAC and the k_2 in a reference region, respectively. Thus, a desired receptor density is obtained as the slope of the plot by using tTACs in both target and reference regions. Note that the k_2 in a reference region should be specified explicitly if Eqn (2) is applied. Therefore, the value is concerned with the reliability of the estimated BP_{ND}.

$$\frac{\int_0^t C(u)\,du}{C(t)} = \left(1 + BP_{ND}\right) \frac{\int_0^t C^R(u)\,du + \frac{1}{k_2^R} C^R(t)}{C(t)} + b_2 \left(t > t_2^*\right) \tag{2}$$

Equation (3) is introduced by assuming an equilibrium state between the tTACs in both target and reference tissues. In reality, the assumption is concerned with the initiation time of the line estimation of t_3^*.

$$\frac{\int_0^t C(u)\,du}{C(t)} = \left(1 + BP_{ND}\right) \frac{\int_0^t C^R(u)\,du}{C(t)} + b_3 \left(t > t_3^*\right) \tag{3}$$

SRTM is shown in Eqn (4) (Lammertsma and Hume, 1996).

$$C(t) = RC^{\mathrm{R}}(t) + k_2\left(1 - \frac{R}{\mathrm{BP_{ND}}}\right)\exp\left(-\frac{k_2}{1+\mathrm{BP_{ND}}}\right) \otimes C^{\mathrm{R}}(t) \quad (4)$$

where R is defined as K_1/K_1^{R} and means a convolution operation. Equation (4) presents the relationship between the tTAC in the reference region and one in the target region. When both tTACs are obtained, $\mathrm{BP_{ND}}$ can be estimated with a mathematical estimation algorithm. Although SRTM has no such parameters that control the performance of the algorithm like t^* and k_2^{R} in the Logan graphical analysis, the estimation process is a nonlinear estimation, and it suffers from statistical uncertainty in the estimates.

2.3.5. Comparison of Algorithms of AR Quantification

Here, estimation algorithms of [^{11}C]MPDX PET for imaging $\mathrm{A_1}$Rs are described as an example of AR quantification. Some parts were presented previously (Kimura et al., 2009).

2.3.5.1. Data Acquisition

A total of 17 male normal volunteers were recruited. The ages ranged from 20 to 77 years (45 ± 23 years). A 60-min dynamic PET scan was carried out with arterial blood sampling using SET-2400 PET camera (Shimadzu, Kyoto, Japan) in a two-dimensional mode. The data were reconstructed using a filtered backprojection algorithm with Butterworth filter having a cutoff at 1.25 cycles/cm. The frame arrangement was 10 s × 6, 30 s × 3, 60 s × 5, 150 s × 5, and 300 s × 8. A total of 27 frames were acquired in 60 min. The dose and specific activity were 652 ± 59 MBq and 58 ± 42 MBq/nmol, respectively. Body motion during the scan was adjusted using the coregistration function in Statistical Parametric Mapping version 8 (SPM8). Ten ROIs were set: the cerebellum (Cbl), caudate (CdH), putamen (Pt), thalamus (Thl), frontal cortex (Frt), temporal cortex (Tmp), occipital cortex (Ocp), posterior cingulate (PCng), parietal cortex (Prt), and centrum semiovale (CSo). The cerebellum was the reference region for [^{11}C]MPDX for $\mathrm{A_1}$R measurement.

2.3.5.2. Kinetic Analysis

The Logan graphical analysis with pTAC named Logan graphical analysis with plasma time–activity curve (LGA) was applied to the frames from 10 to 60 min after the administration (Kimura et al., 2004) to acquire distribution volume. $\mathrm{BP_{ND}}$ was then computed as $(V_{\mathrm{T}}^{\mathrm{R}}/V_{\mathrm{T}}) - 1$ where V_{T} and $V_{\mathrm{T}}^{\mathrm{R}}$ are total distribution volumes in the target and reference regions, respectively. Here, $\mathrm{BP_{ND}}$ is defined as the ratio of the distribution volumes in specific binding and nondisplaceable compartments (Innis et al., 2007). The reference region has negligible receptor sites for the administered radioligand, and the total distribution volume is therefore equal to the distribution volume in the nondisplaceable compartment. Moreover, the distribution volume of the nondisplaceable compartment is assumed to be uniform in the whole brain.

The Logan graphical analysis without pTAC assuming the k_2 in a reference region named LGA-k2, as shown in Eqn (2), was applied to the data set. The k_2 was varied within 0.23 ± 40% (1/min), which was the mean value in the cerebellum (Kimura et al., 2004), and the dependence in the estimated BP_{ND} on the k_2 was investigated.

The Logan graphical analysis without pTAC and the k_2 in the reference region presented in Eqn (3) was also investigated, with the algorithm named LGA-Nok2. The algorithm is sensitive to t_3^*. It was varied from 10 to 50 min after the administration.

Moreover, SRTM was also applied to the PET data, and the performance of BP_{ND} estimation among the algorithms was investigated.

2.3.5.3. *pTAC and tTAC*

A typical pTAC is shown in Figure 9(A). The measured data, the fraction of intact form, and the metabolite-corrected time history are presented. At 60 min after the administration of [^{11}C]MPDX, 74% of the radioactivity was in intact. The tTACs are displayed in Figure 9(B). The waveforms present a typical sort of reversible behavior.

In addition, the cerebellum showed the fastest clearance, which implied low receptor density. The centrum semiovale had different kinetics from other tissues.

2.3.5.4. *Algorithm Comparison*

The typical LGA is presented in Figure 10. Linear relationships were found (r^2 = 0.997 ± 0.004). The comparison of estimated BP_{ND} with those determined by LGA is summarized in Table 2. The regression coefficients, coefficient of determination, and bias are presented as computed using 153 ROIs.

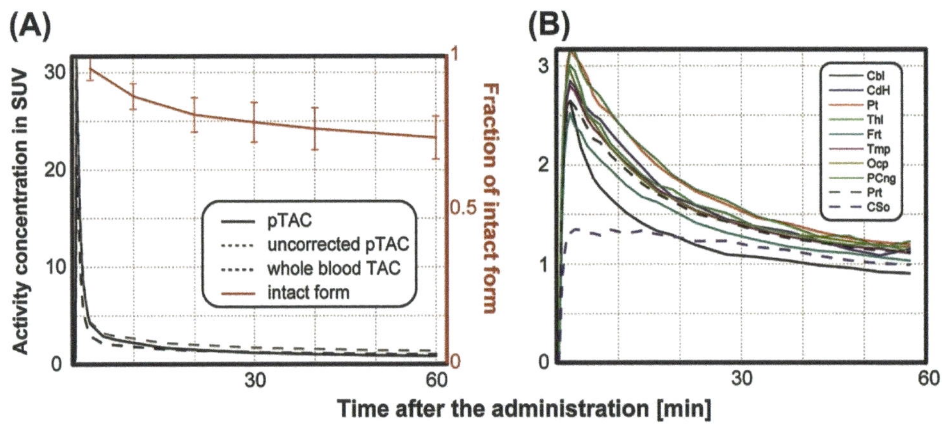

Figure 9 ***Typical pTAC in (A) and tTAC in (B).*** The abscissa and ordinate represent the time after the [^{11}C] MPDX administration in minutes and radioactivity concentration in SUV/ml, respectively. In (A), pTAC, the pTAC without metabolite correction, a time–activity curve in whole blood, and the fraction of an intact form are illustrated with a black solid line, a gray dashed line, a black dashed line, and a red line, respectively. tTACs derived from various regions in the brain are shown in (B).

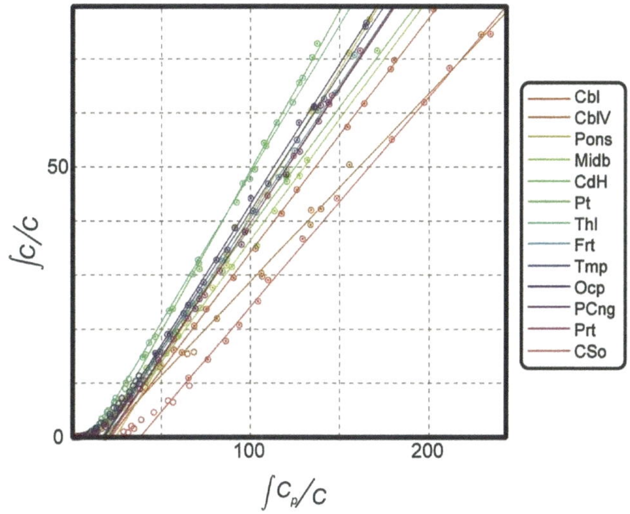

Figure 10 ***A typical logan graphical plot.*** The *x*- and *y*-axes are defined in Eqn (1). Linear relationships are observed 10 min after the administration of [^{11}C]MPDX.

Table 2 Comparison of Algorithms to Estimate BP_{ND}

Algorithms	Regression	Coefficient of determination	% Bias	Note
LGA-k2	$y=1.03x+0.00$	$r^2=0.982$	4.0±6.3	$k_2^R=0.23$ (1/min)
LGA-Nok2	$y=0.98x+0.00$	$r^2=0.954$	−1.1±9.1	$t_3^*=32.5$ (min)
SRTM	$y=1.05x+0.00$	$r^2=0.724$	19±18	

LGA-k2 and LGA-Nok2 means algorithms using no plasma time–activity curve based on the Logan graphical analysis using the k_2 in a reference region and by omitting the k_2. SRTM, Simplified reference region algorithm. The BP_{ND} estimates were compared with those by Logan graphical analysis using pTAC.

The estimates based on the Logan graphical analysis such as LGA-k2 and LGA-Nok2 corresponded to those by LGA. LGA-k2 showed positive bias of less than 10%. LGA-Nok2 was slightly underestimated. SRTM contained large positive bias, and its linear regression was worse than those of the three algorithms based on Logan graphical analysis. SRTM requires a nonlinear estimation algorithm, and it generally suffers from uncertainty in the estimates due to its mathematical framework.

2.3.5.5. k-Dependence

LGA-k2 algorithm requires specification of the k_2 in the reference region before applying the algorithm, and the k_2 is derived from population-based results. It was therefore desirable that the sensitivity of BP_{ND} against k_2^R was low. In Figure 11, mean percentage changes of estimated BP_{ND} are presented to the BP_{ND} at $k_2=0.23$ (1/min). In most regions, the BP_{ND} had positive relationships with k_2^R, and the changes were small (within 5%).

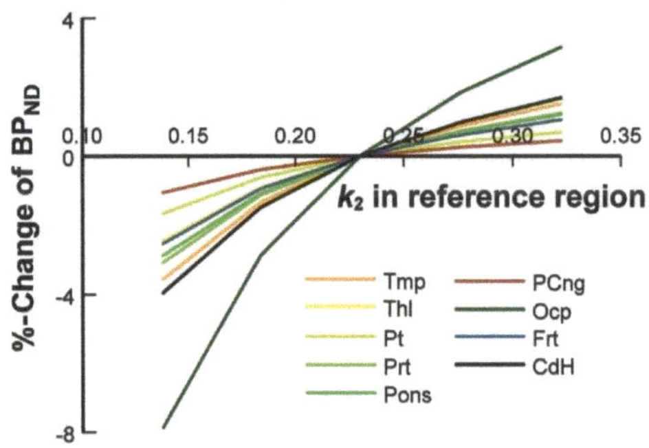

Figure 11 Dependence of the k_2 in a reference region on the estimated binding potential using the logan graphical analysis without a time–activity curve in arterial plasma.

2.3.5.6. t_3^* Dependence

LGA-Nok2 does not require k_2 in a reference region, but the radioactivity concentration in a target tissue has to be in an equilibrium state with that in a reference region. Thus, the starting time to be applied to a line estimation in Logan graphical analysis, t_3^* in Eqn (3), should be given carefully.

The dependence of t_3^* on BP_{ND} estimates is presented in Figure 12. The x- and y-axes denote t_3^* and a mean percentage change of BP_{ND} to the value at $t_3^* = 32.5$ min, respectively. The BP_{ND} was varied according to t_3^*. More than 5% change was found in the putamen, parietal cortex, posterior cingulate, caudate, and thalamus if t_3^* was shifted for 10 min. If t_3^* was placed in the late frames, it made the estimation stable, which was observed as increased percentage bias.

2.3.5.7. Algorithm for A_1R Quantification Using [^{11}C]MPDX

As the results from the comparison, the LGA-k2 was an acceptable choice as an algorithm to quantify the density of A_1R using [^{11}C]MPDX. Such algorithm that does not require pTAC makes the quantification more clinically practical because they omit arterial blood samplings. Moreover, the algorithm also avoids the measurement of metabolite fraction in plasma, and it is helpful to develop the reliability of receptor quantification.

2.3.6. Algorithms and Radioligands Used in A_1R and $A_{2A}R$ Quantification

Regarding A_1R imaging, [^{18}F]CPFPX was applied to human studies in Bauer et al. (2003) in addition to [^{11}C]MPDX. First, the receptor binding of [^{18}F]CPFPX was investigated using a standard two-tissue compartment model compared with the one-tissue compartment model and Logan graphical analysis using pTAC (Meyer et al., 2004). Second, an equilibrium approach was investigated (Meyer et al., 2005a). The binding potential is a function of distribution volume, and distribution volume is

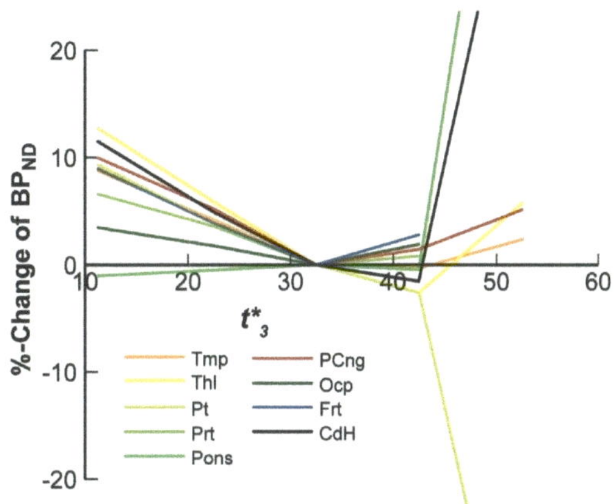

Figure 12 The dependence of t_3^* on the estimated binding potential using the logan graphical analysis without a time–activity curve in arterial plasma.

defined as the ratio of concentrations of administered radioligand in arterial plasma and tissue at equilibrium. If equilibrium states are assumed to be established, distribution volume can be acquired directly from the measured radioligand concentrations in tissues and arterial plasma. This approach was named the bolus/infusion method (Carson et al., 1993), and it enabled monitoring of transient changes of receptor density under a loading condition. Meyer et al. (2005b) also investigated the potential of venous blood sampling, which was a replacement of arterial blood sampling. The implementation of venous blood sampling is easier than that of arterial serial blood sampling clinically. Bauer's group confirmed very low levels of specific binding of [^{18}F]CPFPA in the human cerebellum by in vitro autoradiography (Bauer et al., 2003) and PET in the blocking experiment (Meyer et al., 2006). Later, in order to enhance the applicability of [^{18}F]CPFPA PET for clinical studies, the binding potential of [^{18}F]CPFPA for A_1Rs was estimated noninvasively using the Logan graphical analysis without pTAC using the pons as a reference region (Boy et al., 2008), which showed a similar distribution of A_1Rs as shown by [^{11}C]MPDX PET. In the case of [^{11}C]MPDX PET using the Logan graphical analysis without pTAC, the cerebellum was used as a reference region (Kimura et al., 2004). This is reasonable because a relatively low affinity of [^{11}C]MPDX compared with that of [^{18}F]CPFPA (Table 1) is insensitive for detecting the low A_1R density in the cerebellum.

Regarding A_{2A}R imaging, first, the binding potential of [^{11}C]TMSX was evaluated by a standard 2TCM using metabolite-corrected arterial plasma sampling and the work of Mishina et al. (2007). Second, because [^{11}C]TMSX seemed to bind to atypical A_{2A}Rs in the cerebellum, the centrum semiovale, which belongs to the white matter, was used as the reference region to establish the noninvasive quantification of [^{11}C]TMSX *PET without arterial blood sampling* (Naganawa et al., 2007). The white matter had biological properties different from the gray matter; therefore, the distribution volume in major cortices would differ from that in the white matter. This would cause some bias in $A_{2A}R$ quantification. Further investigations are required to evaluate the performance for $A_{2A}R$ quantification. [^{11}C]SCH442416 was proposed as another radioligand for $A_{2A}R$. The BP_{ND} was computed using a spectrum analysis approach that is a version of the algorithm to estimate a compartment model using pTAC (Ramlackhansingh et al., 2011). The cerebellum was utilized as a reference region, but practical algorithms for A_{2A} quantification have not been reported.

3. CONCLUSIONS

A great deal of basic neuroscience and clinical research revealed evidence for some important roles of ARs in the brain and were applied to various drug developments. Although little information is available for ARs in the living human brain to date, molecular imaging for ARs was successful in several trials for developing PET ligands and is being applied to research on physiology and neurological disorders. Continued efforts to identify high-affinity and selective ligands should lead to PET probes suitable for these binding sites in the near future.

ABBREVIATIONS

AB-MECA N^6-(4-amino-3-iodobenyl)-5′-*N*-methylcarbamoyladenosine
A_1R adenosine A_1 receptor
$A_{2A}R$ adenosine A_{2A} receptor
$A_{2B}R$ adenosine A_{2B} receptor
A_3R adenosine A_3 receptor
AR adenosine receptor
BS-DMPX (*E*)-8-(3-bromostyryl)-3,7-dimethyl-1-propargylxanthine
CHA N^6-cyclohexyladenosine
CGS 21680 2-*p*-(2-carboxyethyl)-phenethylamino-5′-*N*-ethylcarboxamidoadenosine
CNS central nervous system
CPFPX 8-cyclopentyl-3-(3-fluoropropyl)-1-propylxanthine
CPIPX 8-cyclopentyl-3-((*E*)3-iodoprop-2-en-1-yl)-1-propylxanthine
DPCPX 8-cyclopentyl-1,3-dipropylxanthine
EPDX 8-dicyclopropylmethyl-1-ethyl-3-propylxanthine
FE@SUPPY 5-(2-fluoroethyl) 2,4-diethyl-3-(ethylsulfanylcarbonyl)-6-phenylpyridine-5-carboxylate
FRI194921 2-(1-methyl-4-piperidinyl)-6-(2-phenylpyrazolo[1,5-α]pyridin-3-yl)-3(2*H*)-pyridazinone
HPLC high-performance liquid chromatography
IS-DMPX (*E*)-3,7-dimethyl-8-(3-iodostyryl)-1-propargylxanthine
KF15372 8-dicyclopropylmethyl-1,3-dipropylxanthine
KF17837 (*E*)-8-(3,4-dimethoxystyryl)-1,3-dipropyl-7-methylxanthine
LGA Logan graphical analysis with plasma time–activity curve
MPDX 8-dicyclopropylmethyl-1-methyl-3-propylxanthine
PET positron emission tomography
pTAC plasma time–activity curve

***R*-PIA** *R*-N^6-(phenylisopropyl)-adenosine
SCH442416 5-amino-7-(3-(4-methoxyphenyl)propyl)-2-(2-furyl)-pyrazolo[4,3-e]-1,2,4-trizolo[1,5-c] pyrimidine
SCH58261 7-(2-phenylethyl)-5-amino-2-(2-furyl)-pyrazolo[4,3-e]-1,2,4-trizolo[1,5-c]pyrimidine
SPECT single-photon emission computed tomography
SRTM Simplified reference tissue model
TMSX = KF 18446 (*E*)-8-(3,4,5-trimethoxystyryl)-1,3,7-trimethylxanthine
tTAC tissue time–activity curve
ZM 241385 4-(2-[7-amino-2-{2-furyl}{1,2,4}triazolo{2,3,-a}{1,3,5}triazin-5-yl-amino]ethyl)phenol

REFERENCES

Agteresch, H.J., Dagnelie, P.C., van den Berg, J.W., Wilson, J.H., 1999. Adenosine triphosphate: established and potential clinical applications. Drugs 58, 211–232.

Albasanz, J.L., Rodriguez, A., Ferrer, I., Martin, M., 2007. Up-regulation of adenosine A_1 receptors in frontal cortex from Pick's disease cases. Eur. J. Neurosci. 26, 3501–3508.

Albasanz, J.L., Perez, S., Barrachina, M., Ferrer, I., Martin, M., 2008. Up-regulation of adenosine receptors in the frontal cortex in Alzheimer's disease. Brain Pathol. 18, 211–219.

Angelatou, F., Pagonopoulou, O., Maraziotis, T., Olivier, A., Villemeure, J.G., Avoli, M., Kostopoulos, G., 1993. Upregulation of A_1 adenosine receptors in human temporal lobe epilepsy: a quantitative autoradiographic study. Neurosci. Lett. 163, 11–14.

Angulo, E., Casado, V., Mallol, J., Canela, E.I., Vinals, F., Ferrer, I., Lluis, C., Franco, R., 2003. A_1 adenosine receptors accumulate in neurodegenerative structures in Alzheimer disease and mediate both amyloid precursor protein processing and tau phosphorylation and translocation. Brain Pathol. 13, 440–451.

Ascherio, A., Zhang, S.M., Hernan, M.A., Kawachi, I., Colditz, G.A., Speizer, F.E., Willett, W.C., 2001. Prospective study of caffeine consumption and risk of Parkinson's disease in men and women. Ann. Neurol. 50, 56–63.

Ascherio, A., LeWitt, P.A., Xu, K., Eberly, S., Watts, A., Matson, W.R., Marras, C., Kieburtz, K., Rudolph, A., Bogdanov, M.B., Schwid, S.R., Tennis, M., Tanner, C.M., Beal, M.F., Lang, A.E., Oakes, D., Fahn, S., Shoulson, I., Schwarzschild, M.A., 2009. Urate as a predictor of the rate of clinical decline in Parkinson disease. Arch. Neurol. 66, 1460–1468.

Bara-Jimenez, W., Sherzai, A., Dimitrova, T., Favit, A., Bibbiani, F., Gillespie, M., Morris, M.J., Mouradian, M.M., Chase, T.N., 2003. Adenosine A_{2A} receptor antagonist treatment of Parkinson's disease. Neurology 61, 293–296.

Bauer, A., Holschbach, M.H., Meyer, P.T., Boy, C., Herzog, H., Olsson, R.A., Coenen, H.H., Zilles, K., 2003. In vivo imaging of adenosine A_1 receptors in the human brain with [^{18}F]CPFPX and positron emission tomography. NeuroImage 19, 1760–1769.

Bauer, A., Ishiwata, K., 2009. Adenosine receptor ligands and PET imaging of the CNS. Handb. Exp. Pharmacol., 617–642.

Boy, C., Meyer, P.T., Kircheis, G., Holschbach, M.H., Herzog, H., Elmenhorst, D., Kaiser, H.J., Coenen, H.H., Haussinger, D., Zilles, K., Bauer, A., 2008. Cerebral A_1 adenosine receptors (A_1AR) in liver cirrhosis. Eur. J. Nucl. Med. Mol. Imaging 35, 589–597.

Calon, F., Dridi, M., Hornykiewicz, O., Bedard, P.J., Rajput, A.H., Di Paolo, T., 2004. Increased adenosine A_{2A} receptors in the brain of Parkinson's disease patients with dyskinesias. Brain 127, 1075–1084.

Carson, R.E., Channing, M.A., Blasberg, R.G., Dunn, B.B., Cohen, R.M., Rice, K.C., Herscovitch, P., 1993. Comparison of bolus and infusion methods for receptor quantitation: application to [^{18}F]cyclofoxy and positron emission tomography. J. Cereb. Blood Flow Metab. 13, 24–42.

Cross, H.R., Murphy, E., Black, R.G., Auchampach, J., Steenbergen, C., 2002. Overexpression of A_3 adenosine receptors decreases heart rate, preserves energetics, and protects ischemic hearts. Am. J. Physiol. Heart Circ. Physiol. 283, H1562–H1568.

Deckert, J., Abel, F., Kunig, G., Hartmann, J., Senitz, D., Maier, H., Ransmayr, G., Riederer, P., 1998. Loss of human hippocampal adenosine A_1 receptors in dementia: evidence for lack of specificity. Neurosci. Lett. 244, 1–4.

Deckert, J., Brenner, M., Durany, N., Zochling, R., Paulus, W., Ransmayr, G., Tatschner, T., Danielczyk, W., Jellinger, K., Riederer, P., 2003. Up-regulation of striatal adenosine A_{2A} receptors in schizophrenia. Neuroreport 14, 313–316.

Delle Donne, K.T., Sonsalla, P.K., 1994. Protection against methamphetamine-induced neurotoxicity to neostriatal dopaminergic neurons by adenosine receptor activation. J. Pharmacol. Exp. Ther. 271, 1320–1326.

Donoso, M.V., Lopez, R., Miranda, R., Briones, R., Huidobro-Toro, J.P., 2005. A_{2B} adenosine receptor mediates human chorionic vasoconstriction and signals through arachidonic acid cascade. Am. J. Physiol. Heart Circ. Physiol. 288, H2439–H2449.

Dunwiddie, T.V., Masino, S.A., 2001. The role and regulation of adenosine in the central nervous system. Annu. Rev. Neurosci. 24, 31–55.

Elmenhorst, D., Meyer, P.T., Winz, O.H., Matusch, A., Ermert, J., Coenen, H.H., Basheer, R., Haas, H.L., Zilles, K., Bauer, A., 2007. Sleep deprivation increases A_1 adenosine receptor binding in the human brain: a positron emission tomography study. J. Neurosci. 27, 2410–2415.

Eltzschig, H.K., Ibla, J.C., Furuta, G.T., Leonard, M.O., Jacobson, K.A., Enjyoji, K., Robson, S.C., Colgan, S.P., 2003. Coordinated adenine nucleotide phosphohydrolysis and nucleoside signaling in posthypoxic endothelium: role of ectonucleotidases and adenosine A_{2B} receptors. J. Exp. Med. 198, 783–796.

Fastbom, J., Pazos, A., Probst, A., Palacios, J.M., 1987. Adenosine A_1 receptors in the human brain: a quantitative autoradiographic study. Neuroscience 22, 827–839.

Ferre, S., Popoli, P., Gimenez-Llort, L., Finnman, U.B., Martinez, E., Scotti de Carolis, A., Fuxe, K., 1994. Postsynaptic antagonistic interaction between adenosine A_1 and dopamine D_1 receptors. Neuroreport 6, 73–76.

Ferre, S., Fredholm, B.B., Morelli, M., Popoli, P., Fuxe, K., 1997. Adenosine-dopamine receptor-receptor interactions as an integrative mechanism in the basal ganglia. Trends Neurosci. 20, 482–487.

Fredholm, B.B., IJzerman, A.P., Jacobson, K.A., Klotz, K.N., Linden, J., 2001. International Union of Pharmacology. XXV. Nomenclature and classification of adenosine receptors. Pharmacol. Rev. 53, 527–552.

Fredholm, B.B., Svenningsson, P., 2003. Adenosine-dopamine interactions: development of a concept and some comments on therapeutic possibilities. Neurology 61, S5–S9.

Fukumitsu, N., Ishii, K., Kimura, Y., Oda, K., Sasaki, T., Mori, Y., Ishiwata, K., 2003. Imaging of adenosine A_1 receptors in the human brain by positron emission tomography with [^{11}C]MPDX. Ann. Nucl. Med. 17, 511–515.

Fukumitsu, N., Ishii, K., Kimura, Y., Oda, K., Sasaki, T., Mori, Y., Ishiwata, K., 2005. Adenosine A_1 receptor mapping of the human brain by PET with 8-dicyclopropylmethyl-1-^{11}C-methyl-3-propylxanthine. J. Nucl. Med. 46, 32–37.

Fukumitsu, N., Ishii, K., Kimura, Y., Oda, K., Hashimoto, M., Suzuki, M., Ishiwata, K., 2008. Adenosine A_1 receptors using 8-dicyclopropylmethyl-1-[^{11}C]methyl-3-propylxanthine PET in Alzheimer's disease. Ann. Nucl. Med. 22, 841–847.

Giffin, N.J., Kowacs, F., Libri, V., Williams, P., Goadsby, P.J., Kaube, H., 2003. Effect of the adenosine A_1 receptor agonist GR79236 on trigeminal nociception with blink reflex recordings in healthy human subjects. Cephalalgia 23, 287–292.

Glass, M., Faull, R.L., Bullock, J.Y., Jansen, K., Mee, E.W., Walker, E.B., Synek, B.J., Dragunow, M., 1996. Loss of A_1 adenosine receptors in human temporal lobe epilepsy. Brain Res. 710, 56–68.

Glass, M., Dragunow, M., Faull, R.L., 2000. The pattern of neurodegeneration in Huntington's disease: a comparative study of cannabinoid, dopamine, adenosine and $GABA_A$ receptor alterations in the human basal ganglia in Huntington's disease. Neuroscience 97, 505–519.

Gunn, R.N., Lammertsma, A.A., Hume, S.P., Cunningham, V.J., 1997. Parametric imaging of ligand-receptor binding in PET using a simplified reference region model. NeuroImage 6, 279–287.

Guo, Y., Bolli, R., Bao, W., Wu, W.J., Black Jr., R.G., Murphree, S.S., Salvatore, C.A., Jacobson, M.A., Auchampach, J.A., 2001. Targeted deletion of the A_3 adenosine receptor confers resistance to myocardial ischemic injury and does not prevent early preconditioning. J. Mol. Cell. Cardiol. 33, 825–830.

Haas, H.L., Selbach, O., 2000. Functions of neuronal adenosine receptors. Naunyn. Schmiedebergs Arch. Pharmacol. 362, 375–381.

Hendrix, E.M.T., Tóth, B.G., 2010. Nonlinear programming algorithms. In: Du, D.-Z. (Ed.), Introduction to Nonlinear and Global Optimization. Springer Optimization and Its Applications, Vol. 37. Springer, New York, NY, pp. 91–136.

Holschbach, M.H., Olsson, R.A., Bier, D., Wutz, W., Sihver, W., Schuller, M., Palm, B., Coenen, H.H., 2002. Synthesis and evaluation of no-carrier-added 8-cyclopentyl-3-(3-[^{18}F]fluoropropyl)-1-propylxanthine ([^{18}F]CPFPX): a potent and selective A_1-adenosine receptor antagonist for in vivo imaging. J. Med. Chem. 45, 5150–5156.

Hurley, M.J., Mash, D.C., Jenner, P., 2000. Adenosine A_{2A} receptor mRNA expression in Parkinson's disease. Neurosci. Lett. 291, 54–58.

Ichise, M., Toyama, H., Innis, R.B., Carson, R.E., 2002. Strategies to improve neuroreceptor parameter estimation by linear regression analysis. J. Cereb. Blood Flow Metab. 22, 1271–1281.

Ikeda, M., Mackay, K.B., Dewar, D., McCulloch, J., 1993. Differential alterations in adenosine A_1 and kappa 1 opioid receptors in the striatum in Alzheimer's disease. Brain Res. 616, 211–217.

Innis, R.B., Cunningham, V.J., Delforge, J., Fujita, M., Gjedde, A., Gunn, R.N., Holden, J., Houle, S., Huang, S.C., Ichise, M., Iida, H., Ito, H., Kimura, Y., Koeppe, R.A., Knudsen, G.M., Knuuti, J., Lammertsma, A.A., Laruelle, M., Logan, J., Maguire, R.P., Mintun, M.A., Morris, E.D., Parsey, R., Price, J.C., Slifstein, M., Sossi, V., Suhara, T., Votaw, J.R., Wong, D.F., Carson, R.E., 2007. Consensus nomenclature for in vivo imaging of reversibly binding radioligands. J. Cereb. Blood Flow Metab. 27, 1533–1539.

Ishiwata, K., Nariai, T., Kimura, Y., Oda, K., Kawamura, K., Ishii, K., Senda, M., Wakabayashi, S., Shimada, J., 2002a. Preclinical studies on [^{11}C]MPDX for mapping adenosine A_1 receptors by positron emission tomography. Ann. Nucl. Med. 16, 377–382.

Ishiwata, K., Shimada, J., Ishii, K., Suzuki, F., 2002b. Assessment of adenosine A_{2A} receptors with PET as a new diagnostic tool for neurological disorders. Drugs Fut. 27, 569–576.

Ishiwata, K., Mishina, M., Kimura, Y., Oda, K., Sasaki, T., Ishii, K., 2005. First visualization of adenosine A_{2A} receptors in the human brain by positron emission tomography with [^{11}C]TMSX. Synapse 55, 133–136.

Ishiwata, K., Kimura, Y., de Vries, E.F.J., Elsinga, P.H., 2007. PET tracers for mapping adenosine receptors as probes for diagnosis of CNS disorders. CNS Agents Med. Chem. 7, 57–77.

Ishiwata, K., Kimura, Y., Oda, K., Ishii, K., Sakata, M., Kawasaki, K., Nariai, T., Suzuki, Y., Ishibashi, K., Mishina, M., Hashimoto, M., Ishikawa, M., Toyohara, J., 2010. Development of PET radiopharmaceuticals and their clinical applications at the Positron Medical Center. Geriatr. Gerontol. Int. 10 (Suppl. 1), S180–S196.

Jaarsma, D., Sebens, J.B., Korf, J., 1991. Reduction of adenosine A_1-receptors in the perforant pathway terminal zone in Alzheimer hippocampus. Neurosci. Lett. 121, 111–114.

Jacobson, K.A., Gao, Z.G., 2006. Adenosine receptors as therapeutic targets. Nat. Rev. Drug Discov. 5, 247–264.

Jansen, K.L., Faull, R.L., Dragunow, M., Synek, B.L., 1990. Alzheimer's disease: changes in hippocampal *N*-methyl-D-aspartate, quisqualate, neurotensin, adenosine, benzodiazepine, serotonin and opioid receptors – an autoradiographic study. Neuroscience 39, 613–627.

Jenner, P., Mori, A., Hauser, R., Morelli, M., Fredholm, B.B., Chen, J.F., 2009. Adenosine, adenosine A_{2A} antagonists, and Parkinson's disease. Parkinsonism Relat. Disord. 15, 406–413.

Jennings, L.L., Hao, C., Cabrita, M.A., Vickers, M.F., Baldwin, S.A., Young, J.D., Cass, C.E., 2001. Distinct regional distribution of human equilibrative nucleoside transporter proteins 1 and 2 (hENT1 and hENT2) in the central nervous system. Neuropharmacology 40, 722–731.

Joshi, A., Fessler, J.A., Koeppe, R.A., 2008. Improving PET receptor binding estimates from Logan plots using principal component analysis. J. Cereb. Blood Flow Metab. 28, 852–865.

Kachroo, A., Schwarzschild, M.A., 2012. Adenosine A_{2A} receptor gene disruption protects in an α-synuclein model of Parkinson's disease. *Ann. Neurol.*, 71, 278–282.

Kalaria, R.N., Sromek, S., Wilcox, B.J., Unnerstall, J.R., 1990. Hippocampal adenosine A_1 receptors are decreased in Alzheimer's disease. Neurosci. Lett. 118, 257–260.

Kase, H., 2001. New aspects of physiological and pathophysiological functions of adenosine A_{2A} receptor in basal ganglia. Biosci. Biotechnol. Biochem. 65, 1447–1457.

Kelsey, J.E., Langelier, N.A., Oriel, B.S., Reedy, C., 2009. The effects of systemic, intrastriatal, and intrapallidal injections of caffeine and systemic injections of A_{2A} and A_1 antagonists on forepaw stepping in the unilateral 6-OHDA-lesioned rat. Psychopharmacology 201, 529–539.

Kimura, Y., Ishii, K., Fukumitsu, N., Oda, K., Sasaki, T., Kawamura, K., Ishiwata, K., 2004. Quantitative analysis of adenosine A_1 receptors in human brain using positron emission tomography and [1-methyl-^{11}C]8-dicyclopropylmethyl-1-methyl-3-propylxanthine. Nucl. Med. Biol. 31, 975–981.

Kimura, Y., Naganawa, M., Shidahara, M., Ikoma, Y., Watabe, H., 2007. PET kinetic analysis—pitfalls and a solution for the Logan plot. Ann. Nucl. Med. 21, 1–8.

Kimura, Y., Naganawa, M., Mishina, M., Sakata, M., Oda, K., Ishii, K., Ishiwata, K., 2009. Simplified quantification of adenosine A_1 receptor with [^{11}C]MPDX and graphical analysis. In: XXIV International Symposium on Cerebral Blood Flow, Metabolism and Function & IX International Conference on Quantification of Brain Function with PET. , p. 82.

Kostic, V.S., Svetel, M., Sternic, N., Dragasevic, N., Przedborski, S., 1999. Theophylline increases "on" time in advanced Parkinsonian patients. Neurology 52, 1916.

Kulisevsky, J., Barbanoj, M., Gironell, A., Antonijoan, R., Casas, M., Pascual-Sedano, B., 2002. A double-blind crossover, placebo-controlled study of the adenosine A_{2A} antagonist theophylline in Parkinson's disease. Clin. Neuropharmacol. 25, 25–31.

Kurumaji, A., Toru, M., 1998. An increase in [^{3}H] CGS21680 binding in the striatum of postmortem brains of chronic schizophrenics. Brain Res. 808, 320–323.

Lammertsma, A.A., Hume, S.P., 1996. Simplified reference tissue model for PET receptor studies. NeuroImage 4, 153–158.

Latini, S., Pedata, F., 2001. Adenosine in the central nervous system: release mechanisms and extracellular concentrations. J. Neurochem. 79, 463–484.

Lees, A.J., Hardy, J., Revesz, T., 2009. Parkinson's disease. Lancet 373, 2055–2066.

Logan, J., Fowler, J.S., Volkow, N.D., Wolf, A.P., Dewey, S.L., Schlyer, D.J., MacGregor, R.R., Hitzemann, R., Bendriem, B., Gatley, S.J., et al., 1990. Graphical analysis of reversible radioligand binding from time-activity measurements applied to [*N*-^{11}C-methyl]-(-)-cocaine PET studies in human subjects. J. Cereb. Blood Flow Metab. 10, 740–747.

Logan, J., Fowler, J.S., Volkow, N.D., Wang, G.J., Ding, Y.S., Alexoff, D.L., 1996. Distribution volume ratios without blood sampling from graphical analysis of PET data. J. Cereb. Blood Flow Metab. 16, 834–840.

Lopes, L.V., Cunha, R.A., Ribeiro, J.A., 1999. Increase in the number, G protein coupling, and efficiency of facilitatory adenosine A_{2A} receptors in the limbic cortex, but not striatum, of aged rats. J. Neurochem. 73, 1733–1738.

Martinez-Mir, M.I., Probst, A., Palacios, J.M., 1991. Adenosine A_2 receptors: selective localization in the human basal ganglia and alterations with disease. Neuroscience 42, 697–706.

Matsuya, T., Takamatsu, H., Murakami, Y., Noda, A., Ichise, R., Awaga, Y., Nishimura, S., 2005. Synthesis and evaluation of [^{11}C]FR194921 as a nonxanthine-type PET tracer for adenosine A_1 receptors in the brain. Nucl. Med. Biol. 32, 837–844.

Meerlo, P., Roman, V., Farkas, E., Keijser, J.N., Nyakas, C., Luiten, P.G., 2004. Ageing-related decline in adenosine A_1 receptor binding in the rat brain: an autoradiographic study. J. Neurosci. Res. 78, 742–748.

Meyer, P.T., Bier, D., Holschbach, M.H., Boy, C., Olsson, R.A., Coenen, H.H., Zilles, K., Bauer, A., 2004. Quantification of cerebral A_1 adenosine receptors in humans using [^{18}F]CPFPX and PET. J. Cereb. Blood Flow Metab. 24, 323–333.

Meyer, P.T., Elmenhorst, D., Bier, D., Holschbach, M.H., Matusch, A., Coenen, H.H., Zilles, K., Bauer, A., 2005a. Quantification of cerebral A_1 adenosine receptors in humans using [^{18}F]CPFPX and PET: an equilibrium approach. NeuroImage 24, 1192–1204.

Meyer, P.T., Elmenhorst, D., Zilles, K., Bauer, A., 2005b. Simplified quantification of cerebral A_1 adenosine receptors using [^{18}F]CPFPX and PET: analyses based on venous blood sampling. Synapse 55, 212–223.

Meyer, P.T., Elmenhorst, D., Matusch, A., Winz, O., Zilles, K., Bauer, A., 2006. A_1 adenosine receptor PET using [^{18}F]CPFPX: displacement studies in humans. NeuroImage 32, 1100–1105.

Meyer, P.T., Elmenhorst, D., Boy, C., Winz, O., Matusch, A., Zilles, K., Bauer, A., 2007. Effect of aging on cerebral A_1 adenosine receptors: a [^{18}F]CPFPX PET study in humans. Neurobiol. Aging 28, 1914–1924.

Mintun, M.A., Raichle, M.E., Kilbourn, M.R., Wooten, G.F., Welch, M.J., 1984. A quantitative model for the in vivo assessment of drug binding sites with positron emission tomography. Ann. Neurol. 15, 217–227.

Mishina, M., Ishiwata, K., Kimura, Y., Naganawa, M., Oda, K., Kobayashi, S., Katayama, Y., Ishii, K., 2007. Evaluation of distribution of adenosine A_{2A} receptors in normal human brain measured with [^{11}C] TMSX PET. Synapse 61, 778–784.

Mishina, M., Ishiwata, K., Naganawa, M., Kimura, Y., Kitamura, S., Suzuki, M., Hashimoto, M., Ishibashi, K., Oda, K., Sakata, M., Hamamoto, M., Kobayashi, S., Katayama, Y., Ishii, K., 2011. Adenosine A_{2A} receptors measured with [^{11}C]TMSX PET in the striata of Parkinson's disease patients. PLoS One 6, e17338.

Moresco, R.M., Todde, S., Belloli, S., Simonelli, P., Panzacchi, A., Rigamonti, M., Galli-Kienle, M., Fazio, F., 2005. In vivo imaging of adenosine A_{2A} receptors in rat and primate brain using [^{11}C]SCH442416. Eur. J. Nucl. Med. Mol. Imaging 32, 405–413.

Morissette, M., Dridi, M., Calon, F., Tahar, A.H., Meltzer, L.T., Bedard, P.J., Di Paolo, T., 2006. Prevention of dyskinesia by an NMDA receptor antagonist in MPTP monkeys: effect on adenosine A_{2A} receptors. Synapse 60, 239–250.

Naganawa, M., Kimura, Y., Mishina, M., Manabe, Y., Chihara, K., Oda, K., Ishii, K., Ishiwata, K., 2007. Quantification of adenosine A_{2A} receptors in the human brain using [^{11}C]TMSX and positron emission tomography. Eur. J. Nucl. Med. Mol. Imaging 34, 679–687.

Nariai, T., Shimada, Y., Ishiwata, K., Nagaoka, T., Shimada, J., Kuroiwa, T., Ono, K., Ohno, K., Hirakawa, K., Senda, M., 2003. PET imaging of adenosine A_1 receptors with ^{11}C-MPDX as an indicator of severe cerebral ischemic insult. J. Nucl. Med. 44, 1839–1844.

Noguchi, J., Ishiwata, K., Furuta, R., Simada, J., Kiyosawa, M., Ishii, S., Endo, K., Suzuki, F., Senda, M., 1997. Evaluation of carbon-11 labeled KF15372 and its ethyl and methyl derivatives as a potential CNS adenosine A_1 receptor ligand. Nucl. Med. Biol. 24, 53–59.

O'Reilly, E.J., Gao, X., Weisskopf, M.G., Chen, H., Schwarzschild, M.A., Spiegelman, D., Ascherio, A., 2010. Plasma urate and Parkinson's disease in women. Am. J. Epidemiol. 172, 666–670.

Ogden, R.T., 2003. Estimation of kinetic parameters in graphical analysis of PET imaging data. Stat. Med. 22, 3557–3568.

Oishi, Y., Huang, Z.L., Fredholm, B.B., Urade, Y., Hayaishi, O., 2008. Adenosine in the tuberomammillary nucleus inhibits the histaminergic system via A_1 receptors and promotes non-rapid eye movement sleep. Proc. Natl. Acad. Sci. U. S. A. 105, 19992–19997.

Palmer, T.M., Stiles, G.L., 1995. Adenosine receptors. Neuropharmacology 34, 683–694.

Rahman, A., 2009. The role of adenosine in Alzheimer's disease. Curr. Neuropharmacol. 7, 207–216.

Ramlackhansingh, A.F., Bose, S.K., Ahmed, I., Turkheimer, F.E., Pavese, N., Brooks, D.J., 2011. Adenosine 2A receptor availability in dyskinetic and nondyskinetic patients with Parkinson disease. Neurology 76, 1811–1816.

Rodriguez, A., Martin, M., Albasanz, J.L., Barrachina, M., Espinosa, J.C., Torres, J.M., Ferrer, I., 2006. Adenosine A_1 receptor protein levels and activity is increased in the cerebral cortex in Creutzfeldt-Jakob disease and in bovine spongiform encephalopathy-infected bovine-PrP mice. J. Neuropathol. Exp. Neurol. 65, 964–975.

Ross, G.W., Abbott, R.D., Petrovitch, H., Morens, D.M., Grandinetti, A., Tung, K.H., Tanner, C.M., Masaki, K.H., Blanchette, P.L., Curb, J.D., Popper, J.S., White, L.R., 2000. Association of coffee and caffeine intake with the risk of Parkinson disease. JAMA 283, 2674–2679.

Sawynok, J., 1998. Adenosine receptor activation and nociception. Eur. J. Pharmacol. 347, 1–11.

Schwarzschild, M.A., Chen, J.F., Ascherio, A., 2002. Caffeinated clues and the promise of adenosine A_{2A} antagonists in PD. Neurology 58, 1154–1160.

Schwarzschild, M.A., Schwid, S.R., Marek, K., Watts, A., Lang, A.E., Oakes, D., Shoulson, I., Ascherio, A., Hyson, C., Gorbold, E., Rudolph, A., Kieburtz, K., Fahn, S., Gauger, L., Goetz, C., Seibyl, J., Forrest, M., Ondrasik, J., 2008. Serum urate as a predictor of clinical and radiographic progression in Parkinson disease. Arch. Neurol. 65, 716–723.

Schwarzschild, M.A., Marek, K., Eberly, S., Oakes, D., Shoulson, I., Jennings, D., Seibyl, J., Ascherio, A., 2011. Serum urate and probability of dopaminergic deficit in early "Parkinson's disease". Mov. Disord. 26, 1864–1868.

Shah, R.H., Frishman, W.H., 2009. Adenosine$_1$ receptor antagonism: a new therapeutic approach for the treatment of decompensated heart failure. Cardiol. Rev. 17, 125–131.

Shidahara, M., Seki, C., Naganawa, M., Sakata, M., Ishikawa, M., Ito, H., Kanno, I., Ishiwata, K., Kimura, Y., 2009. Improvement of likelihood estimation in Logan graphical analysis using maximum a posteriori for neuroreceptor PET imaging. Ann. Nucl. Med. 23, 163–171.

Slifstein, M., Laruelle, M., 2000. Effects of statistical noise on graphic analysis of PET neuroreceptor studies. J. Nucl. Med. 41, 2083–2088.

Svenningsson, P., Hall, H., Sedvall, G., Fredholm, B.B., 1997. Distribution of adenosine receptors in the postmortem human brain: an extended autoradiographic study. Synapse 27, 322–335.

Todde, S., Moresco, R.M., Simonelli, P., Baraldi, P.G., Cacciari, B., Spalluto, G., Varani, K., Monopoli, A., Matarrese, M., Carpinelli, A., Magni, F., Kienle, M.G., Fazio, F., 2000. Design, radiosynthesis, and biodistribution of a new potent and selective ligand for in vivo imaging of the adenosine A_{2A} receptor system using positron emission tomography. J. Med. Chem. 43, 4359–4362.

Ulas, J., Brunner, L.C., Nguyen, L., Cotman, C.W., 1993. Reduced density of adenosine A_1 receptors and preserved coupling of adenosine A_1 receptors to G proteins in Alzheimer hippocampus: a quantitative autoradiographic study. Neuroscience 52, 843–854.

Urade, Y., Eguchi, N., Qu, W.M., Sakata, M., Huang, Z.L., Chen, J.F., Schwarzschild, M.A., Fink, J.S., Hayaishi, O., 2003. Sleep regulation in adenosine A_{2A} receptor-deficient mice. Neurology 61, S94–S96.

Watabe, H., Ikoma, Y., Kimura, Y., Naganawa, M., Shidahara, M., 2006. PET kinetic analysis—compartmental model. Ann. Nucl. Med. 20, 583–588.

Weisskopf, M.G., O'Reilly, E., Chen, H., Schwarzschild, M.A., Ascherio, A., 2007. Plasma urate and risk of Parkinson's disease. Am. J. Epidemiol. 166, 561–567.

Xiao, D., Cassin, J.J., Healy, B., Burdett, T.C., Chen, J.F., Fredholm, B.B., Schwarzschild, M.A., 2011. Deletion of adenosine A_1 or A_{2A} receptors reduces L-3,4-dihydroxyphenylalanine-induced dyskinesia in a model of Parkinson's disease. Brain Res. 1367, 310–318.

Yabuuchi, K., Kuroiwa, M., Shuto, T., Sotogaku, N., Snyder, G.L., Higashi, H., Tanaka, M., Greengard, P., Nishi, A., 2006. Role of adenosine A_1 receptors in the modulation of dopamine D_1 and adenosine A_{2A} receptor signaling in the neostriatum. Neuroscience 141, 19–25.

Zambrowicz, B.P., Turner, C.A., Sands, A.T., 2003. Predicting drug efficacy: knockouts model pipeline drugs of the pharmaceutical industry. Curr. Opin. Pharmacol. 3, 563–570.

CHAPTER EIGHT

Human Brain Imaging of Dopamine D_1 Receptors

Hironobu Fujiwara[1], Hidehiko Takahashi[1,2], Hitoshi Shimada[1], Yoshiro Okubo[3] and Tetsuya Suhara[1]

[1]Molecular Imaging Center, Department of Molecular Neuroimaging, National Institute of Radiological Sciences, Inage, Chiba, Japan
[2]Department of Psychiatry, Kyoto University Graduate School of Medicine, Sakyo-ku, Kyoto, Japan
[3]Department of Neuropsychiatry, Nippon Medical School, Bunkyo-ku, Tokyo, Japan

1. INTRODUCTION

Dopamine (DA), one of the catecholamines, is considered to play a major role in affective, cognitive, and motor functions such as reward processing and executive function as well as the regulation of involuntary movement. DA cells are located in the midbrain, ventral tegmental area, and substantia nigra and project to the striatum, which is the main input structure of the basal ganglia and cortical regions mainly through the DA system, A9, and A10 (Daunstrom and Fuxe, 1964). More specifically, A9 projects to the dorsal part of the striatum, which may be related more to motor functions than to cognitive functions, and A10 projects to the ventral striatum, which is considered to be related to affective and cognitive functions, for example, pleasure and reward processing. A10 also has direct projections to cortical regions. The physiological function of DA is mediated through its G-protein-coupled receptor family. DA receptors consist of a receptor family that is divided into two major groups, D_1 (D_1 and D_5) and D_2 (D_2, D_3, and D_4), which provide excitatory and inhibitory inputs to the striatum, respectively.

Recent functional imaging studies with neuropsychological measures suggest that cortical regions including the prefrontal cortex (PFC) and the striatum would be included among candidates of the major neural underpinnings of cognitive functions, and ample literature has suggested the relationship between the deficits of these brain regions and several neuropsychiatric illnesses related to the dysregulation of DA-related neurotransmission such as schizophrenia and Parkinson disease (PD).

In human positron emission tomographic (PET) studies, the striatum, the most DA-rich region, has been intensively investigated in terms of association with cognitive functions as well as pathologies of neuropsychiatric illnesses. In addition, the relationship between D_1 receptors (D_1Rs) and PFC functions has been widely investigated because D_1Rs in the cortex (especially in PFC) are several times more abundant than D_2 receptors (D_2Rs) (Hall et al., 1994). The role of D_1R in PFC function including working

Imaging of the Human Brain in Health and Disease
http://dx.doi.org/10.1016/B978-0-12-418677-4.00008-7

memory is well acknowledged, although relatively little is known about its role in other cognitive or emotional functions.

In this chapter, we review the evidence of in vivo D_1R function, focusing on its associations with cognitive functions and pathologies of psychiatric illnesses while introducing our D_1R PET imaging studies.

2.1. AFFECTIVE AND COGNITIVE FUNCTIONS AND D_1 RECEPTORS IN HUMANS

2.1.1. D_1 Receptors and Working Memory

The role of D_1R in cognitive processing has been repeatedly implicated in both animal and human studies these past few decades. In nonhuman primates, Sawaguchi and Goldman-Rakic (1994) demonstrated that local administration of D_1R antagonists, but not D_2R ones, into PFC induced impairment in working memory task. In humans, Müller et al. (1998) reported that systemic administration of a mixed D_1/D_2 agonist facilitated working memory, whereas a selective D_2 agonist had no effect, indicating that the dopaminergic modulation of working memory processes is mediated primarily via D_1R. Moreover, regarding D_1R, a body of animal studies has indicated that stimulation of D_1R in PFC produces an inverted U-shaped dose–response curve, such that too little or too much D_1R stimulation impairs PFC functions (Goldman-Rakic et al., 2000; Williams and Castner, 2006). On the other hand, some human studies reported that systemic administration of D_2 agonist or antagonist modulated cognitive functions including working memory, which are subserved by PFC (McDowell et al., 1998; Mehta et al., 1999). As well, in an animal study, mice lacking D_2R were reported to have a working memory deficit (Glickstein et al., 2002). These studies, however, did not reveal the regions most responsible for these effects. Moreover, although the involvement of D_1R in working memory is widely recognized, as mentioned above, it was not clear whether D_1R stimulation alone or the combination of D_1R and D_2R stimulation is most effective.

Noninvasive molecular neuroimaging, i.e. PET scan, allows us to quantify DA receptors in vivo. Takahashi et al. (2008) has investigated the association between D_1R/D_2R expression and executive functions measured by Wisconsin Card Sorting Test (WCST) in healthy volunteers. The results showed that prefrontal D_1R was more related to executive functions than prefrontal D_2R, and an inverted U-shaped relation between prefrontal D_1R binding and prefrontal function was observed under normal physiological conditions.

More specifically, quadratic regression analysis revealed a significant "U-shaped" relation between the values of binding potentials (BP_{ND}) of [^{11}C]SCH23390 (a selective D_1R antagonist) in PFC and total errors (TEs) and categories achieved (CA) in WCST (Figure 1), i.e. too high or too low level of D_1R expression in PFC leads to high errors

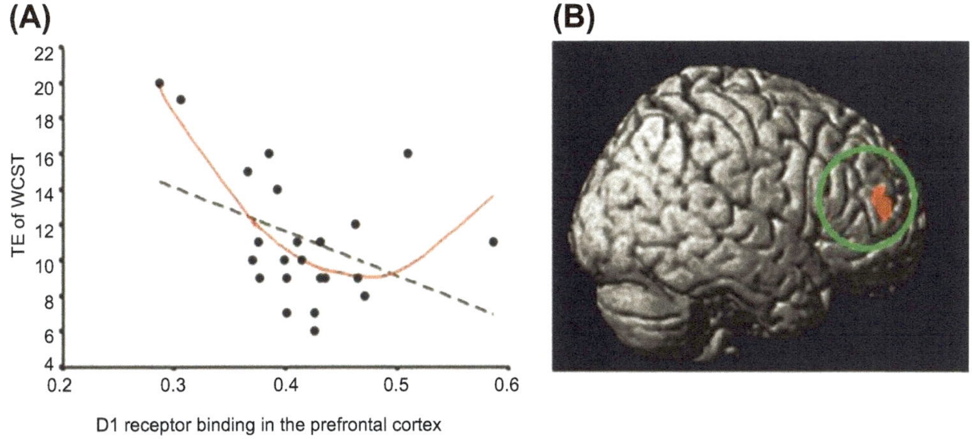

Figure 1 Quadratic (inverted U-shaped) relation between D_1R binding in PFC and performance of WCST. (A) Region of Interest (ROI) analysis revealed a significant quadratic regression between D_1R binding in PFC and TE of WCST. Red solid line, quadratic regression; black broken line, linear regression. (B) Statistical Parametric Mapping (SPM) analysis also revealed a significant quadratic regression between prefrontal D_1R binding and TE of WCST. The results were essentially the same in a subscale "categories achieved" as those in TE.

and a low number of CA. On the other hand, no relation was found between the BP_{ND} value of [^{11}C]FLB457 (a selective D_2 antagonist) in PFC and any neuropsychological measures. In the hippocampus (HPC), not D_1R, but D_2R binding showed positive linear correlations not only with memory function but also with cognitive functions. HPC D_2R contributes to local HPC functions (long-term memory) and to remote cognitive brain functions, which are mainly subserved by PFC, via the HPC–PFC pathway. The findings of the study by Takahashi et al. (2008) might suggest that orchestration of prefrontal D_1R and D_2R in other regions might be necessary for human executive function including working memory.

2.1.2. PET Imaging of D_1 Receptors and D_2 Receptors and Amygdala Function

The role of the amygdala in emotional processing is well known, and especially negative emotional stimuli such as fear and threatening (LeDoux, 2000). The structure receives moderate innervation of dopaminergic fibers (Asan, 1998), and D_1R and D_2R are moderately expressed in this region (Ito et al., 2008). In an animal study, it was demonstrated that DA enhances the response of the amygdala by augmenting excitatory sensory input via D_2R stimulation and attenuating inhibitory prefrontal input to the amygdala through D_1R stimulation, that is, both D_1R and D_2R stimulations would contribute to enhancement of the excitability of amygdala projection neurons via postsynaptic mechanism. In human studies, an functional magnetic resonance imaging (fMRI) study reported that dopaminergic drug therapy such as levodopa or DA agonists partially restored amygdala activation due to emotional task in patients with PD who showed no significant amygdala activation during drug-off states (Tessitore et al., 2002). In addition, another fMRI study of healthy volunteers has demonstrated that amphetamine potentiated the response of the amygdala during an emotional task. More recently, Kienast et al. (2008) reported that DA storage capacity in the human amygdala, measured with 6-[^{18}F]fluoro-L-dopa PET, was positively correlated with fMRI signal changes in the amygdala. However, the contribution of DA D_1R and D_2R to amygdala activation in response to affective stimuli is unknown in humans.

To investigate the relation between amygdala activation and DA receptor subtype, Takahashi et al. (2010a) conducted a multimodal neuroimaging study in which D_1R and D_2R bindings in the amygdala were measured with PET and amygdala activation in response to fearful stimuli was assessed by fMRI. Healthy male subjects underwent fMRI for measuring the amygdala response to fearful face stimuli, and then both D_1R and D_2R in the amygdala were measured using PET with [^{11}C]SCH23390 and [^{11}C]FLB457, respectively.

As a result, bilateral amygdala activations induced by fearful faces were identified in a group analysis. Both voxelwise SPM analysis and ROI analysis revealed that BOLD signals in the amygdala induced by fearful faces were positively correlated with D_1R binding in the amygdala but not with D_2R binding in the amygdala (Figure 8.2). That

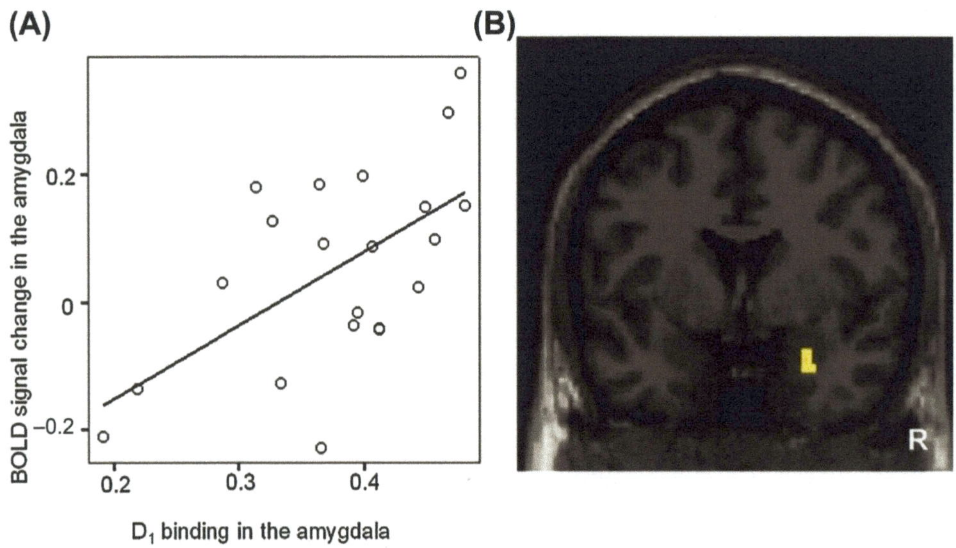

Figure 2 (A) ROIs correlation analysis revealed significant positive correlations between D_1R binding in the amygdala and the degree of amygdala activation. (B) SPM correlation analysis also revealed similar correlations. R indicates right.

is, individuals with high D_1R density in the amygdala tend to show greater amygdala activation in response to fearful stimuli.

It might require some caution to conclude which receptor, i.e. D_1R or D_2R, is more efficient in regulating the amygdala response to emotional stimuli. In our previous pharmacological fMRI study, the involvement of D_2R in amygdala activation was suggested (Takahashi et al., 2005). Also, D_1R antagonist for clinical use is at present not available.

Nevertheless, our multimodal neuroimaging study using PET with dual radioligands and fMRI, i.e. D_1R and D_2R comparison in terms of their association with amygdala activation evoked by fearful condition, suggested that D_1R plays a major role in the overall potentiation of the amygdala response. At a behavioral level, previous animal studies repeatedly reported that systemic and local applications into the amygdala of D_1 agonist and antagonist potentiate and decrease fear response in animal, respectively. Although some studies reported that applications of D_2 agonist and antagonist induced similar effects, the results were less consistent compared to D_1-mediated effects (for review, see de la Mora et al., 2010; Pezze and Feldon, 2004). Thus, our finding could be regarded as being consistent with previous behavioral pharmacological studies. The combination of PET molecular imaging and fMRI seems to represent a powerful approach for understanding molecular functions in affective responses.

2.1.3. D_1 Receptors for Pleasure, Reward, and Substance Dependence

The mesolimbic DA system is considered to be attributable to the processing of pleasure, and it is known that acute use of drugs such as amphetamine, cocaine, etc., induces "dependence". The sensation of "lift", intoxication, and euphoria due to activation of the mesolimbic DA system also reinforces the desire to take more excessive amounts of these drugs, and this repetitive drug taking is considered to affect the DA system itself, leading to psychological (mental) dependence.

The major psychoactive component of tobacco, nicotine, has pharmacologic characteristics common to other drugs of abuse. It causes the desire to take the drug (craving), which contributes to the development and maintenance of drug dependence.

Nicotine is reported to produce its reinforcing and addictive properties by activating the dopaminergic pathway (Corrigall et al., 1992; Koob and Le Moal, 1997; Rice and Cragg, 2004; Takahashi et al., 2007).

Regarding the regional molecular mechanism underlying cigarette craving, in an imaging study using PET, smokers had greater increases in glucose metabolism in the paralimbic system in response to the cigarette cue than did nonsmokers (Brody et al., 2002). Likewise, in an fMRI study of nicotine-deprived smokers, higher activation was observed in the mesolimbic DA system after exposure to smoking-related images compared with neutral images (David et al., 2005; McClernon et al., 2005). In contrast, reduction of D_1R binding in the ventral striatum was reported in cigarette smokers in a PET study (Dagher et al., 2001). These studies suggested that adaptive change of the mesolimbic DA system to chronic cigarette smoking contributed to the development of craving.

Yasuno et al. (2007) directly investigated the association of D_1R binding and regional brain activities related to cigarette craving in a multimodal imaging study. Furthermore, D_1R binding was examined successively to determine the effect of abstinence over 6 months. The results showed cue-induced activation in the left ventral striatum in smokers. D_1R binding in the region showed a negative relationship with cue-induced craving and changes in regional cerebral blood flow (rCBF) (Figure 3). D_1R binding was significantly low in smokers, and then an increasing trend was observed after smoking abstinence

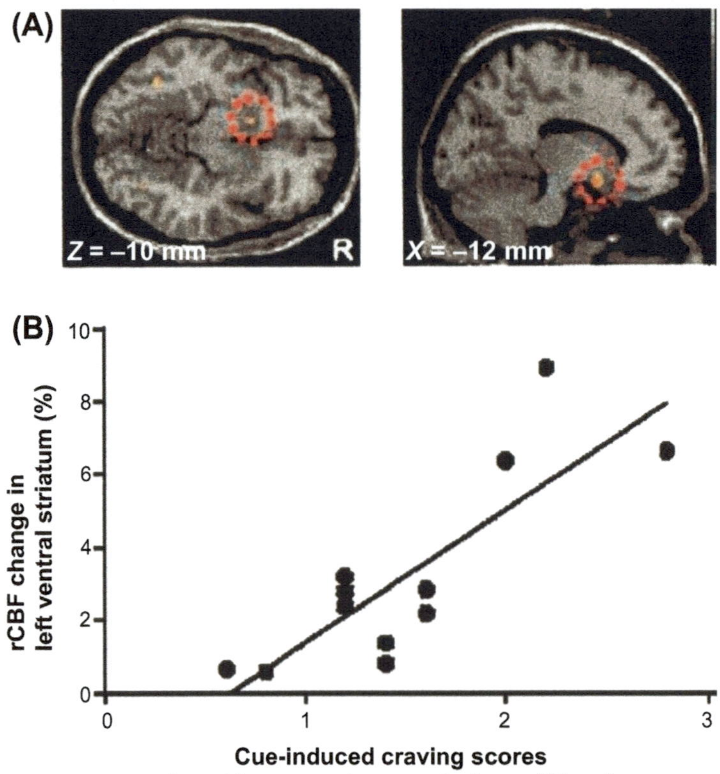

Figure 3 (A) Site of left ventral striatum in $[O^{15}]H_2O$ activation study. SPM projections superimposed on representative transaxial and sagittal magnetic resonance images in voxel-based analysis of regional cerebral blood flow (rCBF) changes induced by smoking cues ($p < 0.001$, uncorrected). R indicates right; coordinates in standard space are indicated at lower left. (B) Scatter plot of the correlation between cue-induced craving scores and rCBF changes in the cluster of the left ventral striatum in 12 smokers.

for 6 month. Thus, D_1R binding and cue-induced rCBF changes in the ventral striatum suggest an important role of D_1R in this region in cigarette craving. Considering the relationship between the role of the ventral striatum in pleasure- and reward-related processing and D_1R functions in the region, D_1R functions of the mesolimbic DA system might be generally due to various kinds of "dependence" such as nicotine, alcohol, and cannabis.

2.1.4. D_1 Receptors Decision Making Under Risk, and Dependence

People often overestimate low probabilities, e.g. in case of playing lotteries, and underestimate high probabilities. In this sense, misestimating risk could lead to disadvantaged choices such as initiation of drug use (or gambling) and transition to their regular use. Although the normative theory in decision making under risks assumes that people typically take the probability-weighted expectation over possible utilities, experimental studies of choices among risks suggest that outcome probabilities are transformed nonlinearly into subjective decision weights by a nonlinear weighting function that overweights low probabilities and underweights high probabilities. Recent studies have revealed the neurocognitive mechanism of decision making under risk. However, the role of modulatory neurotransmission in this process remains unclear.

Using PET imaging, Takahashi et al. (2010b) directly investigated whether DA D_1R and D_2R in the brain are associated with transformation of probabilities into decision weights in healthy volunteers. In their study, gambles were presented to the participants on a computer monitor outside the scanner. Each gamble contained two options with different combinations of objective probabilities and outcomes, and the participants made their choices. By those gambles, it was possible to estimate "certainty equivalent" (Tversky and Kahneman, 1992; Paulus and Frank, 2006), i.e. the amount of a sure payoff at which a player is indifferent between the sure payoff and the gamble. On the basis of this certainty equivalents estimation experiment, probability weighting was estimated using the one-parameter function derived axiomatically by Prelec (1998), $w(p) = \exp\{-(\ln(1/p))^{\alpha}\}$ with $0 < \alpha < 1$. This $w(p)$ function has an inverted S shape with a fixed inflection point at $p = 1/e = 0.37$ (at that point the probability $1/e$ also receives decision weight $1/e$). Nonlinearity is fully captured by a single parameter α. A smaller value of α (closer to 0) means a more nonlinear inflected weighting function and a higher value (closer to 1) means a more linear weighting function. At $\alpha = 1$ the function is linear.

As a result, the binding of striatal D_1R is negatively correlated with the degree of nonlinearity of weighting function, i.e. individuals with lower striatal D_1R density tend to be more risk seeking for low-probability gambles and more risk averse for high-probability gambles. This finding should contribute to a better understanding of the molecular mechanism of risky choice and extreme or impaired decision making observed in drug and gambling addiction (Figure 4). This finding also led us to the intuitive conjecture that D_1 agonist might be useful for easing misestimation of risk, and consequently be beneficial for pathological gambling.

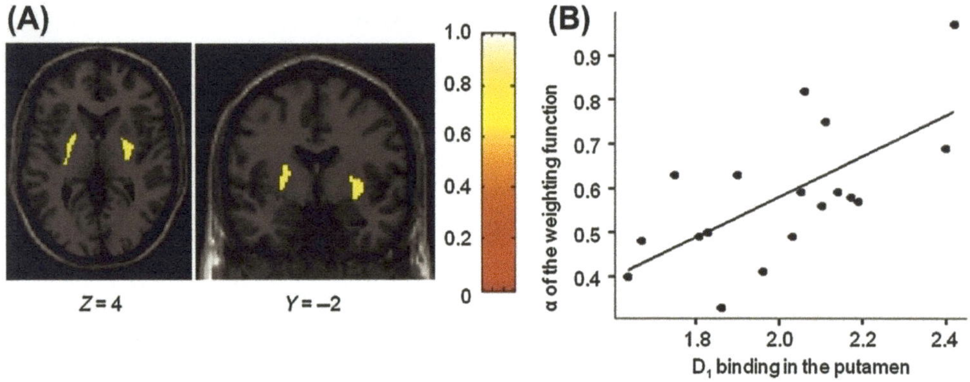

Figure 4 Correlation between nonlinearity of probabilities weighting and D_1 binding in the striatum ($N=18$). (A) Image showing regions of correlation between nonlinearity parameter of weighting function and D_1 binding in the striatum. The bar shows the range of the correlation coefficient. (B) Plots and regression line of correlation between nonlinearity parameter and binding potential of the putamen ($r=0.66$, $p=0.003$).

On the other hand, clinical reports indicated the association between DA agonist medication and the emergence of pathological gambling in PD patients (Gallagher et al., 2007). It has been reported that mixed D_1R/D_2R agonists can promote pathological gambling (Lu et al., 2006). Thus, the role of D_1R in DA agonist-induced pathological gambling remains unclear, and little is known about the complex mechanism of pathological gambling in general. However, we venture to propose a tentative hypothesis: low-level striatal D_1 binding (which might in part be determined by genetic factors) is linked to a risk-seeking trait. Risk-seeking trait has been shown to be linked to enhanced activation and DA release in the striatum during risk-seeking behavior (Leyton et al., 2002; St Onge and Floresco, 2009). Chronic exposure to unusually high release of DA by risk-seeking behavior might induce downregulation of D_1R (Moore et al., 1998; Yasuno et al., 2007). The further decrease in D_1R binding then leads to further risk seeking. Low-level striatal D_1 binding could therefore be a gateway to the vicious cycle, thus creating a predisposition to drug addiction as well as addictions of nonsubstance type, that is, process addiction including pathological gambling, excessive sex, and shopping.

Considering our previous study regarding tobacco and risky choice and their association with D_1R function (Yasuno et al., 2007; Takahashi et al., 2010b), we believe that understanding extreme or impaired decision making can contribute to the assessment and prevention of drug addiction and process dependence including pathological gambling and the development of novel pharmacological therapies for these substance- and nonsubstance-type addictions.

2.1.5. Social Pain and Dopamine Neurotransmission

We often evaluate the self and others from social comparisons. We feel envy when the target person has superior and self-relevant characteristics. When envy is evoked, we often have a desire to possess the same advantage or may wish that the other lacks it. When misfortune occurs to others, we can sympathize and have feelings of concern and sorrow for the other person, although we can also experience schadenfreude, a feeling of pleasure at another's suffering. Schadenfreude is closely related to envy, and it is more likely to arise when misfortune happens to a person who is advantaged and self-relevant than to someone who is neither advantaged nor self-relevant. In a word, schadenfreude occurs when envied persons fall from grace. To elucidate the neurocognitive mechanisms of envy and schadenfreude, Takahashi et al. (2008) conducted two fMRI studies. In the first study, the participants read information concerning target persons characterized by levels of possession and self-relevance of comparison domains. When the target person's possession was superior and self-relevant, stronger envy and stronger anterior cingulate cortex (ACC) activation were induced. In the second study, stronger schadenfreude and stronger striatum activation were induced when misfortunes happened to envied persons. ACC activation in the first study predicted the ventral striatum activation in the second study, indicating that the ACC–ventral striatum connection might play an important role in the processing of envy and schadenfreude (Figure 8.5(A), (B)).

Although it is unclear whether a neural mechanism of envy is associated with that of risky choice (namely, decision making) or not, at least, it would be plausible that the ACC–striatum connection is involved in various types of processing. In this context, it should be noted that subregions of the most DA-rich region, the striatum, would have their own different neural connections with extrastriatal regions and possibly play different roles from one subregion to another, namely, the concept of "striatal functional subdivisions", i.e. associative, sensorimotor, and limbic subdivisions (Alexander et al., 1986; Joel and Weiner, 2000). Regarding the relationship between these distinct functions of each striatal subdivision and decision making, considerable connections between ACC and the striatum across mammalian species would be one of the critical neural networks for the processing of decision making. Although the original model of the segregated parallel loops proposed that the ACC was mainly linked with the nucleus accumbens (NAc) (Alexander et al., 1986), recent studies have shown that the ACC is also connected with the sensorimotor striatum (Ferry et al., 2000; Voorn et al., 2004). In addition, it has been shown that the lower and upper tiers of ACC have their own specific basal ganglia connections. The lower tier is linked to the ventromedial striatum (i.e. NAc and ventral putamen), while the upper tier to the dorsolateral striatum (i.e. the dorsal caudate and putamen) (Ferry et al., 2000). Thus, if the processing of envy and decision making are both independent of each other (alternatively, with overlap to some extent), the difference might be explained by a different neural basis in terms of different striatal functional subdivisions (i.e. ventral striatum for envy and dorsal striatum for risky choice (decision making)) and their differential connections with ventral and dorsal ACC subregions.

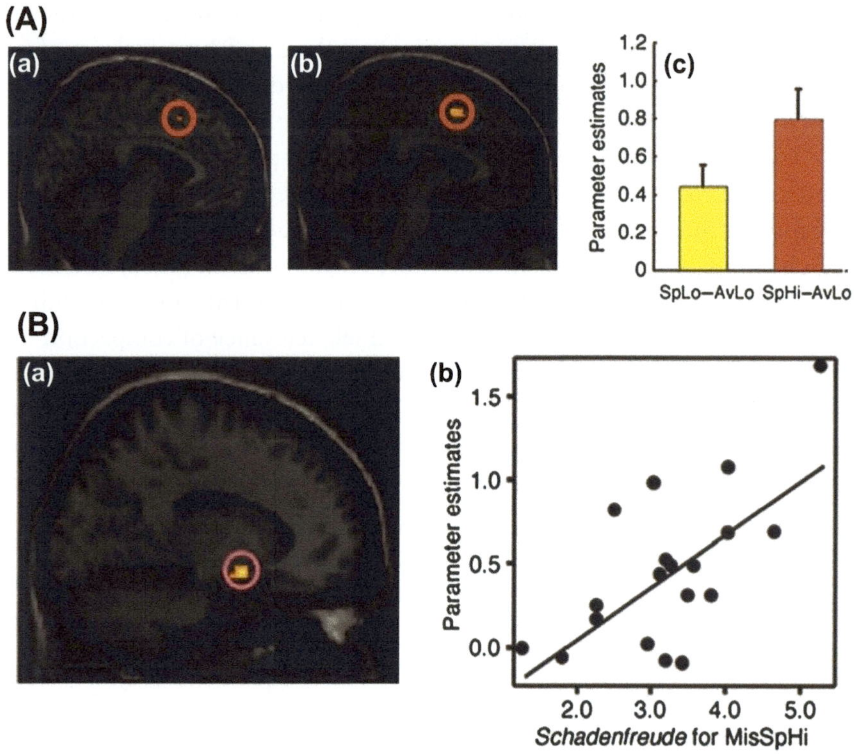

Figure 5 (A) Brain activation in dorsal ACC was modulated by relevance of comparison domain. Brain activations in response to (a) SpLo (superior and low relevance) minus AvLo (average and low relevance) condition. (b) SpHi (superior and high relevance) minus AvLo condition. (c) Mean of parameter estimates at the peak of dorsal Anterior Cingulate Cortex (dACC) activation for SpHi-AvLo contrast (red) was greater than that for SpLo–AvLo contrast (yellow) ($t=2.56$, $P=0.02$). (B) Correlation between self-rating of schadenfreude and ventral striatum activation across participants. (a) Image showing correlation between mean rating of schadenfreude for misfortunes on SpHi, superior and high relevance (MisSpHi) and the ventral striatum in MisSpHi-misfortunes on AvLo (MisAvLo) contrast across participants. (b) Plots and regression line of correlation ($r=0.65$, $P=0.002$) between schadenfreude and parameter estimates of the ventral striatum activation for MisSpHi–MisAvLo contrast at a peak voxel (−14, 2, −12).

The findings of the report regarding schadenfreude document mechanisms of painful emotion, envy, and a rewarding reaction, schadenfreude. Although it is difficult to clarify which type of DA receptor would contribute the neural mechanism of this psychological phenomenon, D_1R might be critical for the processing of schadenfreude considering the relationship among D_1R functions in ACC and the ventral striatum with reward processing. From a clinical point of view, the findings of the study would lead to a better understanding of the mechanisms of pain disorder and improvement of the symptoms, i.e. pain relief, the pathology of which might have a similar background as that of the painful emotion "envy", i.e. social pain (Lieberman et al., 2008) and its relief, schadenfreude.

2.2. D_1 RECEPTOR AND PSYCHIATRIC DISORDERS

2.2.1. D_1 Receptors and Schizophrenia

Schizophrenia is believed to involve an altered DA system (DA hypothesis) because of the action of its therapeutic agents, i.e. DA D_2 antagonist such as haloperidol. Although few studies demonstrated D_2R change in the most DA-rich region, the striatum, Suhara et al. (2002) suggested that D_2R in ACC was decreased in schizophrenia.

Previous reports repeatedly suggested cognitive dysfunction in schizophrenia (Weinberger et al., 1997; McCarley et al., 1999) as well as an important role of another major DA receptor subtype, such as D_1R in cognitive processing mentioned above (Takahashi et al., 2008, 2010a, b), and a PET study using [^{11}C]SCH23390 (Okubo et al., 1997) reported decreased prefrontal D_1R binding in schizophrenia (Figure 6). Furthermore, decreased D_1R binding in the region was associated with negative symptoms as well as executive dysfunctions revealed by WCST in patients with schizophrenia.

This D_1R-related, rather than D_2R, hypodopaminergic function in PFC would be critical for poor performance in cognitive function as well as negative symptomatology in the disorder, which would be consistent with the concept of "hypofrontality in schizophrenia" (Andreasen et al., 1997). Although the degree of D_1R blockade is not correlated with antipsychotic effects, D_1 agonistic agents might possibly be helpful for the recovery of cognitive dysfunctions and negative symptoms such as affective flattening and psychomotor retardation, whereas D_2R blockade would be necessary for the treatment of positive symptoms like hallucination and delusion in patients with schizophrenia.

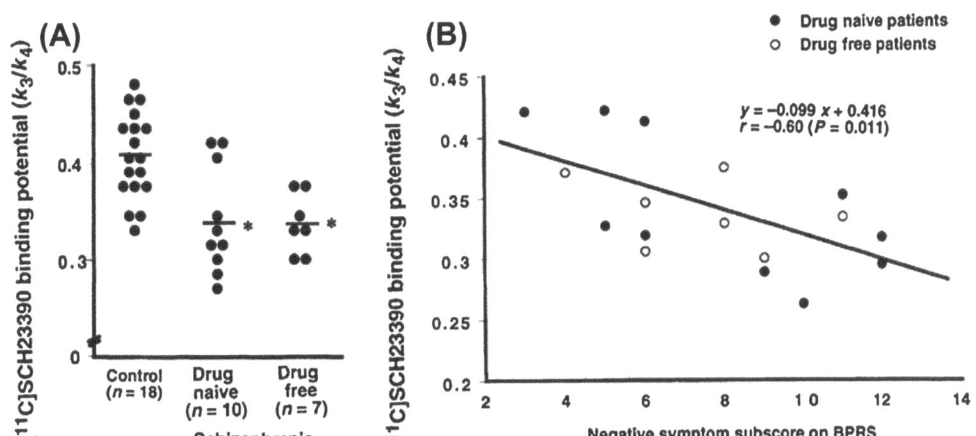

Figure 6 (A) Group comparison in binding potentials (k_3/k_4) of prefrontal D_1R: patient groups had decreased D_1R binding in PFC compared to control ($P < 0.05$). (B) Association between prefrontal D_1R binding and negative symptom subscore on Brief Psychiatric Rating Scale.

Meanwhile, in contrast to the result of the study by Okubo et al. (1997a), a PET study using [^{11}C]NNC112, another radioligand for D_1R, reported that increased D_1R in PFC was associated with working memory deficits in schizophrenia (Abi-Dargham et al., 2002). It has been discussed that these inconsistent results might stem from several factors, including differences in radioligands, but a more recent PET study measuring cortical D_1R with both [^{11}C]SCH23390 and [^{11}C]NNC112 in the same schizophrenia population demonstrated that prefrontal D_1R was decreased in chronic schizophrenia regardless of radioligands (Kosaka et al., 2010). Still, the reasons for these inconsistent results need to be clarified in the future, as an inverted U-shaped response can account for working memory deficits in patients with schizophrenia whether D_1R in PFC is increased or decreased in patients.

As mentioned above, the central profile of most antipsychotics is the D_2R blockade property. Antipsychotics are reasonably effective in ameliorating positive symptoms in schizophrenia. However, negative symptoms and cognitive impairments of schizophrenia are typically not always responsive to antipsychotic therapy. This fact has led to the investigation of alternative agents for the treatment of cognitive impairments in schizophrenia, and a body of data from animal and human studies supports the utility of the D_1 agonist (Buchanan et al., 2007; Okubo et al., 1997b). However, the efficacy of D_1 agonists on cognitive impairments has not been proved so far due to several practical issues of drug development. In addition, we need to take into account the fact that schizophrenia is a heterogeneous disorder. D_1R density might be different according to the type of schizophrenia, changeable even in a single patient according to its stage (prodromal phase, first episode phase, and chronic phase). The inverted U-shaped property of D_1R stimulation might lead to a bidirectional effect of D_1 agonist depending on type or stage. Anhedonia and blunted affect are two of the central features of negative symptoms. Some neuroimaging studies have suggested that reduced amygdala activation was associated with these symptoms (Dowd and Barch, 2010; Takahashi et al., 2004). Therefore, similar to the strategy for cognitive impairment, D_1 agonist might be useful for restoring amygdala activation and consequently improve these negative symptoms.

2.2.2. D_1 Receptor and Other Psychiatric Disorders

It should be mentioned that, even if altered D_1R-related neurotransmission in schizophrenia truly reflects the pathology, including cognitive dysfunction, it is not always disease specific. In other words, deficits in a brain region, for example, the frontal cortex, in terms of its D_1R expression possibly include pathologies of other neuropsychiatric disorders that are considered to have similar neural underpinnings.

In the study of mood disorder, the frontal cortex, and especially the orbitofrontal cortex and subgenual ACC (these regions would be included in the frontal cortex for easier understanding here), has repeatedly been reported as the location of the core neural underpinnings of the disorder, starting with a PET study by Drevets et al. (1997).

Suhara et al. (1992) has suggested decreased D_1R binding in the frontal cortex in bipolar mood disorder. In Huntington disease, 40% of its population has depressive symptoms and has also been reported to have decreased frontal D_1R expression (Sedvall et al., 1994). Lawrence et al. (1998) has reported that cognitive performance such as working memory was correlated with the D_1R binding level as well as D_2R binding level in Huntington's disease. These findings could lead to the clarification of the role of frontal D_1R function in the mechanisms of depressive state and cognitive functions.

Patients with Alzheimer disease (AD) often suffer from depressive symptoms. Kemppainen et al. (2000) has reported that D_1R binding as estimated using [^{11}C]NNC756 was reduced by 14% from the mean control value in both the putamen and caudate nucleus in AD, although there was no significant reduction of D_2R binding as estimated using [^{11}C]raclopride in either of these areas in AD compared to control. In this study, there was no correlation between D_1R or D_2R binding and clinical features; however, the mini–mental state examination consisted only of performed psychological battery, and therefore we could not conclude that the correlation between D_1R binding and depressive symptoms or cognitive dysfunction was significant or not in AD.

Depressive symptoms and frontostriatal cognitive dysfunction are also common in PD. Shinotoh et al. (1993), however, has reported that there was no significant alteration of striatal DIR binding as assessed by [^{11}C]SCH23390 PET in PD patients regardless of whether treated or untreated. Another study measuring D_1R binding using [^{11}C]NNC112 has also shown no regional differences in D_1R density between controls and PD patients (Cropley et al., 2008). These results would indicate that striatal D_1R binding is not related to depressive symptoms or frontostriatal cognitive dysfunction in PD patients.

It is still unclear whether deficits of D_1R functions measured in some neuropsychiatric disorders indicate the cause or result of their symptoms, i.e. whether they are state related or trait related. Thus, further investigations will be needed to clarify this point by D_1R PET imaging based on a longitudinal follow-up design.

3. CONCLUSIONS

In this chapter, we reviewed D_1R function in terms of cognitive functions and its relationships to pathophysiologies of neuropsychiatric disorders. Regarding the relationships between altered D_1R function and the mechanisms of neuropsychiatric disorders, D_1R function would not be regarded as a direct cause of the disorders. That is to say, it would not be disease specific, but rather might be related to the pathophysiologies by influencing cognitive functions of the individuals. In this sense, it might provide a new insight into the mechanism of the symptoms of neuropsychiatric disorders and lead to the development of new ways of dimensional classifications of behavior and cognitive functions. At present, such attempts, such as dimension-based classification of neuropsychiatric disorders, have been made for research purposes as, for example, the Research Domain Criteria project of the National Institute of Mental Health.

On the basis of such concepts, namely, the multidimensional diagnostic approach, "D_1R function for neuropsychiatric disorders" would be clarified further and lead to its clinical use, considering other factors such as genetic background of D_1R expression, the association of the expression, and symptomatology.

REFERENCES

Abi-Dargham, A., Mawlawi, O., Lombardo, I., Gil, R., Martinez, D., Huang, Y., Hwang, D.R., Keilp, J., Kochan, L., Van Heertum, R., Gorman, J.M., Laruelle, M., 2002. Prefrontal dopamine D_1 receptors and working memory in schizophrenia. J. Neurosci. 22, 3708–3719.

Alexander, G.E., DeLong, M.R., Strick, P.L., 1986. Parallel organization of functionally segregated circuits linking basal ganglia and cortex. Annu. Rev. Neurosci. 9, 357–381.

Andreasen, N.C., O'Leary, D.S., Flaum, M., Nopoulos, P., Watkins, G.L., Boles Ponto, L.L., Hichwa, R.D., 1997. Hypofrontality in schizophrenia: distributed dysfunctional circuits in neuroleptic-naïve patients. Lancet 349 (9067), 1730–1734.

Asan, E., 1998. The catecholaminergic innervation of the rat amygdala. Adv. Anat. Embryol. Cell Biol. 142, 1–118.

Brody, A.L., Mandelkern, M.A., London, E.D., Childress, A.R., Lee, G.S., Bota, R.G., Ho, M.L., Saxena, S., Baxter Jr, L.R., Madsen, D., Jarvik, M.E., 2002. Brain metabolic changes during cigarette craving. Arch. Gen. Psychiatry 59, 1162–1172.

Buchanan, R.W., Freedman, R., Javitt, D.C., Abi-Dargham, A., Lieberman, J.A., 2007. Recent advances in the development of novel pharmacological agents for the treatment of cognitive impairments in schizophrenia. Schizophr. Bull. 33, 1120–1130.

Corrigall, W., Franklin, K., Coen, K., Clarke, P., 1992. The mesolimbic dopaminergic system is implicated in the reinforcing effects of nicotine. Psychopharmacology (Berl) 107, 285–289.

Cropley, V.L., Fujita, M., Bara-Jimenez, W., Brown, A.K., Zhang, X.Y., Sangare, J., Herscovitch, P., Pike, V.W., Hallett, M., Nathan, P.J., Innis, R.B., 2008. Pre- and post-synaptic dopamine imaging and its relation with frontostriatal cognitive function in Parkinson disease: PET studies with [11C]NNC 112 and [18F] FDOPA. Psychiatry Res. 163, 171–182.

Dagher, A., Bleicher, C., Aston, J.A., Gunn, R.N., Clarke, P.B., Cumming, P., 2001. Reduced dopamine D_1 receptor binding in the ventral striatum of cigarette smokers. Synapse 42, 48–53.

David, S.P., Munafo, M.R., Johansen-Berg, H., Smith, S.M., Rogers, R.D., Matthews, P.M., Walton, R.T., 2005. Ventral striatum/nucleus accumbens activation to smoking-related pictorial cues in smokers and nonsmokers: a functional magnetic resonance imaging study. Biol. Psychiatry 58, 488–494.

de la Mora, M.P., Gallegos-Cari, A., Arizmendi-Garcia, Y., Marcellino, D., Fuxe, K., 2010. Role of dopamine receptor mechanisms in the amygdaloid modulation of fear and anxiety: structural and functional analysis. Prog. Neurobiol. 90, 198–216.

Dahlström, A., Fuxe, K., 1964. A method for the demonstration of monoamine-containing nerve fibres in the central nervous system. Acta Physiol. Scand. 60, 293–294.

Dowd, E.C., Barch, D.M., 2010. Anhedonia and emotional experience in schizophrenia: neural and behavioral indicators. Biol. Psychiatry 67, 902–911.

Drevets, W.C., Price, J.L., Simpson Jr, J.R., Todd, R.D., Reich, T., Vannier, M., Raichle, M.E., 1997. Subgenual prefrontal cortex abnormalities in mood disorders. Nature 386 (6627), 824–827.

Ferry, A.T., Ongur, D., An, X., Price, J.L., 2000. Prefrontal cortical projections to the striatum in macaque monkeys: evidence for an organization related to prefrontal networks. J. Comp. Neurol. 425, 447–470.

Gallagher, D.A., O'Sullivan, S.S., Evans, A.H., Lees, A.J., Schrag, A., 2007. Pathological gambling in Parkinson's disease: risk factors and differences from dopamine dysregulation. An analysis of published case series. Mov. Disord. 22, 1757–1763.

Glickstein, S.B., Hof, P.R., Schmauss, C., 2002. Mice lacking dopamine D_2 and D_3 receptors have spatial working memory deficits. J. Neurosci. 22, 5619–5629.

Goldman-Rakic, P.S., Muly 3rd, E.C., Williams, G.V., 2000. D(1) receptors in prefrontal cells and circuits. Brain Res. Brain Res. Rev. 31, 295–301.
Hall, H., Sedvall, G., Magnusson, O., Kopp, J., Halldin, C., Farde, L., 1994. Distribution of D_1- and D_2-dopamine receptors, and dopamine and its metabolites in the human brain. Neuropsychopharmacology 11, 245–256.
Ito, H., Takahashi, H., Arakawa, R., Takano, H., Suhara, T., 2008. Normal database of dopaminergic neurotransmission system in human brain measured by positron emission tomography. Neuroimage 39, 555–565.
Joel, D., Weiner, I., 2000. The connections of the dopaminergic system with the striatum in rats and primates: an analysis with respect to the functional and compartmental organization of the striatum. Neuroscience 96, 451–474.
Kemppainen, N., Ruottinen, H., Nâgren, K., Rinne, J.O., 2000. PET shows that striatal dopamine D_1 and D_2 receptors are differentially affected in AD. Neurology 55, 205–209.
Kienast, T., Hariri, A.R., Schlagenhauf, F., Wrase, J., Sterzer, P., Buchholz, H.G., Smolka, M.N., Grunder, G., Cumming, P., Kumakura, Y., Bartenstein, P., Dolan, R.J., Heinz, A., 2008. Dopamine in amygdala gates limbic processing of aversive stimuli in humans. Nat. Neurosci. 11, 1381–1382.
Koob, G.F., Le Moal, M., 1997. Drug abuse: hedonic homeostatic dysregulation. Science 278, 52–58.
Kosaka, J., Takahashi, H., Ito, H., Takano, A., Fujimura, Y., Matsumoto, R., Nozaki, S., Yasuno, F., Okubo, Y., Kishimoto, T., 2010. Decreased binding of [11C] NNC112 and [11C] SCH23390 in patients with chronic schizophrenia. Life Sci. 86, 814–818.
Lawrence, A.D., Weeks, R.A., Brooks, D.J., Andrews, T.C., Watkins, L.H., Harding, A.E., Robbins, T.W., Sahakian, B.J., 1998. The relationship between striatal dopamine receptor binding and cognitive performance in Huntington's disease. Brain 121, 1343–1355.
LeDoux, J.E., 2000. Emotion circuits in the brain. Annu. Rev. Neurosci. 23, 155–184.
Leyton, M., Boileau, I., Benkelfat, C., Diksic, M., Baker, G., Dagher, A., 2002. Amphetamine-induced increases in extracellular dopamine, drug wanting, and novelty seeking: a PET/[11C] raclopride study in healthy men. Neuropsychopharmacology 27, 1027.
Lieberman, M., Eisenberger, N., 2008. The pains and pleasures of social life: a social cognitive neuroscience approach. Science 323, 890–891.
Lu, C., Bharmal, A., Suchowersky, O., 2006. Gambling and Parkinson disease. Arch. Neurol. 63, 298.
McCarley, R.W., Niznikiewicz, M.A., Salisbury, D.F., Nestor, P.G., O'Donnell, B.F., Hirayasu, Y., Grunze, H., Greene, R.W., Shenton, M.E., 1999. Cognitive dysfunction in schizophrenia: unifying basic research and clinical aspects. Eur. Arch. Psychiatry Clin. Neurosci. 249 (Suppl. 4), 69–82.
McClernon, F., Hiott, F., Huettel, S., Rose, J., 2005. Abstinence-induced changes in self-report craving correlate with event-related FMRI responses to smoking cues. Neuropsychopharmacology 30, 1940–1947.
McDowell, S., Whyte, J., D'Esposito, M., 1998. Differential effect of a dopaminergic agonist on prefrontal function in traumatic brain injury patients. Brain 121, 1155–1164.
Mehta, M.A., Sahakian, B.J., McKenna, P.J., Robbins, T.W., 1999. Systemic sulpiride in young adult volunteers simulates the profile of cognitive deficits in Parkinson's disease. Psychopharmacology (Berl) 146, 162–174.
Moore, R., Vinsant, S., Nader, M., Porrino, L., Friedman, D., 1998. Effect of cocaine self-administration on dopamine D_2 receptors in rhesus monkeys. Synapse 30, 88–96.
Müller, U., von Cramon, D.Y., Pollmann, S., 1998. D_1- versus D_2-receptor modulation of visuospatial working memory in humans. J. Neurosci. 18, 2720–2728.
Okubo, Y., Suhara, T., Suzuki, K., Kobayashi, K., Inoue, O., Terasaki, O., Someya, Y., Sassa, T., Sudo, Y., Matsushima, E., Iyo, M., Tateno, Y., Toru, M., 1997a. Decreased prefrontal dopamine D_1 receptors in schizophrenia revealed by PET. Nature 385, 634–636.
Okubo, Y., Suhara, T., Sudo, Y., Toru, M., 1997b. Possible role of dopamine D_1 receptors in schizophrenia. Mol. Psychiatry 2, 291–292.
Paulus, M.P., Frank, L.R., 2006. Anterior cingulate activity modulates nonlinear decision weight function of uncertain prospects. Neuroimage 30 (2), 668–677.
Pezze, M.A., Feldon, J., 2004. Mesolimbic dopaminergic pathways in fear conditioning. Prog. Neurobiol. 74, 301–320.
Prelec, D., 1998. The probability weighting function. Econometrica 66, 497–527.

Rice, M.E., Cragg, S.J., 2004. Nicotine amplifies reward-related dopamine signals in striatum. Nat. Neurosci. 7, 583–584.

Sawaguchi, T., Goldman-Rakic, P.S., 1994. The role of D_1-dopamine receptor in working memory: local injections of dopamine antagonists into the prefrontal cortex of rhesus monkeys performing an oculomotor delayed-response task. J. Neurophysiol. 71, 515–528.

Sedvall, G., Karlsson, P., Lundin, A., Anvret, M., Suhara, T., Halldin, C., Farde, L., 1994. Dopamine D_1 receptor number – a sensitive PET marker for early brain degeneration in Huntington's disease. Eur. Arch. Psychiatry Clin. Neurosci. 243 (5), 249–255.

Shinotoh, H., Inoue, O., Hirayama, K., Aotsuka, A., Asahina, M., Suhara, T., Yamazaki, T., Tateno, Y., 1993. Dopamine D_1 receptors in Parkinson's disease and striatonigral degeneration: a positron emission tomography study. J. Neurol. Neurosurg. Psychiatry 56, 467–472.

St Onge, J.R., Floresco, S.B., 2009. Dopaminergic modulation of risk-based decision making. Neuropsychopharmacology 34, 681–697.

Suhara, T., Okubo, Y., Yasuno, F., Sudo, Y., Inoue, M., Ichimiya, T., Nakashima, Y., Nakayama, K., Tanada, S., Suzuki, K., Halldin, C., Farde, L., 2002. Decreased dopamine D_2 receptor binding in the anterior cingulate cortex in schizophrenia. Arch. Gen. Psychiatry 59 (1), 25–30.

Suhara, T., Nakayama, K., Inoue, O., Fukuda, H., Shimizu, M., Mori, A., Tateno, Y., 1992. D_1 dopamine receptor binding in mood disorders measured by positron emission tomography. Psychopharmacology (Berl) 106 (1), 14–18.

Takahashi, H., Koeda, M., Oda, K., Matsuda, T., Matsushima, E., Matsuura, M., Asai, K., Okubo, Y., 2004. An fMRI study of differential neural response to affective pictures in schizophrenia. Neuroimage 22, 1247–1254.

Takahashi, H., Yahata, N., Koeda, M., Takano, A., Asai, K., Suhara, T., Okubo, Y., 2005. Effects of dopaminergic and serotonergic manipulation on emotional processing: a pharmacological fMRI study. Neuroimage 27, 991–1001.

Takahashi, H., Fujimura, Y., Hayashi, M., Takano, H., Kato, M., Okubo, Y., Kanno, I., Ito, H., Suhara, T., 2007. Enhanced dopamine release by nicotine in cigarette smokers: a double-blind, randomized, placebo-controlled pilot study. Int. J. Neuropsychopharmacol. 11, 413–417.

Takahashi, H., Kato, M., Takano, H., Arakawa, R., Okumura, M., Otsuka, T., Kodaka, F., Hayashi, M., Okubo, Y., Ito, H., Suhara, T., 2008. Differential contributions of prefrontal and hippocampal dopamine D(1) and D(2) receptors in human cognitive functions. J. Neurosci. 28, 12032–12038.

Takahashi, H., Takano, H., Kodaka, F., Arakawa, R., Yamada, M., Otsuka, T., Hirano, Y., Kikyo, H., Okubo, Y., Kato, M., Obata, T., Ito, H., Suhara, T., 2010a. Contribution of dopamine D_1 and D_2 receptors to amygdala activity in human. J. Neurosci. 30, 3043–3047.

Takahashi, H., Matsui, H., Camerer, C., Takano, H., Kodaka, F., Ideno, T., Okubo, S., Takemura, K., Arakawa, R., Eguchi, Y., Murai, T., Okubo, Y., Kato, M., Ito, H., Suhara, T., 2010b. Dopamine D receptors and nonlinear probability weighting in risky choice. J. Neurosci. 30, 16567–16572.

Tessitore, A., Hariri, A.R., Fera, F., Smith, W.G., Chase, T.N., Hyde, T.M., Weinberger, D.R., Mattay, V.S., 2002. Dopamine modulates the response of the human amygdala: a study in Parkinson's disease. J. Neurosci. 22, 9099–9103.

Tversky, A., Kahneman, D., 1992. Advances in prospect theory: cumulative representation of uncertainty. J. Risk Uncertain. 5, 297–323.

Voorn, P., Vanderschren, L.J., Groenewegen, H.J., Robbins, T.W., Pennartz, C.M., 2004. Putting a spin on the dorsal-ventral divide of the striatum. Trends Neurosci. 27, 468–474.

Weinberger, D.R., Gallhofer, B., 1997. Cognitive function in schizophrenia. Int. Clin. Psychopharmacol. 12 (Suppl. 4), S29–S36.

Williams, G.V., Castner, S.A., 2006. Under the curve: critical issues for elucidating D_1 receptor function in working memory. Neuroscience 139, 263–276.

Yasuno, F., Ota, M., Ando, K., Ando, T., Maeda, J., Ichimiya, T., Takano, A., Doronbekov, T.K., Fujimura, Y., Nozaki, S., Suhara, T., 2007. Role of ventral striatal dopamine D_1 receptor in cigarette craving. Biol. Psychiatry 61, 1252–1259.

CHAPTER NINE

Human Brain Imaging of Dopamine Transporters

Andrea Varrone and Christer Halldin
Department of Clinical Neuroscience, Karolinska Institutet, Centre for Psychiatry Research, Stockholm, Sweden

1. INTRODUCTION

1.1. Structure and Function of the Dopamine Transporter

The neurotransmitter dopamine (DA) is one of the main monoamine transmitters in the brain and is involved in the regulation of important brain functions, such as locomotor activity, reward, and cognition (Girault and Greengard, 2004). The DA transporter (DAT) is a plasma membrane protein expressed exclusively in DA neurons where it acts by clearing the DA released into the extracellular space, thereby regulating the amplitude and duration of DA signaling (for reviews, see also Bannon, 2005; Mortensen and Amara, 2003; Uhl, 2003).

The DAT belongs to a large family of Na^+- and Cl^-- dependent transporters, which include monoamine transporters, such as the norepinephrine (NET) and the serotonin transporter (SERT), and carriers of other molecules such as gamma-aminobutyric acid (GAT 1–4), glycine (GLYT1a, 1b, 1c, and 2), proline (PROT), and taurine (TAUT) (Masson et al., 1999). There is a 69–80% homology among DAT, NET, and SERT (Amara and Kuhar, 1993). The DAT consists of 12 transmembrane domains with the C- and N-terminus within the cytoplasm (Giros and Caron, 1993). Within the N-terminus, there are phosphorylated residues (Granas et al., 2003; Lin et al., 2003) that are substrates for protein kinase C and other protein kinases, enzymes implicated in the regulation of the activity of DAT.

The transport process of the DAT involves the translocation of the substrate DA along with two sodium ions and one chloride ion across the cell membrane (Krueger, 1990; McElvain and Schenk, 1992). The major physiologic role of DAT is the termination of the neurotransmission by the rapid reuptake of DA from the synaptic cleft into the presynaptic terminal. It is believed that DAT controls the intensity and duration of DAergic neurotransmission by setting the concentration of DA in the extracellular space. Depending upon cellular conditions, DAT is able to transport DA bidirectionally, but at a normal membrane potential and Na^+ gradient, the inward transport is greater than the outward transport (Hitri et al., 1994).

Imaging of the Human Brain in Health and Disease
http://dx.doi.org/10.1016/B978-0-12-418677-4.00009-9

The activity of the DAT can be regulated by different mechanisms including the direct action of substrates or inhibitors or the changes of the number of transporters available on the plasma membrane as a result of the activation of intracellular pathways or protein-to-protein interactions (Mortensen and Amara, 2003). Phosphorylation of DAT can regulate the activity of the transporter. It has been shown that the phosphorylation of the N-terminus by Protein Kinase C (PKC) activation does not alter internalization or DA uptake (Granas et al., 2003), but it makes the transporter almost insensitive to the effect of amphetamine (Khoshbouei et al., 2004). Kinase activators, phosphatase inhibitors, and transported substrates modulate DAT phosphorylation and activity, but the underlying mechanisms and the role of phosphorylation in these processes are poorly understood. The activity of the transporter can be also modulated by the formation of homodimers or heterodimers with other proteins. In particular, it has been shown that oligomerization has a role in facilitating the trafficking of the transporters to the surface (Hastrup et al., 2001; Sorkina et al., 2003; Torres et al., 2003). Finally, internalization, recycling, and degradation can change the number of transporters on the plasma membrane and thus regulate transport activity.

1.2. Pharmacological Profile of the DAT

Several compounds are substrates of the DAT and can inhibit the activity of the transporter. Inhibitors include drugs such as cocaine, mazindol, methylphenidate, or GBR12909 that act by inhibiting the transporter function, increasing the synaptic DA levels, and the dopaminergic transmission (Schmitt et al., 2008). DA transport is also inhibited by the nonclassical antidepressant bupropion and by the muscarinic antagonist benztropine (Dwoskin et al., 2006; Schmitt et al., 2008). Amphetamine is a substrate of the DAT and of the vesicular monoamine transporter type 2 (VMAT2); it acts by reversing the transport of DA, thereby causing the release of DA and a several-fold increase of its intracellular and extracellular concentrations (Robertson et al., 2009). The substrate of DAT is also the neurotoxin 1-methyl-4-pyridium (MPP^+), which kills the DA neurones and induces parkinsonism. MPP^+ is an active metabolite of 1-methyl-4-phenyl-1,2,3,6-tetrahydropyridine and induces DAergic toxicity by the inhibition of the mitochondrial complex I activity (Vila and Przedborski, 2003).

In vivo positron emission tomography (PET) studies have contributed to the understanding of the pharmacology of some of the above inhibitors, particularly with regard to their occupancy of the DAT. Cocaine and methylphenidate have been labeled with ^{11}C and used as PET radioligands to study their pharmacokinetics, and the relation between DAT occupancy and their reinforcing effects. [^{11}C]cocaine enters the brain rapidly with a peak at ~4 min and a fast clearance of approximately 25 min. The overall time course parallels the previously documented time course of the euphoria experienced after intravenous cocaine administration (Fowler et al., 1989) and the reinforcing effect of cocaine seems to be related also to the speed by which the drug reaches the brain (Volkow et al., 2000). On the other hand, the clearance of [^{11}C]methylphenidate from the striatum (90 min) is slower than that of cocaine. For both drugs, the fast uptake in the striatum paralleled the experience of the "high". For cocaine, the decline in the "high" paralleled its fast rate of clearance from the brain, while for methylphenidate, the "high" decreased very rapidly despite significant binding of the drug in the brain (Volkow et al., 1995). Intravenous cocaine induces a significant level of DAT blockade (ED_{50} = 0.13 mg/kg), which has been associated with the intensity of self-reports of "high" in cocaine abusers. Although intravenous methylphenidate shows a similar potency to the DAT as cocaine does (ED_{50} = 0.07 mg/kg), its DAT occupancy is less well correlated to self-reports of "high". Methylphenidate induces a longer lasting increase of blood pressure and heart rate; therefore, the long-lasting side effects might counterbalance its reinforcing effect (Volkow et al., 1999). The high potency of methylphenidate has been shown also after oral administration, with an ED_{50} of 0.25 mg/kg, which is lower than the dose used therapeutically, suggesting that at clinical doses methylphenidate occupies >50% of striatal DAT (Volkow et al., 1998).

1.3. Drug-Induced Regulation of the DAT

The DAT is a protein undergoing regulation by several drugs acting directly or indirectly on the transporter. These drugs include amphetamine and cocaine and also other drugs acting on other monoamine systems, such as the serotonin-reuptake inhibitors (SSRIs), or inducing changes in the levels of the neurotransmitter DA. The administration of amphetamine and cocaine induces changes in the expression and function of the DAT. Amphetamine administration determines the short- and long-term downregulation of the transporter. In vitro studies in cell cultures have shown that acute amphetamine administration induces an immediate (within seconds) trafficking of DAT to the cell surface, corresponding to the effect on DA efflux (Furman et al., 2009), followed (within minutes) by the downregulation of the DAT molecules that are translocated from the plasma membrane to the intracellular site (Kahlig and Galli, 2003). Chronic administration of amphetamine derivatives can be neurotoxic. Human postmortem and PET studies have shown that in chronic metamphetamine users the number of DAT-binding sites is reduced (McCann et al., 1998; Wilson et al., 1996). The decrease of other neuronal DA markers suggests that chronic metamphetamine use induces an irreversible loss of DA terminals.

A different regulation of the DAT is induced by cocaine administration. Acute cocaine challenge increases the number of the DAT expressed on the plasma membrane as a compensatory mechanism to rapidly clear synaptic DA (Little et al., 2002). This short-term effect explains the behavioral tolerance to the effects of cocaine. Chronic cocaine administration has been shown to be associated with increased DAT-binding sites. This increase has been demonstrated in the striatum and the nucleus accumbens of chronic cocaine users in human autopsy samples (Little et al., 1999; Staley et al., 1994) and in the striatum of cocaine abusers during acute cocaine abstinence in a [^{123}I]β-CIT single-photon emission computed tomography (SPECT) study (Malison et al., 1998a). In detoxified cocaine users, while the [^{11}C]cocaine brain uptake is decreased, the estimated density of the DAT is unchanged (Volkow et al., 1996b).

Antidepressants and SSRIs can also induce changes in DAT density. Preclinical and clinical studies have shown that clomipramine, citalopram, and paroxetine determine an increase in DAT density (Booij et al., 2007; Fujita et al., 1997; Kugaya et al., 2003). These effects on the striatal DAT should be taken under consideration in the clinical and research setting, since often patients are treated with different drugs, including SSRIs, which can affect the imaging results and the quantitative outcome measures of the studies.

Finally, the cell surface expression of the DAT is sensitive to changes in DA release. Preclinical studies in rodents have shown that DA depletion induced by treatment with reserpine or alpha-methyl-para-tyrosine determines a decrease in DAT-binding sites in the striatum (Ikawa et al., 1993; Kilbourn et al., 1992). On the other hand, DA release induced by amantadine treatment is associated with an increase of DAT-binding sites in the rat striatum (Gordon et al., 1996). These data indicate that DAT expression decreases when DA levels are reduced, while it increases when extracellular DA concentration is increased. These findings are relevant with regard to the pathophysiology of Parkinson's disease (PD), which is characterized by DA cell loss, and to the relationship between the decrease of DA content and the quantification of the DA deficit using DAT radioligands.

1.4. Brain Distribution of the DAT

In mammalian brain, DAT mRNA colocalizes with tyrosine hydroxylase immunostaining in the cell body of DA neurones, and it has been detected predominantly in the substantia nigra pars compacta (SNpc) and in the ventral tegmental area (VTA) (Lorang et al., 1994; Usdin et al., 1991). The DAT protein is mainly distributed in the somatodendritic and terminal fields of mesencephalic DA neurones of SNpc and VTA, and is particularly enriched in the striatum and nucleus accumbens (Ciliax et al., 1999). Other regions with established DA circuitry, such as the medial VTA, the hypothalamic neurones, and the medial prefrontal cortex, contain little or no DAT protein (Ciliax et al., 1995). Electron microscopy studies have contributed to show the subcellular distribution of the DAT and to provide insights into the role of the transporter at each site of location (Nirenberg et al., 1996). For instance, the DAT localizes on the plasma membrane and the smooth endoplasmic reticulum of the dendrites in the substantia nigra (SN), where it may regulate DA release and may also be a potential site of entry of neurotoxins. The DAT is also localized in cytoplasmic tubulovesicles in the soma of SN; this localization indicates that the soma is a potential site of synthesis, transport, and recycling of the DAT. In the striatum, the DAT is localized in the plasma membrane of the axons, where it regulates synaptic transmission and can be another potential entry site for neurotoxins. Overall, the highest levels of the DAT are found in the axons of the striatum (caudate, putamen, and nucleus accumbens), while intermediate levels are found in the dendrites of the SN, and the lowest levels in the dendrites of the VTA. Autoradiography studies on postmortem human brain tissue have reported that the density (B_{max}) of the DAT in the striatum is approximately 140–200 pmol/g of tissue (Little et al., 1995; Madras et al., 1998; Staley et al., 1995). The specific binding to the DAT in the nucleus accumbens is approximately 80% of that in the striatum (Little et al., 1995; Staley et al., 1995), whereas in the SN, it is approximately 25–50% of the binding in the striatum (Hall et al., 1999; Staley et al., 1995). Figure 1 shows the distribution of the DAT-binding sites in the human brain.

The amygdala, hippocampus, thalamus, neocortex, and hypothalamus are regions with almost negligible density of axonal DAT (<1% of the striatal DAT density) (Staley et al., 1995).

Since the DAT is responsible for the rapid reuptake of DA from the extracellular space and the termination of the DA signal, the topography of its distribution appears to be directly related to the temporal precision of the dopaminergic transmission (Ciliax et al., 1999). The high concentration of the DAT in mesostriatal/mesolimbic pathways suggests that DA can quickly regulate inputs to the striatum and outputs from the basal ganglia. These DA systems involved in the motor control and behavioral reinforcement can thus make rapid adjustments to novel stimuli. On the other hand, the negligible density of the DAT in the cortex suggests that different mechanisms are involved in the regulation of the extracellular DA levels and on the temporal setting of the DA transmission.

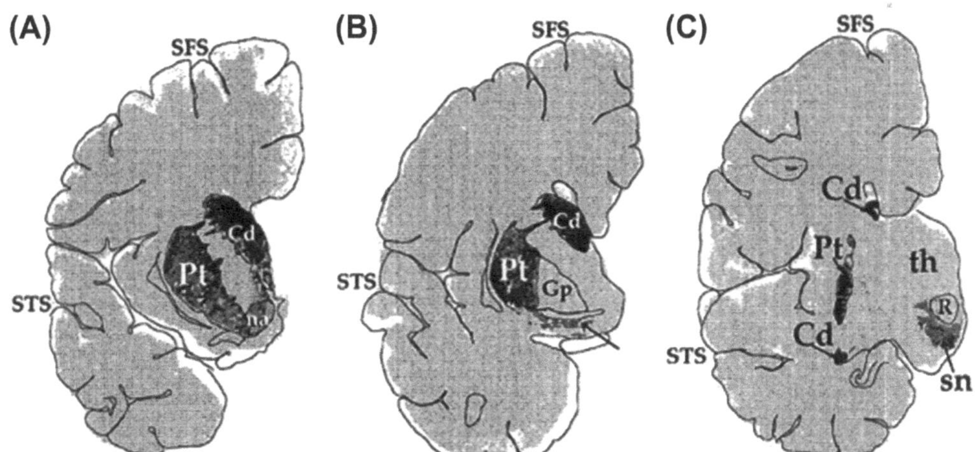

Figure 1 Distribution of DAT-binding sites with [^{123}I]RTI-121 in the human brain. Cd, caudate; Gp, globus pallidus; Pt, Putamen; na, nucleus accumbens; R, red nucleus; SFS, superior frontal sulcus; STS, superior temporal sulcus; sn, substantia nigra; and th, thalamus. *Reproduced from Staley et al. (1995), with the permission of John Wiley and Sons.*

1.5. Genetic Polymorphism and DAT Density. Age and Sex Differences

The coding sequence for the DAT covers about 2 kb corresponding to 617–630 amino acid residues, resulting in the molecular weight of the protein ranging from 60 to 97 kDa depending on glycosylation (Bannon et al., 2001). A polymorphism involving a variable number of tandem repeats (VNTR) has been described in the 3′ untranslated region of the gene (SLC6A3) coding for the DAT (Fuke et al., 2001). This polymorphism has two common alleles with either 9 (*9R) or 10 (*10R) repeats of a 40-basepair sequence, which may influence the expression of the DAT protein (Miller and Madras, 2002). Conflicting results have been obtained on the relationship between VNTR and DAT density. Two SPECT studies with [^{123}I]β-CIT from the same group found an increased DAT availability in 9R-carriers (Jacobsen et al., 2000; van Dyck et al., 2005). On the other hand, two additional studies from different groups using the same radioligand found either a decrease or no change of DAT in 9R-carriers (Heinz et al., 2000; Martinez et al., 2001). A more recent SPECT study with [^{123}I]β-CIT from another group has replicated the findings of increased DAT availability in 9R-carriers as compared with 10R-carriers (van de Giessen et al., 2009). A recent meta-analysis using the data reported in the above studies, although it found a trend toward a higher DAT availability in 9R-carriers, could not demonstrate a statistically significant association between the SLC6A3 polymorphism and striatal DAT availability (Costa et al., 2011). A positive association between SLC6A3 genotype and DAT binding was found in the two studies including the largest group of subjects (van de Giessen et al., 2009; van Dyck et al., 2005) and was also reported in a PET study with [^{11}C]cocaine in 95 control subjects of different ethnicities (Shumay et al., 2011). Therefore, it seems that the sample size is the most important factor to consider in the studies examining the association between genetic polymorphism of SLC6A3 gene and in vivo DAT imaging.

The investigation of the effect of age on DAT availability has provided more consistent results indicating a 6–8% decline of DAT per decade, using various SPECT or PET radioligands (Table 1, Fischman et al., 1998; Ishibashi et al., 2009; Ishikawa et al., 1996; Kazumata et al., 1998; Kuikka et al., 1999; Lavalaye et al., 2000a; Leroy et al., 2007; Mozley et al., 1999; Pirker et al., 2000; Troiano et al., 2009; van Dyck et al., 1995, 2002b; Varrone et al., 2013; Volkow et al., 1994, 1996a). Two SPECT studies have examined a large group of control subjects across a wide age range. The first study with [^{123}I]β-CIT was conducted in one center and included 126 subjects (van Dyck et al., 2002b). The second study with [^{123}I]FP-CIT was conducted in 13 different centers and included 139 subjects (Varrone et al., 2013). A decline of approximately 7% per decade (van Dyck et al., 2002b) and 5.5% per decade (Varrone et al., 2013) has been reported. Despite the different design of the two SPECT studies (one center versus multicenter), a similar age-related decline of DAT availability was found, in agreement with the previous findings. The DAT decline with aging seems to be associated with the deterioration in episodic

Table 1 In Vivo Studies Evaluating the Age-Related Decline of Striatal DAT Density

Radioligand	Modality	No. of Subjects	Age Range	% Decline per Decade	References
[^{123}I]β-CIT	SPECT	28 (14M, 14F)	18–83 y	8	van Dyck et al. (1995)
[^{123}I]FP-CIT	SPECT	15 (11M, 4F)	21–79 y	3.3	Ishikawa et al. (1996)
[^{123}I]altropane	SPECT	7 (5M, 2F)	26–75 y	7.6	Fischman et al. (1998)
[^{123}I]PE2I	SPECT	16 (16M)	23–75 y	4	Kuikka et al. (1999)
[^{123}I]β-CIT	SPECT	35 (14M, 21F)	21–75 y	6.6	Pirker et al. (2000)
[^{123}I]FP-CIT	SPECT	45 (23M, 22F)	18–83 y	4.1	Lavalaye et al. (2000a)
[^{123}I]FP-CIT	SPECT	139 (74M, 65F)	20–83 y	5.5	Varrone et al. (2013)
[^{123}I]β-CIT	SPECT	126 (70M, 56F)	18–88 y	6.6	van Dyck et al. (2002b)
[^{99m}Tc]TRODAT-1	SPECT	55 (27M, 28F)	19–74 y	6.2	Mozley et al. (1999)
[^{11}C]cocaine	PET	26 (26M)	21–63 y	7	Volkow et al. (1994)
[^{11}C]d-*threo*-methylphenidate	PET	23 (14M, 9F)	20–74 y	6.6	Volkow et al. (1996a)
[^{18}F]FP-CIT	PET	7 (2M, 5F)	23–73 y	Caudate: 7.7 Putamen: 6.4	Kazumata et al. (1998)
[^{11}C]PE2I	PET	15 with HRRT 11 with HR$^+$	19–47 y 35–57 y	8.1 11.9	Leroy et al. (2007)
[^{11}C]d-*threo*-methylphenidate	PET	33	27–77 y	Caudate: 11.2 Putamen: 10.5	Troiano et al. (2009)
[^{11}C]β-CFT	PET	16 (15M, 3F)	21–74 y	Caudate: 6.1 Putamen: 5.5	Ishibashi et al. (2009)

HRRT = High-resolution research tomograph; HR$^+$ = ECAT EXACT HR$^+$ (Siemens molecular imaging).
Source: Table partially readapted from van Dyck et al. 2002b.

memory and executive function. Therefore, the DAT as a regulator of DA transmission seems to be an important mediator of cognitive functioning and age-related cognitive changes (Erixon-Lindroth et al., 2005). Gender also has an effect on the DAT availability, which seems to be higher in females than in males (Lavalaye et al., 2000b; Staley et al., 2001; Varrone et al., 2013). Smoking does not seem to have a significant effect on the DAT (Lavalaye et al., 2000b; Staley et al., 2001; Thomsen et al., 2013). Age and gender are important factors to take into consideration in clinical studies evaluating group differences in DAT availability.

2.1. IN VIVO IMAGING OF THE DAT - PART I

2.1.1. Radioligands for the DAT

At present, several PET and SPECT radioligands for DAT are available, as seen in Figures 2 and 3.

All tropane analogs described in this chapter are in the beta form. For simplicity, we will use the most common form in which the radioligands have been conventionally reported in the literature. A state-of-art review has been published in 2001 by Halldin et al. (Halldin et al., 2001); since then, additional PET radioligands have been produced. A more recent review on the imaging of the DAT has been published in 2010 by Varrone and Halldin (Varrone and Halldin, 2010). SPECT radioligands such as $[^{123}\text{I}]$FP-CIT (also referred to as $[^{123}\text{I}]\beta$-CIT-FP, or $[^{123}\text{I}]$ioflupane, commercial name: DaTSCAN, GE Healthcare) and $[^{123}\text{I}]\beta$-CIT (MAP Medical Technologies) are commercially available and have been extensively used in central nervous system (CNS)-research studies and in various neuropsychiatric disorders. $[^{123}\text{I}]$FP-CIT is widely used in Europe for diagnostic purposes, and its approved indications are the differential diagnosis between essential tremor and degenerative parkinsonism and the differential diagnosis between dementia with Lewy bodies (DLB) and Alzheimer's disease (AD) (Darcourt et al., 2009). More recently, $[^{123}\text{I}]$FP-CIT has been also registered by the US Food and Drug Administration for the differential diagnosis between parkinsonism and essential tremor. $[^{123}\text{I}]\beta$-CIT has been for many years one of the DAT tracers most frequently used. It has been used in several studies examining the status of the DA system in parkinsonian and psychiatric

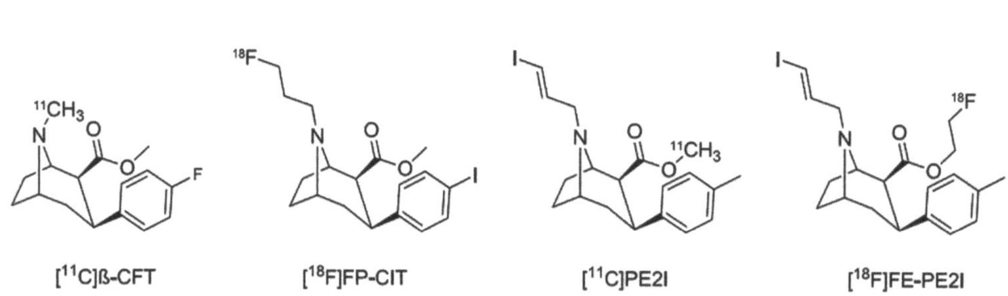

Figure 2 Chemical structure and labeling position of some SPECT and PET radioligands for DAT.

disorders, and the effects of age, gender, and genetic polymorphism on the DAT availability. Among the other SPECT radioligands, [^{123}I]altropane (also known as [^{123}]IACFT), [^{123}I]IPT, and [^{123}I]PE2I (Fischman et al., 1998; Mozley et al., 1996; Pinborg et al., 2002) have been developed and used in some imaging centers. [^{99m}Tc]TRODAT-1 (Mozley et al., 2000) is the only available technetium-labeled DAT radioligand. Because of the advantageous imaging properties of technetium, [^{99m}Tc]TRODAT-1 was initially developed as a promising radioligand for routine clinical use. However, [^{99m}Tc]TRODAT-1 shows a lower target-to-background ratio as compared with other ^{123}I-labeled radioligands. Therefore, the use of the tracer might be more appropriate to examine disease conditions in which there is less decrease of specific signal loss in the striatum, as compared with the advanced stages of dopaminergic degeneration such as in advanced PD (Van Laere et al., 2004).

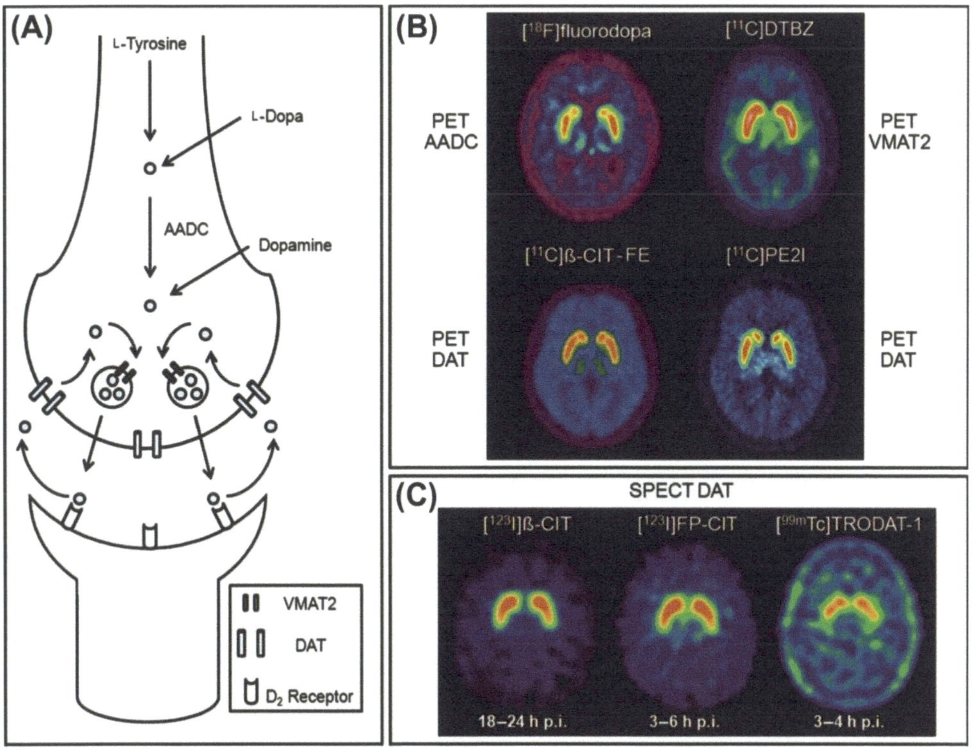

Figure 3 Schematic representation of the dopamine synapse with (A) relevant presynaptic and postsynaptic targets for SPECT or PET imaging. Representative images of different radioligands for the imaging of the dopaminergic terminals. (B) [^{18}F]fluorodopa (AADC=aromatic amino acid decarboxylase), [^{11}C]DTBZ (VMAT2), and PET radioligands for the DAT. (C) SPECT radioligands for the DAT. Images of [^{18}F]fluorodopa, [^{123}I]β-CIT, [^{123}I]FP-CIT, and [^{99m}Tc]TRODAT-1 were kindly provided by Prof. Klaus Tatsch, Municipal Hospital Karlsruhe Inc., Germany. Images of [^{11}C]DTBZ and [^{11}C]β-CIT-FE were kindly provided by Prof. Kirk Frey, Ann Arbor, MI, USA and Prof. Angelo Antonini, Istituti Clinici di Perfezionamento, Milano, Italy.

In the recent years, PET has become a more widely available Nuclear Medicine technique, due to the wider availability of PET/computed tomography (CT) systems. Therefore, the development of new DAT radioligands for PET might provide useful diagnostic tools for clinical use, considering also the advantage of the higher resolution and sensitivity of PET as compared with that of SPECT. PET radioligands for the DAT labeled with ^{11}C and applied in human studies include [^{11}C]PE2I (Halldin et al., 2003; Hirvonen et al., 2008; Jucaite et al., 2006; Leroy et al., 2007; Ribeiro et al., 2007; Sureau et al., 2008), [^{11}C]β-CFT (also referred to as [^{11}C]WIN 35,428) (Frost et al., 1993; Rinne et al., 1995, 1998), [^{11}C]β-CIT (Farde et al., 1994; Ginovart et al., 1997; Laihinen et al., 1995), and [^{11}C]β-CIT-FE (also referred to as [^{11}C]FE-CIT) (Antonini et al., 2001; Erixon-Lindroth et al., 2005; Halldin et al., 1996). The affinity and selectivity for the DAT are important requisites for a successful PET radioligand. These factors contribute to a high target-to-background ratio, which provides not only a high image contrast but also the possibility to visualize the DAT in the midbrain or SN. This property has been reported for [^{18}F]FP-CIT (Kazumata et al., 1998) and [^{11}C]PE2I (Halldin et al., 2003) (Figure 4), and more recently for [^{18}F]FE-PE2I (Varrone et al., 2009b, 2011c) (Figure 5).

It is particularly interesting for the DAT quantification in PD, since the SN is the site where the cell bodies of the DA neurons are located and where the pathology of PD is found. In addition, the combination of a high target-to-background ratio and high resolution provided by new pet systems, such as the high-resolution research tomograph (HRRT), together with advanced methods of image reconstruction including modeling of the scanner resolution (Leroy et al., 2007; Varrone et al., 2009a), give additional opportunities for the improved quantification of the DAT in the midbrain. [^{11}C]PE2I is one example of such radioligands successfully used for high-resolution PET imaging of the DAT (Hirvonen et al., 2008; Leroy et al., 2007; Sureau et al., 2008).

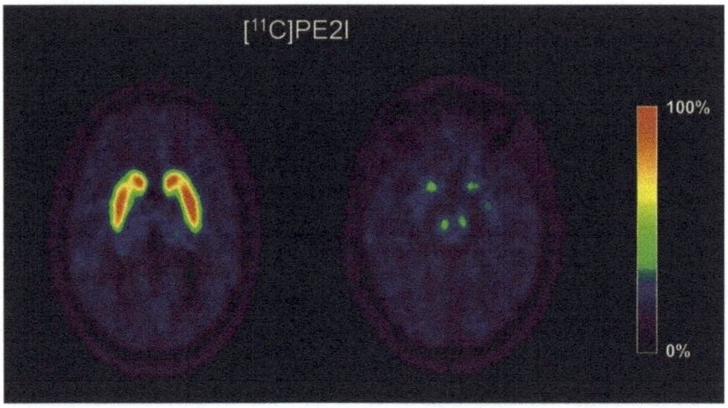

Figure 4 Dopamine transporter PET imaging with [^{11}C]PE2I in a healthy control. The image was acquired in the ECAT EXACT HR system. Not only the striatum (left) but also the substantia nigra (right) can be visualized.

Another important aspect to consider in the development of new PET radioligands for DAT is the labeling with ^{18}F. ^{18}F-labeled radioligands can be of advantage since ^{18}F has lower energy, resulting in a shorter positron range and theoretically a higher intrinsic resolution. This might provide a better imaging quality. In addition, the labeling with ^{18}F makes the tracer suitable for distribution to centers without the cyclotron and the radiochemistry facility, thereby facilitating a wider clinical use, particularly in those center having only PET/CT systems. DAT radioligands labeled with ^{18}F and applied to human subjects include [^{18}F]β-CFT (Laakso et al., 1998; Rinne et al., 1999a,b), [^{18}F]FP-CIT (also referred to as [^{18}F]β-CIT-FP) (Figure 6) (Khan et al., 2005; Lundkvist et al., 1997; Ma et al., 2002; Robeson et al., 2003; Yaqub et al., 2007), and [^{18}F] 2beta-carbomethoxy-3beta-(4-chlorophenyl)-8-(2-fluoroethyl)nort ropane (FECNT) (Davis et al., 2003).

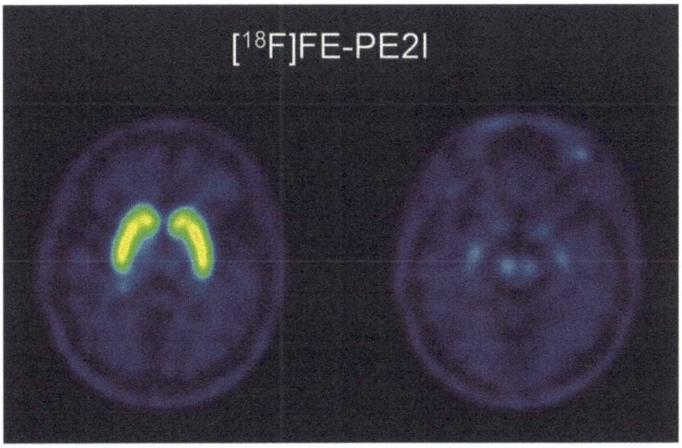

Figure 5 Dopamine transporter PET imaging with [^{18}F]FE-PE2I in a healthy control. The image was acquired in the ECAT EXACT HR system. Clear visualization of the caudate, putamen, and the substantia nigra can be seen. *Courtesy of the National Institute of Radiological Sciences, Chiba, Japan (Drs Hiroshi Ito and Tetsuya Suhara).*

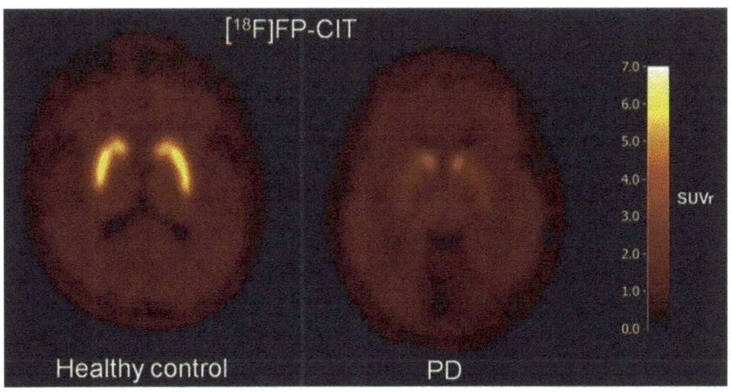

Figure 6 Representative images of [^{18}F]FP-CIT in a healthy control and a patient with parkinson's disease. Images are expressed as the standardized uptake value ratio (SUVr) relative to the mean radioactivity in the cerebellum. *Courtesy of the VU University Medical Centre, Amsterdam, The Netherlands.*

Although promising, one limitation of FECNT is the production of polar metabolites, fluoroacetaldehyde, and fluoroethanol, which have been found to enter the brain and affect the quantification of the DAT (Zoghbi et al., 2006). Second-generation ^{18}F radioligands for the DAT are LBT-999 (Chalon et al., 2006) and FE-PE2I (Schou et al., 2009). The ^{11}C analog of LBT-999 has been studied in baboons and seems to be a suitable radioligand for the in vivo quantification of the DAT in the striatum and extrastriatal regions (Saba et al., 2007). [^{18}F]LBT-999 has been studied in rhesus monkeys. Similarly to the ^{11}C analog, [^{18}F]LBT-999 shows very slow kinetics and the quantification of DAT in the striatum would require at least 4 h of imaging (Varrone et al., 2011b). Therefore, this radioligand seems more suitable for the quantification of the DAT in extrastriatal regions, such as in the midbrain and the thalamus (Varrone et al., 2011b). [^{18}F]FE-PE2I has been initially evaluated in preclinical studies and more recently in humans and seemed to be a promising radioligand for the DAT due to its fast kinetics and favorable metabolism (Varrone et al., 2009b). Because of the fast kinetics, peak equilibrium of [^{18}F]FE-PE2I in the striatum and the midbrain is reached between 20 and 40 min after injection (Varrone et al., 2011c), a property that seems rather unique for a ^{18}F-labeled radioligand. Therefore, the accurate quantification of the DAT availability with [^{18}F]FE-PE2I is possible with only a 60 min duration of the PET measurements (Varrone et al., 2011a). [^{18}F]FE-PE2I also shows a favorable metabolism with less production in the plasma of a metabolite peak (Varrone et al., 2011c), which in the case of [^{11}C]PE2I has been shown to correspond to a lipophilic metabolite accumulating in the rat striatum (Shetty et al., 2007). This property of [^{18}F]FE-PE2I contributes to less or negligible bias in the estimation of the binding potential as compared with [^{11}C]PE2I (Varrone et al., 2011c).

Another requisite for a useful PET radioligand for the DAT is the possibility of applying noninvasive methods of quantification with limited imaging time, which is particularly important to applied studies in parkinsonian patients as has been demonstrated for the two main SPECT radioligands, [^{123}I]β-CIT and [^{123}I]FP-CIT (see below). In this regard, [^{18}F]FE-PE2I appears to be particularly promising (see below) based on the initial studies in nonhuman primates (Varrone et al., 2011a,b,c) and in human subjects (Sasaki et al., 2012).

2.1.2. Quantification of DAT Availability

The use of DAT imaging as a biomarker of nigrostriatal integrity and as a tool to evaluate the status of the dopaminergic system requires the estimation of quantitative outcome measures such as the binding potential (BP_{ND}), which is equal to $B_{max}/K_D \star f_{ND}$ (where B_{max} is the density of transporters, $1/K_D$ is the radioligand affinity for the DAT and f_{ND} is the free fraction of the radioligand in the nondisplaceable tissue) (Innis et al., 2007) and proportional to the density of the DAT. The most widely used approach to estimate the BP relies on reference region methods in which either the cerebellum or the occipital cortex, regions with a negligible density of DAT-binding sites, are used as regions to estimate the nondisplaceable binding of the radioligand in the brain. Reference region methods such as the Logan noninvasive method (Logan et al., 1996), the simplified reference tissue method (SRTM) (Hirvonen et al., 2008; Jucaite et al., 2006; Leroy et al., 2007), and a simple ratio between the target and the reference regions at late time points (Frost et al., 1993; Rinne et al., 1995; Yaqub et al., 2007) or at transient equilibrium (Laakso et al., 1998; Rinne et al., 1999a) have been used for the quantification of PET radioligands such as [^{11}C]d-*threo*-methylphenidate (Logan et al., 1996), [^{11}C]PE2I (Hirvonen et al., 2008; Jucaite et al., 2006; Leroy et al., 2007), [^{18}F]FP-CIT (Yaqub et al., 2007), [^{11}C]β-CIT (Rinne et al., 1995), [^{11}C]β-CFT (Frost et al., 1993; Rinne et al., 1995), and [^{18}F]β-CFT (Laakso et al., 1998; Rinne et al., 1999a). In the case of [^{18}F]FE-PE2I, in nonhuman primates, reference tissue methods, such as SRTM, the noninvasive Logan graphical analysis, and the multilinear reference tissue model, have all been found to provide BP_{ND} values within 5% accuracy as compared with the full quantification using the arterial input function (Varrone et al., 2011c). In humans, SRTM has been shown to provide BP_{ND} values in excellent agreement with the kinetic analysis and compartmental modeling (Sasaki et al., 2012).

Simplified methods are particularly useful in the case of DAT radioligands used for routine clinical applications. Noninvasive approaches and acquisition protocols with short imaging time are desirable to improve patient comfort and optimize patient throughput. [^{123}I]β-CIT and [^{123}I]FP-CIT are good examples of such radioligands. [^{123}I]β-CIT is characterized by very slow kinetics; between 18 and 24 h after injection, the tracer reaches a steady-state concentration in the striatum and the occipital cortex and the BP can be measured by the ratio (Striatum/Occipital cortex—1) (van Dyck et al., 1995). In the case of [^{123}I]FP-CIT, it has been shown that the tracer reaches transient equilibrium within 3–4 h after injection yielding a stable ratio between the striatum and occipital cortex (Booij et al., 1997a,b). The time window within 3–4 h can be used to estimate the BP. The advantage of [^{123}I]FP-CIT over [^{123}I]β-CIT is the possibility of completing SPECT imaging within the same day (Booij et al., 1997b).

With regard to other SPECT radioligands, similar or alternative methods of quantification have been reported. The peak equilibrium method, in which the BP is estimated by the ratio between the specific binding and the reference region at the time of peak equilibrium (Farde et al., 1989), has been applied to [^{123}I]altropane showing a good correlation with the Logan invasive plot, the Logan noninvasive plot and the SRTM (Bonab et al., 2000; Fischman et al., 1998). As for [^{99m}Tc]TRODAT-1, 4–5 h postinjection has been found to be a suitable imaging time for DAT quantification, using the occipital cortex (Kao et al., 2001) or the whole supratentorial brain as reference regions (Acton et al., 2000). In the case of [^{123}I]PE2I, a bolus-to-infusion method with a 3 h-duration of infusion and a bolus-to-infusion ratio of 2.7 h has been proposed as an alternative method for the quantification of the DAT to implement in the clinical setting (Pinborg et al., 2005).

2.2. IN VIVO IMAGING OF THE DAT - PART II

2.2.1. Diagnostic Biomarker

DAT imaging has been successfully applied to several neurodegenerative disorders. The main results of the various studies are summarized in Table 2. The pathology of PD consists of the loss of DA neurones in the SN and the reduction of DA projections to the striatum. Therefore, the main application of DAT imaging has been the quantification of the loss of nigrostriatal dopaminergic projections in PD. The ventrolateral part of the SN shows the most pronounced pathological changes. Since the ventrolateral part of the nigra mainly projects to the posterior putamen, this is the region most affected in PD. [^{123}I]β-CIT and [^{123}I]FP-CIT SPECT have a high discriminative power to differentiate PD from normal aging as shown in Figure 7, particularly when the DAT availability in the putamen is evaluated (Seibyl et al., 1995; Tissingh et al., 1998). Similar results with a sensitivity and specificity between 95 and 100% have been obtained with [^{99m}Tc]TRODAT-1 (Weng et al., 2004), suggesting that the three radioligands are equally sensitive in detecting the DA deficit at an early stage of PD. This property might be related to the fact that in early PD, the DAT seems to be downregulated as opposed to other markers of DA function such as [^{18}F]F-dopa and [^{11}C]DTBZ, radioligands for the decarboxylase enzyme and the VMAT2 (Lee et al., 2000). In addition, α-synuclein has been found to directly interact with the DAT and functional changes, such as increased trafficking or reduced surface expression of the DAT could occur as a result of the accumulation of α-synuclein or reduced DA levels (Kilbourn et al., 1992; Oaks and Sidhu, 2011).

In the routine clinical setting, the main clinical application of DAT imaging is to differentiate degenerative parkinsonism from other conditions such as essential tremor, drug-induced, vascular, or psychogenic parkinsonism that do not present a DA deficit. [^{123}I]β-CIT and [^{123}I]FP-CIT have both shown a high sensitivity and specificity in the differentiation between essential tremor and PD. In the DOPASCAN study conducted in the United States and in the FP-CIT study conducted in Europe, for the differential diagnosis between essential tremor and PD the sensitivity and specificity of [^{123}I]β-CIT were 98% and 83% and those of [^{123}I]FP-CIT were 95% and 94% (2000; Benamer et al., 2000). SPECT imaging of the DAT is clinically useful to prove the presence of the DA deficit in those conditions in which the diagnosis of degenerative parkinsonism is uncertain and an appropriate treatment is warranted. In a multicenter study with [^{123}I]FP-CIT SPECT, it has been shown that in ~50% of patients, a change in diagnosis was made after DAT imaging, with an increased number of correct diagnoses of degenerative parkinsonism. In ~70% of patients, DAT imaging also contributed to changes of clinical management, consisting of the initiation of the correct treatment or the change of the appropriate treatment (Catafau and Tolosa, 2004). Atypical parkinsonian disorders, such as multiple system atrophy (MSA) and progressive supranuclear palsy (PSP), present with a degree and pattern of DA deficits similar to PD (Plotkin et al., 2005; Varrone et al., 2001) (Figure 8).

Table 2 DAT Imaging in Parkinsonism, Related Movement Disorders, and Other Neurodegenerative Disorders

Disease	Radioligand	DAT Density	Reference
Parkinson's disease[1]	[^{123}I]β-CIT	↓↓/↓↓↓[2]	Marek et al. (1996); Seibyl et al. (1995); Varrone et al. (2001); Winogrodzka et al. (2003)
	[^{123}I]FP-CIT	↓↓/↓↓↓[2]	Booij et al. (1997a,b); Ishikawa et al. (1996)
	[^{123}I]altropane	↓↓/↓↓↓[2]	Fischman et al. (1998)
	[^{123}I]IPT	↓↓/↓↓↓[2]	Chouker et al. (2001)
	[^{99m}Tc]TRODAT-1	↓↓/↓↓↓[2]	Mozley et al. (2000); Weng et al. (2004)
	[^{11}C]β-CFT	↓↓/↓↓↓[2]	Frost et al. (1993); Rinne et al. (1995)
	[^{11}C]β-CIT	↓↓/↓↓↓[2]	Laihinen et al. (1995); Rinne et al. (1995)
	[^{11}C]β-CIT-FE	↓↓/↓↓↓[2]	Antonini et al. (2001)
	[^{18}F]β-CFT	↓↓/↓↓↓[2]	Rinne et al. (1999a,b)
	[^{18}F]FP-CIT	↓↓/↓↓↓[2]	Ma et al. (2002)
Atypical parkinsonism (MSA, PSP)	[^{123}I]β-CIT	↓↓/↓↓↓[2]	Messa et al. (1998); Scherfler et al. (2005); Varrone et al. (2001)
	[^{123}I]FP-CIT	↓↓/↓↓↓[2]	Plotkin et al. (2005)
	[^{11}C]β-CFT	↓↓/↓↓↓[2]	Ilgin et al. (1999)
Essential tremor	[^{123}I]β-CIT	↔[3]	2000
	[^{123}I]FP-CIT	↔	Benamer et al. (2000)
	[^{11}C]β-CIT-FE	↔	Antonini et al. (2001)
Drug-induced or psychogenic parkinsonism	[^{123}I]FP-CIT	↔	Marshall and Grosset (2003)
Vascular parkinsonism	[^{123}I]β-CIT	↔	Gerschlager et al. (2002)
	[^{123}I]FP-CIT	↓	Lorberboym et al. (2004)
	[^{99m}Tc]TRODAT-1	↔	Tzen et al. (2001)
Huntington's disease	[^{11}C]β-CIT	↓↓	Ginovart et al. (1997)
Dementia with Lewy bodies	[^{123}I]FP-CIT	↓↓/↓↓↓[2]	McKeith et al. (2007); Walker et al. (2002, 2004)
Alzheimer's disease	[^{123}I]FP-CIT	↔	Walker et al. (2002)
	[^{11}C]β-CFT	↓	Rinne et al. (1998)
Frontotemporal dementia	[^{123}I]FP-CIT	↓↓	Sedaghat et al. (2007)
	[^{11}C]β-CFT	↓	Rinne et al. (2002)

↓ = Mild reduction of DAT density; ↓↓ = Moderate reduction of DAT density; ↓↓↓ = Severe reduction of DAT density; ↔ = No change in the DAT density; ↑ = Increase of the DAT density.

[1]In Parkinson's disease, for simplicity, some of the most representative publications have been selected.

[2]Moderate to severe reduction depending upon the stage of the disease.

[3]In the DOPASCAN study, a reduction of the DAT density has been reported in five patients with essential tremor.

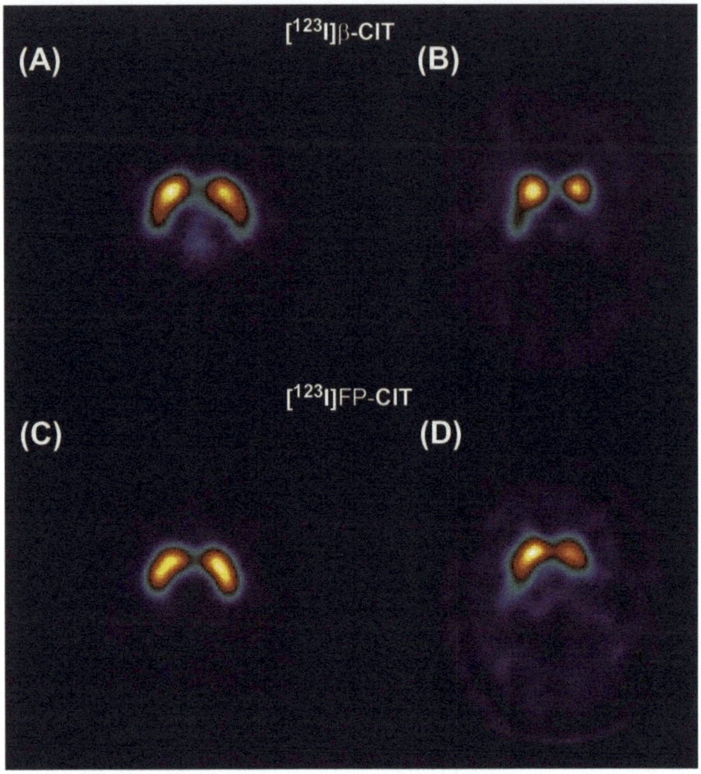

Figure 7 Dopamine transporter SPECT imaging with [^{123}I]β-CIT and [^{123}I]FP-CIT in two healthy controls (A, C) and two patients with Parkinson's disease (B, D). Images of [^{123}I]β-CIT were kindly provided by Dr John P. Seibyl, Molecular Neuroimaging, New Haven, CT, USA. [^{123}I]FP-CIT SPECT images were acquired at the Biostructure and Bioimaging Institute, Napoli, Italy.

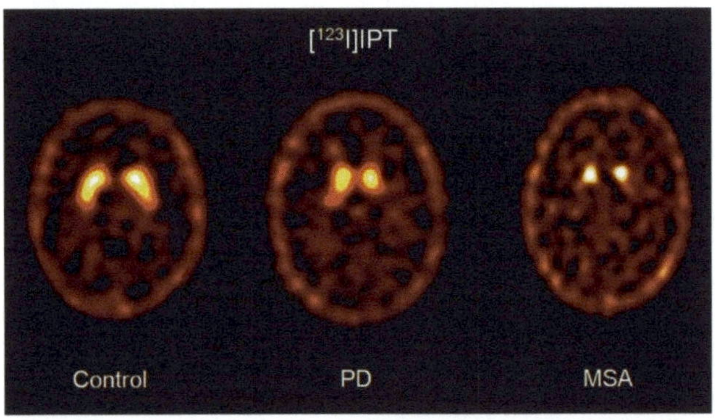

Figure 8 Dopamine transporter SPECT images with [^{123}I]IPT in a healthy control, a Parkinson's disease patient and a patient with multiple system atrophy. *Courtesy of Prof. Klaus Tatsch, Municipal Hospital Karlsruhe Inc., Germany.*

Although a specific decrease of [^{123}I]β-CIT binding in the midbrain has been reported in MSA patients (Scherfler et al., 2005) and a uniform loss of DAT has been found in PSP patients (Ilgin et al., 1999; Messa et al., 1998), DAT imaging cannot be used on a patient basis to distinguish these two conditions from PD. Complementary imaging of the D_2 receptor availability in the striatum could help differentiate PD from atypical parkinsonism symptoms as seen is Figure 9 (Koch et al., 2007).

Disorders characterized by Lewy-body pathology, including PD dementia and DLB also show a loss of DAT-binding sites. In DLB, parkinsonism usually follows the onset of the dementia. Therefore, the early identification of the DAT deficit is clinically relevant for the management of the patients. AD pathology is not characterized by the loss of DA terminals and DAT imaging can be used to differentiate DLB from AD (Walker et al., 2002) as shown in Figure 10.

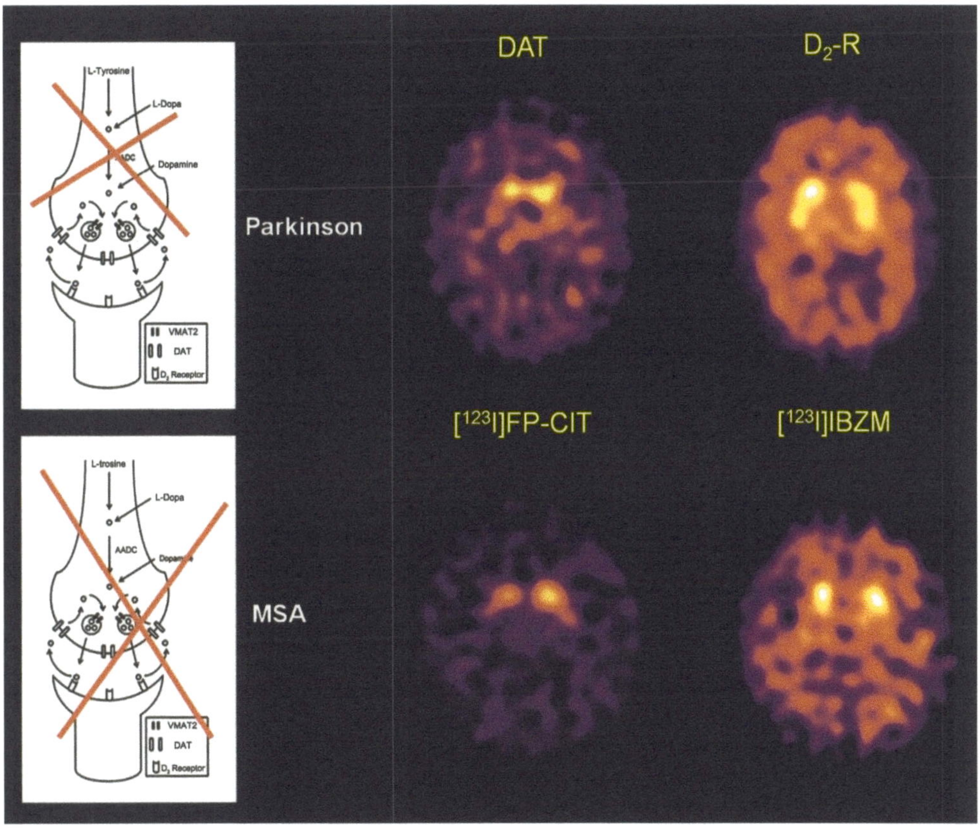

Figure 9 Schematic representation of the loss of presynaptic terminals in Parkinson's disease and presynaptic and postsynaptic binding sites in multiple system atrophy (MSA). SPECT images of DAT and D_2 receptors with [^{123}I]FP-CIT and [^{123}I]IBZM in a Parkinson's disease patient and in an MSA patient. [^{123}I]FP-CIT and [^{123}I]IBZM SPECT images were acquired at the Biostructure and Bioimaging Institute, Napoli, Italy.

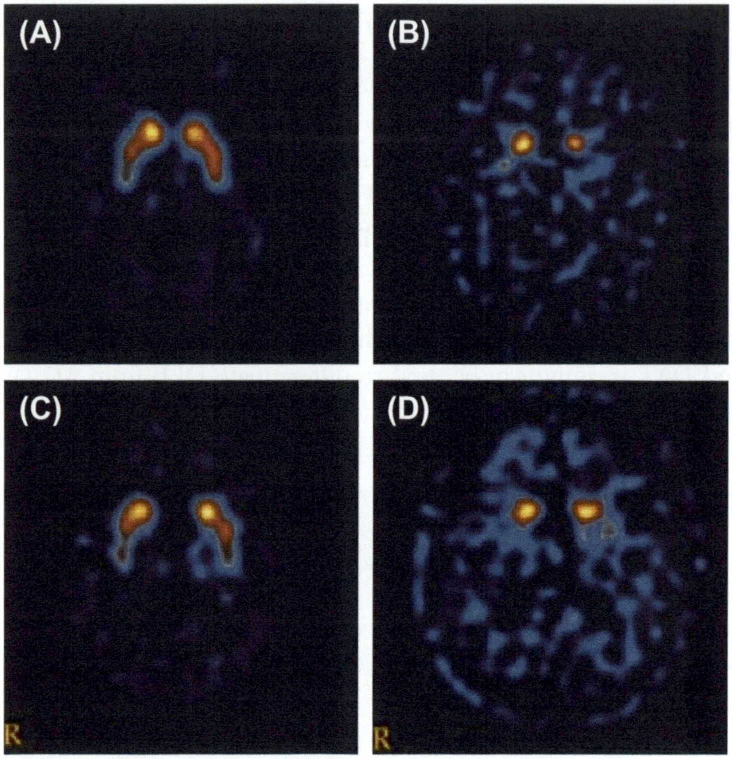

Figure 10 [^{123}I]FP-CIT SPECT images of a healthy control (A) and patients with (B) Parkinson's disease, (C) Alzheimer's disease, or (D) dementia with Lewy bodies. *Reproduced from Walker et al. (2002), with the permission of the BMJ Publishing Group, Ltd.*

A phase III clinical trial using [^{123}I]FP-CIT SPECT has examined 94 patients with probable DLB, 57 patients with possible DLB, and 147 patients with non-DLB dementia. The mean sensitivity for detecting clinical probable DLB was 77.7%. The specificity for excluding non-DLB dementia was 90.4%. The diagnostic accuracy and the positive and the negative predictive values were 85.7%, 82.4%, and 87.5% (McKeith et al., 2007). In DLB, the DAT loss in the caudate appears to be more severe as compared with the one observed in PD, whereas a similar impairment is found in the putamen. Therefore, the pattern of DAT deficit observed in DLB seems to be slightly different from the one typically described in PD (Walker et al., 2004).

2.3. IN VIVO IMAGING OF THE DAT - PART III

2.3.1. Early Biomarker in PD

The onset of symptoms in PD usually occurs when approximately 50% of DA terminals are lost. However, in patients with hemiparkinsonism, the striatum contralateral to the clinically unaffected side already shows an approximately 30% decrease in DAT binding (Booij et al., 1997a; Marek et al., 1996). This finding indicates that the loss of DAT-binding sites can be detected before the onset of the symptoms. Therefore, DAT imaging could be used as a tool for the preclinical assessment of the DA deficit in selected conditions carrying an increased risk of developing symptomatic parkinsonism. Genetic mutations have been found to be responsible for autosomal or recessive forms of parkinsonism. Hyposmia and rapid-eye movement behavioral disorder (RBD) are also common conditions that can precede the onset of PD. Among the various genetic mutations associated with parkinsonism, mutations in the Parkin gene are the cause of autosomal recessive early-onset PD. It has been shown that Parkin patients present with a more severe DA dysfunction as compared with non-Parkin PD patients (Scherfler et al., 2004; Varrone et al., 2004). In vitro and preclinical studies in Parkin −/− mice have shown that the DAT function is strictly linked with the parkin protein and that parkin deficit is associated with the reduction of the DAT and a variety of abnormalities of the DA function (Goldberg et al., 2003; Itier et al., 2003; Jiang et al., 2004). Heterozygous and homozygous carriers of Parkin mutations show a dopaminergic deficit already at a presymptomatic stage (Khan et al., 2005). The severity of the dopaminergic deficit in those carriers seems to be closely related to the number of mutated alleles (Hilker et al., 2002; Pellecchia et al., 2007). Other frequent genetic mutations associated with familial and sporadic forms of PD are those in the leucine-rich repeat kinase 2 (LRRK2) gene (PARK8). A PET study in asymptomatic members of a parkinsonian LRRK2 kindred has shown a longitudinal decline of PET markers with reduced DAT binding being the earliest marker of dopaminergic dysfunction and suggesting that DAT imaging could be the most sensitive indicator of subclinical nigrostriatal deficit (Nandhagopal et al., 2008). Hyposmia is one of the most common sign that can be present years before the onset of PD (Berendse et al., 2011). Approximately 75% of PD patients have a decreased sense of smell (Haehner et al., 2009) and they are frequently unaware of this abnormality. Some simple tests such as the University of Pennsylvania Smell Identification Test (Doty et al., 1984) or the Sniffin' sticks test (Hummel et al., 1997) can be used to objectively assess the presence of hyposmia. In a study of a large cohort of healthy subjects having first-degree relatives with parkinsonism, it was found that approximately 10% of hyposmic subjects showed a decrease in the DAT binding. These subjects developed parkinsonism within two years of follow-up. An additional 12% of hyposmic individuals showed a DAT decline over two years of follow-up (Ponsen et al., 2004). With regard to RBD, some studies have shown that patients with "idiopathic RBD" show a decrease in the DAT availability or DA function of a similar degree as in PD patients (Albin et al., 2000; Eisensehr et al., 2000, 2003; Stiasny-Kolster et al., 2005). Longitudinal DAT imaging in RBD patients shows a progressive decline of nigrostriatal dopaminergic function (Iranzo et al., 2011) and the clinical follow-up in these patients suggests that idiopathic RBD can be a prodromal condition of a Lewy-body disorder (Iranzo et al., 2013).

The common finding of these studies is that the loss of DAT-binding sites can be identified before the onset of parkinsonian symptoms. Therefore, DAT imaging is a very sensitive marker for early or presymptomatic diagnosis of degenerative parkinsonism. The combination of DAT imaging with the identification of early clinical signs of PD, such as hyposmia or RBD, and of known genetic mutations can be a strategy to identify in a presymptomatic stage those individuals carrying an increased risk for developing PD (Haehner et al., 2009; Marek and Jennings, 2009; Siderowf and Lang, 2012).

2.3.2. Biomarker of Disease Progression in PD

The neurodegenerative loss of DA terminals in PD is a progressive process that can be followed in vivo through the imaging of the DAT. Recent longitudinal studies conducted over several years have shown that the process follows an exponential decline (Nandhagopal et al., 2009; Schwarz et al., 2004). The three markers of DA integrity, namely, fluorodopa, DAT, and VMAT2 seem to provide similar results with regard to the exponential decline. As displayed in Table 3, there is a large number of data on the DAT decline. The annual decrease of DAT has been found to be approximately 8–10% per year (Table 3, Chouker et al., 2001; Marek et al., 2001; Nandhagopal et al., 2009; Nurmi et al., 2000, 2003; Pirker et al., 2002, 2003; Schwarz et al., 2004; Staffen et al., 2000; Winogrodzka et al., 2003), which is approximately 10 times higher than the reported annual decline of DAT in normal aging. Markers of DA terminal integrity should in principle be suitable as biomarkers of disease progression in PD. At least, fluorodopa and DAT imaging would fulfill the requirements as biomarkers of disease progression, since fluorodopa and DAT imaging have both been found to correlate with the number of DA neurons and with the severity and stage of PD (Booij et al., 1997a; Brooks et al., 2003; Seibyl et al., 1995; Snow et al., 1993). When using a biomarker to assess the changes of putative neuroprotective or disease-modifying drugs on the DA terminal integrity, it is important to evaluate whether the imaging signal is influenced by the type of treatment. In other words, it is important to understand whether the observed change in the imaging outcome measure is related to the effect of the treatment on the imaging target (e.g., competition or downregulation) or specifically on the integrity of the DA terminals. To some extent, the issue is not yet resolved. The Comparison of the agonist pramipexole with levodopa on motor complications of Parkinson's disease (CALM-PD) and the REAL-PET studies (2002; Whone et al., 2003) were two independent clinical trials with [^{123}I]β-CIT SPECT and [^{18}F]fluorodopa PET in which PD patients were treated with either L-dopa or a D_2 agonist (pramipexole or ropinirole). These two independent studies have shown that PD patients treated with a D_2 agonist had a slower decline of DAT availability or [^{18}F]fluorodopa influx constant as compared with patients treated with L-dopa. It is yet to be determined whether the observed difference between the two groups was related to the possible neuroprotective effects of D_2 agonists or neurotoxic effects of L-dopa, or perhaps to L-dopa interference with DAT expression in the case of the CALM-PD study (Morrish, 2003). Although preclinical and clinical studies

Table 3 Longitudinal Imaging Studies on the Progression of DAT Loss in PD Patients

Radioligand	Modality	No of Patients/ Controls	No of Scans	Mean Imaging Interval (months)	DAT Loss per Year	References
[^{123}I]β-CIT	SPECT	15 PD patients	2	15	5.6%	Staffen et al. (2000)
[^{123}I]IPT	SPECT	8	3	12 (Scan 1–2) 24 (Scan 2–3)	1st year: 6.6% 2nd year: 5.3%	Chouker et al. (2001)
[^{123}I]β-CIT	SPECT	32 PD patients 24 Healthy controls	2 2	12–48	11.2% 0.8%	Marek et al. (2001)
[^{123}I]β-CIT	SPECT	24 PD patients 10 APS patients	2 2	13–63	4.3%[2] 9.6%[2]	Pirker et al. (2002)
[^{123}I]β-CIT	SPECT	21 PD patients	3	26 (Scan 1–2) 38 (Scan 2–3)	Scan 1–2: 4.5%[2] Scan 2–3: 3.0%[2]	Pirker et al. (2003)
[^{123}I]β-CIT	SPECT	50 PD patients	2	12	Caudate: 4% Putamen: 8%	Winogrodzka et al. (2003)
[^{123}I]IPT	SPECT	6 PD patients	5	90	Exponential fit[3]	Schwarz et al. (2004)
[^{18}F]β-CFT	PET	8 PD patients 7 Healthy controls	2 2	24	Caudate: 12.5%[1] Putamen: 13.1%[1] Caudate: 2.9%[1] Putamen: 2.1%[1]	Nurmi et al. (2000)
[^{18}F]β-CFT	PET	12 PD patients 11 Healthy controls	2	26.4	Caudate: 5.6%[2] Anterior Putamen: 5.3%[2] Posterior Putamen: 3.3%[2]	Nurmi et al. (2003)
[^{11}C]d-*threo*-methylphenidate	PET	76 PD patients	3	48 and 96	Exponential fit[3]	Nandhagopal et al. (2009)

H&Y = Hoehn and Yahr; PD = Parkinson's disease; APS = Atypical parkinsonian syndrome.
[1]The rate of annual decline was calculated compared to the baseline mean.
[2]The rate of annual decline was calculated compared to the normal control mean (age-expected normal values).
[3]The data were fitted with an exponential model and corresponding equation. The percent decline per year was not reported.

in a small number of subjects have shown small or not significant effects of L-dopa or D_2 agonists on the DAT (Ahlskog et al., 1999; Guttman et al., 2001; Lavalaye et al., 2000b), some of the above questions still remain unanswered (Marek et al., 2002). A randomized, double-blind, placebo-controlled trial (Earlier versus Later Levodopa Therapy in Parkinson Disease (ELLDOPA)) was conducted in 361 PD patients. Patients received daily doses of L-dopa of 150, 300, 600 mg, or placebo. There was a significant clinical benefit of L-dopa versus placebo (Fahn et al., 2004), but in the [^{123}I]β-CIT SPECT substudy of 142 patients from the same cohort, the L-dopa groups showed a more significant DAT decline versus the placebo group (−6%, −4%, and −7.2% in the patients receiving L-dopa at 150, 300, or 600 mg daily, as compared with −1.4% in the placebo group). A recent study using micro-PET in the 6-OH DA rodent model of PD has shown that the binding of both [^{11}C]methylphenidate (DAT tracer) and [^{11}C]DTBZ (VMAT2 tracer) can be affected by treatment with L-dopa or pramipexole (Sossi et al., 2010). It is not clear whether the effect of dopaminergic treatment on the imaging target could depend on the type or radioligand used, by its affinity to the target, its kinetic properties (i.e., low or high dissociation constant k_4), or by the method of quantification used. However, at least for [^{123}I]β-CIT SPECT, which is the tracer with the largest experience in clinical trials of PD, the ELLDOPA study has indeed raised questions on the interpretation of the imaging results in the presence of dopaminergic agents (Fahn, 2005). Perhaps, for new imaging probes, preclinical or clinical studies would be required to assess the potential effect of dopaminergic treatment on the imaging outcome measures. With regard to VMAT2 as a third potential target for the assessment of the DA terminal integrity, along with DAT and fluorodopa, the experience from imaging studies and clinical trials is more limited. Although [^{18}F]FP-DTBZ (or AV-133) has been developed for potential clinical use (Okamura et al., 2010), data on the suitability of VMAT2 as marker of disease progression in PD patients have not been provided yet.

In conclusion, bearing in mind all considerations about the usefulness of DAT imaging or fluorodopa PET as surrogate biomarkers of disease progression, these imaging modalities can be viewed as useful, and at present probably the only, molecular biomarkers to assess the effects of restorative therapies on the primary pathological dysfunction (i.e., DA deficit) in PD. For instance, DAT imaging and fluorodopa PET have been used to assess the effect of fetal mesenchephalic cell transplants or glial-derived nerve factor (GDNF) in preclinical models of PD (Inaji et al., 2005; Kordower et al., 2000). They have also been used to assess the effect on the DA system in PD patients transplanted with mesencephalic cells (Piccini et al., 2005; Pogarell et al., 2006), and in patients treated with GDNF (Gill et al., 2003) or with deep brain stimulation (Hesse et al., 2008).

2.3.3. Other CNS Disorders. Attention-Deficit Hyperactivity Disorder, Tourette's Syndrome, Depression, and Schizophrenia

DAT imaging has been also used to study the dopaminergic system in disorders such as attention-deficit hyperactivity disorders (ADHD), Tourette's syndrome (TS), depression, and schizophrenia (Table 4). ADHD is a psychiatric condition of the childhood

Table 4 DAT Imaging in Other CNS Disorders

Disease	Radioligand	DAT Density	References
ADHD	[^{123}I]altropane	↑	Dougherty et al. (1999); Spencer et al. (2005)
	[^{123}I]IPT	↑	Cheon et al. (2003); Spencer et al. (2005)
	[^{99m}Tc]TRODAT-1	↑	Dresel et al. (2000); Spencer et al. (2005)
	[^{11}C]altropane	↑	Spencer et al. (2007)
	[^{123}I]β-CIT	↔	van Dyck et al. (2002a)
	[^{123}I]FP-CIT	↓	Hesse et al. (2009)
	[^{11}C]PE2I	↓[1]	Jucaite et al. (2005)
	[^{11}C]cocaine	↓	Volkow et al. (2007, 2009)
Tourette's syndrome	[^{99m}Tc]TRODAT-1	↔	Yeh et al. (2006)
	[^{99m}Tc]TRODAT-1	↔	Hwang et al. (2008)
	[^{11}C]Methylphenidate	↔[2]	Albin et al. (2009)
	[^{123}I]β-CIT	↑	Malison et al. (1995)
	[^{123}I]FP-CIT	↑	Serra-Mestres et al. (2004)
	[^{123}I]IPT	↑	(Cheon et al., 2004)
Depression	[^{123}I]β-CIT	↑	Laasonen-Balk et al. (1999)
	[^{99m}Tc]TRODAT-1	↑	Brunswick et al. (2003); Yang et al. (2008)
	[^{123}I]β-CIT	↔	Malison et al. (1998b)
	[^{11}C]RTI-32	↓	Meyer et al. (2001)
Schizophrenia	[^{123}I]β-CIT	↔	Laruelle et al. (2000)
	[^{99m}Tc]TRODAT-1	↔	Hsiao et al. (2003)
	[^{18}F]β-CFT	↔	Laakso et al. (2000) #100;
	[^{11}C]PE2I	↑	Arakawa et al. (2009)

↑ = increase of the DAT density, ↔ = no change in the DAT density, ↓ = reduction of the DAT density.
[1]In the study by Jucaite et al. (2005), a decrease of the DAT availability was found in the midbrain, but not in the striatum.
[2]In the study of Albin et al. (2009), [^{11}C]DTBZ was also used as a marker of the VMAT2 availability, and no differences were found between controls and patients with Tourette's syndrome.

characterized by the presence of inattention, hyperactivity, and impulsivity. The symptoms can be alleviated by treatment with methylphenidate, and the disorder can also manifest in adulthood. Contradictory results have been reported in previous imaging studies in ADHD using DAT imaging. Some studies in children and adults with ADHD have reported an increased striatal DAT density (Cheon et al., 2003; Dougherty et al., 1999; Spencer et al., 2005, 2007), while other studies have indicated no changes (van Dyck et al., 2002a) or rather a decreased DAT density in adults with ADHD (Hesse et al., 2009; Volkow et al., 2007, 2009). The different findings could have been related to differences in patient populations, imaging modalities, or radioligands used. In adolescents with ADHD, a decrease of the DAT availability in the midbrain has been reported with the high-affinity radioligand [^{11}C]PE2I (Jucaite et al., 2005). In the same study, no differences in DAT availability were found in the striatum (Jucaite et al., 2005). More recently, a study in a relatively large group of unmedicated adults with ADHD also found a decreased DAT availability in the midbrain, nucleus accumbens and caudate, using voxel-based and region of interest (ROI)-based analyses (Volkow et al., 2009). These findings suggest that the SN is also involved in the DA dysfunction associated with ADHD, together with other areas (such as the nucleus accumbens) involved in the DA reward pathway. In adults with ADHD, the striatal DAT availability, measured with [^{99m}Tc]TRODAT-1 SPECT, was found to be closely related to the response to methylphenidate, being higher than that for controls in the responders and lower than that for controls in the nonresponders (Krause et al., 2000; Krause et al., 2003). It has been also shown that nicotine use and treatment with methylphenidate are associated with a decreased DAT availability in the striatum of ADHD patients (Dresel et al., 2000; Krause, 2008). Therefore, in ADHD, it seems that DAT imaging could be used to examine the functional status of the DA system and to monitor the effect of treatment with methylphenidate. The duration of methylphenidate treatment and the time of DAT imaging in relation to the administration of the drug are very important to understand the regulatory effects of methylphenidate on the DAT. A recent PET study with [^{11}C]cocaine in 18 never-medicated adult ADHD patients and 11 control subjects has shown an increase of DAT availability of approximately 24% in the ventral striatum of ADHD patients after 12 months of treatment with methylphenidate (Wang et al., 2013). No differences in striatal DAT availability between ADHD patients and control subjects were present at baseline, but a statistically significant higher striatal DAT availability was found in ADHD patients compared with that in control subjects after 12 months. Since the PET measurements were conducted 24 h after the last pharmacological dose of methylphenidate, when the DAT occupancy by the drug is negligible, the results have been interpreted as a compensatory increase of DAT after a long-term exposure to methylphenidate and increased synaptic DA concentration. The suggestion of the authors was that if the levels of synaptic DA modulate the DAT levels, the discrepancy in the results reported in the literature with regard to the DAT and ADHD might also depend on the variability of the treatment history of the patients included in the different studies.

TS is a common neurodevelopmental disorder characterized by tics and behavioral comorbidities. In vivo PET and SPECT studies have shown no differences in the DAT availability between TS patients and controls suggesting no specific abnormalities of the integrity of the dopaminergic system in this disorder (Albin et al., 2009; Hwang et al., 2008; Yeh et al., 2006). However, three SPECT studies in drug-naive children or adults with TS have shown an increased DAT density in the striatum (Cheon et al., 2004; Malison et al., 1995; Serra-Mestres et al., 2004); thus, at present, it is unclear whether the controversial reports are due to differences in methodology, patient groups, or treatment effect. Nevertheless, it has reported that TS patients have an increased release of DA by amphetamine (Wong et al., 2008) suggesting that increased phasic DA release is a distinctive alteration of the DA system in these patients.

In major depression, SPECT studies with [^{123}I]β-CIT or [^{99m}Tc]TRODAT-1 have mainly shown an increase in the DAT availability in drug-free or drug-naive patients as compared with controls (Brunswick et al., 2003; Laasonen-Balk et al., 1999; Yang et al., 2008). This finding might seem unexpected since the hypothesis in depression would be a decreased dopaminergic activity, which would be associated with decreased DAT availability. Possible explanations for increased DAT availability have been either a primary upregulation of the DAT, which could in turn lead to a decreased intrasynaptic DA concentration and hence to a lower DA transmission (Laasonen-Balk et al., 1999), or a compensatory mechanism secondary to an increased dopaminergic "tone" in major depression (Brunswick et al., 2003). It must be said that the above studies have shown a slight increase in the DAT availability (~20%). In addition, a [^{123}I]β-CIT SPECT study in 15 drug-free patients with major depression and 15 controls showed a decrease of the 5-HTT availability in the brain stem, but did not report any differences in striatal DAT availability between patients and controls (Malison et al., 1998b). Finally, a PET study with the radioligand [^{11}C]RTI-32 in 23 healthy controls and 9 patients with a major depressive episode secondary to major depression (5 drug-naive and 4 drug-free) reported an average 14% decrease of striatal DAT availability in the patients compared with the controls (Meyer et al., 2001). This finding has been interpreted in line with the hypothesis of reduced DA neurotransmission in depression.

Schizophrenia is a psychiatric disorder associated with the dysregulation of the DA system. Previous studies have not reported any abnormalities of the DAT in the striatum of schizophrenic patients (Hsiao et al., 2003; Laakso et al., 2000; Laruelle et al., 2000). However, a recent study has reported an increase in the thalamic DAT density in schizophrenic patients with a significant correlation with both positive and negative symptoms (Arakawa et al., 2009). These findings, along with the concomitant evidence of a decreased D_2 receptor density in thalamic subregions of schizophrenic patients (Talvik et al., 2003; Yasuno et al., 2004), suggest that the thalamus is a brain region likely involved in the pathophysiology of schizophrenia.

3. CONCLUSIONS

The DAT is a transmembrane protein involved in the regulation of the extracellular levels of DA and of the temporal precision of DA transmission. Many psychotropic drugs act on the DAT and have been used to study the pharmacology of the DA transmission. Several radioligands for the DAT have been developed as biomarkers of DA function and integrity in physiological processes such as aging, in neurodegenerative disorders such as PD, and in other psychiatric disorders associated with the changes in DA transmission. SPECT radioligands for the DAT as [^{123}I]FP-CIT, [^{123}I]β-CIT, and [^{99m}Tc]TRODAT-1 have been widely applied in the clinical setting and nowadays represent molecular imaging tools available for daily practice. The development of ^{18}F radioligands for the DAT, such as [^{18}F]FE-PE2I, may provide new opportunities for a more widespread use of PET or PET/CT in the clinical setting for the evaluation of patients with PD or related movement disorders.

REFERENCES

Acton, P.D., Meyer, P.T., Mozley, P.D., Plossl, K., Kung, H.F., 2000. Simplified quantification of dopamine transporters in humans using [^{99m}Tc]TRODAT-1 and single-photon emission tomography. Eur. J. Nucl. Med. 27, 1714–1718.

Ahlskog, J.E., Uitti, R.J., O'Connor, M.K., Maraganore, D.M., Matsumoto, J.Y., Stark, K.F., Turk, M.F., Burnett, O.L., 1999. The effect of dopamine agonist therapy on dopamine transporter imaging in Parkinson's disease. Mov. Disord. 14, 940–946.

Albin, R.L., Koeppe, R.A., Chervin, R.D., Consens, F.B., Wernette, K., Frey, K.A., Aldrich, M.S., 2000. Decreased striatal dopaminergic innervation in REM sleep behavior disorder. Neurology 55, 1410–1412.

Albin, R.L., Koeppe, R.A., Wernette, K., Zhuang, W., Nichols, T., Kilbourn, M.R., Frey, K.A., 2009. Striatal [^{11}C]dihydrotetrabenazine and [^{11}C]methylphenidate binding in Tourette syndrome. Neurology 72, 1390–1396.

Amara, S.G., Kuhar, M.J., 1993. Neurotransmitter transporters: recent progress. Annu. Rev. Neurosci. 16, 73–93.

A multicenter assessment of dopamine transporter imaging with DOPASCAN/SPECT in parkinsonism. Parkinson Study Group. 2000. Neurology 55, 1540–1547.

Dopamine transporter brain imaging to assess the effects of pramipexole vs levodopa on Parkinson disease progression. 2002. JAMA 287, 1653–1661.

Antonini, A., Moresco, R.M., Gobbo, C., De Notaris, R., Panzacchi, A., Barone, P., Calzetti, S., Negrotti, A., Pezzoli, G., Fazio, F., 2001. The status of dopamine nerve terminals in Parkinson's disease and essential tremor: a PET study with the tracer [11-C]FE-CIT. Neurol. Sci. 22, 47–48.

Arakawa, R., Ichimiya, T., Ito, H., Takano, A., Okumura, M., Takahashi, H., Takano, H., Yasuno, F., Kato, M., Okubo, Y., Suhara, T., 2009. Increase in thalamic binding of [(11)C]PE2I in patients with schizophrenia: a positron emission tomography study of dopamine transporter. J. Psychiatr. Res. 43, 1219–1223.

Bannon, M.J., 2005. The dopamine transporter: role in neurotoxicity and human disease. Toxicol. Appl. Pharmacol. 204, 355–360.

Bannon, M.J., Michelhaugh, S.K., Wang, J., Sacchetti, P., 2001. The human dopamine transporter gene: gene organization, transcriptional regulation, and potential involvement in neuropsychiatric disorders. Eur. Neuropsychopharmacol. 11, 449–455.

Benamer, T.S., Patterson, J., Grosset, D.G., Booij, J., de Bruin, K., van Royen, E., Speelman, J.D., Horstink, M.H., Sips, H.J., Dierckx, R.A., Versijpt, J., Decoo, D., Van Der Linden, C., Hadley, D.M., Doder, M., Lees, A.J., Costa, D.C., Gacinovic, S., Oertel, W.H., Pogarell, O., Hoeffken, H., Joseph, K., Tatsch, K., Schwarz, J., Ries, V., 2000. Accurate differentiation of parkinsonism and essential tremor using visual assessment of [^{123}I]-FP-CIT SPECT imaging: the [^{123}I]-FP-CIT study group. Mov. Disord. 15, 503–510.

Berendse, H.W., Roos, D.S., Raijmakers, P., Doty, R.L., 2011. Motor and non-motor correlates of olfactory dysfunction in Parkinson's disease. J. Neurol. Sci. 310, 21–24.

Bonab, A.A., Fischman, A.J., Alpert, N.M., 2000. Comparison of 4 methods for quantification of dopamine transporters by SPECT with [^{123}I]IACFT. J. Nucl. Med. 41, 1086–1092.

Booij, J., Tissingh, G., Boer, G.J., Speelman, J.D., Stoof, J.C., Janssen, A.G., Wolters, E.C., van Royen, E.A., 1997a. [^{123}I]FP-CIT SPECT shows a pronounced decline of striatal dopamine transporter labelling in early and advanced Parkinson's disease. J. Neurol. Neurosurg. Psychiatry 62, 133–140.

Booij, J., Tissingh, G., Winogrodzka, A., Boer, G.J., Stoof, J.C., Wolters, E.C., van Royen, E.A., 1997b. Practical benefit of [^{123}I]FP-CIT SPET in the demonstration of the dopaminergic deficit in Parkinson's disease. Eur. J. Nucl. Med. 24, 68–71.

Booij, J., de Jong, J., de Bruin, K., Knol, R., de Win, M.M., van Eck-Smit, B.L., 2007. Quantification of striatal dopamine transporters with ^{123}I-FP-CIT SPECT is influenced by the selective serotonin reuptake inhibitor paroxetine: a double-blind, placebo-controlled, crossover study in healthy control subjects. J. Nucl. Med. 48, 359–366.

Brooks, D.J., Frey, K.A., Marek, K.L., Oakes, D., Paty, D., Prentice, R., Shults, C.W., Stoessl, A.J., 2003. Assessment of neuroimaging techniques as biomarkers of the progression of Parkinson's disease. Exp. Neurol. 184 (Suppl. 1), S68–S79.

Brunswick, D.J., Amsterdam, J.D., Mozley, P.D., Newberg, A., 2003. Greater availability of brain dopamine transporters in major depression shown by [^{99m}Tc]TRODAT-1 SPECT imaging. Am. J. Psychiatry 160, 1836–1841.

Catafau, A.M., Tolosa, E., 2004. Impact of dopamine transporter SPECT using ^{123}I-Ioflupane on diagnosis and management of patients with clinically uncertain Parkinsonian syndromes. Mov. Disord. 19, 1175–1182.

Chalon, S., Hall, H., Saba, W., Garreau, L., Dolle, F., Halldin, C., Emond, P., Bottlaender, M., Deloye, J.B., Helfenbein, J., Madelmont, J.C., Bodard, S., Mincheva, Z., Besnard, J.C., Guilloteau, D., 2006. Pharmacological characterization of (E)-N-(4-fluorobut-2-enyl)-2beta-carbomethoxy-3beta-(4'-tolyl)nortropane (LBT-999) as a highly promising fluorinated ligand for the dopamine transporter. J. Pharmacol. Exp. Ther. 317, 147–152.

Cheon, K.A., Ryu, Y.H., Kim, Y.K., Namkoong, K., Kim, C.H., Lee, J.D., 2003. Dopamine transporter density in the basal ganglia assessed with [^{123}I]IPT SPET in children with attention deficit hyperactivity disorder. Eur. J. Nucl. Med. Mol. Imaging 30, 306–311.

Cheon, K.A., Ryu, Y.H., Namkoong, K., Kim, C.H., Kim, J.J., Lee, J.D., 2004. Dopamine transporter density of the basal ganglia assessed with [^{123}I]IPT SPECT in drug-naive children with Tourette's disorder. Psychiatry Res. 130, 85–95.

Chouker, M., Tatsch, K., Linke, R., Pogarell, O., Hahn, K., Schwarz, J., 2001. Striatal dopamine transporter binding in early to moderately advanced Parkinson's disease: monitoring of disease progression over 20 years. Nucl. Med. Commun. 22, 721–725.

Ciliax, B.J., Heilman, C., Demchyshyn, L.L., Pristupa, Z.B., Ince, E., Hersch, S.M., Niznik, H.B., Levey, A.I., 1995. The dopamine transporter: immunochemical characterization and localization in brain. J. Neurosci. 15, 1714–1723.

Ciliax, B.J., Drash, G.W., Staley, J.K., Haber, S., Mobley, C.J., Miller, G.W., Mufson, E.J., Mash, D.C., Levey, A.I., 1999. Immunocytochemical localization of the dopamine transporter in human brain. J. Comp. Neurol. 409, 38–56.

Costa, A., Riedel, M., Muller, U., Moller, H.J., Ettinger, U., 2011. Relationship between SLC6A3 genotype and striatal dopamine transporter availability: a meta-analysis of human single photon emission computed tomography studies. Synapse 65, 998–1005.

Darcourt, J., Booij, J., Tatsch, K., Varrone, A., Vander Borght, T., Kapucu, O.L., Nagren, K., Nobili, F., Walker, Z., Van Laere, K., 2009. EANM procedure guidelines for brain neurotransmission SPECT using (^{123}) I-labelled dopamine transporter ligands, version 2. Eur. J. Nucl. Med. Mol. Imaging.

Davis, M.R., Votaw, J.R., Bremner, J.D., Byas-Smith, M.G., Faber, T.L., Voll, R.J., Hoffman, J.M., Grafton, S.T., Kilts, C.D., Goodman, M.M., 2003. Initial human PET imaging studies with the dopamine transporter ligand ^{18}F-FECNT. J. Nucl. Med. 44, 855–861.

Doty, R.L., Shaman, P., Kimmelman, C.P., Dann, M.S., 1984. University of Pennsylvania Smell Identification Test: a rapid quantitative olfactory function test for the clinic. Laryngoscope 94, 176–178.

Dougherty, D.D., Bonab, A.A., Spencer, T.J., Rauch, S.L., Madras, B.K., Fischman, A.J., 1999. Dopamine transporter density in patients with attention deficit hyperactivity disorder. Lancet 354, 2132–2133.

Dresel, S., Krause, J., Krause, K.H., LaFougere, C., Brinkbaumer, K., Kung, H.F., Hahn, K., Tatsch, K., 2000. Attention deficit hyperactivity disorder: binding of [^{99m}Tc]TRODAT-1 to the dopamine transporter before and after methylphenidate treatment. Eur. J. Nucl. Med. 27, 1518–1524.

Dwoskin, L.P., Rauhut, A.S., King-Pospisil, K.A., Bardo, M.T., 2006. Review of the pharmacology and clinical profile of bupropion, an antidepressant and tobacco use cessation agent. CNS Drug Rev. 12, 178–207.

van Dyck, C.H., Seibyl, J.P., Malison, R.T., Laruelle, M., Wallace, E., Zoghbi, S.S., Zea-Ponce, Y., Baldwin, R.M., Charney, D.S., Hoffer, P.B., 1995. Age-related decline in striatal dopamine transporter binding with iodine-123-beta-CITSPECT. J. Nucl. Med. 36, 1175–1181.

van Dyck, C.H., Quinlan, D.M., Cretella, L.M., Staley, J.K., Malison, R.T., Baldwin, R.M., Seibyl, J.P., Innis, R.B., 2002a. Unaltered dopamine transporter availability in adult attention deficit hyperactivity disorder. Am. J. Psychiatry 159, 309–312.

van Dyck, C.H., Seibyl, J.P., Malison, R.T., Laruelle, M., Zoghbi, S.S., Baldwin, R.M., Innis, R.B., 2002b. Age-related decline in dopamine transporters: analysis of striatal subregions, nonlinear effects, and hemispheric asymmetries. Am. J. Geriatr. Psychiatry 10, 36–43.

van Dyck, C.H., Malison, R.T., Jacobsen, L.K., Seibyl, J.P., Staley, J.K., Laruelle, M., Baldwin, R.M., Innis, R.B., Gelernter, J., 2005. Increased dopamine transporter availability associated with the 9-repeat allele of the SLC6A3 gene. J. Nucl. Med. 46, 745–751.

Eisensehr, I., Linke, R., Noachtar, S., Schwarz, J., Gildehaus, F.J., Tatsch, K., 2000. Reduced striatal dopamine transporters in idiopathic rapid eye movement sleep behaviour disorder. Comparison with Parkinson's disease and controls. Brain 123 (Pt 6), 1155–1160.

Eisensehr, I., Linke, R., Tatsch, K., Kharraz, B., Gildehaus, J.F., Wetter, C.T., Trenkwalder, C., Schwarz, J., Noachtar, S., 2003. Increased muscle activity during rapid eye movement sleep correlates with decrease of striatal presynaptic dopamine transporters. IPT and IBZM SPECT imaging in subclinical and clinically manifest idiopathic REM sleep behavior disorder, Parkinson's disease, and controls. Sleep 26, 507–512.

Erixon-Lindroth, N., Farde, L., Wahlin, T.B., Sovago, J., Halldin, C., Backman, L., 2005. The role of the striatal dopamine transporter in cognitive aging. Psychiatry Res. 138, 1–12.

Fahn, S., 2005. Does levodopa slow or hasten the rate of progression of Parkinson's disease? J. Neurol. 252 (Suppl. 4), IV37–IV42.

Fahn, S., Oakes, D., Shoulson, I., Kieburtz, K., Rudolph, A., Lang, A., Olanow, C.W., Tanner, C., Marek, K., 2004. Levodopa and the progression of Parkinson's disease. N. Engl. J. Med. 351, 2498–2508.

Farde, L., Eriksson, L., Blomquist, G., Halldin, C., 1989. Kinetic analysis of central [^{11}C]raclopride binding to D_2-dopamine receptors studied by PET—a comparison to the equilibrium analysis. J. Cereb. Blood Flow Metab. 9, 696–708.

Farde, L., Halldin, C., Muller, L., Suhara, T., Karlsson, P., Hall, H., 1994. PET study of [^{11}C]beta-CIT binding to monoamine transporters in the monkey and human brain. Synapse 16, 93–103.

Fischman, A.J., Bonab, A.A., Babich, J.W., Palmer, E.P., Alpert, N.M., Elmaleh, D.R., Callahan, R.J., Barrow, S.A., Graham, W., Meltzer, P.C., Hanson, R.N., Madras, B.K., 1998. Rapid detection of Parkinson's disease by SPECT with altropane: a selective ligand for dopamine transporters. Synapse 29, 128–141.

Fowler, J.S., Volkow, N.D., Wolf, A.P., Dewey, S.L., Schlyer, D.J., Macgregor, R.R., Hitzemann, R., Logan, J., Bendriem, B., Gatley, S.J., et al., 1989. Mapping cocaine binding sites in human and baboon brain in vivo. Synapse 4, 371–377.

Frost, J.J., Rosier, A.J., Reich, S.G., Smith, J.S., Ehlers, M.D., Snyder, S.H., Ravert, H.T., Dannals, R.F., 1993. Positron emission tomographic imaging of the dopamine transporter with ^{11}C-WIN 35,428 reveals marked declines in mild Parkinson's disease. Ann. Neurol. 34, 423–431.

Fujita, M., Takatoku, K., Matoba, Y., Nishiura, M., Kobayashi, K., Inoue, O., Nishimura, T., 1997. Enhancement of [^{123}I]beta-CIT binding in the striatum with clomipramine: is there a serotonin-dopamine interaction? Eur. J. Nucl. Med. 24, 403–408.

Fuke, S., Suo, S., Takahashi, N., Koike, H., Sasagawa, N., Ishiura, S., 2001. The VNTR polymorphism of the human dopamine transporter (DAT1) gene affects gene expression. Pharmacogenomics J. 1, 152–156.

Furman, C.A., Chen, R., Guptaroy, B., Zhang, M., Holz, R.W., Gnegy, M., 2009. Dopamine and amphetamine rapidly increase dopamine transporter trafficking to the surface: live-cell imaging using total internal reflection fluorescence microscopy. J. Neurosci. 29, 3328–3336.

Gerschlager, W., Bencsits, G., Pirker, W., Bloem, B.R., Asenbaum, S., Prayer, D., Zijlmans, J.C., Hoffmann, M., Brucke, T., 2002. [^{123}I]beta-CIT SPECT distinguishes vascular parkinsonism from Parkinson's disease. Mov. Disord. 17, 518–523.

van de Giessen, E., de Win, M.M., Tanck, M.W., van den Brink, W., Baas, F., Booij, J., 2009. Striatal dopamine transporter availability associated with polymorphisms in the dopamine transporter gene SLC6A3. J. Nucl. Med. 50, 45–52.

Gill, S.S., Patel, N.K., Hotton, G.R., O'Sullivan, K., McCarter, R., Bunnage, M., Brooks, D.J., Svendsen, C.N., Heywood, P., 2003. Direct brain infusion of glial cell line-derived neurotrophic factor in Parkinson disease. Nat. Med. 9, 589–595.

Ginovart, N., Lundin, A., Farde, L., Halldin, C., Backman, L., Swahn, C.G., Pauli, S., Sedvall, G., 1997. PET study of the pre- and post-synaptic dopaminergic markers for the neurodegenerative process in Huntington's disease. Brain 120 (Pt 3), 503–514.

Girault, J.A., Greengard, P., 2004. The neurobiology of dopamine signaling. Arch. Neurol. 61, 641–644.
Giros, B., Caron, M.G., 1993. Molecular characterization of the dopamine transporter. Trends Pharmacol. Sci. 14, 43–49.
Goldberg, M.S., Fleming, S.M., Palacino, J.J., Cepeda, C., Lam, H.A., Bhatnagar, A., Meloni, E.G., Wu, N., Ackerson, L.C., Klapstein, G.J., Gajendiran, M., Roth, B.L., Chesselet, M.F., Maidment, N.T., Levine, M.S., Shen, J., 2003. Parkin-deficient mice exhibit nigrostriatal deficits but not loss of dopaminergic neurons. J. Biol. Chem. 278, 43628–43635.
Gordon, I., Weizman, R., Rehavi, M., 1996. Modulatory effect of agents active in the presynaptic dopaminergic system on the striatal dopamine transporter. Eur. J. Pharmacol. 298, 27–30.
Granas, C., Ferrer, J., Loland, C.J., Javitch, J.A., Gether, U., 2003. N-terminal truncation of the dopamine transporter abolishes phorbol ester- and substance P receptor-stimulated phosphorylation without impairing transporter internalization. J. Biol. Chem. 278, 4990–5000.
Guttman, M., Stewart, D., Hussey, D., Wilson, A., Houle, S., Kish, S., 2001. Influence of L-dopa and pramipexole on striatal dopamine transporter in early PD. Neurology 56, 1559–1564.
Haehner, A., Boesveldt, S., Berendse, H.W., Mackay-Sim, A., Fleischmann, J., Silburn, P.A., Johnston, A.N., Mellick, G.D., Herting, B., Reichmann, H., Hummel, T., 2009. Prevalence of smell loss in Parkinson's disease–a multicenter study. Parkinsonism Relat. Disord. 15, 490–494.
Hall, H., Halldin, C., Guilloteau, D., Chalon, S., Emond, P., Besnard, J., Farde, L., Sedvall, G., 1999. Visualization of the dopamine transporter in the human brain postmortem with the new selective ligand [^{125}I] PE2I. Neuroimage 9, 108–116.
Halldin, C., Farde, L., Lundkvist, C., Ginovart, N., Nakashima, Y., Karlsson, P., Swahn, C.G., 1996. [^{11}C]beta-CIT-FE, a radioligand for quantitation of the dopamine transporter in the living brain using positron emission tomography. Synapse 22, 386–390.
Halldin, C., Gulyas, B., Langer, O., Farde, L., 2001. Brain radioligands—state of the art and new trends. Q. J. Nucl. Med. 45, 139–152.
Halldin, C., Erixon-Lindroth, N., Pauli, S., Chou, Y.H., Okubo, Y., Karlsson, P., Lundkvist, C., Olsson, H., Guilloteau, D., Emond, P., Farde, L., 2003. [(11)C]PE2I: a highly selective radioligand for PET examination of the dopamine transporter in monkey and human brain. Eur. J. Nucl. Med. Mol. Imaging 30, 1220–1230.
Hastrup, H., Karlin, A., Javitch, J.A., 2001. Symmetrical dimer of the human dopamine transporter revealed by cross-linking Cys-306 at the extracellular end of the sixth transmembrane segment. Proc. Natl. Acad. Sci. U.S.A. 98, 10055–10060.
Heinz, A., Goldman, D., Jones, D.W., Palmour, R., Hommer, D., Gorey, J.G., Lee, K.S., Linnoila, M., Weinberger, D.R., 2000. Genotype influences in vivo dopamine transporter availability in human striatum. Neuropsychopharmacology 22, 133–139.
Hesse, S., Strecker, K., Winkler, D., Luthardt, J., Scherfler, C., Reupert, A., Oehlwein, C., Barthel, H., Schneider, J.P., Wegner, F., Meyer, P., Meixensberger, J., Sabri, O., Schwarz, J., 2008. Effects of subthalamic nucleus stimulation on striatal dopaminergic transmission in patients with Parkinson's disease within one-year follow-up. J. Neurol. 255, 1059–1066.
Hesse, S., Ballaschke, O., Barthel, H., Sabri, O., 2009. Dopamine transporter imaging in adult patients with attention-deficit/hyperactivity disorder. Psychiatry Res. 171, 120–128.
Hilker, R., Klein, C., Hedrich, K., Ozelius, L.J., Vieregge, P., Herholz, K., Pramstaller, P.P., Heiss, W.D., 2002. The striatal dopaminergic deficit is dependent on the number of mutant alleles in a family with mutations in the parkin gene: evidence for enzymatic parkin function in humans. Neurosci. Lett. 323, 50–54.
Hirvonen, J., Johansson, J., Teras, M., Oikonen, V., Lumme, V., Virsu, P., Roivainen, A., Nagren, K., Halldin, C., Farde, L., Hietala, J., 2008. Measurement of striatal and extrastriatal dopamine transporter binding with high-resolution PET and [^{11}C]PE2I: quantitative modeling and test-retest reproducibility. J. Cereb. Blood Flow Metab. 28, 1059–1069.
Hitri, A., Hurd, Y.L., Wyatt, R.J., Deutsch, S.I., 1994. Molecular, functional and biochemical characteristics of the dopamine transporter: regional differences and clinical relevance. Clin. Neuropharmacol. 17, 1–22.
Hsiao, M.C., Lin, K.J., Liu, C.Y., Tzen, K.Y., Yen, T.C., 2003. Dopamine transporter change in drug-naive schizophrenia: an imaging study with ^{99m}Tc-TRODAT-1. Schizophr. Res. 65, 39–46.
Hummel, T., Sekinger, B., Wolf, S.R., Pauli, E., Kobal, G., 1997. 'Sniffin' sticks': olfactory performance assessed by the combined testing of odor identification, odor discrimination and olfactory threshold. Chem. Senses 22, 39–52.

Hwang, W.J., Yao, W.J., Fu, Y.K., Yang, A.S., 2008. [^{99m}Tc]TRODAT-1/[^{123}I]IBZM SPECT studies of the dopaminergic system in Tourette syndrome. Psychiatry Res. 162, 159–166.

Ikawa, K., Watanabe, A., Kaneno, S., Toru, M., 1993. Modulation of [^{3}H]mazindol binding sites in rat striatum by dopaminergic agents. Eur. J. Pharmacol. 250, 261–266.

Ilgin, N., Zubieta, J., Reich, S.G., Dannals, R.F., Ravert, H.T., Frost, J.J., 1999. PET imaging of the dopamine transporter in progressive supranuclear palsy and Parkinson's disease. Neurology 52, 1221–1226.

Inaji, M., Yoshizaki, T., Okauchi, T., Maeda, J., Nagai, Y., Nariai, T., Ohno, K., Ando, K., Okano, H., Obayashi, S., Suhara, T., 2005. In vivo PET measurements with [^{11}C]PE2I to evaluate fetal mesencephalic transplantations to unilateral 6-OHDA-lesioned rats. Cell Transplant. 14, 655–663.

Innis, R.B., Cunningham, V.J., Delforge, J., Fujita, M., Gjedde, A., Gunn, R.N., Holden, J., Houle, S., Huang, S.C., Ichise, M., Iida, H., Ito, H., Kimura, Y., Koeppe, R.A., Knudsen, G.M., Knuuti, J., Lammertsma, A.A., Laruelle, M., Logan, J., Maguire, R.P., Mintun, M.A., Morris, E.D., Parsey, R., Price, J.C., Slifstein, M., Sossi, V., Suhara, T., Votaw, J.R., Wong, D.F., Carson, R.E., 2007. Consensus nomenclature for in vivo imaging of reversibly binding radioligands. J. Cereb. Blood Flow Metab. 27, 1533–1539.

Iranzo, A., Valldeoriola, F., Lomena, F., Molinuevo, J.L., Serradell, M., Salamero, M., Cot, A., Ros, D., Pavia, J., Santamaria, J., Tolosa, E., 2011. Serial dopamine transporter imaging of nigrostriatal function in patients with idiopathic rapid-eye-movement sleep behaviour disorder: a prospective study. Lancet Neurol. 10, 797–805.

Iranzo, A., Tolosa, E., Gelpi, E., Molinuevo, J.L., Valldeoriola, F., Serradell, M., Sanchez-Valle, R., Vilaseca, I., Lomena, F., Vilas, D., Llado, A., Gaig, C., Santamaria, J., 2013. Neurodegenerative disease status and post-mortem pathology in idiopathic rapid-eye-movement sleep behaviour disorder: an observational cohort study. Lancet Neurol. 12, 443–453.

Ishibashi, K., Ishii, K., Oda, K., Kawasaki, K., Mizusawa, H., Ishiwata, K., 2009. Regional analysis of age-related decline in dopamine transporters and dopamine D_2-like receptors in human striatum. Synapse 63, 282–290.

Ishikawa, T., Dhawan, V., Kazumata, K., Chaly, T., Mandel, F., Neumeyer, J., Margouleff, C., Babchyck, B., Zanzi, I., Eidelberg, D., 1996. Comparative nigrostriatal dopaminergic imaging with iodine-123-beta CIT-FP/SPECT and fluorine-18-FDOPA/PET. J. Nucl. Med. 37, 1760–1765.

Itier, J.M., Ibanez, P., Mena, M.A., Abbas, N., Cohen-Salmon, C., Bohme, G.A., Laville, M., Pratt, J., Corti, O., Pradier, L., Ret, G., Joubert, C., Periquet, M., Araujo, F., Negroni, J., Casarejos, M.J., Canals, S., Solano, R., Serrano, A., Gallego, E., Sanchez, M., Denefle, P., Benavides, J., Tremp, G., Rooney, T.A., Brice, A., Garcia de Yebenes, J., 2003. Parkin gene inactivation alters behaviour and dopamine neurotransmission in the mouse. Hum. Mol. Genet. 12, 2277–2291.

Jacobsen, L.K., Staley, J.K., Zoghbi, S.S., Seibyl, J.P., Kosten, T.R., Innis, R.B., Gelernter, J., 2000. Prediction of dopamine transporter binding availability by genotype: a preliminary report. Am. J. Psychiatry 157, 1700–1703.

Jiang, H., Jiang, Q., Feng, J., 2004. Parkin increases dopamine uptake by enhancing the cell surface expression of dopamine transporter. J. Biol. Chem. 279, 54380–54386.

Jucaite, A., Fernell, E., Halldin, C., Forssberg, H., Farde, L., 2005. Reduced midbrain dopamine transporter binding in male adolescents with attention-deficit/hyperactivity disorder: association between striatal dopamine markers and motor hyperactivity. Biol. Psychiatry 57, 229–238.

Jucaite, A., Odano, I., Olsson, H., Pauli, S., Halldin, C., Farde, L., 2006. Quantitative analyses of regional [^{11}C]PE2I binding to the dopamine transporter in the human brain: a PET study. Eur. J. Nucl. Med. Mol. Imaging 33, 657–668.

Kahlig, K.M., Galli, A., 2003. Regulation of dopamine transporter function and plasma membrane expression by dopamine, amphetamine, and cocaine. Eur. J. Pharmacol. 479, 153–158.

Kao, P.F., Tzen, K.Y., Yen, T.C., Lu, C.S., Weng, Y.H., Wey, S.P., Ting, G., 2001. The optimal imaging time for [^{99}Tcm]TRODAT-1/SPET in normal subjects and patients with Parkinson's disease. Nucl. Med. Commun. 22, 151–154.

Kazumata, K., Dhawan, V., Chaly, T., Antonini, A., Margouleff, C., Belakhlef, A., Neumeyer, J., Eidelberg, D., 1998. Dopamine transporter imaging with fluorine-18-FPCIT and PET. J. Nucl. Med. 39, 1521–1530.

Khan, N.L., Scherfler, C., Graham, E., Bhatia, K.P., Quinn, N., Lees, A.J., Brooks, D.J., Wood, N.W., Piccini, P., 2005. Dopaminergic dysfunction in unrelated, asymptomatic carriers of a single parkin mutation. Neurology 64, 134–136.

Khoshbouei, H., Sen, N., Guptaroy, B., Johnson, L., Lund, D., Gnegy, M.E., Galli, A., Javitch, J.A., 2004. N-terminal phosphorylation of the dopamine transporter is required for amphetamine-induced efflux. PLoS Biol. 2, E78.

Kilbourn, M.R., Sherman, P.S., Pisani, T., 1992. Repeated reserpine administration reduces in vivo [^{18}F] GBR 13119 binding to the dopamine uptake site. Eur. J. Pharmacol. 216, 109–112.

Koch, W., Hamann, C., Radau, P.E., Tatsch, K., 2007. Does combined imaging of the pre- and postsynaptic dopaminergic system increase the diagnostic accuracy in the differential diagnosis of parkinsonism? Eur. J. Nucl. Med. Mol. Imaging 34, 1265–1273.

Kordower, J.H., Emborg, M.E., Bloch, J., Ma, S.Y., Chu, Y., Leventhal, L., McBride, J., Chen, E.Y., Palfi, S., Roitberg, B.Z., Brown, W.D., Holden, J.E., Pyzalski, R., Taylor, M.D., Carvey, P., Ling, Z., Trono, D., Hantraye, P., Deglon, N., Aebischer, P., 2000. Neurodegeneration prevented by lentiviral vector delivery of GDNF in primate models of Parkinson's disease. Science 290, 767–773.

Krause, J., 2008. SPECT and PET of the dopamine transporter in attention-deficit/hyperactivity disorder. Expert Rev. Neurother. 8, 611–625.

Krause, K.H., Dresel, S.H., Krause, J., Kung, H.F., Tatsch, K., 2000. Increased striatal dopamine transporter in adult patients with attention deficit hyperactivity disorder: effects of methylphenidate as measured by single photon emission computed tomography. Neurosci. Lett. 285, 107–110.

Krause, K.H., Dresel, S.H., Krause, J., la Fougere, C., Ackenheil, M., 2003. The dopamine transporter and neuroimaging in attention deficit hyperactivity disorder. Neurosci. Biobehav. Rev. 27, 605–613.

Krueger, B.K., 1990. Kinetics and block of dopamine uptake in synaptosomes from rat caudate nucleus. J. Neurochem. 55, 260–267.

Kugaya, A., Seneca, N.M., Snyder, P.J., Williams, S.A., Malison, R.T., Baldwin, R.M., Seibyl, J.P., Innis, R.B., 2003. Changes in human in vivo serotonin and dopamine transporter availabilities during chronic antidepressant administration. Neuropsychopharmacology 28, 413–420.

Kuikka, J.T., Tupala, E., Bergstrom, K.A., Hiltunen, J., Tiihonen, J., 1999. Iodine-123 labelled PE2I for dopamine transporter imaging: influence of age in healthy subjects. Eur. J. Nucl. Med. 26, 1486–1488.

Laakso, A., Bergman, J., Haaparanta, M., Vilkman, H., Solin, O., Hietala, J., 1998. [^{18}F]CFT [(^{18}F)WIN 35,428], a radioligand to study the dopamine transporter with PET: characterization in human subjects. Synapse 28, 244–250.

Laakso, A., Vilkman, H., Alakare, B., Haaparanta, M., Bergman, J., Solin, O., Peurasaari, J., Rakkolainen, V., Syvalahti, E., Hietala, J., 2000. Striatal dopamine transporter binding in neuroleptic-naive patients with schizophrenia studied with positron emission tomography. Am. J. Psychiatry 157, 269–271.

Laasonen-Balk, T., Kuikka, J., Viinamaki, H., Husso-Saastamoinen, M., Lehtonen, J., Tiihonen, J., 1999. Striatal dopamine transporter density in major depression. Psychopharmacology (Berl) 144, 282–285.

Van Laere, K., De Ceuninck, L., Dom, R., Van den Eynden, J., Vanbilloen, H., Cleynhens, J., Dupont, P., Bormans, G., Verbruggen, A., Mortelmans, L., 2004. Dopamine transporter SPECT using fast kinetic ligands: ^{123}I-FP-beta-CIT versus ^{99m}Tc-TRODAT-1. Eur. J. Nucl. Med. Mol. Imaging 31, 1119–1127.

Laihinen, A.O., Rinne, J.O., Nagren, K.A., Lehikoinen, P.K., Oikonen, V.J., Ruotsalainen, U.H., Ruottinen, H.M., Rinne, U.K., 1995. PET studies on brain monoamine transporters with carbon-11-beta-CIT in Parkinson's disease. J. Nucl. Med. 36, 1263–1267.

Laruelle, M., Abi-Dargham, A., van Dyck, C., Gil, R., D'Souza, D.C., Krystal, J., Seibyl, J., Baldwin, R., Innis, R., 2000. Dopamine and serotonin transporters in patients with schizophrenia: an imaging study with [(123)I]beta-CIT. Biol. Psychiatry 47, 371–379.

Lavalaye, J., Booij, J., Reneman, L., Habraken, J.B., van Royen, E.A., 2000a. Effect of age and gender on dopamine transporter imaging with [^{123}I]FP-CIT SPECT in healthy volunteers. Eur. J. Nucl. Med. 27, 867–869.

Lavalaye, J., Knol, R.J., de Bruin, K., Reneman, L., Janssen, A.G., Booij, J., 2000b. [^{123}I]FP-CIT binding in rat brain after acute and sub-chronic administration of dopaminergic medication. Eur. J. Nucl. Med. 27, 346–349.

Lee, C.S., Samii, A., Sossi, V., Ruth, T.J., Schulzer, M., Holden, J.E., Wudel, J., Pal, P.K., de la Fuente-Fernandez, R., Calne, D.B., Stoessl, A.J., 2000. In vivo positron emission tomographic evidence for compensatory changes in presynaptic dopaminergic nerve terminals in Parkinson's disease. Ann. Neurol. 47, 493–503.

Leroy, C., Comtat, C., Trebossen, R., Syrota, A., Martinot, J.L., Ribeiro, M.J., 2007. Assessment of ^{11}C-PE2I binding to the neuronal dopamine transporter in humans with the high-spatial-resolution PET scanner HRRT. J. Nucl. Med. 48, 538–546.

Lin, Z., Zhang, P.W., Zhu, X., Melgari, J.M., Huff, R., Spieldoch, R.L., Uhl, G.R., 2003. Phosphatidylinositol 3-kinase, protein kinase C, and MEK1/2 kinase regulation of dopamine transporters (DAT) require N-terminal DAT phosphoacceptor sites. J. Biol. Chem. 278, 20162–20170.

Little, K.Y., Carroll, F.I., Cassin, B.J., 1995. Characterization and localization of [^{125}I]RTI-121 binding sites in human striatum and medial temporal lobe. J. Pharmacol. Exp. Ther. 274, 1473–1483.

Little, K.Y., Zhang, L., Desmond, T., Frey, K.A., Dalack, G.W., Cassin, B.J., 1999. Striatal dopaminergic abnormalities in human cocaine users. Am. J. Psychiatry 156, 238–245.

Little, K.Y., Elmer, L.W., Zhong, H., Scheys, J.O., Zhang, L., 2002. Cocaine induction of dopamine transporter trafficking to the plasma membrane. Mol. Pharmacol. 61, 436–445.

Logan, J., Fowler, J.S., Volkow, N.D., Wang, G.J., Ding, Y.S., Alexoff, D.L., 1996. Distribution volume ratios without blood sampling from graphical analysis of PET data. J. Cereb. Blood Flow Metab. 16, 834–840.

Lorang, D., Amara, S.G., Simerly, R.B., 1994. Cell-type-specific expression of catecholamine transporters in the rat brain. J. Neurosci. 14, 4903–4914.

Lorberboym, M., Djaldetti, R., Melamed, E., Sadeh, M., Lampl, Y., 2004. ^{123}I-FP-CIT SPECT imaging of dopamine transporters in patients with cerebrovascular disease and clinical diagnosis of vascular parkinsonism. J. Nucl. Med. 45, 1688–1693.

Lundkvist, C., Halldin, C., Ginovart, N., Swahn, C.G., Farde, L., 1997. [^{18}F] beta-CIT-FP is superior to [^{11}C] beta-CIT-FP for quantitation of the dopamine transporter. Nucl. Med. Biol. 24, 621–627.

Ma, Y., Dhawan, V., Mentis, M., Chaly, T., Spetsieris, P.G., Eidelberg, D., 2002. Parametric mapping of [^{18}F] FPCIT binding in early stage Parkinson's disease: a PET study. Synapse 45, 125–133.

Madras, B.K., Gracz, L.M., Fahey, M.A., Elmaleh, D., Meltzer, P.C., Liang, A.Y., Stopa, E.G., Babich, J., Fischman, A.J., 1998. Altropane, a SPECT or PET imaging probe for dopamine neurons: III. Human dopamine transporter in postmortem normal and Parkinson's diseased brain. Synapse 29, 116–127.

Malison, R.T., McDougle, C.J., van Dyck, C.H., Scahill, L., Baldwin, R.M., Seibyl, J.P., Price, L.H., Leckman, J.F., Innis, R.B., 1995. [^{123}I]beta-CIT SPECT imaging of striatal dopamine transporter binding in Tourette's disorder. Am. J. Psychiatry 152, 1359–1361.

Malison, R.T., Best, S.E., van Dyck, C.H., McCance, E.F., Wallace, E.A., Laruelle, M., Baldwin, R.M., Seibyl, J.P., Price, L.H., Kosten, T.R., Innis, R.B., 1998a. Elevated striatal dopamine transporters during acute cocaine abstinence as measured by [^{123}I] beta-CIT SPECT. Am. J. Psychiatry 155, 832–834.

Malison, R.T., Price, L.H., Berman, R., van Dyck, C.H., Pelton, G.H., Carpenter, L., Sanacora, G., Owens, M.J., Nemeroff, C.B., Rajeevan, N., Baldwin, R.M., Seibyl, J.P., Innis, R.B., Charney, D.S., 1998b. Reduced brain serotonin transporter availability in major depression as measured by [^{123}I]-2 beta-carbomethoxy-3 beta-(4-iodophenyl)tropane and single photon emission computed tomography. Biol. Psychiatry 44, 1090–1098.

Marek, K., Jennings, D., 2009. Can we image premotor Parkinson disease? Neurology 72, S21–S26.

Marek, K.L., Seibyl, J.P., Zoghbi, S.S., Zea-Ponce, Y., Baldwin, R.M., Fussell, B., Charney, D.S., van Dyck, C., Hoffer, P.B., Innis, R.P., 1996. [^{123}I] beta-CIT/SPECT imaging demonstrates bilateral loss of dopamine transporters in hemi-Parkinson's disease. Neurology 46, 231–237.

Marek, K., Innis, R., van Dyck, C., Fussell, B., Early, M., Eberly, S., Oakes, D., Seibyl, J., 2001. [^{123}I]beta-CIT SPECT imaging assessment of the rate of Parkinson's disease progression. Neurology 57, 2089–2094.

Marek, K., Jennings, D., Seibyl, J., 2002. Do dopamine agonists or levodopa modify Parkinson's disease progression? Eur. J. Neurol. 9 (Suppl. 3), 15–22.

Marshall, V., Grosset, D., 2003. Role of dopamine transporter imaging in routine clinical practice. Mov. Disord. 18, 1415–1423.

Martinez, D., Gelernter, J., Abi-Dargham, A., van Dyck, C.H., Kegeles, L., Innis, R.B., Laruelle, M., 2001. The variable number of tandem repeats polymorphism of the dopamine transporter gene is not associated with significant change in dopamine transporter phenotype in humans. Neuropsychopharmacology 24, 553–560.

Masson, J., Sagne, C., Hamon, M., El Mestikawy, S., 1999. Neurotransmitter transporters in the central nervous system. Pharmacol. Rev. 51, 439–464.

McCann, U.D., Wong, D.F., Yokoi, F., Villemagne, V., Dannals, R.F., Ricaurte, G.A., 1998. Reduced striatal dopamine transporter density in abstinent methamphetamine and methcathinone users: evidence from positron emission tomography studies with [^{11}C]WIN-35,428. J. Neurosci. 18, 8417–8422.

McElvain, J.S., Schenk, J.O., 1992. A multisubstrate mechanism of striatal dopamine uptake and its inhibition by cocaine. Biochem. Pharmacol. 43, 2189–2199.

McKeith, I., O'Brien, J., Walker, Z., Tatsch, K., Booij, J., Darcourt, J., Padovani, A., Giubbini, R., Bonuccelli, U., Volterrani, D., Holmes, C., Kemp, P., Tabet, N., Meyer, I., Reininger, C., 2007. Sensitivity and specificity of dopamine transporter imaging with ^{123}I-FP-CIT SPECT in dementia with Lewy bodies: a phase III, multicentre study. Lancet Neurol. 6, 305–313.

Messa, C., Volonte, M.A., Fazio, F., Zito, F., Carpinelli, A., d'Amico, A., Rizzo, G., Moresco, R.M., Paulesu, E., Franceschi, M., Lucignani, G., 1998. Differential distribution of striatal [^{123}I]beta-CIT in Parkinson's disease and progressive supranuclear palsy, evaluated with single-photon emission tomography. Eur. J. Nucl. Med. 25, 1270–1276.

Meyer, J.H., Kruger, S., Wilson, A.A., Christensen, B.K., Goulding, V.S., Schaffer, A., Minifie, C., Houle, S., Hussey, D., Kennedy, S.H., 2001. Lower dopamine transporter binding potential in striatum during depression. Neuroreport 12, 4121–4125.

Miller, G.M., Madras, B.K., 2002. Polymorphisms in the 3′-untranslated region of human and monkey dopamine transporter genes affect reporter gene expression. Mol. Psychiatry 7, 44–55.

Morrish, P.K., 2003. How valid is dopamine transporter imaging as a surrogate marker in research trials in Parkinson's disease? Mov. Disord. 18 (Suppl. 7), S63–S70.

Mortensen, O.V., Amara, S.G., 2003. Dynamic regulation of the dopamine transporter. Eur. J. Pharmacol. 479, 159–170.

Mozley, P.D., Kim, H.J., Gur, R.C., Tatsch, K., Muenz, L.R., McElgin, W.T., Kung, M.P., Mu, M., Myers, A.M., Kung, H.F., 1996. Iodine-123-IPT SPECT imaging of CNS dopamine transporters: nonlinear effects of normal aging on striatal uptake values. J. Nucl. Med. 37, 1965–1970.

Mozley, P.D., Acton, P.D., Barraclough, E.D., Plossl, K., Gur, R.C., Alavi, A., Mathur, A., Saffer, J., Kung, H.F., 1999. Effects of age on dopamine transporters in healthy humans. J. Nucl. Med. 40, 1812–1817.

Mozley, P.D., Schneider, J.S., Acton, P.D., Plossl, K., Stern, M.B., Siderowf, A., Leopold, N.A., Li, P.Y., Alavi, A., Kung, H.F., 2000. Binding of [^{99m}Tc]TRODAT-1 to dopamine transporters in patients with Parkinson's disease and in healthy volunteers. J. Nucl. Med. 41, 584–589.

Nandhagopal, R., Mak, E., Schulzer, M., McKenzie, J., McCormick, S., Sossi, V., Ruth, T.J., Strongosky, A., Farrer, M.J., Wszolek, Z.K., Stoessl, A.J., 2008. Progression of dopaminergic dysfunction in a LRRK2 kindred: a multitracer PET study. Neurology 71, 1790–1795.

Nandhagopal, R., Kuramoto, L., Schulzer, M., Mak, E., Cragg, J., Lee, C.S., McKenzie, J., McCormick, S., Samii, A., Troiano, A., Ruth, T.J., Sossi, V., de la Fuente-Fernandez, R., Calne, D.B., Stoessl, A.J., 2009. Longitudinal progression of sporadic Parkinson's disease: a multi-tracer positron emission tomography study. Brain 132, 2970–2979.

Nirenberg, M.J., Vaughan, R.A., Uhl, G.R., Kuhar, M.J., Pickel, V.M., 1996. The dopamine transporter is localized to dendritic and axonal plasma membranes of nigrostriatal dopaminergic neurons. J. Neurosci. 16, 436–447.

Nurmi, E., Ruottinen, H.M., Kaasinen, V., Bergman, J., Haaparanta, M., Solin, O., Rinne, J.O., 2000. Progression in Parkinson's disease: a positron emission tomography study with a dopamine transporter ligand [^{18}F]CFT. Ann. Neurol. 47, 804–808.

Nurmi, E., Bergman, J., Eskola, O., Solin, O., Vahlberg, T., Sonninen, P., Rinne, J.O., 2003. Progression of dopaminergic hypofunction in striatal subregions in Parkinson's disease using [^{18}F]CFT PET. Synapse 48, 109–115.

Oaks, A.W., Sidhu, A., 2011. Synuclein modulation of monoamine transporters. FEBS Lett. 585, 1001–1006.

Okamura, N., Villemagne, V.L., Drago, J., Pejoska, S., Dhamija, R.K., Mulligan, R.S., Ellis, J.R., Ackermann, U., O'Keefe, G., Jones, G., Kung, H.F., Pontecorvo, M.J., Skovronsky, D., Rowe, C.C., 2010. In vivo measurement of vesicular monoamine transporter type 2 density in Parkinson disease with (18)F-AV-133. J. Nucl. Med. 51, 223–228.

Pellecchia, M.T., Varrone, A., Annesi, G., Amboni, M., Cicarelli, G., Sansone, V., Annesi, F., Rocca, F.E., Vitale, C., Pappata, S., Quattrone, A., Barone, P., 2007. Parkinsonism and essential tremor in a family with pseudo-dominant inheritance of PARK2: an FP-CIT SPECT study. Mov. Disord. 22, 559–563.

Piccini, P., Pavese, N., Hagell, P., Reimer, J., Bjorklund, A., Oertel, W.H., Quinn, N.P., Brooks, D.J., Lindvall, O., 2005. Factors affecting the clinical outcome after neural transplantation in Parkinson's disease. Brain 128, 2977–2986.

Pinborg, L.H., Videbaek, C., Svarer, C., Yndgaard, S., Paulson, O.B., Knudsen, G.M., 2002. Quantification of [(123)I]PE2I binding to dopamine transporters with SPET. Eur. J. Nucl. Med. Mol. Imaging 29, 623–631.

Pinborg, L.H., Ziebell, M., Frokjaer, V.G., de Nijs, R., Svarer, C., Haugbol, S., Yndgaard, S., Knudsen, G.M., 2005. Quantification of ^{123}I-PE2I binding to dopamine transporter with SPECT after bolus and bolus/infusion. J. Nucl. Med. 46, 1119–1127.

Pirker, W., Asenbaum, S., Hauk, M., Kandlhofer, S., Tauscher, J., Willeit, M., Neumeister, A., Praschak-Rieder, N., Angelberger, P., Brucke, T., 2000. Imaging serotonin and dopamine transporters with ^{123}I-beta-CIT SPECT: binding kinetics and effects of normal aging. J. Nucl. Med. 41, 36–44.

Pirker, W., Djamshidian, S., Asenbaum, S., Gerschlager, W., Tribl, G., Hoffmann, M., Brucke, T., 2002. Progression of dopaminergic degeneration in Parkinson's disease and atypical parkinsonism: a longitudinal beta-CIT SPECT study. Mov. Disord. 17, 45–53.

Pirker, W., Holler, I., Gerschlager, W., Asenbaum, S., Zettinig, G., Brucke, T., 2003. Measuring the rate of progression of Parkinson's disease over a 5-year period with beta-CIT SPECT. Mov. Disord. 18, 1266–1272.

Plotkin, M., Amthauer, H., Klaffke, S., Kuhn, A., Ludemann, L., Arnold, G., Wernecke, K.D., Kupsch, A., Felix, R., Venz, S., 2005. Combined ^{123}I-FP-CIT and ^{123}I-IBZM SPECT for the diagnosis of parkinsonian syndromes: study on 72 patients. J. Neural Transm. 112, 677–692.

Pogarell, O., Koch, W., Gildehaus, F.J., Kupsch, A., Lindvall, O., Oertel, W.H., Tatsch, K., 2006. Long-term assessment of striatal dopamine transporters in Parkinsonian patients with intrastriatal embryonic mesencephalic grafts. Eur. J. Nucl. Med. Mol. Imaging 33, 407–411.

Ponsen, M.M., Stoffers, D., Booij, J., van Eck-Smit, B.L., Wolters, E., Berendse, H.W., 2004. Idiopathic hyposmia as a preclinical sign of Parkinson's disease. Ann. Neurol. 56, 173–181.

Ribeiro, M.J., Ricard, M., Lievre, M.A., Bourgeois, S., Emond, P., Gervais, P., Dolle, F., Syrota, A., 2007. Whole-body distribution and radiation dosimetry of the dopamine transporter radioligand [(11)C]PE2I in healthy volunteers. Nucl. Med. Biol. 34, 465–470.

Rinne, J.O., Laihinen, A., Nagren, K., Ruottinen, H., Ruotsalainen, U., Rinne, U.K., 1995. PET examination of the monoamine transporter with [^{11}C]beta-CIT and [^{11}C]beta-CFT in early Parkinson's disease. Synapse 21, 97–103.

Rinne, J.O., Sahlberg, N., Ruottinen, H., Nagren, K., Lehikoinen, P., 1998. Striatal uptake of the dopamine reuptake ligand [^{11}C]beta-CFT is reduced in Alzheimer's disease assessed by positron emission tomography. Neurology 50, 152–156.

Rinne, J.O., Bergman, J., Ruottinen, H., Haaparanta, M., Eronen, E., Oikonen, V., Sonninen, P., Solin, O., 1999a. Striatal uptake of a novel PET ligand, [^{18}F]beta-CFT, is reduced in early Parkinson's disease. Synapse 31, 119–124.

Rinne, J.O., Ruottinen, H., Bergman, J., Haaparanta, M., Sonninen, P., Solin, O., 1999b. Usefulness of a dopamine transporter PET ligand [(18)F]beta-CFT in assessing disability in Parkinson's disease. J. Neurol. Neurosurg. Psychiatry 67, 737–741.

Rinne, J.O., Laine, M., Kaasinen, V., Norvasuo-Heila, M.K., Nagren, K., Helenius, H., 2002. Striatal dopamine transporter and extrapyramidal symptoms in frontotemporal dementia. Neurology 58, 1489–1493.

Robertson, S.D., Matthies, H.J., Galli, A., 2009. A closer look at amphetamine-induced reverse transport and trafficking of the dopamine and norepinephrine transporters. Mol. Neurobiol. 39, 73–80.

Robeson, W., Dhawan, V., Belakhlef, A., Ma, Y., Pillai, V., Chaly, T., Margouleff, C., Bjelke, D., Eidelberg, D., 2003. Dosimetry of the dopamine transporter radioligand ^{18}F-FPCIT in human subjects. J. Nucl. Med. 44, 961–966.

Saba, W., Valette, H., Schollhorn-Peyronneau, M.A., Coulon, C., Ottaviani, M., Chalon, S., Dolle, F., Emond, P., Halldin, C., Helfenbein, J., Madelmont, J.C., Deloye, J.B., Guilloteau, D., Bottlaender, M., 2007. [^{11}C]LBT-999: a suitable radioligand for investigation of extra-striatal dopamine transporter with PET. Synapse 61, 17–23.

Sasaki, T., Ito, H., Kimura, Y., Arakawa, R., Takano, H., Seki, C., Kodaka, F., Fujie, S., Takahata, K., Nogami, T., Suzuki, M., Fujiwara, H., Takahashi, H., Nakao, R., Fukumura, T., Varrone, A., Halldin, C., Nishikawa, T., Suhara, T., 2012. Quantification of dopamine transporter in human brain using PET with ^{18}F-FE-PE2I. J. Nucl. Med. 53, 1065–1073.

Scherfler, C., Khan, N.L., Pavese, N., Eunson, L., Graham, E., Lees, A.J., Quinn, N.P., Wood, N.W., Brooks, D.J., Piccini, P.P., 2004. Striatal and cortical pre- and postsynaptic dopaminergic dysfunction in sporadic parkin-linked parkinsonism. Brain 127, 1332–1342.

Scherfler, C., Seppi, K., Donnemiller, E., Goebel, G., Brenneis, C., Virgolini, I., Wenning, G.K., Poewe, W., 2005. Voxel-wise analysis of [^{123}I]beta-CIT SPECT differentiates the Parkinson variant of multiple system atrophy from idiopathic Parkinson's disease. Brain 128, 1605–1612.

Schmitt, K.C., Zhen, J., Kharkar, P., Mishra, M., Chen, N., Dutta, A.K., Reith, M.E., 2008. Interaction of cocaine-, benztropine-, and GBR12909-like compounds with wild-type and mutant human dopamine transporters: molecular features that differentially determine antagonist-binding properties. J. Neurochem. 107, 928–940.

Schou, M., Steiger, C., Varrone, A., Guilloteau, D., Halldin, C., 2009. Synthesis, radiolabeling and preliminary in vivo evaluation of [^{18}F]FE-PE2I, a new probe for the dopamine transporter. Bioorg. Med. Chem. Lett. 19, 4843–4845.

Schwarz, J., Storch, A., Koch, W., Pogarell, O., Radau, P.E., Tatsch, K., 2004. Loss of dopamine transporter binding in Parkinson's disease follows a single exponential rather than linear decline. J. Nucl. Med. 45, 1694–1697.

Sedaghat, F., Gotzamani-Psarrakou, A., Dedousi, E., Arnaoutoglou, M., Psarrakos, K., Baloyannis, I., Dimitriadis, A.S., Baloyannis, S.J., 2007. Evaluation of dopaminergic function in frontotemporal dementia using I-FP-CIT single photon emission computed tomography. Neurodegener. Dis. 4, 382–385.

Seibyl, J.P., Marek, K.L., Quinlan, D., Sheff, K., Zoghbi, S., Zea-Ponce, Y., Baldwin, R.M., Fussell, B., Smith, E.O., Charney, D.S., et al., 1995. Decreased single-photon emission computed tomographic [^{123}I]beta-CIT striatal uptake correlates with symptom severity in Parkinson's disease. Ann. Neurol. 38, 589–598.

Serra-Mestres, J., Ring, H.A., Costa, D.C., Gacinovic, S., Walker, Z., Lees, A.J., Robertson, M.M., Trimble, M.R., 2004. Dopamine transporter binding in Gilles de la Tourette syndrome: a [^{123}I]FP-CIT/SPECT study. Acta Psychiatr. Scand. 109, 140–146.

Shetty, H.U., Zoghbi, S.S., Liow, J.S., Ichise, M., Hong, J., Musachio, J.L., Halldin, C., Seidel, J., Innis, R.B., Pike, V.W., 2007. Identification and regional distribution in rat brain of radiometabolites of the dopamine transporter PET radioligand [^{11}C]PE2I. Eur. J. Nucl. Med. Mol. Imaging 34, 667–678.

Shumay, E., Chen, J., Fowler, J.S., Volkow, N.D., 2011. Genotype and ancestry modulate brain's DAT availability in healthy humans. PLoS One 6, e22754.

Siderowf, A., Lang, A.E., 2012. Premotor Parkinson's disease: concepts and definitions. Mov. Disord. 27, 608–616.

Snow, B.J., Tooyama, I., McGeer, E.G., Yamada, T., Calne, D.B., Takahashi, H., Kimura, H., 1993. Human positron emission tomographic [^{18}F]fluorodopa studies correlate with dopamine cell counts and levels. Ann. Neurol. 34, 324–330.

Sorkina, T., Doolen, S., Galperin, E., Zahniser, N.R., Sorkin, A., 2003. Oligomerization of dopamine transporters visualized in living cells by fluorescence resonance energy transfer microscopy. J. Biol. Chem. 278, 28274–28283.

Sossi, V., Dinelle, K., Schulzer, M., Mak, E., Doudet, D.J., de la Fuente-Fernandez, R., 2010. Levodopa and pramipexole effects on presynaptic dopamine PET markers and estimated dopamine release. Eur. J. Nucl. Med. Mol. Imaging 37, 2364–2370.

Spencer, T.J., Biederman, J., Madras, B.K., Faraone, S.V., Dougherty, D.D., Bonab, A.A., Fischman, A.J., 2005. In vivo neuroreceptor imaging in attention-deficit/hyperactivity disorder: a focus on the dopamine transporter. Biol. Psychiatry 57, 1293–1300.

Spencer, T.J., Biederman, J., Madras, B.K., Dougherty, D.D., Bonab, A.A., Livni, E., Meltzer, P.C., Martin, J., Rauch, S., Fischman, A.J., 2007. Further evidence of dopamine transporter dysregulation in ADHD: a controlled PET imaging study using altropane. Biol. Psychiatry 62, 1059–1061.

Staffen, W., Mair, A., Unterrainer, J., Trinka, E., Ladurner, G., 2000. Measuring the progression of idiopathic Parkinson's disease with [^{123}I] beta-CIT SPECT. J. Neural Transm. 107, 543–552.

Staley, J.K., Hearn, W.L., Ruttenber, A.J., Wetli, C.V., Mash, D.C., 1994. High affinity cocaine recognition sites on the dopamine transporter are elevated in fatal cocaine overdose victims. J. Pharmacol. Exp. Ther. 271, 1678–1685.

Staley, J.K., Boja, J.W., Carroll, F.I., Seltzman, H.H., Wyrick, C.D., Lewin, A.H., Abraham, P., Mash, D.C., 1995. Mapping dopamine transporters in the human brain with novel selective cocaine analog [^{125}I] RTI-121. Synapse 21, 364–372.

Staley, J.K., Krishnan-Sarin, S., Zoghbi, S., Tamagnan, G., Fujita, M., Seibyl, J.P., Maciejewski, P.K., O'Malley, S., Innis, R.B., 2001. Sex differences in [^{123}I]beta-CIT SPECT measures of dopamine and serotonin transporter availability in healthy smokers and nonsmokers. Synapse 41, 275–284.

Stiasny-Kolster, K., Doerr, Y., Moller, J.C., Hoffken, H., Behr, T.M., Oertel, W.H., Mayer, G., 2005. Combination of 'idiopathic' REM sleep behaviour disorder and olfactory dysfunction as possible indicator for alpha-synucleinopathy demonstrated by dopamine transporter FP-CIT-SPECT. Brain 128, 126–137.

Sureau, F.C., Reader, A.J., Comtat, C., Leroy, C., Ribeiro, M.J., Buvat, I., Trebossen, R., 2008. Impact of image-space resolution modeling for studies with the high-resolution research tomograph. J. Nucl. Med. 49, 1000–1008.

Talvik, M., Nordstrom, A.L., Olsson, H., Halldin, C., Farde, L., 2003. Decreased thalamic D_2/D_3 receptor binding in drug-naive patients with schizophrenia: a PET study with [^{11}C]FLB 457. Int. J. Neuropsychopharmacol. 6, 361–370.

Thomsen, G., Knudsen, G.M., Jensen, P.S., Ziebell, M., Holst, K.K., Asenbaum, S., Booij, J., Darcourt, J., Dickson, J.C., Kapucu, O.L., Nobili, F., Sabri, O., Sera, T., Tatsch, K., Tossici-Bolt, L., Laere, K.V., Borght, T.V., Varrone, A., Pagani, M., Pinborg, L.H., 2013. No difference in striatal dopamine transporter availability between active smokers, ex-smokers and non-smokers using [^{123}I]FP-CIT (DaTSCAN) and SPECT. EJNMMI Res. 3, 39.

Tissingh, G., Booij, J., Bergmans, P., Winogrodzka, A., Janssen, A.G., van Royen, E.A., Stoof, J.C., Wolters, E.C., 1998. Iodine-123-N-omega-fluoropropyl-2beta-carbomethoxy-3beta-(4-iodophenyl) tropane SPECT in healthy controls and early-stage, drug-naive Parkinson's disease. J. Nucl. Med. 39, 1143–1148.

Torres, G.E., Carneiro, A., Seamans, K., Fiorentini, C., Sweeney, A., Yao, W.D., Caron, M.G., 2003. Oligomerization and trafficking of the human dopamine transporter. Mutational analysis identifies critical domains important for the functional expression of the transporter. J. Biol. Chem. 278, 2731–2739.

Troiano, A.R., Schulzer, M., Fuente-Fernandez, R.D., Mak, E., McKenzie, J., Sossi, V., McCormick, S., Ruth, T.J., Stoessl, A.J., 2009. Dopamine transporter PET in normal aging: dopamine transporter decline and its possible role in preservation of motor function. Synapse 64, 146–151.

Tzen, K.Y., Lu, C.S., Yen, T.C., Wey, S.P., Ting, G., 2001. Differential diagnosis of Parkinson's disease and vascular parkinsonism by (^{99m})Tc-TRODAT-1. J. Nucl. Med. 42, 408–413.

Uhl, G.R., 2003. Dopamine transporter: basic science and human variation of a key molecule for dopaminergic function, locomotion, and parkinsonism. Mov. Disord. 18 (Suppl. 7), S71–S80.

Usdin, T.B., Mezey, E., Chen, C., Brownstein, M.J., Hoffman, B.J., 1991. Cloning of the cocaine-sensitive bovine dopamine transporter. Proc. Natl. Acad. Sci. U.S.A. 88, 11168–11171.

Varrone, A., Halldin, C., 2010. Molecular imaging of the dopamine transporter. J. Nucl. Med. 51, 1331–1334.

Varrone, A., Marek, K.L., Jennings, D., Innis, R.B., Seibyl, J.P., 2001. [(123)I]beta-CIT SPECT imaging demonstrates reduced density of striatal dopamine transporters in Parkinson's disease and multiple system atrophy. Mov. Disord. 16, 1023–1032.

Varrone, A., Pellecchia, M.T., Amboni, M., Sansone, V., Salvatore, E., Ghezzi, D., Garavaglia, B., Brice, A., Brunetti, A., Bonavita, V., De Michele, G., Salvatore, M., Pappata, S., Barone, P., 2004. Imaging of dopaminergic dysfunction with [^{123}I]FP-CIT SPECT in early-onset parkin disease. Neurology 63, 2097–2103.

Varrone, A., Sjoholm, N., Eriksson, L., Gulyas, B., Halldin, C., Farde, L., 2009a. Advancement in PET quantification using 3D-OP-OSEM point spread function reconstruction with the HRRT. Eur. J. Nucl. Med. Mol. Imaging 36, 1639–1650.

Varrone, A., Steiger, C., Schou, M., Takano, A., Finnema, S.J., Guilloteau, D., Gulyas, B., Halldin, C., 2009b. In vitro autoradiography and in vivo evaluation in cynomolgus monkey of [^{18}F]FE-PE2I, a new dopamine transporter PET radioligand. Synapse 63, 871–880.

Varrone, A., Gulyas, B., Takano, A., Stabin, M.G., Jonsson, C., Halldin, C., 2011a. Simplified quantification and whole-body distribution of [(18)F]FE-PE2I in nonhuman primates: prediction for human studies. Nucl. Med. Biol.

Varrone, A., Stepanov, V., Nakao, R., Toth, M., Gulyas, B., Emond, P., Deloye, J.B., Vercouillie, J., Stabin, M.G., Jonsson, C., Guilloteau, D., Halldin, C., 2011b. Imaging of the striatal and extrastriatal dopamine transporter with (18)F-LBT-999: quantification, biodistribution, and radiation dosimetry in nonhuman primates. J. Nucl. Med. 52, 1313–1321.

Varrone, A., Toth, M., Steiger, C., Takano, A., Guilloteau, D., Ichise, M., Gulyas, B., Halldin, C., 2011c. Kinetic analysis and quantification of the dopamine transporter in the nonhuman primate brain with ^{11}C-PE2I and ^{18}F-FE-PE2I. J. Nucl. Med. 52, 132–139.

Varrone, A., Dickson, J.C., Tossici-Bolt, L., Sera, T., Asenbaum, S., Booij, J., Kapucu, O.L., Kluge, A., Knudsen, G.M., Koulibaly, P.M., Nobili, F., Pagani, M., Sabri, O., Vander Borght, T., Van Laere, K., Tatsch, K., 2013. European multicentre database of healthy controls for [^{123}I]FP-CIT SPECT (ENC-DAT): age-related effects, gender differences and evaluation of different methods of analysis. Eur. J. Nucl. Med. Mol. Imaging 40, 213–227.

Vila, M., Przedborski, S., 2003. Targeting programmed cell death in neurodegenerative diseases. Nat. Rev. Neurosci. 4, 365–375.

Volkow, N.D., Fowler, J.S., Wang, G.J., Logan, J., Schlyer, D., MacGregor, R., Hitzemann, R., Wolf, A.P., 1994. Decreased dopamine transporters with age in health human subjects. Ann. Neurol. 36, 237–239.

Volkow, N.D., Ding, Y.S., Fowler, J.S., Wang, G.J., Logan, J., Gatley, J.S., Dewey, S., Ashby, C., Liebermann, J., Hitzemann, R., et al., 1995. Is methylphenidate like cocaine? Studies on their pharmacokinetics and distribution in the human brain. Arch. Gen. Psychiatry 52, 456–463.

Volkow, N.D., Ding, Y.S., Fowler, J.S., Wang, G.J., Logan, J., Gatley, S.J., Hitzemann, R., Smith, G., Fields, S.D., Gur, R., 1996a. Dopamine transporters decrease with age. J. Nucl. Med. 37, 554–559.

Volkow, N.D., Wang, G.J., Fowler, J.S., Logan, J., Hitzemannn, R., Gatley, S.J., MacGregor, R.R., Wolf, A.P., 1996b. Cocaine uptake is decreased in the brain of detoxified cocaine abusers. Neuropsychopharmacology 14, 159–168.

Volkow, N.D., Wang, G.J., Fowler, J.S., Gatley, S.J., Logan, J., Ding, Y.S., Hitzemann, R., Pappas, N., 1998. Dopamine transporter occupancies in the human brain induced by therapeutic doses of oral methylphenidate. Am. J. Psychiatry 155, 1325–1331.

Volkow, N.D., Wang, G.J., Fowler, J.S., Fischman, M., Foltin, R., Abumrad, N.N., Gatley, S.J., Logan, J., Wong, C., Gifford, A., Ding, Y.S., Hitzemann, R., Pappas, N., 1999. Methylphenidate and cocaine have a similar in vivo potency to block dopamine transporters in the human brain. Life Sci. 65, PL7–12.

Volkow, N.D., Wang, G.J., Fischman, M.W., Foltin, R., Fowler, J.S., Franceschi, D., Franceschi, M., Logan, J., Gatley, S.J., Wong, C., Ding, Y.S., Hitzemann, R., Pappas, N., 2000. Effects of route of administration on cocaine induced dopamine transporter blockade in the human brain. Life Sci. 67, 1507–1515.

Volkow, N.D., Wang, G.J., Newcorn, J., Fowler, J.S., Telang, F., Solanto, M.V., Logan, J., Wong, C., Ma, Y., Swanson, J.M., Schulz, K., Pradhan, K., 2007. Brain dopamine transporter levels in treatment and drug naive adults with ADHD. Neuroimage 34, 1182–1190.

Volkow, N.D., Wang, G.J., Kollins, S.H., Wigal, T.L., Newcorn, J.H., Telang, F., Fowler, J.S., Zhu, W., Logan, J., Ma, Y., Pradhan, K., Wong, C., Swanson, J.M., 2009. Evaluating dopamine reward pathway in ADHD: clinical implications. JAMA 302, 1084–1091.

Walker, Z., Costa, D.C., Walker, R.W., Shaw, K., Gacinovic, S., Stevens, T., Livingston, G., Ince, P., McKeith, I.G., Katona, C.L., 2002. Differentiation of dementia with Lewy bodies from Alzheimer's disease using a dopaminergic presynaptic ligand. J. Neurol. Neurosurg. Psychiatry 73, 134–140.

Walker, Z., Costa, D.C., Walker, R.W., Lee, L., Livingston, G., Jaros, E., Perry, R., McKeith, I., Katona, C.L., 2004. Striatal dopamine transporter in dementia with Lewy bodies and Parkinson disease: a comparison. Neurology 62, 1568–1572.

Wang, G.J., Volkow, N.D., Wigal, T., Kollins, S.H., Newcorn, J.H., Telang, F., Logan, J., Jayne, M., Wong, C.T., Han, H., Fowler, J.S., Zhu, W., Swanson, J.M., 2013. Long-term stimulant treatment affects brain dopamine transporter level in patients with attention deficit hyperactive disorder. PLoS One 8, e63023.

Weng, Y.H., Yen, T.C., Chen, M.C., Kao, P.F., Tzen, K.Y., Chen, R.S., Wey, S.P., Ting, G., Lu, C.S., 2004. Sensitivity and specificity of ^{99m}Tc-TRODAT-1 SPECT imaging in differentiating patients with idiopathic Parkinson's disease from healthy subjects. J. Nucl. Med. 45, 393–401.

Whone, A.L., Watts, R.L., Stoessl, A.J., Davis, M., Reske, S., Nahmias, C., Lang, A.E., Rascol, O., Ribeiro, M.J., Remy, P., Poewe, W.H., Hauser, R.A., Brooks, D.J., 2003. Slower progression of Parkinson's disease with ropinirole versus levodopa: the REAL-PET study. Ann. Neurol. 54, 93–101.

Wilson, J.M., Kalasinsky, K.S., Levey, A.I., Bergeron, C., Reiber, G., Anthony, R.M., Schmunk, G.A., Shannak, K., Haycock, J.W., Kish, S.J., 1996. Striatal dopamine nerve terminal markers in human, chronic methamphetamine users. Nat. Med. 2, 699–703.

Winogrodzka, A., Bergmans, P., Booij, J., van Royen, E.A., Stoof, J.C., Wolters, E.C., 2003. [(123)I]beta-CIT SPECT is a useful method for monitoring dopaminergic degeneration in early stage Parkinson's disease. J. Neurol. Neurosurg. Psychiatry 74, 294–298.

Wong, D.F., Brasic, J.R., Singer, H.S., Schretlen, D.J., Kuwabara, H., Zhou, Y., Nandi, A., Maris, M.A., Alexander, M., Ye, W., Rousset, O., Kumar, A., Szabo, Z., Gjedde, A., Grace, A.A., 2008. Mechanisms of dopaminergic and serotonergic neurotransmission in Tourette syndrome: clues from an in vivo neurochemistry study with PET. Neuropsychopharmacology 33, 1239–1251.

Yang, Y.K., Yeh, T.L., Yao, W.J., Lee, I.H., Chen, P.S., Chiu, N.T., Lu, R.B., 2008. Greater availability of dopamine transporters in patients with major depression–a dual-isotope SPECT study. Psychiatry Res. 162, 230–235.

Yaqub, M., Boellaard, R., van Berckel, B.N., Ponsen, M.M., Lubberink, M., Windhorst, A.D., Berendse, H.W., Lammertsma, A.A., 2007. Quantification of dopamine transporter binding using [^{18}F]FP-beta-CIT and positron emission tomography. J. Cereb. Blood Flow Metab. 27, 1397–1406.

Yasuno, F., Suhara, T., Okubo, Y., Sudo, Y., Inoue, M., Ichimiya, T., Takano, A., Nakayama, K., Halldin, C., Farde, L., 2004. Low dopamine d(2) receptor binding in subregions of the thalamus in schizophrenia. Am. J. Psychiatry 161, 1016–1022.

Yeh, C.B., Lee, C.H., Chou, Y.H., Chang, C.J., Ma, K.H., Huang, W.S., 2006. Evaluating dopamine transporter activity with ^{99m}Tc-TRODAT-1 SPECT in drug-naive Tourette's adults. Nucl. Med. Commun. 27, 779–784.

Zoghbi, S.S., Shetty, H.U., Ichise, M., Fujita, M., Imaizumi, M., Liow, J.S., Shah, J., Musachio, J.L., Pike, V.W., Innis, R.B., 2006. PET imaging of the dopamine transporter with ^{18}F-FECNT: a polar radiometabolite confounds brain radioligand measurements. J. Nucl. Med. 47, 520–527.

CHAPTER TEN

Imaging of Dopamine and Serotonin Receptors and Transporters

Andre C. Felicio and A. Jon Stoessl
Pacific Parkinson's Research Centre, Vancouver Hospital and Health Sciences Centre, University of British Columbia, Vancouver, BC, Canada

1. INTRODUCTION

Human brain imaging as assessed by molecular neuroimaging techniques provides a unique opportunity not only to understand different in vivo pathophysiological mechanisms, but also to aid in the diagnosis of different neurodegenerative disorders (Shen et al., 2011).

In this chapter we will address the basic principles of modern neuroimaging with a focus on the use of Single-Photon Computed Tomography (SPECT) or Positron Emission Tomography (PET) for the molecular imaging of Parkinson's disease (PD). Other neurodegenerative disorders will be considered to a lesser extent when appropriate. Additionally, our focus will be on molecular imaging techniques, with much less emphasis placed on structural and functional magnetic resonance imaging (MRI) studies, an area in which there have been several recent advances.

Molecular imaging enables the analysis of molecules or proteins (transporters or receptors) at picomolar concentrations (Hammoud et al., 2007). However, it is important to remember that what is being measured is just radioactivity. Thus, the biological interpretation of the results depends upon the specificity of the particular radioligand or substrate, a knowledge of the radiolabeled metabolites and whether they cross the blood-brain barrier and if so whether they are active at the site of interest or are cleared, and the application of biological models to estimate the binding/uptake/metabolism of the radiotracer based on radioactivity counts.

SPECT is dependent upon the detection of single photons (gamma rays) emitted typically by 123I or 99mTc. PET is in contrast based upon the concurrent emission of 2 gamma rays traveling in opposite directions following the annihilation reaction that occurs when a positron collides with an electron. Positron emitters tend to be much less stable, with radioactive half-lives ranging from 2 minutes (15O) to approximately 2 hours (18F) in most biological applications. Modification of the molecule of interest to incorporate the radioactive gamma or positron emitter may change its biological properties. A radiotracer with a longer half-life permits a more leisurely pace from tracer production to patient injection and is also preferable if ligand binding takes a longer

Imaging of the Human Brain in Health and Disease
http://dx.doi.org/10.1016/B978-0-12-418677-4.00010-5

Table 1 Differences Regarding SPECT and PET Techniques

	PET	SPECT
Cost	Higher	Lower
Radioisotopes	^{11}C, ^{18}F, ^{13}N and ^{15}O	^{123}I and ^{99m}Tc
Irradiation	Positron emission	Single-photon emission
Image resolution	4 mm	8 mm
Radioisotope half-life	Shorter	Longer
Cyclotron dependent	Yes	No

time to reach equilibrium (thereby resulting in reduced radioactive counts). In contrast, tracers with shorter half-lives can be injected repeatedly thereby allowing multi-tracer studies or repeated studies with the same tracer before and after a variety of interventions. Differences between PET and SPECT are summarized below (Table 1).

2.1. PARKINSON'S DISEASE AS A MODEL FOR UNDERSTANDING MOLECULAR IMAGING

Parkinson's disease (PD) is the second most frequent neurodegenerative disorder after Alzheimer disease (AD), and to date, the diagnosis is based primarily on clinical grounds (Marsden, 1994; Jankovic, 2008). At the time motor signs are identified (bradykinesia, rest tremor, rigidity or postural imbalance), it is estimated that the degenerative process has already affected 30–50% of nigral dopaminergic neurons (Bernheimer et al., 1973; Fearnley and Lees, 1991), and that 80% of striatal dopamine has been lost. Therefore, an important issue is whether one can identify individuals at a risk of developing PD, prior to the emergence of motor symptoms. Second, based on the currently most widely accepted guidelines from the United Kingdom Parkinson's Disease Society Brain Bank, misdiagnosis is evident in 10–20% of individuals followed over time, higher when people are seen only once in relatively early stages of disease (Hughes et al., 1992, 2001).

While misdiagnosis may be considered to be primarily of academic interest, it has implications for prognosis, for the appropriate use of symptomatic therapy and may lead to the inappropriate inclusion of patients in clinical trials, thereby resulting in misleading findings.

As is the case for all neurodegenerative disorders, PD progresses relentlessly over time, and standardized measures of physical disability or quality of life may not be sufficient to determine the extent of disease progression. This becomes of particular interest if one is attempting to assess neurorestorative or neuroprotective treatment strategies.

Molecular imaging may be particularly useful in PD for all these reasons (Scherfler et al., 2007). Using these modern imaging techniques, it is possible to identify dopamine deficiency sometimes long before the initial presentation of motor symptoms in individuals with at-risk manifestations. Some of the current known risk factors for the

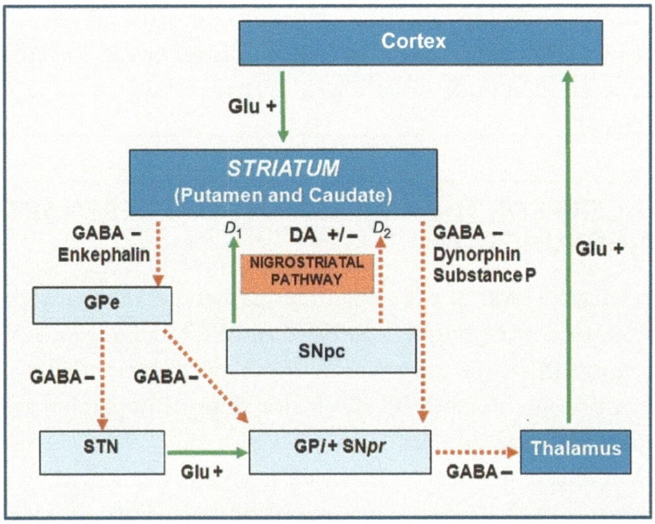

Figure 1 Basal ganglia circuitry. + = excitatory (represented as a green line); − = inhibitory (represented as a red line); Glu = glutamate; GPe = globus pallidus externus; GPi = globus pallidus internus; STN = subthalamic nucleus; SNpc = substantia nigra pars compacta; SNpr = substantia nigra pars reticulata. *Adapted from Obeso et al. (2008) and Wichmann et al. (2011).*

future development of PD that can be studied with molecular imaging are hyposmia (Ponsen et al., 2004), late-onset depression (Schuurman et al., 2002; Leentjens et al., 2003; Alonso et al., 2009), or rapid eye movement (REM) sleep behavior disorder (RBD) (Schenck et al., 1996; Albin et al., 2000; Hickey et al., 2007; Postuma et al., 2009; Iranzo et al., 2011). It is also feasible to differentiate patients with degenerative parkinsonism from those who do not have dopaminergic dysfunction as assessed by different radioligands (Poewe and Scherfler, 2003; Felicio et al., 2009, 2010a). Finally, using molecular imaging as a biomarker to assess dopaminergic dysfunction, it is possible to monitor disease progression and treatment strategies.

Before discussing the various single photon emission computed tomography (SPECT) or positron emission tomography (PET) radioligands, it is desirable to understand the basic aspects of the basal ganglia circuitry, its main neurotransmitter systems, and the several related input and output pathways (Alexander and Crutcher, 1990) as shown in Figure 1. Dopamine denervation in PD leads to widespread alterations in synaptic activity through the cortico–striato–pallido–thalamo–cortical loops and beyond. Overall, the basal ganglia circuitry involves direct and indirect excitatory and inhibitory striatal output pathways originating from the sensory and motor cortex, but eventually this outflow is somatotopically funneled into different thalamic relay nuclei before projecting back to virtually all the areas of the frontal lobes. The functional organization of basal ganglia circuitry reflects a highly organized network, where different parts are activated depending on specific functions and circumstances. In PD, with dopaminergic degeneration particularly at the level of the striatum, loss of GPi inhibition results in the inhibition of the thalamus, with a consequent reduction in excitatory impulses to cortex (Obeso et al., 2008; Wichmann et al., 2011) (Table 1).

2.2. TRACERS FOR THE ASSESSMENT OF PRESYNAPTIC DOPAMINE FUNCTION

At the presynaptic level of the dopaminergic neuron there are several important sites that can be assessed using different SPECT and PET radioligands (Figure 2). These tracers provide complementary information about the function of the dopaminergic neuron, since it is possible to study all the major steps of dopamine synthesis, storage, release, and reuptake (Stoessl, 2011).

Dopamine synthesis depends on a complex machinery that starts with the hydroxylation of tyrosine to L-3,4-dihydroxyphenylalanine (L-dopa). L-Dopa is thereafter decarboxylated by L-aromatic acid decarboxylase (L-AADC) to dopamine. Both newly synthesized dopamine and dopamine that has been taken back up into the presynaptic nerve terminal following its release are stored in synaptic vesicles via the Vesicular Monoamine Transporter type 2 (VMAT2). This prevents the further catabolism of dopamine, which would otherwise lead to toxic oxidative species. This represents another target for evaluation using PET tracers. Finally, following depolarization, dopamine is released into the synaptic cleft where it can act on both presynaptic and postsynaptic

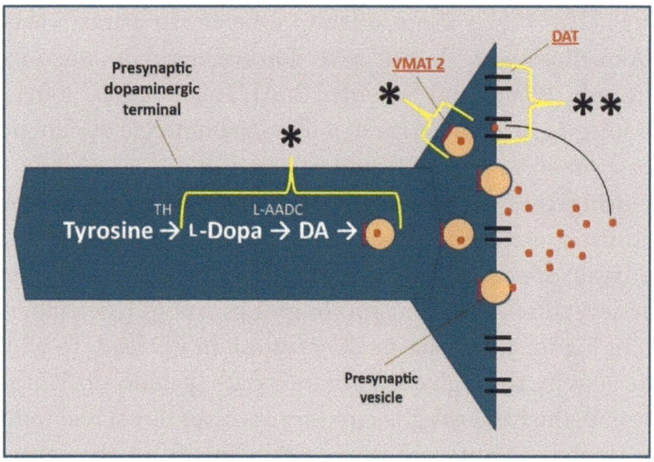

Figure 2 Diagram highlighting SPECT and PET sites at the presynaptic dopaminergic neuron. * = PET radiotracers such as [^{18}F]-fluorodopa and [^{11}C]-dihydrotetrabenazine ([^{11}C]-DTBZ); ** = PET and SPECT radiotracers (Table 2). TH = tyrosine-hydroxylase; L-dopa = L-3,4-dihydroxyphenylalanine; L-AADC = L-aromatic acid decarboxylase; DA = dopamine; VMAT2 = vesicular monoamine transporter type 2; DAT = dopamine transporter. *Adapted from Stoessl (2011).*

dopamine receptors, following which its action is terminated primarily (in the healthy brain) by reuptake through the membrane Dopamine Transporter (DAT). DAT binding can be assessed using a variety of PET and SPECT dopamine tracers.

In the following section, we discuss in greater detail each of these steps, focusing not only on SPECT and PET-related tracers but also on their clinical relevance in PD.

2.3. TRANSPORTERS

2.3.1. L-Aromatic Acid Decarboxylase

The assessment of L-AADC through PET tracers such as [^{18}F]-fluorodopa allows one to estimate dopamine synthesis and storage. Information derived from activity measured over the first 90–120 min of radiotracer administration provides the evaluation of the uptake and conversion of fluorodopa to fluorodopamine by L-AAAD (Martin et al., 1989). However, the uptake or influx constant so derived is based on an assumption of unidirectional transport with irreversible trapping of fluorodopamine in the nerve terminal. Over time, fluorodopamine leaks out of the nerve terminal and is then subject to enzymatic degradation. By measuring the activity over a prolonged time (4 h following tracer administration), one can estimate the distribution volume and its inverse, dopamine turnover (Sossi et al., 2002). Interestingly, dopamine turnover assessed using this technique is inversely related to DAT expression (Sossi et al., 2007), reinforcing the importance of transmitter reuptake for the normal physiology of dopamine function.

It is worth noting that AADC is subject to compensatory changes. For example, in early PD (Lee et al., 2000; Sossi et al., 2002), fluorodopa uptake is increased relative to nerve terminal density as assessed by VMAT2 binding, probably reflecting compensatory upregulation of AADC activity. Hence, fluorodopa uptake may underestimate the degree of dopaminergic denervation, particularly in early disease, and this should be taken into consideration. Additionally, AADC is not specific to the dopaminergic neuron, but is also expressed by serotonergic neurons. The ability of serotonergic neurons to synthesize dopamine has been recently postulated as a potential mechanism underlying levodopa-induced dyskinesias in PD (Munoz et al., 2009; Bjorklund et al., 2010).

2.3.2. Vesicular Monoamine Transporter Type 2

VMAT2 allows the transport of biogenic monoamines, such as serotonin and dopamine, from the cytoplasm into the vesicles. It is assessed by PET with [^{11}C]-DTBZ. While the VMAT2 is expressed by all monoaminergic neurons, striatal binding of DTBZ reflects mainly dopaminergic projections and DTBZ PET is accordingly very useful to assess the density of dopamine nerve terminals. Although DTBZ binding is thought to be less subject to pharmacological effects (Vander Borght et al., 1995) or compensatory changes than either DAT ligands or fluorodopa (Sections 2.1 and 2.3), it is nonetheless subject to competition from endogenous intravesicular dopamine and may accordingly be increased in states of profound DA deficiency such as dopa-responsive dystonia (de La Fuente-Fernandez et al., 2003) and early methamphetamine abstinence (Boileau et al., 2008a), and is reduced following levodopa administration in patients with advanced PD (de la Fuente-Fernandez et al., 2009). Figure 3 shows [^{11}C]-DTBZ PET in a healthy subject and a PD patient.

2.3.3. Dopamine Transporter

The plasmalemmal DAT is located on the presynaptic terminals of dopaminergic neurons and is responsible for the reuptake of up to 70% of the dopamine released into the synaptic cleft (Eisenhofer, 2001). In the healthy brain, this is the leading mechanism responsible for terminating the actions of DA (Varrone and Halldin, 2010). DAT imaging can be assessed using both PET and SPECT tracers as presented in Table 2. DAT scan, for instance, is a commercially available iodine-labeled SPECT tracer previously approved in the European Union for the differentiation of PD and Essential Tremor, and recently

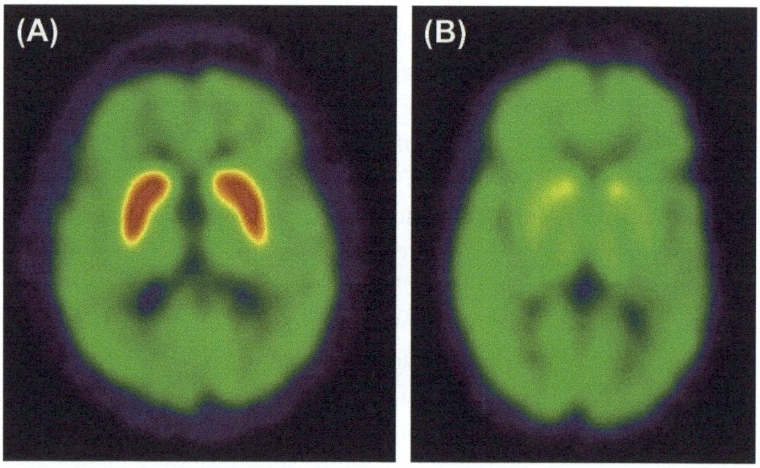

Figure 3 Vesicular monoamine transporter type 2 imaging scan using [^{11}C]-DTBZ, in a healthy subject versus that in a matched Parkinson's disease patient. On the left (A): Axial slice through the striatum showing normal radiotracer uptake. On the right (B): Abnormal radiotracer uptake showing an asymmetric decrease of [^{11}C]-DTBZ binding affecting the posterior striatum more than the anterior striatum. *Images are courtesy of the Pacific Parkinson's Research Centre, University of British Columbia, Canada.*

Table 2 SPECT and PET DAT Ligands

	SPECT	PET
DAT ligand	[^{99m}T c]-TRODAT-l	[^{11}C]-dMP (d-threo methylphenidate)
	[^{123}I]-Altropane	[^{11}C]-Cocaine
	[^{123}I]-FP-CIT	[^{11}C]-Altropane
	[^{123}I]-PE2I	[^{11}C]/[^{18}F]-β-CFT
	[^{123}I]-β-CIT	[^{11}C]-FE-CIT

approved by the Food and Drug Administration (FDA) to be used in the United States. While the majority of DAT SPECT tracers are ^{123}I or ^{99m}Tc-labeled, PET tracers for the DAT tracers are generally ^{11}C- or ^{18}F-labeled. The development of technetium-99m labeled DAT tracers, such as [^{99m}Tc]-TRODAT-1, provided a useful and less expensive alternative in molecular neuroimaging studies (Mozley et al., 2000) as compared to iodine-123 labeled DAT tracers such as [^{123}I]-FP-CIT (DATscan) (Booij et al., 1997) or [^{123}I]-β-CIT (Marek et al., 1996). The cost of cyclotron-produced ^{123}I, for instance, is relatively higher than the more widely available and generator-produced ^{99m}Tc. Moreover, [^{99m}Tc]-TRODAT-1 also shows favorable pharmacokinetics and allows for similar semiquantification. Figure 4 shows a healthy subject submitted to brain SPECT imaging using the technetium-ligand TRODAT-1.

Although DAT imaging is frequently used as a surrogate marker to assess dopaminergic dysfunction, one should take care of the potential caveats of this methodology (Piccini, 2003). First, age-associated physiological decline in DAT is estimated to range from 3% to 8% per decade, and corresponds to nigrostriatal cell loss. This is substantially lower than the approximately 8% per year pathological loss of DAT seen in PD (Volkow

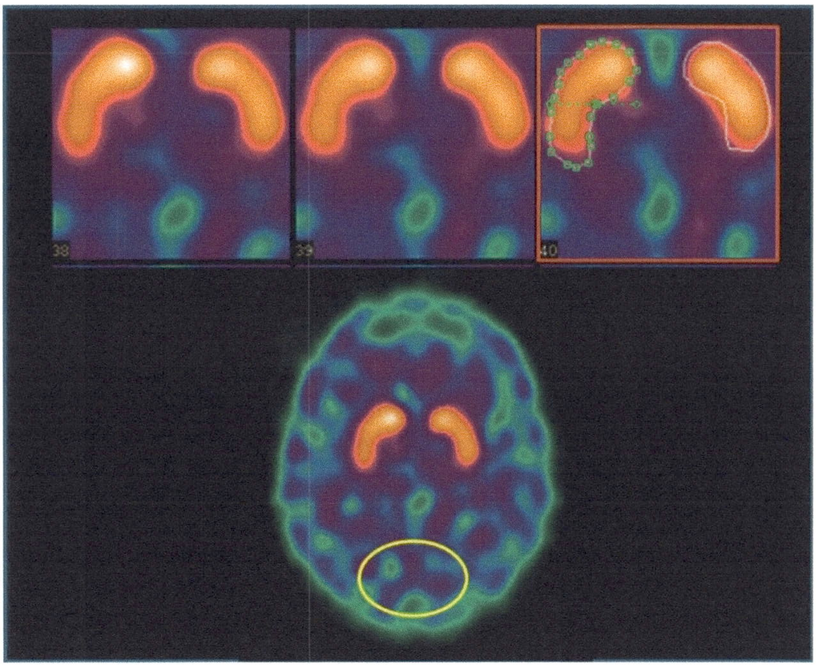

Figure 4 Dopamine transporter imaging scan using a technetium-based SPECT ligand, TRODAT-1, in a healthy subject. At the top level: Axial slice through the striatum showing the radiotracer uptake, and the region of interest (ROI) drawn in the area with specifically bound radioligand. At the bottom level: ROI drawn in the occipital area or nondisplaceable radioligand tissue. *Images are courtesy of the Laboratório Interdisciplicar de Neurociências Clínicas, Universidade Federal de São Paulo, Brazil.*

et al., 1994). Another major potential confounder is that the DAT may be subject to pharmacological regulation arising from medications such as dopamine reuptake inhibitors (bupropion or methylphenidate) or other dopaminergic medications (Kagi et al., 2010). Finally, DAT is also subject to compensatory changes, such as may occur in the early stages of PD. The downregulation of DAT should allow dopamine to interact for a longer time with presynaptic and postsynaptic receptors, by preventing its rapid reuptake (Lee et al., 2000). While several tracers are available for assessing DAT binding, it should be noted that their specificity for the DAT (as opposed to other plasmalemmal monoamine transporters) is variable, as are their pharmacokinetic properties. The latter point may be quite important as tracers that are slow to reach equilibrium may require longer scan times to provide accurate estimates of binding potential and are accordingly better labeled with longer half-life emitters (e.g., F-18 rather than C-11 for PET).

2.4. TRACERS FOR THE ASSESSMENT OF POSTSYNAPTIC DOPAMINE FUNCTION

Dopamine receptors are the primary targets in the treatment of many disorders such as PD, Huntington's chorea, and schizophrenia (Seeman and van Tol, 1994). There are several radiotracers that can be used to access D_1 and D_2 dopamine receptors. [^{18}F]-fallypride and [^{11}C]-FLB-457, which are PET ligands, and [^{123}I]-epidepride, an SPECT ligand, are ultra-high-affinity dopamine receptor antagonist radioligands, which allow the quantification and visualization of low-density dopamine extrastriatal D_2/D_3 receptors as well as striatal receptors (de Paulis, 2003). On the other hand, the radioligands [^{11}C]-raclopride (or [^{11}C]-RAC) for PET, and [^{123}I]-IBZM for SPECT are widely employed to assess striatal DA receptor availability. In contrast to the former, these ligands have lower D_2/D_3 receptor affinity; hence, the quantification of extrastriatal receptors is not possible (Pinborg et al., 2007). However, RAC and IBZM are subject to competition from endogenous DA for in vivo binding to D_2 receptors, and changes in binding can therefore be used to infer alterations in synaptic DA concentration.

Similar to DAT ligands, dopamine receptor binding is also influenced by age and, to a lesser extent, the stage of PD and therapy. While increased tracer binding is observed in the more affected putamen in early PD (Antonini et al., 1997; Kaasinen et al., 2000), advanced PD and chronic treatment result in the normalization of binding in the putamen and decreased binding in the caudate (Antonini et al., 1997; Thobois et al., 2004). In early PD, the loss of midbrain dopamine is accompanied by a matching loss in the DAT and a rise in D_1 and D_2 receptor densities (Seeman and Niznik, 1990); the latter is also demonstrated using [^{11}C]-N-methylspiperone PET (Kaasinen et al., 2000). Unlike RAC, this ligand is not thought to be subject to displacement by endogenous dopamine and the findings therefore suggest that the increased binding of D_2 ligands to putaminal D_2 receptors in early PD reflects receptor upregulation and not just increased receptor occupancy due to endogenous dopamine deficiency. In contrast, D_1 binding as assessed by [^{11}C]-SCH23390 and PET is normal in PD (Rinne et al., 1990), but may be decreased in conditions characterized by the loss of striatal neurons, such as multiple system atrophy (MSA; Shinotoh et al., 1993). Since dopamine receptors are an important site for antipsychotic drugs, it is important to take the patient's drug record into consideration at the time of a dopamine receptor scans.

2.5. OTHER TRACERS: SEROTONERGIC BINDING LIGANDS

Dopamine dysfunction does not account for the entire spectrum of disability in PD. Indeed, several other neurotransmitters are involved as well, including the serotonergic system (Barone, 2010).

Serotonergic innervation is a potential additional source of dopamine in patients with PD, as serotonin neurons have a neurotransmitter-release machinery closely resembling that of dopaminergic neurons (Ugrumov, 2009). Both serotonergic and dopaminergic neurons express L-aromatic amino acid decarboxylase (a key step in the synthesis of both serotonin and dopamine) and the VMAT2 (responsible for the storage of newly synthesized monoamines). Presumably, dopamine released as a false neurotransmitter by serotonergic neurons may to some extent be implicated in both motor and perhaps nonmotor complications of PD (Carta et al., 2008; Politis et al., 2010a). One possible explanation is that without the regulatory control of DATs or D_2 autoreceptors, levodopa-derived dopamine is released into the synapse at higher than desired concentrations, but in the absence of a normal reuptake mechanism, it may be rapidly cleared, resulting in the pulsatile stimulation of dopamine receptors. Furthermore, there is evidence from both 6-hydroxydopamine-lesioned rats and 1-methyl-4-phenyl-1,2,3,6-tetrahydropyridine(MPTP)-treated macaques that serotonin receptor (5-HT_{1A} and 5-HT_{1B}) agonists may prevent or reduce the development of levodopa-induced dyskinesia (Munoz et al., 2008).

From the perspective of molecular imaging studies, the serotonergic system can be assessed using several selective or nonselective radiotracers that bind to the serotonin transporter (5-HTT) or some of the 14 subtypes of serotonin receptors. 5-HT synthesis rate can also be evaluated, but will not be considered further here (Paterson et al., 2011).

The currently available radioligands for the in vivo brain imaging of the 5-HT system in humans include antagonists for the 5-HT_{1A}, 5-HT_{1B}, 5-HT_{2A}, and 5-HT_4 receptors, and for 5-HTT (Paterson et al., 2011). The latter can be evaluated with both selective and nonselective tracers. Nonselective radiotracers for the 5-HTT include [^{123}I]-FP-CIT, and [^{123}I]-β-CIT, both of which bind also to the DAT. On the other hand, [^{11}C]-3-amino-4-[2-[(di(methyl)amino)methyl]phenyl]sulfanylbenzonitrile (DASB) (PET) and [^{123}I]-ADAM (SPECT) are selective 5-HTT radioligands. PET studies of 5-HT receptors in PD have included the use of [^{11}C]-WAY-100635 (5-HT_{1A} receptors) (Doder et al., 2003), and [^{18}F]-setoperone (5-HT_{2A} receptors) (Ballanger et al., 2010).

A recent study using [^{11}C]-DASB PET and [^{123}I]-FP-CIT SPECT (used to determine the DAT binding in the striatum) showed preserved 5-HTT binding in early PD patients and a negative correlation with DAT, suggesting that the serotonergic system is remarkably spared at least at this disease stage (Strecker et al., 2011), in line with a previous SPECT study using [^{123}I]-β-CIT (Kim et al., 2003). On the other hand, if one considers all PD stages, 5-HTT binding as assessed by [^{11}C]-DASB PET progressively reduces in a nonlinear pattern, regardless of disease severity, and without influence of dopamine replacement therapy (Politis et al., 2010a).

Interestingly, 5-HTT binding has also been used as a method to help distinguish different forms of degenerative parkinsonism, as shown in a [^{123}I]-FP-CIT SPECT study using the midbrain as the ROI. In this study, the authors found a greater reduction of 5-HTT binding in dementia with Lewy bodies (DLBs) as compared to PD (Roselli et al., 2010).

The drawbacks of serotonin imaging are related to fast systemic metabolism, difficult kinetic modeling requiring arterial input function, and the presence of radioactive metabolites in plasma. Also, similar to SPECT ligands for the 5-HTT, some serotonin receptor ligands are nonselective, such as [^{18}F]-setoperone, which binds to D_2 receptors as well (Paterson et al., 2011).

2.6. CLINICAL RELEVANCE

2.6.1. PD Diagnosis, Disease Severity, and Disease Progression

The diagnosis of PD is essentially clinical (Marsden, 1994; Jankovic, 2008), and we believe that the clinical evaluation of patients will not be replaced by modern imaging techniques. Nevertheless, as pointed out earlier, misdiagnosis can approach 20% of cases (Hughes et al., 1992). Recently, in line with the European Union, the U.S. FDA approved DAT scan as a diagnostic tool to help differentiate between PD and Essential Tremor. This is the only current FDA-approved subsidiary examination to aid in PD diagnosis.

The role of magnetic resonance imaging (MRI) in the diagnosis of PD is still not fully established. Traditionally, it has been seen as a means to potentially exclude alternate diagnoses such as vascular parkinsonism, or more importantly (but also more problematically), atypical forms of degenerative parkinsonism such as MSA or progressive supranuclear palsy (PSP). Traditional structural MRI findings of these latter diagnoses such as increased pallidal iron, rim of putaminal hyperintensity, "hot cross bun" sign and/or cerebellar atrophy in MSA (Schrag et al., 2000) or the hummingbird or penguin sign in PSP (Kato et al., 2003) may not be reliable, particularly in early disease.

Diffusion-weighted studies may permit differentiation between these conditions (Seppi et al., 2003; Nicoletti et al., 2006, 2008) and detailed measurements of the midbrain and superior cerebellar peduncles may help identify PSP (Morelli et al., 2011), but these approaches are not yet in widespread use. Multimodal imaging, combining measurements of iron, microstructural damage (mean diffusivity), and white matter integrity (fractional anisotropy), appear to differentiate PD patients from healthy controls (Cherubini et al., 2010; Peran et al., 2010), but the role of such techniques in routine diagnostic practice is similarly not established. Figure 5 shows an illustrative multimodal MRI between patients with PD and controls.

Transcranial sonography reveals increased nigral echogenicity, thought to reflect increased iron deposition, in approximately 80% of patients with PD. Nigral hyperechogenicity is not generally seen in patients with MSA or PSP, and this may accordingly be a helpful if not entirely reliable diagnostic tool (Gaenslen et al., 2008). However, nigral echogenicity is increased in about 10% of putatively normal individuals, possibly reflecting presymptomatic Lewy body disease, and the degree of echogenicity does not change with disease progression (Berg et al., 2005).

Owing to the pattern of nigrostriatal degeneration that occurs in PD (Au et al., 2005), the loss of dopamine is most prominent in the posterior part of the putamen (anteroposterior gradient). Therefore, early PD is associated with the reduced binding of DAT, VMAT2 and fluorodopa uptake affecting the posterior putamen contralateral to the most affected limb first and most prominently throughout the course of the disease. In keeping with the need to lose 80% of striatal dopamine prior to symptom onset, it is perhaps not surprising that the clinically unaffected striatum shows reduced DAT binding even in patients with Hoehn & Yahr (H&Y) Stage 1 (clinically unilateral) disease (Marek et al., 1996). With disease progression, dopaminergic dysfunction is seen in the more anterior and ventral regions of the striatum, but while the differences between the

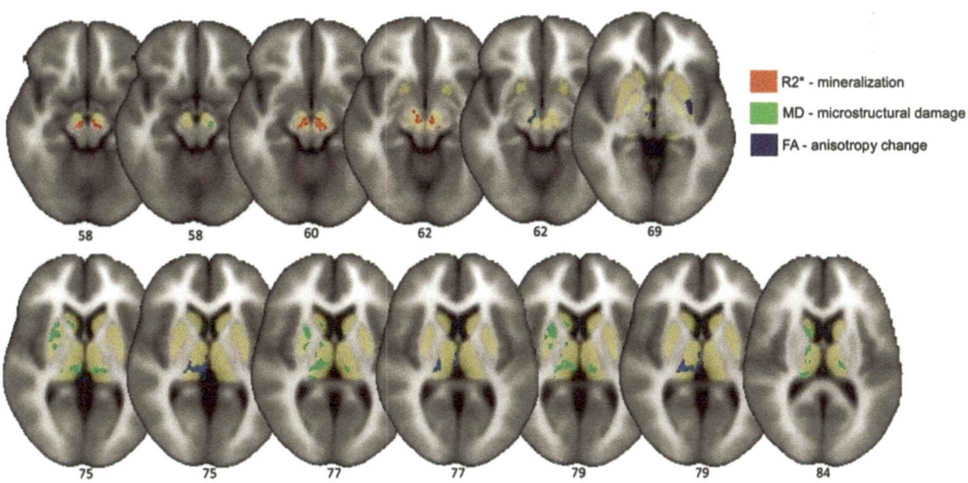

Figure 5 Multimodal MRI in PD. Differences are noted between patients with PD and controls through voxel-based analysis of R2*, mean diffusivity, and fractional anisotropy maps. *Reproduced from Peran et al. (2010) by permission of Oxford University Press.*

two sides become less marked over time, the posterior striatum remains more severely affected (Nandhagopal et al., 2009). This pattern of nigrostriatal dopamine dysfunction detected on functional imaging is in keeping with the findings derived from *postmortem* studies (Kish et al., 1988; Gibb and Lees, 1991). Furthermore, an age-specific progression of nigrostriatal dysfunction has been recently reported suggesting that, as compared to older patients, young-onset PD patients have a slower disease progression, and are able to endure more damage to the dopaminergic system (de la Fuente-Fernandez et al., 2011) in spite of a greater nigrostriatal dysfunction at symptom onset (Shih et al., 2007b).

The reduction of dopaminergic tracer uptake correlates with the scales routinely used to access PD disease severity. Hence, there is in general a significant inverse correlation between reductions in DAT or VMAT2 binding or fluorodopa uptake and measures of motor dysfunction such as the Unified Parkinson's disease Rating scale part III or the H&Y scale (Pirker, 2003). In patients with tremor predominant PD, this correlation may be lacking since it is believed that bradykinesia rather than tremor best correlates with functional imaging dopamine loss, perhaps because the underlying cause of tremor is much more serotonergic than dopaminergic (Vingerhoets et al., 1997). This is in line with a molecular imaging study that showed a better correlation of tremor in PD with brain stem serotonergic innervation as assessed by $[^{11}C]$-WAY 100635 (a 5-HT_{1A} ligand, whose binding in the brain stem is predominantly to somatodendritic 5-HT autoreceptors) PET binding (Doder et al., 2003).

Radiotracer imaging has been used as an in vivo biomarker to assess the effect of treatment on disease progression in various clinical trials. These studies include the CALM-PD study, which used $[^{123}I]$-β-CIT SPECT to compare the effects of L-dopa versus pramipexole in early PD (Parkinson Study Group, 2000), the REAL-PET study, which compared the use of ropinirole and L-dopa in de novo PD patients using $[^{18}F]$-fluorodopa PET as the primary outcome measure (Whone et al., 2003), the earlier versus later levodopa in Parkinson disease (ELLDOPA) study, in which the effects of L-dopa on clinical progression of PD were studied in a subgroup of patients (Fahn et al., 2004), among others. Overall, the functional imaging findings in these studies have raised many questions. Thus, both the Comparison of the agonist pramipexole with levodopa on motor complications of Parkinson's disease (CALM-PD) and REAL-PET studies found a lower rate of decline in tracer uptake in patients treated with dopamine agonists as compared to levodopa, but this did not translate into clinical benefit in either trial. The ELLDOPA study compared three doses of levodopa against placebo and in the subset of patients studied with $[^{123}I]$-β-CIT SPECT, there was a greater rate of decline in DAT binding in patients taking levodopa, yet clinical performance following medication withdrawal was better. It should also be noted that while there is usually a general agreement between clinical and imaging measures of disease severity, most studies have failed to find a correlation between *rate of change* in the two measures. This discordance between clinical progression and radiotracer imaging findings could reflect many factors

(Brooks et al., 2003; Morrish, 2003; Pavese et al., 2009b), including potential effects of the therapies on the surrogate markers rather than on the disease process itself. Moreover, clinical progression was measured using Unified Parkinson's Disease Rating Scale (UPRDS), which reflects a composite of dopaminergic and nondopaminergic dysfunction in PD; the clinical sign that best reflects the severity of the nigrostriatal lesion is bradykinesia. Finally, clinical performance will be affected by compensatory changes in the brain that may serve to preserve function particularly in early disease, but fail with disease progression. Thus, the assessment of the nigrostriatal DA system alone may be inadequate to determine the overall disease progression in PD. Proper study design and analysis are needed, and the imaging data must be interpreted with caution and in the context of the clinical outcome.

2.6.2. PD Motor and Nonmotor Symptoms Disease Related and Treatment Related

2.6.2.1. Fluctuations and Dyskinesias

Fluctuations in motor function and/or dyskinesias are the major generally long-term complications of dopamine replacement therapy. Most PD patients will eventually develop these motor side effects at varying degrees of severity. Although preclinical and clinical findings suggest the pulsatile stimulation of striatal postsynaptic receptors as a key mechanism underlying levodopa-induced dyskinesias, their pathogenesis is still unclear (Calabresi et al., 2010). Using PET with [^{11}C]-RAC to estimate DA release following the oral administration of levodopa (de la Fuente-Fernandez et al., 2001), it was shown that altered patterns of dopamine release preceded the emergence of motor complications. In PD patients who went on to develop motor fluctuations but still had a stable response to medication at the time of the PET scan, DA release was of a comparatively greater magnitude, but shorter duration compared to that in those subjects who maintained a stable response to medication. Levodopa-derived DA release increases with disease duration and has a similarly greater magnitude but a shorter duration in patients with dyskinesias (de la Fuente-Fernandez et al., 2004), in keeping with impaired regulation that may secondarily result in abnormally pulsatile stimulation of striatal DA receptors and subsequent changes downstream.

One of the factors regulating the temporal profile of DA activity in the synapse is the DAT. DAT expression is downregulated in early PD (Lee et al., 2000), presumably in an effort to maintain the levels of DA in the synapse. However, using [^{18}F]-fluorodopa PET with prolonged scanning times (Sossi et al., 2007), it was shown that dopamine turnover is inversely related to DAT expression—thus, for a subject with a comparable degree of DA denervation, reduced DAT expression should be accompanied by increased DA turnover (and consequently more pulsatile stimulation of DA receptors), presumably because rather than being taken back up into the presynaptic neuron and reused, synaptic DA is subject to enzymatic degradation. Another study using the PET ligands [^{11}C]-MP and [^{11}C]-DTBZ in 36 PD patients with motor fluctuations showed that putaminal MP–DTBZ ratio was significantly decreased in individuals with dyskinesia (Troiano et al., 2009).

In addition to changes that occur downstream to dopamine receptors, nondopaminergic mechanisms may also take place at the presynaptic level and contribute to dyskinesias in PD. Thus, serotonergic neurons may release dopamine but without the regulatory control of either DAT or presynaptic DA autoreceptors (Munoz et al., 2009; Bjorklund et al., 2010). It is therefore of interest that recent PET studies have demonstrated increased serotonin transporter binding within the grafted tissue of PD patients who developed dyskinesias following fetal transplantation (Politis et al., 2010b).

2.6.2.2. Cognitive Impairment and Dementia

The point prevalence of dementia in PD is close to 30%, and the incidence rate is increased 4–6 times as compared to controls. The cumulative prevalence is very high, and at least 75% of PD patients who survive for >10 years will develop dementia (Aarsland and Kurz, 2010). With the contribution of molecular imaging, a further step in understanding the pathophysiology of dementia in PD dementia has been possible (Johansen et al., 2010). Thus, studies using the PET tracer Pittsburgh Compound-B (PIB) have shown a greater β-amyloid protein load in patients with DLBs compared to those with PD dementia, suggesting an Alzheimer-like pathology in the former (Brooks, 2009). However, the results are variable, and some authors have demonstrated increased amyloid deposition in a subset of patients with both DLB and Parkinson's disease dementia (PDD) seemingly linked to the APOE-E4 allele (Maetzler et al., 2009). This is in keeping with postmortem studies that show dementia in PD is associated with amyloid deposition, but not necessarily with neurofibrillary pathology (Burack et al., 2010).

PET can also be used to demonstrate glucose metabolic abnormalities among patients with PD dementia and with DLB. In both cases, there is a reduction of parietooccipital glucose metabolism, distinctive from the pattern seen in AD (Minoshima et al., 2001). Interestingly, the neuronal acetylcholine esterase PET tracer MP4A showed topographic abnormalities similar to those seen on fluorodeoxyglucose (FDG) PET, but significantly greater in patients with PD and dementia than in DLB or PD without dementia (Klein et al., 2010). However, reductions in cholinergic activity as assessed by a different cholinesterase substrate methylpiperidin-4-yl propionate (PMP) are widespread in PD dementia and greater than those seen in AD (Bohnen et al., 2003).

Finally, regardless of the imaging findings in studies of glucose metabolism, β-amyloid deposition, and cholinergic activity, the pattern of dopaminergic involvement seems to be relatively equal in PD with or without dementia and DLB, suggesting that parkinsonism or motor involvement in these subjects is independent of the degree of cognitive impairment (Hilker et al., 2005; Klein et al., 2010). The corollary is that the presence of significant dopamine dysfunction as measured by the radiotracer imaging in a patient with dementia suggests a diagnosis of DLBs as opposed to AD.

2.6.2.3. Depression

Depression is a frequent nonmotor symptom in PD, affecting approximately 45% of PD patients at all stages (Rickards, 2005). Interestingly, depression can also anticipate by many years the onset of PD motor symptoms (Shiba et al., 2000; Ishihara and Brayne, 2006). Although molecular imaging findings using different PET or SPECT tracers to assess dopaminergic cell loss are somewhat variable, it is possible to implicate dopamine dysfunction as a potential underlying mechanism related to depression in PD. One study found a higher DAT uptake as assessed by the SPECT tracer TRODAT-1 (Felicio et al., 2010b), while others have shown a decreased DAT uptake using different less selective DAT tracers, such as the DAT/SERT tracer β-CIT with SPECT (Murai et al., 2001) or the DAT/NAT tracer RTI-32 with PET (Remy et al., 2005) or even selective ones such as TRODAT-1 (Weintraub et al., 2005) and FP-CIT for SPECT (Hesse et al., 2009). Using the PET tracer [^{11}C]-DASB, a highly selective serotonin transporter (5-HTT) ligand, it was shown that depressed-PD patients displayed a widespread increase (8–68%) in DASB-specific binding outside of the striatum, which was significant in dorsolateral (37%) and prefrontal (68%) cortices (Boileau et al., 2008b). This is in line with a more recent study that also showed elevated DASB binding in the raphe nuclei and limbic structures of depressed-PD antidepressant-naive patients, possibly reflecting lower extracellular serotonin levels (Politis et al., 2010c).

2.6.2.4. Impulse Control Disorder

A number of different impulse control disorders (ICDs) may arise during the course of PD, particularly in those patients taking dopamine agonists. Among these ICDs, the main subtypes are pathological gambling, hypersexuality, and compulsive eating and buying. Dopamine dysregulation syndrome and punding are other forms of impulsive or compulsive disorders frequently reported, that may be seen in association with levodopa use (Wu et al., 2009). Although there is a growing body of evidence toward the understanding of these pathological behaviors, we are still far from an adequate management (Weintraub, 2008; Ceravolo et al., 2009). Molecular imaging studies may provide insights into their neurobiological basis. Indeed, the brain reward systems have been implicated as abnormal in a recent fMRI study (Cilia et al., 2011). Additionally, an RAC-PET study comparing dopaminergic function in PD patients with and without pathological gambling demonstrated greater dopamine release in the ventral striatum of PD-gambling subjects (13.9%) than in PD patients without this problem (8.1%) during the performance of a video gambling task. Although prior imaging studies suggested that abnormalities in dopaminergic binding and dopamine release may be markers of vulnerability to addiction, this study presented the first evidence of these phenomena in pathological gambling (Steeves et al., 2009). RAC-PET has also been used to demonstrate increased DA release in response to visual cues associated with ICDs in PD patients compared to those without these behavioral complications (O'Sullivan et al., 2011). ICDs induced by dopamine agonists likely reflect dysfunction in extrastriatal regions as well. Thus, dopamine agonist administration is associated with the loss of the normal relationship between reward prediction error and the activation of the orbitofrontal cortex, mainly reflecting impaired deactivation in response to negative reward prediction errors (van Eimeren et al., 2009).

2.6.3. Individuals at an Increased Risk to Develop PD

Individuals at an increased risk to develop PD are a group of particular interest, not only because of the possibility to better understand the pathological progression in PD but also as a window of opportunity to study neuroprotective treatment strategies. A growing body of evidence has shown some preclinical signs to be predictors or risk factors for the future development of PD (Siderowf and Stern, 2006). Among these the most important are hyposmia, RBD, and late-onset depression. Constipation is another risk factor, but as the majority of people with constipation will not go on to develop PD, this alone is clearly not sufficient to justify detailed prospective imaging studies. Therefore, functional imaging can be used to detect preclinical evidence of dopamine deficiency in people deemed to be at increased risk of PD (Stoessl, 2007), for instance, asymptomatic relatives of PD patients who have a decreased sense of smell (Berendse et al., 2001). The addition of transcranial sonography to the combination of radiotracer imaging and olfactory testing may increase the yield even further (Sommer et al., 2004).

In a prospective study in 361 asymptomatic first-degree relatives of PD patients (Ponsen et al., 2010), the authors studied the value of a two-step approach, combining olfactory testing and DAT SPECT imaging, in detecting patients in the premotor phase of PD. Unexplained hyposmia alone was associated with a 12.5% risk of developing PD within a five-year period. Furthermore, all relatives who later developed PD had both hyposmia, and abnormally reduced striatal DAT binding at baseline.

As >50% of patients with RBD will go on to develop a neurodegenerative disease within 12 years, mostly PD, but also MSA, DLB, or Alzheimer's disease (Postuma et al., 2009; Yoritaka et al., 2009), it is important to recognize this condition and to ensure that patients receive adequate follow-up. In a recent serial DAT scan study on patients with idiopathic RBD, those with a reduced DAT uptake at baseline had a greater risk to develop parkinsonian signs within 3 years (Iranzo et al., 2011). Interestingly, the combination of "idiopathic" RBD and olfactory dysfunction has been considered an early indicator for possible alpha-synucleinopathy as demonstrated by a DAT SPECT study using FP-CIT, and corresponding to Braak stage 2 neuropathological staging (Stiasny-Kolster et al., 2005).

In the majority of patients with classic PD, the etiology is unknown. However, a clear genetic component has been determined in a minority. Mutations in five causative genes combined (*α-Synuclein, Parkin, PTEN-induced kinase 1 (PINK1), DJ-1, Leucine-rich repeat kinase 2 (LRRK2)*) account for 2–3% of all cases with classical parkinsonism, often clinically indistinguishable from idiopathic PD (Klein and Lohmann-Hedrich, 2007). Molecular imaging has demonstrated incipient nigrostriatal dysfunction in subjects who carry mutations of genes associated with dominantly inherited PD but who do not yet manifest motor signs (Shih et al., 2007a). One multitracer PET study of asymptomatic LRRK2 mutation carriers revealed dopaminergic dysfunction that progressed over time, highlighting the importance of this group not only to study the natural history of early disease but also for potential neuroprotective interventions (Nandhagopal et al., 2008). Changes in fluorodopa uptake have also been reported in asymptomatic heterozygotes for mutations in genes associated with recessively inherited parkinsonism, such as parkin and PINK1 (Khan et al., 2002, 2005). The significance of these abnormalities is not clear, and they do not appear to be associated with significant progression or the development of clinical parkinsonism (Pavese et al., 2009a).

3. CONCLUSIONS

In this chapter, we reviewed the basic and clinical principles of molecular imaging techniques as a key to understanding how human brain works in vivo, both in health and in PD, where a profound dopamine deficiency is found.

Using a combination of tracers, it is possible to evaluate different presynaptic and postsynaptic dopaminergic as well as nondopaminergic (such as serotonergic) sites, and the resulting findings may provide numerous insights. Tracers of dopamine function can help, for instance, to differentiate between parkinsonism and other disorders resulting in tremor where this is clinically unclear and radiotracer binding can be used as an independent measure of PD severity. The pathophysiology of motor and nonmotor complications of PD such as fluctuations and dyskinesias, cognitive impairment, dementia, depression, or ICDs can be readily investigated using molecular imaging. Finally, a growing and increasingly rewarding field of interest is the identification of individuals at a risk to develop PD. In this context, molecular imaging can provide a unique opportunity to detect premotor abnormalities of dopamine function in these individuals and to monitor them over time so we can better understand disease progression, the compensatory changes that take place in response to dopamine cell loss, and to assess the effects of future neuroprotective treatment strategies.

ACKNOWLEDGMENTS

AJS is supported by the Canada Research Chairs program and by grants from the Canadian Institutes of Health Research, the Michael J. Fox Foundation and the Pacific Alzheimer Research Foundation.

REFERENCES

Aarsland, D., Kurz, M.W., 2010. The epidemiology of dementia associated with Parkinson disease. J. Neurol. Sci. 289, 18–22.

Albin, R.L., Koeppe, R.A., Chervin, R.D., Consens, F.B., Wernette, K., Frey, K.A., Aldrich, M.S., 2000. Decreased striatal dopaminergic innervation in REM sleep behavior disorder. Neurology 55, 1410–1412.

Alexander, G.E., Crutcher, M.D., 1990. Functional architecture of basal ganglia circuits: neural substrates of parallel processing. Trends Neurosci. 13, 266–271.

Alonso, A., Rodriguez, L.A., Logroscino, G., Hernan, M.A., 2009. Use of antidepressants and the risk of Parkinson's disease: a prospective study. J. Neurol. Neurosurg. Psychiatry 80, 671–674.

Antonini, A., Schwarz, J., Oertel, W.H., Pogarell, O., Leenders, K.L., 1997. Long-term changes of striatal dopamine D_2 receptors in patients with Parkinson's disease: a study with positron emission tomography and [^{11}C]raclopride. Mov. Disord. 12, 33–38.

Au, W.L., Adams, J.R., Troiano, A.R., Stoessl, A.J., 2005. Parkinson's disease: in vivo assessment of disease progression using positron emission tomography. Brain Res. Mol. Brain Res. 134, 24–33.

Ballanger, B., Strafella, A.P., van Eimeren, T., Zurowski, M., Rusjan, P.M., Houle, S., Fox, S.H., 2010. Serotonin 2A receptors and visual hallucinations in Parkinson disease. Arch. Neurol. 67, 416–421.

Barone, P., 2010. Neurotransmission in Parkinson's disease: beyond dopamine. Eur. J. Neurol. 17, 364–376.

Berendse, H.W., Booij, J., Francot, C.M., Bergmans, P.L., Hijman, R., Stoof, J.C., Wolters, E.C., 2001. Subclinical dopaminergic dysfunction in asymptomatic Parkinson's disease patients' relatives with a decreased sense of smell. Ann. Neurol. 50, 34–41.

Berg, D., Merz, B., Reiners, K., Naumann, M., Becker, G., 2005. Five-year follow-up study of hyperechogenicity of the substantia nigra in Parkinson's disease. Mov. Disord. 20, 383–385.

Bernheimer, H., Birkmayer, W., Hornykiewicz, O., Jellinger, K., Seitelberger, F., 1973. Brain dopamine and the syndromes of Parkinson and Huntington. Clinical, morphological and neurochemical correlations. J. Neurol. Sci. 20, 415–455.

Bjorklund, T., Carlsson, T., Cederfjall, E.A., Carta, M., Kirik, D., 2010. Optimized adeno-associated viral vector-mediated striatal DOPA delivery restores sensorimotor function and prevents dyskinesias in a model of advanced Parkinson's disease. Brain 133, 496–511.

Bohnen, N.I., Kaufer, D.I., Ivanco, L.S., Lopresti, B., Koeppe, R.A., Davis, J.G., Mathis, C.A., Moore, R.Y., DeKosky, S.T., 2003. Cortical cholinergic function is more severely affected in parkinsonian dementia than in Alzheimer disease: an in vivo positron emission tomographic study. Arch. Neurol. 60, 1745–1748.

Boileau, I., Rusjan, P., Houle, S., Wilkins, D., Tong, J., Selby, P., Guttman, M., Saint-Cyr, J.A., Wilson, A.A., Kish, S.J., 2008a. Increased vesicular monoamine transporter binding during early abstinence in human methamphetamine users: is VMAT2 a stable dopamine neuron biomarker? J. Neurosci. 28, 9850–9856.

Boileau, I., Warsh, J.J., Guttman, M., Saint-Cyr, J.A., McCluskey, T., Rusjan, P., Houle, S., Wilson, A.A., Meyer, J.H., Kish, S.J., 2008b. Elevated serotonin transporter binding in depressed patients with Parkinson's disease: a preliminary PET study with [^{11}C]-DASB. Mov. Disord. 23, 1776–1780.

Booij, J., Tissingh, G., Winogrodzka, A., Boer, G.J., Stoof, J.C., Wolters, E.C., van Royen, E.A., 1997. Practical benefit of [^{123}I]FP-CIT SPET in the demonstration of the dopaminergic deficit in Parkinson's disease. Eur. J. Nucl. Med. 24, 68–71.

Brooks, D.J., 2009. Imaging amyloid in Parkinson's disease dementia and dementia with Lewy bodies with positron emission tomography. Mov. Disord. 24 (Suppl. 2), S742–S747.

Brooks, D.J., Frey, K.A., Marek, K.L., Oakes, D., Paty, D., Prentice, R., Shults, C.W., Stoessl, A.J., 2003. Assessment of neuroimaging techniques as biomarkers of the progression of Parkinson's disease. Exp. Neurol. 184 (Suppl. 1), S68–S79.

Burack, M.A., Hartlein, J., Flores, H.P., Taylor-Reinwald, L., Perlmutter, J.S., Cairns, N.J., 2010. In vivo amyloid imaging in autopsy-confirmed Parkinson disease with dementia. Neurology 74, 77–84.

Calabresi, P., Di Filippo, M., Ghiglieri, V., Tambasco, N., Picconi, B., 2010. Levodopa-induced dyskinesias in patients with Parkinson's disease: filling the bench-to-bedside gap. Lancet Neurol. 9, 1106–1117.

Carta, M., Carlsson, T., Munoz, A., Kirik, D., Bjorklund, A., 2008. Serotonin-dopamine interaction in the induction and maintenance of L-DOPA-induced dyskinesias. Prog. Brain Res. 172, 465–478.

Ceravolo, R., Frosini, D., Rossi, C., Bonuccelli, U., 2009. Impulse control disorders in Parkinson's disease: definition, epidemiology, risk factors, neurobiology and management. Parkinsonism Relat. Disord. 15 (Suppl. 4), S111–S115.

Cherubini, A., Spoletini, I., Peran, P., Luccichenti, G., Di Paola, M., Sancesario, G., Gianni, W., Giubilei, F., Bossu, P., Sabatini, U., Caltagirone, C., Spalletta, G., 2010. A multimodal MRI investigation of the subventricular zone in mild cognitive impairment and Alzheimer's disease patients. Neurosci. Lett. 469, 214–218.

Cilia, R., Cho, S.S., van Eimeren, T., Marotta, G., Siri, C., Ko, J.H., Pellecchia, G., Pezzoli, G., Antonini, A., Strafella, A.P., 2011. Pathological gambling in patients with Parkinson's disease is associated with fronto-striatal disconnection: a path modeling analysis. Mov. Disord. 26, 225–233.

Doder, M., Rabiner, E.A., Turjanski, N., Lees, A.J., Brooks, D.J., 2003. Tremor in Parkinson's disease and serotonergic dysfunction: an 11C-WAY 100635 PET study. Neurology 60, 601–605.

Eisenhofer, G., 2001. The role of neuronal and extraneuronal plasma membrane transporters in the inactivation of peripheral catecholamines. Pharmacol. Ther. 91, 35–62.

van Eimeren, T., Ballanger, B., Pellecchia, G., Miyasaki, J.M., Lang, A.E., Strafella, A.P., 2009. Dopamine agonists diminish value sensitivity of the orbitofrontal cortex: a trigger for pathological gambling in Parkinson's disease? Neuropsychopharmacology 34, 2758–2766.

Fahn, S., Oakes, D., Shoulson, I., Kieburtz, K., Rudolph, A., Lang, A., Olanow, C.W., Tanner, C., Marek, K., 2004. Levodopa and the progression of Parkinson's disease. N. Engl. J. Med. 351, 2498–2508.

Fearnley, J.M., Lees, A.J., 1991. Ageing and Parkinson's disease: substantia nigra regional selectivity. Brain 114 (Pt 5), 2283–2301.

Felicio, A.C., Godeiro-Junior, C., Shih, M.C., Borges, V., Silva, S.M., Aguiar Pde, C., Hoexter, M.Q., Barsottini, O.G., Andrade, L.A., Bressan, R.A., Ferraz, H.B., 2010a. Evaluation of patients with clinically unclear Parkinsonian syndromes submitted to brain SPECT imaging using the technetium-99m labeled tracer TRODAT-1. J. Neurol. Sci. 291, 64–68.

Felicio, A.C., Moriyama, T.S., Godeiro-Junior, C., Shih, M.C., Hoexter, M.Q., Borges, V., Silva, S.M., Amaro-Junior, E., Andrade, L.A., Ferraz, H.B., Bressan, R.A., 2010b. Higher dopamine transporter density in Parkinson's disease patients with depression. Psychopharmacology (Berl) 211, 27–31.

Felicio, A.C., Shih, M.C., Godeiro-Junior, C., Andrade, L.A., Bressan, R.A., Ferraz, H.B., 2009. Molecular imaging studies in Parkinson disease: reducing diagnostic uncertainty. Neurologist 15, 6–16.

de la Fuente-Fernandez, R., Furtado, S., Guttman, M., Furukawa, Y., Lee, C.S., Calne, D.B., Ruth, T.J., Stoessl, A.J., 2003. VMAT2 binding is elevated in dopa-responsive dystonia: visualizing empty vesicles by PET. Synapse 49, 20–28.

de la Fuente-Fernandez, R., Lu, J.Q., Sossi, V., Jivan, S., Schulzer, M., Holden, J.E., Lee, C.S., Ruth, T.J., Calne, D.B., Stoessl, A.J., 2001. Biochemical variations in the synaptic level of dopamine precede motor fluctuations in Parkinson's disease: PET evidence of increased dopamine turnover. Ann. Neurol. 49, 298–303.

de la Fuente-Fernandez, R., Schulzer, M., Kuramoto, L., Cragg, J., Ramachandiran, N., Au, W.L., Mak, E., McKenzie, J., McCormick, S., Sossi, V., Ruth, T.J., Lee, C.S., Calne, D.B., Stoessl, A.J., 2011. Age-specific progression of nigrostriatal dysfunction in Parkinson's disease. Ann. Neurol.

de la Fuente-Fernandez, R., Schulzer, M., Mak, E., Calne, D.B., Stoessl, A.J., 2004. Presynaptic mechanisms of motor fluctuations in Parkinson's disease: a probabilistic model. Brain 127, 888–899.

de la Fuente-Fernandez, R., Sossi, V., McCormick, S., Schulzer, M., Ruth, T.J., Stoessl, A.J., 2009. Visualizing vesicular dopamine dynamics in Parkinson's disease. Synapse 63, 713–716.

Gaenslen, A., Unmuth, B., Godau, J., Liepelt, I., Di Santo, A., Schweitzer, K.J., Gasser, T., Machulla, H.J., Reimold, M., Marek, K., Berg, D., 2008. The specificity and sensitivity of transcranial ultrasound in the differential diagnosis of Parkinson's disease: a prospective blinded study. Lancet Neurol. 7, 417–424.

Gibb, W.R., Lees, A.J., 1991. Anatomy, pigmentation, ventral and dorsal subpopulations of the substantia nigra, and differential cell death in Parkinson's disease. J. Neurol. Neurosurg. Psychiatry 54, 388–396.

Hammoud, D.A., Hoffman, J.M., Pomper, M.G., 2007. Molecular neuroimaging: from conventional to emerging techniques. Radiology, 245, 21–42.

Hesse, S., Meyer, P.M., Strecker, K., Barthel, H., Wegner, F., Oehlwein, C., Isaias, I.U., Schwarz, J., Sabri, O., 2009. Monoamine transporter availability in Parkinson's disease patients with or without depression. Eur. J. Nucl. Med. Mol. Imaging 36, 428–435.

Hickey, M.G., Demaerschalk, B.M., Caselli, R.J., Parish, J.M., Wingerchuk, D.M., 2007. "Idiopathic" rapid-eye-movement (REM) sleep behavior disorder is associated with future development of neurodegenerative diseases. Neurologist 13, 98–101.

Hilker, R., Thomas, A.V., Klein, J.C., Weisenbach, S., Kalbe, E., Burghaus, L., Jacobs, A.H., Herholz, K., Heiss, W.D., 2005. Dementia in Parkinson disease: functional imaging of cholinergic and dopaminergic pathways. Neurology 65, 1716–1722.

Hughes, A.J., Daniel, S.E., Kilford, L., Lees, A.J., 1992. Accuracy of clinical diagnosis of idiopathic Parkinson's disease: a clinico-pathological study of 100 cases. J. Neurol. Neurosurg. Psychiatry 55, 181–184.

Hughes, A.J., Daniel, S.E., Lees, A.J., 2001. Improved accuracy of clinical diagnosis of Lewy body Parkinson's disease. Neurology 57, 1497–1499.

Iranzo, A., Valldeoriola, F., Lomena, F., Molinuevo, J.L., Serradell, M., Salamero, M., Cot, A., Ros, D., Pavia, J., Santamaria, J., Tolosa, E., 2011. Serial dopamine transporter imaging of nigrostriatal function in patients with idiopathic rapid-eye-movement sleep behaviour disorder: a prospective study. Lancet Neurol. 10, 797–805.

Ishihara, L., Brayne, C., 2006. A systematic review of depression and mental illness preceding Parkinson's disease. Acta Neurol. Scand. 113, 211–220.

Jankovic, J., 2008. Parkinson's disease: clinical features and diagnosis. J. Neurol. Neurosurg. Psychiatry 79, 368–376.

Johansen, K.K., White, L.R., Sando, S.B., Aasly, J.O., 2010. Biomarkers: Parkinson disease with dementia and dementia with Lewy bodies. Parkinsonism Relat. Disord. 16, 307–315.

Kaasinen, V., Ruottinen, H.M., Nagren, K., Lehikoinen, P., Oikonen, V., Rinne, J.O., 2000. Upregulation of putaminal dopamine D_2 receptors in early Parkinson's disease: a comparative PET study with [^{11}C] raclopride and [^{11}C]N-methylspiperone. J. Nucl. Med. 41, 65–70.

Kagi, G., Bhatia, K.P., Tolosa, E., 2010. The role of DAT-SPECT in movement disorders. J. Neurol. Neurosurg. Psychiatry 81, 5–12.

Kato, N., Arai, K., Hattori, T., 2003. Study of the rostral midbrain atrophy in progressive supranuclear palsy. J. Neurol. Sci. 210, 57–60.

Khan, N.L., Scherfler, C., Graham, E., Bhatia, K.P., Quinn, N., Lees, A.J., Brooks, D.J., Wood, N.W., Piccini, P., 2005. Dopaminergic dysfunction in unrelated, asymptomatic carriers of a single Parkin mutation. Neurology 64, 134–136.

Khan, N.L., Valente, E.M., Bentivoglio, A.R., Wood, N.W., Albanese, A., Brooks, D.J., Piccini, P., 2002. Clinical and subclinical dopaminergic dysfunction in PARK6-linked parkinsonism: an ^{18}F-dopa PET study. Ann. Neurol. 52, 849–853.

Kim, S.E., Choi, J.Y., Choe, Y.S., Choi, Y., Lee, W.Y., 2003. Serotonin transporters in the midbrain of Parkinson's disease patients: a study with ^{123}I-beta-CIT SPECT. J. Nucl. Med. 44, 870–876.

Kish, S.J., Shannak, K., Hornykiewicz, O., 1988. Uneven pattern of dopamine loss in the striatum of patients with idiopathic Parkinson's disease. Pathophysiologic and clinical implications. N. Engl. J. Med. 318, 876–880.

Klein, C., Lohmann-Hedrich, K., 2007. Impact of recent genetic findings in Parkinson's disease. Curr. Opin. Neurol. 20, 453–464.

Klein, J.C., Eggers, C., Kalbe, E., Weisenbach, S., Hohmann, C., Vollmar, S., Baudrexel, S., Diederich, N.J., Heiss, W.D., Hilker, R., 2010. Neurotransmitter changes in dementia with Lewy bodies and Parkinson disease dementia in vivo. Neurology 74, 885–892.

Lee, C.S., Samii, A., Sossi, V., Ruth, T.J., Schulzer, M., Holden, J.E., Wudel, J., Pal, P.K., de la Fuente-Fernandez, R., Calne, D.B., Stoessl, A.J., 2000. In vivo positron emission tomographic evidence for compensatory changes in presynaptic dopaminergic nerve terminals in Parkinson's disease. Ann. Neurol. 47, 493–503.

Leentjens, A.F., Van den Akker, M., Metsemakers, J.F., Lousberg, R., Verhey, F.R., 2003. Higher incidence of depression preceding the onset of Parkinson's disease: a register study. Mov. Disord. 18, 414–418.

Maetzler, W., Liepelt, I., Reimold, M., Reischl, G., Solbach, C., Becker, C., Schulte, C., Leyhe, T., Keller, S., Melms, A., Gasser, T., Berg, D., 2009. Cortical PIB binding in Lewy body disease is associated with Alzheimer-like characteristics. Neurobiol. Dis. 34, 107–112.

Marek, K.L., Seibyl, J.P., Zoghbi, S.S., Zea-Ponce, Y., Baldwin, R.M., Fussell, B., Charney, D.S., van Dyck, C., Hoffer, P.B., Innis, R.P., 1996. [^{123}I] beta-CIT/SPECT imaging demonstrates bilateral loss of dopamine transporters in hemi-Parkinson's disease. Neurology 46, 231–237.

Marsden, C.D., 1994. Parkinson's disease. J. Neurol. Neurosurg. Psychiatry 57, 672–681.

Martin, W.R., Palmer, M.R., Patlak, C.S., Calne, D.B., 1989. Nigrostriatal function in humans studied with positron emission tomography. Ann. Neurol. 26, 535–542.

Minoshima, S., Foster, N.L., Sima, A.A., Frey, K.A., Albin, R.L., Kuhl, D.E., 2001. Alzheimer's disease versus dementia with Lewy bodies: cerebral metabolic distinction with autopsy confirmation. Ann. Neurol. 50, 358–365.

Morelli, M., Arabia, G., Novellino, F., Salsone, M., Giofre, L., Condino, F., Messina, D., Quattrone, A., 2011. MRI measurements predict PSP in unclassifiable parkinsonisms: a cohort study. Neurology 77, 1042–1047.

Morrish, P.K., 2003. REAL and CALM: what have we learned? Mov. Disord. 18, 839–840.

Mozley, P.D., Schneider, J.S., Acton, P.D., Plossl, K., Stern, M.B., Siderowf, A., Leopold, N.A., Li, P.Y., Alavi, A., Kung, H.F., 2000. Binding of [^{99m}Tc]TRODAT-1 to dopamine transporters in patients with Parkinson's disease and in healthy volunteers. J. Nucl. Med. 41, 584–589.

Munoz, A., Carlsson, T., Tronci, E., Kirik, D., Bjorklund, A., Carta, M., 2009. Serotonin neuron-dependent and -independent reduction of dyskinesia by 5-HT1A and 5-HT1B receptor agonists in the rat Parkinson model. Exp. Neurol. 219, 298–307.

Munoz, A., Li, Q., Gardoni, F., Marcello, E., Qin, C., Carlsson, T., Kirik, D., Di Luca, M., Bjorklund, A., Bezard, E., Carta, M., 2008. Combined 5-HT1A and 5-HT1B receptor agonists for the treatment of L-DOPA-induced dyskinesia. Brain 131, 3380–3394.

Murai, T., Muller, U., Werheid, K., Sorger, D., Reuter, M., Becker, T., von Cramon, D.Y., Barthel, H., 2001. In vivo evidence for differential association of striatal dopamine and midbrain serotonin systems with neuropsychiatric symptoms in Parkinson's disease. J. Neuropsychiatry Clin. Neurosci. 13, 222–228.

Nandhagopal, R., Kuramoto, L., Schulzer, M., Mak, E., Cragg, J., Lee, C.S., McKenzie, J., McCormick, S., Samii, A., Troiano, A., Ruth, T.J., Sossi, V., de la Fuente-Fernandez, R., Calne, D.B., Stoessl, A.J., 2009. Longitudinal progression of sporadic Parkinson's disease: a multi-tracer positron emission tomography study. Brain 132, 2970–2979.

Nandhagopal, R., Mak, E., Schulzer, M., McKenzie, J., McCormick, S., Sossi, V., Ruth, T.J., Strongosky, A., Farrer, M.J., Wszolek, Z.K., Stoessl, A.J., 2008. Progression of dopaminergic dysfunction in a LRRK2 kindred: a multitracer PET study. Neurology 71, 1790–1795.

Nicoletti, G., Lodi, R., Condino, F., Tonon, C., Fera, F., Malucelli, E., Manners, D., Zappia, M., Morgante, L., Barone, P., Barbiroli, B., Quattrone, A., 2006. Apparent diffusion coefficient measurements of the middle cerebellar peduncle differentiate the Parkinson variant of MSA from Parkinson's disease and progressive supranuclear palsy. Brain 129, 2679–2687.

Nicoletti, G., Tonon, C., Lodi, R., Condino, F., Manners, D., Malucelli, E., Morelli, M., Novellino, F., Paglionico, S., Lanza, P., Messina, D., Barone, P., Morgante, L., Zappia, M., Barbiroli, B., Quattrone, A., 2008. Apparent diffusion coefficient of the superior cerebellar peduncle differentiates progressive supranuclear palsy from Parkinson's disease. Mov. Disord. 23, 2370–2376.

O'Sullivan, S.S., Wu, K., Politis, M., Lawrence, A.D., Evans, A.H., Bose, S.K., Djamshidian, A., Lees, A.J., Piccini, P., 2011. Cue-induced striatal dopamine release in Parkinson's disease-associated impulsive-compulsive behaviours. Brain 134, 969–978.

Obeso, J.A., Rodriguez-Oroz, M.C., Benitez-Temino, B., Blesa, F.J., Guridi, J., Marin, C., Rodriguez, M., 2008. Functional organization of the basal ganglia: therapeutic implications for Parkinson's disease. Mov. Disord. 23 (Suppl. 3), S548–S559.

Parkinson Study Group, 2000. A randomized controlled trial comparing pramipexole with levodopa in early Parkinson's disease: design and methods of the CALM-PD study. Clin. Neuropharmacol 23, 34–44.

Paterson, L.M., Kornum, B.R., Nutt, D.J., Pike, V.W., Knudsen, G.M., 2011. 5-HT radioligands for human brain imaging with PET and SPECT. Med. Res. Rev.

Pavese, N., Khan, N.L., Scherfler, C., Cohen, L., Brooks, D.J., Wood, N.W., Bhatia, K.P., Quinn, N.P., Lees, A.J., Piccini, P., 2009a. Nigrostriatal dysfunction in homozygous and heterozygous parkin gene carriers: an ^{18}F-dopa PET progression study. Mov. Disord. 24, 2260–2266.

Pavese, N., Kiferle, L., Piccini, P., 2009b. Neuroprotection and imaging studies in Parkinson's disease. Parkinsonism Relat. Disord. 15 (Suppl. 4), S33–S37.

Peran, P., Cherubini, A., Assogna, F., Piras, F., Quattrocchi, C., Peppe, A., Celsis, P., Rascol, O., Demonet, J.F., Stefani, A., Pierantozzi, M., Pontieri, F.E., Caltagirone, C., Spalletta, G., Sabatini, U., 2010. Magnetic resonance imaging markers of Parkinson's disease nigrostriatal signature. Brain 133, 3423–3433.

Piccini, P.P., 2003. Dopamine transporter: basic aspects and neuroimaging. Mov. Disord. 18 (Suppl. 7), S3–S8.

Pinborg, L.H., Videbaek, C., Ziebell, M., Mackeprang, T., Friberg, L., Rasmussen, H., Knudsen, G.M., Glenthoj, B.Y., 2007. [^{123}I]epidepride binding to cerebellar dopamine D_2/D_3 receptors is displaceable: implications for the use of cerebellum as a reference region. Neuroimage 34, 1450–1453.

Pirker, W., 2003. Correlation of dopamine transporter imaging with parkinsonian motor handicap: how close is it? Mov. Disord. 18 (Suppl. 7), S43–S51.

Poewe, W., Scherfler, C., 2003. Role of dopamine transporter imaging in investigation of parkinsonian syndromes in routine clinical practice. Mov. Disord. 18 (Suppl. 7), S16–S21.

Politis, M., Wu, K., Loane, C., Kiferle, L., Molloy, S., Brooks, D.J., Piccini, P., 2010a. Staging of serotonergic dysfunction in Parkinson's disease: an in vivo ^{11}C-DASB PET study. Neurobiol. Dis. 40, 216–221.

Politis, M., Wu, K., Loane, C., Quinn, N.P., Brooks, D.J., Rehncrona, S., Bjorklund, A., Lindvall, O., Piccini, P., 2010b. Serotonergic neurons mediate dyskinesia side effects in Parkinson's patients with neural transplants. Sci. Transl. Med. 2, 38ra46.

Politis, M., Wu, K., Loane, C., Turkheimer, F.E., Molloy, S., Brooks, D.J., Piccini, P., 2010c. Depressive symptoms in PD correlate with higher 5-HTT binding in raphe and limbic structures. Neurology 75, 1920–1927.

Ponsen, M.M., Stoffers, D., Booij, J., van Eck-Smit, B.L., Wolters, E., Berendse, H.W., 2004. Idiopathic hyposmia as a preclinical sign of Parkinson's disease. Ann. Neurol. 56, 173–181.

Ponsen, M.M., Stoffers, D., Wolters, E., Booij, J., Berendse, H.W., 2010. Olfactory testing combined with dopamine transporter imaging as a method to detect prodromal Parkinson's disease. J. Neurol. Neurosurg. Psychiatry 81, 396–399.

Postuma, R.B., Gagnon, J.F., Vendette, M., Fantini, M.L., Massicotte-Marquez, J., Montplaisir, J., 2009. Quantifying the risk of neurodegenerative disease in idiopathic REM sleep behavior disorder. Neurology 72, 1296–1300.

de Paulis, T., 2003. The discovery of epidepride and its analogs as high-affinity radioligands for imaging extrastriatal dopamine D(2) receptors in human brain. Curr. Pharm. Des. 9, 673–696.

Remy, P., Doder, M., Lees, A., Turjanski, N., Brooks, D., 2005. Depression in Parkinson's disease: loss of dopamine and noradrenaline innervation in the limbic system. Brain 128, 1314–1322.

Rickards, H., 2005. Depression in neurological disorders: Parkinson's disease, multiple sclerosis, and stroke. J. Neurol. Neurosurg. Psychiatry 76 (Suppl. 1), i48–52.

Rinne, J.O., Laihinen, A., Nagren, K., Bergman, J., Solin, O., Haaparanta, M., Ruotsalainen, U., Rinne, U.K., 1990. PET demonstrates different behaviour of striatal dopamine D_1 and D_2 receptors in early Parkinson's disease. J. Neurosci. Res. 27, 494–499.

Roselli, F., Pisciotta, N.M., Pennelli, M., Aniello, M.S., Gigante, A., De Caro, M.F., Ferrannini, E., Tartaglione, B., Niccoli-Asabella, A., Defazio, G., Livrea, P., Rubini, G., 2010. Midbrain SERT in degenerative parkinsonisms: a ^{123}I-FP-CIT SPECT study. Mov. Disord. 25, 1853–1859.

Schenck, C.H., Bundlie, S.R., Mahowald, M.W., 1996. Delayed emergence of a parkinsonian disorder in 38% of 29 older men initially diagnosed with idiopathic rapid eye movement sleep behaviour disorder. Neurology 46, 388–393.

Scherfler, C., Schwarz, J., Antonini, A., Grosset, D., Valldeoriola, F., Marek, K., Oertel, W., Tolosa, E., Lees, A.J., Poewe, W., 2007. Role of DAT-SPECT in the diagnostic work up of parkinsonism. Mov. Disord. 22, 1229–1238.

Schrag, A., Good, C.D., Miszkiel, K., Morris, H.R., Mathias, C.J., Lees, A.J., Quinn, N.P., 2000. Differentiation of atypical parkinsonian syndromes with routine MRI. Neurology 54, 697–702.

Schuurman, A.G., van den Akker, M., Ensinck, K.T., Metsemakers, J.F., Knottnerus, J.A., Leentjens, A.F., Buntinx, F., 2002. Increased risk of Parkinson's disease after depression: a retrospective cohort study. Neurology 58, 1501–1504.

Seeman, P., Niznik, H.B., 1990. Dopamine receptors and transporters in Parkinson's disease and schizophrenia. FASEB J. 4, 2737–2744.

Seeman, P., Van Tol, H.H., 1994. Dopamine receptor pharmacology. Trends Pharmacol. Sci. 15, 264–270.

Seppi, K., Schocke, M.F., Esterhammer, R., Kremser, C., Brenneis, C., Mueller, J., Boesch, S., Jaschke, W., Poewe, W., Wenning, G.K., 2003. Diffusion-weighted imaging discriminates progressive supranuclear palsy from PD, but not from the Parkinson variant of multiple system atrophy. Neurology 60, 922–927.

Shen, L.H., Tseng, Y.C., Liao, M.H., Fu, Y.K., 2011. The role of molecular imaging in the diagnosis and management of neuropsychiatric disorders. J. Biomed. Biotechnol., 439397.

Shiba, M., Bower, J.H., Maraganore, D.M., McDonnell, S.K., Peterson, B.J., Ahlskog, J.E., Schaid, D.J., Rocca, W.A., 2000. Anxiety disorders and depressive disorders preceding Parkinson's disease: a case-control study. Mov. Disord. 15, 669–677.

Shih, M.C., Felicio, A.C., de Oliveira Godeiro-Junior, C., de Carvalho Aguiar, P., de Andrade, L.A., Ferraz, H.B., Bressan, R.A., 2007a. Molecular imaging in hereditary forms of parkinsonism. Eur. J. Neurol. 14, 359–368.

Shih, M.C., Franco de Andrade, L.A., Amaro Jr., E., Felicio, A.C., Ferraz, H.B., Wagner, J., Hoexter, M.Q., Lin, L.F., Fu, Y.K., Mari, J.J., Tufik, S., Bressan, R.A., 2007b. Higher nigrostriatal dopamine neuron loss in early than late onset Parkinson's disease? – a [^{99m}Tc]-TRODAT-1 SPECT study. Mov. Disord. 22, 863–866.

Shinotoh, H., Inoue, O., Hirayama, K., Aotsuka, A., Asahina, M., Suhara, T., Yamazaki, T., Tateno, Y., 1993. Dopamine D_1 receptors in Parkinson's disease and striatonigral degeneration: a positron emission tomography study. J. Neurol. Neurosurg. Psychiatry 56, 467–472.

Siderowf, A., Stern, M.B., 2006. Preclinical diagnosis of Parkinson's disease: are we there yet? Curr. Neurol. Neurosci. Rep. 6, 295–301.

Sommer, U., Hummel, T., Cormann, K., Mueller, A., Frasnelli, J., Kropp, J., Reichmann, H., 2004. Detection of presymptomatic Parkinson's disease: combining smell tests, transcranial sonography, and SPECT. Mov. Disord. 19, 1196–1202.

Sossi, V., de La Fuente-Fernandez, R., Holden, J.E., Doudet, D.J., McKenzie, J., Stoessl, A.J., Ruth, T.J., 2002. Increase in dopamine turnover occurs early in Parkinson's disease: evidence from a new modeling approach to PET ^{18}F-fluorodopa data. J. Cereb. Blood Flow Metab. 22, 232–239.

Sossi, V., de la Fuente-Fernandez, R., Schulzer, M., Troiano, A.R., Ruth, T.J., Stoessl, A.J., 2007. Dopamine transporter relation to dopamine turnover in Parkinson's disease: a positron emission tomography study. Ann. Neurol. 62, 468–474.

Steeves, T.D., Miyasaki, J., Zurowski, M., Lang, A.E., Pellecchia, G., Van Eimeren, T., Rusjan, P., Houle, S., Strafella, A.P., 2009. Increased striatal dopamine release in Parkinsonian patients with pathological gambling: a [^{11}C] raclopride PET study. Brain 132, 1376–1385.

Stiasny-Kolster, K., Doerr, Y., Moller, J.C., Hoffken, H., Behr, T.M., Oertel, W.H., Mayer, G., 2005. Combination of 'idiopathic' REM sleep behaviour disorder and olfactory dysfunction as possible indicator for alpha-synucleinopathy demonstrated by dopamine transporter FP-CIT-SPECT. Brain 128, 126–137.

Stoessl, A.J., 2007. Positron emission tomography in premotor Parkinson's disease. Parkinsonism Relat. Disord. 13 (Suppl. 3), S421–S424.

Stoessl, A.J., 2011. Neuroimaging in Parkinson's disease. Neurotherapeutics 8, 72–81.

Strecker, K., Wegner, F., Hesse, S., Becker, G.A., Patt, M., Meyer, P.M., Lobsien, D., Schwarz, J., Sabri, O., 2011. Preserved serotonin transporter binding in de novo Parkinson's disease: negative correlation with the dopamine transporter. J. Neurol. 258, 19–26.

Thobois, S., Vingerhoets, F., Fraix, V., Xie-Brustolin, J., Mollion, H., Costes, N., Mertens, P., Benabid, A.L., Pollak, P., Broussolle, E., 2004. Role of dopaminergic treatment in dopamine receptor down-regulation in advanced Parkinson disease: a positron emission tomographic study. Arch. Neurol. 61, 1705–1709.

Troiano, A.R., de la Fuente-Fernandez, R., Sossi, V., Schulzer, M., Mak, E., Ruth, T.J., Stoessl, A.J., 2009. PET demonstrates reduced dopamine transporter expression in PD with dyskinesias. Neurology 72, 1211–1216.

Ugrumov, M.V., 2009. Non-dopaminergic neurons partly expressing dopaminergic phenotype: distribution in the brain, development and functional significance. J. Chem. Neuroanat. 38, 241–256.

Vander Borght, T., Kilbourn, M., Desmond, T., Kuhl, D., Frey, K., 1995. The vesicular monoamine transporter is not regulated by dopaminergic drug treatments. Eur. J. Pharmacol. 294, 577–583.

Varrone, A., Halldin, C., 2010. Molecular imaging of the dopamine transporter. J. Nucl. Med. 51, 1331–1334.

Vingerhoets, F.J., Schulzer, M., Calne, D.B., Snow, B.J., 1997. Which clinical sign of Parkinson's disease best reflects the nigrostriatal lesion? Ann. Neurol. 41, 58–64.

Volkow, N.D., Fowler, J.S., Wang, G.J., Logan, J., Schlyer, D., MacGregor, R., Hitzemann, R., Wolf, A.P., 1994. Decreased dopamine transporters with age in health human subjects. Ann. Neurol. 36, 237–239.

Weintraub, D., 2008. Dopamine and impulse control disorders in Parkinson's disease. Ann. Neurol. 64 (Suppl. 2), S93–S100.
Weintraub, D., Newberg, A.B., Cary, M.S., Siderowf, A.D., Moberg, P.J., Kleiner-Fisman, G., Duda, J.E., Stern, M.B., Mozley, D., Katz, I.R., 2005. Striatal dopamine transporter imaging correlates with anxiety and depression symptoms in Parkinson's disease. J. Nucl. Med. 46, 227–232.
Whone, A.L., Watts, R.L., Stoessl, A.J., Davis, M., Reske, S., Nahmias, C., Lang, A.E., Rascol, O., Ribeiro, M.J., Remy, P., Poewe, W.H., Hauser, R.A., Brooks, D.J., 2003. Slower progression of Parkinson's disease with ropinirole versus levodopa: the REAL-PET study. Ann. Neurol. 54, 93–101.
Wichmann, T., DeLong, M.R., Guridi, J., Obeso, J.A., 2011. Milestones in research on the pathophysiology of Parkinson's disease. Mov. Disord. 26, 1032–1041.
Wu, K., Politis, M., Piccini, P., 2009. Parkinson disease and impulse control disorders: a review of clinical features, pathophysiology and management. Postgrad. Med. J. 85, 590–596.
Yoritaka, A., Ohizumi, H., Tanaka, S., Hattori, N., 2009. Parkinson's disease with and without REM sleep behaviour disorder: are there any clinical differences? Eur. Neurol. 61, 164–170.

CHAPTER ELEVEN

Imaging the Dopamine D_3 Receptor In Vivo

Mark Slifstein[1,2], Eugenii A. Rabiner[3,4] and Roger N. Gunn[3,5,6]
[1]Department of Psychiatry, Columbia University, New York, NY, USA
[2]Division of Translational Imaging, New York State Psychiatric Institute, New York, NY, USA
[3]Imanova Limited, London, UK
[4]Institute of Psychiatry, Kings College, London, UK
[5]Department of Medicine, Imperial College, London, UK
[6]Department of Engineering Science, University of Oxford, UK

1. INTRODUCTION

Dopamine receptors in the brain are broadly divided into two classes of G-protein-coupled receptors. D_1-like receptors (D_1, D_5) couple to $G_{\alpha s}$ and G_{olf} and, when stimulated by dopamine, increase cyclic adenosine monophosphate (AMP) production through the activation of adenylate cyclase, whereas D_2-like receptors (D_2, D_3, and D_4) couple to $G_{\alpha i}$ and $G_{\alpha o}$ and inhibit adenylate cyclase activity and cyclic AMP production (Neve et al., 2004). The D_3 receptor was first recognized as a distinct molecular subtype within the D_2-like family by Sokoloff et al. who cloned D_3 from rat (Sokoloff et al., 1990) and human (Giros et al., 1990) DNA. D_3 has a high sequence homology with D_2, and thus, many ligands for D_2 also readily bind to D_3 (Table 1, Levant, 1997; Table 1, Sokoloff et al., 2006). The anatomic distribution of D_3, however, is more limited than that of D_2. The discovery that [^{3}H](+)7-[^{3}H]hydroxy-*N*,*N*-di-*n*-propyl-2-aminotetralin (7-OH-DPAT) is 100-fold more selective for D_3 than for D_2 in transfected CHO cells enabled autoradiography to be performed in rat brain, where it was shown that D_3 was confined to the islands of Calleja, lobules 9 and 10 of the cerebellum, nucleus accumbens, and olfactory tubercle (Levesque et al., 1992). D_3 binding in the dorsal striatum, where D_2 is abundant, was low.

In human postmortem brains, Murray et al. used unlabeled 7-OH-DPAT as a blocking agent in conjunction with the $D_{2/3}$ ligand [^{125}I]epidipride to characterize the distribution of D_3 (Murray et al., 1994). 7-OH-DPAT blocked [^{125}I]epidipride binding in islands of Calleja, nucleus accumbens, the extended amygdala, ventral tegmental area, and globus pallidus, but not in the dorsal striatum. Using another D_3 selective compound, [^{125}I]*trans* 7-OH-PIPAT ([^{125}I](R)-*trans*-7-hydroxy-2-[*N*-propyl-*N*-(3'-iodo-2'-propenyl)-amino]tetralin) (Foulon et al., 1993), Gurevich et al. examined postmortem tissue from patients with schizophrenia (age = 78 ± 5 years) and matched healthy controls (Gurevich et al., 1997). A binding pattern similar to that reported by Murray et al. was observed. Intriguingly, patients with

Imaging of the Human Brain in Health and Disease
http://dx.doi.org/10.1016/B978-0-12-418677-4.00011-7

Table 1 BP_{ND} for [^{11}C]Raclopride and [^{11}C]PHNO for BP897 Studies ($n=3$, Baboons)

	[^{11}C]Raclopride			[^{11}C]-(+)-PHNO		
Regions of Interest	**Baseline BP_{ND}**	**Post-BP897 BP_{ND}**	**Difference (%)**	**Baseline BP_{ND}**	**Post-BP897 BP_{ND}**	**Difference (%)**
Ventral striatum	2.05 ± 0.84	1.71 ± 0.88	-19 ± 8	2.81 ± 0.79	1.91 ± 0.34	-30 ± 11*
Dorsal striatum	2.57 ± 1.12	2.33 ± 1.08	-10 ± 7	2.10 ± 0.52	1.79 ± 0.30	-13 ± 8*
Globus pallidus	1.57 ± 0.45	1.13 ± 0.46	-29 ± 9*	4.05 ± 1.42	1.67 ± 0.37	-57 ± 11*
Thalamus				1.18 ± 0.21	0.47 ± 0.14	-60 ± 9*
SN/VTA				0.60 ± 0.02	0.06 ± 0.08	-90 ± 14*

SN/VTA = substantia nigra/ventral tegmental area.
*represents "$p < 0.05$, one sample t test".

schizophrenia who were antipsychotic drug-free for an extended period of time prior to death had elevated D_3 in the ventral striatum (VST) compared to both controls and patients who had been treated up to 72h before death, leading the authors to speculate that there was an upregulation of D_3 in schizophrenia that had been normalized by antipsychotic drug treatment.

Using [^{3}H](+)7-OH-DPAT, Staley and Mash performed membrane binding and autoradiography in postmortem tissue from subjects who had died from cocaine overdose, subjects who had died from excited delirium following cocaine ingestion, and matched controls (Staley and Mash, 1996). Again, the same binding pattern, with a high binding in the VST and nucleus accumbens, was observed. It was also shown that the subjects who died from cocaine overdose had elevated binding in most regions compared to that in the other groups, further confirmation of which was provided by the analysis of mRNA (Segal et al., 1997). Thus, during the 1990s, it was well established that in humans, D_3 receptors were mostly localized to the "limbic" striatum, and furthermore, studies hinted that D_3 expression was altered in psychiatric illness.

In vivo imaging studies that could distinguish D_3 from D_2 in human subjects, however, were not possible for many years. This was due to the lack of a selective or D_3 preferring radiotracer for positron emission tomography (PET) or single photon emission tomography (SPECT). Available D_2-like tracers such as [^{11}C]raclopride, [^{123}I]IBZM ((S)-(-)-3-[^{123}I]iodo-2-hydroxy-6-methoxy-*N*-[(1-ethyl-2-pyrrolidinyl) methyl]benzamide), or [^{18}F]fallypride have shown a similar or lower preference for D_3 compared to D_2 receptors, either when the affinity is measured (Malmberg et al., 1996; Sokoloff et al., 1990; Vanhauwe et al., 1999; Videbaek et al., 2000), or as detected by the visual inspection of the binding pattern in scans, which always display a higher binding in the dorsal striatum than in the VST or other known loci of D_3 expression.

This situation changed serendipitously with the introduction of the D_2/D_3 agonist PET radiotracer [^{11}C]-(+)-PHNO in 2005 (Wilson et al., 2005). Developed originally as a therapy for Parkinson's disease (Ahlskog et al., 1991; Dykstra et al., 1985; Jones et al., 1984), it was first labeled with tritium for binding and autoradiography experiments (Seeman et al., 1993). (+)-PHNO was labeled for PET with ^{11}C in 1997 (Brown et al., 1997), but no follow-up imaging studies were reported and the use of the compound as a PET tracer was not developed at that time. Wilson et al. published their radiolabeling procedure in 2005 and rapidly followed that with several studies characterizing the in vivo imaging properties of the tracer. It was thought at the time that an agonist tracer might provide more information about D_2/D_3 receptor affinity states and possibly be more sensitive to competition from endogenous dopamine than antagonist radiotracers.

[^{11}C]-(+)-PHNO does act as an agonist at D_2 receptors and the question as to whether agonist tracers confer any advantages over antagonist tracers in detecting receptor affinity states in the in vivo setting is an area of ongoing investigation (Finnema et al., 2009; McCormick et al., 2008; Narendran et al., 2010; Seeman, 2012; Shotbolt et al., 2012; Skinbjerg et al., 2012), but the high binding in the globus pallidus observed in the initial [^{11}C]-(+)-PHNO scans performed in humans (Willeit et al., 2006) strongly suggested the possibility that [^{11}C]-(+)-PHNO is D_3-preferring, as the globus pallidus is one of the brain regions identified as having a high D_3 expression in human postmortem tissue (Table 1, Murray et al., 1994).

Some (Freedman et al., 1994; van Vliet et al., 2000) but not all (Seeman et al., 2005) in vitro assays suggested the D_3-preferring nature of (+)-PHNO; subsequent in vivo investigations (described below) definitively demonstrated that [^{11}C]-(+)-PHNO is D_3-preferring, although it exhibits specific binding to D_2 receptors as well. [^{11}C]-(+)-PHNO displays a mixed D_2/D_3 signal which, with careful interpretation, can provide information about D_3 receptors in vivo that was not previously accessible. Despite ongoing research efforts to develop other D_3-preferring or D_3 selective radiotracers with suitable pharmacokinetic properties for in vivo imaging with PET or SPECT (Bennacef et al., 2009; Chu et al., 2005; Hocke et al., 2010; Hofling et al., 2010; Mach et al., 2004; Tu et al., 2011), to date, only [^{11}C]-(+)-PHNO has proven to be viable for this purpose.

Thus, the story of in vivo D_3 imaging is largely the story of [^{11}C]-(+)-PHNO imaging. In the following sections, we describe the initial studies with [^{11}C]-(+)-PHNO in preclinical species and humans, the studies that characterized the tracer's D_3-preferring properties, the development of quantitative methods for uncoupling the D_3 signal from the D_2 signal, and finally, several studies in populations with psychiatric or neurological illnesses that have already been performed with [^{11}C]-(+)-PHNO.

2.1. PET IMAGING WITH [^{11}C]-(+)-PHNO

2.1.1. Early PET Studies, Recognition of the D_3-Preferring Nature of [^{11}C]-(+)-PHNO

Following the initial publication describing the radiosynthesis of [^{11}C]-(+)-PHNO (Wilson et al., 2005), one of the earliest in vivo studies utilized the tracer to image human volunteers (Willeit et al., 2006). These investigators scanned four healthy volunteers with [^{11}C]-(+)-PHNO at baseline and then again under several comparison conditions. Two subjects underwent retest scans with [^{11}C]-(+)-PHNO, two were scanned again with [^{11}C]-(+)-PHNO following 2 mg of haloperidol, and two were scanned with [^{11}C]raclopride at baseline. Ninety minutes of dynamic data were acquired for each scan, and the simplified reference tissue method (SRTM, Lammertsma and Hume, 1996) with the cerebellum as the reference tissue was applied to the regions of interest in the caudate, putamen, and globus pallidus to derive the outcome measure BP_{ND} (Binding Potential relative to the "Nondisplaceable" tracer pool in brain; Innis et al., 2007). It was observed that [^{11}C]-(+)-PHNO BP_{ND} in the globus pallidus was 33% larger than in the caudate or putamen, whereas [^{11}C] raclopride BP_{ND} in the globus pallidus was less than half the dorsal striatum values. It was also observed that, on average, 2 mg of haloperidol reduced [^{11}C]-(+)-PHNO BP_{ND} by approximately 30% compared to baseline in all the three regions.

The authors also commented on the slower uptake and washout in the globus pallidus compared to that in the dorsal striatum, and the fact that there was detectable specific binding in the substantia nigra/ventral tegmental area, though they did not quantify this. The ability of the D_2/D_3 antagonist haloperidol to block [^{11}C]-(+)-PHNO binding was evidence that the detected signal was due to specific binding to D_2-like receptors. The authors discussed possible reasons for the elevated binding in the globus pallidus, including a high fraction of D_2 receptors configured in the high affinity state for agonists as well as possible binding to D_3 receptors, but reached no conclusion.

At the same time, this group of authors also published a study using in vivo imaging with [^{11}C]-(+)-PHNO in anesthetized cats to examine a number of the pharmacological and pharmacokinetic properties of the tracer (Ginovart et al., 2006). Interestingly, one of the experiments performed used pretreatment with a highly selective D_3 antagonist SB-277011 (Stemp et al., 2000), but no effect of D_3 receptor blockade on tracer binding parameters was detected, because quantification was only performed in dorsal striatum where D_3 density is negligible.

Also, Narendran et al. performed an imaging study in anesthetized baboons to test the degree to which the high signal of [^{11}C]-(+)-PHNO in the globus pallidus was due to specific binding to D_3 receptors (Narendran et al., 2006). Animals were scanned with [^{11}C]-(+)-PHNO and [^{11}C]raclopride at baseline and following a single intravenous dose of 0.25 mg/kg of BP897, a D_3 partial agonist that is 70-fold selective for D_3 compared to D_2 in vitro (Pilla et al., 1999). As in the previous study from Willett et al., baseline BP_{ND} of [^{11}C]raclopride in the globus pallidus was lower than BP_{ND} in the dorsal striatum (1.48 ± 0.41 in the globus pallidus vs 2.56 ± 0.91 in the dorsal striatum in $n=4$ subjects, Table 1), whereas BP_{ND} of [^{11}C]-(+)-PHNO was higher in the globus pallidus than in the dorsal striatum (3.88 ± 1.15 vs 2.07 ± 0.43 in the same subjects, Table 1 and Figures 1 and 2).

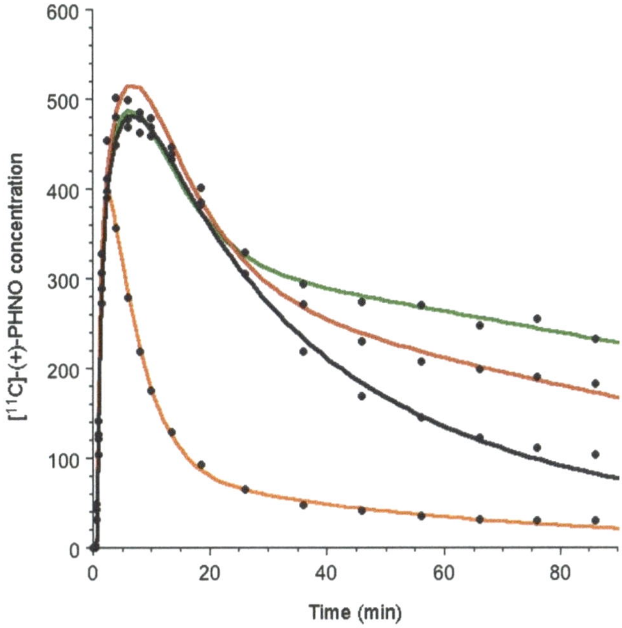

Figure 1 Time activity curves from one representative [^{11}C]-(+)-PHNO scan in an anesthetized baboon as presented in Narendran et al. (2006). Markers are measured data; continuous curves are model fits. Regions are the globus pallidus (green), ventral striatum (red), dorsal striatum (black), and cerebellum (orange). *Figure reproduced with permission from Narendran (2006).*

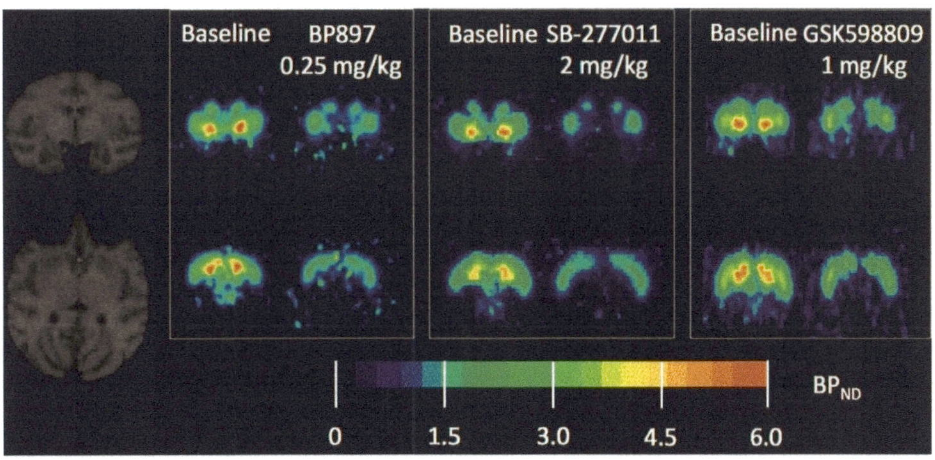

Figure 2 [^{11}C]-(+)-PHNO BP_{ND} maps in individual anesthetized baboons at baseline and following the administration of D_3-selective blocking compounds. The representative magnetic resonance imaging (MRI) images (left) show the levels of the coronal (top) and transverse (bottom) images, all of which pass through the globus pallidus and the dorsal striatum. In each case, binding in the globus pallidus is the highest during the baseline condition (red) but is nearly zero following the administration of the blocking compound. Dorsal striatum (green) is mostly unaffected by the blocking compound. The BP897 data were presented in Narendran et al. (2006); the SB-277,011 data were presented in Rabiner et al. (2009); the GSK598809 data have not been previously published. *Figure reproduced with permission from Narendran (2006) and Rabiner (2009).*

$[^{11}C]$-(+)-PHNO BP_{ND} was also quantifiable in the substantia nigra/ventral tegmental area and the thalamus, in contrast to $[^{11}C]$raclopride. Following BP897, large decreases in $[^{11}C]$-(+)-PHNO BP_{ND} and $[^{11}C]$raclopride BP_{ND} were observed in all measured brain regions except in the dorsal striatum (dorsal striatum decrease = 13% and 10% for $[^{11}C]$raclopride and $[^{11}C]$-(+)-PHNO respectively, Table 1). The $[^{11}C]$raclopride decrease in the globus pallidus was 30% on average. Given the nearly equal affinity of raclopride for D_2 and D_3 receptors, this would suggest that 30% of D_2-like receptors in the globus pallidus are D_3, provided that BP897 blocked nearly all D_3 and no D_2, an assumption that is a rough idealization at best. But the much larger binding decrease for $[^{11}C]$-(+)-PHNO in the globus pallidus (57%, Figure 2) provided strong evidence of the D_3-preferring nature of the tracer. Additionally, 90% blockade of $[^{11}C]$-(+)-PHNO binding was achieved in the substantia nigra/ventral tegmental area, suggesting that nearly all $[^{11}C]$-(+)-PHNO BP_{ND} in that region was due to binding to D_3.

The estimated proportions of D_3 and D_2 receptors in the globus pallidus based on $[^{11}C]$raclopride binding (30 and 70%, respectively) are in reasonably good agreement with those of a binding study performed around the same time by Seeman et al. using tissue homogenates from the globus pallidus of human postmortem brains, $[^{3}H]$raclopride, $[^{3}H]$(+)-PHNO, and the D_3-selective compound FAUC 365 (Seeman et al., 2006), who estimated the D_3 fraction in the globus pallidus to be 43 ± 11%. The study of Narendran et al. provided the first direct evidence that, in vivo, $[^{11}C]$-(+)-PHNO is D_3-preferring. But the characterization was not complete. A single dose of BP897 was used, and this did not allow for the estimation of the relative selectivity of BP897 for D_3 and D_2 in vivo, which might have been different than in the in vitro setting, and therefore, the portion of blocking associated with each receptor type could not be determined precisely. Also, as BP897 is a partial agonist, it may have induced unknown pharmacological effects that influenced the results.

A confirmatory and more comprehensive study was performed by Rabiner et al., also in anesthetized baboons (Rabiner et al., 2009). The subjects were scanned at baseline with $[^{11}C]$-(+)-PHNO and then again during a constant infusion of SB-277011, a compound shown to have a 100-fold selectivity for D_3 compared to D_2 in CHO cells transfected with human receptors (Reavill et al., 2000). Three doses of SB-277011 were used (0.5, 2, and 5 mg/kg) leading to a range of receptor occupancies, and a nonlinear regression model was formulated to assess both the occupancy by SB-277011 of D_2 and D_3 receptors, as well as the relative fraction of $[^{11}C]$-(+)-PHNO binding potential associated with the different receptor types. The model divides $[^{11}C]$-(+)-PHNO BP_{ND} into the portion due to D_3 binding (the D_3 fraction, or f_{D_3}) and the portion due to D_2 binding ($1 - f_{D_3}$). This can also be represented as

$$BP_{ND}\,(\text{baseline}) = f_{ND}\left(\frac{[D_3]}{K_D\,(D_3)} + \frac{[D_2]}{K_D\,(D_2)}\right) \quad (1)$$

where f_{ND} is the free fraction of free plus nonspecifically bound [^{11}C]-(+)-PHNO in brain tissue, $[D_3]$ and $[D_2]$ are the concentrations of those receptors available to bind to the tracer and $K_D(D_3)$, $K_D(D_2)$ are the equilibrium dissociation constants of the tracer for those receptors, with the first term on the right equal to f_{D_3} times the observed BP_{ND} and the second term equal to $1 - f_{D_3}$ times the observed BP_{ND}. An assumption of this model is additivity, that is, that the observed binding potential, which is measured using a kinetic model with a single specific binding site, is equal to the sum of the binding potentials associated with the separate receptor types. If a blocking compound with an affinity for both D_2 and D_3 is present, then

$$BP_{ND}\,(\text{with competitor}) = f_{ND}\left(\frac{[D_3]}{K_D(D_3)\left(1+\frac{[F]}{K_i(D_3)}\right)} + \frac{[D_2]}{K_D(D_2)\left(1+\frac{[F]}{K_i(D_2)}\right)}\right) \tag{2}$$

where [F] is the free concentration of the competitor and $K_i(D_3)$, $K_i(D_2)$ are its inhibition constants for D_3 and D_2 receptors, respectively. This can be written equivalently as

$$BP_{ND}\,(\text{with competitor}) = f_{ND}\left(\frac{[D_3]}{K_D(D_3)}\,(1-\text{occupancy at } D_3) + \frac{[D_2]}{K_D(D_2)}\,(1-\text{occupancy at } D_2)\right) \tag{3}$$

The fractional change in BP_{ND} between conditions is then given by

$$\frac{BP_{ND}\,(\text{baseline}) - BP_{ND}\,(\text{with competitor})}{BP_{ND}\,(\text{baseline})} = f_{D3} \times (\text{occupancy at } D_3) + \left(1-f_{D_3}\right) \times (\text{occupancy at } D_2) \tag{4}$$

Here, f_{D_3} will vary across brain regions but be independent of competitor concentration, whereas occupancy by the competitor at D_2 and D_3 will be independent of brain region, but vary with the blocker concentration. By simultaneously regressing the data across all scans for all doses of SB-277011 and all brain regions where [^{11}C]-(+)-PHNO was quantifiable onto Eqn 4), Rabiner et al. were able to form estimates of the mean f_{D_3} in each region and the mean percentages of D_2 and D_3 receptors occupied by SB-277011 at each dose across subjects. This additionally allowed the in vivo selectivity of SB-277011 to be estimated. The results were that f_{D_3} followed the following rank order: substantia nigra/ventral tegmental area (95%) > thalamus (88%) > globus pallidus (72%) >VST (49%) > caudate (23%) > putamen (8%). The estimated inhibition constants of SB-277011 for [^{11}C]-(+)-PHNO binding at D_3 and D_2, as functions of injected doses, were 0.13 and 21.16 mg/kg, corresponding to a selectivity ratio of 162. This was somewhat higher, but within the order of magnitude agreement of the in vitro estimate of 100. Qualitatively, the results were in strong agreement with those of the 2006 study of Narendran et al., in that in anesthetized baboons, compounds known a priori to be D_3 selective (BP897 or SB-277011) could be administered at a dose that inhibited nearly all of [^{11}C]-(+)-PHNO binding in the substantia nigra/ventral tegmental area and a sizable fraction of binding in the globus pallidus, VST, and thalamus, while leaving the dorsal striatum binding nearly unchanged (Figure 2).

The 2009 Rabiner study provided additional evidence that PHNO is D_3-preferring in vivo from a mutant mouse model in which autoradiography was performed on brains of animals that had been injected with [^{3}H]-(+)-PHNO and either vehicle or SB-277011 prior to sacrifice. In vehicle-treated wild-type animals, high binding of [^{3}H]-(+)-PHNO was detected in the substantia nigra, ventral pallidum, habenula, cerebellar lobes X and IX, VST, and caudate-putamen; D_2-knockout mice had a similar binding in all regions except in the caudate putamen, where it was undetectable, whereas D_3-knockout mice had moderate binding in the caudate putamen but negligible binding in all extrastriatal areas. In animals that were pretreated with SB-277011, wild types showed a nearly complete blockade in all extrastriatal regions but only approximately 30% blockade in the caudate putamen. Binding was blocked in all regions in D_2-knockout mice where it had previously been detected in vehicle-treated animals, and binding was unaffected in the caudate putamen of D_3-knockouts.

Evidence that the observed pattern in nonhuman primates is reflective of the behavior of [^{11}C]-(+)-PHNO in humans was provided subsequently in a study from Searle et al. (Searle et al., 2010), in which subjects were scanned with [^{11}C]-(+)-PHNO at baseline and following the administration of GSK598809, a D_3-selective antagonist under development and approved for administration to humans for research use (120-fold selective for D_3 over D_2 in CHO cells expressing human D_2 and D_3 receptors, unpublished data). Nineteen healthy human volunteers were scanned with [^{11}C]-(+)-PHNO at baseline (Figure 3) and following a range of orally administered doses of GSK598809 (5–175 mg).

The model developed in the Rabiner et al. study was used to uncouple the D_2 and D_3 components of [^{11}C]-(+)-PHNO BP_{ND}. In the model, the binding of GSK598809 was constrained to a one-site fit (D_3 only), as this was more statistically parsimonious than a two-site fit and the inhibition constant of GSK598809 for [^{11}C]-(+)-PHNO D_3 binding was estimated in terms of the plasma concentration of GSK598809. The f_{D_3} results were similar to those observed in anesthetized baboons: substantia nigra/ventral tegmental area (100%) > globus pallidus (67%) > thalamus (46%) >VST (26%) > caudate (1%) > putamen (0%). A subsequent reanalysis of these data in which brain regions were more carefully parsed also showed that there was specific binding in the hypothalamus due almost exclusively to D_3 (Tziortzi et al., 2011).

Most recently, a study was performed in anesthetized rhesus monkeys in which the in vivo saturation binding technique was used to estimate the in vivo affinity of the

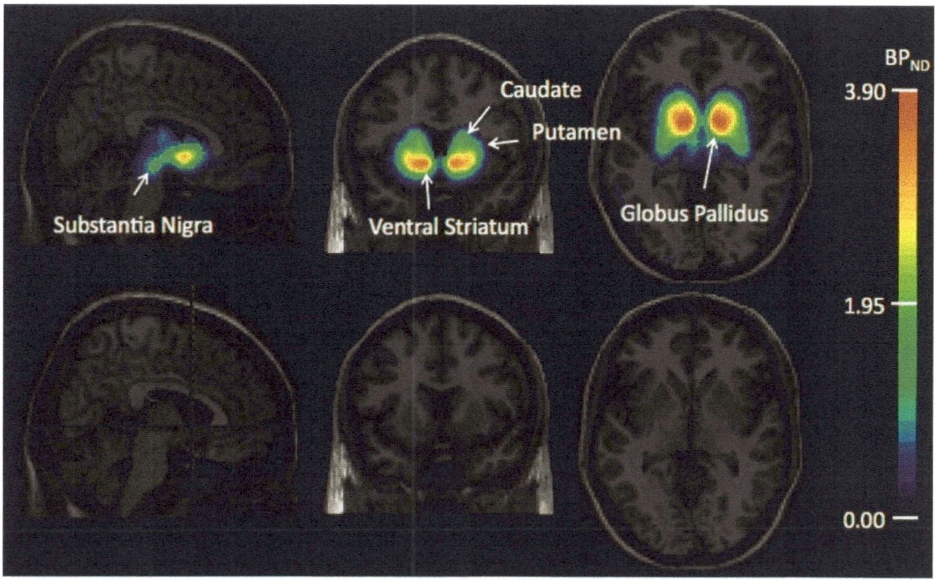

Figure 3 Average [^{11}C]-(+)-PHNO BP_{ND} map across $n = 19$ healthy human volunteers (top Row, superposed on MRI). The data are from the study presented in Searle et al. (2010) and reanalyzed in Tziortzi et al. (2011). Parametric BP_{ND} maps were computed for each subject using the SRTM algorithm and then spatially normalized into a common stereotaxic space. Lines in the sagittal MRI image (bottom left) show the levels of the coronal (center) and transverse (right) images. *Figure reproduced with permission from Searle (2010).*

tracer (Gallezot et al., 2011). Subjects were scanned using a bolus plus constant infusion imaging design. Mass of the tracer ranged from 0.36 to 5.61 μg/kg. The same set of brain regions was analyzed as above. Data were fitted to Eqn 2), except that in this case the "competitor" was (+)-PHNO itself. Equilibrium dissociation constants were assumed to be the same across brain regions, whereas the receptor availability of both receptor types was allowed to vary across regions within subject. The free concentration [F] was estimated from the cerebellum concentration of the tracer during the equilibrium phase, based on the assumed negligible receptor concentration in that region. Data were analyzed both by kinetic analysis (SRTM) and by concentration ratios measured during the equilibrium phase. Depending on which method of analysis was used, [^{11}C]-(+)-PHNO was estimated to be 25- to 48-fold more selective for D_3 than for D_2 and the regional fractions f_{D_3} were highly in accord with that in the previous studies (Figure 4).

Thus, based on these studies, several general conclusions can be drawn. First, in vivo, [^{11}C]-(+)-PHNO is D_3-preferring. Second, there is a considerable homology between primate species in terms of the decomposition of [^{11}C]-(+)-PHNO binding across brain regions. Figure 4 compares the f_{D_3} results from the studies described here as well as additional unpublished data acquired in rhesus monkeys. The similarities are clear. It is also evident that nearly all [^{11}C]-(+)-PHNO specific binding detected in the substantia

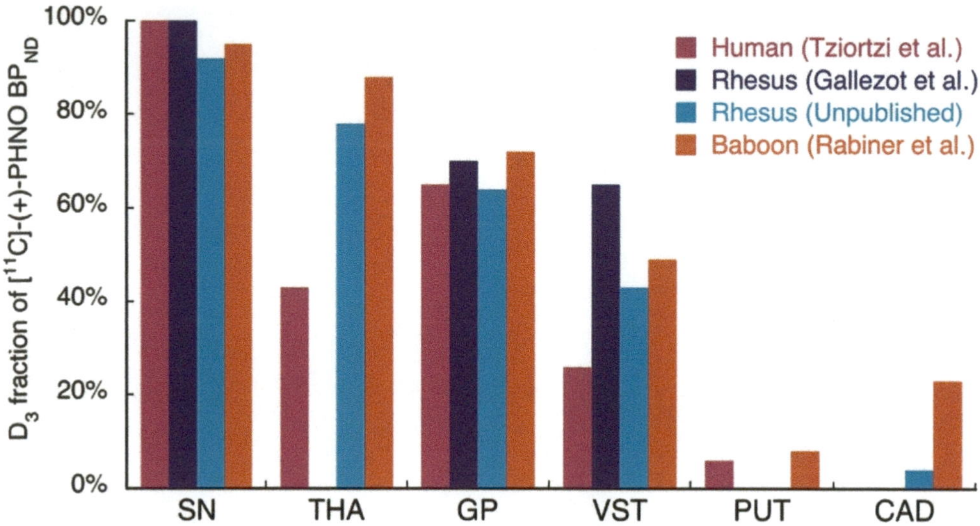

Figure 4 Homology. Graphical depiction of the D_3-associated fraction of [^{11}C]-(+)-PHNO BP_{ND} in humans, rhesus monkeys, and baboons. The human data are the same as those in the Searle et al. study (Searle et al., 2010) following a reanalysis in which the region boundaries were refined (Tziortzi et al., 2011). The rhesus monkey data are from one published study (Gallezot et al., 2011) and an independent unpublished study in $n=3$ subjects. The baboon data are from the study of Rabiner et al. (Rabiner et al., 2009). Data in the thalamus were not available for the rhesus study of Gallezot et al. Very high similarity across species can be seen in the substantia nigra/ventral tegmental area, globus pallidus, and dorsal striatum (caudate and putamen). The thalamus and VST are more heterogeneous; possible reasons are species differences in receptor expression or region definitions, or less statistical precision in these regions due to the small size (ventral striatum) or low signal (thalamus). Abbreviations: SN = substantia nigra/ventral tegmental area, THA = thalamus, GP = globus pallidus, VST = ventral striatum, PUT = putamen, CAD = caudate. *Figure reproduced with permission from Searle et al. (2010), Tziortzi et al. (2011), Gallezot et al. (2011), and Rabiner et al. (2009).*

nigra/ventral tegmental area is to D_3 receptors, whereas nearly all binding in the putamen is to D_2 receptors.

Other regions, including the globus pallidus, have more mixed signals. The putamen signal yields a robust measure of D_2 binding and provides a setting for the investigation of the relationship between agonist tracers and D_2 receptor affinity states. The signal from the dopamine midbrain nuclei, however, while due almost exclusively to D_3 binding, does not provide an extremely precise measure, due to the overall low specific binding and the small size of that brain region. Thus, an accurate estimation of D_3 vs D_2 occupancy or selectivity of an exogenously administered drug is likely to require the application of a comprehensive model that utilizes all the available information, as in the studies described above.

One methodological issue that warrants comment relates to the affinity of [^{11}C]-(+)-PHNO for D_3 receptors. Gallezot et al. estimated the affinity, in rhesus monkeys, as the parameter K_D/f_{ND} that was found to be 0.23–0.56 nM for D_3 (Gallezot et al., 2011).

In a separate set of 19 scans acquired in rhesus monkeys, average f_{ND} (computed under the assumption of passive diffusion as fp/V_{ND}, where fp was measured, fp = 0.±0.06, and cerebellum VT was taken as an estimate of V_{ND}, V_{ND} = 6.10±0.97) was f_{ND} = 0.098 (unpublished data). This gives a rough estimate of K_D in the range of 0.02–0.05 nM, suggesting that [^{11}C]-(+)-PHNO is a very high affinity tracer at the D_3 receptor.

This implies both that it is challenging to synthesize [^{11}C]-(+)-PHNO with a high enough specific activity to maintain tracer dose conditions (i.e., that peak D_3 receptor occupancy by the tracer does not exceed 5%) and that mass carryover between back to back scans, where unlabeled (+)-PHNO from the first scan leads to quantifiable receptor occupancy in the subsequent scan, is a possibility. In a set of test-retest scans performed in three anesthetized baboons where the retest scan began 172±36 min after the start of the test scan, a substantial decrease of BP_{ND} was detected in D_3-rich regions, but not in dorsal striatum, consistent with a mass-carryover effect (Girgis et al., 2011).

A final comment on methodology pertains to the suitability of the cerebellum as a reference tissue. Small decreases in cerebellar VT following high doses of SB-277011 in baboons (Rabiner et al., 2009), in cerebellar standard uptake value (SUV) following GSK598809 in humans (Searle et al., 2010), as well preliminary data in humans showing decreased cerebellar VT following the administration of aripiprazole (Shotbolt et al., 2011) all point to the existence of a small but quantifiable amount of [^{11}C]-(+)-PHNO specific binding in the cerebellum. This suggests that a corrected BP_{ND} as presented in Gunn et al. (2011) might be appropriate for this tracer. Thus, [^{11}C]-(+)-PHNO imaging presents many challenges in radiosynthesis, pharmacokinetic quantification, and experimental design.

2.2. D_3 RECEPTOR IMAGING IN NEUROPSYCHIATRIC ILLNESS

2.2.1. Schizophrenia

Since the discovery of D_3, several lines of evidence have suggested that D_3 might play some role in schizophrenia. The most obvious of these is that D_3 is in the D_2-like family and it has long been known that antipsychotic drug action involves the blockade of D_2-like receptors (Seeman, 1980). Most antipsychotic drugs in current use bind to both D_2 and D_3 receptors, though they tend to have a twofold to fivefold higher affinity for D_2 (Levant, 1997; Schotte et al., 1993). Analysis based on this reported selectivity suggests that, at generally accepted therapeutic dose levels and associated D_2 occupancy, moderate binding to D_3 receptors should occur as well (Girgis et al., 2011 and Table 2), implying a role for D_3 in the action of antipsychotics.

As described above, the study of Gurevich et al. suggested that D_3 receptors are upregulated in schizophrenia, but that antipsychotic treatment normalizes their expression. There have been some intriguing genetic results as well. A polymorphism involving a glycine to serine substitution on exon 1 of the D_3 gene was discovered for which the homozygosity of either allele was slightly more prevalent in patients with schizophrenia

Table 2 Average D_2:D_3 selectivity for some commonly used antipsychotic drugs and predicted occupancy at D_3, given concentrations associated with conventionally accepted values for D_2 occupancy leading to therapeutic efficacy

Antipsychotic	# Studies	D_2:D_3 Selectivity ± SD	D_2 Occupancy	Predicted D_3 Occupancy
Clozapine	9	2.82 ± 2.01	60%	35%
Haloperidol	11	3.03 ± 4.42	75%	50%
Risperidone	3	4.9 ± 3.71	75%	38%
Olanzapine	1	3.18	75%	49%
Amisulpride lo	3	2.29 ± 0.58	67%	47%
Amisulpride hi	3	2.29 ± 0.58	80%	64%
Raclopride	3	1.16 ± 0.53	*	*

*indicates that no occupancy is being estimated for raclopride.

than in control subjects (Crocq et al., 1992). Several meta-analyses of the large number of replicating studies that followed confirmed a small but significant increase of homozygosity in schizophrenia (Dubertret et al., 1998; Jonsson et al., 2004; Williams et al., 1998).

The functional significance, if any, of this polymorphism has not been clear, though there is some evidence that patients who are glycine carriers are more susceptible to tardive dyskinesia than to serine homozygotes (Lerer et al., 2002). Animal models have also been suggestive. For example, it was shown that the locomotor hyperactivity induced in mice by the *N*-methyl-D-aspartate (NMDA) receptor antagonist MK-801, a psychotomimetic, was not present in D_3 knockout mice (Leriche et al., 2003). In summary, while there was no direct evidence implicating a role for D_3 in schizophrenia, several lines of inference existed.

The first study to use [^{11}C]-(+)-PHNO imaging in schizophrenia came from Graff-Guerrero et al. (Graff-Guerrero et al., 2009b). Thirteen patients with schizophrenia (age = 26 ± 6 years) who were actively psychotic (mean positive and negative symptom scale (PANSS) total scores = 75, mean PANSS positive score = 21) and had been drug-free for at least 2 weeks were scanned along with 13 healthy comparison subjects. SRTM analysis was applied to the same set of brain regions as in the earlier studies of Searle et al., Rabiner et al., and Narendran et al. No significant differences in BP_{ND} were observed in any region; in fact, the group means were nearly identical in every region. In particular, no differences were observed in the substantia nigra where BP_{ND} is due almost entirely to the D_3 binding, nor in the VST or the globus pallidus where Gurevich et al. had observed the upregulation of D_3 in postmortem tissue from older drug free patients. This was strongly suggestive of no upregulation in vivo, though not conclusive due to the mixed D_2/D_3 nature of the [^{11}C]-(+)-PHNO signal in the relevant regions.

Other factors that may have contributed to the difference in observations are the age differences between the subjects in the Graff-Guerrero and Gurevich studies, as well as the possibility that elevated baseline dopamine at D_2-like receptors in patients with schizophrenia, which has been observed in several previous studies (Abi-Dargham et al.,

2000; Abi-Dargham et al., 2009; Kegeles et al., 2010), may have masked underlying differences. To date, no study attempting to replicate this result has been published. In 2009, Graff-Guerrero et al. published a study looking at the effects of antipsychotic drugs on [^{11}C]-(+)-PHNO binding (Graff-Guerrero et al., 2009a). Twenty-three patients with schizophrenia (age 38 ± 7 years) who were on stable therapeutic doses of clozapine, olanzapine, or risperidone were scanned with [^{11}C]-(+)-PHNO and [^{11}C]raclopride, along with 23 matched healthy control volunteers. In the dorsal striatum, BP_{ND} was decreased compared to controls by a similar amount for all three drugs, though slightly less for clozapine. The averages were 53% in the caudate and 42% in the putamen for [^{11}C]-(+)-PHNO, 71% in the caudate and 69% in the putamen for [^{11}C]raclopride. These results were consistent with a picture of high occupancy by drugs at D_2 receptors, and possibly less occupancy at D_3, as embodied in the difference between the two tracers. However, in the VST, [^{11}C]-(+)-PHNO binding was only 17% lower than controls and in the globus pallidus, a paradoxical increase of 71% was observed. Inspection of the scatter plot of the binding data (Figure 2, Graff-Guerrero et al., 2009a) suggests that this large mean increase may have been driven by a single subject, but even without that subject it is evident that there was no decrease or some increase in binding in the globus pallidus in patients compared to that in controls. [^{11}C]raclopride binding was decreased compared to that in controls to a similar extent as in other regions (72% in the VST, 59% in the globus pallidus).

Previous in vitro assays with these drugs (Table 2) suggested that measurable occupancy should have occurred at D_3, given the observed D_2 occupancy in the dorsal striatum, whereas the results of this study suggested that either the drugs were behaving differently in vivo than predicted in vitro, or that D_3, but not D_2 receptor upregulation had occurred, or a combination of these. A methodological aside is that the estimation of occupancy through the relative change in BP_{ND}, as embodied in Eqn (4), requires the implicit assumption that receptor pool available to both the tracer and the competitor is the same during the baseline scan and the scan with competitor, so that the receptor concentration is normalized out of the equation of relative change. When this is not the case, as in upregulation, the competitor may be bound to some fraction of the receptors, but the increased total pool of available receptors can also lead to an increase in the concentration of unoccupied receptors, which would be recorded as an increase in BP_{ND}.

To test the ability of their method to detect D_3 binding of a competitor, Graff-Guerrero et al. scanned three additional patients on stable doses of risperidone following the administration of 0.5 mg of pramipexole, a D_3-preferring agonist at D_2-like receptors (Levant et al., 1999; Seeman et al., 2005). BP_{ND} in the globus pallidus was reduced by 45%, and while quantitative results were not presented for the substantia nigra, images suggested that binding was reduced in that region as well. Because this study suggested the possibility that antipsychotic drugs were not binding to D_3 receptors in vivo,

Girgis et al. performed a study with $[^{11}C]$-(+)-PHNO in anesthetized baboons to test the binding of haloperidol and clozapine following acute dosing. Four baboons were scanned at baseline and following intravenous doses of drugs designed to lead to D_2 occupancy comparable to therapeutic levels (0.0109 mg/kg for haloperidol, 0.553 mg/kg for clozapine). Using a regression model similar to those of the Rabiner and Searle studies, the relative selectivity of each drug for D_2 and D_3 was estimated. Both drugs were found to bind to D_3 receptors following single intravenous doses. Haloperidol was estimated to be 2.38-fold more selective for D_2 than for D_3 (the average over 11 in vitro studies was 3.03, Table 2) and clozapine was estimated to be 5.25-fold more selective for D_2 (the average over nine in vitro studies was 2.82, Table 2). These results suggested that antipsychotic drugs did in fact bind to D_3 receptors in vivo, suggesting in turn that Graff-Guerrero et al. had an observed D_3 upregulation following stable antipsychotic treatment.

A limitation of the Graff-Guerrero study was the use of healthy volunteers for comparison rather than a within-subject design, though the earlier study of baseline binding from these authors suggested that $[^{11}C]$-(+)-PHNO BP_{ND} is the same in schizophrenia as in healthy volunteers. To address this limitation, researchers from the same laboratory conducted another study in patients with schizophrenia utilizing a within-subject design (Mizrahi et al., 2011). Eight patients who were antipsychotic drug-naive were scanned at baseline and then following 2.5 weeks of antipsychotic drug therapy. Seven subjects were administered risperidone (2.5 ± 0.1 mg/day), and one subject received 10 mg of olanzapine per day. Following treatment, $[^{11}C]$-(+)-PHNO BP_{ND} was reduced by approximately 40% in the dorsal striatum, but increased in the globus pallidus and substantia nigra by approximately 50%.

Thus, the in vivo imaging studies performed so far suggest a scenario in which commonly used antipsychotic drugs do bind to D_3 receptors, with some preference for D_2 as predicted from in vitro assays, but that there is then a relatively rapid (on the order of weeks or less) upregulation of D_3. Clearly, all these results require expansion and replication, but two, nearly opposite clinical interpretations could be drawn based on the currently available data—either that D_3 receptor blockade is not important for antipsychotic efficacy, or that D_3 represents an underexploited target for antipsychotic therapy that awaits the development of more selective agents.

2.2.2. Parkinson's Disease

Idiopathic Parkinson's disease, an age-related neurodegenerative disease, is characterized by the loss of dopaminergic neurons in the substantia nigra pars compacta, leading, in particular, to the progressive loss of dopaminergic transmission in the striatum (Hornykiewicz, 1966). There are several lines of evidence that D_3 receptors may play a special role. In terms of receptor expression in humans, some (Piggott et al., 1999; Ryoo and Joyce, 1994), but not all (Hurley et al., 1996), postmortem autoradiography studies have shown both increased D_2 density, presumably as a homeostatic response to reduced dopaminergic transmission, and paradoxically decreased D_3 receptor density in Parkinson's patients compared to matched controls, an effect which has also been detected in animal models of Parkinson's disease in rodents (Levesque et al., 1995) and primates (Morissette et al., 1998).

The traditional therapy has been to augment dopamine production with levodopa, but other drugs, alone or in combination with levodopa, including dopamine agonists such as the D_3-preferring agonist pramipexole or the mixed D_2/D_3 agonist ropinirole are used as well (Diamond and Jankovic, 2006); in fact (+)-PHNO itself was originally tested as a Parkinson's disease therapy (Ahlskog et al., 1991; Dykstra et al., 1985; Jones et al., 1984). There is evidence from several animal models of Parkinson's disease that D3 agonists such as pramipexole may be neuroprotective or even neuroregenerative (reviewed in Joyce and Millan, 2007). Data from an [^{18}F] 3,4-dihydroxyphenylalanine (DOPA) PET study in Parkinson's patients supports a neuroprotective role for ropinirole (Whone et al., 2003), and an [^{123}I]b-CIT SPECT study in Parkinson's patients supports a neuroprotective role for pramipexole (Parkinson Study Group, 2002), though there is also some indication that these medications may contribute to the development of undesirable consequences of dopaminergic augmentation therapy such as pathological gambling (Lader, 2008).

Thus, a reasonably large body of evidence exists to warrant interest in D_3 imaging in Parkinson's disease. To date, one such study has been performed. Boileau et al. performed a PET study in which 10 drug-naive, early stage Parkinson's patients and 10 demographically matched control subjects were imaged with [^{11}C]-(+)-PHNO and [^{11}C] raclopride under baseline conditions (Boileau et al., 2009). It was observed that BP_{ND} in the putamen was elevated by 25% in Parkinson's patients with both tracers, and also that this effect was more pronounced on the side of the brain contralateral to symptoms. In the globus pallidus, however, whereas [^{11}C]raclopride binding was not significantly different between groups, [^{11}C]-(+)-PHNO BP_{ND} was reduced by 42% compared with that in controls, and to a lesser extent (11%) in the VST. An interesting additional result was that when the ratio of [^{11}C]-(+)-PHNO BP_{ND} to [^{11}C]raclopride BP_{ND} in the globus pallidus of patients was plotted against performance on a motor skill task, a positive correlation was observed.

These results are consistent with those of the postmortem and animal studies that found elevated D_2 and decreased D_3 density. The former is demonstrated by the increase in both [^{11}C]raclopride BP_{ND} and [^{11}C]-(+)-PHNO BP_{ND} in the putamen (both [^{11}C] raclopride BP_{ND} and [^{11}C]-(+)-PHNO BP_{ND} in the putamen are almost exclusively D_2 binding), and the latter demonstrated by a decreased [^{11}C]-(+)-PHNO, but not [^{11}C] raclopride binding in the globus pallidus. This observation can be interpreted in terms of the fact that while the [^{11}C]-(+)-PHNO signal in the globus pallidus is dominated by D_3 binding, D_2 density is still much greater than the D_3 density in this structure so that [^{11}C] raclopride, which binds with similar affinities to both receptor types, is less sensitive to moderate D_3 decreases. The intriguing relationship observed between the binding ratios of the tracers and preservation of motor skills suggests that symptoms in this domain may be related to the balance between D_2 and D_3 signaling, an area that warrants further study.

2.2.3. Substance Abuse

In the area of PET imaging studies examining D_2-like receptors in cohorts of substance abusing or dependent subjects, a relatively large body of work exists (Table 3; also see chapter by Volkow et al., in this volume). The majority of these have been performed with [^{11}C]raclopride; several have also used [^{18}F]fallypride or the SPECT tracer [^{123}I]IBZM. While there were some exceptions, the vast majority of these found a decreased receptor availability and blunted responses to dopaminergic stimulants. These studies all used tracers that have similar affinities for D_2 and D_3, and hence cannot inform the question as to the separate contributions of the receptor subtypes to the phenotype.

There has also been extensive literature in animal models of substance abuse, including studies showing that D_3 selective antagonists, such as SB-277011 or NGB 2904 (Yuan et al., 1998), disrupt or attenuate behaviors thought to be models of dependence (Heidbreder and Newman, 2010). Similar effects are seen with the D_3

Table 3 Previous Studies with D_2-Like Receptor Tracers in Substance-Dependent Populations

Study	Drug	Tracer	Baseline BP_{ND}	Pharmacological Challenge
Hietala et al. (1994)	ETOH	[^{11}C]raclopride	Down	NA
Volkow et al. (1996)	ETOH	[^{11}C]raclopride	Down	NA
Wang et al. (1997)	Opiates	[^{11}C]raclopride	Down	NA
Volkow et al. (1997)	Cocaine	[^{11}C]raclopride	Down	MP blunted
Volkow et al. (2001)	METH	[^{11}C]raclopride	Down	NA
Martinez et al. (2005)	ETOH	[^{11}C]raclopride	Down	AMPH blunted VST only
Martinez et al. (2007)	Cocaine	[^{11}C]raclopride	Down	AMPH blunted
Volkow et al. (2007)	ETOH	[^{11}C]raclopride	Down VST only	MP
Zijlstra et al. (2008)	Heroin	[^{123}I]IBZM	Down in caudate	Increased in putamen
Fehr et al. (2008)	Nicotine	[^{18}F]fallypride	Down	NA
Lee et al. (2009)	METH	[^{18}F]fallypride	Down	NA
Martinez et al. (2012)	Heroin	[^{11}C]raclopride	Down	MP blunted
Urban et al. (2012)	Cannabis	[^{11}C]raclopride	Normal	AMPH normal

D_2-like imaging studies of drug-dependent subjects. ETOH = alcohol, Opiates = heroin and/or methadone, METH = methamphetamine, NA = not applicable (no stimulant challenge in the study), MP = methylphenidate, AMPH = D-amphetamine.

partial agonist BP897 (Table 2, Sokoloff et al., 2006). There is some evidence, again, from animal models, that chronic exposure to cocaine (Le Foll et al., 2002) or nicotine (Le Foll et al., 2003) induces the upregulation of D_3, but not that of the D_2 or D_1 receptors.

One study has also shown that indirect dopaminergic agonism (with levodopa administration) in conjunction with D_1 receptor stimulation induces ectopic expression of D_3 receptors in the caudate putamen of 6-OHDA-treated rats (Bordet et al., 1997). Taken together, these studies suggest that D_3 may be upregulated in stimulant abusers even in the presence of global blunting of D_2-like receptors. To date, only one $[^{11}C]$-(+)-PHNO study with substance-dependent subjects has been published (however, there have been several recent conference abstracts utilizing $[^{11}C]$-(+)-PHNO to study alcohol- or cocaine-dependent subjects, suggesting more peer reviewed studies may arrive presently). Boileau et al. imaged 16 methamphetamine users and 16 matched control subjects with $[^{11}C]$-(+)-PHNO under baseline conditions. They did in fact observe significantly higher BP_{ND} in methamphetamine users compared to controls in the substantia nigra (46%) and globus pallidus (9–11%), but observed no significant differences in the dorsal striatum, a result that is reasonably consistent with predictions from the animal literature, and a scenario in which D_3 upregulation is detected in the normally D_3-rich regions, but D_2 blunting in the dorsal striatum, observed in previous studies, is not detected due to offsetting effects of increased D_3 in those structures as well. A within-subject dual tracer study (e.g., $[^{11}C]$-(+)-PHNO combined with $[^{11}C]$ raclopride) in this population could provide more definitive evidence for the plausibility of this explanation.

3. CONCLUSIONS

Despite a large body of suggestive studies in animal models and postmortem data, the function of the dopamine D_3 receptor in healthy subjects and its role in the pathophysiology of psychiatric illness is still not well understood. Until recently, it was not possible to use in vivo imaging methodology to study D_3, due to its structural similarity to D_2, the more extensive expression of D_2, and the propensity of most radiotracers to bind to both receptor types with similar affinity. The recent development of $[^{11}C]$-(+)-PHNO, which is strongly D_3-preferring, has allowed D_3 to become a target of inquiry in PET imaging, and several intriguing results from psychiatric populations have already been published. But D_3 imaging with $[^{11}C]$-(+)-PHNO has limitations, due to both the mixed D_2/D_3 signal and the high affinity of the tracer at D_3. Despite multiple attempts by radiochemists, no other D_3-selective or D_3-preferring radiotracer with suitable pharmacokinetic properties has been developed, so that in the foreseeable future, $[^{11}C]$-(+)-PHNO imaging will be the only route to elucidate the in vivo properties of this receptor.

REFERENCES

Abi-Dargham, A., Rodenhiser, J., Printz, D., Zea-Ponce, Y., Gil, R., Kegeles, L., Weiss, R., Cooper, T., Mann, J.J., Van Heertum, R., Gorman, J.M., Laruelle, M., 2000. Increased baseline occupancy of D_2 receptors by dopamine in schizophrenia. Proc. Natl. Acad. Sci. U.S.A. 97, 8104–8109.

Abi-Dargham, A., van de Giessen, E., Slifstein, M., Kegeles, L.S., Laruelle, M., 2009. Baseline and amphetamine-stimulated dopamine activity are related in drug-naive schizophrenic subjects. Biol. Psychiatry 65, 1091–1093.

Ahlskog, J.E., Muenter, M.D., Bailey, P.A., Miller, P.M., 1991. Parkinson's-disease monotherapy with controlled-release Mk-458 (Phno) – double-blind-study and comparison to carbidopa levodopa. Clin. Neuropharmacol. 14, 214–227.

Bennacef, I., Salinas, C.A., Bonasera, T.A., Gunn, R.N., Audrain, H., Jakobsen, S., Nabulsi, N., Weinzimmer, D., Carson, R.E., Huang, Y., Holmes, I., Micheli, F., Heidbreder, C., Gentile, G., Rossi, T., Laruelle, M., 2009. Dopamine D_3 receptor antagonists: the quest for a potentially selective PET ligand. Part 3: radiosynthesis and in vivo studies. Bioorg. Med. Chem. Lett. 19, 5056–5059.

Boileau, I., Guttman, M., Rusjan, P., Adams, J.R., Houle, S., Tong, J.C., Hornykiewicz, O., Furukawa, Y., Wilson, A.A., Kapur, S., Kish, S.J., 2009. Decreased binding of the D_3 dopamine receptor-preferring ligand [^{11}C]-(+)-PHNO in drug-naive Parkinson's disease. Brain 132, 1366–1375.

Bordet, R., Ridray, S., Carbon, S., Diaz, J., Sokoloff, P., Schwartz, J.C., 1997. Induction of dopamine D_3 receptor expression as a mechanism of behavioral sensitization to levodopa. Proc. Natl. Acad. Sci. U.S.A. 94, 3363–3367.

Brown, D., Luthra, S., Brady, F., Prenant, C., Dijkstra, D., Wikstrom, H., Brooks, D., 1997. Labelling of the D_2 agonist (1)-PHNO using [^{11}C]-propionyl chloride. In: Proceedings of the XIIth International Symposium on Radiopharmaceutical Chemistry. Uppsala, Sweden. Wiley, Chichester, UK, pp. 565–566.

Chu, W., Tu, Z., McElveen, E., Xu, J., Taylor, M., Luedtke, R.R., Mach, R.H., 2005. Synthesis and in vitro binding of *N*-phenyl piperazine analogs as potential dopamine D_3 receptor ligands. Bioorg. Med. Chem. 13, 77–87.

Crocq, M.A., Mant, R., Asherson, P., Williams, J., Hode, Y., Mayerova, A., Collier, D., Lannfelt, L., Sokoloff, P., Schwartz, J.C., et al., 1992. Association between schizophrenia and homozygosity at the dopamine D_3 receptor gene. J. Med. Genet. 29, 858–860.

Diamond, A., Jankovic, J., 2006. Treatment of advanced Parkinson's disease. Expert Rev. Neurother. 6, 1181–1197.

Dubertret, C., Gorwood, P., Ades, J., Feingold, J., Schwartz, J.C., Sokoloff, P., 1998. Meta-analysis of DRD3 gene and schizophrenia: ethnic heterogeneity and significant association in Caucasians. Am. J. Med. Genet. 81, 318–322.

Dykstra, D., Hazelhoff, B., Mulder, T.B.A., Devries, J.B., Wynberg, H., Horn, A.S., 1985. Synthesis and pharmacological activity of the hexahydro-4h-naphth[1,2b][1,4]-oxazines—a new series of potent dopamine receptor agonists. Eur. J. Med. Chem. 20, 247–250.

Fehr, C., Yakushev, I., Hohmann, N., Buchholz, H.G., Landvogt, C., Deckers, H., Eberhardt, A., Klager, M., Smolka, M.N., Scheurich, A., Dielentheis, T., Schmidt, L.G., Rosch, F., Bartenstein, P., Grunder, G., Schreckenberger, M., 2008. Association of low striatal dopamine D_2 receptor availability with nicotine dependence similar to that seen with other drugs of abuse. Am. J. Psychiatry 165, 507–514.

Finnema, S.J., Halldin, C., Bang-Andersen, B., Gulyas, B., Bundgaard, C., Wikstrom, H.V., Farde, L., 2009. Dopamine $D_{(2/3)}$ receptor occupancy of apomorphine in the nonhuman primate brain—a comparative PET study with [^{11}C]raclopride and [^{11}C]MNPA. Synapse 63, 378–389.

Foulon, C., Kung, M.P., Kung, H.F., 1993. Synthesis of (R, S)-2'-trans-7-hydroxy-2-[*N*-*n*-propyl-*N*-(3'-iodo-2'- propenyl)-amino]tetralin (trans-7-OH-PIPAT): a new D_3 dopamine receptor ligand. J. Med. Chem. 36, 1499–1500.

Freedman, S.B., Patel, S., Marwood, R., Emms, F., Seabrook, G.R., Knowles, M.R., Mcallister, G., 1994. Expression and pharmacological characterization of the human D-3 dopamine-receptor. J. Pharmacol. Exp. Ther. 268, 417–426.

Gallezot, J.D., Beaver, J.D., Gunn, R.N., Nabulsi, N., Weinzimmer, D., Singhal, T., Slifstein, M., Fowles, K., Ding, Y.S., Huang, Y., Laruelle, M., Carson, R.E., Rabiner, E.A., 2011. Affinity and selectivity of [^{11}C]-(+)-PHNO for the D_3 and D_2 receptors in the rhesus monkey brain in vivo. Synapse 24 Feb 2012. http://dx.doi.org/10.1002/syn.21535.

Ginovart, N., Galineau, L., Willeit, M., Mizrahi, R., Bloomfield, P.M., Seeman, P., Houle, S., Kapur, S., Wilson, A.A., 2006. Binding characteristics and sensitivity to endogenous dopamine of [^{11}C]-(+)-PHNO, a new agonist radiotracer for imaging the high-affinity state of D_2 receptors in vivo using positron emission tomography. J. Neurochem. 97, 1089–1103.
Girgis, R., Xu, X., Miyake, N., Easwaramoorthy, B., Gunn, R., Rabiner, E., Abi-Dargham, A., Slifstein, M., 2011. In vivo binding of antipsychotics to D_3 and D_2 receptors: a PET study in baboons with [^{11}C]-(+)-PHNO. Neuropsychopharmacology 36, 887–895.
Giros, B., Martres, M.P., Sokoloff, P., Schwartz, J.C., 1990. Gene cloning of human dopaminergic D_3 receptor and identification of its chromosome. C. R. Acad. Sci. 311, 501–508.
Graff-Guerrero, A., Mamo, D., Shammi, C.M., Mizrahi, R., Marcon, H., Barsoum, P., Rusjan, P., Houle, S., Wilson, A.A., Kapur, S., 2009a. The effect of antipsychotics on the high-affinity state of D_2 and D_3 receptors: a positron emission tomography study with [^{11}C]-(+)-PHNO. Arch. Gen. Psychiatry 66, 606–615.
Graff-Guerrero, A., Mizrahi, R., Agid, O., Marcon, H., Penny, B., Rusjan, P., Wilson, A.A., Zipursky, R.B., Kapur, S., 2009b. The dopamine D_2 receptors in high-affinity state and D_3 receptors in schizophrenia: a clinical [^{11}C]-(+)-PHNO PET study. Neuropsychopharmacology 34, 1078–1086.
Group, P.S., 2002. Dopamine transporter brain imaging to assess the effects of pramipexole vs levodopa on parkinson disease progression. JAMA 287, 1653–1661.
Gunn, R.N., Murthy, V., Catafau, A.M., Searle, G., Bullich, S., Slifstein, M., Ouellet, D., Zamuner, S., Herance, R., Salinas, C., Pardo-Lozano, R., Rabiner, E.A., Farre, M., Laruelle, M., 2011. Translational characterization of [^{11}C] GSK931145, a PET ligand for the glycine transporter type 1. Synapse 65, 1319–1332.
Gurevich, E.V., Bordelon, Y., Shapiro, R.M., Arnold, S.E., Gur, R.E., Joyce, J.N., 1997. Mesolimbic dopamine D_3 receptors and use of antipsychotics in patients with schizophrenia—a postmortem study. Arch. Gen. Psychiatry 54, 225–232.
Heidbreder, C.A., Newman, A.H., 2010. Current perspectives on selective dopamine D_3 receptor antagonists as pharmacotherapeutics for addictions and related disorders. Ann. N.Y. Acad. Sci. 1187, 4–34.
Hietala, J., West, C., Syvälahti, E., Nagren, K., Lehikoinene, P., Sonninen, P., Ruotsalainen, U., 1994. Striatal D_2 dopamine receptor binding characteristics in vivo in patients with alcohol dependence. Psychopharmacology 116, 285–290.
Hocke, C., Maschauer, S., Hubner, H., Lober, S., Utz, W., Kuwert, T., Gmeiner, P., Prante, O., 2010. A series of ^{18}F-labelled pyridinylphenyl amides as subtype-selective radioligands for the dopamine D_3 receptor. Chem. Med. Chem. 5, 941–948.
Hofling, S.B., Maschauer, S., Hubner, H., Gmeiner, P., Wester, H.J., Prante, O., Heinrich, M.R., 2010. Synthesis, biological evaluation and radiolabelling by [^{18}F]-fluoroarylation of a dopamine D_3-selective ligand as prospective imaging probe for PET. Bioorg. Med. Chem. Lett. 20, 6933–6937.
Hornykiewicz, O., 1966. Dopamine (3-hydroxytyramine) and brain function. Pharmacol. Rev. 18, 925–964.
Hurley, M.J., Stubbs, C.M., Jenner, P., Marsden, C.D., 1996. D_3 receptor expression within the basal ganglia is not affected by Parkinson's disease. Neurosci. Lett. 214, 75–78.
Innis, R.B., Cunningham, V.J., Delforge, J., Fujita, M., Gjedde, A., Gunn, R.N., Holden, J., Houle, S., Huang, S.C., Ichise, M., Iida, H., Ito, H., Kimura, Y., Koeppe, R.A., Knudsen, G.M., Knuuti, J., Lammertsma, A.A., Laruelle, M., Logan, J., Maguire, R.P., Mintun, M.A., Morris, E.D., Parsey, R., Price, J.C., Slifstein, M., Sossi, V., Suhara, T., Votaw, J.R., Wong, D.F., Carson, R.E., 2007. Consensus nomenclature for in vivo imaging of reversibly binding radioligands. J. Cereb. Blood Flow Metab. 27, 1533–1539.
Jones, J.H., Anderson, P.S., Baldwin, J.J., Clineschmidt, B.V., McClure, D.E., Lundell, G.F., Randall, W.C., Martin, G.E., Williams, M., Hirshfield, J.M., et al., 1984. Synthesis of 4-substituted ^{2}H-naphth[1,2-b]-1,4-oxazines, a new class of dopamine agonists. J. Med. Chem. 27, 1607–1613.
Jonsson, E.G., Kaiser, R., Brockmoller, J., Nimgaonkar, V.L., Crocq, M.A., 2004. Meta-analysis of the dopamine D_3 receptor gene (DRD3) Ser9Gly variant and schizophrenia. Psychiatr. Genet. 14, 9–12.
Joyce, J.N., Millan, M.J., 2007. Dopamine D_3 receptor agonists for protection and repair in Parkinson's disease. Curr. Opin. Pharmacol. 7, 100–105.
Kegeles, L., Abi-Dargham, A., Frankle, W.G., Gil, R., Cooper, T., Slifstein, M., Hwang, D.-R., Huang, Y., Haber, S., Laruelle, M., 2010. Increased synaptic dopamine in associative regions of the striatum in schizophrenia. Arch. Gen. Psychiatry 67, 231–239.
Lader, M., 2008. Antiparkinsonian medication and pathological gambling. CNS Drugs 22, 407–416.
Lammertsma, A.A., Hume, S.P., 1996. Simplified reference tissue model for PET receptor studies. Neuroimage 4, 153–158.

Le Foll, B., Diaz, J., Sokoloff, P., 2003. Increased dopamine D_3 receptor expression accompanying behavioral sensitization to nicotine in rats. Synapse 47, 176–183.

Le Foll, B., Frances, H., Diaz, J., Schwartz, J.C., Sokoloff, P., 2002. Role of the dopamine D_3 receptor in reactivity to cocaine-associated cues in mice. Eur. J. Neurosci. 15, 2016–2026.

Lee, B., London, E.D., Poldrack, R.A., Farahi, J., Nacca, A., Monterosso, J.R., Mumford, J.A., Bokarius, A.V., Dahlbom, M., Mukherjee, J., Bilder, R.M., Brody, A.L., Mandelkern, M.A., 2009. Striatal dopamine D_2/D_3 receptor availability is reduced in methamphetamine dependence and is linked to impulsivity. J. Neurosci. 29, 14734–14740.

Lerer, B., Segman, R.H., Fangerau, H., Daly, A.K., Basile, V.S., Cavallaro, R., Aschauer, H.N., McCreadie, R.G., Ohlraun, S., Ferrier, N., Masellis, M., Verga, M., Scharfetter, J., Rietschel, M., Lovlie, R., Levy, U.H., Meltzer, H.Y., Kennedy, J.L., Steen, V.M., Macciardi, F., 2002. Pharmacogenetics of tardive dyskinesia: combined analysis of 780 patients supports association with dopamine D_3 receptor gene Ser9Gly polymorphism. Neuropsychopharmacology 27, 105–119.

Leriche, L., Schwartz, J.C., Sokoloff, P., 2003. The dopamine D_3 receptor mediates locomotor hyperactivity induced by NMDA receptor blockade. Neuropharmacology 45, 174–181.

Levant, B., 1997. The D_3 dopamine receptor: neurobiology and potential clinical relevance. Pharmacol. Rev. 49, 231–252.

Levant, B., Ling, Z.D., Carvey, P.M., 1999. Dopamine D_3 receptors—relevance for the drug treatment of Parkinson's disease. CNS Drugs 12, 391–402.

Levesque, D., Diaz, J., Pilon, C., Martres, M.P., Giros, B., Souil, E., Schott, D., Morgat, J.L., Schwartz, J.C., Sokoloff, P., 1992. Identification, characterization, and localization of the dopamine D_3 receptor in rat brain using 7-[^{3}H]hydroxy-*N*, *N*-di-*n*-propyl-2-aminotetralin. Proc. Natl. Acad. Sci. U.S.A. 89, 8155–8159.

Levesque, D., Martres, M.P., Diaz, J., Griffon, N., Lammers, C.H., Sokoloff, P., Schwartz, J.C., 1995. A paradoxical regulation of the dopamine D_3 receptor expression suggests the involvement of an anterograde factor from dopamine neurons. Proc. Natl. Acad. Sci. U.S.A. 92, 1719–1723.

Mach, R.H., Huang, Y., Freeman, R.A., Wu, L., Vangveravong, S., Luedtke, R.R., 2004. Conformationally-flexible benzamide analogues as dopamine D_3 and sigma 2 receptor ligands. Bioorg. Med. Chem. Lett. 14, 195–202.

Malmberg, A., Jerning, E., Mohell, N., 1996. Critical reevaluation of spiperone and benzamide binding to D_2 receptors: evidence for identical binding sites. Eur. J. Pharmacol. 303, 123–128.

Martinez, D., Gil, R., Slifstein, M., Hwang, D.R., Huang, Y.Y., Perez, A., Kegeles, L., Talbot, P., Evans, S., Krystal, J., Laruelle, M., Abi-Dargham, A., 2005. Alcohol dependence is associated with blunted dopamine transmission in the ventral striatum. Biol. Psychiatry 58, 779–786.

Martinez, D., Narendran, R., Foltin, R.W., Slifstein, M., Hwang, D.R., Broft, A., Huang, Y.Y., Cooper, T.B., Fischman, M.W., Kleber, H.D., Laruelle, M., 2007. Amphetamine-induced dopamine release: markedly blunted in cocaine dependence and predictive of the choice to self-administer cocaine. Am. J. Psychiatry 164, 622–629.

Martinez, D., Saccone, P.A., Liu, F., Slifstein, M., Orlowska, D., Grassetti, A., Cook, S., Broft, A., Van Heertum, R., Comer, S.D., 2012. Deficits in dopamine D_2 receptors and presynaptic dopamine in heroin dependence: commonalities and differences with other types of addiction. Biol. Psychiatry 71, 192–198.

McCormick, P.N., Kapur, S., Seeman, P., Wilson, A.A., 2008. Dopamine D_2 receptor radiotracers [^{11}C](+)-PHNO and [^{3}H]raclopride are indistinguishably inhibited by D_2 agonists and antagonists ex vivo. Nucl. Med. Biol. 35, 11–17.

Mizrahi, R., Agid, O., Borlido, C., Suridjan, I., Rusjan, P., Houle, S., Remington, G., Wilson, A.A., Kapur, S., 2011. Effects of antipsychotics on D_3 receptors: a clinical PET study in first episode antipsychotic naive patients with schizophrenia using [^{11}C]-(+)-PHNO. Schizophr. Res. 131, 63–68.

Morissette, M., Goulet, M., Grondin, R., Blanchet, P., Bedard, P.J., Di Paolo, T., Levesque, D., 1998. Associative and limbic regions of monkey striatum express high levels of dopamine D_3 receptors: effects of MPTP and dopamine agonist replacement therapies. Eur. J. Neurosci. 10, 2565–2573.

Murray, A., Ryoo, H.L., Gurevich, E., Joyce, J.N., 1994. Localization of dopamine D_3 receptors to mesolimbic and D_2 receptors to mesofrontal regions of human forebrain. Proc. Natl. Acad. Sci. 91, 11271–11275.

Narendran, R., Mason, N.S., Laymon, C.M., Lopresti, B.J., Velasquez, N.D., May, M.A., Kendro, S., Martinez, D., Mathis, C.A., Frankle, W.G., 2010. A comparative evaluation of the dopamine $D_{(2/3)}$ agonist radiotracer [^{11}C](−)-*N*-propyl-norapomorphine and antagonist [^{11}C]raclopride to measure amphetamine-induced dopamine release in the human striatum. J. Pharmacol. Exp. Ther. 333, 533–539.

Narendran, R., Slifstein, M., Guillin, O., Hwang, Y.Y., Hwang, D.R., Scher, E., Reeder, S., Rabiner, E., Laruelle, M., 2006. The dopamine ($D_{2/3}$) receptor agonist positron emission tomography radiotracer [^{11}C]-(+)-PHNO is a D_3 receptor preferring agonist in vivo. Synapse 60, 485–495.

Neve, K.A., Seamans, J.K., Trantham-Davidson, H., 2004. Dopamine receptor signaling. J. Recept. Signal. Transduct. Res. 24, 165–205.

Piggott, M.A., Marshall, E.F., Thomas, N., Lloyd, S., Court, J.A., Jaros, E., Burn, D., Johnson, M., Perry, R.H., McKeith, I.G., Ballard, C., Perry, E.K., 1999. Striatal dopaminergic markers in dementia with Lewy bodies, Alzheimer's and Parkinson's diseases: rostrocaudal distribution. Brain 122, 1449–1468.

Pilla, M., Perachon, S., Sautel, F., Garrido, F., Mann, A., Wermuth, C., Schwartz, J.C., Everitt, B.J., Sokoloff, P., 1999. Selective inhibition of cocaine-seeking behaviour by a partial dopamine D_3 receptor agonist. Nature 400, 371–375.

Rabiner, E., Slifstein, M., Nobrega, J., Plisson, C., Huiban, M., Raymond, R., Diwan, M., Wilson, A.A., McCormick, P., Gentile, G., Gunn, R.N., Laruelle, M., 2009. In vivo quantification of regional dopamine-D_3 receptor binding potential of (+)PHNO: studies in non-human primates and transgenic mice. Synapse 63, 782–793.

Reavill, C., Taylor, S.G., Wood, M.D., Ashmeade, T., Austin, N.E., Avenell, K.Y., Boyfield, I., Branch, C.L., Cilia, J., Coldwell, M.C., Hadley, M.S., Hunter, A.J., Jeffrey, P., Jewitt, F., Johnson, C.N., Jones, D.N., Medhurst, A.D., Middlemiss, D.N., Nash, D.J., Riley, G.J., Routledge, C., Stemp, G., Thewlis, K.M., Trail, B., Vong, A.K., Hagan, J.J., 2000. Pharmacological actions of a novel, high-affinity, and selective human dopamine D_3 receptor antagonist, SB-277011-A. J. Pharmacol. Exp. Ther. 294, 1154–1165.

Ryoo, H.L., Joyce, J.N., 1994. Loss of dopamine D_2 receptors varies along the rostrocaudal axis of the hippocampal complex in Alzheimer's disease. J. Comp. Neurol. 348, 94–110.

Schotte, A., Janssen, P.F.M., Megens, A.A.H.P., Leysen, J.E., 1993. Occupancy of central neurotransmitter receptors by risperidone, clozapine and haloperidol, measured ex vivo by autoradiography. Brain Res. 631, 191–202.

Searle, G., Beaver, J.D., Comley, R.A., Bani, M., Tziortzi, A., Slifstein, M., Mugnaini, M., Griffante, C., Wilson, A.A., Merlo-Pich, E., Houle, S., Gunn, R., Rabiner, E.A., Laruelle, M., 2010. Imaging dopamine D_3 receptors in the human brain with positron emission tomography, [^{11}C]PHNO, and a selective D_3 receptor antagonist. Biol. Psychiatry 68, 392–399.

Seeman, P., 1980. Brain dopamine receptors. Pharmacol. Rev. 32, 229–313.

Seeman, P., 2012. Dopamine agonist radioligand binds to both D_2 high and D_2 low receptors, explaining why alterations in D_2high are not detected in human brain scans. Synapse 66, 88–93.

Seeman, P., Ko, F., Willeit, M., McCormick, P., Ginovart, N., 2005. Antiparkinson concentrations of pramipexole and PHNO occupy dopamine D_2(high) and D_3(high) receptors. Synapse 58, 122–128.

Seeman, P., Ulpian, C., Larsen, R.D., Anderson, P.S., 1993. Dopamine receptors labelled by PHNO. Synapse 14, 254–262.

Seeman, P., Wilson, A., Gmeiner, P., Kapur, S., 2006. Dopamine D_2 and D_3 receptors in human putamen, caudate nucleus, and globus pallidus. Synapse 60, 205–211.

Segal, D.M., Moraes, C.T., Mash, D.C., 1997. Up-regulation of D_3 dopamine receptor mRNA in the nucleus accumbens of human cocaine fatalities. Brain Res. Mol. Brain Res. 45, 335–339.

Shotbolt, P., Tziortzi, A.C., Searle, G., Plisson, C., Huiban, M., Gunn, R.N., Rabiner, E.A., 2011. Quantification of [^{11}C]-(+)-PHNO Binding in the Human Brain: Mass Effects and Suitability of Reference Region (Abstract 476). ISCBFM 2011 (BrainPET). International Society for Cerebral Blood Flow and Metabolism, Barcelona, Spain.

Shotbolt, P., Tziortzi, A.C., Searle, G.E., Colasanti, A., van der Aart, J., Abanades, S., Plisson, C., Miller, S.R., Huiban, M., Beaver, J.D., Gunn, R.N., Laruelle, M., Rabiner, E.A., 2012. Within-subject comparison of [^{11}C]-(+)-PHNO and [^{11}C]raclopride sensitivity to acute amphetamine challenge in healthy humans. J. Cereb. Blood Flow Metab. 32, 127–136.

Skinbjerg, M., Sibley, D.R., Javitch, J.A., Abi-Dargham, A., 2012. Imaging the high-affinity state of the dopamine D_2 receptor in vivo: fact or fiction? Biochem. Pharmacol. 83, 193–198.

Sokoloff, P., Diaz, J., LeFoll, B., Guillin, O., Leriche, L., Bezard, E., Gross, C., 2006. The dopamine D_3 receptor: a therapeutic target for the treatment of neuropsychiatric disorders. CNS Neurol. Disord. Drug Targets 5, 25–43.

Sokoloff, P., Giros, B., Martres, M.-P., Bouthenet, M.-L., Schwartz, J.-C., 1990. Molecular cloning and characterization of a novel dopamine receptor D_3 as a target for neuroleptics. Nature 347, 146–151.

Staley, J.K., Mash, D.C., 1996. Adaptive increase in D_3 dopamine receptors in the brain reward circuits of human cocaine fatalities. J. Neurosci. 16, 6100–6106.

Stemp, G., Ashmeade, T., Branch, C.L., Hadley, M.S., Hunter, A.J., Johnson, C.N., Nash, D.J., Thewlis, K.M., Vong, A.K., Austin, N.E., Jeffrey, P., Avenell, K.Y., Boyfield, I., Hagan, J.J., Middlemiss, D.N., Reavill, C., Riley, G.J., Routledge, C., Wood, M., 2000. Design and synthesis of trans-*N*-[4-[2-(6-cyano-1,2,3, 4-tetrahydroisoquinolin-2-yl)ethyl]cyclohexyl]-4-quinolinecarboxamide (SB-277011): a potent and selective dopamine D_3 receptor antagonist with high oral bioavailability and CNS penetration in the rat. J. Med. Chem. 43, 1878–1885.

Tu, Z., Li, S., Cui, J., Xu, J., Taylor, M., Ho, D., Luedtke, R.R., Mach, R.H., 2011. Synthesis and pharmacological evaluation of fluorine-containing D dopamine receptor ligands. J. Med. Chem. 54, 1555–1564.

Tziortzi, A.C., Searle, G.E., Tzimopoulou, S., Salinas, C., Beaver, J.D., Jenkinson, M., Laruelle, M., Rabiner, E.A., Gunn, R.N., 2011. Imaging dopamine receptors in humans with [^{11}C]-(+)-PHNO: dissection of D_3 signal and anatomy. NeuroImage 54, 264–277.

Urban, N.B., Slifstein, M., Thompson, J.L., Xu, X., Girgis, R.R., Raheja, S., Haney, M., Abi-Dargham, A., 2012. Dopamine release in chronic cannabis users: a [^{11}C]raclopride positron emission tomography study. Biol. Psychiatry http://dx.doi.org/10.1016/j.biopsych.2011.12.018.

van Vliet, L.A., Rodenhuis, N., Dijkstra, D., Wikstrom, H., Pugsley, T.A., Serpa, K.A., Meltzer, L.T., Heffner, T.G., Wise, L.D., Lajiness, M.E., Huff, R.M., Svensson, K., Sundell, S., Lundmark, M., 2000. Synthesis and pharmacological evaluation of thiopyran analogues of the dopamine D_3 receptor-selective agonist (4aR,10bR)-(+)-trans-3,4,4a,10b-tetrahydro-4-n-propyl-2H,5H [1]b enzopyrano[4,3-b]-1,4-oxazin-9-ol (PD 128907). J. Med. Chem. 43, 2871–2882.

Vanhauwe, J.F., Fraeyman, N., Francken, B.J., Luyten, W.H., Leysen, J.E., 1999. Comparison of the ligand binding and signaling properties of human dopamine D_2 and D_3 receptors in Chinese hamster ovary cells. J. Pharmacol. Exp. Ther. 290, 908–916.

Videbaek, C., Toska, K., Scheideler, M.A., Paulson, O.B., Moos Knudsen, G., 2000. SPECT tracer [^{123}I] IBZM has similar affinity to dopamine D_2 and D_3 receptors. Synapse 38, 338–342.

Volkow, N.D., Chang, L., Wang, G.J., Fowler, J.S., Ding, Y.S., Sedler, M., Logan, J., Franceschi, D., Gatley, J., Hitzemann, R., Gifford, A., Wong, C., Pappas, N., 2001. Low level of brain dopamine D_2 receptors in methamphetamine abusers: association with metabolism in the orbitofrontal cortex. Am. J. Psychiatry 158, 2015–2021.

Volkow, N.D., Wang, G.J., Fowler, J.S., Logan, J., Gatley, S.J., Hitzemann, R., Chen, A.D., Dewey, S.L., Pappas, N., 1997. Decreased striatal dopaminergic responsiveness in detoxified cocaine-dependent subjects. Nature 386, 830–833.

Volkow, N.D., Wang, G.J., Fowler, J.S., Logan, J., Hitzemann, R., Ding, Y.S., Pappas, N., Shea, C., Piscani, K., 1996. Decreases in dopamine receptors but not in dopamine transporters in alcoholics. Alcohol. Clin. Exp. Res. 20, 1594–1598.

Volkow, N.D., Wang, G.J., Telang, F., Fowler, J.S., Logan, J., Jayne, M., Ma, Y., Pradhan, K., Wong, C., 2007. Profound decreases in dopamine release in striatum in detoxified alcoholics: possible orbitofrontal involvement. J. Neurosci. 27, 12700–12706.

Wang, G.J., Volkow, N.D., Fowler, J.S., Logan, J., Abumrad, N.N., Hitzemann, R.J., Pappas, N.S., Pascani, K., 1997. Dopamine D_2 receptor availability in opiate-dependent subjects before and after naloxone-precipitated withdrawal. Neuropsychopharmacology 16, 174–182.

Whone, A.L., Watts, R.L., Stoessl, A.J., Davis, M., Reske, S., Nahmias, C., Lang, A.E., Rascol, O., Ribeiro, M.J., Remy, P., Poewe, W.H., Hauser, R.A., Brooks, D.J., 2003. Slower progression of Parkinson's disease with ropinirole versus levodopa: the REAL-PET study. Ann. Neurol. 54, 93–101.

Willeit, M., Ginovart, N., Kapur, S., Houle, S., Hussey, D., Seeman, P., Wilson, A.A., 2006. High-affinity states of human brain dopamine $D_{2/3}$ receptors imaged by the agonist [^{11}C]-(+)-PHNO. Biol. Psychiatry 59, 389–394.

Williams, J., Spurlock, G., Holmans, P., Mant, R., Murphy, K., Jones, L., Cardno, A., Asherson, P., Blackwood, D., Muir, W., Meszaros, K., Aschauer, H., Mallet, J., Laurent, C., Pekkarinen, P., Seppala, J., Stefanis, C.N., Papadimitriou, G.N., Macciardi, F., Verga, M., Pato, C., Azevedo, H., Crocq, M.A., Gurling, H., Owen, M.J., et al., 1998. A meta-analysis and transmission disequilibrium study of association between the dopamine D_3 receptor gene and schizophrenia. Mol. Psychiatry 3, 141–149.

Wilson, A.A., McCormick, P., Kapur, S., Willeit, M., Garcia, A., Hussey, D., Houle, S., Seeman, P., Ginovart, N., 2005. Radiosynthesis and evaluation of [^{11}C]-(+)-4-propyl-3,4,4a,5,6,10b-hexahydro-2H-naphtho[1,2-b][1,4]oxazin -9-ol as a potential radiotracer for in vivo Imaging of the dopamine D_2 high-affinity state with positron emission tomography. J. Med. Chem. 48, 4153–4160.

Yuan, J., Chen, X., Brodbeck, R., Primus, R., Braun, J., Wasley, J.W., Thurkauf, A., 1998. NGB 2904 and NGB 2849: two highly selective dopamine D_3 receptor antagonists. Bioorg. Med. Chem. Lett. 8, 2715–2718.

Zijlstra, F., Booij, J., van den Brink, W., Franken, I.H., 2008. Striatal dopamine D_2 receptor binding and dopamine release during cue-elicited craving in recently abstinent opiate-dependent males. Eur. Neuropsychopharmacol. 18, 262–270.

CHAPTER TWELVE

Dopamine Receptors and Dopamine Release

Robert M. Kessler
Department of Radiology and Radiological Sciences, Vanderbilt University School of Medicine, Nashville, TN, USA

1. INTRODUCTION

Over the past three decades, positron emission tomographic (PET) and single-photon emission computed tomographic (SPECT) studies of dopamine (DA) receptors have been used to study a number of disorders in which DA neurotransmission is believed to play an important role and in examining the effects of drugs which affect DA neurotransmission. The purpose of this chapter is to review the methods used to image DA D_1- and DA D_2-type receptors and to touch upon some of the important applications of these methods. A comprehensive review of applications of DA D_1 and DA $D_{2/3}$ receptor imaging is beyond the scope of this chapter. In addition, DA $D_{2/3}$ receptor ligands have been used to estimate releasable DA and baseline extracellular DA levels. These methods are reviewed as well. Imaging of DA D_1-, DA D_2-type receptors and studies of extracellular DA levels will be examined separately.

2.1. DOPAMINE D_1 FAMILY RECEPTORS

2.1.1. Overview

DA D_1-type receptors include the DA D_1 and DA D_5 receptors. Both signal intracellularly through Gαs proteins to activate adenyl cyclase which leads to the formation of cyclic adenosine monophosphate (cAMP), activation of protein kinase A, and phosphorylation of the Thr34 site on DARRP32 (Beaulieu and Gainetdinov, 2011). The DA D_1 and D_5 receptors are structurally homologous, but the D_5 receptor has higher affinity for DA than the D_1 receptor, is more constitutively active in the absence of agonist, has differing affinities for agonists and antagonists, and has a somewhat different regional distribution in brain (Civelli and Borrelli, 2010; Tiberi and Caron, 1994; Bergson et al., 1995). At a functional level, the DA D_1 receptor promotes locomotor activity, while the DA D_5 receptor appears to inhibit locomotor activity in animals which have been unilaterally lesioned with 6-hydroxydopamine (Dziewczapolski et al., 1998).

Imaging of the Human Brain in Health and Disease
http://dx.doi.org/10.1016/B978-0-12-418677-4.00012-9

2.1.2. Distribution of Dopamine D_1-Type Receptors in Brain

With regard to regional localization, the highest levels DA D_1 receptors are found in the striatum with a rostrocaudal gradient seen in the putamen but not in the caudate; levels of DA D_1 receptors similar to those seen in the dorsal striatum are seen in the nucleus accumbens (Piggott et al., 1999). There are lower levels in the globus pallidus, zona reticulata of the substantia nigra, and anterior cingulate—about 25% of striatal levels. In cortical regions and amygdala, DA D_1 receptor levels are about 20% of striatal levels; still lower levels are present in the hippocampus, thalamus, and basal forebrain (Bergson et al., 1995; Cortés et al., 1989; Abi-Dargham et al., 2000a). DA D_5 receptors occur in a different regional distribution and are seen at lower levels than DA D_1 receptors in most regions. The highest levels of DA D_5 receptors are in the frontal and temporal cortices. In contrast to DA D_1 receptors which are localized in the zona reticulata of the substantia nigra, relatively high levels of DA D_5 receptors are seen in the zona compacta of the substantia nigra where they are predominantly localized on dopaminergic neurons. DA D_5 receptors are also seen in the striatum, nucleus accumbens, olfactory tubercle, hippocampus, gray matter layers of the superior and inferior colliculus, oculomotor nucleus, hypothalamus, thalamus, and globus pallidus; low levels are present in the cerebellum (Khan et al., 2000; Civelli and Borrelli, 2010; Bergson et al., 1995).

2.1.3. Dopamine D_1-Type PET and SPECT Radioligands

A number of PET and some SPECT radioligands for the DA D_1 receptor have been developed including [^{11}C]SCH23390; [^{11}C]SCH39166; [^{11}C]NNC687; [^{11}C]NNC756; [^{11}C]NNC112; [^{11}C]NNC22-0010; [^{11}C]NNC22-0215; [^{11}C], [^{76}Br], and [^{123}I]A-69024; [^{11}C]SKF82957, a DA D_1 agonist; [^{123}I]TISCH; and [^{123}I]FISCH (Halldin et al., 1986, 1998; Karlsson et al., 1993, 1995; Foged et al., 1996, 1998; Billings et al., 1989, 1992; Kassiou et al., 2001, 2002; Besret et al., 2008; Palner et al., 2010). Of these, [^{11}C]NNC112 and [^{11}C]NNC756 have the highest striatal to cerebellar ratios in humans. The most widely used DA D_1 radioligands are [^{11}C]NNC112 (Figure 1) and [^{11}C]SCH23390, which was the first PET DA D_1 radioligand available for studying DA D_1 receptors in humans.

[^{11}C]NNC112 was preferred to [^{11}C]NNC756 as initial studies suggested that 24–28% of cortical uptake of [^{11}C]NNC756 was due to binding at 5-hydroxytryptamine $(HT)_2$ receptors, while little displacement of [^{11}C]NNC112 cortical uptake was seen after administration of the 5-HT_2 receptor antagonist ketanserin in monkeys (Karlsson et al., 1993; Halldin et al., 1998). However, subsequent studies using MDL100907, a specific 5-HT_{2A} antagonist, to displace cortical uptake have shown that about 24–29% of [^{11}C]NNC112 cortical uptake in baboons was due to binding at the 5-HT_{2A} receptor (Ekelund et al., 2007). Similarly in humans 20–30% of

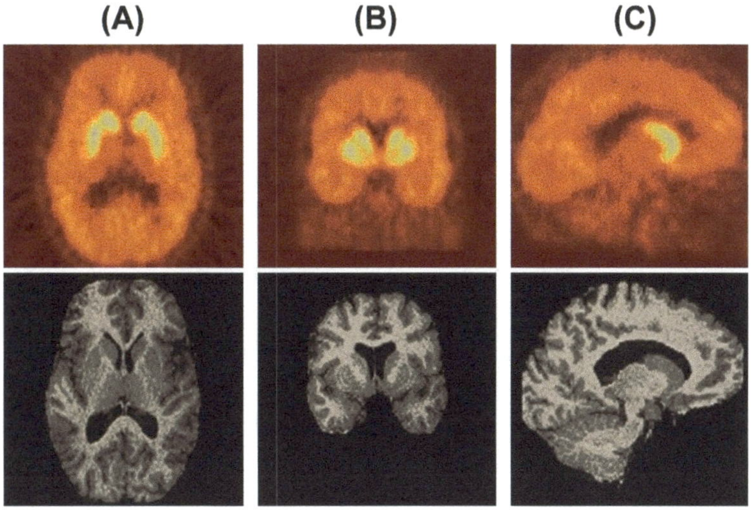

Figure 1 Axial (A), coronal (B), and sagittal (C) images of [^{11}C]NNC112 demonstrating highest levels of DA D_1 receptors in the striatum with lower levels of uptake in cortical regions. Corresponding MRI scans are below each image. Specific [^{11}C]NNC112 uptake in striatum is to DA D_1 receptors, while uptake in extrastriatal regions has a significant 5-HT_{2A} component, about 20–30%. There is very low uptake in cerebellum. *Images courtesy of Dr Anissa Abi Dargham, Columbia University.*

[^{11}C]NNC112 cortical uptake was blocked by pretreatment with 2 mg risperidone (Slifstein et al., 2007). A second study in humans using ketanserin to block 5-HT_{2A} receptors found that 30% of cortical uptake and 20% of uptake in limbic regions was due to [^{11}C]NNC112 binding to 5-HT_{2A} receptors (Catafau et al., 2010). No significant displacement or blockade of striatal [^{11}C]NNC112 uptake was seen following administration of 5-HT_{2A} antagonists in these studies. Similar findings have been reported for [^{11}C]SCH23390; a 21% blockade of cortical [^{11}C]SCH23390 uptake was seen with MDL100907 pretreatment. There was a suggestion of some displacement of striatal [^{11}C]SCH23390 uptake—about 11–12%—which did not reach statistical significance. In vivo Scatchard plots of [^{11}C]NNC112 uptake in baboons indicate that the ratio of in vivo DA D_1 to 5-HT_{2A} affinities is between 6 and 14, not the 100-fold difference reported in vitro (Ekelund et al., 2007). Similar studies are not currently available for the other DA D_1 radioligands listed above. These results suggest that both [^{11}C]NNC112 and [^{11}C]SCH23390 are specific DA D_1 radioligands in the striatum but not in extrastriatal regions. Currently, there are no DA D_1 radioligands which have been demonstrated to be specific for studying cortical and limbic DA D_1 receptors in humans. It has been recommended that if studies of extrastriatal DA D_1 receptors are performed, then it may be appropriate to pretreat subjects with a 5-HT_{2A} antagonist (Catafau et al., 2010).

2.1.4. Effects of Extracellular Dopamine Levels on Dopamine D_1 Receptor Radioligand Binding

The effects of altered extracellular DA levels on in vivo specific brain uptake of DA D_1 receptor radioligands are considerably different from those seen with DA $D_{2/3}$ benzamide antagonist and $D_{2/3}$ agonist radioligands. Amphetamine doses which in anesthetized primates produced 20–40% decreases in apparent striatal DA D_2 receptor levels as measured using benzamide or agonist radioligands (see below) did not affect in vivo uptake of DA D_1 antagonists, i.e. [^{11}C]NNC756, [^{11}C]NNC112, or [^{11}C]SCH23390, in primates (Chou et al., 1999; Abi-Dargham et al., 1999; Laruelle, 2000; Palner et al., 2010). Acute amphetamine administration to primates produced equivocal effects on striatal uptake of a DA D_1 agonist, [^{11}C]SKF82957; some animals showed no effect, while others showed a slight increase in uptake (Palner et al., 2010; Wilson, personal communication). Acute DA depletion which produces increased striatal binding of benzamide DA D_2 radioligands (see below) produced no change in [^{11}C]NNC112 striatal binding but decreased [^{11}C]SCH23390 striatal binding in rodents (Guo et al., 2003). These differing effects of DA depletion on uptake of DA D_1 radioligands may reflect the ability of these radioligands to bind to internalized receptors with similar affinity to membrane-externalized receptors or in the case of [^{11}C]SCH23390 perhaps with even greater affinity. However, in humans, DA depletion by alpha-methyl-para-tyrosine (AMPT) produced no change in [^{11}C]SCH23390 striatal uptake (Verhoeff et al., 2002). Differences in intracellular receptor trafficking of DA D_1 vs D_2 receptors may be a factor in these differing effects. DA D_1 receptors have more rapid intracellular cycling (20 min) than DA D_2 receptors and do not undergo lysosomal degradation as do DA D_2 receptors as they are not a substrate for GASP which sorts intracellular receptors for degradation (Martin-Negrier et al., 2006; Thompson et al., 2007).

While longer term DA depletion would be expected to produce increased DA D_1 receptor levels due to receptor upregulation, DA depletion studies in primates using reserpine produced 13–20% decreases in in vivo [^{11}C]NNC112 B_{max} without affecting the K_d in both striatal and extrastriatal regions at 3–28 days after reserpine administration (Chou et al., 1999). In rats, chronic DA depletion produced decreased [^{11}C]SCH23390 binding, but, in contrast to primate studies, rodent studies reported increased [^{11}C]NNC112 striatal and extrastriatal binding (Guo et al., 2003). While no long-term interventional manipulations of extracellular DA levels have been reported in humans, a study of the val158met catechol-*O*-methyltransferase polymorphism using [^{11}C]NNC112 demonstrated that individuals with the val/val polymorphism, which would be expected to have lower extracellular DA levels, particularly in cortical and limbic regions, had significantly higher DA D_1 receptor levels in cortex and limbic regions than in subjects with the met/met polymorphism. No significant difference was seen in the striatum (Slifstein et al., 2008). These results again suggest that apparent changes in DA

D_1 levels due to changes in extracellular DA levels vary depending on the radioligand used, the species examined, and the duration of changes in extracellular DA levels.

2.2. CLINICAL STUDIES OF THE DOPAMINE D_1 RECEPTOR

Studies of changes in DA D_1 receptor levels with normal aging have demonstrated differing rates of receptor loss depending on the region and age. During adolescence and early adulthood there is a rapid loss of cortical DA D_1 receptors consistent with postmortem studies of DA D_1 receptors (Seeman et al., 1987; Rinne et al., 1990; Montague et al., 1999), i.e. an approximately 3% decrease per year in the dorsolateral prefrontal, parietal, temporal, occipital, and anterior cingulate cortices but only a 1% decline per year in the caudate and putamen (Jucaite et al., 2010). During adulthood, from ages 20–74 years, there is a slower rate of loss, a 9% decrease per decade in prefrontal and occipital cortical receptors and a 7–8% decrease per decade in striatal receptors (Suhara et al., 1991; Wang et al., 1998).

Studies of schizophrenia, depression, drug abuse, movement disorders, and the relationship of cognitive function to DA D_1 receptor levels have been performed. These will be briefly examined with a view to examining methodological issues. Studies of schizophrenia in both drug-naïve and drug-free subjects using [^{11}C]SCH23390 have reported decreased frontal cortical levels which were correlated with the level of negative symptoms and related to poor performance on the Wisconsin Card Sorting Test (Okubo et al., 1997). However, a second study using [^{11}C]SCH23390 reported no change in B_{max} or K_d in the cortex or striatum in drug-naïve schizophrenic subjects (Karlsson et al., 2002). In a study of monozygotic and dizygotic twins discordant for schizophrenia using [^{11}C]SCH23390, increasing genetic risk for schizophrenia was associated with increasing apparent DA D_1 receptor levels in medial prefrontal cortex, superior temporal gyrus, and the angular gyrus in the inferior parietal lobule; chronically medicated schizophrenic subjects demonstrated decreased cortical and striatal DA D_1 receptor levels consistent with studies in animals (Hirvonen et al., 2006; Kosaka et al., 2010). Two studies of drug-naïve and drug-free schizophrenic subjects using [^{11}C]NNC112 have been reported (Abi-Dargham et al., 2002, 2012). The first reported increased dorsolateral prefrontal cortical (DLPFC) DA D_1 receptor levels in a combined cohort of drug-naïve and drug-free subjects; the increased DLPFC levels were negatively correlated with the *N*-back test at the 3-back level. The second study found increased levels only in drug-naïve schizophrenics and no correlation with the *N*-back test. The length of drug-free interval was positively correlated with DLPFC DA D_1 levels consistent with the findings in treated schizophrenic subjects. In addition, the second study found that the cerebellar volume of distribution for [^{11}C]NNC112 differed for schizophrenic and healthy subjects which resulted in a loss of significant findings when the BP_{ND} was used as a metric of DA D_1 levels, suggesting that reference regions methods are inappropriate for

[^{11}C]NNC112 studies in schizophrenics. While the Okubo study (Okubo et al., 1997) reported decreased frontal DA D_1 receptor levels, the available studies suggest increased cortical DA D_1 receptor levels in untreated schizophrenic subjects but decreased levels in chronically treated subjects.

Studies of the DA D_1 receptor have been performed in depression, obsessive-compulsive disorder, and drug abuse. There have been two PET studies of the DA D_1 receptor in depression (Dougherty et al., 2006; Cannon et al., 2009). The first study utilized [^{11}C]SCH23390 and reported decreased DA D_1 receptor levels in the striatum bilaterally in 10 subjects with major depression and anger attacks (Dougherty et al., 2006). The second study utilized [^{11}C]NNC112 and found decreased DA D_1 receptor levels in the left anterior caudate (Cannon et al., 2009) which were negatively correlated with both duration of illness and anhedonia. There have been two small studies (each $N=7$) of the DA D_1 receptor using [^{11}C]SCH23390 in obsessive-compulsive disorder which reported decreased caudate and anterior cingulate DA D_1 receptor levels bilaterally; the anterior cingulate but not caudate receptor levels were significantly and negatively correlated with scores on the Yale-Brown Obsessive-Compulsive Rating Scale (Olver et al., 2009, 2010). Studies of DA D_1 receptor levels using [^{11}C]SCH23390 in tobacco smokers have reported decreased ventral striatal DA D_1 receptor levels which were negatively correlated with both cue-induced activation in the ventral striatum and the probability of making risky choices (Yasuno et al., 2007; Takahashi et al., 2010). In a study of DA D_1 receptor levels in cocaine abusers using [^{11}C]NNC112, no difference in receptor availability was seen in any striatal or cortical region despite primate studies which have reported increased DA D_1 receptor levels in cocaine self-administering nonhuman primates (Nader et al., 2002; Martinez et al., 2009a). However, a significant negative correlation was seen in humans between limbic striatum, i.e. ventral striatum, and the choice to self-administer cocaine vs accepting a monetary reward. This correlation was interpreted as suggesting that lower ventral striatal DA D_1 receptor availability was associated with an increased probability of cocaine relapse in cocaine-dependent subjects. In depression, obsessive-compulsive disorder, and substance abuse decreased DA D_1 receptor levels were seen and these decreases were significantly and negatively correlated with symptom levels or abnormal behaviors in these disorders.

Ketamine in subanesthetic doses is a psychotomimetic drug which produces a state in normal subjects similar to that seen in schizophrenic subjects. Ketamine abusers demonstrated a significant 24% increase in dorsolateral prefrontal DA D_1 receptor levels similar to the increased prefrontal DA D_1 receptor levels seen in schizophrenic subjects; increases were seen in other cortical regions but did not reach significance (Narendran et al., 2005a). Unlike some studies in schizophrenic subjects, the increased DLFPC DA D_1 levels were not significantly correlated with working memory (semantic, visuospatial, and auditory), executive functions, attention, reaction time, verbal learning and memory, verbal fluency, motor function, and intellectual functioning.

There have been a number of PET studies using [^{11}C]SCH23390 to examine DA D_1 receptor levels in movement disorders. In Huntington disease 35–75% reductions in striatal DA D_1 receptor levels have been reported (Sedvall et al., 1994; Turjanski et al., 1995; Ginovart et al., 1997a). The magnitude of this decrement is highly correlated with years of symptomatic illness reflecting a high rate of striatal DA D_1 receptor loss in symptomatic subjects, about 5% per year (Andrews et al., 1999); asymptomatic carriers of the Huntington gene showed a mean loss of DA D_1 receptors of about 2% per year with a subgroup showing a more rapid loss of 4.5% per year. These observations suggest an increasing rate of loss of DA D_1 receptors in asymptomatic gene carriers with the highest rate of decline in the symptomatic period. In symptomatic Huntington subjects, DA D_1 receptor levels in the caudate and putamen were significantly correlated with spatial span, a test of spatial sequencing, and sequence generation (Lawrence et al., 1998). In contrast, studies of DA D_1 receptor levels using both [^{11}C]NNC112 and [^{11}C]SCH23390 in Parkinson disease patients have failed to show a significant change in both striatal and extrastriatal regions (Cropley et al., 2008a; Shinotoh et al., 1993); while not reaching significance, decreases in striatal DA D_1 levels were significantly correlated with duration of illness and L-dopa treatment (Turjanski et al., 1997). Patients with striatonigral degeneration have been reported to have decreased striatal DA D_1 receptor levels with the largest decrement in the posterior putamen of 31%, a 21% decrement in the anterior putamen, and the smallest decrements in the caudate of 12% (Shinotoh et al., 1993).

2.3. DOPAMINE D_2 FAMILY RECEPTORS

2.3.1. Overview

The DA D_2 family of receptors includes the D_2, D_3, and D_4 receptors. DA D_2 receptors are further divided into short and long forms, i.e. DA D_{2S} and DA D_{2L}. The DA D_{2S} variant is predominantly a presynaptic autoreceptor, while the DA D_{2L} is predominantly a postsynaptic receptor (Khan et al., 1998). DA D_2-type receptors are G-protein-coupled receptors which signal intracellularly via a number of intracellular signaling pathways. DARRP32 is an integrator of DA D_1 and DA D_2 cAMP-mediated intracellular signaling as well as mediating intracellular signaling of a number of other neurotransmitters which signal via G-protein-coupled receptors (Svenningsson et al., 2004). DA D_2 family receptors signal through Gαi to decrease phosphorylation of the Thr32 site on DARRP32 (Svenningsson et al., 2004). In addition, DA D_2-type receptors are coupled directly to phospholipase C leading to the production of inositol triphosphate and activation of the Ca^{2+}/calmodulin-dependent protein phosphatase, calcineurin (also known as PP-2B) which dephosphorylates Thr32 on DARRP32 (Nishi et al., 1997). These actions all lead to a decrease in the phosphorylation of the Thr32 site on DARRP32 and oppose the actions of DA D_1 receptor stimulation (Nairn et al., 2004). Internalized DA D_2 and D_3 but apparently not D_4-type receptors form a complex with β-arrestin and PP2A which negatively modulates Akt activity (Beaulieu et al., 2007a,b). Akt is modulated by a number of intracellular signaling pathways including BDNF, serotonin, and insulin signaling and regulates both GSK3 and mTOR. Akt is involved in the action of antidepressants, antipsychotic drugs, and DA-mediated behaviors (Beaulieu et al., 2007b; Allen et al., 2011). In addition to these intracellular signaling pathways, direct receptor/receptor protein interactions have been described for DA D_2-type receptors including heteromers with the A_{2A} receptor, mGluR5 receptor, H_3, and CB_1 receptors (Fuxe et al., 2010). At present, imaging techniques are largely restricted to cell surface receptors.

2.3.2. Distribution of Dopamine D_2-Type Receptors in Brain

The levels and distribution of DA D_2, D_3, and D_4 receptors varies across the brain (Kessler et al., 1993a; Primus et al., 1997; Gurevich and Joyce, 1999; Rieck et al., 2004; Tziortzi et al., 2011). The DA D_2-type receptor has highest levels in the striatum, moderate levels in the external segment of the globus pallidus—about one-fourth that in the dorsal striatum, and approximately 10-fold lower levels in the anterior paraventricular, centromedian, parafascicular, and dorsomedial thalamic nuclei; substantia nigra; hypothalamus; and amygdala. Within the amygdala, the highest levels of DA D_2 receptors are in the basolateral nuclei and the amygdylostriatal transition zone. Much lower levels are seen in cortical and other limbic regions—approximately 25- to 100-fold lower than in the striatum. In the human brain, low levels of DA D_2 receptors are seen throughout the cortex with the highest levels in the inferomedial temporal cortex and low levels in the prefrontal cortex, particularly the dorsolateral prefrontal cortex. This differs from both the rat and rhesus monkey brain where considerably higher levels, both absolute and relative, are seen in frontal regions. Outside the amygdala the highest levels of DA D_2 receptors in limbic regions are in the cingulate, particularly the anterior cingulate, and in the hippocampus. Due to partial voluming in PET imaging studies, apparent hippocampal DA D_2 receptor levels may also reflect a contribution from the tail of the caudate which is located immediately superior to the temporal horn of the lateral ventricle, while the hippocampus sits immediately inferior to the temporal horn.

The distribution of DA D_3 receptors in the human brain is considerably more restricted and lower in level than that seen with the DA D_2 receptor (Hall et al., 1996; Gurevich and Joyce, 1999; Tziortzi et al., 2011). The highest levels of DA D_3 receptors are in the ventral striatum/ventral pallidum where DA D_3 levels are about 25–30% that of dorsal striatal DA D_2 receptor levels. Unlike the rodent brain where DA D_3 receptors in the nucleus accumbens are concentrated in the islands of Calleja, in humans there is a more diffuse expression of DA D_3 receptors in the ventral striatum with highest levels in the more posterior portions of the ventral striatum just under the anterior commissure extending posteriorly into the ventral pallidum with some extension superiorly to just above anterior commissure. Other regions with appreciable levels of DA D_3 receptors include the hypothalamus, in particular the mammillary bodies, the substantia nigra, and the anterior nucleus of the thalamus. High levels of DA D_3 receptors can be seen with autoradiography in the mammillothalamic tract which projects from the mammillary bodies to the anterior nucleus of thalamus. In the dorsal striatum, postmortem studies find that DA D_3 receptor levels are 10–20% of DA D_2 receptor levels (Gurevich and Joyce, 1999; Kessler, unpublished results), but imaging studies report that the available dorsal striatal DA D_3 levels are less than 1% of dorsal striatal DA D_2 levels (Tziortzi et al., 2011). The divergence between imaging and postmortem studies may be due to the high levels of extracellular DA in the striatum and the higher affinity of DA D_3 receptors for DA compared to DA D_2 receptors which may produce high levels of DA D_3 receptor occupancy by DA in the striatum.

The distribution of the DA D_4 receptor in human brain differs from that of the DA D_2 and D_3 receptors. The density of DA D_4 receptors is significantly lower than that of either DA D_2 or D_3 receptors (Primus et al., 1997). There are no detectable levels in the striatum—dorsal or ventral. Highest levels of DA D_4 receptors are seen in the lateral septal nucleus with somewhat lower levels, about 30% lower, in the entorhinal cortex and dorsomedial thalamus. DA D_4 receptors are also seen in the hippocampus, hypothalamus, and prefrontal cortex where the levels of D_4 receptors are about one-third those in the lateral septum.

2.3.3. Dopamine D_2, D_3, and D_4 PET and SPECT Radioligands

2.3.3.1. Dopamine D_2-Type Antagonist Radioligands

DA D_2-type receptors have been a target of radioligand development for nearly three decades. As a comprehensive review of all proposed DA D_2 type radioligands is beyond the scope of this section, the focus is on DA D_2 radioligands which have been used in humans most frequently. The majority of radioligands developed for the DA D_2-type receptor are not selective for DA D_2 vs D_3 receptors. Most DA $D_{2/3}$ radioligands are antagonists and are either substituted benzamides or butyrophenones. Substituted benzamide and butyrophenone DA D_2 radioligands bind to overlapping but nonidentical sites on the DA D_2 receptor resulting in differing ionic and pH requirements for optimal binding to the DA D_2 receptor, differing effects of point mutations on affinity for the DA D_2 receptor, differing B_{max} values for the DA D_2 receptor levels in rat striatal membranes, and differences in the mode of inhibition of binding to the DA D_2 receptor (Hall et al., 1990; Kessler et al., 1993b). Inhibition of raclopride, a benzamide, binding to DA D_2 receptors by *N*-methylspiperone, a butyrophenone, is competitive, while inhibition of *N*-methylspiperone binding by raclopride is both competitive and noncompetitive (Hall et al., 1990).

These differences in binding sites can lead to differing findings in clinical studies. [^{11}C]Raclopride studies of quetiapine occupancy of striatal DA D_2 receptors at clinically effective doses demonstrated 51% occupancy of the DA D_2 receptor at the 750 mg dose level; however, [^{11}C]*N*-methylspiperone studies failed to detect any occupancy of

DA D_2 receptors by quetiapine at any dose studied (Hagberg et al., 1998). In addition, as discussed below, there are significant differences between benzamides and butyrophenone DA D_2 radioligands with regard to sensitivity to extracellular DA levels. These differences need to be kept in mind when interpreting the results of studies with these radioligands.

2.3.3.2. Dopamine D_2 Receptor Antagonist Radioligands for Striatal Regions

Butyrophenone DA D_2-type radioligands include [^{11}C]*N*-methylspiperone, [^{18}F]3-*N*-(2-fluoroethyl)spiperone, and [^{18}F]*N*-methylbenperidol ([^{18}F]NMB). [^{11}C]*N*-methylspiperone was the first PET radioligand for the DA D_2 receptor (Burns et al., 1984; Wagner et al., 1984). [^{11}C]*N*-methylspiperone has high affinity for the DA D_2-type receptor (K_d = 0.11 nM in human caudate) and for 5-HT_2 receptor (K_d = 0.44 nM in human frontal cortex) (Lyon et al., 1986). As a result, specific uptake in the striatum, particularly the dorsal striatum, is principally to DA D_2-type receptors, but 91% of frontal cortical specific uptake is to 5-HT_2 receptors (Frost et al., 1987). As a result, [^{11}C]*N*-methylspiperone cannot be used to study DA D_2 receptors in extrastriatal regions. [^{18}F]3-*N*-(2-fluoroethyl)spiperone has nearly identical affinities to those seen with [^{11}C]*N*-methylspiperone for the DA D_2 and 5-HT_2 receptors as well as low affinity for the α_1 receptor and negligible affinity for other neuroreceptor sites (Goffinet et al., 1990). Striatal uptake of [^{11}C]*N*-methylspiperone appears to be nearly irreversible in vivo.

Wong and Gjedde have validated a strategy for estimating striatal DA D_2 receptor levels by varying the k_3, i.e. the on rate of [^{11}C]*N*-methylspiperone to the DA D_2 receptor, by administration of a 6 mg dose of haloperidol (Wong et al., 1986a,b). This method requires arterial sampling, plasma metabolite analysis, as well as measurement of plasma haloperidol levels. Comparison to simple striatal:cerebellar ratios have shown that such ratios are significantly less sensitive to changes in striatal DA D_2 receptor levels (Wong et al., 1997). Modeling studies of [^{18}F]3-*N*-(2-fluoroethyl)spiperone have similarly shown nearly irreversible striatal binding for tracer studies performed up to 90 min. Due to its very high affinity for the DA D_2 receptor and slow dissociation from the DA D_2 receptor, estimation of [^{18}F]3-*N*-(2-fluoroethyl)spiperone binding potentials requires either two studies, one at tracer doses and one with a radioligand having medium to low specific activity producing partial receptor saturation—similar to the use of haloperidol with [^{11}C]*N*-methylspiperone, or longer modeling studies lasting 3 h using a single dose of a lower specific activity radiotracer producing partial receptor saturation (Huang et al., 1989; Bahn et al., 1989). The need for partial receptor saturation and an arterial metabolite-corrected plasma input function for accurate kinetic modeling has restricted the use of these radioligands in the past decade. Most recent studies of cerebral DA D_2 receptors have used benzamide radioligands. [^{11}C]*N*-methylspiperone has also been used by several groups to provide an estimate of cortical 5-HT_2 receptor levels (Nordström et al., 1993; Nyberg et al., 1999; Gefvert et al., 2001).

[^{18}F]NMB is a butyrophenone with somewhat lower affinity for the DA D_2 receptor than spiperone analogs (in vitro $K_d = 0.58$ nM), but about twofold higher affinity for the DA D_2 receptor than raclopride. [^{18}F]NMB has low affinity for the 5-HT_{2A} receptor and is selective for the DA D_2-type receptor (Moerlein et al., 1997). Unlike other available DA D_2 radioligands, [^{18}F]NMB has been reported to have greater than 100-fold selectivity for the DA D_2 vs the DA D_3 receptor (K_d of 0.58 vs 114 nM) (Karimi et al., 2011). [^{18}F]NMB has only an eightfold greater affinity for the D_2 vs the D_4 receptor. Given the paucity of D_4 receptors in the striatum, [^{18}F]NMB may be useful in separating changes in striatal D_2 vs D_3 levels. In humans, PET [^{18}F]NMB studies have shown that it is a reversible DA D_2 radioligand with striatal:cerebellar ratios of about 5:1 and caudate and putamenal binding potentials in the range of 3.5 (Antenor-Dorsey et al., 2008a). Unlike the spiperone analogs, reference region methods can be used to quantitate [^{18}F]NMB binding potentials. Although a time/stability analysis of estimated receptor level has not been reported, about 2 h of data collection appears to be sufficient. Initial studies suggest that [^{18}F]NMB binding potentials may be insensitive to extracellular DA levels (Moerlein et al., 1997). [^{11}C]NMB has also been reported in animals and appears to be a promising radioligand with lower radiation dosimetry than [^{18}F]NMB (Suehiro et al., 1990; Antenor-Dorsey et al., 2008b).

Substituted benzamide radioligands have been the DA $D_{2/3}$-type radioligands most frequently used in humans since the mid-1990s. These radioligands include [^{123}I]IBZM, [^{11}C]raclopride, [^{123}I]epidepride, [^{18}F]fallypride, [^{18}F]desmethylfallypride, as well as [^{11}C] and [^{75}Br]FLB457. As a group, these substituted benzamides do not have high affinity for the DA D_4 receptor but have similar affinities for D_{2S}, D_{2L}, and D_3 receptors. Substituted benzamide DA D_2-type radioligands are the most widely used in humans due to a number of factors including (1) their high selectivity for DA $D_{2/3}$ type receptors, (2) their reversible in vivo binding kinetics allowing straightforward tracer kinetic modeling, and (3) the sensitivity of a number of benzamides to extracellular DA levels allowing estimation of psychostimulant-induced DA release and baseline extracellular DA levels. Benzamide DA $D_{2/3}$ radioligands with moderately high affinity for the DA D_2 receptor, i.e. [^{123}I]IBZM (in vitro $K_d = 0.43$ nM), for SPECT studies and [^{11}C]raclopride ($K_d = 1.2$ nM) and [^{18}F]desmethylfallypride ($K_d = 0.34$ nM) for PET studies, have adequate specific uptake sufficient to quantify DA $D_{2/3}$ receptor levels in the striatum and, in some studies, in the anterior thalamus, but not in other extrastriatal regions where there are lower DA $D_{2/3}$ receptor levels (Köhler et al., 1985; Schonwetter et al., 1989; Mukherjee et al., 1996; Farde et al., 1985; Kung et al., 1989; Seibyl et al., 1996; Gründer et al., 2003; Hirvonen et al., 2008).

Both [^{11}C]raclopride and [^{18}F]desmethylfallypride have reported striatal: cerebellar ratios of 3-4; the specific uptake in the anterior thalamus for both radioligands is about 20% of that seen in the striatum. Specific striatal binding ratios for [^{123}I]IBZM are somewhat lower, about 1.75. With regard to modeling of substituted benzamide DA

D_2 radioligands, a number of approaches have been validated. Two-tissue compartmental tracer kinetic modeling with a metabolite-corrected arterial input function provides excellent fits of time–activity curves of these radioligands with moderately high affinity. Given the lack of specific cerebellar binding by radioligands with moderately high affinity for the DA D_2 receptor (Farde et al., 1988), modeling of striatal DA $D_{2/3}$ receptor levels has been validated for reference region methods, bolus/infusion methods, and Logan graphical plots with a cerebellar input function (Seibyl et al., 1996; Lammertsma et al., 1996; Logan et al., 1996, 2001; Carson et al., 1997; Gründer et al., 2003).

2.3.4. Dopamine D_2-Type Antagonist Radioligands for Extrastriatal Regions

Benzamide radioligands with very high affinity and selectivity for the DA $D_{2/3}$ receptors, i.e. [^{123}I]epidepride ($K_d = 24$ pM), [^{18}F]fallypride ($K_d = 31$ pM), as well as [^{11}C] and [^{76}Br] FLB457 ($K_d = 20$ pM) are the principal benzamide antagonist radioligands used to study the low to very low levels of DA $D_{2/3}$ receptors in extrastriatal brain regions (Figure 2). Compared to striatum, extrastriatal brain regions have DA $D_{2/3}$ receptor levels which vary from about 10-fold lower, i.e. medial thalamus, to 100-fold lower, i.e. dorsolateral prefrontal cortex (Kessler et al., 1991, 1993a–c, 2000; Halldin et al., 1995). Given the very high affinity of these radioligands for the DA $D_{2/3}$ receptor, the mass dose of tracer is an issue. Olsson has shown that for [^{11}C]FLB457, tracer dose mandates a mass of less than 0.5 μg which corresponds to a specific activity of 6000 Ci/mmol for an 8.1 mCi dose (Olsson et al., 2004). A mass dose of 1.6 μg, corresponding to a specific activity of 1875/mmol for an 8.1 mCi dose, leads to 23% occupancy of extrastriatal DA D_2 receptors. For [^{18}F]fallypride, a dose of 5 mCi with a specific activity of 2000 Ci/mmol corresponding to about 1 μg produces a receptor occupancy of 5% or less in extrastriatal regions (Kessler et al., 2000). For [^{123}I]epidepride, specific activities are typically very high due to use of no-carrier-added ^{123}I and so mass dose is not usually a concern.

The relatively slow in vivo dissociation of [^{123}I]epidepride, and [^{11}C] and [^{76}Br] FLB457 from the DA $D_{2/3}$ receptor in vivo makes quantitation of the high levels of DA $D_{2/3}$ in striatum difficult for these radioligands (Kessler et al., 1993b; Fujita et al., 1999; Delforge et al., 1999). After bolus administration of tracer doses to humans, [^{123}I]epidepride has an apparent $t_{½}$ of dissociation from striatal DA $D_{2/3}$ receptors of approximately 4.3 h (Fujita et al., 1999). Although a formal time/stability of estimates analysis was not presented by Fujita for two-compartment tracer kinetic analysis following a bolus injection, the data presented for [^{123}I]epidepride suggests that at least 12 h of serial scanning is needed to estimate striatal binding potentials (Fujita et al., 1999). Pinborg (Pinborg et al., 2000) has reported that extrastriatal DA $D_{2/3}$ receptor levels can be estimated using a 7-h bolus infusion paradigm. Both Pinborg and Fujita have examined the use of a bolus infusion paradigm for estimating striatal DA $D_{2/3}$ receptor levels and have shown that at least a 24-h infusion is needed for striatal DA $D_{2/3}$ receptor level quantitation.

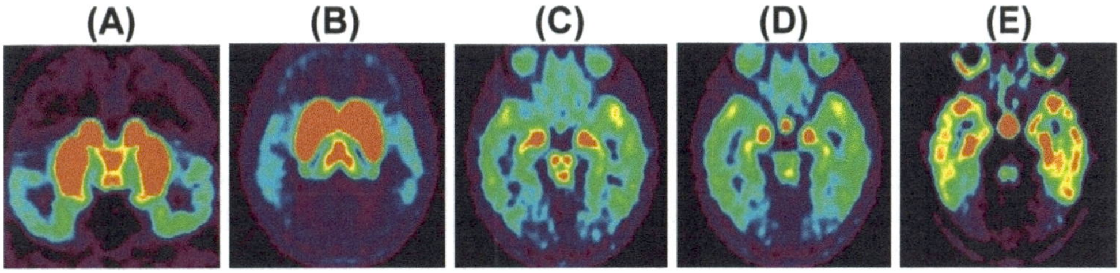

Figure 2 Coronal and axial images of [^{18}F]fallypride uptake in brain windowed to demonstrate extrastriatal DA $D_{2/3}$ receptors. (A) Coronal uptake through the level of the anterior thalamus. Relatively high uptake is seen in the in the midline in the anterior intralaminar nuclei and anterior nucleus of the thalamus. Just inferior to the anterior thalamus, the paired mammillary bodies of the hypothalamus are seen. Lower uptake is seen in the temporal cortex with very low uptake in the frontal lobes. (B–E) Axial images. (B) Image through the dorsal striatum at the level of the midthalamus and demonstrating high midline uptake in the anterior intralaminar and midline thalamic nuclei, and posteriorly and laterally in the centromedian/parafascicular nuclei. (C) Image at the level of the midbrain and demonstrating uptake in the substantia nigra in the ventral midbrain and in the colliculi in the dorsal midbrain as well as moderate uptake in the temporal cortex. (D) Amygdala DA $D_{2/3}$ receptor binding as well some uptake in the dorsal pons in the region of the dorsal raphe and temporal cortex. (E) Image shows that the highest levels of cortical DA $D_{2/3}$ receptors are in the inferior temporal cortex, particularly in the region of the collateral sulcus. Typically, there are higher levels of DA $D_{2/3}$ receptors in left temporal cortex as shown in (E). High uptake is also seen in the pituitary.

Modeling studies of tracer kinetics following a bolus injection indicate that 5 h are need for [^{76}Br]FLB457 to achieve an equilibrium state in the putamen which makes [^{11}C]FLB457 unsuitable for quantitating striatal DA $D_{2/3}$ receptor levels (Delforge et al., 1999). As a result, both [^{123}I]epidepride and [^{11}C]FLB457 are generally used only to quantitate extrastriatal DA $D_{2/3}$ receptor levels. If quantitation of both striatal and extrastriatal DA $D_{2/3}$ is needed, then a second study with radioligands such as [^{11}C] raclopride must be performed. [^{18}F]Fallypride, while having a slightly lower in vitro affinity for the DA D_2 receptor (a K_d of about 31 pM), has a much faster off rate in vivo than [^{123}I]epidepride or [^{11}C]FLB457; following displacement by large doses of haloperidol or raclopride, the in vivo $t_{½}$ of displacement is about 14 min for [^{18}F]fallypride vs 55 min for [^{123}I]epidepride and 33 min for [^{11}C]FLB457 (Kessler, unpublished data; Kessler et al., 1993b; Delforge et al., 1999). The faster off rate seen with [^{18}F]fallypride as well as its somewhat higher lipophilicity allows quantitation of striatal and extrastriatal regions with scan times of 3–3.5 h (Kessler et al., 2000; Schmidt et al., 1994; Vernaleken et al., 2011).

While the reference region methods can be used to quantitate striatal and extrastriatal DA $D_{2/3}$ receptor levels when using PET [^{18}F]fallypride studies (Kessler et al., 2000, 2005; Siessmeier et al., 2005; Kegeles et al., 2008), caution should be exercised when using reference region methods to quantify extrastriatal DA D_2 receptor levels with [^{123}I] epidepride and [^{11}C] or [^{76}Br]FLB457. There are very low cerebellar DA $D_{2/3}$ receptor levels which these radioligands can detect due to their low lipophilicity and very high affinity for the DA $D_{2/3}$ receptors (Kessler et al., 1991; Pinborg et al., 2007; Vandehey et al., 2010). For [^{123}I]epidepride, Pinborg reported that the cerebellar volume of distribution was decreased by 22% in schizophrenic subjects after months of risperidone treatment (mean dose = 3.8 mg) or zuclopenthixol therapy (mean dose = 9.6 mg). [^{123}I] Epidepride studies performed at baseline and 4 h following a 6 mg oral dose of haloperidol produced a 27% decrement in cerebellar uptake (Kessler, unpublished data). Narendran has reported that a single 15 mg dose of aripiprazole produced a 79% occupancy of DA $D_{2/3}$ receptors which led to a 17% decrease in the cerebellar V_T for [^{11}C]FLB457 in healthy subjects suggesting that about 21% of cerebellar uptake is to cerebellar DA $D_{2/3}$ receptors (Narendran et al., 2011a).

Studies in baboons demonstrate that 77% of cerebellar uptake of [^{76}Br]FLB457 is due to specific binding, while studies in rodents indicate that 60% of cerebellar [^{11}C]FLB457 uptake is specific (Delforge et al., 1999; Asselin et al., 2007). Modeling studies demonstrate that considerable caution is required in the use of a cerebellar reference region method for quantitation of extrastriatal DA D_2 receptor levels when using [^{11}C]FLB457 (Delforge et al., 1999; Asselin et al., 2007; Vandehey et al., 2010). If reference region methods are used, the specific cerebellar binding seen with [^{123}I] epidepride and [^{11}C] and [^{75}Br]FLB457 may lead to overestimates of extrastriatal DA D_2 receptor occupancy by antipsychotic drugs and underestimation of DA D_2 receptor levels in regions with very low levels of specific uptake such as cortex (Asselin et al., 2007; Pinborg et al., 2007; Vandehey et al., 2010). In addition, nonuniform changes in regional cerebral blood flow may lead to appreciable changes in regional estimates of DA D_2-type receptor density when reference tissue methods are used with these radioligands (Asselin et al., 2007). This sensitivity to regional cerebral blood flow may confound the interpretation of changes in regional binding with cognitive activations which are postulated to produce DA release and the effect of drugs which produce regionally specific changes in cerebral blood flow (Aalto et al., 2005). In contrast, [^{18}F] fallypride which has also been used to quantify extrastriatal receptors appears to have very low levels of specific cerebellar uptake in humans and primates. The low level of specific cerebellar uptake for [^{18}F]fallypride is likely due to its higher lipophilicity than epidepride and FLB457 (i.e. logK_W at pH 7.4 = 2.48 vs 2.05 and 1.82) leading to higher nonspecific binding and slightly lower affinity for the DA D_2 receptor both in vitro and in vivo (Schmidt et al., 1994; Olsson et al., 2004; Slifstein et al., 2004a,b; Kessler et al., 2005; Vandehey et al., 2010). As a result, [^{18}F]fallypride does not have appreciable specific cerebellar binding and so reference region methods are appropriate for modeling regional DA D_2 receptor levels (Kessler et al., 2000, 2005; Kegeles et al., 2008; Vandehey et al., 2010).

Of the available butyrophenone PET radioligands for the DA D_2 receptor, only [^{18}F] NMB studies has demonstrated an ability to quantitate DA D_2 receptor levels in some extrastriatal regions (Eisenstein et al., 2012). An analysis of extrastriatal binding of PET [^{18}F]NMB indicates that binding potentials in thalamus and amygdala are high enough to allow studies in these regions but groups of 35–50 subjects are needed to demonstrate between group effects. Uptake in cortex, hippocampus, and midbrain was so low that large groups of subjects would be required to adequately quantitate such levels.

The available antagonist radioligands for quantitation of extrastriatal DA $D_{2/3}$ receptors all have strengths and limitations. [^{18}F]NMB is limited by low binding potentials in extrastriatal brain regions. [^{18}F]Fallypride does not require arterial sampling for quantitation of DA $D_{2/3}$ receptor levels and can estimate DA $D_{2/3}$ receptor levels in both striatal and extrastriatal regions, but requires about 3–3.5 h of scanning. The mass injected should be less than 1 μg of fallypride requiring a specific activity of greater than 2000 Ci/mmol; this is an achievable specific activity for ^{18}F-labeled compounds. [^{18}F]Fallypride has somewhat lesser affinity and higher lipophilicity than [^{123}I]epidepride and [^{11}C]FLB457 and so has somewhat lower BP's in cortical regions; [^{11}C]FLB457 has 30–50% higher BP's (Narendran et al., 2009a). However, [^{11}C]FLB457 studies require an arterial input function for adequate quantitation as well as very high specific activity, at least 5000–6000 Ci/mmol which can be difficult to achieve. In addition, a second tracer injection is needed if striatal DA D_2 receptor levels are to be quantitated. However, if quantitation of frontal cortical DA D_2 receptor levels and a within subject same day test–retest paradigm is needed, then [^{11}C]FLB457 may be the tracer of choice. [^{123}I]Epidepride is the only available SPECT radioligand which has been validated for quantitation of extrastriatal DA D_2 receptor levels; it requires either a long bolus infusion (7 h) or an arterial input function with tracer kinetic modeling.

2.3.5. Dopamine $D_{2/3}$ Agonist Radioligands for Striatal Regions

High-affinity DA $D_{2/3}$ agonist radioligands for imaging striatal DA $D_{2/3}$ receptors include [^{11}C]NPA, [^{11}C]MNPA, and [^{18}F]MCL-524 (Narendran et al., 2004; Finnema et al., 2005, 2012). These ligands can be used to estimate DA $D_{2/3}$ receptor levels in the striatum but not in extrastriatal regions. Their affinities for the high-affinity agonist state of the DA D_2 receptor are in the range of 0.1–0.4 nM (Narendran et al., 2004; Skinbjerg et al., 2009) with 50-fold lower affinity for the low-affinity agonist state of the DA D_2 receptor in vitro. All appear to have reversible kinetics with 60–90 min of data collection leading to stable estimates of binding potentials and are appropriately modeled by reference region methods (Otsuka et al., 2009; Narendran et al., 2009b, 2010). Mass doses of less than 2 μg of [^{11}C]NPA are needed to maintain a tracer dose, i.e. less than 5% receptor occupancy; while similar analyses have not been published for [^{11}C]MNPA and [^{18}F]MCL-524, it is likely that similar mass dose limitations apply. The motivation for development of DA $D_{2/3}$ agonist radioligands is at least twofold. First the high-affinity state of the DA D_2 receptor is believed to be the functional state of the receptor. Second, if as suggested by in vitro membrane binding studies, only a fraction of DA D_2 receptors are in the high-affinity state, then agonist radioligands may be more sensitive to extracellular DA levels than antagonist radioligands.

While in vitro studies have clearly shown the presence of two affinity states of the DA D_2 receptor, primate PET studies have not shown convincing evidence that PET imaging with DA $D_{2/3}$ agonist ligands can successfully detect the high-affinity vs low-affinity state of the DA $D_{2/3}$ receptor in vivo (Skinbjerg et al., 2012). PET studies in primates comparing competitive displacement of [^{11}C]MNPA and [^{11}C]raclopride by apomorphine, a DA agonist, over a broad range of doses have failed to show a difference in displacement curves (Finnema et al., 2009). A primate PET study of (+)PD128907 displacement of [^{11}C]raclopride showed that agonist-induced displacement showed no evidence of two affinity states (Kortekaas et al., 2004). Ex vivo studies in conscious rats using [^{3}H]raclopride and [^{11}C]PHNO similarly showed no differences in displacement curves following administration of DA D_2 antagonists, partial or full agonists (McCormick et al., 2008). The reasons for the discrepancy between in vitro and in vivo studies may relate to the limited temporal resolution of PET studies vs the half-life of the high-affinity agonist state. It is unclear whether there are large numbers of stable receptor G protein complexes in vivo; there is considerable evidence to suggest rapid coupling of receptors to G proteins within the membrane after binding of the ligand to the receptor.

PET imaging may have insufficient temporal resolution to distinguish these states (Skinbjerg et al., 2012). Another explanation has been put forward by Seeman (2012). Seeman notes that [^{11}C]NPA and [^{11}C]PHNO have in vitro half-lives of dissociation of about 100–600 s (22 °C in vitro), while the transition from high- to low-affinity states may occur in a second; thus these radioligands will remain on the receptor after its transition to a low-affinity state. While Narendran has performed an in vivo saturation binding analysis of [^{11}C]NPA and [^{11}C]raclopride in baboons which suggests that 79% of striatal DA $D_{2/3}$ receptors were in the high-affinity state (Narendran et al., 2005a,b), the above studies suggest that the ability of agonist radioligands to distinguish the high- from the low-affinity agonist state of the DA D_2 receptor is limited. The issue of greater sensitivity of DA $D_{2/3}$ agonist radioligands to extracellular DA levels will be discussed below.

2.3.6. Dopamine D_3 Radioligands

While a number of radioligands for the DA D_3 receptor have been synthesized (Sóvágó et al., 2004; Kuhnast et al., 2006; Turolla et al., 2005; Hocke et al., 2008; Bennacef et al., 2009), to date only [^{11}C]PHNO appears capable of reliably imaging DA D_3 receptors in vivo (Wilson et al., 2005; Narendran et al., 2006; Graff-Guerrero et al., 2009a; Searle et al., 2010) (Figure 3). In vitro PHNO has been reported to have a $K_d = 0.16$ nM for the DA D_3 and a $K_d = 8.5$ nM for the high-affinity state of the DA D_2 receptor. This is similar to calculated in vivo affinities (K_d/f_{ND}) of [^{11}C]PHNO for the DA D_3 receptor vs DA D_2 receptor, i.e. in the range of 0.23–0.56 nM vs 11–14 nM (Gallezot et al., 2012). Both estimates suggest a 50-fold selectivity of PHNO for the DA D_3

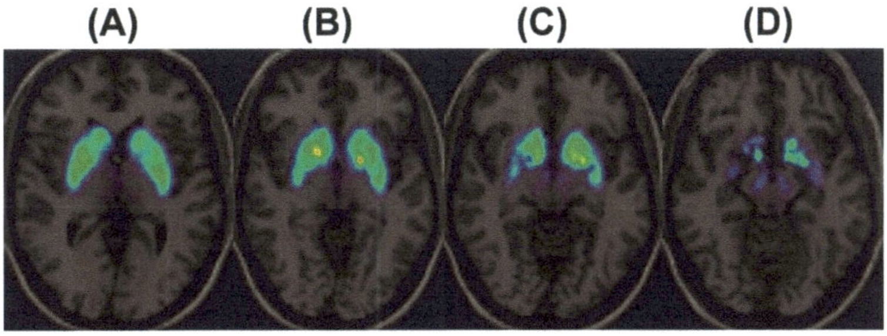

Figure 3 Sequential images of [^{11}C]PHNO uptake in human brain. Highest uptake is seen in the ventral striatum/ventral pallidal region (B,C). Uptake in the dorsal striatum (A) is due almost entirely to binding to DA D_2-type receptors, while binding in the ventral midbrain, i.e. substantia nigra, (D) is almost entirely due to binding to DA D_3 receptors. *Images courtesy of Dr Romina Mizrahi, University of Toronto.*

receptor. [^{11}C]PHNO demonstrates a different pattern of regional localization from that seen with high-affinity DA $D_{2/3}$ radioligands and consistent with the known distribution of DA D_3 receptors. While different publications report slightly differing regional localizations depending on how regions of interest are delineated, the highest uptake is observed in the region of the ventral pallidum/substantia innominata (BP_{ND} = 3.2, fraction of uptake due to D_3 receptors (f_{D_3}) = 0.69 – 0.75), globus pallidus (BP_{ND} = 2.6, f_{D_3} = 0.57–0.67), ventral striatum (BP_{ND} = 2.48–2.94, f_{D_3} = 0.19–0.26), substantia nigra (BP_{ND} = 1–1.39, f_{D_3} = 0.92–1.0), hypothalamus (BP_{ND} = 0.96–1.1, f_{D_3} = 1.0), and thalamus (BP_{ND} = 0.26–0.39, f_{D_3} = 0.43–0.46) (Searle et al., 2010; Tziortzi et al., 2011).

DA D_3 binding in the hypothalamus likely is largely due to the high levels of DA D_3 receptors in the mammillary bodies (Gurevich and Joyce, 1999). DA D_3 uptake in the thalamus appears to be due to the high levels of DA D_3 receptors in the anterior thalamic nucleus. The f_{D_3} reported for dorsal striatal regions in humans is very low, 0.01 or less, in contrast to postmortem studies which find that 10–20% of total DA D_2-type dorsal striatal binding is due to the DA D_3 receptor (Gurevich and Joyce, 1999). The divergence between imaging and postmortem studies may be due to the high levels of extracellular DA in the striatum and the higher affinity of DA D_3 receptors for DA compared to DA D_2 receptors which may produce higher levels of DA D_3 receptor occupancy by DA in the striatum than in extrastriatal regions (Freedman, 1993). This explanation is consistent with studies of [^{18}F]5, a very selective, high-affinity DA D_3 radioligand which demonstrated no selective striatal uptake in rhesus monkeys at baseline but some specific uptake after administration of lorazepam which decreases extracellular DA levels (Mach et al., 2011). [^{11}C]PHNO binding in the substantia nigra and hypothalamus is almost entirely due to binding to the DA D_3 receptor, while uptake in the dorsal striatum is almost entirely due to binding to the DA D_2 receptor. In the ventral striatum, globus pallidus, and thalamus, [^{11}C]PHNO uptake reflects binding to both DA D_3 and D_2 receptors.

With regard to quantitation of DA D_3 receptor levels with [^{11}C]PHNO, there are caveats regarding both the use of reference region methods and mass doses. Regional kinetics of [^{11}C]PHNO uptake depend on the fraction of uptake due to the DA D_3 receptor; in DA D_3-rich regions such as the ventral pallidum/ventral striatum and substantia nigra, there is slower washout of radioactivity reflecting the higher affinity of [^{11}C]PHNO for the DA D_3 than D_2 receptor. A two-tissue compartment kinetic model with an arterial input function, provides excellent fits of regional [^{11}C]PHNO time–activity curves and stable estimates of binding potentials; however, in the primate cerebellum, there is appreciable specific uptake, i.e. 17–30%, due to DA D_3 receptor binding (Rabiner et al., 2009; Searle et al., 2010). This specific cerebellar uptake may explain differences in published studies regarding occupancy of DA D_3 receptors by antipsychotic drugs.

Primate [^{11}C]PHNO PET studies of occupancy of DA D_2 and D_3 receptors in the caudate, putamen, and globus pallidus by haloperidol and clozapine demonstrated regional occupancies which corresponded to expected levels based on the affinities of these antipsychotic drugs to DA D_2 and D_3 receptors (Girgis et al., 2011); a two-tissue compartment model with a metabolite-corrected arterial input function was used in this primate study. In contrast, studies of DA D_2 and D_3 receptor occupancy using [^{11}C] PHNO PET studies in drug-naïve schizophrenic subjects prior to and following 2.5 weeks of antipsychotic drug therapy demonstrated unexpected increases in globus pallidus and substantia nigra binding potentials suggesting DA D_3 receptor upregulation (Mizrahi et al., 2011); similar findings with [^{11}C]PHNO PET studies were also seen in chronically treated schizophrenic subjects (Graff-Guerrero et al., 2009b). A reference region method was used in the human studies. It is unclear whether the higher DA D_3 receptor levels in substantia nigra and globus pallidus are due to upregulation of receptors with antipsychotic drug therapy or whether some of the differences in these studies reflect methodological issues related to the use of the reference region method (Girgis et al., 2011).

A second caveat regarding [^{11}C]PHNO studies relates to the issue of acceptable mass doses. Mass doses above 3 μg are associated with side effects including nausea and vomiting (Mizrahi et al., 2010). Studies of [^{11}C]PHNO affinity for the DA D_3 receptor suggest that in humans a dose of 0.01 μg/kg, corresponding to an average mass dose of about 0.7 μg, is needed to maintain a receptor occupancy of under 10% (Rabiner and Laruelle, 2010; Gallezot et al., 2012). Clinical studies, however, have administered mass doses of 2–3 μg which produces significant DA D_3 receptor occupancy (Mizrahi et al., 2011; Tziortzi et al., 2011); this may be a confounding factor in evaluating DA D_3 receptor levels particularly in studies of receptor occupancy by antipsychotic drugs. The available data suggest that the presence of significant levels of DA D_3 receptors in the cerebellum makes the cerebellum inappropriate as a reference region. Further advances in radiochemistry are needed to achieve tracer-level mass doses of [^{11}C]PHNO.

2.3.7. Dopamine D_4 Receptor Radioligands

Currently, there are no PET or SPECT radioligands for imaging the DA D_4 receptor.

2.4. EFFECTS OF EXTRACELLULAR DA LEVELS ON DOPAMINE $D_{2/3}$ RECEPTOR BINDING

2.4.1. Studieso of Striatal DA Release

The specific brain uptake of many but not all substituted benzamide DA D_2 radioligands is modulated by extracellular DA levels (Innis et al., 1992; Dewey et al., 1992; Kessler et al., 1993c; al-Tikriti et al., 1994). Butyrophenone DA D_2 radioligands such as spiperone derivatives or radiolabeled benperidol analogs have not reliably shown such modulation or in some studies have shown paradoxical changes such as increased specific binding following elevation of extracellular DA levels (Laruelle, 2000; Moerlein et al., 1997). Both amphetamine and methylphenidate have been used to increase extracellular DA levels leading to decreased specific striatal binding of benzamides such as [^{123}I]IBZM, [^{11}C]raclopride, [^{18}F]fallypride, [18F]desmethoxyfallypride, and decreased specific extrastriatal binding of [^{18}F]fallypride and [^{11}C]FLB457 (Laruelle et al., 1995; Breier et al., 1997; Mukherjee et al., 1996; Riccardi et al., 2006; Narendran et al., 2009a). DA D_2- and D_3-type agonist radioligands such as [^{11}C]NPA, [^{11}C]PHNO, [^{18}F]MCL-524, and [^{11}C]MNPA may have greater sensitivity to extracellular DA levels than benzamide DA $D_{2/3}$ antagonist radioligands (Narendran et al., 2010; Ginovart et al., 2006; Shotbolt et al., 2012; Seneca et al., 2006; Finnema et al., 2012).

Amphetamine releases intracellular DA by reversing the transport of DA at both the level of the vesicular monoamine transporter and the DA transporter; amphetamine releases not only DA but also norepinephrine and serotonin. In contrast to amphetamine, methylphenidate raises extracellular DA levels by blocking the reuptake of extracellular DA into presynaptic neurons by the DA transporter; methylphenidate also blocks the norepinephrine transporter. Laruelle and Breier demonstrated that the amphetamine-induced decreases in striatal uptake of [^{123}I]IBZM and [^{11}C]raclopride were linearly related to the peak level of extracellular DA following amphetamine administration. Extracellular DA levels increased linearly up to a maximum intravenous (IV) amphetamine dose of about 0.75–1.0 mg/kg in primates; these doses produced a maximal decrease in striatal [^{123}I]IBZM uptake of about 50% (Laruelle et al., 1997a,b; Laruelle, 2000; Breier et al., 1997). In primates, combined PET and microdialysis studies suggest that the ratio of peak amphetamine-induced increases in extracellular striatal DA levels to decreases in striatal [^{11}C]raclopride and [^{123}I]IBZM binding potentials/V_3's is about 44:1 (Breier et al., 1997; Laruelle et al., 1997a,b). Modeling studies of the effects of amphetamine-induced increases in extracellular DA levels on [^{11}C]raclopride striatal binding potentials demonstrated that the reduction in specific binding is related to the integral of released DA (Endres et al., 1997).

Interestingly, combined animal imaging/microdialysis studies of striatal DA release demonstrate that amphetamine and methylphenidate doses which produce comparable displacement of striatal [^{11}C]raclopride specific binding show fourfold differences in psychostimulant-induced extracellular DA levels, being fourfold lower following methylphenidate administration than after amphetamine administration (Schiffer et al., 2006). Similarly, scopolamine has been shown to produce decreased striatal [^{11}C]raclopride binding potentials and increased striatal DA synthesis, but produced no change in extracellular DA levels as assessed by microdialysis (Tsukada et al., 2000). In the striatum there is a predominantly synaptic mode of dopaminergic neurotransmission (Garris and Wightman, 1994). As striatal DA transporters are believed to be arrayed at the edge of the synaptic cleft, different drugs may produce comparable intrasynaptic DA levels but with varying levels of extrasynaptic spillover where DA is detected by microdialysis (Sesack et al., 1998). The effect of increased striatal DA release on benzamide DA D_2 receptor radioligands appears to be predominantly mediated by intrasynaptic DA which may not be accurately reflected by microdialysis (Laruelle, 2000).

While psychostimulant-induced DA release leads to decreased specific uptake of benzamide radioligands such as [^{11}C]raclopride and [^{123}I]IBZM, the mechanism of this decrease is not completely delineated (Laruelle et al., 1997a,b; Laruelle, 2000). Microdialysis studies in primates have shown that IV amphetamine rapidly increases extracellular DA to peak levels by 20 min with a subsequent exponential decrease in extracellular DA levels with a half-life of about 40 min. However, PET and SPECT primate studies have shown no change in the decrements in striatal binding of [^{123}I]IBZM, [^{11}C]raclopride, and [^{11}C]NPA at 6 h after IV amphetamine administration, partial normalization of specific binding by 24 h, and complete normalization by 48 h (Laruelle et al., 1997a,b; Carson et al., 2001; Narnedran et al., 2007). Studies in mice have reported similar findings for [^{11}C]MNPA and [^{18}F]fallypride (Skinbjerg et al., 2010). An in vivo Scatchard analysis of amphetamine-induced changes in B_{max} and K_d reported that amphetamine induces decrements in both B_{max} and K_d (Ginovart et al., 2004).

These findings argue against a simple competitive inhibition mechanism being responsible for amphetamine-induced decrements in striatal-specific binding. As benzamides require a high Na^+ environment with a pH > 7 for optimal binding to the DA D_2 receptor, it has been proposed that DA-induced internalization of DA $D_{2/3}$ receptors would produce a decrease in affinity of benzamide DA $D_{2/3}$ radioligands for the DA $D_{2/3}$ receptor leading to dissociation from the internalized receptor and a decrease in specific binding (Kessler et al., 1993b; Laruelle, 2000). Studies of the duration of amphetamine-induced changes in specific striatal binding of [^{11}C]MNPA and [^{18}F]fallypride in arrestin-3-knockout mice have shown that arrestin-3-knockout mice who cannot internalize DA D_2 receptors have a similar decrement in specific striatal binding as wild-type mice at 30 min following amphetamine administration, but a much shorter duration of such decrements (Skinbjerg et al., 2010). Internalization is a factor in the prolonged duration of decrements of specific binding.

Internalization does not appear to be the principal factor mediating the magnitude of the amphetamine-induced decrement in specific binding. In vitro fluorescent microscopic studies of cultured cells expressing the DA D_2 receptor have shown lower affinity for the internalized DA D_2 receptor than the surface-expressed receptor across a group of benzamides and agonists, but no correlation was seen between these shifts in affinities and the magnitudes of D-amphetamine-induced decreases in specific binding in vivo (Guo et al., 2010). It has been argued that the magnitude of the amphetamine effect is not a function of radioligand affinity if studies are conducted using a tracer dose of a benzamide or agonist radioligand sensitive to extracellular DA levels and if an equilibrium state is achieved during the course of the measurement (Laruelle, 2000; Slifstein et al., 2004a,b); such arguments are based on a competitive inhibition model. Nearly all studies in humans use a pretreatment/blockade model with amphetamine being administered at varying times prior to the administration of radioligand, either as a slow IV bolus given 5–45 min prior to the administration of radioligand or orally about 3 h prior to radioligand administration (Abi-Dargham et al., 1998; Yoder et al., 2004; Riccardi et al., 2006).

Psychostimulant administration produces time-dependent changes in DA release. Thus even if an equilibrium is achieved with regard to the radioligand kinetics, it is convolved with the time-dependent DA release pattern. If one assumes competitive binding kinetics, then the magnitude of the decrement in specific radioligand binding resulting from a time-dependent increase in extracellular DA is determined not by the equilibrium K_d of the radioligand for the receptor but by the net transit time of the radioligand out of the brain which is determined by both the k_4, the transfer constant describing the transition from receptor-bound to unbound ligand in brain, and the k_2, the brain to plasma transit constant (Kessler et al., 1993c; Yoder et al., 2004; Morris and Yoder, 2007). The k_4 in the striatum reflects not only the k_{off} from the receptor but also the ability to diffuse from the synapse prior to rebinding to synaptic DA D_2 receptors which is a function of both lipophilicity and the concentration of receptors at the synapse (Kessler et al., 1993c; Votaw et al., 1993). The in vivo k_{off} in the striatum for benzamide radioligands determined by displacement using large doses of antagonist can be 5- to 10-fold larger than the modeled k_4 when tracer doses of radioligand are used. As the specific activity of the radioligand is decreased and the number of available synaptic receptors decrease, the modeled k_4 becomes larger (Kessler et al., 1993a,c; Votaw et al., 1993; Fujita et al., 1999). The value of k_2 is largely determined by the lipophilicity of the radioligand.

Thus radioligands with low lipophilicity and/or slow dissociation from the DA D_2 receptor will demonstrate lesser decrements in specific binding when D-amphetamine or methylphenidate displacement/blockade studies are performed (Kessler et al., 1993c; Votaw et al., 1993). These factors explain why [^{18}F]fallypride which has fivefold higher equilibrium affinity for the DA $D_{2/3}$ receptor than [^{18}F]5-fluoropropylepidepride but a more optimal lipophilicity for membrane transit and a similar k_{off} has greater sensitivity to D-amphetamine-induced striatal DA release (Kessler et al., 1993a,c; Slifstein et al., 2004a,b; Mukherjee et al., 2005). Although the K_d and k_{off} are usually correlated, fallypride is an exception as its k_{off} is considerably larger than expected for its affinity for the DA $D_{2/3}$ receptor. For benzamide DA $D_{2/3}$ radioligands, the transfer constants determining the rate at which a radioligand can move from being receptor bound to being transferred to plasma, k_2 and k_4, are important determinants of the magnitude of amphetamine- and methylphenidate-induced decrements in specific striatal binding. Similar factors may be operative for DA $D_{2/3}$ agonist radioligands.

One of the motivations for development of DA agonist radioligands has been the belief that DA $D_{2/3}$ agonist radioligands may be more sensitive to extracellular striatal DA levels given their selectivity for the high-affinity agonist state of the DA D_2 receptor. Studies of [^{11}C]NPA, [^{11}C]PHNO, and [^{11}C]MNPA displacement by amphetamine-induced DA release in anesthetized primates and cats reported approximately 40–60% higher decrements in striatal agonist binding potentials compared to [^{11}C]raclopride (Narendran et al., 2004; Seneca et al., 2006; Ginovart et al., 2006). Studies of conscious vs anesthetized rats have reported that anesthesia induces increases in specific striatal binding of [^{11}C]NPA and [^{11}C]PHNO but not [^{11}C]raclopride; these increases in agonist binding led to much greater amphetamine-induced decrements in striatal specific uptake of these agonist radioligands when animals were studied in the anesthetized state; however, no significant difference in amphetamine effects was seen between agonist and antagonist DA $D_{2/3}$ radioligands when animals were studied in the conscious state (McCormick et al., 2011). Similarly, studies in conscious and anesthetized monkeys demonstrated that ketamine/xylazine anesthesia produced increased baseline [^{11}C]MNPA binding potentials and 2.5-fold greater methamphetamine-induced decrements in striatal binding potentials in the anesthetized compared to the conscious state (Ohba et al., 2009). In contrast to studies in rats and monkeys, studies in humans comparing amphetamine-induced decrements in striatal [^{11}C]NPA and [^{11}C]PHNO binding potentials to those seen with [^{11}C]raclopride have reported approximately 50–60% greater decrements in BP_{ND} for these agonist tracers; only a 20% decrease was seen in BP_P suggesting that BP_{ND} is a more sensitive measure of striatal DA release (Narendran et al., 2010; Shotbolt et al., 2012).

It should be noted that in the dorsal striatum [^{11}C]PHNO binds almost completely to DA D_2 as opposed to D_3 receptors and that between 65% and 79% of striatal DA D_2 receptors may be in the high-affinity agonist state (Narendran et al., 2005a,b; Ginovart et al., 2006; Shotbolt et al., 2012). The available literature indicates that agonist DA $D_{2/3}$ radioligands are more sensitive to increased extracellular DA levels than antagonist radioligands despite the questions which have been raised regarding their ability to distinguish agonist from antagonist states of the DA D_2 receptor in vivo.

While doses of up to 1–2 mg/kg of amphetamine have been used in primate studies, typical amphetamine doses in humans are 0.3 mg/kg IV or 0.43–0.50 mg/kg orally

(Laruelle et al., 1995; Martinez et al., 2007; Cárdenas et al., 2004; Riccardi et al., 2006; Cropley et al., 2008b); these doses produce decreases of 6–18% in humans in striatal subregion benzamide binding potentials ([^{11}C]raclopride, [^{123}I]IBZM, and [^{18}F]fallypride) with lowest decrements in the caudate. Oral doses of 60 mg of methylphenidate and IV doses of 0.5 mg/kg of methylphenidate produce similar decrements in striatal binding potentials (Volkow et al., 1997, 2002; Martinez et al., 2011). While oral psychostimulants produce levels of DA release similar to those seen with IV administration, oral administration has a significantly lower incidence of side effects (Volkow et al., 1998; Cardenas et al., 2004).

Studies in humans of the test–retest reliability of amphetamine-induced decreases in striatal [^{123}I]IBZM binding potentials have shown a test–retest intraclass correlation coefficient of 0.89 and no significant difference between the test and retest studies; these indicate excellent reproducibility of this measurement (Kegeles et al., 1999). Although not always statistically significant, in humans, amphetamine-induced decrements in striatal DA $D_{2/3}$ agonist binding potentials, particularly BP_{ND}, were larger than those seen with benzamide antagonist radioligands (Narendran et al., 2010; Shotbolt et al., 2012). Given these greater effects, DA $D_{2/3}$ agonists such as [^{11}C]NPA and [^{11}C]PHNO may be a more sensitive indicator of psychostimulant-induced striatal DA release.

2.4.2. Studies of Extrastriatal Dopamine Release

[^{18}F]Fallypride, [^{11}C]FLB457, and [^{11}C]PHNO are the DA $D_{2/3}$ and D_3 preferring radioligands which have been reported to be sensitive to amphetamine- and methylphenidate-induced DA release in extrastriatal regions (Riccardi et al., 2006; Montgomery et al., 2007; Cropley et al., 2008b; Narendran et al., 2009a; Shotbolt, 2012). Narendran has recently validated the use of [^{11}C]FLB457 PET studies to estimate frontal cortical DA release; a linear relationship between amphetamine-induced frontal cortical DA release in primates and decrements in [^{11}C]FLB457 binding potentials was demonstrated using microdialysis and PET studies (Narendran et al., 2012). While amphetamine produces 10-fold or greater increases in absolute levels of extracellular DA release in striatum vs extrastriatal regions in primates (Laruelle et al., 1997a,b; Gerhardt et al., 2002; Narendran et al., 2012), amphetamine-induced decrements in striatal vs extrastriatal regions are similar, i.e. in the range of 7–13%, for many regions when using [^{18}F]fallypride (Riccardi et al., 2006; Cropley et al., 2008b).

The explanation for this unexpected similarity in level of decrements despite a marked difference in DA release is likely related to differences in modes of DA neurotransmission in striatal and extrastriatal regions. Unlike striatum where DA neurotransmission is largely synaptic, there is considerable evidence that in extrastriatal regions it operates predominantly in a volume/extrasynaptic mode (Garris and Wightman, 1994; Lewis et al., 2001; Pinto and Sesack, 2008). There are fewer DA transporters at extrastriatal DA synapses allowing diffusion of DA out of the synapse (Smiley et al., 1994; Sesack et al., 1998; Melchitzky et al., 2006). Only 39% of cortical DA fiber varicosities are seen at synapses suggesting predominantly nonsynaptic DA release (Smiley and Goldman-Rakic, 1993). In cortex the norepinephrine transporter and not the DA transporter appears to be the primary reuptake mechanism for extracellular DA which is also consistent with an extrasynaptic/volume mode of neurotransmission (Yamamoto and Novotney, 1998). As a result of these factors, extracellular DA has a much longer half-life than striatal extracellular DA, about 2 s in the frontal cortex and amygdala vs less than 60 ms in caudate, putamen, and nucleus accumbens (Garris and Wightman, 1994). In cortical and limbic regions, both DA D_2 and DA D_1 receptors are predominantly located extrasynaptically (Smiley et al., 1994; Negyessy and Goldman-Rakic, 2005; Pinto and Sesack, 2008).

These differences in striatal and extrastriatal DA neurotransmission produce differing DA D_2 radioligand kinetics in response to an amphetamine challenge. The lower levels of DA D_2 receptors and the absence of synaptic membranes to constrain diffusion away from the DA D_2 receptor markedly decreases the probability of radioligand rebinding to the DA D_2 receptor; this, for example, results in a larger k_4 which increases the effect of increased DA levels. Fujita has shown that following a tracer dose of [^{123}I] epidepride, the modeled k_4 in temporal cortex is 4.6 times larger than that in striatum likely due to the decreased probability of rebinding (Fujita et al., 1999). The decreased probability of rebinding following dissociation from the receptor is also likely the reason why amphetamine-induced displacements of very high affinity benzamide DA radioligands in extrastriatal regions are higher than expected. The available data regarding D-amphetamine-induced displacement of [^{11}C]PHNO demonstrates relatively high amphetamine-induced decrements in BP_{ND} in the thalamus (14%), ventral pallidum (24%), substantia nigra (27%), and hypothalamus (28%). These relatively high decrements likely result from the combination of high DA D_3 vs D_2 levels in these regions, the 20-fold higher affinity of DA for the DA D_3 vs D_2 receptor, and the volume/extrasynaptic mode of DA neurotransmission in extrastriatal regions (Shotbolt et al., 2012).

One issue regarding DA release in extrastriatal regions is the role of noradrenergic vs dopaminergic neurons as the source of extracellular DA. There is a denser noradrenergic than dopaminergic innervation in many extrastriatal regions, particularly in cortex and hippocampus where there is sparse DA innervation (Devoto et al., 2003; Smith and Greene, 2012). Rodent studies have shown that selective inactivation of noradrenergic innervation in cortex and hippocampus leads to approximately 75% decreases in extracellular DA levels in the occipital cortex and 50% decreases in frontal cortex, and prevents amphetamine-induced increases in extracellular DA in the hippocampus (Devoto et al., 2001, 2003; Smith and Greene, 2012). Electrical stimulation of the locus coeruleus significantly increases extracellular DA levels in both medial prefrontal cortex and occipital cortex (Devoto et al., 2005). The 6-OHDA lesions of VTA DA neurons reduced extracellular DA levels in frontal cortex by only 20%, but decreased nucleus accumbens levels by 92% (Devoto et al., 2008).

Disulfiram administration which blocks the conversion of DA to norepinephrine led to marked decreases in extracellular noradrenalin levels, being reduced by 60% and 80% in occipital and medial prefrontal cortex respectively, but led to 250% and 300% increases in extracellular DA levels in occipital and medial prefrontal cortex, respectively. In addition, disulfiram treatment markedly increased cocaine-induced DA release in the prefrontal cortex (Devoto et al., 2012). No studies in humans or primates to date have examined the role of noradrenergic vs dopaminergic neuronal origin of amphetamine-induced increases in extracellular DA in cortex and the hippocampus.

The available radioligands for studying extrastriatal DA release have differing strengths. It appears likely that to detect amphetamine-induced DA release in cortex a within-subject design is needed (Montgomery et al., 2007; Aalto et al., 2009; Narendran et al., 2009b). [^{18}F]Fallypride can be used to estimate DA release in striatum and extrastriatal regions without arterial sampling, but amphetamine administration requires a second day of scanning. Baseline and postamphetamine [^{11}C]FLB457 studies can be performed in a single day and there has been one publication suggesting that its sensitivity to cortical DA release is greater than that seen with [^{11}C]fallypride (Narendran et al., 2009a). However, no comparison was made to [^{18}F]fallypride which will have greater signal to noise at later time points which are critical for determination of binding potentials.

As discussed above, about 21% of [^{11}C]FLB457 cerebellar uptake appears to be specific; and so arterial sampling and metabolite analyses appear to be required for accurate quantitation of DA release if [^{11}C]FLB457 is used (Montgomery et al., 2007; Narendran et al., 2011a). Estimation of striatal DA release is not possible with [^{11}C]FLB457. [^{11}C]PHNO also appears to require arterial sampling with metabolite analysis given its specific cerebellar binding. However, its major advantages are the larger amphetamine effects in striatum as well as in extrastriatal regions which have a significant level of DA D_3 receptors, i.e. substantia nigra, hypothalamus, and thalamus. The choice of radioligand used to estimate extrastriatal DA release will depend on the regions which need to be examined and whether both baseline and postpsychostimulant studies need to be performed within a single day.

2.4.3. Studies of Baseline Extracellular Dopamine Levels

Depletion of cerebral DA levels by inhibition of DA synthesis using multiple doses of AMPT (5–8 g/day over 24–48 h) leads to depletion of DA stores and decreased levels of extracellular DA which produces increased binding potentials of radiolabeled DA $D_{2/3}$ benzamide antagonist and agonist radioligands due to increased radioligand affinity for the DA $D_{2/3}$ receptor (Ross and Jackson, 1989; Ginovart et al., 1997b; Laruelle et al., 1997b; Fujita et al., 2000; Verhoeff et al., 2002, 2003; Riccardi et al., 2008; Cropley et al., 2008b). AMPT dosing schedules of 5 g over 26 h, 5.25 g over 29 h, 5.5 g over 37 h, and 6–8 g over 48 h have reported a 64–70% decrease in plasma HVA levels, and plasma AMPT levels of about 20–29 μg/ml have been calculated to produce about a 75% inhibition of tyrosine hydroxylase, the rate-limiting enzyme for DA synthesis (Laruelle et al., 1997a,b; Fujita et al., 2000; Verhoeff et al., 2002; Riccardi et al., 2008; Cropley et al., 2008b). These depletion schedules have been reported to produce increases in [^{11}C]raclopride striatal binding potentials of 11–13% with 10, 9, and 14% increases in the ventral striatum, anterior caudate, and putamen, respectively (Verhoeff et al., 2002; Kegeles et al., 2001, 2010a,b); a broad range of changes in [^{123}I]IBZM striatal binding potentials have been reported, i.e. from about 9% to 28%; the 9% change is more consistent with results from [^{11}C]raclopride studies (Laruelle et al., 1997a,b; Abi-Dargham et al., 2000b).

There have been two studies of AMPT effects on [^{18}F]fallypride binding potentials (Riccardi et al., 2008; Cropley et al., 2008b) and single studies using [^{123}I]epidepride (Fujita et al., 2000) or [^{11}C]FLB457 (Frankle et al., 2010); all studies which examined changes in cortical binding potentials failed to detect consistent significant changes in cortical regions outside the temporal lobe. In one study (Fujita et al., 2000) a significant effect in temporal cortex was reported and a second study found a significant effect in temporal cortex only when using BP_{ND} as the outcome measure but not when using BP_P (Frankle et al., 2010). The low level of baseline extracellular DA seen in cortical regions likely leads to a small effect which may not be easily detectable (Narendran et al., 2012).

There is a divergence in results for the two [^{18}F]fallypride studies. The first study (Riccardi et al., 2008) reported effects in the striatum similar to that seen with [^{11}C]raclopride studies, a 12.7% change in substantia nigra, and a parametric image analysis which showed significant changes in the hypothalamus as well; the second study did not find significant effects (Cropley et al., 2008b). There are methodological differences in these studies. In the first study, the AMPT PET studies were performed within 2 days following the initial PET study, while subjects in the second study had AMPT PET studies about 10 weeks following the baseline studies which would be expected to lead to greater variability between baseline and post-AMPT studies. A second difference is the lack of AMPT side effects in the Cropley study, while the Riccardi study reported a very high incidence of dysphoria and akathisia which required treatment with either diphenhydramine or benzodiazepines in order to complete the imaging studies. No dropouts were seen in the Cropley study due to side effects; other studies have observed dropout rates of 30–45% due to AMPT side effects (Fujita et al., 2000; Frankle et al., 2010; Kessler, unpublished data; Kegeles, personal communication). Given the expected 75% DA depletion with the AMPT plasma level reported by Cropley, the lack of side effects is puzzling and suggests that DA depletion with the 3 g/day schedule has been insufficient to adequately deplete DA.

Baseline extracellular striatal DA levels can be estimated using [^{11}C]raclopride, [^{123}I]IBZM, and likely [^{18}F]fallypride using dose schedules of 5–8 g for a 70-kg subject administered over 26–48 h. In addition, it appears that baseline extracellular DA levels in the substantia nigra can be estimated when using [^{18}F]fallypride if the pre- and post-AMPT studies are conducted within a few days of each other and adequate doses of AMPT are given.

The high incidence of akathisia and dysphoria and high dropout rates with AMPT administration schedules of 5–8 g over 26–48 h raises the issue of whether dose schedules exist which allow adequate depletion without such a high incidence of side effects. There are only two studies which have examined this issue. The first reported a within-subject comparison of 4 g administered over 24 h (1 g every 6 h for 24 h) vs 8 g administered over 48 h using [^{11}C]raclopride PET studies in seven healthy subjects (Kegeles et al., 2001); 4 g of AMPT administered over 24 h produced a mean increase of 8.4% in striatal [^{11}C]raclopride V_3s, while 8 g administered over 48 h resulted in an 11.4% increase. Comparable changes were seen in all striatal subregions. However, due to the large standard deviations, no significant changes were seen between 4 g over 24 h and 8 g over 48 h. The second study examined oral administration of 1500 mg AMPT in six subjects over a 4-h period but found no change in striatal [^{123}I]IBZM BP_{ND}s; a 40 mg/kg dose corresponding to 2.8 g over 24 h for a 70-kg subject produced an 8.8 ± 9.0% (SD) increase which is remarkably close to the value reported by Kegeles for the 4-g schedule over 24 h (Boot et al., 2008). The incidence of akathisia was not reported but there was a significant drop in the subjective well-being of the subjects studied at the 40 mg/kg dose. While the changes seen with 2.8–4 g doses over 24 h are significant, greater increases of 11–13% in striatal binding potentials have been reported with dosing schedules of 5–8 g over 26–48 h (Kegeles et al., 2001; Verhoeff et al., 2002, 2003; Riccardi et al., 2006). Given the noise of these determinations, extra signal translates to an important increase in power in comparing patient groups to controls.

2.5. CLINICAL STUDIES USING DOPAMINE D_2 TYPE RECEPTOR LIGANDS

2.5.1. Overview

While an extensive review of all clinical applications of DA $D_{2/3}$ radioligands is beyond the scope of this chapter, imaging studies of DA $D_{2/3}$ receptors in schizophrenia and drug abuse will be reviewed as these are disorders in which DA $D_{2/3}$-mediated neurotransmission plays an important role and which have been the subject of a number of studies. As with DA D_1 receptors, there is a significant age-related loss of striatal and extrastriatal DA $D_{2/3}$ receptors. The mean rate of striatal and thalamic DA $D_{2/3}$ receptor loss is about 5–6% per decade in adults; the decrements in cortical DA $D_{2/3}$ receptor loss are higher, i.e. 9–12% per decade, in frontal than temporal cortex while, decrements in amygdala and hippocampus are 7 and 10% per decade, respectively (Rinne et al., 1993; Antonini et al., 1993; Kaasinen et al., 2000, 2002).

2.5.2. Schizophrenia

In the 1980s, imaging studies of DA D_2 receptors in schizophrenia were motivated by the observations that all clinical antipsychotic drugs were DA D_2 receptor antagonists, that the clinically effective dose of antipsychotic drug was directly correlated with its potency at the DA D_2 receptor, and that postmortem studies demonstrated increased striatal DA D_2 receptor levels in at least a subgroup of schizophrenic subjects (Seeman et al., 1975, 1984). While early studies of striatal DA $D_{2/3}$ receptor levels in schizophrenia using [^{11}C]*N*-methylspiperone demonstrated increased DA $D_{2/3}$ levels in drug-free schizophrenic subjects (Wong et al., 1986c), many subsequent studies using substituted benzamide radioligands, i.e. [^{11}C]raclopride and [^{123}I]IBZM, failed to find significant elevations in striatal DA $D_{2/3}$ receptor density in drug-naïve and drug-free schizophrenic subjects, although there was greater variability in DA D_2 receptor levels in schizophrenic subjects (Farde et al., 1990; Hietala et al., 1994a,b; Talvik et al., 2006; Kegeles et al., 2010b). However, a meta-analysis of striatal DA $D_{2/3}$ receptor levels in schizophrenia concluded that DA $D_{2/3}$ receptor levels are mildly elevated in schizophrenia (Laruelle, 1998).

Consistent with this finding are recent [^{123}I]IBZM studies (Corripio et al., 2011; Pogarell et al., 2012) which have reported increased DA $D_{2/3}$ levels in unmedicated schizophrenics. Interestingly, a study of AMPT-induced DA depletion in schizophrenics demonstrated that striatal [^{123}I]IBZM binding potentials were significantly higher in schizophrenic subjects than in controls in the DA-depleted state suggesting that higher DA $D_{2/3}$ receptor levels in schizophrenic subjects may be masked by higher extracellular DA levels (Abi-Dargham et al., 2000b). This study is an important reminder that binding potentials are a function of both radioligand affinity, which for many radioligands is affected by endogenous neurotransmitter levels, and the levels of receptors. A study of monozygotic twins discordant for schizophrenia has reported increased caudate DA $D_{2/3}$ receptor levels in the unaffected twin suggesting that increased DA $D_{2/3}$ receptor levels may be a risk factor for schizophrenia (Hirvonen et al., 2005).

While several studies of striatal DA $D_{2/3}$ receptor levels have not reported significant correlations of striatal DA $D_{2/3}$ receptor levels with symptoms in unmedicated schizophrenic subjects, two recent [^{123}I]IBZM studies have reported significant correlations of positive symptoms with striatal DA $D_{2/3}$ receptor levels (Schmitt et al., 2009; Pogarell et al., 2012). A [^{18}F]fallypride PET study reported a significant correlation of hallucinations with ventral striatal receptor levels (Kessler et al., 2009). It appears that only positive symptoms have been correlated with striatal DA $D_{2/3}$ receptor levels. There appears to be mildly increased striatal levels of DA $D_{2/3}$ receptors and greater variability in striatal receptor levels in schizophrenic subjects reflecting greater within-subject lability of DA $D_{2/3}$ levels over time in schizophrenic than in healthy subjects (Syvälahti et al., 2000).

Most studies of extrastriatal DA D_2 receptors using both [^{11}C]FLB457 and [^{18}F]fallypride have reported decreased levels of DA $D_{2/3}$ receptors in the thalamus, particularly in the region of the dorsomedial nucleus and pulvinar (Talvik et al., 2003, 2006; Yasuno et al., 2004; Buchsbaum et al., 2006; Kessler et al., 2009; Lehrer et al., 2010; Kegeles et al., 2010a,b). While a number of studies have failed to find changes in cortical DA

$D_{2/3}$ receptor levels (Talvik et al., 2003; Glenthoj et al., 2006; Kessler et al., 2009; Kegeles et al., 2010a,b), other studies have reported decreased levels in the anterior cingulate, frontal, and temporal cortices with the specific regional decrements varying from study to study (Suhara et al., 2002; Buchsbaum et al., 2006; Lehrer et al., 2010). One study has reported increased DA $D_{2/3}$ receptor levels in the substantia nigra (Kessler et al., 2009). There has also been one PET [^{11}C]PHNO study in acutely ill schizophrenic subjects which did not find altered levels in any striatal subregion, globus pallidus, substantia nigra, or thalamus suggesting that there are no altered levels of D_{2high} or D_3 receptors in schizophrenic subjects (Graff-Guerrero et al., 2009a,b).

There have been a number of studies correlating symptom complexes with regional DA $D_{2/3}$ levels in extrastriatal regions. While these studies have varying results, only positive symptoms were significantly correlated with regional DA $D_{2/3}$ receptor levels, including correlations with thalamic, frontal, and temporal cortical DA $D_{2/3}$ receptor levels (Talvik et al., 2003, 2006; Yasuno et al., 2004; Glenthoj et al., 2006; Kessler et al., 2009). In contrast to striatal DA $D_{2/3}$ levels in schizophrenia which appear to be increased, in most extrastriatal regions there are decreased DA $D_{2/3}$ levels, particularly in the medial thalamus, and possibly in frontal and temporal cortex. There appears to be considerable variability in findings across studies likely reflecting both heterogeneity across subjects in alterations in dopaminergic neurotransmission in schizophrenia and within-subject lability of DA $D_{2/3}$ receptor levels (Syvälahti et al., 2000). Only positive symptoms were correlated with changes in extrastriatal DA $D_{2/3}$ receptor levels.

Unlike the variability seen across studies of DA $D_{2/3}$ receptor levels in schizophrenia, studies of releasable DA using amphetamine and of baseline extracellular DA levels using AMPT depletion in schizophrenic subjects have shown consistent changes. All four studies of amphetamine-induced striatal DA release have shown increased DA release in unmedicated schizophrenic subjects compared to healthy subjects (Laruelle et al., 1996; Breier et al., 1997; Abi-Dargham et al., 1998; Pogarell et al., 2012). Acutely ill schizophrenic subjects demonstrated the highest levels of DA release, being more than twice as high as seen in healthy subjects; schizophrenic patients in remission demonstrated significantly higher DA release than healthy subjects (40% increased) but significantly lower DA release than seen in the acutely ill group (Laruelle et al., 1999). Interestingly, striatal DA release was strongly correlated with a transient increase in positive symptoms following amphetamine administration ($r = 0.54$, $P = 0.0009$), although the baseline level of positive symptoms was not correlated with amphetamine-induced DA release. Although 2 of 34 subjects had decreased negative symptoms following amphetamine administration, the remaining 32 subjects showed no relationship between negative symptoms and DA release (Laruelle et al., 1999).

Mental stress has been reported to produce increased striatal DA release in drug-naïve schizophrenic subjects. This was greatest in the associative striatum; somewhat lesser increases were also seen in high-risk subjects (Mizrahi et al., 2012). Increased amphetamine-induced DA release has been reported in subjects with schizotypal personality disorder, the increase being similar to that seen in remitted schizophrenic subjects (Abi-Dargham et al., 2004). In a moderately large study in normal subjects ($N=63$), schizotypal traits were correlated with DA release in the associative striatum (Woodward et al., 2011). Increased striatal DA release, particularly in the associative striatum, is seen in schizophrenia which appears to be related to positive symptoms. Increased striatal DA release in schizophrenia appears to be associated with increased positive symptoms.

Two studies of baseline levels of extracellular DA levels in unmedicated schizophrenic subjects have shown higher and more variable baseline extracellular DA levels in striatum, particularly in the anterior/associative striatum using a 48-h schedule of AMPT administration (Abi-Dargham et al., 2000a,b; Kegeles et al., 2010a). Baseline extracellular DA levels were predictive of response of positive symptoms to antipsychotic drug therapy but did not predict the level of baseline positive symptoms, suggesting that some subjects may have a DA-driven psychosis, while in other subjects DA may not be the driver of the psychotic state. It is noteworthy that schizophrenic subjects have both increased baseline DA levels and increased amphetamine-induced DA release, and a significant positive correlation of baseline DA levels with amphetamine-induced DA release which is not seen in healthy subjects which indicates increased and dysregulated striatal DA function in schizophrenia (Abi-Dargham et al., 2009).

In summary, studies of DA $D_{2/3}$ receptor levels, amphetamine-induced DA release, and baseline extracellular DA levels demonstrate (1) dysregulated, increased striatal presynaptic DA function in schizophrenia which varies with the clinical state of the patient but remains increased in remission; (2) mildly increased striatal DA $D_{2/3}$ levels but decreased DA $D_{2/3}$ receptor levels in other portions of frontal cortical/striatal/thalamic circuits; (3) the changes in striatal and extrastriatal DA $D_{2/3}$ neurotransmission are associated only with positive symptoms; and (4) lability of changes in DA neurotransmission producing greater variance than seen in healthy subjects. The etiology of these regional changes in DA neurotransmission is unclear and there have been a number of hypotheses regarding the pathophysiology of these changes, i.e. decreased DA neurotransmission in thalamic, cortical, and limbic regions leading to dysregulation of striatal DA neurotransmission (Davis et al., 1991); increased striatal DA neurotransmission leading to dysfunction of cortical and limbic region DA neurotransmission (Li et al., 2011); and altered glutamatergic regulation of DA neurotransmission (Krystal et al., 2003; Javitt, 2007, 2010; Adell et al., 2012).

2.5.3. Substance Abuse

Studies in both animals and humans have shown that the development of substance addiction is accompanied by both decreased levels of striatal DA $D_{2/3}$ receptors and decreased DA release. Decreased DA $D_{2/3}$ receptor levels are seen in cocaine, methamphetamine, opiate, alcohol, and tobacco addiction in humans (Volkow et al., 1993, 1996, 1997, 2001, 2002; Wang et al., 1997; Martinez et al., 2004, 2005, 2007, 2012; Lee et al., 2009; Brown et al., 2012). Obesity has also been compared to drug addiction; decreased DA $D_{2/3}$ is seen in extreme obesity, body mass index of 45 or greater (Wang et al., 2001; de Weijer et al., 2011). In animals, the development of decreased striatal DA D_2 receptor levels appears to mediate the development of compulsive behaviors which are resistant to punishment, i.e. addiction; conversely, increasing ventral striatal DA D_2 receptor levels in addicted animals leads to decreased intake of abused substances (Johnson and Kenny, 2010; Thanos et al., 2005).

In humans, the magnitude of decreases in striatal DA $D_{2/3}$ receptors in addiction is generally in the range of about 10–20%; in human cocaine abusers, little change has been seen in DA $D_{2/3}$ receptor levels with up to 4 months of abstinence (Volkow et al., 1993). No significant correlations were seen between ventral striatal or other striatal subregion DA $D_{2/3}$ levels and positive effects of cocaine, primed cocaine-seeking behavior, or years of exposure (Martinez et al., 2004). In alcohol-dependent human subjects decreased DA $D_{2/3}$ levels have been observed in extrastriatal regions, i.e. thalamus, hippocampus, insula, and temporal cortex.

Most of the studies used to evaluate DA $D_{2/3}$ receptor levels in substance abuse have utilized benzamide radioligands which are sensitive to extracellular DA level. AMPT depletion studies in cocaine abusers have shown decreased baseline levels of extracellular DA indicating that decreased apparent DA $D_{2/3}$ receptor levels in cocaine abusers are due to decreased receptor levels and not to increased extracellular DA levels (Martinez et al., 2009b). While, as discussed above, there is some question as to whether DA $D_{2/3}$ agonist radioligands can identify the agonist state of the receptor in vivo, [^{11}C]NPA PET studies in cocaine-dependent subjects found no change of the fraction of DA $D_{2/3}$ receptors in the high-affinity agonist state (Narendran et al., 2011b). Decreased striatal DA $D_{2/3}$ receptor levels are seen in across a number of substance abuse states and the development of decreased DA D_2 receptor levels appears to mediate at least in part the development of compulsive alcohol abuse and likely other substance abuse. There is, however, no apparent relationship between cocaine abuse behaviors and the decrements in striatal DA $D_{2/3}$ receptor levels seen in cocaine abusers.

In humans there have been fewer studies of DA release than DA $D_{2/3}$ receptor levels in substance abuse. However, studies of psychostimulant-induced striatal DA release in cocaine, methamphetamine, morphine, and alcohol-abusing subjects demonstrate 50–90% decreases in amphetamine- or methylphenidate-induced DA release depending on the substance and striatal regions examined in contrast to 10–20% reductions in DA $D_{2/3}$ receptor levels (Volkow et al., 1997, 2007a,b; Martinez et al., 2005, 2007, 2012; Zijlstra et al., 2008; Wang et al., 2012). Unlike other substance abuse-dependent states, cannabis-dependent subjects did not demonstrate either decreased baselines striatal DA $D_{2/3}$ receptor levels or decreased striatal DA release (Urban et al., 2012); however, an inverse correlation of anterior striatal DA release with age of onset of cannabis use was seen suggesting that individuals who began using cannabis in adolescence to early adulthood had greater impairment in dopaminergic neurotransmission. In methamphetamine and cocaine addiction, the changes in anteroventral striatal striatal DA release are predictive of both altered reward function, i.e. preference of abused drug over monetary rewards, and the response to therapy (Martinez et al., 2007, 2011; Wang et al., 2012). Cocaine-abusing subjects who were able to maintain an abstinent state, i.e. therapy responders, had baseline methylphenidate-induced ventral DA release levels nearly that of control subjects, i.e. 12.1% in ventral striatum compared to 13.7% in healthy subjects, while cocaine-abusing subjects who were therapy nonresponders had markedly decreased DA release, 1.3%. Similar findings were seen in methamphetamine abusers (Wang et al., 2012). Presentation of cocaine cues to cocaine abusers produces decreases in [^{11}C]raclopride binding potentials presumably due to DA release; the magnitude of the changes in [^{11}C]raclopride binding potentials in the dorsal striatum was correlated with self-reports of craving (Volkow et al., 2006a,b). This is consistent with the role of the dorsal striatum in compulsive, addictive behaviors. Altered DA release appears to predict the changes in reward function and therapeutic response in cocaine and methamphetamine abuse.

In morphine addiction, decreased methylphenidate-induced DA release similar to that seen with psychostimulants has been reported; however, the magnitude of DA release was not related to preference for abused substance over monetary reward, as was reported for cocaine (Martinez et al., 2012). Opiate cue induced craving-induced striatal DA release but unlike studies in cocaine abuse no significant correlation was seen between cue-induced craving and DA release. However, significant correlations were seen between cue-induced DA release and baseline chronic craving and anhedonia (Zijlstra et al., 2008). Opiate addiction also differs from psychostimulant addiction in that impulsivity does not predict escalation of opiate self-administration in rodents. Opiate addiction appears to have a different relationship to DA neurotransmission than psychostimulant abuse which directly targets the DA system.

In alcohol-dependent human subjects, amphetamine- and methylphenidate-induced DA release was significantly decreased in ventral striatum, not in caudate, and to a lesser extent in putamen (Martinez et al., 2005; Volkow et al., 2007a,b). Alcohol-dependent subjects have more localized decrements in striatal DA release, being principally in ventral striatum, in contrast to psychostimulant abusers who have more widespread striatal decrements. The decreased ventral striatal DA release seen in alcohol-dependent subjects was associated with a decrease in the rewarding effects of methylphenidate (Volkow et al., 2007a,b). In healthy subjects, oral alcohol administration produced ventral striatal DA release (Boileau et al., 2003).

A second, larger study of alcohol-induced DA release reported significant ventral striatal DA release, i.e. decreases in [^{11}C]raclopride binding potentials, with smaller changes in other striatal subregions; a significant sex difference was observed with men having greater DA release than women (Urban et al., 2010). DA release in men but not women was significantly correlated with the subjective effects of alcohol. Using [^{11}C]raclopride PET studies, healthy carriers of the A118G polymorphism of the mu opioid receptor were shown to have greater alcohol-induced striatal, particularly ventral striatal, DA release than subjects with the AA polymorphism (Ramchandani et al., 2011); microdialysis studies in mice bred to express the AA and GG polymorphisms showed increased ventral striatal alcohol-induced DA release following an alcohol challenge in the GG- vs AA-bearing mice. Alcohol dependence appears to have more selective effects on DA release in the ventral striatum in comparison to psychostimulant abuse.

DA $D_{2/3}$ receptor levels and DA release appear to mediate risk factors for substance abuse. Studies in humans demonstrate that high striatal DA $D_{2/3}$ receptor levels predict an aversive response to psychostimulants while lower levels predict a pleasurable response (Volkow et al., 1999). Consistent with these results is a study of unaffected members of alcoholic families which found increased striatal DA $D_{2/3}$ levels in these individuals (Volkow et al., 2006a). Other studies of novelty-seeking behaviors have reported an inverse correlation with insular DA D_2 levels and an inverted-U-shaped correlation of sensation seeking with striatal DA D_2 levels which predicts an inverse relationship between sensation seeking and DA $D_{2/3}$ levels (Suhara et al., 2001; Gjedde et al., 2010). These studies suggest that high striatal DA $D_{2/3}$ receptor levels may be protective for psychostimulant drug abuse. Studies of DA release suggest that there may be an inverted-U-shaped curve of risk for psychostimulant drug abuse vs striatal DA release.

Novelty seeking behaviors have been postulated to be a risk factor for substance use; in animals impulsivity appears to be the factor most predictive of the development of cocaine but not morphine addiction (Piazza et al., 1989; Belin et al., 2008, 2011). In rats novelty seeking behavior has been related to decreased midbrain DA D_2 autoreceptor function, increased midbrain DA neuronal firing rates, and increased nucleus accumbens DA release (Bradberry et al., 1991; Marinelli and White, 2000); impulsivity in rats has been related to decreased ventral striatal DA $D_{2/3}$ receptors levels (Dalley et al., 2007). In humans both novelty seeking behaviors and impulsivity have been inversely correlated with ventral midbrain DA $D_{2/3}$ receptor levels and positively correlated with ventral and anteroventral striatal amphetamine-induced DA release (Leyton et al., 2002; Zald et al., 2008; Buckholtz et al., 2010). Increased amphetamine induced ventral striatal DA release in healthy subjects also predicts greater pleasurable responses to amphetamine administration (Drevets et al., 2001). These findings suggest that increased anteroventral striatal DA release predisposes to greater novelty seeking, impulsivity, and increased rewarding properties of psychostimulants leading to a greater risk of substance abuse.

While increased ventral striatal DA release may be a risk factor for substance abuse, never-treated subjects with attention deficit/hyperactivity disorder, which is associated with a high risk of substance abuse, demonstrate decreased striatal, particularly caudate,

DA release and increased rewarding effects of methylphenidate administration; an inverse correlation of methylphenidate's rewarding properties with striatal DA release was seen (Volkow et al., 2007a,b). These studies suggest an inverted-U-shaped curve of striatal, particularly ventral striatal, DA release with risk for substance abuse. Striatal DA $D_{2/3}$ receptor-mediated neurotransmission appears to play an important role in the risk of substance abuse.

In summary, while there are differences in the relationship of DA $D_{2/3}$ neurotransmission to the development of psychostimulant vs opiate vs alcohol abuse, decreased DA $D_{2/3}$ neurotransmission is seen in many types of substance abuse. A recent study of cannabis abuse suggests that earlier onset of substance use, i.e. in adolescence, may be associated with greater impairments in striatal DA neurotransmission.

3. CONCLUSIONS

Over the past three decades a great deal of progress has been made in the development of PET and SPECT imaging methods for studying DA neurotransmission in the living human brain. There are now well-validated radioligands for studying striatal DA D_1 receptor levels; striatal and extrastriatal DA $D_{2/3}$ receptor levels; DA D_3 receptor levels, psychostimulant-induced DA release in striatum, ventral midbrain, limbic, and cortical regions; as well as baseline extracellular DA levels in striatum and ventral midbrain. To date there are no radioligands for the DA D_5 and DA D_4 receptors suitable for imaging studies in humans and there is a need for more specific DA D_1 radioligands for studying extrastriatal DA D_1 receptors.

There remain questions regarding the mechanisms underlying the psychostimulant-induced reductions in benzamide DA D_2 antagonist, DA $D_{2/3}$ agonist, and DA D_3-preferring agonist radioligands used to estimate releasable DA. It appears that internalization of DA D_2- and DA D_3-type receptors is responsible for the protracted reductions in binding potentials of radioligands following psychostimulant administration. The mechanism underlying the magnitude of these decrements is not completely resolved but likely relates at least in part to the combination of the time course of psychostimulant-induced DA release and the kinetics of radioligand binding in brain; care must be taken to utilize tracer doses of radioligand and have robust methods of quantitation which for some radioligands involves the use of a metabolite-corrected arterial input function so as not to have changes in regional cerebral blood flow affect the results.

An unresolved question is the source of extracellular DA in cortex and limbic regions; there is evidence in animals to suggest that a considerable fraction arises from noradrenergic and not dopaminergic neurons. These methods have provided important insights into the pathophysiology of a number of disorders including schizophrenia, drug addiction, extreme obesity, and movement disorders. Additional methodological developments are needed for studying DA D_4 receptors, DA D_5 receptors, extrastriatal DA D_1 receptors, and intracellular DA signaling pathways.

REFERENCES

Aalto, S., Brück, A., Laine, M., Någren, K., Rinne, J.O., March 9, 2005. Frontal and temporal dopamine release during working memory and attention tasks in healthy humans: a positron emission tomography study using the high-affinity dopamine D_2 receptor ligand [^{11}C]FLB 457. J. Neurosci. 25 (10), 2471–2477.

Aalto, S., Hirvonen, J., Kaasinen, V., Hagelberg, N., Kajander, J., Någren, K., Seppälä, T., Rinne, J.O., Scheinin, H., Hietala, J., March 2009. The effects of D-amphetamine on extrastriatal dopamine D_2/D_3 receptors: a randomized, double-blind, placebo-controlled PET study with [^{11}C]FLB 457 in healthy subjects. Eur. J. Nucl. Med. Mol. Imaging 36 (3), 475–483.

Abi-Dargham, A., Gil, R., Krystal, J., Baldwin, R.M., Seibyl, J.P., Bowers, M., van Dyck, C.H., Charney, D.S., Innis, R.B., Laruelle, M., June 1998. Increased striatal dopamine transmission in schizophrenia: confirmation in a second cohort. Am. J. Psychiatry 155 (6), 761–767.

Abi-Dargham, A., Simpson, N., Kegeles, L., Parsey, R., Hwang, D.R., Anjilvel, S., Zea-Ponce, Y., Lombardo, I., Van Heertum, R., Mann, J.J., Foged, C., Halldin, C., Laruelle, M., May 1999. PET studies of binding competition between endogenous dopamine and the D_1 radiotracer [^{11}C]NNC 756. Synapse 32 (2), 93–109.

Abi-Dargham, A., Martinez, D., Mawlawi, O., Simpson, N., Hwang, D.R., Slifstein, M., Anjilvel, S., Pidcock, J., Guo, N.N., Lombardo, I., Mann, J.J., Van Heertum, R., Foged, C., Halldin, C., Laruelle, M., February 2000a. Measurement of striatal and extrastriatal dopamine D_1 receptor binding potential with [^{11}C] NNC 112 in humans: validation and reproducibility. J. Cereb. Blood Flow Metab. 20 (2), 225–243.

Abi-Dargham, A., Rodenhiser, J., Printz, D., Zea-Ponce, Y., Gil, R., Kegeles, L.S., Weiss, R., Cooper, T.B., Mann, J.J., Van Heertum, R.L., Gorman, J.M., Laruelle, M., July 5, 2000b. Increased baseline occupancy of D_2 receptors by dopamine in schizophrenia. Proc. Natl. Acad. Sci. USA 97 (14), 8104–8109.

Abi-Dargham, A., Mawlawi, O., Lombardo, I., Gil, R., Martinez, D., Huang, Y., Hwang, D.R., Keilp, J., Kochan, L., Van Heertum, R., Gorman, J.M., Laruelle, M., May 1, 2002. Prefrontal dopamine D_1 receptors and working memory in schizophrenia. J. Neurosci. 22 (9), 3708–3719.

Abi-Dargham, A., Kegeles, L.S., Zea-Ponce, Y., Mawlawi, O., Martinez, D., Mitropoulou, V., O'Flynn, K., Koenigsberg, H.W., Van Heertum, R., Cooper, T., Laruelle, M., Siever, L.J., May 15, 2004. Striatal amphetamine-induced dopamine release in patients with schizotypal personality disorder studied with single photon emission computed tomography and [^{123}I]iodobenzamide. Biol. Psychiatry 55 (10), 1001–1006.

Abi-Dargham, A., van de Giessen, E., Slifstein, M., Kegeles, L.S., Laruelle, M., June 15, 2009. Baseline and amphetamine-stimulated dopamine activity are related in drug-naïve schizophrenic subjects. Biol. Psychiatry 65 (12), 1091–1093.

Abi-Dargham, A., Xu, X., Thompson, J.L., Gil, R., Kegeles, L.S., Urban, N., Narendran, R., Hwang, D.R., Laruelle, M., Slifstein, M., June 2012. Increased prefrontal cortical D_1 receptors in drug naive patients with schizophrenia: a PET study with [^{11}C]NNC112. J. Psychopharmacol. 26 (6), 794–805.

Adell, A., Jiménez-Sánchez, L., López-Gil, X., Romón, T., January 2012. Is the acute NMDA receptor hypofunction a valid model of schizophrenia? Schizophr. Bull. 38 (1), 9–14.

Allen, J.A., Yost, J.M., Setola, V., Chen, X., Sassano, M.F., Chen, M., Peterson, S., Yadav, P.N., Huang, X.P., Feng, B., Jensen, N.H., Che, X., Bai, X., Frye, S.V., Wetsel, W.C., Caron, M.G., Javitch, J.A., Roth, B.L., Jin, J., November 8, 2011. Discovery of β-arrestin-biased dopamine D_2 ligands for probing signal transduction pathways essential for antipsychotic efficacy. Proc. Natl. Acad. Sci. USA 108 (45), 18488–18493.

al-Tikriti, M.S., Baldwin, R.M., Zea-Ponce, Y., Sybirska, E., Zoghbi, S.S., Laruelle, M., Malison, R.T., Kung, H.F., Kessler, R.M., Charney, D.S., et al., February 1994. Comparison of three high affinity SPECT radiotracers for the dopamine D_2 receptor. Nucl. Med. Biol. 21 (2), 179–188.

Andrews, T.C., Weeks, R.A., Turjanski, N., Gunn, R.N., Watkins, L.H., Sahakian, B., Hodges, J.R., Rosser, A.E., Wood, N.W., Brooks, D.J., December 1999. Huntington's disease progression. PET and clinical observations. Brain 122 (Pt 12), 2353–2363.

Antenor-Dorsey, J.A., Markham, J., Moerlein, S.M., Videen, T.O., Perlmutter, J.S., April 2008a. Validation of the reference tissue model for estimation of dopaminergic D_2-like receptor binding with [^{18}F](*N*-methyl)benperidol in humans. Nucl. Med. Biol. 35 (3), 335–341.

Antenor-Dorsey, J.A., Laforest, R., Moerlein, S.M., Videen, T.O., Perlmutter, J.S., April 2008b. Radiation dosimetry of *N*-([^{11}C]methyl)benperidol as determined by whole-body PET imaging of primates. Eur. J. Nucl. Med. Mol. Imaging 35 (4), 771–778.

Antonini, A., Leenders, K.L., Reist, H., Thomann, R., Beer, H.F., Locher, J., May 1993. Effect of age on D_2 dopamine receptors in normal human brain measured by positron emission tomography and ^{11}C-raclopride. Arch. Neurol. 50 (5), 474–480.

Asselin, M.C., Montgomery, A.J., Grasby, P.M., Hume, S.P., February 2007. Quantification of PET studies with the very high-affinity dopamine D_2/D_3 receptor ligand [^{11}C]FLB 457: re-evaluation of the validity of using a cerebellar reference region. J. Cereb. Blood Flow Metab. 27 (2), 378–392.

Bahn, M.M., Huang, S.C., Hawkins, R.A., Satyamurthy, N., Hoffman, J.M., Barrio, J.R., Mazziotta, J.C., Phelps, M.E., December 1989. Models for in vivo kinetic interactions of dopamine D_2-neuroreceptors and 3-(2′-[^{18}F]fluoroethyl)spiperone examined with positron emission tomography. J. Cereb. Blood Flow Metab. 9 (6), 840–849.

Beaulieu, J.M., Tirotta, E., Sotnikova, T.D., Masri, B., Salahpour, A., Gainetdinov, R.R., Borrelli, E., Caron, M.G., January 24, 2007a. Regulation of Akt signaling by D_2 and D_3 dopamine receptors in vivo. J. Neurosci. 27 (4), 881–885.

Beaulieu, J.M., Gainetdinov, R.R., Caron, M.G., April 2007b. The Akt-GSK-3 signaling cascade in the actions of dopamine. Trends Pharmacol. Sci. 28 (4), 166–172.

Beaulieu, J.M., Gainetdinov, R.R., March 2011. The physiology, signaling, and pharmacology of dopamine receptors. Pharmacol. Rev. 63 (1), 182–217.

Belin, D., Mar, A.C., Dalley, J.W., Robbins, T.W., Everitt, B.J., June 6, 2008. High impulsivity predicts the switch to compulsive cocaine-taking. Science 320 (5881), 1352–1355.

Belin, D., Berson, N., Balado, E., Piazza, P.V., Deroche-Gamonet, V., February 2011. High-novelty-preference rats are predisposed to compulsive cocaine self-administration. Neuropsychopharmacology 36 (3), 569–579.

Bennacef, I., Salinas, C.A., Bonasera, T.A., Gunn, R.N., Audrain, H., Jakobsen, S., Nabulsi, N., Weinzimmer, D., Carson, R.E., Huang, Y., Holmes, I., Micheli, F., Heidbreder, C., Gentile, G., Rossi, T., Laruelle, M., September 1, 2009. Dopamine D_3 receptor antagonists: the quest for a potentially selective PET ligand. Part 3: radiosynthesis and in vivo studies. Bioorg. Med. Chem. Lett. 19 (17), 5056–5059.

Bergson, C., Mrzljak, L., Smiley, J.F., Pappy, M., Levenson, R., Goldman-Rakic, P.S., December 1995. Regional, cellular, and subcellular variations in the distribution of D_1 and D_5 dopamine receptors in primate brain. J. Neurosci. 15 (12), 7821–7836.

Besret, L., Dollé, F., Hérard, A.S., Guillermier, M., Demphel, S., Hinnen, F., Coulon, C., Ottaviani, M., Bottlaender, M., Hantraye, P., Kassiou, M., July 2008. Dopamine D_1 receptor imaging in the rodent and primate brain using the isoquinoline +-[11C]A-69024 and positron emission tomography. J. Pharm. Sci. 97 (7), 2811–2819.

Billings, J., Kung, M.P., Chumpradit, S., Pan, S., Kung, H.F., 1989. [^{125}I](+/−)FISCH: a new CNS D-1 dopamine receptor imaging ligand. Life Sci. 45 (8), 711–718.

Billings, J.J., Kung, M.P., Chumpradit, S., Mozley, D., Alavi, A., Kung, H.F., January 1992. Characterization of radioiodinated TISCH: a high-affinity and selective ligand for mapping CNS D_1 dopamine receptor. J. Neurochem. 58 (1), 227–236.

Boileau, I., Assaad, J.M., Pihl, R.O., Benkelfat, C., Leyton, M., Diksic, M., Tremblay, R.E., Dagher, A., September 15, 2003. Alcohol promotes dopamine release in the human nucleus accumbens. Synapse 49 (4), 226–231.

Boot, E., Booij, J., Hasler, G., Zinkstok, J.R., de Haan, L., Linszen, D.H., van Amelsvoort, T.A., July 2008. AMPT-induced monoamine depletion in humans: evaluation of two alternative [^{123}I]IBZM SPECT procedures. Eur. J. Nucl. Med. Mol. Imaging 35 (7), 1350–1356.

Bradberry, C.W., Gruen, R.J., Berridge, C.W., Roth, R.H., August 1991. Individual differences in behavioral measures: correlations with nucleus accumbens dopamine measured by microdialysis. Pharmacol. Biochem. Behav. 39 (4), 877–882.

Breier, A., Su, T.P., Saunders, R., Carson, R.E., Kolachana, B.S., de Bartolomeis, A., Weinberger, D.R., Weisenfeld, N., Malhotra, A.K., Eckelman, W.C., Pickar, D., March 18, 1997. Schizophrenia is associated with elevated amphetamine-induced synaptic dopamine concentrations: evidence from a novel positron emission tomography method. Proc. Natl. Acad. Sci. USA 94 (6), 2569–2574.

Brown, A.K., Mandelkern, M.A., Farahi, J., Robertson, C., Ghahremani, D.G., Sumerel, B., Moallem, N., London, E.D., August 2012. Sex differences in striatal dopamine D_2/D_3 receptor availability in smokers and non-smokers. Int. J. Neuropsychopharmacol. 15 (7), 989–994.

Buchsbaum, M.S., Christian, B.T., Lehrer, D.S., Narayanan, T.K., Shi, B., Mantil, J., Kemether, E., Oakes, T.R., Mukherjee, J., July 2006. D_2/D_3 dopamine receptor binding with [F-18]fallypride in thalamus and cortex of patients with schizophrenia. Schizophr. Res. 85 (1–3), 232–244.

Buckholtz, J.W., Treadway, M.T., Cowan, R.L., Woodward, N.D., Shelby, E.S., Smith, C.E., Kessler, R.M., Zald, D.H., July 30, 2010. Dopaminergic network differences in human impulsivity. Science 329 (5991), 532.

Burns, H.D., Dannals, R.F., Langström, B., Ravert, H.T., Zemyan, S.E., Duelfer, T., Wong, D.F., Frost, J.J., Kuhar, M.J., Wagner, H.N., November 1984. (3-*N*-[^{11}C]methyl)spiperone, a ligand binding to dopamine receptors: radiochemical synthesis and biodistribution studies in mice. J. Nucl. Med. 25 (11), 1222–1227.

Cannon, D.M., Klaver, J.M., Peck, S.A., Rallis-Voak, D., Erickson, K., Drevets, W.C., April 2009. Dopamine type-1 receptor binding in major depressive disorder assessed using positron emission tomography and [^{11}C]NNC-112. Neuropsychopharmacology 34 (5), 1277–1287.

Cárdenas, L., Houle, S., Kapur, S., Busto, U.E., January 2004. Oral D-amphetamine causes prolonged displacement of [^{11}C]raclopride as measured by PET. Synapse 51 (1), 27–31.

Carson, R.E., Breier, A., de Bartolomeis, A., Saunders, R.C., Su, T.P., Schmall, B., Der, M.G., Pickar, D., Eckelman, W.C., April 1997. Quantification of amphetamine-induced changes in [^{11}C]raclopride binding with continuous infusion. J. Cereb. Blood Flow Metab. 17 (4), 437–447.

Carson, R.E., Vuong, B.-K., Watabe, H., Herscovitch, P., Eckelman, W., 2001. Amphetamine Induced Dopamine Release: Duration of Action Assessed with [^{11}C]raclopride in Anesthetized Monkeys. Physiological Imaging of the Brain with PET. Academic Press, San Diego, pp. 205–209.

Catafau, A.M., Searle, G.E., Bullich, S., Gunn, R.N., Rabiner, E.A., Herance, R., Radua, J., Farre, M., Laruelle, M., May 2010. Imaging cortical dopamine D_1 receptors using [^{11}C]NNC112 and ketanserin blockade of the 5-HT_{2A} receptors. J. Cereb. Blood Flow Metab. 30 (5), 985–993.

Chou, Y.H., Karlsson, P., Halldin, C., Olsson, H., Farde, L., September 1999. A PET study of D_1-like dopamine receptor ligand binding during altered endogenous dopamine levels in the primate brain. Psychopharmacology (Berlin) 146 (2), 220–227.

Civelli, O., Borrelli, E., 2010. The Dopamine Receptors. Encyclopedia of Life Sciences, Wiley on Line Library.

Corripio, I., Escartí, M.J., Portella, M.J., Pérez, V., Grasa, E., Sauras, R.B., Alonso, A., Safont, G., Camacho, M.V., Dueñas, R., Arranz, B., San, L., Catafau, A.M., Carrió, I., Alvarez, E., December 2011. Density of striatal D_2 receptors in untreated first-episode psychosis: an I^{123}-IBZM SPECT study. Eur. Neuropsychopharmacol. 21 (12), 861–866.

Cortés, R., Gueye, B., Pazos, A., Probst, A., Palacios, J.M., 1989. Dopamine receptors in human brain: autoradiographic distribution of D_1 sites. Neuroscience 28 (2), 263–273.

Cropley, V.L., Fujita, M., Bara-Jimenez, W., Brown, A.K., Zhang, X.Y., Sangare, J., Herscovitch, P., Pike, V.W., Hallett, M., Nathan, P.J., Innis, R.B., July 15, 2008a. Pre- and post-synaptic dopamine imaging and its relation with frontostriatal cognitive function in Parkinson disease: PET studies with [^{11}C]NNC 112 and [^{18}F]FDOPA. Psychiatry Res. 163 (2), 171–182.

Cropley, V.L., Innis, R.B., Nathan, P.J., Brown, A.K., Sangare, J.L., Lerner, A., Ryu, Y.H., Sprague, K.E., Pike, V.W., Fujita, M., June 2008b. Small effect of dopamine release and no effect of dopamine depletion on [^{18}F]fallypride binding in healthy humans. Synapse 62 (6), 399–408.

Dalley, J.W., Fryer, T.D., Brichard, L., Robinson, E.S., Theobald, D.E., Lääne, K., Peña, Y., Murphy, E.R., Shah, Y., Probst, K., Abakumova, I., Aigbirhio, F.I., Richards, H.K., Hong, Y., Baron, J.C., Everitt, B.J., Robbins, T.W., March 2, 2007. Nucleus accumbens $D_{2/3}$ receptors predict trait impulsivity and cocaine reinforcement. Science 315 (5816), 1267–1270.

Davis, K.L., Kahn, R.S., Ko, G., Davidson, M., November 1991. Dopamine in schizophrenia: a review and reconceptualization. Am. J. Psychiatry 148 (11), 1474–1486.

Delforge, J., Bottlaender, M., Loc'h, C., Guenther, I., Fuseau, C., Bendriem, B., Syrota, A., Mazière, B., May 1999. Quantitation of extrastriatal D_2 receptors using a very high-affinity ligand (FLB 457) and the multi-injection approach. J. Cereb. Blood Flow Metab. 19 (5), 533–546.

Devoto, P., Flore, G., Pani, L., Gessa, G.L., November 2001. Evidence for co-release of noradrenaline and dopamine from noradrenergic neurons in the cerebral cortex. Mol. Psychiatry 6 (6), 657–664.

Devoto, P., Flore, G., Longu, G., Pira, L., Gessa, G.L., December 1, 2003. Origin of extracellular dopamine from dopamine and noradrenaline neurons in the medial prefrontal and occipital cortex. Synapse 50 (3), 200–205.

Devoto, P., Flore, G., Saba, P., Fà, M., Gessa, G.L., May 2, 2005. Co-release of noradrenaline and dopamine in the cerebral cortex elicited by single train and repeated train stimulation of the locus coeruleus. BMC Neurosci. 6, 31.

Devoto, P., Flore, G., Saba, P., Castelli, M.P., Piras, A.P., Luesu, W., Viaggi, M.C., Ennas, M.G., Gessa, G.L., May 15, 2008. 6-Hydroxydopamine lesion in the ventral tegmental area fails to reduce extracellular dopamine in the cerebral cortex. J. Neurosci. Res. 86 (7), 1647–1658.

Devoto, P., Flore, G., Saba, P., Cadeddu, R., Gessa, G.L., February 2012. Disulfiram stimulates dopamine release from noradrenergic terminals and potentiates cocaine-induced dopamine release in the prefrontal cortex. Psychopharmacology (Berlin) 219 (4), 1153–1164.

Dewey, S.L., Smith, G.S., Logan, J., Brodie, J.D., Yu, D.W., Ferrieri, R.A., King, P.T., MacGregor, R.R., Martin, T.P., Wolf, A.P., et al., October 1992. GABAergic inhibition of endogenous dopamine release measured in vivo with ^{11}C-raclopride and positron emission tomography. J. Neurosci. 12 (10), 3773–3780.

de Weijer, B.A., van de Giessen, E., van Amelsvoort, T.A., Boot, E., Braak, B., Janssen, I.M., van de Laar, A., Fliers, E., Serlie, M.J., Booij, J., December 16, 2011. Lower striatal dopamine $D_{2/3}$ receptor availability in obese compared with non-obese subjects. EJNMMI Res. 1 (1), 37.

Dougherty, D.D., Bonab, A.A., Ottowitz, W.E., Livni, E., Alpert, N.M., Rauch, S.L., Fava, M., Fischman, A.J., 2006. Decreased striatal D_1 binding as measured using PET and [^{11}C]SCH 23,390 in patients with major depression with anger attacks. Depress Anxiety 23 (3), 175–177.

Drevets, W.C., Gautier, C., Price, J.C., Kupfer, D.J., Kinahan, P.E., Grace, A.A., Price, J.L., Mathis, C.A., January 15, 2001. Amphetamine-induced dopamine release in human ventral striatum correlates with euphoria. Biol. Psychiatry 49 (2), 81–96.

Dziewczapolski, G., Menalled, L.B., García, M.C., Mora, M.A., Gershanik, O.S., Rubinstein, M., January 5, 1998. Opposite roles of D_1 and D_5 dopamine receptors in locomotion revealed by selective antisense oligonucleotides. Neuroreport 9 (1), 1–5.

Eisenstein, S.A., Koller, J.M., Piccirillo, M., Kim, A., Antenor-Dorsey, J.A., Videen, T.O., Snyder, A.Z., Karimi, M., Moerlein, S.M., Black, K.J., Perlmutter, J.S., Hershey, T., September 2012. Characterization of extrastriatal D_2 in vivo specific binding of [^{18}F](*N*-methyl)benperidol using PET. Synapse 66 (9), 770–780.

Ekelund, J., Slifstein, M., Narendran, R., Guillin, O., Belani, H., Guo, N.N., Hwang, Y., Hwang, D.R., Abi-Dargham, A., Laruelle, M., May–June 2007. In vivo DA D_1 receptor selectivity of NNC 112 and SCH 23390. Mol. Imaging Biol. 9 (3), 117–125.

Endres, C.J., Kolachana, B.S., Saunders, R.C., Su, T., Weinberger, D., Breier, A., Eckelman, W.C., Carson, R.E., September 1997. Kinetic modeling of [^{11}C]raclopride: combined PET-microdialysis studies. J. Cereb. Blood Flow Metab. 17 (9), 932–942.

Farde, L., Ehrin, E., Eriksson, L., Greitz, T., Hall, H., Hedström, C.G., Litton, J.E., Sedvall, G., June 1985. Substituted benzamides as ligands for visualization of dopamine receptor binding in the human brain by positron emission tomography. Proc. Natl. Acad. Sci. USA 82 (11), 3863–3867.

Farde, L., Pauli, S., Hall, H., Eriksson, L., Halldin, C., Högberg, T., Nilsson, L., Sjögren, I., Stone-Elander, S., 1988. Stereoselective binding of ^{11}C-raclopride in living human brain–a search for extrastriatal central D_2-dopamine receptors by PET. Psychopharmacology (Berlin) 94 (4), 471–478.

Farde, L., Wiesel, F.A., Stone-Elander, S., Halldin, C., Nordström, A.L., Hall, H., Sedvall, G., March 1990. D_2 dopamine receptors in neuroleptic-naive schizophrenic patients. A positron emission tomography study with [^{11}C]raclopride. Arch. Gen. Psychiatry 47 (3), 213–219.

Finnema, S.J., Seneca, N., Farde, L., Shchukin, E., Sóvágó, J., Gulyás, B., Wikström, H.V., Innis, R.B., Neumeyer, J.L., Halldin, C., May 2005. A preliminary PET evaluation of the new dopamine D_2 receptor agonist [^{11}C]MNPA in cynomolgus monkey. Nucl. Med. Biol. 32 (4), 353–360.

Finnema, S.J., Halldin, C., Bang-Andersen, B., Gulyás, B., Bundgaard, C., Wikström, H.V., Farde, L., May 2009. Dopamine $D_{2/3}$ receptor occupancy of apomorphine in the nonhuman primate brain–a comparative PET study with [^{11}C]raclopride and [^{11}C]MNPA. Synapse 63 (5), 378–389.

Finnema, S.J., Vladimir, S., Nakao, R., Amini, N., Sromek, A., Neumeyer, J., Seeman, P., Farde, L., Halldin, C., August 2012. The novel dopamine D_2/D_3 receptor agonist radioligand [^{18}F]MCL-524 is sensitive to dopamine levels. J. Cereb. Blood Flow Metab. 32 (Suppl. 1), SPO27.

Foged, C., Halldin, C., Loc'h, C., Mazière, B., Karlsson, P., Mazière, M., Swahn, C.G., Farde, L., August 1996. ^{11}C- and ^{76}Br-labelled NNC 22-0010, selective dopamine D_1 receptor radioligands for PET. Nucl. Med. Biol. 23 (6), 837–844.

Foged, C., Halldin, C., Swahn, C.G., Ginovart, N., Karlsson, P., Lundkvist, C., Farde, L., July 1998. [^{11}C]NNC 22-0215, a metabolically stable dopamine D_1 radioligand for PET. Nucl. Med. Biol. 25 (5), 503–508.

Frankle, W.G., Mason, N.S., Rabiner, E.A., Ridler, K., May, M.A., Asmonga, D., Chen, C.M., Kendro, S., Cooper, T.B., Mathis, C.A., Narendran, R., December 2010. No effect of dopamine depletion on the binding of the high-affinity $D_{2/3}$ radiotracer [^{11}C]FLB 457 in the human cortex. Synapse 64 (12), 879–885.

Frost, J.J., Smith, A.C., Kuhar, M.J., Dannals, R.F., Wagner Jr., H.N., March 9, 1987. In vivo binding of ^{3}H-*N*-methylspiperone to dopamine and serotonin receptors. Life Sci. 40 (10), 987–995.

Fujita, M., Seibyl, J.P., Verhoeff, N.P., Ichise, M., Baldwin, R.M., Zoghbi, S.S., Burger, C., Staley, J.K., Rajeevan, N., Charney, D.S., Innis, R.B., December 15, 1999. Kinetic and equilibrium analyses of [(^{123})I] epidepride binding to striatal and extrastriatal dopamine D_2 receptors. Synapse 34 (4), 290–304.

Fujita, M., Verhoeff, N.P., Varrone, A., Zoghbi, S.S., Baldwin, R.M., Jatlow, P.A., Anderson, G.M., Seibyl, J.P., Innis, R.B., January 10, 2000. Imaging extrastriatal dopamine D_2 receptor occupancy by endogenous dopamine in healthy humans. Eur. J. Pharmacol. 387 (2), 179–188.

Fuxe, K., Marcellino, D., Leo, G., Agnati, L.F., February 2010. Molecular integration via allosteric interactions in receptor heteromers. A working hypothesis. Curr. Opin. Pharmacol. 10 (1), 14–22.

Gallezot, J.D., Beaver, J.D., Gunn, R.N., Nabulsi, N., Weinzimmer, D., Singhal, T., Slifstein, M., Fowles, K., Ding, Y.S., Huang, Y., Laruelle, M., Carson, R.E., Rabiner, E.A., February 2012. Affinity and selectivity of [^{11}C]-(+)-PHNO for the D_3 and D_2 receptors in the rhesus monkey brain in vivo. Synapse 66 (6), 489–500.

Garris, P.A., Wightman, R.M., January 1994. Different kinetics govern dopaminergic transmission in the amygdala, prefrontal cortex, and striatum: an in vivo voltammetric study. J. Neurosci. 14 (1), 442–450.

Gefvert, O., Lundberg, T., Wieselgren, I.M., Bergström, M., Långström, B., Wiesel, F., Lindström, L., April 2001. D_2 and $5HT_{2A}$ receptor occupancy of different doses of quetiapine in schizophrenia: a PET study. Eur. Neuropsychopharmacol. 11 (2), 105–110.

Gerhardt, G.A., Cass, W.A., Yi, A., Zhang, Z., Gash, D.M., January 2002. Changes in somatodendritic but not terminal dopamine regulation in aged rhesus monkeys. J. Neurochem. 80 (1), 168–177.

Ginovart, N., Lundin, A., Farde, L., Halldin, C., Bäckman, L., Swahn, C.G., Pauli, S., Sedvall, G., March 1997a. PET study of the pre- and post-synaptic dopaminergic markers for the neurodegenerative process in Huntington's disease. Brain 120 (Pt 3), 503–514.

Ginovart, N., Farde, L., Halldin, C., Swahn, C.G., April 1997b. Effect of reserpine-induced depletion of synaptic dopamine on [^{11}C]raclopride binding to D_2-dopamine receptors in the monkey brain. Synapse 25 (4), 321–325.

Ginovart, N., Wilson, A.A., Houle, S., Kapur, S., June 15, 2004. Amphetamine pretreatment induces a change in both D_2-receptor density and apparent affinity: a [^{11}C]raclopride positron emission tomography study in cats. Biol. Psychiatry 55 (12), 1188–1194.

Ginovart, N., Galineau, L., Willeit, M., Mizrahi, R., Bloomfield, P.M., Seeman, P., Houle, S., Kapur, S., Wilson, A.A., May 2006. Binding characteristics and sensitivity to endogenous dopamine of [^{11}C]-(+)-PHNO, a new agonist radiotracer for imaging the high-affinity state of D_2 receptors in vivo using positron emission tomography. J. Neurochem. 97 (4), 1089–1103.

Girgis, R.R., Xu, X., Miyake, N., Easwaramoorthy, B., Gunn, R.N., Rabiner, E.A., Abi-Dargham, A., Slifstein, M., March 2011. In vivo binding of antipsychotics to D_3 and D_2 receptors: a PET study in baboons with [^{11}C]-(+)-PHNO. Neuropsychopharmacology 36 (4), 887–895.

Gjedde, A., Kumakura, Y., Cumming, P., Linnet, J., Møller, A., February 23, 2010. Inverted-U-shaped correlation between dopamine receptor availability in striatum and sensation seeking. Proc. Natl. Acad. Sci. USA 107 (8), 3870–3875.

Glenthoj, B.Y., Mackeprang, T., Svarer, C., Rasmussen, H., Pinborg, L.H., Friberg, L., Baaré, W., Hemmingsen, R., Videbaek, C., September 15, 2006. Frontal dopamine $D_{2/3}$ receptor binding in drug-naive first-episode schizophrenic patients correlates with positive psychotic symptoms and gender. Biol. Psychiatry 60 (6), 621–629. 3.

Goffinet, A.M., Leysen, J., Labar, D., January 1990. In vitro pharmacological profile of 3-*N*-(2-fluoroethyl) spiperone. J. Cereb. Blood Flow Metab. 10 (1), 140–142.

Graff-Guerrero, A., Mizrahi, R., Agid, O., Marcon, H., Barsoum, P., Rusjan, P., Wilson, A.A., Zipursky, R., Kapur, S., March 2009a. The dopamine D_2 receptors in high-affinity state and D_3 receptors in schizophrenia: a clinical [^{11}C]-(+)-PHNO PET study. Neuropsychopharmacology 34 (4), 1078–1086.

Graff-Guerrero, A., Mamo, D., Shammi, C.M., Mizrahi, R., Marcon, H., Barsoum, P., Rusjan, P., Houle, S., Wilson, A.A., Kapur, S., June 2009b. The effect of antipsychotics on the high-affinity state of D_2 and D_3 receptors: a positron emission tomography study with [^{11}C]-(+)-PHNO. Arch. Gen. Psychiatry 66 (6), 606–615.

Gründer, G., Siessmeier, T., Piel, M., Vernaleken, I., Buchholz, H.G., Zhou, Y., Hiemke, C., Wong, D.F., Rösch, F., Bartenstein, P., January 2003. Quantification of D_2-like dopamine receptors in the human brain with ^{18}F-desmethoxyfallypride. J. Nucl. Med. 44 (1), 109–116.

Guo, N., Hwang, D.R., Lo, E.S., Huang, Y.Y., Laruelle, M., Abi-Dargham, A., September 2003. Dopamine depletion and in vivo binding of PET D_1 receptor radioligands: implications for imaging studies in schizophrenia. Neuropsychopharmacology 28 (9), 1703–1711.

Guo, N., Guo, W., Kralikova, M., Jiang, M., Schieren, I., Narendran, R., Slifstein, M., Abi-Dargham, A., Laruelle, M., Javitch, J.A., Rayport, S., February 2010. Impact of D_2 receptor internalization on binding affinity of neuroimaging radiotracers. Neuropsychopharmacology 35 (3), 806–817.

Gurevich, E.V., Joyce, J.N., January 1999. Distribution of dopamine D_3 receptor expressing neurons in the human forebrain: comparison with D_2 receptor expressing neurons. Neuropsychopharmacology 20 (1), 60–80.

Hagberg, G., Gefvert, O., Bergström, M., Wieselgren, I.M., Lindström, L., Wiesel, F.A., Långström, B., June 30, 1998. *N*-[^{11}C]methylspiperone PET, in contrast to [^{11}C]raclopride, fails to detect D_2 receptor occupancy by an atypical neuroleptic. Psychiatry Res. 82 (3), 147–160.

Hall, H., Wedel, I., Halldin, C., Kopp, J., Farde, L., December 1990. Comparison of the in vitro receptor binding properties of *N*-[^{3}H]methylspiperone and [^{3}H]raclopride to rat and human brain membranes. J. Neurochem. 55 (6), 2048–2057.

Hall, H., Halldin, C., Dijkstra, D., Wikström, H., Wise, L.D., Pugsley, T.A., Sokoloff, P., Pauli, S., Farde, L., Sedvall, G., December 1996. Autoradiographic localisation of D_3-dopamine receptors in the human brain using the selective D_3-dopamine receptor agonist (+)-[^{3}H]PD 128907. Psychopharmacology (Berlin) 128 (3), 240–247.

Halldin, C., Stone-Elander, S., Farde, L., Ehrin, E., Fasth, K.J., Långström, B., Sedvall, G., 1986. Preparation of ^{11}C-labelled SCH 23390 for the in vivo study of dopamine D-1 receptors using positron emission tomography. Int. J. Rad. Appl. Instrum. A 37 (10), 1039–1043.

Halldin, C., Farde, L., Högberg, T., Mohell, N., Hall, H., Suhara, T., Karlsson, P., Nakashima, Y., Swahn, C.G., July 1995. Carbon-11-FLB 457: a radioligand for extrastriatal D_2 dopamine receptors. J. Nucl. Med. 36 (7), 1275–1281.

Halldin, C., Foged, C., Chou, Y.H., Karlsson, P., Swahn, C.G., Sandell, J., Sedvall, G., Farde, L., December 1998. Carbon-11-NNC 112: a radioligand for PET examination of striatal and neocortical D_1-dopamine receptors. J. Nucl. Med. 39 (12), 2061–2068.

Hietala, J., West, C., Syvälahti, E., Någren, K., Lehikoinen, P., Sonninen, P., Ruotsalainen, U., November 1994. Striatal D_2 dopamine receptor binding characteristics in vivo in patients with alcohol dependence. Psychopharmacology (Berlin) 116 (3), 285–290.

Hietala, J., Syvälahti, E., Vuorio, K., Någren, K., Lehikoinen, P., Ruotsalainen, U., Räkköläinen, V., Lehtinen, V., Wegelius, U., February 1994. Striatal D_2 dopamine receptor characteristics in neuroleptic-naive schizophrenic patients studied with positron emission tomography. Arch. Gen. Psychiatry 51 (2), 116–123.

Hirvonen, J., van Erp, T.G., Huttunen, J., Aalto, S., Någren, K., Huttunen, M., Lönnqvist, J., Kaprio, J., Hietala, J., Cannon, T.D., April 2005. Increased caudate dopamine D_2 receptor availability as a genetic marker for schizophrenia. Arch. Gen. Psychiatry 62 (4), 371–378.

Hirvonen, J., van Erp, T.G., Huttunen, J., Aalto, S., Någren, K., Huttunen, M., Lönnqvist, J., Kaprio, J., Cannon, T.D., Hietala, J., October 2006. Brain dopamine D_1 receptors in twins discordant for schizophrenia. Am. J. Psychiatry 163 (10), 1747–1753.
Hirvonen, J., Karlsson, H., Kajander, J., Markkula, J., Rasi-Hakala, H., Någren, K., Salminen, J.K., Hietala, J., May 2008. Striatal dopamine D_2 receptors in medication-naive patients with major depressive disorder as assessed with [^{11}C]raclopride PET. Psychopharmacology (Berlin) 197 (4), 581–590 5.
Hocke, C., Prante, O., Salama, I., Hübner, H., Löber, S., Kuwert, T., Gmeiner, P., May 2008. ^{18}F-Labeled FAUC 346 and BP 897 derivatives as subtype-selective potential PET radioligands for the dopamine D_3 receptor. ChemMedChem 3 (5), 788–793.
Huang, S.C., Bahn, M.M., Barrio, J.R., Hoffman, J.M., Satyamurthy, N., Hawkins, R.A., Mazziotta, J.C., Phelps, M.E., December 1989. A double-injection technique for in vivo measurement of dopamine D_2-receptor density in monkeys with 3-(2′-[^{18}F]fluoroethyl)spiperone and dynamic positron emission tomography. J. Cereb. Blood Flow Metab. 9 (6), 850–858.
Innis, R.B., Malison, R.T., al-Tikriti, M., Hoffer, P.B., Sybirska, E.H., Seibyl, J.P., Zoghbi, S.S., Baldwin, R.M., Laruelle, M., Smith, E.O., et al., December 1992. Amphetamine-stimulated dopamine release competes in vivo for [^{123}I]IBZM binding to the D_2 receptor in nonhuman primates. Synapse 10 (3), 177–184.
Javitt, D.C., 2007. Glutamate and schizophrenia: phencyclidine, *N*-methyl-D-aspartate receptors, and dopamine-glutamate interactions. Int. Rev. Neurobiol. 78, 69–108.
Javitt, D.C., 2010. Glutamatergic theories of schizophrenia. Isr. J. Psychiatry Relat. Sci. 47 (1), 4–16.
Johnson, P.M., Kenny, P.J., May 2010. Dopamine D_2 receptors in addiction-like reward dysfunction and compulsive eating in obese rats. Nat. Neurosci. 13 (5), 635–641.
Jucaite, A., Forssberg, H., Karlsson, P., Halldin, C., Farde, L., April 28, 2010. Age-related reduction in dopamine D_1 receptors in the human brain: from late childhood to adulthood, a positron emission tomography study. Neuroscience 167 (1), 104–110.
Kaasinen, V., Vilkman, H., Hietala, J., Någren, K., Helenius, H., Olsson, H., Farde, L., Rinne, J., September–October 2000. Age-related dopamine D_2/D_3 receptor loss in extrastriatal regions of the human brain. Neurobiol. Aging 21 (5), 683–688.
Kaasinen, V., Kemppainen, N., Någren, K., Helenius, H., Kurki, T., Rinne, J.O., June 2002. Age-related loss of extrastriatal dopamine D_2-like receptors in women. J. Neurochem. 81 (5), 1005–1010.
Karimi, M., Moerlein, S.M., Videen, T.O., Luedtke, R.R., Taylor, M., Mach, R.H., Perlmutter, J.S., January 2011. Decreased striatal dopamine receptor binding in primary focal dystonia: a D_2 or D_3 defect? Mov. Disord. 26 (1), 100–106.
Karlsson, P., Farde, L., Halldin, C., Swahn, C.G., Sedvall, G., Foged, C., Hansen, K.T., Skrumsager, B., 1993. PET examination of [^{11}C]NNC 687 and [^{11}C]NNC 756 as new radioligands for the D_1-dopamine receptor. Psychopharmacology (Berlin) 113 (2), 149–156.
Karlsson, P., Sedvall, G., Halldin, C., Swahn, C.G., Farde, L., October 1995. Evaluation of SCH 39166 as PET ligand for central D_1 dopamine receptor binding and occupancy in man. Psychopharmacology (Berlin) 121 (3), 300–308.
Karlsson, P., Farde, L., Halldin, C., Sedvall, G., May 2002. PET study of D_1 dopamine receptor binding in neuroleptic-naive patients with schizophrenia. Am. J. Psychiatry 159 (5), 761–767.
Kassiou, M., Mardon, K., Mattner, F., Katsifis, A., Dikic, B., June 29, 2001. Pharmacological evaluation of (+)-2-[^{123}I]A-69024: a radioligand for in vivo studies of dopamine D_1 receptors. Life Sci. 69 (6), 669–675.
Kassiou, M., Loc'h, C., Bottlaender, M., Mardon, K., Ottaviani, M., Coulon, C., Katsifis, A., Maziere, B., April 2002. (+)-[^{76}Br]A-69024: a non-benzazepine radioligand for studies of dopamine D_1 receptors using PET. Nucl. Med. Biol. 29 (3), 295–302.
Kegeles, L.S., Zea-Ponce, Y., Abi-Dargham, A., Rodenhiser, J., Wang, T., Weiss, R., Van Heertum, R.L., Mann, J.J., Laruelle, M., March 15, 1999. Stability of [^{123}I]IBZM SPECT measurement of amphetamine-induced striatal dopamine release in humans. Synapse 31 (4), 302–308.
Kegeles, L.S., Mawlawi, O., Rodenhiser, J., Martienez, D., Broft, A., Fullerton, C., Perey, A., Cooper, T., Van Heertum, R., Abi-Dargham, A., Laruelle, M., 2001. Baseline dopamine activity in striatal subregions determined by [^{11}C] raclopride PET in healthy humans. J. Nucl. Med. 42, 16.

Kegeles, L.S., Slifstein, M., Frankle, W.G., Xu, X., Hackett, E., Bae, S.A., Gonzales, R., Kim, J.H., Alvarez, B., Gil, R., Laruelle, M., Abi-Dargham, A., December 2008. Dose-occupancy study of striatal and extrastriatal dopamine D_2 receptors by aripiprazole in schizophrenia with PET and [^{18}F]fallypride. Neuropsychopharmacology 33 (13), 3111–3125.

Kegeles, L.S., Abi-Dargham, A., Frankle, W.G., Gil, R., Cooper, T.B., Slifstein, M., Hwang, D.R., Huang, Y., Haber, S.N., Laruelle, M., March 2010a. Increased synaptic dopamine function in associative regions of the striatum in schizophrenia. Arch. Gen. Psychiatry 67 (3), 231–239.

Kegeles, L.S., Slifstein, M., Xu, X., Urban, N., Thompson, J.L., Moadel, T., Harkavy-Friedman, J.M., Gil, R., Laruelle, M., Abi-Dargham, A., October 1, 2010b. Striatal and extrastriatal dopamine D_2/D_3 receptors in schizophrenia evaluated with [^{18}F]fallypride positron emission tomography. Biol. Psychiatry 68 (7), 634–641.

Kessler, R.M., Ansari, M.S., Schmidt, D.E., de Paulis, T., Clanton, J.A., Innis, R., al-Tikriti, M., Manning, R.G., Gillespie, D., 1991. High affinity dopamine D_2 receptor radioligands. 2. [^{125}I]epidepride, a potent and specific radioligand for the characterization of striatal and extrastriatal dopamine D_2 receptors. Life Sci. 49 (8), 617–628.

Kessler, R.M., Whetsell, W.O., Ansari, M.S., Votaw, J.R., de Paulis, T., Clanton, J.A., Schmidt, D.E., Mason, N.S., Manning, R.G., April 23, 1993a. Identification of extrastriatal dopamine D_2 receptors in post mortem human brain with [^{125}I]epidepride. Brain Res. 609 (1–2), 237–243.

Kessler, R.M., Votaw, J.R., Schmidt, D.E., Ansari, M.S., Holdeman, K.P., de Paulis, T., Clanton, J.A., Pfeffer, R., Manning, R.G., Ebert, M.H., 1993b. High affinity dopamine D_2 receptor radioligands. 3. [^{123}I] and [^{125}I]epidepride: in vivo studies in rhesus monkey brain and comparison with in vitro pharmacokinetics in rat brain. Life Sci. 53 (3), 241–250.

Kessler, R.M., Mason, N.S., Votow, J.R., Ansari, M.S., Depaulis, T., Schimidt, D.E., Manning, R.G., 1993c. Factors underlying the D-amphetamine induced displacement of dopamine D_2 radioligands in rhesus-monkeys. J. Nucl. Med. 34 (5), 132.

Kessler, R.M., Mason, N.S., Jonesc, C., Ansari, M.S., Manning, R.G., Price, R.R., 2000. [^{18}F]*N*-ally-5-fluoro proplepidepride (fallypride): radiation dosimetry, quantification of striatal and extrastriatal dopamine receptors in man. NeuroImage 11, S32.

Kessler, R.M., Ansari, M.S., Riccardi, P., Li, R., Jayathilake, K., Dawant, B., Meltzer, H.Y., December 2005. Occupancy of striatal and extrastriatal dopamine D_2/D_3 receptors by olanzapine and haloperidol. Neuropsychopharmacology 30 (12), 2283–2289.

Kessler, R.M., Woodward, N.D., Riccardi, P., Li, R., Ansari, M.S., Anderson, S., Dawant, B., Zald, D., Meltzer, H.Y., June 15, 2009. Dopamine D_2 receptor levels in striatum, thalamus, substantia nigra, limbic regions, and cortex in schizophrenic subjects. Biol. Psychiatry 65 (12), 1024–1031.

Khan, Z.U., Mrzljak, L., Gutierrez, A., de la Calle, A., Goldman-Rakic, P.S., June 23, 1998. Prominence of the dopamine D_2 short isoform in dopaminergic pathways. Proc. Natl. Acad. Sci. USA 95 (13), 7731–7736.

Khan, Z.U., Gutiérrez, A., Martín, R., Peñafiel, A., Rivera, A., de la Calle, A., 2000. Dopamine D_5 receptors of rat and human brain. Neuroscience 100 (4), 689–699.

Köhler, C., Hall, H., Ogren, S.O., Gawell, L., July 1, 1985. Specific in vitro and in vivo binding of ^{3}H-raclopride. A potent substituted benzamide drug with high affinity for dopamine D-2 receptors in the rat brain. Biochem. Pharmacol. 34 (13), 2251–2259.

Kortekaas, R., Maguire, R.P., Cremers, T.I., Dijkstra, D., van Waarde, A., Leenders, K.L., May 2004. In vivo binding behavior of dopamine receptor agonist (+)-PD 128907 and implications for the "ceiling effect" in endogenous competition studies with [(^{11})C]raclopride-a positron emission tomography study in *Macaca mulatta*. J. Cereb. Blood Flow Metab. 24 (5), 531–535.

Kosaka, J., Takahashi, H., Ito, H., Takano, A., Fujimura, Y., Matsumoto, R., Nozaki, S., Yasuno, F., Okubo, Y., Kishimoto, T., Suhara, T., May 22, 2010. Decreased binding of [^{11}C]NNC112 and [^{11}C]SCH23390 in patients with chronic schizophrenia. Life Sci. 86 (21–22), 814–818.

Krystal, J.H., D'Souza, D.C., Mathalon, D., Perry, E., Belger, A., Hoffman, R., September 2003. NMDA receptor antagonist effects, cortical glutamatergic function, and schizophrenia: toward a paradigm shift in medication development. Psychopharmacology (Berlin) 169 (3–4), 215–233.

Kuhnast, B., Valette, H., Besret, L., Demphel, S., Coulon, C., Ottaviani, M., Guillermier, M., Bottlaender, M., Dollé, F., August 2006. Synthesis and radiolabeling of *N*-[4-[4-(2-[^{11}C]methoxyphenyl)piperazin-1-yl] butyl]benzo[*b*]thiophene-2-carboxamide – a potential radiotracer for D_3 receptor imaging with PET. Nucl. Med. Biol. 33 (6), 785–795.

Kung, H.F., Pan, S., Kung, M.P., Billings, J., Kasliwal, R., Reilley, J., Alavi, A., January 1989. In vitro and in vivo evaluation of [^{123}I]IBZM: a potential CNS D-2 dopamine receptor imaging agent. J. Nucl. Med. 30 (1), 88–92.
Lammertsma, A.A., Bench, C.J., Hume, S.P., Osman, S., Gunn, K., Brooks, D.J., Frackowiak, R.S., January 1996. Comparison of methods for analysis of clinical [^{11}C]raclopride studies. J. Cereb. Blood Flow Metab. 16 (1), 42–52.
Laruelle, M., Abi-Dargham, A., van Dyck, C.H., Rosenblatt, W., Zea-Ponce, Y., Zoghbi, S.S., Baldwin, R.M., Charney, D.S., Hoffer, P.B., Kung, H.F., et al., July 1995. SPECT imaging of striatal dopamine release after amphetamine challenge. J. Nucl. Med. 36 (7), 1182–1190.
Laruelle, M., Abi-Dargham, A., van Dyck, C.H., Gil, R., D'Souza, C.D., Erdos, J., McCance, E., Rosenblatt, W., Fingado, C., Zoghbi, S.S., Baldwin, R.M., Seibyl, J.P., Krystal, J.H., Charney, D.S., Innis, R.B., August 20, 1996. Single photon emission computerized tomography imaging of amphetamine-induced dopamine release in drug-free schizophrenic subjects. Proc. Natl. Acad. Sci. USA 93 (17), 9235–9240.
Laruelle, M., Iyer, R.N., al-Tikriti, M.S., Zea-Ponce, Y., Malison, R., Zoghbi, S.S., Baldwin, R.M., Kung, H.F., Charney, D.S., Hoffer, P.B., Innis, R.B., Bradberry, C.W., January 1997a. Microdialysis and SPECT measurements of amphetamine-induced dopamine release in nonhuman primates. Synapse 25 (1), 1–14.
Laruelle, M., D'Souza, C.D., Baldwin, R.M., Abi-Dargham, A., Kanes, S.J., Fingado, C.L., Seibyl, J.P., Zoghbi, S.S., Bowers, M.B., Jatlow, P., Charney, D.S., Innis, R.B., September 1997b. Imaging D_2 receptor occupancy by endogenous dopamine in humans. Neuropsychopharmacology 17 (3), 162–174.
Laruelle, M., September 1998. Imaging dopamine transmission in schizophrenia. A review and meta-analysis. Q. J. Nucl. Med. 42 (3), 211–221.
Laruelle, M., Abi-Dargham, A., Gil, R., Kegeles, L., Innis, R., July 1, 1999. Increased dopamine transmission in schizophrenia: relationship to illness phases. Biol. Psychiatry 46 (1), 56–72.
Laruelle, M., March 2000. Imaging synaptic neurotransmission with in vivo binding competition techniques: a critical review. J. Cereb. Blood Flow Metab. 20 (3), 423–451.
Lawrence, A.D., Weeks, R.A., Brooks, D.J., Andrews, T.C., Watkins, L.H., Harding, A.E., Robbins, T.W., Sahakian, B.J., July 1998. The relationship between striatal dopamine receptor binding and cognitive performance in Huntington's disease. Brain 121 (Pt 7), 1343–1355.
Lee, B., London, E.D., Poldrack, R.A., Farahi, J., Nacca, A., Monterosso, J.R., Mumford, J.A., Bokarius, A.V., Dahlbom, M., Mukherjee, J., Bilder, R.M., Brody, A.L., Mandelkern, M.A., November 25, 2009. Striatal dopamine D_2/D_3 receptor availability is reduced in methamphetamine dependence and is linked to impulsivity. J. Neurosci. 29 (47), 14734–14740.
Lehrer, D.S., Christian, B.T., Kirbas, C., Chiang, M., Sidhu, S., Short, H., Wang, B., Shi, B., Chu, K.W., Merrill, B., Buchsbaum, M.S., September 2010. ^{18}F-fallypride binding potential in patients with schizophrenia compared to healthy controls. Schizophr. Res. 122 (1–3), 43–52.
Lewis, D.A., Melchitzky, D.S., Sesack, S.R., Whitehead, R.E., Auh, S., Sampson, A., March 26, 2001. Dopamine transporter immunoreactivity in monkey cerebral cortex: regional, laminar, and ultrastructural localization. J. Comp. Neurol. 432 (1), 119–136.
Leyton, M., Boileau, I., Benkelfat, C., Diksic, M., Baker, G., Dagher, A., December 2002. Amphetamine-induced increases in extracellular dopamine, drug wanting, and novelty seeking: a PET/[^{11}C]raclopride study in healthy men. Neuropsychopharmacology 27 (6), 1027–1035.
Li, Y.C., Kellendonk, C., Simpson, E.H., Kandel, E.R., Gao, W.J., July 19, 2011. D_2 receptor overexpression in the striatum leads to a deficit in inhibitory transmission and dopamine sensitivity in mouse prefrontal cortex. Proc. Natl. Acad. Sci. USA 108 (29), 12107–12112.
Logan, J., Fowler, J.S., Volkow, N.D., Wang, G.J., Ding, Y.S., Alexoff, D.L., September 1996. Distribution volume ratios without blood sampling from graphical analysis of PET data. J. Cereb. Blood Flow Metab. 16 (5), 834–840.
Logan, J., Fowler, J.S., Volkow, N.D., Ding, Y.S., Wang, G.J., Alexoff, D.L., March 2001. A strategy for removing the bias in the graphical analysis method. J. Cereb. Blood Flow Metab. 21 (3), 307–320.
Lyon, R.A., Titeler, M., Frost, J.J., Whitehouse, P.J., Wong, D.F., Wagner Jr., H.N., Dannals, R.F., Links, J.M., Kuhar, M.J., October 1986. ^{3}H-3-*N*-methylspiperone labels D_2 dopamine receptors in basal ganglia and S_2 serotonin receptors in cerebral cortex. J. Neurosci. 6 (10), 2941–2949.
Mach, R.H., Tu, Z., Xu, J., Li, S., Jones, L.A., Taylor, M., Luedtke, R.R., Derdeyn, C.P., Perlmutter, J.S., Mintun, M.A., August 2011. Endogenous dopamine (DA) competes with the binding of a radiolabeled D_3 receptor partial agonist in vivo: a positron emission tomography study. Synapse 65 (8), 724–732.

Marinelli, M., White, F.J., December 1, 2000. Enhanced vulnerability to cocaine self-administration is associated with elevated impulse activity of midbrain dopamine neurons. J. Neurosci. 20 (23), 8876–8885.

Martin-Negrier, M.L., Charron, G., Bloch, B., September 1, 2006. Receptor recycling mediates plasma membrane recovery of dopamine D_1 receptors in dendrites and axons after agonist-induced endocytosis in primary cultures of striatal neurons. Synapse 60 (3), 194–204.

Martinez, D., Broft, A., Foltin, R.W., Slifstein, M., Hwang, D.R., Huang, Y., Perez, A., Frankle, W.G., Cooper, T., Kleber, H.D., Fischman, M.W., Laruelle, M., June 2004. Cocaine dependence and D_2 receptor availability in the functional subdivisions of the striatum: relationship with cocaine-seeking behavior. Neuropsychopharmacology 29 (6), 1190–1202.

Martinez, D., Gil, R., Slifstein, M., Hwang, D.R., Huang, Y., Perez, A., Kegeles, L., Talbot, P., Evans, S., Krystal, J., Laruelle, M., Abi-Dargham, A., November 15, 2005. Alcohol dependence is associated with blunted dopamine transmission in the ventral striatum. Biol. Psychiatry 58 (10), 779–786.

Martinez, D., Narendran, R., Foltin, R.W., Slifstein, M., Hwang, D.R., Broft, A., Huang, Y., Cooper, T.B., Fischman, M.W., Kleber, H.D., Laruelle, M., April 2007. Amphetamine-induced dopamine release: markedly blunted in cocaine dependence and predictive of the choice to self-administer cocaine. Am. J. Psychiatry 164 (4), 622–629.

Martinez, D., Slifstein, M., Narendran, R., Foltin, R.W., Broft, A., Hwang, D.R., Perez, A., Abi-Dargham, A., Fischman, M.W., Kleber, H.D., Laruelle, M., June 2009a. Dopamine D_1 receptors in cocaine dependence measured with PET and the choice to self-administer cocaine. Neuropsychopharmacology 34 (7), 1774–1782.

Martinez, D., Greene, K., Broft, A., Kumar, D., Liu, F., Narendran, R., Slifstein, M., Van Heertum, R., Kleber, H.D., October 2009b. Lower level of endogenous dopamine in patients with cocaine dependence: findings from PET imaging of D_2/D_3 receptors following acute dopamine depletion. Am. J. Psychiatry 166 (10), 1170–1177.

Martinez, D., Carpenter, K.M., Liu, F., Slifstein, M., Broft, A., Friedman, A.C., Kumar, D., Van Heertum, R., Kleber, H.D., Nunes, E., June 2011. Imaging dopamine transmission in cocaine dependence: link between neurochemistry and response to treatment. Am. J. Psychiatry 168 (6), 634–641.

Martinez, D., Saccone, P.A., Liu, F., Slifstein, M., Orlowska, D., Grassetti, A., Cook, S., Broft, A., Van Heertum, R., Comer, S.D., February 1, 2012. Deficits in dopamine D_2 receptors and presynaptic dopamine in heroin dependence: commonalities and differences with other types of addiction. Biol. Psychiatry 71 (3), 192–198.

McCormick, P.N., Kapur, S., Seeman, P., Wilson, A.A., January 2008. Dopamine D_2 receptor radiotracers $[(^{11})C](+)$-PHNO and $[(^{3})H]$raclopride are indistinguishably inhibited by D_2 agonists and antagonists ex vivo. Nucl. Med. Biol. 35 (1), 11–17.

McCormick, P.N., Ginovart, N., Wilson, A.A., August 2011. Isoflurane anaesthesia differentially affects the amphetamine sensitivity of agonist and antagonist D_2/D_3 positron emission tomography radiotracers: implications for in vivo imaging of dopamine release. Mol. Imaging Biol. 13 (4), 737–746.

Melchitzky, D.S., Erickson, S.L., Lewis, D.A., December 28, 2006. Dopamine innervation of the monkey mediodorsal thalamus: location of projection neurons and ultrastructural characteristics of axon terminals. Neuroscience 143 (4), 1021–1030.

Mizrahi, R., Houle, S., Vitcu, I., Ng, A., Wilson, A.A., March 2010. Side effects profile in humans of $(^{11})$C-(+)-PHNO, a dopamine $D_{2/3}$ agonist ligand for PET. J. Nucl. Med. 51 (3), 496–497.

Mizrahi, R., Agid, O., Borlido, C., Suridjan, I., Rusjan, P., Houle, S., Remington, G., Wilson, A.A., Kapur, S., September 2011. Effects of antipsychotics on D_3 receptors: a clinical PET study in first episode antipsychotic naive patients with schizophrenia using $[^{11}C]$-(+)-PHNO. Schizophr. Res. 131 (1–3), 63–68.

Mizrahi, R., Addington, J., Rusjan, P.M., Suridjan, I., Ng, A., Boileau, I., Pruessner, J.C., Remington, G., Houle, S., Wilson, A.A., March 15, 2012. Increased stress-induced dopamine release in psychosis. Biol. Psychiatry 71 (6), 561.

Moerlein, S.M., Perlmutter, J.S., Markham, J., Welch, M.J., August 1997. In vivo kinetics of $[^{18}F]$(*N*-methyl) benperidol: a novel PET tracer for assessment of dopaminergic D_2-like receptor binding. J. Cereb. Blood Flow Metab. 17 (8), 833–845.

Montague, D.M., Lawler, C.P., Mailman, R.B., Gilmore, J.H., November 1999. Developmental regulation of the dopamine D_1 receptor in human caudate and putamen. Neuropsychopharmacology 21 (5), 641–649.

Montgomery, A.J., Asselin, M.C., Farde, L., Grasby, P.M., February 2007. Measurement of methylphenidate-induced change in extrastriatal dopamine concentration using [^{11}C]FLB 457 PET. J. Cereb. Blood Flow Metab. 27 (2), 369–377.

Morris, E.D., Yoder, K.K., March 2007. Positron emission tomography displacement sensitivity: predicting binding potential change for positron emission tomography tracers based on their kinetic characteristics. J. Cereb. Blood Flow Metab. 27 (3), 606–617.

Mukherjee, J., Yang, Z.Y., Brown, T., Roemer, J., Cooper, M., 1996. ^{18}F-desmethoxyfallypride: a fluorine-18 labeled radiotracer with properties similar to carbon-11 raclopride for PET imaging studies of dopamine D_2 receptors. Life Sci. 59 (8), 669–678.

Mukherjee, J., Christian, B.T., Narayanan, T.K., Shi, B., Collins, D., January 25, 2005. Measurement of D-amphetamine-induced effects on the binding of dopamine D_2/D_3 receptor radioligand, ^{18}F-fallypride in extrastriatal brain regions in non-human primates using PET. Brain Res. 1032 (1–2), 77–84.

Nader, M.A., Daunais, J.B., Moore, T., Nader, S.H., Moore, R.J., Smith, H.R., Friedman, D.P., Porrino, L.J., July 2002. Effects of cocaine self-administration on striatal dopamine systems in rhesus monkeys: initial and chronic exposure. Neuropsychopharmacology 27 (1), 35–46.

Nairn, A.C., Svenningsson, P., Nishi, A., Fisone, G., Girault, J.A., Greengard, P., 2004. The role of DARPP-32 in the actions of drugs of abuse. Neuropharmacology 47 (Suppl. 1), 14–23 Review.

Narendran, R., Hwang, D.R., Slifstein, M., Talbot, P.S., Erritzoe, D., Huang, Y., Cooper, T.B., Martinez, D., Kegeles, L.S., Abi-Dargham, A., Laruelle, M., January 1, 2004. In vivo vulnerability to competition by endogenous dopamine: comparison of the D_2 receptor agonist radiotracer (−)-*N*-[^{11}C]propyl-norapomorphine ([^{11}C]NPA) with the D_2 receptor antagonist radiotracer [^{11}C]-raclopride. Synapse 52 (3), 188–208.

Narendran, R., Frankle, W.G., Keefe, R., Gil, R., Martinez, D., Slifstein, M., Kegeles, L.S., Talbot, P.S., Huang, Y., Hwang, D.R., Khenissi, L., Cooper, T.B., Laruelle, M., Abi-Dargham, A., December 2005a. Altered prefrontal dopaminergic function in chronic recreational ketamine users. Am. J. Psychiatry 162 (12), 2352–2359.

Narendran, R., Hwang, D.R., Slifstein, M., Hwang, Y., Huang, Y., Ekelund, J., Guillin, O., Scher, E., Martinez, D., Laruelle, M., October 2005b. Measurement of the proportion of D_2 receptors configured in state of high affinity for agonists in vivo: a positron emission tomography study using [^{11}C]*N*-propyl-norapomorphine and [^{11}C]raclopride in baboons. J. Pharmacol. Exp. Ther. 315 (1), 80–90.

Narendran, R., Slifstein, M., Guillin, O., Hwang, Y., Hwang, D.R., Scher, E., Reeder, S., Rabiner, E., Laruelle, M., December 1, 2006. Dopamine ($D_{2/3}$) receptor agonist positron emission tomography radiotracer [^{11}C]-(+)-PHNO is a D_3 receptor preferring agonist in vivo. Synapse 60 (7), 485–495.

Narendran, R., Slifstein, M., Hwang, D.R., Hwang, Y., Scher, E., Reeder, S., Martinez, D., Laruelle, M., February 2007. Amphetamine-induced dopamine release: duration of action as assessed with the $D_{2/3}$ receptor agonist radiotracer (−)-*N*-[(^{11})C]propyl-norapomorphine ([^{11}C]NPA) in an anesthetized non-human primate. Synapse 61 (2), 106–109.

Narendran, R., Frankle, W.G., Mason, N.S., Rabiner, E.A., Gunn, R.N., Searle, G.E., Vora, S., Litschge, M., Kendro, S., Cooper, T.B., Mathis, C.A., Laruelle, M., June 2009a. Positron emission tomography imaging of amphetamine-induced dopamine release in the human cortex: a comparative evaluation of the high affinity dopamine $D_{2/3}$ radiotracers [^{11}C]FLB 457 and [^{11}C]fallypride. Synapse 63 (6), 447–461.

Narendran, R., Frankle, W.G., Mason, N.S., Laymon, C.M., Lopresti, B.J., Price, J.C., Kendro, S., Vora, S., Litschge, M., Mountz, J.M., Mathis, C.A., July 2009b. Positron emission tomography imaging of $D_{2/3}$ agonist binding in healthy human subjects with the radiotracer [(^{11})C]-*N*-propyl-norapomorphine: preliminary evaluation and reproducibility studies. Synapse 63 (7), 574–584.

Narendran, R., Mason, N.S., Laymon, C.M., Lopresti, B.J., Velasquez, N.D., May, M.A., Kendro, S., Martinez, D., Mathis, C.A., Frankle, W.G., May 2010. A comparative evaluation of the dopamine $D_{2/3}$ agonist radiotracer [^{11}C](−)-*N*-propyl-norapomorphine and antagonist [^{11}C]raclopride to measure amphetamine-induced dopamine release in the human striatum. J. Pharmacol. Exp. Ther. 333 (2), 533–539.

Narendran, R., Mason, N.S., Chen, C.M., Himes, M., Keating, P., May, M.A., Rabiner, E.A., Laruelle, M., Mathis, C.A., Frankle, W.G., May 2011a. Evaluation of dopamine $D_{2/3}$ specific binding in the cerebellum for the positron emission tomography radiotracer [^{11}C]FLB 457: implications for measuring cortical dopamine release. Synapse 65 (10), 991–997.

Narendran, R., Mason, N.S., May, M.A., Chen, C.M., Kendro, S., Ridler, K., Rabiner, E.A., Laruelle, M., Mathis, C.A., Frankle, W.G., January 2011b. Positron emission tomography imaging of dopamine $D_{2/3}$ receptors in the human cortex with [^{11}C]FLB 457: reproducibility studies. Synapse 65 (1), 35–40.

Narendran, R., Hank, J., Lopresti, B., Mason, N., Gurnsey, K., Ruskiewicz, J., Chen, C.M., Mathis, C., Frankle, W., Bradberry, C., 2012. Imaging dopamine transmission in the prefrontal cortex: a combined microdialysis and [^{11}C]FLB 457PET study. J. Cerb. Blood Flow Metab. 32, S1–S157.

Negyessy, L., Goldman-Rakic, P.S., August 8, 2005. Subcellular localization of the dopamine D_2 receptor and coexistence with the calcium-binding protein neuronal calcium sensor-1 in the primate prefrontal cortex. J. Comp. Neurol. 488 (4), 464–475.

Nishi, A., Snyder, G.L., Greengard, P., November 1, 1997. Bidirectional regulation of DARPP-32 phosphorylation by dopamine. J. Neurosci. 17 (21), 8147–8155.

Nordström, A.L., Farde, L., Halldin, C., 1993. High 5-HT_2 receptor occupancy in clozapine treated patients demonstrated by PET. Psychopharmacology (Berlin) 110 (3), 365–367.

Nyberg, S., Eriksson, B., Oxenstierna, G., Halldin, C., Farde, L., June 1999. Suggested minimal effective dose of risperidone based on PET-measured D_2 and 5-HT_{2A} receptor occupancy in schizophrenic patients. Am. J. Psychiatry 156 (6), 869–875.

Ohba, H., Harada, N., Nishiyama, S., Kakiuchi, T., Tsukada, H., June 2009. Ketamine/xylazine anesthesia alters [^{11}C]MNPA binding to dopamine D_2 receptors and response to methamphetamine challenge in monkey brain. Synapse 63 (6), 534–537.

Okubo, Y., Suhara, T., Suzuki, K., Kobayashi, K., Inoue, O., Terasaki, O., Someya, Y., Sassa, T., Sudo, Y., Matsushima, E., Iyo, M., Tateno, Y., Toru, M., February 13, 1997. Decreased prefrontal dopamine D_1 receptors in schizophrenia revealed by PET. Nature 385 (6617), 634–636.

Olsson, H., Halldin, C., Farde, L., June 2004. Differentiation of extrastriatal dopamine D_2 receptor density and affinity in the human brain using PET. NeuroImage 22 (2), 794–803.

Olver, J.S., O'Keefe, G., Jones, G.R., Burrows, G.D., Tochon-Danguy, H.J., Ackermann, U., Scott, A., Norman, T.R., April 2009. Dopamine D_1 receptor binding in the striatum of patients with obsessive-compulsive disorder. J. Affect Disord. 114 (1–3), 321–326.

Olver, J.S., O'Keefe, G., Jones, G.R., Burrows, G.D., Tochon-Danguy, H.J., Ackermann, U., Scott, A.M., Norman, T.R., July 30, 2010. Dopamine D_1 receptor binding in the anterior cingulate cortex of patients with obsessive-compulsive disorder. Psychiatry Res. 183 (1), 85–88.

Otsuka, T., Ito, H., Halldin, C., Takahashi, H., Takano, H., Arakawa, R., Okumura, M., Kodaka, F., Miyoshi, M., Sekine, M., Seki, C., Nakao, R., Suzuki, K., Finnema, S., Hirayasu, Y., Suhara, T., Farde, L., May 2009. Quantitative PET analysis of the dopamine D_2 receptor agonist radioligand ^{11}C-(*R*)-2-CH_3O-*N*-n-propylnorapomorphine in the human brain. J. Nucl. Med. 50 (5), 703–710.

Palner, M., McCormick, P., Parkes, J., Knudsen, G.M., Wilson, A.A., May 2010. Systemic catechol-*O*-methyl transferase inhibition enables the D_1 agonist radiotracer *R*-[^{11}C]SKF 82957. Nucl. Med. Biol. 37 (7), 837–843.

Piazza, P.V., Deminière, J.M., Le Moal, M., Simon, H., September 29, 1989. Factors that predict individual vulnerability to amphetamine self-administration. Science 245 (4925), 1511–1513.

Piggott, M.A., Marshall, E.F., Thomas, N., Lloyd, S., Court, J.A., Jaros, E., Costa, D., Perry, R.H., Perry, E.K., May 1999. Dopaminergic activities in the human striatum: rostrocaudal gradients of uptake sites and of D_1 and D_2 but not of D_3 receptor binding or dopamine. Neuroscience 90 (2), 433–445.

Pinborg, L.H., Videbaek, C., Knudsen, G.M., Swahn, C.G., Halldin, C., Friberg, L., Paulson, O.B., Lassen, N.A., June 15, 2000. Dopamine D_2 receptor quantification in extrastriatal brain regions using [(^{123})I] epidepride with bolus/infusion. Synapse 36 (4), 322–329.

Pinborg, L.H., Videbaek, C., Ziebell, M., Mackeprang, T., Friberg, L., Rasmussen, H., Knudsen, G.M., Glenthoj, B.Y., February 15, 2007. [^{123}I]epidepride binding to cerebellar dopamine D_2/D_3 receptors is displaceable: implications for the use of cerebellum as a reference region. NeuroImage 34 (4), 1450–1453.

Pinto, A., Sesack, S.R., September 2008. Ultrastructural analysis of prefrontal cortical inputs to the rat amygdala: spatial relationships to presumed dopamine axons and D_1 and D_2 receptors. Brain Struct. Funct. 213 (1–2), 159–175.

Pogarell, O., Koch, W., Karch, S., Dehning, S., Müller, N., Tatsch, K., Poepperl, G., Möller, H.J., May 2012. Dopaminergic neurotransmission in patients with schizophrenia in relation to positive and negative symptoms. Pharmacopsychiatry 45 (Suppl. 1), S36–S41.

Primus, R.J., Thurkauf, A., Xu, J., Yevich, E., McInerney, S., Shaw, K., Tallman, J.F., Gallagher, D.W., August 1997. II. Localization and characterization of dopamine D_4 binding sites in rat and human brain by use of the novel, D_4 receptor-selective ligand [^{3}H]NGD 94-1. J. Pharmacol. Exp. Ther. 282 (2), 1020–1027.

Rabiner, E.A., Slifstein, M., Nobrega, J., Plisson, C., Huiban, M., Raymond, R., Diwan, M., Wilson, A.A., McCormick, P., Gentile, G., Gunn, R.N., Laruelle, M.A., September 2009. In vivo quantification of regional dopamine-D_3 receptor binding potential of (+)-PHNO: studies in non-human primates and transgenic mice. Synapse 63 (9), 782–793.

Rabiner, E.A., Laruelle, M., April 2010. Imaging the D_3 receptor in humans in vivo using [^{11}C](+)-PHNO positron emission tomography (PET). Int. J. Neuropsychopharmacol. 13 (3), 289–290.

Ramchandani, V.A., Umhau, J., Pavon, F.J., Ruiz-Velasco, V., Margas, W., Sun, H., Damadzic, R., Eskay, R., Schoor, M., Thorsell, A., Schwandt, M.L., Sommer, W.H., George, D.T., Parsons, L.H., Herscovitch, P., Hommer, D., Heilig, M., August 2011. A genetic determinant of the striatal dopamine response to alcohol in men. Mol. Psychiatry 16 (8), 809–817.

Riccardi, P., Li, R., Ansari, M.S., Zald, D., Park, S., Dawant, B., Anderson, S., Doop, M., Woodward, N., Schoenberg, E., Schmidt, D., Baldwin, R., Kessler, R., September 2006. Amphetamine-induced displacement of [^{18}F] fallypride in striatum and extrastriatal regions in humans. Neuropsychopharmacology 31 (5), 1016–1026.

Riccardi, P., Baldwin, R., Salomon, R., Anderson, S., Ansari, M.S., Li, R., Dawant, B., Bauernfeind, A., Schmidt, D., Kessler, R., January 15, 2008. Estimation of baseline dopamine D_2 receptor occupancy in striatum and extrastriatal regions in humans with positron emission tomography with [^{18}F] fallypride. Biol. Psychiatry 63 (2), 241–244.

Rieck, R.W., Ansari, M.S., Whetsell Jr., W.O., Deutch, A.Y., Kessler, R.M., February 2004. Distribution of dopamine D_2-like receptors in the human thalamus: autoradiographic and PET studies. Neuropsychopharmacology 29 (2), 362–372.

Rinne, J.O., Lönnberg, P., Marjamäki, P., February 5, 1990. Age-dependent decline in human brain dopamine D_1 and D_2 receptors. Brain Res. 508 (2), 349–352.

Rinne, J.O., Hietala, J., Ruotsalainen, U., Säkö, E., Laihinen, A., Någren, K., Lehikoinen, P., Oikonen, V., Syvälahti, E., March 1993. Decrease in human striatal dopamine D_2 receptor density with age: a PET study with [^{11}C]raclopride. J. Cereb. Blood Flow Metab. 13 (2), 310–314.

Ross, S.B., Jackson, D.M., July 1989. Kinetic properties of the in vivo accumulation of ^{3}H-(−)-*N*-n-propylnorapomorphine in mouse brain. Naunyn Schmiedebergs Arch. Pharmacol. 340 (1), 13–20.

Schiffer, W.K., Volkow, N.D., Fowler, J.S., Alexoff, D.L., Logan, J., Dewey, S.L., March 15, 2006. Therapeutic doses of amphetamine or methylphenidate differentially increase synaptic and extracellular dopamine. Synapse 59 (4), 243–251.

Schmidt, D.E., Votaw, J.R., Kessler, R.M., de Paulis, T., March 1994. Aromatic and amine substituent effects on the apparent lipophilicities of *N*-[(2-pyrrolidinyl)methyl]-substituted benzamides. J. Pharm. Sci. 83 (3), 305–315.

Schmitt, G.J., Meisenzahl, E.M., Frodl, T., La Fougère, C., Hahn, K., Möller, H.J., Dresel, S., September 30, 2009. Increase of striatal dopamine transmission in first episode drug-naive schizophrenic patients as demonstrated by [(^{123})I]IBZM SPECT. Psychiatry Res. 173 (3), 183–189.

Schonwetter, B.S., Luedtke, R.R., Kung, M.P., Billings, J., Kung, H.F., Molinoff, P.B., July 1989. Characterization of membrane-bound and soluble D_2 receptors in canine caudate using [^{125}I]IBZM. J. Pharmacol. Exp. Ther. 250 (1), 110–116.

Searle, G., Beaver, J.D., Comley, R.A., Bani, M., Tziortzi, A., Slifstein, M., Mugnaini, M., Griffante, C., Wilson, A.A., Merlo-Pich, E., Houle, S., Gunn, R., Rabiner, E.A., Laruelle, M., August 15, 2010. Imaging dopamine D_3 receptors in the human brain with positron emission tomography, [^{11}C]PHNO, and a selective D_3 receptor antagonist. Biol. Psychiatry 68 (4), 392–399.

Sedvall, G., Karlsson, P., Lundin, A., Anvret, M., Suhara, T., Halldin, C., Farde, L., 1994. Dopamine D_1 receptor number—a sensitive PET marker for early brain degeneration in Huntington's disease. Eur. Arch. Psychiatry Clin. Neurosci. 243 (5), 249–255.

Seeman, P., Chau-Wong, M., Tedesco, J., Wong, K., November 1975. Brain receptors for antipsychotic drugs and dopamine: direct binding assays. Proc. Natl. Acad. Sci. USA 72 (11), 4376–4380.

Seeman, P., Ulpian, C., Bergeron, C., Riederer, P., Jellinger, K., Gabriel, E., Reynolds, G.P., Tourtellotte, W.W., August 17, 1984. Bimodal distribution of dopamine receptor densities in brains of schizophrenics. Science 225 (4663), 728–731.

Seeman, P., Bzowej, N.H., Guan, H.C., Bergeron, C., Reynolds, G.P., Bird, E.D., Riederer, P., Jellinger, K., Tourtellotte, W.W., December 1987. Human brain D_1 and D_2 dopamine receptors in schizophrenia, Alzheimer's, Parkinson's, and Huntington's diseases. Neuropsychopharmacology 1 (1), 5–15.

Seeman, P., January 2012. Dopamine agonist radioligand binds to both D_2High and D_2Low receptors, explaining why alterations in D_2High are not detected in human brain scans. Synapse 66 (1), 88–93.

Seibyl, J.P., Zea-Ponce, Y., Brenner, L., Baldwin, R.M., Krystal, J.H., Offord, S.J., Mochoviak, S., Charney, D.S., Hoffer, P.B., Innis, R.B., January 1996. Continuous intravenous infusion of iodine-123-IBZM for SPECT determination of human brain dopamine receptor occupancy by antipsychotic agent RWJ-37796. J. Nucl. Med. 37 (1), 11–15.

Seneca, N., Finnema, S.J., Farde, L., Gulyás, B., Wikström, H.V., Halldin, C., Innis, R.B., April 2006. Effect of amphetamine on dopamine D_2 receptor binding in nonhuman primate brain: a comparison of the agonist radioligand [^{11}C]MNPA and antagonist [^{11}C]raclopride. Synapse 59 (5), 260–269.

Sesack, S.R., Hawrylak, V.A., Matus, C., Guido, M.A., Levey, A.I., April 1, 1998. Dopamine axon varicosities in the prelimbic division of the rat prefrontal cortex exhibit sparse immunoreactivity for the dopamine transporter. J. Neurosci. 18 (7), 2697–2708.

Shinotoh, H., Inoue, O., Hirayama, K., Aotsuka, A., Asahina, M., Suhara, T., Yamazaki, T., Tateno, Y., May 1993. Dopamine D_1 receptors in Parkinson's disease and striatonigral degeneration: a positron emission tomography study. J. Neurol. Neurosurg. Psychiatry 56 (5), 467–472.

Shotbolt, P., Tziortzi, A.C., Searle, G.E., Colasanti, A., van der Aart, J., Abanades, S., Plisson, C., Miller, S.R., Huiban, M., Beaver, J.D., Gunn, R.N., Laruelle, M., Rabiner, E.A., January 2012. Within-subject comparison of [(^{11})C]-(+)-PHNO and [(^{11})C]raclopride sensitivity to acute amphetamine challenge in healthy humans. J. Cereb. Blood Flow Metab. 32 (1), 127–136.

Siessmeier, T., Zhou, Y., Buchholz, H.G., Landvogt, C., Vernaleken, I., Piel, M., Schirrmacher, R., Rösch, F., Schreckenberger, M., Wong, D.F., Cumming, P., Gründer, G., Bartenstein, P., June 2005. Parametric mapping of binding in human brain of D_2 receptor ligands of different affinities. J. Nucl. Med. 46 (6), 964–972.

Skinbjerg, M., Namkung, Y., Halldin, C., Innis, R.B., Sibley, D.R., June 2009. Pharmacological characterization of 2-methoxy-*N*-propylnorapomorphine's interactions with D_2 and D_3 dopamine receptors. Synapse 63 (6), 462–475.

Skinbjerg, M., Liow, J.S., Seneca, N., Hong, J., Lu, S., Thorsell, A., Heilig, M., Pike, V.W., Halldin, C., Sibley, D.R., Innis, R.B., May 1, 2010. D_2 dopamine receptor internalization prolongs the decrease of radioligand binding after amphetamine: a PET study in a receptor internalization-deficient mouse model. NeuroImage 50 (4), 1402–1407.

Skinbjerg, M., Sibley, D.R., Javitch, J.A., Abi-Dargham, A., January 15, 2012. Imaging the high-affinity state of the dopamine D_2 receptor in vivo: fact or fiction? Biochem. Pharmacol. 83 (2), 193–198.

Slifstein, M., Hwang, D.R., Huang, Y., Guo, N., Sudo, Y., Narendran, R., Talbot, P., Laruelle, M., September 2004. In vivo affinity of [^{18}F]fallypride for striatal and extrastriatal dopamine D_2 receptors in nonhuman primates. Psychopharmacology (Berlin) 175 (3), 274–286.

Slifstein, M., Narendran, R., Hwang, D.R., Sudo, Y., Talbot, P.S., Huang, Y., Laruelle, M., October 2004. Effect of amphetamine on [(^{18})F]fallypride in vivo binding to D_2 receptors in striatal and extrastriatal regions of the primate brain: single bolus and bolus plus constant infusion studies. Synapse 54 (1), 46–63.

Slifstein, M., Kegeles, L.S., Gonzales, R., Frankle, W.G., Xu, X., Laruelle, M., Abi-Dargham, A., October 2007. [^{11}C]NNC 112 selectivity for dopamine D_1 and serotonin 5-HT_{2A} receptors: a PET study in healthy human subjects. J. Cereb. Blood Flow Metab. 27 (10), 1733–1741.

Slifstein, M., Kolachana, B., Simpson, E.H., Tabares, P., Cheng, B., Duvall, M., Frankle, W.G., Weinberger, D.R., Laruelle, M., Abi-Dargham, A., August 2008. COMT genotype predicts cortical-limbic D_1 receptor availability measured with [^{11}C]NNC112 and PET. Mol. Psychiatry 13 (8), 821–827.

Smiley, J.F., Goldman-Rakic, P.S., May–June 1993. Heterogeneous targets of dopamine synapses in monkey prefrontal cortex demonstrated by serial section electron microscopy: a laminar analysis using the silver-enhanced diaminobenzidine sulfide (SEDS) immunolabeling technique. Cereb. Cortex 3 (3), 223–238.

Smiley, J.F., Levey, A.I., Ciliax, B.J., Goldman-Rakic, P.S., June 7, 1994. D_1 dopamine receptor immunoreactivity in human and monkey cerebral cortex: predominant and extrasynaptic localization in dendritic spines. Proc. Natl. Acad. Sci. USA 91 (12), 5720–5724.

Smith, C.C., Greene, R.W., May 2, 2012. CNS dopamine transmission mediated by noradrenergic innervation. J. Neurosci. 32 (18), 6072–6080.

Sóvágó, J., Farde, L., Halldin, C., Langer, O., Laszlovszky, I., Kiss, B., Gulyás, B., October 2004. Positron emission tomographic evaluation of the putative dopamine-D_3 receptor ligand, [^{11}C]RGH-1756 in the monkey brain. Neurochem. Int. 45 (5), 609–617.

Suehiro, M., Dannals, R.F., Scheffel, U., Stathis, M., Wilson, A.A., Ravert, H.T., Villemagne, V.L., Sanchez-Roa, P.M., Wagner Jr., H.N., December 1990. In vivo labeling of the dopamine D_2 receptor with *N*-^{11}C-methyl-benperidol. J. Nucl. Med. 31 (12), 2015–2021.

Suhara, T., Fukuda, H., Inoue, O., Itoh, T., Suzuki, K., Yamasaki, T., Tateno, Y., 1991. Age-related changes in human D_1 dopamine receptors measured by positron emission tomography. Psychopharmacology (Berlin) 103 (1), 41–45.

Suhara, T., Yasuno, F., Sudo, Y., Yamamoto, M., Inoue, M., Okubo, Y., Suzuki, K., May 2001. Dopamine D_2 receptors in the insular cortex and the personality trait of novelty seeking. NeuroImage 13 (5), 891–895.

Suhara, T., Okubo, Y., Yasuno, F., Sudo, Y., Inoue, M., Ichimiya, T., Nakashima, Y., Nakayama, K., Tanada, S., Suzuki, K., Halldin, C., Farde, L., January 2002. Decreased dopamine D_2 receptor binding in the anterior cingulate cortex in schizophrenia. Arch. Gen. Psychiatry 59 (1), 25–30.

Svenningsson, P., Nishi, A., Fisone, G., Girault, J.A., Nairn, A.C., Greengard, P., 2004. DARPP-32: an integrator of neurotransmission. Annu. Rev. Pharmacol. Toxicol. 44, 269–296.

Syvälahti, E.K., Räkköläinen, V., Aaltonen, J., Lehtinen, V., Hietala, J., June 16, 2000. Striatal D_2 dopamine receptor density and psychotic symptoms in schizophrenia: a longitudinal study. Schizophr. Res. 43 (2–3), 159–161.

Takahashi, H., Matsui, H., Camerer, C., Takano, H., Kodaka, F., Ideno, T., Okubo, S., Takemura, K., Arakawa, R., Eguchi, Y., Murai, T., Okubo, Y., Kato, M., Ito, H., Suhara, T., December 8, 2010. Dopamine D_1 receptors and nonlinear probability weighting in risky choice. J. Neurosci. 30 (49), 16567–16572.

Talvik, M., Nordström, A.L., Olsson, H., Halldin, C., Farde, L., December 2003. Decreased thalamic D_2/D_3 receptor binding in drug-naive patients with schizophrenia: a PET study with [^{11}C]FLB 457. Int. J. Neuropsychopharmacol. 6 (4), 361–370.

Talvik, M., Nordström, A.L., Okubo, Y., Olsson, H., Borg, J., Halldin, C., Farde, L., December 1, 2006. Dopamine D_2 receptor binding in drug-naïve patients with schizophrenia examined with raclopride-C11 and positron emission tomography. Psychiatry Res. 148 (2–3), 165–173.

Thanos, P.K., Rivera, S.N., Weaver, K., Grandy, D.K., Rubinstein, M., Umegaki, H., Wang, G.J., Hitzemann, R., Volkow, N.D., May 27, 2005. Dopamine D_2R DNA transfer in dopamine D_2 receptor-deficient mice: effects on ethanol drinking. Life Sci. 77 (2), 130–139.

Thompson, D., Pusch, M., Whistler, J.L., October 5, 2007. Changes in G protein-coupled receptor sorting protein affinity regulate postendocytic targeting of G protein-coupled receptors. J. Biol. Chem. 282 (40), 29178–29185.

Tiberi, M., Caron, M.G., November 11, 1994. High agonist-independent activity is a distinguishing feature of the dopamine D_1B receptor subtype. J. Biol. Chem. 269 (45), 27925–27931.

Tsukada, H., Harada, N., Nishiyama, S., Ohba, H., Kakiuchi, T., September 15, 2000. Cholinergic neuronal modulation alters dopamine D_2 receptor availability in vivo by regulating receptor affinity induced by facilitated synaptic dopamine turnover: positron emission tomography studies with microdialysis in the conscious monkey brain. J. Neurosci. 20 (18), 7067–7073.

Turjanski, N., Weeks, R., Dolan, R., Harding, A.E., Brooks, D.J., June 1995. Striatal D_1 and D_2 receptor binding in patients with Huntington's disease and other choreas. A PET study. Brain 118 (Pt 3), 689–696.

Turjanski, N., Lees, A.J., Brooks, D.J., September 1997. In vivo studies on striatal dopamine D_1 and D_2 site binding in L-dopa-treated Parkinson's disease patients with and without dyskinesias. Neurology 49 (3), 717–723.

Turolla, E.A., Matarrese, M., Belloli, S., Moresco, R.M., Simonelli, P., Todde, S., Fazio, F., Magni, F., Kienle, M.G., Leopoldo, M., Berardi, F., Colabufo, N.A., Lacivita, E., Perrone, R., November 3, 2005. ^{11}C-labeling of n-[4-[4-(2,3-dichlorophenyl)piperazin-1-yl]butyl]arylcarboxamide derivatives and evaluation as potential radioligands for PET imaging of dopamine D_3 receptors. J. Med. Chem. 48 (22), 7018–7023.

Tziortzi, A.C., Searle, G.E., Tzimopoulou, S., Salinas, C., Beaver, J.D., Jenkinson, M., Laruelle, M., Rabiner, E.A., Gunn, R.N., January 1, 2011. Imaging dopamine receptors in humans with [^{11}C]-(+)-PHNO: dissection of D_3 signal and anatomy. NeuroImage 54 (1), 264–277.

Urban, N.B., Kegeles, L.S., Slifstein, M., Xu, X., Martinez, D., Sakr, E., Castillo, F., Moadel, T., O'Malley, S.S., Krystal, J.H., Abi-Dargham, A., October 15, 2010. Sex differences in striatal dopamine release in young adults after oral alcohol challenge: a positron emission tomography imaging study with [^{11}C]raclopride. Biol. Psychiatry 68 (8), 689–696.

Urban, N.B., Slifstein, M., Thompson, J.L., Xu, X., Girgis, R.R., Raheja, S., Haney, M., Abi-Dargham, A., April 15, 2012. Dopamine release in chronic cannabis users: a [^{11}C]raclopride positron emission tomography study. Biol. Psychiatry 71 (8), 677–683.

Vandehey, N.T., Moirano, J.M., Converse, A.K., Holden, J.E., Mukherjee, J., Murali, D., Nickles, R.J., Davidson, R.J., Schneider, M.L., Christian, B.T., May 2010. High-affinity dopamine D_2/D_3 PET radioligands ^{18}F-fallypride and ^{11}C-FLB457: a comparison of kinetics in extrastriatal regions using a multiple-injection protocol. J. Cereb. Blood Flow Metab. 30 (5), 994–1007.

Verhoeff, N.P., Hussey, D., Lee, M., Tauscher, J., Papatheodorou, G., Wilson, A.A., Houle, S., Kapur, S., 2002. Dopamine depletion results in increased neostriatal D_2, but not D_1, receptor binding in humans. Mol. Psychiatry 7 (3) 233, 322–328.

Verhoeff, N.P., Christensen, B.K., Hussey, D., Lee, M., Papatheodorou, G., Kopala, L., Rui, Q., Zipursky, R.B., Kapur, S., January 2003. Effects of catecholamine depletion on D_2 receptor binding, mood, and attentiveness in humans: a replication study. Pharmacol. Biochem. Behav. 74 (2), 425–432.

Vernaleken, I., Peters, L., Raptis, M., Lin, R., Buchholz, H.G., Zhou, Y., Winz, O., Rösch, F., Bartenstein, P., Wong, D.F., Schäfer, W.M., Gründer, G., September 2011. The applicability of SRTM in [(^{18})F]fallypride PET investigations: impact of scan durations. J. Cereb. Blood Flow Metab. 31 (9), 1958–1966.

Volkow, N.D., Fowler, J.S., Wang, G.J., Hitzemann, R., Logan, J., Schlyer, D.J., Dewey, S.L., Wolf, A.P., June 1993. Decreased dopamine D_2 receptor availability is associated with reduced frontal metabolism in cocaine abusers. Synapse 14 (2), 169–177.

Volkow, N.D., Wang, G.J., Fowler, J.S., Logan, J., Hitzemann, R., Ding, Y.S., Pappas, N., Shea, C., Piscani, K., December 1996. Decreases in dopamine receptors but not in dopamine transporters in alcoholics. Alcohol. Clin. Exp. Res. 20 (9), 1594–1598.

Volkow, N.D., Wang, G.J., Fowler, J.S., Logan, J., Gatley, S.J., Hitzemann, R., Chen, A.D., Dewey, S.L., Pappas, N., April 24, 1997. Decreased striatal dopaminergic responsiveness in detoxified cocaine-dependent subjects. Nature 386 (6627), 830–833.

Volkow, N.D., Wang, G.J., Fowler, J.S., Gatley, S.J., Logan, J., Ding, Y.S., Hitzemann, R., Pappas, N., October 1998. Dopamine transporter occupancies in the human brain induced by therapeutic doses of oral methylphenidate. Am. J. Psychiatry 155 (10), 1325–1331.

Volkow, N.D., Wang, G.J., Fowler, J.S., Logan, J., Gatley, S.J., Gifford, A., Hitzemann, R., Ding, Y.S., Pappas, N., September 1999. Prediction of reinforcing responses to psychostimulants in humans by brain dopamine D_2 receptor levels. Am. J. Psychiatry 156 (9), 1440–1443.

Volkow, N.D., Chang, L., Wang, G.J., Fowler, J.S., Ding, Y.S., Sedler, M., Logan, J., Franceschi, D., Gatley, J., Hitzemann, R., Gifford, A., Wong, C., Pappas, N., December 2001. Low level of brain dopamine D_2 receptors in methamphetamine abusers: association with metabolism in the orbitofrontal cortex. Am. J. Psychiatry 158 (12), 2015–2021.

Volkow, N.D., Wang, G.J., Maynard, L., Fowler, J.S., Jayne, B., Telang, F., Logan, J., Ding, Y.S., Gatley, S.J., Hitzemann, R., Wong, C., Pappas, N., December 30, 2002. Effects of alcohol detoxification on dopamine D_2 receptors in alcoholics: a preliminary study. Psychiatry Res. 116 (3), 163–172.

Volkow, N.D., Wang, G.J., Begleiter, H., Porjesz, B., Fowler, J.S., Telang, F., Wong, C., Ma, Y., Logan, J., Goldstein, R., Alexoff, D., Thanos, P.K., September 2006a. High levels of dopamine D_2 receptors in unaffected members of alcoholic families: possible protective factors. Arch. Gen. Psychiatry 63 (9), 999–1008.

Volkow, N.D., Wang, G.J., Telang, F., Fowler, J.S., Logan, J., Childress, A.R., Jayne, M., Ma, Y., Wong, C., June 14, 2006b. Cocaine cues and dopamine in dorsal striatum: mechanism of craving in cocaine addiction. J. Neurosci. 26 (24), 6583–6588.

Volkow, N.D., Wang, G.J., Telang, F., Fowler, J.S., Logan, J., Jayne, M., Ma, Y., Pradhan, K., Wong, C., November 14, 2007a. Profound decreases in dopamine release in striatum in detoxified alcoholics: possible orbitofrontal involvement. J. Neurosci. 27 (46), 12700–12706.

Volkow, N.D., Wang, G.J., Newcorn, J., Telang, F., Solanto, M.V., Fowler, J.S., Logan, J., Ma, Y., Schulz, K., Pradhan, K., Wong, C., Swanson, J.M., August 2007b. Depressed dopamine activity in caudate and preliminary evidence of limbic involvement in adults with attention-deficit/hyperactivity disorder. Arch. Gen. Psychiatry 64 (8), 932–940.

Votaw, J.R., Kessler, R.M., de Paulis, T., November 1993. Failure of the three compartment model to describe the pharmacokinetics in brain of a high affinity substituted benzamide. Synapse 15 (3), 177–190.

Wagner Jr., H.N., Burns, H.D., Dannals, R.F., Wong, D.F., Langström, B., Duelfer, T., Frost, J.J., Ravert, H.T., Links, J.M., Rosenbloom, S.B., et al., 1984. Assessment of dopamine receptor densities in the human brain with carbon-11-labeled *N*-methylspiperone. Ann. Neurol. 15 (Suppl.), S79–S84.

Wang, G.J., Volkow, N.D., Fowler, J.S., Logan, J., Abumrad, N.N., Hitzemann, R.J., Pappas, N.S., Pascani, K., February 1997. Dopamine D_2 receptor availability in opiate-dependent subjects before and after naloxone-precipitated withdrawal. Neuropsychopharmacology 16 (2), 174–182.

Wang, Y., Chan, G.L., Holden, J.E., Dobko, T., Mak, E., Schulzer, M., Huser, J.M., Snow, B.J., Ruth, T.J., Calne, D.B., Stoessl, A.J., September 1998. Age-dependent decline of dopamine D_1 receptors in human brain: a PET study. Synapse 30 (1), 56–61.

Wang, G.J., Volkow, N.D., Logan, J., Pappas, N.R., Wong, C.T., Zhu, W., Netusil, N., Fowler, J.S., February 3, 2001. Brain dopamine and obesity. Lancet 357 (9253), 354–357.

Wang, G.J., Smith, L., Volkow, N.D., Telang, F., Logan, J., Tomasi, D., Wong, C.T., Hoffman, W., Jayne, M., Alia-Klein, N., Thanos, P., Fowler, J.S., September 2012. Decreased dopamine activity predicts relapse in methamphetamine abusers. Mol. Psychiatry 17 (9), 918–925.

Wilson, A.A., McCormick, P., Kapur, S., Willeit, M., Garcia, A., Hussey, D., Houle, S., Seeman, P., Ginovart, N., June 16, 2005. Radiosynthesis and evaluation of [^{11}C]-(+)-4-propyl-3,4,4a,5,6,10b-hexahydro-2*H*-naphtho[1,2-*b*][1,4]oxazin-9-ol as a potential radiotracer for in vivo imaging of the dopamine D_2 high-affinity state with positron emission tomography. J. Med. Chem. 48 (12), 4153–4160.

Wong, D.F., Gjedde, A., Wagner Jr., H.N., April 1986a. Quantification of neuroreceptors in the living human brain. I. Irreversible binding of ligands. J. Cereb. Blood Flow Metab. 6 (2), 137–146.

Wong, D.F., Gjedde, A., Wagner Jr., H.N., Dannals, R.F., Douglass, K.H., Links, J.M., Kuhar, M.J., April 1986b. Quantification of Neuroreceptors in the living human brain. II. Inhibition studies of receptor density and affinity. J. Cereb. Blood Flow Metab. 6 (2), 147–153.

Wong, D.F., Wagner Jr., H.N., Tune, L.E., Dannals, R.F., Pearlson, G.D., Links, J.M., Tamminga, C.A., Broussolle, E.P., Ravert, H.T., Wilson, A.A., Toung, J.K., Malat, J., Williams, J.A., O'Tuama, L.A., Snyder, S.H., Kuhar, M.J., Gjedde, A., December 19, 1986c. Positron emission tomography reveals elevated D_2 dopamine receptors in drug-naive schizophrenics. Science 234 (4783), 1558–1563.

Wong, D.F., Singer, H.S., Brandt, J., Shaya, E., Chen, C., Brown, J., Kimball, A.W., Gjedde, A., Dannals, R.F., Ravert, H.T., Wilson, P.D., Wagner Jr., H.N., August 1997. D_2-like dopamine receptor density in Tourette syndrome measured by PET. J. Nucl. Med. 38 (8), 1243–1247.

Woodward, N.D., Cowan, R.L., Park, S., Ansari, M.S., Baldwin, R.M., Li, R., Doop, M., Kessler, R.M., Zald, D.H., April 2011. Correlation of individual differences in schizotypal personality traits with amphetamine-induced dopamine release in striatal and extrastriatal brain regions. Am. J. Psychiatry 168 (4), 418–426.

Yamamoto, B.K., Novotney, S., July 1998. Regulation of extracellular dopamine by the norepinephrine transporter. J. Neurochem. 71 (1), 274–280.

Yasuno, F., Suhara, T., Okubo, Y., Sudo, Y., Inoue, M., Ichimiya, T., Takano, A., Nakayama, K., Halldin, C., Farde, L., June 2004. Low dopamine D_2 receptor binding in subregions of the thalamus in schizophrenia. Am. J. Psychiatry 161 (6), 1016–1022.

Yasuno, F., Ota, M., Ando, K., Ando, T., Maeda, J., Ichimiya, T., Takano, A., Doronbekov, T.K., Fujimura, Y., Nozaki, S., Suhara, T., June 1, 2007. Role of ventral striatal dopamine D_1 receptor in cigarette craving. Biol. Psychiatry 61 (11), 1252–1259.

Yoder, K.K., Wang, C., Morris, E.D., May 2004. Change in binding potential as a quantitative index of neurotransmitter release is highly sensitive to relative timing and kinetics of the tracer and the endogenous ligand. J. Nucl. Med. 45 (5), 903–911.

Zald, D.H., Cowan, R.L., Riccardi, P., Baldwin, R.M., Ansari, M.S., Li, R., Shelby, E.S., Smith, C.E., McHugo, M., Kessler, R.M., December 31, 2008. Midbrain dopamine receptor availability is inversely associated with novelty-seeking traits in humans. J. Neurosci. 28 (53), 14372–14378.

Zijlstra, F., Booij, J., van den Brink, W., Franken, I.H., April 2008. Striatal dopamine D_2 receptor binding and dopamine release during cue-elicited craving in recently abstinent opiate-dependent males. Eur. Neuropsychopharmacol. 18 (4), 262–270.

CHAPTER THIRTEEN

Dopamine Receptor Imaging in Schizophrenia: Focus on Genetic Vulnerability

Jussi Hirvonen[1,2] and Jarmo Hietala[2,3]
[1]Department of Radiology, Turku University Hospital and University of Turku, Turku, Finland
[2]Turku PET Centre, Turku University Hospital and University of Turku, Turku, Finland
[3]Department of Psychiatry, University of Turku, Turku, Finland

1. INTRODUCTION

Schizophrenia is a severe mental illness with a worldwide prevalence of about 1–2% of the population. The exact pathophysiology of schizophrenia has remained elusive, but the disease is considered to be multifactorial with strong genetic contributions as well as prominent gene–environment interaction (van Os et al., 2008). In the brain, schizophrenia is associated with subtle structural and functional abnormalities of various brain regions, such as striatum, hippocampus, and prefrontal cortex (Harrison and Weinberger, 2005; Lisman et al., 2008). At the molecular level, multiple neurotransmitter systems have been hypothesized to be abnormal: glutamate, dopamine, γ-aminobutyric acid (GABA), and serotonin (Carlsson et al., 2001).

Among the neurotransmitter systems, dopamine has been most strongly linked to schizophrenia, owing to the propensity of dopamine D_2 receptor antagonist to alleviate and of dopamine-releasing drugs to exacerbate positive symptoms of schizophrenia (Howes and Kapur, 2009; Davis et al., 1991). Briefly, the dopamine hypothesis of schizophrenia suggests that subcortical dopamine overactivity is a final common pathway brought on by multiple factors from genes and environment and contributed by dysfunction of many brain circuits (Howes and Kapur, 2009; Tost et al., 2010). Also, the cortical dopamine neurotransmission is suggested to be dysfunctional in schizophrenia (Goldman-Rakic et al., 2000). Dopamine abnormalities are tightly connected to glutamate and GABA deficits that are thought to be relevant in the pathophysiology in schizophrenia (Laruelle et al., 2003; Lisman et al., 2008). More specifically, glutamatergic corticostriatal afferents regulate the activity of mesostriatal and mesolimbic dopamine neurons, both at the level of basal ganglia and the midbrain, and both through direct pathways and indirectly via GABA interneurons (Pycock et al., 1980; Sesack and Carr, 2002).

Subcortical dopamine function is thought to occur via two distinct mechanisms: tonic dopamine release setting the background dopamine level (extrasynaptic, nanomolar) and

Imaging of the Human Brain in Health and Disease
http://dx.doi.org/10.1016/B978-0-12-418677-4.00013-0

phasic dopamine release associated with dopamine neuron firing (synaptic, millimolar). In a pathophysiological model advanced by Grace (Grace, 1991, 1993), the prefrontal cortical pathology in schizophrenia impairs appropriate stimulation of subcortical tonic dopamine release, which leads to upregulation of postsynaptic dopamine receptors and disinhibition of dopamine neurons via terminal autoreceptors as compensatory phenomena. Together, these changes render the subcortical dopamine system hypersensitive to phasic dopamine release in schizophrenia (Grace, 1991, 1993; Moore et al., 1999).

In general, brain imaging studies in patients with schizophrenia do not distinguish between state and trait phenomena. That is, is it possible that brain abnormalities are associated with vulnerability to develop schizophrenia rather than the clinical phenotype itself? Are these abnormalities also present in the brains of individuals at risk of developing the illness? These questions are relevant because epidemiological studies suggest that schizophrenia is highly heritable (Sullivan et al., 2003; Tsuang, 2000). A meta-analysis of twin studies suggests that the heritability of schizophrenia is about 80%, reflecting the extent to which variation in susceptibility to schizophrenia may be attributed to genetic variation in the human population (Sullivan et al., 2003).

Schizophrenia is not a single-gene disorder but rather likely results from a combination of both rare and common genetic variants (Owen et al., 2010), environmental risk factors (Cannon et al., 2003), and their interaction (van Os et al., 2008). Because of high clinical heterogeneity, research into intermediate phenotypes has become a popular strategy in identifying the neurobiological factors responsible for transmitting the increased risk from the genes to the phenotype. This strategy is based on the premise that intermediate phenotypes are more proximal to the underlying genes than complex clinical phenotypes by having higher penetrance and more simple genetic architecture. These studies have shown that genetic risk variants for schizophrenia modulate brain structure, function, and connectivity in a way reminiscent of pathophysiology of schizophrenia (Meyer-Lindenberg and Weinberger, 2006).

Furthermore, similar alterations have been found in individuals with genetic risk for schizophrenia but who do not express the illness (Cannon, 2005; Whalley et al., 2005; Kaymaz and van Os, 2009). The strategy of studying people at genetic risk is attractive because findings in these individuals are not confounded by antipsychotic medication and chronic illness.

This chapter intends to review the extant literature pertaining to dopamine receptor imaging in schizophrenia by means of molecular imaging techniques. To complement previous scholarly reviews on the topic (Laruelle, 1998), we will focus on the effects of genetic vulnerability to schizophrenia.

2.1. OVERVIEW OF MOLECULAR IMAGING TECHNIQUES

Molecular imaging techniques such as single-photon emission computed tomography (SPECT) and positron emission tomography (PET) are based on short-lived radioactive isotopes that are attached to a molecule of interest (e.g. water, glucose, or receptor ligand). These techniques have extremely high sensitivity and specificity to measure different proteins in the human body. High sensitivity stems from the ability of the scanner detectors to measure extremely small amounts of radiation from the radionuclide (e.g. ^{11}C); therefore, radioligands only need to be given at doses in the microgram range. High specificity is a pharmacological property of the ligand; for example, the dopamine D_2 receptor antagonist raclopride binds specifically and selectively to D_2 receptors in the human caudate nucleus. Thus, the radioligand (e.g. [^{11}C]raclopride) combines both high sensitivity and high specificity. To quantify receptor binding, the principles of in vitro receptor binding techniques are applied to the in vivo situation—with obvious limitations. For example, radioligands are typically given at very small doses, or "tracer" doses, which are assumed to occupy only a minimal proportion of the target molecules.

This technique of using one small radiotracer concentration measures the ratio of receptor density and affinity, which is often referred to as the binding potential (Innis et al., 2007). Separate measurement of receptor density and affinity requires the simultaneous injection of multiple, increasingly saturating doses of the unlabeled ligand (Farde et al., 1986), but this approach has the disadvantage of potentially significant pharmacological side effects. Developing new PET radioligands is a challenging task: for example, an optimal radioligand should only bind to the target in question, have good brain penetration, have sufficiently rapid kinetics to allow imaging in a few hours, and should not create radioactive metabolites that may interfere with quantification (Pike, 2009). Given these potential complications, radioligands are typically carefully validated first in animals and then in healthy human subjects before applying them in patient populations.

Molecular imaging techniques are typically used to measure the availability of the target (e.g. receptor, transporter, or enzyme) in a baseline, or resting, condition. Bearing in mind the limitations outlined above, the result is then inferred to reflect the density of the target in brain tissue. Another technique is based on the principle that endogenous neurotransmitters compete with the radioligand in binding to the target. Changes in synaptic neurotransmitter concentrations can thus be indirectly measured by imaging the same subject before and after a pharmacological challenge. Binding of the dopamine D_2 receptor antagonist radioligands [^{11}C]raclopride and [^{123}I] (S(-)-N-[(1-ethyl-2-pyrrolidinyl)-methyl]-2-hydroxy-3-iodo-6-methoxy- benzamide (IBZM) has been shown to be sensitive to increases and decreases in endogenous dopamine concentrations (Laruelle, 2000), although factors other than competitive inhibition may be involved (Skinbjerg et al., 2010).

These molecular imaging techniques have been extensively used in the past years to provide evidence for dysregulated striatal presynaptic dopamine system. Studies using radiolabeled analogs of dopamine precursors have shown increased presynaptic dopamine synthesis capacity in schizophrenia (Hietala et al., 1995, 1999; Reith et al., 1994;

Dao-Castellana et al., 1997; McGowan et al., 2004; Elkashef et al., 2000; Lindstrom et al., 1999; Meyer-Lindenberg et al., 2002) and that this increase in striatal dopamine function correlates with poor task-related engagement of the prefrontal cortex (Meyer-Lindenberg et al., 2002), consistent with the idea that striatal dopamine hyperactivity is connected to cortical pathology (Howes and Kapur, 2009). Studies using the "challenge paradigm" have shown increased amphetamine-induced striatal dopamine release in patients with schizophrenia (Laruelle et al., 1996; Breier et al., 1997; Abi-Dargham et al., 1998), which is present only in active ill but not in remitted patients and correlates with amphetamine-induced positive symptoms (Laruelle et al., 1999).

Recently, increased stress-induced dopamine release was also demonstrated in patients with schizophrenia (Mizrahi et al., 2012). Interestingly, increased dopamine release is also seen in patients with schizotypal personality disorder (Abi-Dargham et al., 2004), suggesting additional genetic contributions, given that this condition is genetically related to schizophrenia (Siever and Davis, 2004). Studies using a dopamine depletion paradigm have shown increased baseline occupancy of striatal D_2 receptors by dopamine (Abi-Dargham et al., 2000a,b; Kegeles et al., 2010a,b), which is predictive of decrease in positive symptoms after subsequent antipsychotic medication (Abi-Dargham et al., 2000a,b).

2.2. DOPAMINE D_2 RECEPTOR IMAGING IN SCHIZOPHRENIA

The highest density of dopamine D_2 receptors is in the striatum, followed by the thalamus and the cortex (Hall et al., 1988). In the striatum, dopamine D_2 receptors modulate the function of cortico–striato–thalamo–cortical loops (Gerfen et al., 1990) mainly via inhibition of adenylyl cyclase (Neve et al., 2004), and are a major target of antipsychotic drugs (van Rossum, 1966; Farde et al., 1992). Various SPECT and PET tracers are available for imaging D_2 receptors. The spiperone derivative [^{11}C]*N*-methyl-spiperone (NMSP) was the first PET radioligand to be used for imaging striatal D_2 receptors (Wagner et al., 1983). The first radioligand to show appropriate selectivity for D_2-like receptors was the benzazepine derivative [^{11}C]raclopride (Farde et al., 1986, 1987). Compared with the striatum, the D_2 receptor density in the cerebral cortex is low (Hall et al., 1994), and reliable imaging of these receptors requires ligands with higher (picomolar) affinity such as [^{11}C] FLB (a substituted benzamide) 457 (Halldin et al., 1995), which has high signal in extrastriatal regions in the human brain (Farde et al., 1997).

2.2.1. Dopamine D_2 Receptor Abnormalities in Patients with Schizophrenia

Early studies on striatal dopamine D_2 receptors found increased binding in patients with schizophrenia (Wong et al., 1986; Crawley et al., 1986; Tune et al., 1993). These studies used the spiperone radioligand [^{11}C]NMSP and documented up to twofold higher ligand binding in patients than in healthy subjects. However, later studies conducted with a different class of radioligands, notably [^{11}C]raclopride, failed to confirm this finding (Farde et al., 1990; Hietala et al., 1994). To date, 21 reports have been published on striatal dopamine D_2 receptor imaging in schizophrenia. A meta-analysis of all published studies (Wong et al., 1986; Crawley et al., 1986; Farde et al., 1990; Martinot et al., 1991, 1994, 1990; Tune et al., 1993; Hietala et al., 1994; Pilowsky et al., 1994; Nordström et al., 1995; Laruelle et al., 1996; Okubo et al., 1997; Breier et al., 1997; Knable et al., 1997; Abi-Dargham et al., 1998, 2000a,b; Yang et al., 2004; Corripio et al., 2006; Kessler et al., 2009; Kegeles et al., 2010a,b) suggests a small but significant increase in striatal D_2 receptors in schizophrenia (see Figure 1) (Laruelle, 1998).

Only a few studies have found significant differences between patients and healthy subjects (Wong et al., 1986; Crawley et al., 1986; Corripio et al., 2006), although most studies have found positive effect sizes (effect size is calculated as the difference between the means of patients and controls, divided by the standard deviation in the control group). Extending the previous work by Laruelle (Laruelle, 1998), when we pool the

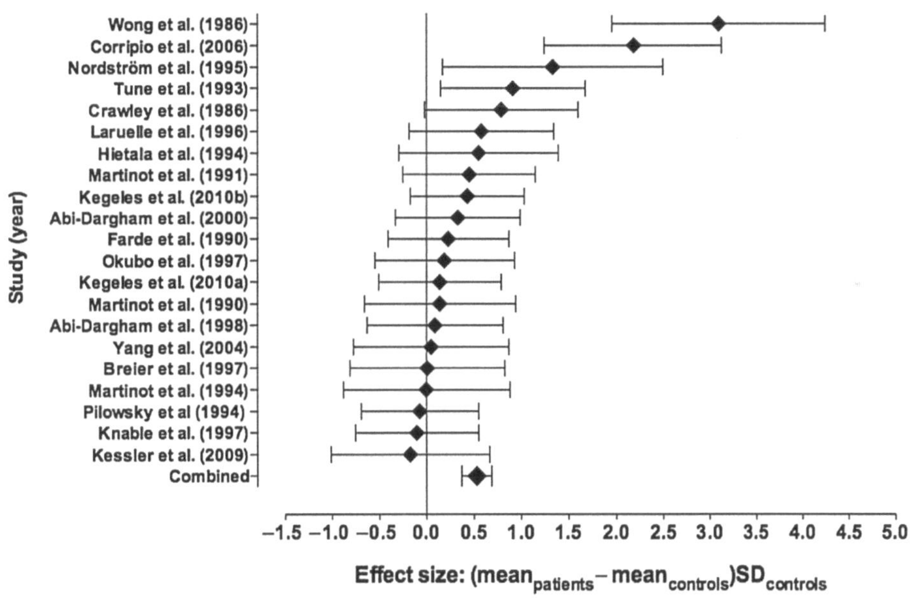

Figure 1 Meta-analysis combines all 21 published studies on striatal dopamine D_2 receptor binding in patients with schizophrenia (total of 294 patients) in comparisons with healthy subjects (total of 315 subjects). Effect size is the difference between the means of patients and controls divided by the standard deviation in the control group. Values are effect sizes, and error bars are lower and upper bounds for the 95% confidence interval of the effect size. All diamonds are the same size and not proportional to any study characteristic. Solid horizontal line represents zero (that is, no difference between patients and healthy subjects). As is evident, the distribution of effect sizes is significantly different from that expected based on the null hypothesis of no overall differences between groups (one-sample *t*-test, $p=0.004$). That is, under the null hypothesis, effect sizes should be distributed evenly around zero.

data from all published studies (N = 21), we get an estimate of a pooled effect size of 0.53 (standard deviation 0.80, 95% confidence interval 0.37–0.69), which is of modest magnitude. The distribution of effect sizes is significantly different from that expected based on the null hypothesis of no overall differences between groups (one-sample t-test, $p = 0.004$). This pooled analysis has a total of 294 patients with schizophrenia and 315 healthy subjects.

Thus, there appears to be a moderate overall increase in striatal dopamine D_2 receptor density in a heterogeneous population of patients with schizophrenia. Future studies should be able to form more homogeneous samples of patients for analysis, as exemplified by a recent finding that increased striatal D_2 receptors can be found only in psychotic patients with poor prognosis (Corripio et al., 2006). In addition to clinical heterogeneity, the type of radioligand might also have contributed to the discrepant finding, as indicated by the fact that the pooled effect size for studies using butyrophenone radioligands (such as [^{11}C]NSMP) is higher than that for studies using benzamide radioligands (such as [^{11}C]raclopride or [^{123}I]IBZM) (Laruelle, 1998). This issue has been extensively discussed previously (Seeman, 1987; Seeman and Kapur, 2000; Wong, 2002). Benzamide radioligands appear to be more sensitive to pharmacological competition with endogenous dopamine than butyrophenones (Seeman et al., 1989, 1990; Young et al., 1991) and thus, increased levels of endogenous dopamine in schizophrenia would underestimate benzamide radioligand binding and mask the increased D_2 receptor density. This hypothesis has gained support from recent imaging studies showing increased endogenous dopamine levels at baseline (Abi-Dargham et al., 2000a,b; Kegeles et al., 2010a,b). In other words, patients with schizophrenia exhibited increased occupancy by dopamine at D_2 receptors and higher D_2 receptor binding was revealed only after depleting endogenous dopamine.

Early studies focused on striatal D_2 receptors in schizophrenia, in part owing to technical difficulties in imaging the low density of extrastriatal D_2 receptors. Cortical and thalamic dopamine D_2 receptors contribute to cognitive functions and mechanism of action of antipsychotic drugs (Lidow et al., 1998) and can now be imaged with high-affinity radioligands (Halldin et al., 1995). Most studies have shown decreased extrastriatal D_2 receptor binding in schizophrenia, especially in the thalamus (Suhara et al., 2002; Yasuno et al., 2004; Talvik et al., 2003; Tuppurainen et al., 2003, 2006; Kessler et al., 2009; Kegeles et al., 2010a,b; Lehrer et al., 2010), although not all results have been consistent. Convergence of the results in the thalamus is interesting given the central role of this brain structure in the pathophysiology of schizophrenia (Clinton and Meador-Woodruff, 2004). Thus, lower D_2 receptor density in the thalamus may contribute to sensory gating abnormalities in schizophrenia.

2.2.2. Effects of Genetic Risk on Dopamine D_2 Receptors

Although molecular imaging studies have confirmed dopaminergic abnormalities in patients with schizophrenia, these studies do not distinguish between state and trait phenomena. Is it possible that dopaminergic abnormalities are associated with vulnerability to develop schizophrenia and therefore are present in the brain before the clinical disease begins, or are they only related to the expression of illness? Could we detect dopaminergic abnormalities in people with genetic risk for schizophrenia? We recently sought to answer these questions by employing a powerful study design to investigate genetic vulnerability: a twin study. Nonidentical twins share 50% of their segregating genes, while identical twins share 100%. Therefore, twin studies have the advantage of being able to gradually increase the sharing of genetic similarity, and to closely match for environmental factors. Studies using discordant twins (i.e. only one co-twin expressing the illness) can investigate both the effects of genetic vulnerability by comparing unaffected co-twins with healthy controls and the effects of expressing the illness by comparing affected and unaffected co-twins. The twin design is especially powerful in studying schizophrenia, because nonidentical twins have about 10% risk and identical twins have about 50% risk of developing schizophrenia, both of which greatly exceed the risk of 1–2% in the general population. We sought to sample twins discordant for schizophrenia for PET studies on imaging dopamine D_2 and D_1 receptor binding. We recruited middle-aged twins by cross-referencing the National Twin Registry in Finland (Kaprio and Koskenvuo, 2002) including all same-sex twin pairs born in Finland between 1940 and 1957 ($N = 9692$ pairs) with three national registers related to hospitalization, medications, and pensions (Cannon et al., 1998). We identified 348 twin pairs with either or both co-twins having schizophrenia and 9214 healthy twin pairs. Among these twins, we randomly sampled 11 discordant twin pairs (six identical and five nonidentical) and seven healthy twin pairs (four identical, three nonidentical) for PET studies. Using PET, we measured dopamine D_2 and D_1 receptor binding with the well-established radioligands [^{11}C]raclopride and [^{11}C]SCH 23390, respectively. We did not scan medicated patients for dopamine D_2 receptors because chronic antipsychotic medication affects these receptors and thereby confounds the results.

When we looked at the effects of genetic vulnerability for schizophrenia on dopamine D_2 receptors, we found that unaffected identical co-twins of patients with schizophrenia had higher striatal D_2 receptor density than did healthy twins, specifically in the caudate nucleus (effect size 0.75) (Figure 2) (Hirvonen et al., 2005). We found no significant differences between unaffected nonidentical co-twins and controls and were thus not able to directly demonstrate association with increasing genetic loading in this small sample. High caudate D_2 receptor density predicted poor cognitive performance in the whole sample. In identical healthy twins, we found a high intraclass correlation coefficient (about 0.60) in D_2 receptor density specifically in the caudate, suggesting that striatal dopamine D_2 density is highly heritable. This finding of increased D_2 receptors in the caudate in unaffected co-twins of patients with schizophrenia further supports the hypothesis that dopamine abnormalities in schizophrenia are in part genetically determined. The fact that the caudate nucleus was specifically afflicted and

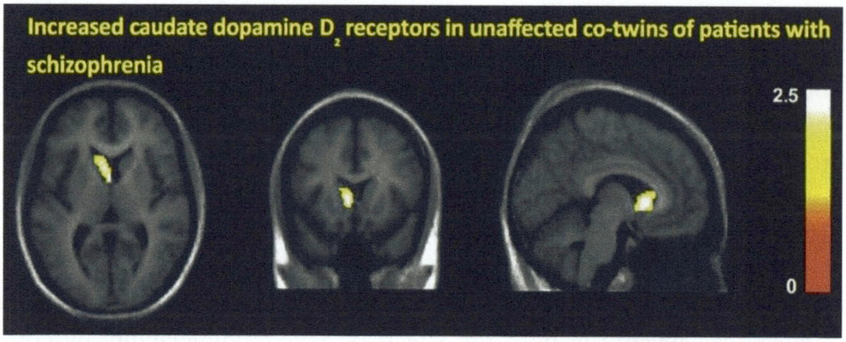

Figure 2 Increased caudate dopamine D_2 receptor binding in unaffected co-twins of patients with schizophrenia (Hirvonen et al., 2005). Results from a voxel-based receptor mapping analysis are visualized on a magnetic resonance imaging template. Color scales represent *T* statistic values at voxel level. The right side of the brain appears on the right side of the images.

that high dopamine D_2 receptor density predicted poor cognitive performance support the view that dopamine function may be particularly dysregulated in the associative striatum in schizophrenia. Two more recent studies in unaffected relatives of patients with schizophrenia have not replicated increased caudate D_2 receptor binding (Lee et al., 2008; Brunelin et al., 2010). The reasons for this discrepancy are not known, but modest genetic loading and small sample sizes in these studies may have contributed, because the interindividual variability of dopamine D_2 measurements is large (Farde et al., 1995).

In summary, we found that striatal dopamine D_2 receptor binding is increased in people at genetic risk for schizophrenia. Could this biomarker represent a primary abnormality in schizophrenia, and could it actually be more readily detected in people at genetic risk than in patients who express the illness? The contemporary version of the dopamine hypothesis of schizophrenia emphasizes abnormal regulation of subcortical dopamine by cortical and hippocampal afferents (Howes and Kapur, 2009; Davis et al., 1991), and it therefore seems likely that increased striatal D_2 receptor binding in the unaffected co-twins represents a secondary rather than a primary pathological change. Consistent with hypothesis of dysregulation networks, reduced markers of pyramidal cell integrity in the prefrontal cortex predict increased dopamine D_2 receptors (Bertolino et al., 1999) and increased amphetamine-induced dopamine release (Bertolino et al., 2000) in patients with schizophrenia, and disturbed prefrontal activity is associated with increased presynaptic striatal dopamine in both patients (Meyer-Lindenberg et al., 2002) and individuals at clinical risk (Fusar-Poli et al., 2010). Cortical glutamatergic control over subcortical dopamine may be important, because glutamate N-methyl-D-aspartate receptor antagonism potentiates amphetamine-induced dopamine release in healthy humans (Kegeles et al., 2000). Reduced prefrontal gray matter volume (Cannon et al., 2002) and deficits in frontally mediated cognitive functioning (Cannon et al., 2000) in unaffected co-twins of patients with schizophrenia support prefrontal pathology as a genetically mediated trait in schizophrenia.

We might speculate that increased striatal D_2 receptor binding is more readily detected in individuals with genetic risk for schizophrenia than in patients who express the illness. Given the link between synaptic dopamine concentrations and positive symptoms of schizophrenia (Abi-Dargham et al., 2000a,b), we might expect people at genetic risk, who are spared of frank psychotic symptoms, to not have higher synaptic dopamine concentration brought on by phasic neuron bursting. These individuals would then be expected to show less underestimation of [^{11}C]raclopride binding by endogenous dopamine than patients with schizophrenia, and we would better be able to detect increased D_2 receptor density secondary to cortical dysregulation of striatal tonic dopamine function (Grace, 1991, 1993). This hypothesis predicts that striatal dopamine D_2 receptor may be a reasonable imaging biomarker of vulnerability to schizophrenia. Whether this biomarker is useful for genetic association studies or clinical studies on early interventions remains to be determined.

2.3. DOPAMINE D_1 RECEPTOR IMAGING IN SCHIZOPHRENIA

Cortical dopamine D_1 receptors are crucial for working memory performance, which is disturbed in schizophrenia (Goldman-Rakic et al., 2000; Sawaguchi and Goldman-Rakic, 1991). These receptors are located in apical dendrites and spines of prefrontal pyramidal neurons and interneurons (Smiley et al., 1994) where they modulate neuronal responses to incoming depolarizing signaling (Goldman-Rakic et al., 2000) and increase the signal-to-noise ratio of prefrontal processing (Abi-Dargham and Moore, 2003). An inverted-"U"-shaped curve for working memory dependence on D_1 receptor stimulation has been proposed, meaning that both too much and too little D_1 receptor stimulation can impair working memory (Lidow et al., 1998). The benzazepine radioligand [^{11}C]SCH 23390 was the first to image D_1 receptors in humans (Halldin et al., 1986; Farde et al., 1987). A more recently discovered radioligand of the same class, [^{11}C]NNC 112, has a higher signal-to-noise ratio in the human brain (Halldin et al., 1998; Abi-Dargham et al., 2000a,b).

2.3.1. Dopamine D_1 Receptor Abnormalities in Patients with Schizophrenia

Okubo and colleagues first described decreased prefrontal cortical D_1 receptor binding in drug-naïve and drug-free patients with schizophrenia using [^{11}C]SCH 23390 as the radioligand (Okubo et al., 1997). Karlsson and colleagues used the same radioligand to study drug-free patients with schizophrenia, but did not find significant differences between patients and healthy controls (Karlsson et al., 2002). In contrast to previous studies, Abi-Dargham used the more recent radioligand with higher signal-to-noise ratio, [^{11}C]NNC 112, and demonstrated increased D_1 receptor binding in patients with schizophrenia, which was correlated with poor cognitive performance (Abi-Dargham et al., 2002). Choice of radioligand ([^{11}C]SCH 23390 vs [^{11}C]NNC 112) was then discussed as a potential reason for the discrepant findings, based on differential responses to changes in endogenous dopamine concentrations in rodents (Abi-Dargham and Moore, 2003). However, a recent study by Kosaka and colleagues found decreased binding of both tracers, [^{11}C]SCH 23390 and [^{11}C]NNC 112, in chronically ill and medicated patients with schizophrenia (Kosaka et al., 2010), suggesting that the choice of radioligand may not be enough to explain the discrepant finding between the early studies. Finally, Abi-Dargham and colleagues recently published a larger sample of patients with schizophrenia scanned with [^{11}C] NNC 112 showing increased D_1 receptor binding only in drug-naïve patients, but not in medicated patients with schizophrenia (Abi-Dargham et al., 2011). Thus, D_1 receptor binding appears be increased in early, nonmedicated stages of the illness, but normalized or even decreased in medicated patients. In addition to antipsychotic medication, a potentially major source of confusion is that a sizable portion of specific binding in these radioligands is to cortical serotonin 5-hydroxytryptamine 2A receptors (Catafau et al., 2010), due to lack of pharmacological selectivity. Yet another potential confusion is genetics: variation in the catechol-*O*-monoamine transporter (*COMT*) gene affects dopamine D_1 receptor binding in humans (Slifstein et al., 2008).

2.3.2. Effects of Genetic Risk on Dopamine D_1 Receptors

Given that dopamine D_1 receptor abnormalities are found only in the early, nonmedicated stage of the illness, could these abnormalities be detected in individuals at genetic risk for schizophrenia? We studied dopamine D_1 receptor binding in the brain with PET and [^{11}C]SCH 23390 using the twin study design outlined below. Specifically, we recruited 11 discordant twin pairs (six identical and five nonidentical) and seven healthy twin pairs (four identical, three nonidentical). The results from this study proved more complex than those for the dopamine D_2 receptor study. We found that unaffected identical co-twins had higher D_1 receptor densities in the prefrontal, temporal, and parietal cortex than did healthy twins (see Figure 3) (Hirvonen et al., 2006a,b). Nonidentical co-twins had binding values that were intermediate between those from identical co-twins and healthy twins, consistent with direct association between D_1 receptor density and degree of genetic loading. In addition, high cortical D_1 receptor binding predicted poor spatial working memory performance in these unaffected co-twins. Among healthy identical twins, cortical D_1 receptors were under tight genetic control as indicated by high intraclass correlation coefficients (in the order of 0.60–0.80 in the cortical regions), reminiscent of the finding in striatal D_2 receptors. We also found high correlations between striatal dopamine D_1 and D_2 receptors among healthy twins (Hirvonen et al., 2006a,b).

Patients with schizophrenia were then compared with their own unaffected co-twins to examine disease-specific contributions. We found that patients with schizophrenia had

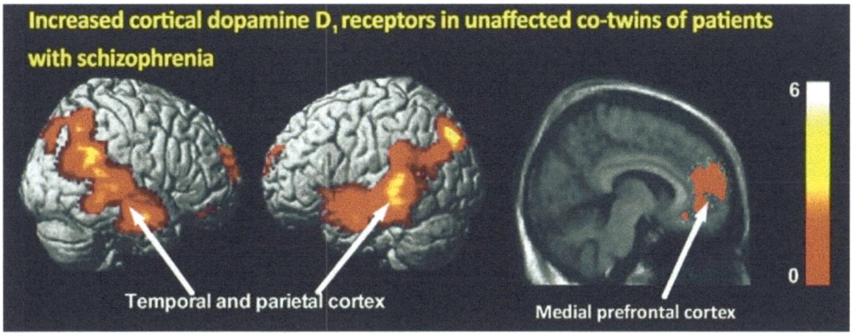

Figure 3 Increased cortical dopamine D_1 receptor binding in unaffected co-twins of patients with schizophrenia (Hirvonen et al., 2006a,b). Results are presented as in Figure 2.

much lower D_1 receptor binding in most regions in the brain than did their unaffected co-twins, consistent with the previous study by Okubo and colleagues in medicated patients (Okubo et al., 1997). Among the patients, dose of antipsychotic medication correlated negatively with dopamine D_1 receptor binding: those with higher drug doses had less D_1 receptor binding than those with lower drug doses, suggesting that antipsychotic drug treatment may have downregulated D_1 receptors in schizophrenia.

In summary, there appears to be distinct vulnerability and disease-related contributions to cortical D_1 receptors: receptors are increased in individuals at genetic risk but decreased in patients with chronic, medicated schizophrenia (Hirvonen et al., 2006a,b). This conclusion from our twin studies is partially supported by other published work showing increased D_1 receptor binding in drug-naïve schizophrenia (Abi-Dargham et al., 2011) and decreased D_1 receptor binding in chronic, medicated patients (Okubo et al., 1997; Kosaka et al., 2010). The twin design has the obvious advantage of very high statistical power to examine disease-specific contributions: an identical but unaffected co-twin is a perfectly matched control subject for a patient with schizophrenia.

2.4. SUMMARY: THE DOPAMINE HYPOTHESIS OF SCHIZOPHRENIA VULNERABILITY

Molecular imaging studies have provided evidence that schizophrenia is associated with dysregulated dopamine system in brain. Presynaptic dopamine systems appears to be overactive, striatal dopamine D_2 receptors may be increased and occupied with excess endogenous dopamine, thalamic D_2 receptors may be decreased, and cortical D_1 receptors may be decreased in chronic, medicated disease. Recent studies suggest that the dopaminergic abnormalities in schizophrenia are shared by individuals at risk for developing the illness, suggesting that they serve as potential intermediate phenotypes that convey risk for schizophrenia.

In addition to receptor binding abnormalities, higher presynaptic dopamine function appears to be associated with genetic and clinical risk and not just clinical phenotype. As reviewed above, increased presynaptic dopamine function has been robustly shown in schizophrenia (Hietala et al., 1995). We recently studied first-degree relatives (children and siblings) of patients with schizophrenia in comparison with healthy subjects using PET and [^{18}F]3,4-dihydroxy-6-(18)F-fluoro-l-phenylalanine (FDOPA), an indicator of presynaptic dopamine synthesis capacity (Huttunen et al., 2008). We found about 20% higher striatal [^{18}F]FDOPA influx values in first-degree relatives than in healthy controls, consistent with increased striatal dopamine synthesis capacity in individuals at genetic risk. The effect size was roughly similar to that we previously found in patients with schizophrenia (Hietala et al., 1995). Increased [^{18}F]FDOPA influx values have also been shown recently in individuals at clinical risk for schizophrenia (Fusar-Poli et al., 2010; Howes and Kapur, 2009; Stone et al., 2010). Recently, increased stress-induced dopamine release was demonstrated in individuals at high clinical risk for schizophrenia (Mizrahi et al., 2012).

The question of cause or consequence remains in studies looking at abnormalities in people at risk of developing a disease. In this context, we cannot definitely answer the question whether dopamine abnormalities in unaffected relatives with patients of schizophrenia reflect vulnerability or protection; that is, do these individuals remain unaffected despite or because of these abnormalities. With regard to our twin studies, we included middle-aged subjects, who are past the typical age of onset of schizophrenia, thus suggesting that protective factors may be involved. Directionality of the abnormalities and correlations with the clinical phenotype might also provide useful hints to this issue. With regard to increased striatal D_2 receptors (Hirvonen et al., 2005), the changes are qualitatively similar to those reported in patients with schizophrenia. Moreover, high striatal D_2 receptors correlated with poor cognitive performance—thus, we might conclude that these changes represent vulnerability rather than protection.

The same logic might be applied to higher presynaptic dopamine function in individuals at risk (Huttunen et al., 2008). The finding of increased cortical D_1 receptors in unaffected co-twins is more complex to interpret, because findings in patients with schizophrenia have not been consistent. Recent data suggest that D_1 receptors may be increased in drug-naïve patients but downregulated in medicated patients (Abi-Dargham et al., 2011), suggesting that increased D_1 receptors in unaffected twins may reflect vulnerability. Alternatively, D_1 receptor upregulation might be a homeostatic response to loss of cortical neuropil (Selemon and Goldman-Rakic, 1999), where D_1 receptors are preferentially localized at the ultrastructural level (Smiley et al., 1994). That is, D_1 receptor upregulation might attempt to compensate for neuropil loss in individuals at genetic risk (Cannon et al., 2002).

The exact genes responsible for dopaminergic abnormalities in schizophrenia are not known. Adding to the complexity, dopaminergic abnormalities are likely mediated by multiple genes and their interactions (van Os et al., 2008). The gene for D_2 receptors has many polymorphisms, and some of these may affect receptor binding in vivo. The C allele of the C957T polymorphism may be overrepresented in schizophrenia (Monakhov et al., 2008), but this allele is associated with lower striatal D_2 receptor binding in healthy subjects (Hirvonen et al., 2004). The cysteine allele of the Ser311Cys polymorphism is associated with schizophrenia (Glatt et al., 2003), but does not affect receptor binding in vivo (Pohjalainen et al., 1997). Effects of D_1 receptor gene variants are not well understood (Wong et al., 2000).

On a larger scale, many risk genes for schizophrenia, such as dysbindin and neuregulin, modulate dopaminergic or glutamatergic neurotransmission (Lisman et al., 2008; Tost et al., 2010). The gene coding for *COMT*, which degrades cortical dopamine, is interesting since it provides a plausible mechanism for cortical dopaminergic dysfunction as a genetically mediated trait for schizophrenia. Variation in this locus is implicated in prefrontal cortical pathology of schizophrenia (Meyer-Lindenberg and Weinberger, 2006; Egan et al., 2001), increased dopamine turnover rate postmortem (Akil et al., 2003) and in vivo (Meyer-Lindenberg et al., 2005), and higher cortical dopamine D_1 receptors in vivo (Slifstein et al., 2008). Thus, variation in the *COMT* gene might contribute to higher striatal D_2 receptors (Hirvonen et al., 2005) and higher cortical D_1 receptors (Hirvonen et al., 2006a,b) in unaffected co-twins of patients with schizophrenia. However, this variant does not directly determine striatal or cortical D_2 receptor binding in the healthy population (Hirvonen et al., 2010), suggesting interactions with other genes or the environment in the population at risk.

3. CONCLUSIONS

In conclusion, molecular imaging studies have provided robust evidence for dopaminergic abnormalities in schizophrenia. Schizophrenia appears to be characterized by increased presynaptic dopamine turnover, increased striatal dopamine D_2 receptors that are in part occupied with excess endogenous dopamine, decreased thalamic D_2 receptors, and decreased cortical D_1 receptors in chronic, medicated disease. Similar alterations are now seen in individuals who are at genetic risk but who do not express the illness, suggesting that dopaminergic mechanisms may serve as intermediate phenotypes, part of the neural substrate that conveys the risk from the genotype to the complex phenotype. Specifically, studies show increased striatal dopamine D_2 receptors, increased cortical D_1 receptors, and increased presynaptic dopamine turnover in the striatum in individuals at risk.

Thus, we propose a novel hypothesis to be tested further, the "dopamine hypothesis of schizophrenia vulnerability" (Hirvonen and Hietala, 2011). Future studies should further characterize the role of dopamine disturbance in the genetic risk for schizophrenia, for example, by trying to distinguish between alterations in tonic (Abi-Dargham et al., 2000a,b) vs phasic (Laruelle et al., 1996) dopamine regulation. Furthermore, dopaminergic abnormalities should be integrated with structural and functional deficits that characterize schizophrenia (Fusar-Poli et al., 2010; Meyer-Lindenberg et al., 2005). Finally, imaging studies should try to examine the gene–environment interaction in schizophrenia in order to elucidate the effects of environmental triggers (e.g. stress and substance abuse) in vulnerable individuals (van Os et al., 2008). In conclusion, molecular imaging is likely able to reveal the intermediate neurobiological phenotypes responsible for transmitting risk for schizophrenia from genes to clinical phenotype. Ultimately, a comprehensive view of these endophenotypes and their interaction with environmental factors will facilitate discovery of novel disease mechanisms and better treatment.

REFERENCES

Abi-Dargham, A., Gil, R., Krystal, J., Baldwin, R.M., Seibyl, J.P., Bowers, M., van Dyck, C.H., Charney, D.S., Innis, R.B., Laruelle, M., 1998. Increased striatal dopamine transmission in schizophrenia: confirmation in a second cohort. Am. J. Psychiatry 155, 761–767.

Abi-Dargham, A., Kegeles, L.S., Zea-Ponce, Y., Mawlawi, O., Martinez, D., Mitropoulou, V., O'Flynn, K., Koenigsberg, H.W., Van Heertum, R., Cooper, T., Laruelle, M., Siever, L.J., 2004. Striatal amphetamine-induced dopamine release in patients with schizotypal personality disorder studied with single photon emission computed tomography and [^{123}I]iodobenzamide. Biol. Psychiatry 55, 1001–1006.

Abi-Dargham, A., Martinez, D., Mawlawi, O., Simpson, N., Hwang, D.R., Slifstein, M., Anjilvel, S., Pidcock, J., Guo, N.N., Lombardo, I., Mann, J.J., Van Heertum, R., Foged, C., Halldin, C., Laruelle, M., 2000a. Measurement of striatal and extrastriatal dopamine D_1 receptor binding potential with [^{11}C]NNC 112 in humans: validation and reproducibility. J. Cereb. Blood Flow Metab. 20, 225–243.

Abi-Dargham, A., Rodenhiser, J., Printz, D., Zea-Ponce, Y., Gil, R., Kegeles, L.S., Weiss, R., Cooper, T.B., Mann, J.J., Van Heertum, R.L., Gorman, J.M., Laruelle, M., 2000b. Increased baseline occupancy of D_2 receptors by dopamine in schizophrenia. Proc. Natl. Acad. Sci. USA 97, 8104–8109.

Abi-Dargham, A., Mawlawi, O., Lombardo, I., Gil, R., Martinez, D., Huang, Y., Hwang, D.R., Keilp, J., Kochan, L., Van Heertum, R., Gorman, J.M., Laruelle, M., 2002. Prefrontal dopamine D_1 receptors and working memory in schizophrenia. J. Neurosci. 22, 3708–3719.

Abi-Dargham, A., Moore, H., 2003. Prefrontal DA transmission at D_1 receptors and the pathology of schizophrenia. Neuroscientist 9, 404–416.

Abi-Dargham, A., Xu, X., Thompson, J.L., Gil, R., Kegeles, L.S., Urban, N.B., Narendran, R., Hwang, D.R., Laruelle, M., Slifstein, M., July 18, 2011. Increased prefrontal cortical D_1 receptors in drug naive patients with schizophrenia: a PET study with [^{11}C]NNC112. J. Psychopharmacol. (Epub ahead of print).

Akil, M., Kolachana, B.S., Rothmond, D.A., Hyde, T.M., Weinberger, D.R., Kleinman, J.E., 2003. Catechol-*O*-methyltransferase genotype and dopamine regulation in the human brain. J. Neurosci. 23, 2008–2013.

Bertolino, A., Breier, A., Callicott, J.H., Adler, C., Mattay, V.S., Shapiro, M., Frank, J.A., Pickar, D., Weinberger, D.R., 2000. The relationship between dorsolateral prefrontal neuronal *N*-acetylaspartate and evoked release of striatal dopamine in schizophrenia. Neuropsychopharmacology 22, 125–132.

Bertolino, A., Knable, M.B., Saunders, R.C., Callicott, J.H., Kolachana, B., Mattay, V.S., Bachevalier, J., Frank, J.A., Egan, M., Weinberger, D.R., 1999. The relationship between dorsolateral prefrontal *N*-acetylaspartate measures and striatal dopamine activity in schizophrenia. Biol. Psychiatry 45, 660–667.

Breier, A., Su, T.P., Saunders, R., Carson, R.E., Kolachana, B.S., de Bartolomeis, A., Weinberger, D.R., Weisenfeld, N., Malhotra, A.K., Eckelman, W.C., Pickar, D., 1997. Schizophrenia is associated with elevated amphetamine-induced synaptic dopamine concentrations: evidence from a novel positron emission tomography method. Proc. Natl. Acad. Sci. USA 94, 2569–2574.

Brunelin, J., d'Amato, T., Van Os, J., Costes, N., Suaud Chagny, M.F., Saoud, M., 2010. Increased left striatal dopamine transmission in unaffected siblings of schizophrenia patients in response to acute metabolic stress. Psychiatry Res. 181, 130–135.

Cannon, T.D., 2005. Clinical and genetic high-risk strategies in understanding vulnerability to psychosis. Schizophr. Res. 79, 35–44.

Cannon, T.D., Huttunen, M.O., Lonnqvist, J., Tuulio-Henriksson, A., Pirkola, T., Glahn, D., Finkelstein, J., Hietanen, M., Kaprio, J., Koskenvuo, M., 2000. The inheritance of neuropsychological dysfunction in twins discordant for schizophrenia. Am. J. Hum. Genet. 67, 369–382.

Cannon, T.D., Kaprio, J., Lonnqvist, J., Huttunen, M., Koskenvuo, M., 1998. The genetic epidemiology of schizophrenia in a Finnish twin cohort. A population-based modeling study. Arch. Gen. Psychiatry 55, 67–74.

Cannon, T.D., Thompson, P.M., van Erp, T.G., Toga, A.W., Poutanen, V.P., Huttunen, M., Lonnqvist, J., Standerskjold-Nordenstam, C.G., Narr, K.L., Khaledy, M., Zoumalan, C.I., Dail, R., Kaprio, J., 2002. Cortex mapping reveals regionally specific patterns of genetic and disease-specific gray-matter deficits in twins discordant for schizophrenia. Proc. Natl. Acad. Sci. USA 99, 3228–3233.

Cannon, T.D., van Erp, T.G., Bearden, C.E., Loewy, R., Thompson, P., Toga, A.W., Huttunen, M.O., Keshavan, M.S., Seidman, L.J., Tsuang, M.T., 2003. Early and late neurodevelopmental influences in the prodrome to schizophrenia: contributions of genes, environment, and their interactions. Schizophr. Bull. 29, 653–669.

Carlsson, A., Waters, N., Holm-Waters, S., Tedroff, J., Nilsson, M., Carlsson, M.L., 2001. Interactions between monoamines, glutamate, and GABA in schizophrenia: new evidence. Annu. Rev. Pharmacol. Toxicol. 41, 237–260.

Catafau, A.M., Searle, G.E., Bullich, S., Gunn, R.N., Rabiner, E.A., Herance, R., Radua, J., Farre, M., Laruelle, M., 2010. Imaging cortical dopamine D_1 receptors using [^{11}C]NNC112 and ketanserin blockade of the 5-HT_{2A} receptors. J. Cereb. Blood Flow Metab. 30, 985–993.

Clinton, S.M., Meador-Woodruff, J.H., 2004. Thalamic dysfunction in schizophrenia: neurochemical, neuropathological, and in vivo imaging abnormalities. Schizophr. Res. 69, 237–253.

Corripio, I., Perez, V., Catafau, A.M., Mena, E., Carrio, I., Alvarez, E., 2006. Striatal D_2 receptor binding as a marker of prognosis and outcome in untreated first-episode psychosis. NeuroImage 29, 662–666.

Crawley, J.C., Owens, D.G., Crow, T.J., Poulter, M., Johnstone, E.C., Smith, T., Oldland, S.R., Veall, N., Owen, F., Zanelli, G.D., 1986. Dopamine D_2 receptors in schizophrenia studied in vivo. Lancet 2, 224–225.

Dao-Castellana, M.H., Paillere-Martinot, M.L., Hantraye, P., Attar-Levy, D., Remy, P., Crouzel, C., Artiges, E., Feline, A., Syrota, A., Martinot, J.L., 1997. Presynaptic dopaminergic function in the striatum of schizophrenic patients. Schizophr. Res. 23, 167–174.

Davis, K.L., Kahn, R.S., Ko, G., Davidson, M., 1991. Dopamine in schizophrenia: a review and reconceptualization. Am. J. Psychiatry 148, 1474–1486.

Egan, M.F., Goldberg, T.E., Kolachana, B.S., Callicott, J.H., Mazzanti, C.M., Straub, R.E., Goldman, D., Weinberger, D.R., 2001. Effect of COMT Val$^{108/158}$ Met genotype on frontal lobe function and risk for schizophrenia. Proc. Natl. Acad. Sci. USA 98, 6917–6922.

Elkashef, A.M., Doudet, D., Bryant, T., Cohen, R.M., Li, S.H., Wyatt, R.J., 2000. 6-(18)F-DOPA PET study in patients with schizophrenia. Positron emission tomography. Psychiatry Res. 100, 1–11.

Farde, L., Hall, H., Ehrin, E., Sedvall, G., 1986. Quantitative analysis of D_2 dopamine receptor binding in the living human brain by PET. Science 231, 258–261.

Farde, L., Hall, H., Pauli, S., Halldin, C., 1995. Variability in D_2-dopamine receptor density and affinity: a PET study with [^{11}C]raclopride in man. Synapse 20, 200–208.

Farde, L., Halldin, C., Stone-Elander, S., Sedvall, G., 1987. PET analysis of human dopamine receptor subtypes using ^{11}C-SCH 23390 and ^{11}C-raclopride. Psychopharmacology (Berlin) 92, 278–284.

Farde, L., Nordstrom, A.L., Wiesel, F.A., Pauli, S., Halldin, C., Sedvall, G., 1992. Positron emission tomographic analysis of central D_1 and D_2 dopamine receptor occupancy in patients treated with classical neuroleptics and clozapine. Relation to extrapyramidal side effects. Arch. Gen. Psychiatry 49, 538–544.

Farde, L., Suhara, T., Nyberg, S., Karlsson, P., Nakashima, Y., Hietala, J., Halldin, C., 1997. A PET-study of [^{11}C]FLB 457 binding to extrastriatal D_2-dopamine receptors in healthy subjects and antipsychotic drug-treated patients. Psychopharmacology (Berlin) 133, 396–404.

Farde, L., Wiesel, F.A., Stone-Elander, S., Halldin, C., Nordstrom, A.L., Hall, H., Sedvall, G., 1990. D_2 dopamine receptors in neuroleptic-naive schizophrenic patients. A positron emission tomography study with [^{11}C]raclopride. Arch. Gen. Psychiatry 47, 213–219.

Fusar-Poli, P., Howes, O.D., Allen, P., Broome, M., Valli, I., Asselin, M.C., Grasby, P.M., McGuire, P.K., 2010. Abnormal frontostriatal interactions in people with prodromal signs of psychosis: a multimodal imaging study. Arch. Gen. Psychiatry 67, 683–691.

Gerfen, C.R., Engber, T.M., Mahan, L.C., Susel, Z., Chase, T.N., Monsma Jr., F.J., Sibley, D.R., 1990. D_1 and D_2 dopamine receptor-regulated gene expression of striatonigral and striatopallidal neurons. Science 250, 1429–1432.

Glatt, S.J., Faraone, S.V., Tsuang, M.T., 2003. Meta-analysis identifies an association between the dopamine D_2 receptor gene and schizophrenia. Mol. Psychiatry 8, 911–915.

Goldman-Rakic, P.S., Muly 3rd, E.C., Williams, G.V., 2000. D_1 receptors in prefrontal cells and circuits. Brain Res. Brain Res. Rev. 31, 295–301.

Grace, A.A., 1991. Phasic versus tonic dopamine release and the modulation of dopamine system responsivity: a hypothesis for the etiology of schizophrenia. Neuroscience 41, 1–24.

Grace, A.A., 1993. Cortical regulation of subcortical dopamine systems and its possible relevance to schizophrenia. J. Neural Transm. Gen. Sect. 91, 111–134.

Hall, H., Farde, L., Sedvall, G., 1988. Human dopamine receptor subtypes–in vitro binding analysis using ^{3}H-SCH 23390 and ^{3}H-raclopride. J. Neural Transm. 73, 7–21.

Hall, H., Sedvall, G., Magnusson, O., Kopp, J., Halldin, C., Farde, L., 1994. Distribution of D_1- and D_2-dopamine receptors, and dopamine and its metabolites in the human brain. Neuropsychopharmacology 11, 245–256.

Halldin, C., Farde, L., Hogberg, T., Mohell, N., Hall, H., Suhara, T., Karlsson, P., Nakashima, Y., Swahn, C.G., 1995. Carbon-11-FLB 457: a radioligand for extrastriatal D_2 dopamine receptors. J. Nucl. Med. 36, 1275–1281.

Halldin, C., Foged, C., Chou, Y.H., Karlsson, P., Swahn, C.G., Sandell, J., Sedvall, G., Farde, L., 1998. Carbon-11-NNC 112: a radioligand for PET examination of striatal and neocortical D_1-dopamine receptors. J. Nucl. Med. 39, 2061–2068.

Halldin, C., Stone-Elander, S., Farde, L., Ehrin, E., Fasth, K.J., Langstrom, B., Sedvall, G., 1986. Preparation of ^{11}C-labelled SCH 23390 for the in vivo study of dopamine D-1 receptors using positron emission tomography. Int. J. Rad. Appl. Instrum. A 37, 1039–1043.

Harrison, P.J., Weinberger, D.R., 2005. Schizophrenia genes, gene expression, and neuropathology: on the matter of their convergence. Mol. Psychiatry 10 (1), 40–68 Errata in Mol. Psychiatry 2005; 10(4), 420 and Mol. Psychiatry 2005; 10(8), 804.

Hietala, J., Syvalahti, E., Vilkman, H., Vuorio, K., Rakkolainen, V., Bergman, J., Haaparanta, M., Solin, O., Kuoppamaki, M., Eronen, E., Ruotsalainen, U., Salokangas, R.K., 1999. Depressive symptoms and presynaptic dopamine function in neuroleptic-naive schizophrenia. Schizophr. Res. 35, 41–50.

Hietala, J., Syvalahti, E., Vuorio, K., Nagren, K., Lehikoinen, P., Ruotsalainen, U., Rakkolainen, V., Lehtinen, V., Wegelius, U., 1994. Striatal D_2 dopamine receptor characteristics in neuroleptic-naive schizophrenic patients studied with positron emission tomography. Arch. Gen. Psychiatry 51, 116–123.

Hietala, J., Syvalahti, E., Vuorio, K., Rakkolainen, V., Bergman, J., Haaparanta, M., Solin, O., Kuoppamaki, M., Kirvela, O., Ruotsalainen, U., et al., 1995. Presynaptic dopamine function in striatum of neuroleptic-naive schizophrenic patients. Lancet 346, 1130–1131.

Hirvonen, J., Hietala, J., 2011. Dysfunctional brain networks and genetic risk for schizophrenia: specific neurotransmitter systems. CNS Neurosci. Ther. 17, 89–96.

Hirvonen, J., van Erp, T.G., Huttunen, J., Aalto, S., Nagren, K., Huttunen, M., Lonnqvist, J., Kaprio, J., Cannon, T.D., Hietala, J., 2006a. Brain dopamine D_1 receptors in twins discordant for schizophrenia. Am. J. Psychiatry 163, 1747–1753.

Hirvonen, J., van Erp, T.G., Huttunen, J., Nagren, K., Huttunen, M., Aalto, S., Lonnqvist, J., Kaprio, J., Cannon, T.D., Hietala, J., 2006b. Striatal dopamine D_1 and D_2 receptor balance in twins at increased genetic risk for schizophrenia. Psychiatry Res. 146, 13–20.

Hirvonen, J., van Erp, T.G., Huttunen, J., Aalto, S., Nagren, K., Huttunen, M., Lonnqvist, J., Kaprio, J., Hietala, J., Cannon, T.D., 2005. Increased caudate dopamine D_2 receptor availability as a genetic marker for schizophrenia. Arch. Gen. Psychiatry 62, 371–378.

Hirvonen, M., Laakso, A., Nagren, K., Rinne, J.O., Pohjalainen, T., Hietala, J., 2004. C957T polymorphism of the dopamine D_2 receptor (DRD2) gene affects striatal DRD2 availability in vivo. Mol. Psychiatry 9, 1060–1061.

Hirvonen, M.M., Nagren, K., Rinne, J.O., Pesonen, U., Vahlberg, T., Hagelberg, N., Hietala, J., 2010. COMT Val^{158}Met genotype does not alter cortical or striatal dopamine D_2 receptor availability in vivo. Mol. Imaging Biol. 12, 192–197.

Howes, O.D., Kapur, S., 2009. The dopamine hypothesis of schizophrenia: version III–the final common pathway. Schizophr. Bull. 35, 549–562.
Huttunen, J., Heinimaa, M., Svirskis, T., Nyman, M., Kajander, J., Forsback, S., Solin, O., Ilonen, T., Korkeila, J., Ristkari, T., McGlashan, T., Salokangas, R.K., Hietala, J., 2008. Striatal dopamine synthesis in first-degree relatives of patients with schizophrenia. Biol. Psychiatry 63, 114–117.
Innis, R.B., Cunningham, V.J., Delforge, J., Fujita, M., Gjedde, A., Gunn, R.N., Holden, J., Houle, S., Huang, S.C., Ichise, M., Iida, H., Ito, H., Kimura, Y., Koeppe, R.A., Knudsen, G.M., Knuuti, J., Lammertsma, A.A., Laruelle, M., Logan, J., Maguire, R.P., Mintun, M.A., Morris, E.D., Parsey, R., Price, J.C., Slifstein, M., Sossi, V., Suhara, T., Votaw, J.R., Wong, D.F., Carson, R.E., 2007. Consensus nomenclature for in vivo imaging of reversibly binding radioligands. J. Cereb. Blood Flow Metab. 27, 1533–1539.
Kaprio, J., Koskenvuo, M., 2002. Genetic and environmental factors in complex diseases: the older Finnish Twin Cohort. Twin Res. 5, 358–365.
Karlsson, P., Farde, L., Halldin, C., Sedvall, G., 2002. PET study of D_1 dopamine receptor binding in neuroleptic-naive patients with schizophrenia. Am. J. Psychiatry 159, 761–767.
Kaymaz, N., van Os, J., 2009. Heritability of structural brain traits: an endophenotype approach to deconstruct schizophrenia. Int. Rev. Neurobiol. 89, 85–130.
Kegeles, L.S., Abi-Dargham, A., Frankle, W.G., Gil, R., Cooper, T.B., Slifstein, M., Hwang, D.R., Huang, Y., Haber, S.N., Laruelle, M., 2010a. Increased synaptic dopamine function in associative regions of the striatum in schizophrenia. Arch. Gen. Psychiatry 67, 231–239.
Kegeles, L.S., Slifstein, M., Xu, X., Urban, N., Thompson, J.L., Moadel, T., Harkavy-Friedman, J.M., Gil, R., Laruelle, M., Abi-Dargham, A., 2010b. Striatal and extrastriatal dopamine D_2/D_3 receptors in schizophrenia evaluated with [^{18}F]fallypride positron emission tomography. Biol. Psychiatry 68 (7), 634–641.
Kegeles, L.S., Abi-Dargham, A., Zea-Ponce, Y., Rodenhiser-Hill, J., Mann, J.J., Van Heertum, R.L., Cooper, T.B., Carlsson, A., Laruelle, M., 2000. Modulation of amphetamine-induced striatal dopamine release by ketamine in humans: implications for schizophrenia. Biol. Psychiatry 48, 627–640.
Kessler, R.M., Woodward, N.D., Riccardi, P., Li, R., Ansari, M.S., Anderson, S., Dawant, B., Zald, D., Meltzer, H.Y., 2009. Dopamine D_2 receptor levels in striatum, thalamus, substantia nigra, limbic regions, and cortex in schizophrenic subjects. Biol. Psychiatry 65, 1024–1031.
Knable, M.B., Egan, M.F., Heinz, A., Gorey, J., Lee, K.S., Coppola, R., Weinberger, D.R., 1997. Altered dopaminergic function and negative symptoms in drug-free patients with schizophrenia. [^{123}I]-iodobenzamide SPECT study. Br. J. Psychiatry 171, 574–577.
Kosaka, J., Takahashi, H., Ito, H., Takano, A., Fujimura, Y., Matsumoto, R., Nozaki, S., Yasuno, F., Okubo, Y., Kishimoto, T., Suhara, T., 2010. Decreased binding of [^{11}C]NNC112 and [^{11}C]SCH23390 in patients with chronic schizophrenia. Life Sci. 86, 814–818.
Laruelle, M., 1998. Imaging dopamine transmission in schizophrenia. A review and meta-analysis. Q. J. Nucl. Med. 42, 211–221.
Laruelle, M., 2000. Imaging synaptic neurotransmission with in vivo binding competition techniques: a critical review. J. Cereb. Blood Flow Metab. 20, 423–451.
Laruelle, M., Abi-Dargham, A., Gil, R., Kegeles, L., Innis, R., 1999. Increased dopamine transmission in schizophrenia: relationship to illness phases. Biol. Psychiatry 46, 56–72.
Laruelle, M., Abi-Dargham, A., van Dyck, C.H., Gil, R., D'Souza, C.D., Erdos, J., McCance, E., Rosenblatt, W., Fingado, C., Zoghbi, S.S., Baldwin, R.M., Seibyl, J.P., Krystal, J.H., Charney, D.S., Innis, R.B., 1996. Single photon emission computerized tomography imaging of amphetamine-induced dopamine release in drug-free schizophrenic subjects. Proc. Natl. Acad. Sci. USA 93, 9235–9240.
Laruelle, M., Kegeles, L.S., Abi-Dargham, A., 2003. Glutamate, dopamine, and schizophrenia: from pathophysiology to treatment. Ann. N.Y. Acad. Sci. 1003, 138–158.
Lee, K.J., Lee, J.S., Kim, S.J., Correll, C.U., Wee, H., Yoo, S.Y., Jeong, J.M., Lee, D.S., Lee, S.I., Kwon, J.S., 2008. Loss of asymmetry in D_2 receptors of putamen in unaffected family members at increased genetic risk for schizophrenia. Acta Psychiatr. Scand. 118, 200–208.
Lehrer, D.S., Christian, B.T., Kirbas, C., Chiang, M., Sidhu, S., Short, H., Wang, B., Shi, B., Chu, K.W., Merrill, B., Buchsbaum, M.S., 2010. ^{18}F-Fallypride binding potential in patients with schizophrenia compared to healthy controls. Schizophr. Res. 122 (1–3), 43–52.
Lidow, M.S., Williams, G.V., Goldman-Rakic, P.S., 1998. The cerebral cortex: a case for a common site of action of antipsychotics. Trends Pharmacol. Sci. 19, 136–140.

Lindstrom, L.H., Gefvert, O., Hagberg, G., Lundberg, T., Bergstrom, M., Hartvig, P., Langstrom, B., 1999. Increased dopamine synthesis rate in medial prefrontal cortex and striatum in schizophrenia indicated by L-(beta-11C) DOPA and PET. Biol. Psychiatry 46, 681–688.

Lisman, J.E., Coyle, J.T., Green, R.W., Javitt, D.C., Benes, F.M., Heckers, S., Grace, A.A., 2008. Circuit-based framework for understanding neurotransmitter and risk gene interactions in schizophrenia. Trends Neurosci. 31, 234–242.

Martinot, J.L., Paillere-Martinot, M.L., Loc'h, C., Hardy, P., Poirier, M.F., Mazoyer, B., Beaufils, B., Maziere, B., Allilaire, J.F., Syrota, A., 1991. The estimated density of D_2 striatal receptors in schizophrenia. A study with positron emission tomography and ^{76}Br-bromolisuride. Br. J. Psychiatry 158, 346–350.

Martinot, J.L., Paillere-Martinot, M.L., Loc'h, C., Lecrubier, Y., Dao-Castellana, M.H., Aubin, F., Allilaire, J.F., Mazoyer, B., Maziere, B., Syrota, A., 1994. Central D_2 receptors and negative symptoms of schizophrenia. Br. J. Psychiatry 164, 27–34.

Martinot, J.L., Peron-Magnan, P., Huret, J.D., Mazoyer, B., Baron, J.C., Boulenger, J.P., Loc'h, C., Maziere, B., Caillard, V., Loo, H., et al., 1990. Striatal D_2 dopaminergic receptors assessed with positron emission tomography and [^{76}Br]bromospiperone in untreated schizophrenic patients. Am. J. Psychiatry 147, 44–50.

McGowan, S., Lawrence, A.D., Sales, T., Quested, D., Grasby, P., 2004. Presynaptic dopaminergic dysfunction in schizophrenia: a positron emission tomographic [^{18}F]fluorodopa study. Arch. Gen. Psychiatry 61, 134–142.

Meyer-Lindenberg, A., Kohn, P.D., Kolachana, B., Kippenhan, S., McInerney-Leo, A., Nussbaum, R., Weinberger, D.R., Berman, K.F., 2005. Midbrain dopamine and prefrontal function in humans: interaction and modulation by COMT genotype. Nat. Neurosci. 8, 594–596.

Meyer-Lindenberg, A., Miletich, R.S., Kohn, P.D., Esposito, G., Carson, R.E., Quarantelli, M., Weinberger, D.R., Berman, K.F., 2002. Reduced prefrontal activity predicts exaggerated striatal dopaminergic function in schizophrenia. Nat. Neurosci. 5, 267–271.

Meyer-Lindenberg, A., Weinberger, D.R., 2006. Intermediate phenotypes and genetic mechanisms of psychiatric disorders. Nat. Rev. Neurosci. 7, 818–827.

Mizrahi, R., Addington, J., Rusjan, P.M., Suridjan, I., Ng, A., Boileau, I., Pruessner, J.C., Remington, G., Houle, S., Wilson, A.A., 2012. Increased stress-induced dopamine release in psychosis. Biol. Psychiatry 71, 561–567.

Monakhov, M., Golimbet, V., Abramova, L., Kaleda, V., Karpov, V., 2008. Association study of three polymorphisms in the dopamine D_2 receptor gene and schizophrenia in the Russian population. Schizophr. Res. 100, 302–307.

Moore, H., West, A.R., Grace, A.A., 1999. The regulation of forebrain dopamine transmission: relevance to the pathophysiology and psychopathology of schizophrenia. Biol. Psychiatry 46, 40–55.

Neve, K.A., Seamans, J.K., Trantham-Davidson, H., 2004. Dopamine receptor signaling. J. Recept. Signal Transduct. Res. 24, 165–205.

Nordström, A.L., Farde, L., Eriksson, L., Halldin, C., 1995. No elevated D_2 dopamine receptors in neuroleptic-naive schizophrenic patients revealed by positron emission tomography and [^{11}C]*N*-methylspiperone. Psychiatry Res. 61, 67–83.

Okubo, Y., Suhara, T., Suzuki, K., Kobayashi, K., Inoue, O., Terasaki, O., Someya, Y., Sassa, T., Sudo, Y., Matsushima, E., Iyo, M., Tateno, Y., Toru, M., 1997. Decreased prefrontal dopamine D_1 receptors in schizophrenia revealed by PET. Nature 385, 634–636.

Owen, M.J., Craddock, N., O'Donovan, M.C., 2010. Suggestion of roles for both common and rare risk variants in genome-wide studies of schizophrenia. Arch. Gen. Psychiatry 67, 667–673.

Pike, V.W., 2009. PET radiotracers: crossing the blood–brain barrier and surviving metabolism. Trends Pharmacol. Sci. 30, 431–440.

Pilowsky, L.S., Costa, D.C., Ell, P.J., Verhoeff, N.P., Murray, R.M., Kerwin, R.W., 1994. D_2 dopamine receptor binding in the basal ganglia of antipsychotic-free schizophrenic patients. An 123I-IBZM single photon emission computerised tomography study. Br. J. Psychiatry 164, 16–26.

Pohjalainen, T., Cravchik, A., Gejman, P.V., Rinne, J., Nagren, K., Syvalahti, E., Hietala, J., 1997. Antagonist binding characteristics of the Ser311→Cys variant of human dopamine D_2 receptor in vivo and in vitro. Biochem. Biophys. Res. Commun. 232, 143–146.

Pycock, C.J., Kerwin, R.W., Carter, C.J., 1980. Effect of lesion of cortical dopamine terminals on subcortical dopamine receptors in rats. Nature 286, 74–76.

Reith, J., Benkelfat, C., Sherwin, A., Yasuhara, Y., Kuwabara, H., Andermann, F., Bachneff, S., Cumming, P., Diksic, M., Dyve, S.E., Etienne, P., Evans, A.C., Lal, S., Shevell, M., Savard, G., Wong, D.F., Chouinard, G., Gjedde, A., 1994. Elevated dopa decarboxylase activity in living brain of patients with psychosis. Proc. Natl. Acad. Sci. USA 91, 11651–11654.

Sawaguchi, T., Goldman-Rakic, P.S., 1991. D_1 dopamine receptors in prefrontal cortex: involvement in working memory. Science 251, 947–950.

Seeman, P., 1987. Dopamine receptors and the dopamine hypothesis of schizophrenia. Synapse 1, 133–152.

Seeman, P., Guan, H.C., Niznik, H.B., 1989. Endogenous dopamine lowers the dopamine D_2 receptor density as measured by [^{3}H]raclopride: implications for positron emission tomography of the human brain. Synapse 3, 96–97.

Seeman, P., Kapur, S., 2000. Schizophrenia: more dopamine, more D_2 receptors. Proc. Natl. Acad. Sci. USA 97, 7673–7675.

Seeman, P., Niznik, H.B., Guan, H.C., 1990. Elevation of dopamine D_2 receptors in schizophrenia is underestimated by radioactive raclopride. Arch. Gen. Psychiatry 47, 1170–1172.

Selemon, L.D., Goldman-Rakic, P.S., 1999. The reduced neuropil hypothesis: a circuit based model of schizophrenia. Biol. Psychiatry 45, 17–25.

Sesack, S.R., Carr, D.B., 2002. Selective prefrontal cortex inputs to dopamine cells: implications for schizophrenia. Physiol. Behav. 77, 513–517.

Siever, L.J., Davis, K.L., 2004. The pathophysiology of schizophrenia disorders: perspectives from the spectrum. Am. J. Psychiatry 161, 398–413.

Skinbjerg, M., Liow, J.S., Seneca, N., Hong, J., Lu, S., Thorsell, A., Heilig, M., Pike, V.W., Halldin, C., Sibley, D.R., Innis, R.B., 2010. D_2 dopamine receptor internalization prolongs the decrease of radioligand binding after amphetamine: a PET study in a receptor internalization-deficient mouse model. NeuroImage 50, 1402–1407.

Slifstein, M., Kolachana, B., Simpson, E.H., Tabares, P., Cheng, B., Duvall, M., Frankle, W.G., Weinberger, D.R., Laruelle, M., Abi-Dargham, A., 2008. COMT genotype predicts cortical-limbic D_1 receptor availability measured with [^{11}C]NNC112 and PET. Mol. Psychiatry 13, 821–827.

Smiley, J.F., Levey, A.I., Ciliax, B.J., Goldman-Rakic, P.S., 1994. D_1 dopamine receptor immunoreactivity in human and monkey cerebral cortex: predominant and extrasynaptic localization in dendritic spines. Proc. Natl. Acad. Sci. USA 91, 5720–5724.

Stone, J.M., Howes, O.D., Egerton, A., Kambeitz, J., Allen, P., Lythgoe, D.J., O'Gorman, R.L., McLean, M.A., Barker, G.J., McGuire, P., 2010. Altered relationship between hippocampal glutamate levels and striatal dopamine function in subjects at ultra high risk of psychosis. Biol. Psychiatry 68 (7), 599–602.

Suhara, T., Okubo, Y., Yasuno, F., Sudo, Y., Inoue, M., Ichimiya, T., Nakashima, Y., Nakayama, K., Tanada, S., Suzuki, K., Halldin, C., Farde, L., 2002. Decreased dopamine D_2 receptor binding in the anterior cingulate cortex in schizophrenia. Arch. Gen. Psychiatry 59, 25–30.

Sullivan, P.F., Kendler, K.S., Neale, M.C., 2003. Schizophrenia as a complex trait: evidence from a meta-analysis of twin studies. Arch. Gen. Psychiatry 60, 1187–1192.

Talvik, M., Nordstrom, A.L., Olsson, H., Halldin, C., Farde, L., 2003. Decreased thalamic D_2/D_3 receptor binding in drug-naive patients with schizophrenia: a PET study with [^{11}C]FLB 457. Int. J. Neuropsychopharmacol. 6, 361–370.

Tost, H., Alam, T., Meyer-Lindenberg, A., 2010. Dopamine and psychosis: theory, pathomechanisms and intermediate phenotypes. Neurosci. Biobehav. Rev. 34, 689–700.

Tsuang, M., 2000. Schizophrenia: genes and environment. Biol. Psychiatry 47, 210–220.

Tune, L.E., Wong, D.F., Pearlson, G., Strauss, M., Young, T., Shaya, E.K., Dannals, R.F., Wilson, A.A., Ravert, H.T., Sapp, J., et al., 1993. Dopamine D_2 receptor density estimates in schizophrenia: a positron emission tomography study with ^{11}C-*N*-methylspiperone. Psychiatry Res. 49, 219–237.

Tuppurainen, H., Kuikka, J., Viinamaki, H., Husso-Saastamoinen, M., Bergstrom, K., Tiihonen, J., 2003. Extrastriatal dopamine $D_{2/3}$ receptor density and distribution in drug-naive schizophrenic patients. Mol. Psychiatry 8, 453–455.

Tuppurainen, H., Kuikka, J.T., Laakso, M.P., Viinamaki, H., Husso, M., Tiihonen, J., 2006. Midbrain dopamine $D_{2/3}$ receptor binding in schizophrenia. Eur. Arch. Psychiatry Clin. Neurosci. 256, 382–387.
van Os, J., Rutten, B.P., Poulton, R., 2008. Gene-environment interactions in schizophrenia: review of epidemiological findings and future directions. Schizophr. Bull. 34, 1066–1082.
van Rossum, J.M., 1966. The significance of dopamine-receptor blockade for the mechanism of action of neuroleptic drugs. Arch. Int. Pharmacodyn. Ther. 160, 492–494.
Wagner Jr., H.N., Burns, H.D., Dannals, R.F., Wong, D.F., Langstrom, B., Duelfer, T., Frost, J.J., Ravert, H.T., Links, J.M., Rosenbloom, S.B., Lukas, S.E., Kramer, A.V., Kuhar, M.J., 1983. Imaging dopamine receptors in the human brain by positron tomography. Science 221, 1264–1266.
Whalley, H.C., Whyte, M.C., Johnstone, E.C., Lawrie, S.M., 2005. Neural correlates of enhanced genetic risk for schizophrenia. Neuroscientist 11, 238–249.
Wong, A.H., Buckle, C.E., Van Tol, H.H., 2000. Polymorphisms in dopamine receptors: what do they tell us? Eur. J. Pharmacol. 410, 183–203.
Wong, D.F., 2002. In vivo imaging of D_2 dopamine receptors in schizophrenia: the ups and downs of neuroimaging research. Arch. Gen. Psychiatry 59, 31–34.
Wong, D.F., Wagner Jr., H.N., Tune, L.E., Dannals, R.F., Pearlson, G.D., Links, J.M., Tamminga, C.A., Broussolle, E.P., Ravert, H.T., Wilson, A.A., Toung, J.K., Malat, J., Williams, J.A., O'Tuama, L.A., Snyder, S.H., Kuhar, M.J., Gjedde, A., 1986. Positron emission tomography reveals elevated D_2 dopamine receptors in drug-naive schizophrenics. Science 234, 1558–1563.
Yang, Y.K., Yu, L., Yeh, T.L., Chiu, N.T., Chen, P.S., Lee, I.H., 2004. Associated alterations of striatal dopamine D_2/D_3 receptor and transporter binding in drug-naive patients with schizophrenia: a dual-isotope SPECT study. Am. J. Psychiatry 161, 1496–1498.
Yasuno, F., Suhara, T., Okubo, Y., Sudo, Y., Inoue, M., Ichimiya, T., Takano, A., Nakayama, K., Halldin, C., Farde, L., 2004. Low dopamine D_2 receptor binding in subregions of the thalamus in schizophrenia. Am. J. Psychiatry 161, 1016–1022.
Young, L.T., Wong, D.F., Goldman, S., Minkin, E., Chen, C., Matsumura, K., Scheffel, U., Wagner Jr., H.N., 1991. Effects of endogenous dopamine on kinetics of [^{3}H]*N*-methylspiperone and [^{3}H]raclopride binding in the rat brain. Synapse 9, 188–194.

CHAPTER FOURTEEN

Human Brain Imaging in Tardive Dyskinesia

Sofia N. Chatziioannou[1] and Emmanouil N. Rizos[2]

[1]Department of Radiology, Nuclear Medicine Section, National and Kapodistrian University of Athens Medical School, Attikon General Hospital, Athens, Greece

[2]Department of Psychiatry, National and Kapodistrian University of Athens Medical School, Attikon General Hospital, Athens, Greece

1. INTRODUCTION

Tardive dyskinesia (TD) is a serious adverse effect, neuroleptic induced, characterized by irregular stereotypical movements of the mouth, face and tongue, and choreathetoid movements of the fingers, arms, legs and trunk with a mean prevalence of 24% approximately. Although it is usually mild, 1% to 8% of patients may develop severe and often irreversible TD (Llorca et al., 2002). Atypical antipsychotics cause TD less often when compared to conventional antipsychotics. Although atypical medications are less likely to induce TD, several cases of TD have been reported in patients treated with atypical antipsychotic drugs (Correl et al. 2004; Lykouras et al. 1999).

The pathogenetic mechanism of TD still remains unknown, and while some patients develop TD, others do not. The known risk factors such as age, gender, ethnicity, psychiatric diagnosis, smoking, diabetes mellitus, organic brain dysfunction, early extrapyramidal side-effects, type of neuroleptic treatment (typical or atypical agent, dosage and duration) and genetic factors predict partly the variance in the incidence of TD. The association of genetic factors in the development of TD comes from clinical observations and animal studies, suggesting that a genetic mechanism might underlie vulnerability (Bakker et al. 2006; Kane et al. 2004).

The dominant hypothesis of tardive dyskinesia's pathogenetic mechanism is the overactivity of the dopaminergic neurotransmission in the basal ganglia; upregulation of the dopamine D_2-like receptors (D_2, D_3 and D_4) have been postulated to play a role in its pathophysiology (Jeste et al., 1993).

All preclinical studies suggest an elevation of D2 receptors in the striatum of antipsychotic-medicated animals and a long-term exposure to these agents generally raise D2 receptors in 35% of cases (Seeman et al., 1988). Moreover a recent preclinical report has demonstrated the reduction of dopamine uptake in the striatum of rats presenting orofacial dyskinesia under haloperidol medication through a probable mechanism of alteration in dopamine transporter function (Fachinetto et al., 2007).

Imaging of the Human Brain in Health and Disease
http://dx.doi.org/10.1016/B978-0-12-418677-4.00014-2

Studies have shown that the density of D_2 receptors is increasaed in the post-mortem brains of schizophrenic patients and this finding may be a result of the medications. Additionally other neuroimaging studies have shown an elevation of D2 receptors in brains of drug-naïve schizophrenic patients. Neuroleptics are thought to be highly surface-active agents and so they expand and invaginate cell membranes in striatal brain regions. Thus, the prolonged administration of an antipsychotic agent could result in a damage of cell membranes with the possible death of nigral cells, via an accumulation mechanism of neuroleptics, through the high affinity of antipsychotic agents to the striatal neuromelanin, within two types of cell neuromelanin granules or deposits and neuroleptic-myelin figures (Seeman et al. 1987, Seeman et al. 2011). Studies have suggested that such neuroleptic – induced nigral cell damage occurs in post-mortem human tissue from patients with tardive dyskinesia. Following the same hypothesis, this type of cell damage in striatal brain regions by long-term neuroleptic exposure is the phenomenon that is called dopaminergic supersensitivity. Moreover, it seems that there are various degrees of dopaminergic denervation supersensitivity in the striatum controlling oral and other types of motor activity.

Except of antipsychotics related TD, another gastro-intestinal pharmaceutical agent, metoclopramide, has been associated with even higher incidence of TD, than that of antipsychotics, despite its very week antipsychotic potency (Ganzini et al., 1993). In a recent post-mortem study of Chen et al. at 2011, the above mentioned accumulated mechanism was confirmed. Specifically metoclopramide had a significant concentration in substantia nigra, a fact that caused nigral nerve cell membrane damage, possibly explaining its toxic action to the striatal brain regions and thus the appearance of early TD symptoms.

Another interesting research area of the pathophysiology of TD is the neurodegenerative hypothesis of TD that is in association with above mentioned data. Brain-derived neurotrophic factor (BDNF), belonging to the neurotrophic family, is critical for the maintenance of functional neurons and has been found to be involved in the neurodevelopment of dopaminergic-related systems, and prevent dopaminergic neuronal damage, which is relevant to the pathophysiology of TD. Studies have found lower BDNF levels in patients with TD, than those without TD. Thus, decreased BDNF may play a role in the pathophysiology of TD as a type of neurodegeneration or an adaptive response to neuroleptic induced supersensitisation (up regulation) of the dopamine systems (Yang et al., 2011).

2.1. NUCLEAR MEDICINE MODALITIES

2.1.1. Dopamine Transporter Scan

The dominant hypothesis about the mechanism of tardive dyskinesia (TD) is an overactivity of dopaminergic neurotransmission in basal ganglia (nigrostriatal pathway) and upregulation of dopamine (DA) receptors due to chronic blockade of dopamine D_2

receptors by antipsychotics (DA hypersensitivity hypothesis) (Kapur and Seeman, 2001). It is possible that the TD phenotype may be heterogeneous with respect to the localization of abnormal involuntary movements. Some patients exhibit only orofacial TD along with other abnormal involuntary movements of their extremities and truncal region. The variety of TD symptomatology may have different underlying pathophysiological mechanism and therefore distinct etiology.

The integrity of the dopaminergic neurons can be marked by the dopamine transporter (DAT) density. The main pathway for the clearance of extracellular DA and the regulation of the duration of DA signaling is DA uptake by DAT (Kahlig and Galli, 2003). The decrease of DAT shows either loss of striatal DA neurons or more likely decrease in DAT expression (Lieberman et al., 1990) because of the excessive postsynaptic dopamine D_2 blockade (Jibson and Tandom, 1998) of the activity of antipsychotics. In order to visualize these DATs in the striatum, single-photon emission computed tomography (SPECT) with *N*-ω-fluoropropyl-2β-carbomethoxy-3β [4-iodophenyl] tropane ([^{123}I] FP-CIT, ioflupane) or DAT scan (DAT imaging) can be used. This nuclear medicine method shows presynaptic DA degeneration (Tatsch, 2001) and is quite sensitive to demonstrate even a small loss of nigrostriatal dopaminergic degeneration.

In patients with nigrostriatal degeneration, DAT imaging with [^{123}I] FP-CIT SPECT was able to show decreased uptake (Booij et al., 1999). The more the decrease in striatal [^{123}I] FP-CIT binding, the more the progression of degeneration. Therefore, [^{123}I] FP-CIT uptake is in accordance with the density of nigrostriatal neurons. However, a recent study by Lavalaye et al. (2001a,b) questioned the previous results, as no significant difference in striatal [^{123}I] FP-CIT uptake between patients with TD and controls was detected. Nevertheless, it must be taken into account that the number of subjects studied in this study was small, and, therefore, no definite conclusions could be drawn.

Some trials have demonstrated that increased DAT uptake in DAT scan, induced by some medications, seems to be associated with the improvement of TD. Rizos et al. (2010a) presented a case of a psychotic patient with TD symptoms. A DAT scan showed decreased DAT uptake in the area of basal ganglia. Quetiapine, a novel atypical antipsychotic agent with a mainly high affinity for 5-hydroxytryptamine (HT) (serotonin) 2A receptors (5-HT_{2A}), histamine (H_1), and noradrenergic (α_1) receptor and a much lower affinity for D_2/D_3 dopaminergic receptors in the striatal regions (Nemeroff et al., 2002), was then initiated. Both psychotic and TD symptoms improved while a second DAT scan showed improvement in the radiotracer uptake by the basal ganglia. The authors concluded that increased DAT uptake seemed to be associated with the improvement of TD.

Another case reports a patient with paranoid schizophrenia who was initially treated with pipamperon, but because of presentation of TD, the patient was switched to ziprasidone. Before the initiation of ziprasidone, a DAT scan showed decreased uptake in the basal ganglia. The initiation of ziprasidone and the discontinuation of pipamperon resulted in remission of TD symptoms in this case, which could be associated with an increase in striatal uptake through the lower DA receptor binding of ziprasidone as compared to the typical agent. The lower ziprasidone DA receptor binding seemed to have caused an increase in striatal dopaminergic neurotransmission and, thus, an increase in uptake as indicated in the second DAT scan (Rizos et al., 2010b).

In a study by Chatziioannou et al. (2011), it was shown that there was a difference in striatal DAT uptake between TD status and non-TD status in a patient with a psychiatric history of paranoid schizophrenia (Figure 1). This patient who was initially on quetiapine was switched to sertindole because of a severe relapse of his mental condition as well as development of TD symptoms. The initiation of sertindole and the discontinuation of quetiapine resulted in an improvement of TD symptoms, which was associated with a normal striatal uptake in the DAT scan through sertindole's low D_2 occupancy in the basal ganglia. Six months later the patient presented with severe syncopal episodes with normal QT interval and hence sertindole was discontinued. The patient started quetiapine again on his own. The discontinuation of sertindole and the reinitiation of quetiapine by the patient resulted in TD relapse. A second DAT scan showed decreased DAT uptake in the area of basal ganglia. The higher quetiapine D_2 occupancy, compared to the low sertindole D_2 occupancy, seemed to have caused a decrease in striatal dopaminergic neurotransmission and, therefore, a decrease in the basal ganglia uptake confirmed by the second DAT scan. The present study shows that the deterioration of TD symptoms was associated with a decrease of striatal DAT levels.

2.1.2. Positron Emission Tomography

One of the most functional tracers of presynaptic dopaminergic neuron terminations in the nigrostriatal system is ^{18}F-fluorodopa (^{18}F-DOPA). For the diagnosis and evaluation of Parkinson's Disease (PD) progression, ^{18}F-DOPA-positron emission tomography (PET) is the most essential tool (Puñal-Riobóo et al., 2009). By measuring aromatic amino acid decarboxylase (AADC) activity, DAT activity, and vesicular monoamine transporter (VMAT2) density, PET has the capability of investigating the proper presynaptic DA terminal functionality. In order to assess AADC, ^{18}F-DOPA-PET can be used in the DA terminals (Garnett et al., 1983), as it is a measure of their functionability, since AADC activity permits the conversion of L-DOPA to DA. This ligand has been used, so as the correlation between ^{18}F-DOPA binding and motor disability is assessed. Based on measurements obtained by the Unified Parkinson's Disease Rating Scale, an inverse correlation has been proved to exist between striatal ^{18}F-DOPA binding and motor disability (Vingerhoets et al., 1997; Broussolle et al., 1999), i.e. loss of DA terminals has as effect on increase in motor disability. D_2 receptor blockade in the associative and limbic subdivisions of the striatum seems to be one of the mental side effects of antipsychotics, which influence cognition and reward motivation (Kim et al., 2011). For the assessment

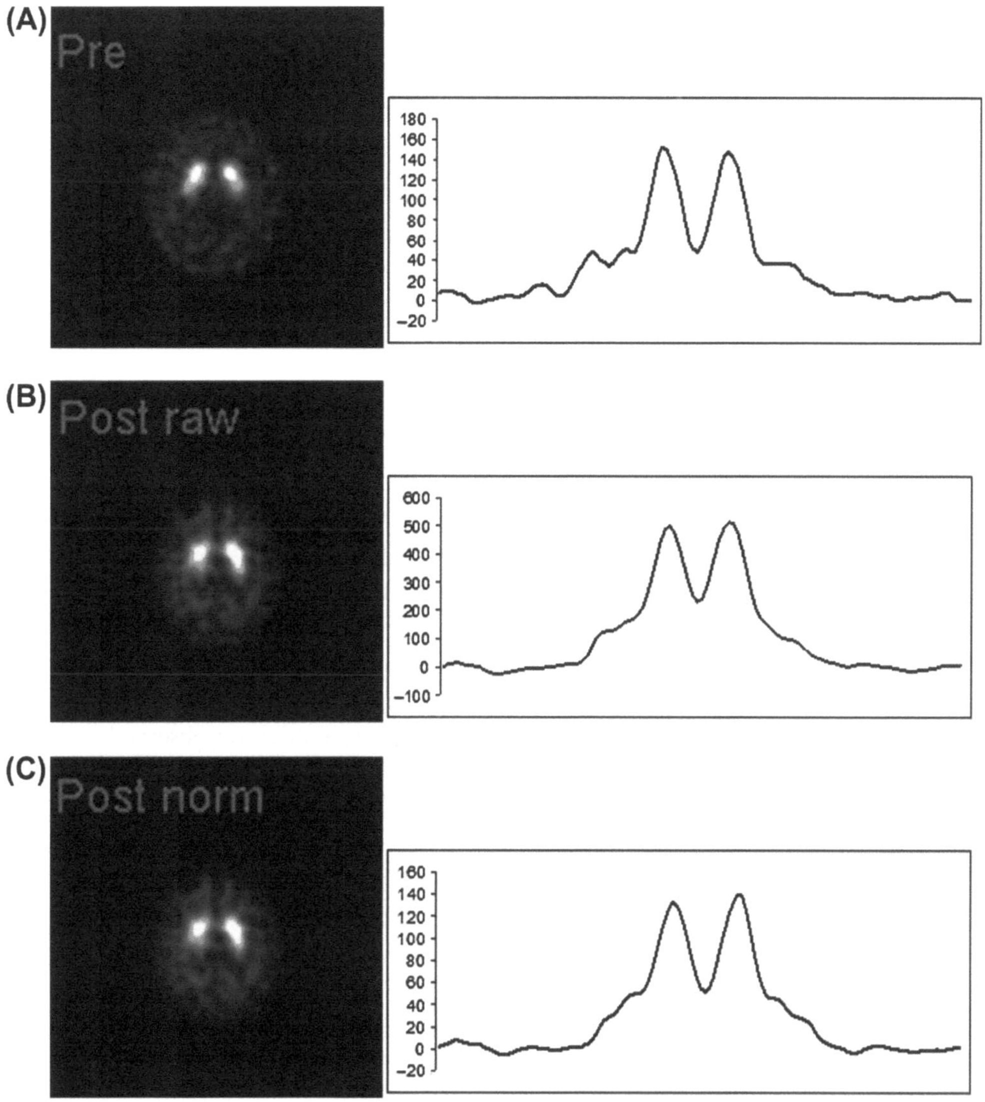

Figure 1 Representative corresponding slices, i.e. slices at approximately the same location or height in the patient without (Pre) and with (Post raw) symptoms of TD, demonstrating worse uptake of the radiopharmaceutical in the basal ganglia in the presence of the symptoms. (A) Image marked as "Pre" presents a transaxial slice from the scan prior to the initiation of quetiapine and the occurrence of TD symptoms and while the patient was on sertindole; the profile next to it shows the pixel values along a horizontal line that passes through the upper part of the ganglia. (B) Image marked as "Post raw" presents a transaxial slice from the scan after the initiation of quetiapine and while the patient had TD symptoms; the profile next to it shows the pixel values along a horizontal line that passes through the upper part of the ganglia. (C) Image marked as "Post norm" is the normalized version of the Post raw image (normalization factor of 0.27) and the profile next to it corresponds to the normalized pixel values. *Figure reproduced with permission from Chatziioannou et al. (2011).*

of striatal D_2 receptors, the radiotracer ^{11}C-raclopride (RAC) has been used (Rinne et al., 1995; Farde et al., 1989).

The dominant hypothesis on the pathophysiology of TD supposes a supersensitivity of DA receptors from prolonged receptor blockade or their upregulation (Lavalaye et al., 2001a,b). The mechanism of TD is poorly understood, even though it was described 50 years ago. TD may result primarily from DA supersensitivity induced by neuroleptics in the nigrostriatal pathway, affecting mostly the D_2 DA receptor. First-generation antipsychotics, which have greater affinity for the D_2 binding site, mainly provoke TD. Evidence of a dose–response relationship, withdrawal effects, studies on D_2 agonists and antagonists, animal studies, and genetic polymorphism research support the D_2 hypersensitivity hypothesis (Hoergen, 2007).

Seeman (1988) demonstrated for the first time increased dopamine D_2 receptor binding after long-term treatment with antipsychotics in humans using an in vivo neuroreceptor imaging technique. It is conceivable that D_2 receptor upregulation has clinical implications in humans. In a PET study using *N*-^{11}C-methylspiperone, the subject with the highest increase in the D_2 binding potential developed severe and persistent symptoms of TD in the upper limbs and neck within the first week after concluding the antipsychotic washout and being started on Seroquel. The same results came up from the observation undertaken in animals that implicate antipsychotic-induced D_2 receptor upregulation in motor hyperactivity. The previously masked symptoms of TD, which are a fundamental aspect of the dopaminergic supersensitivity model, are observed once the antipsychotic treatment is stopped or its dose is reduced.

In a later study, Silvestri et al. showed that dopamine D_2 receptor binding is increased after long-term treatment with antipsychotics in humans using in vivo neuroreceptor imaging techniques (PET scan using RAC). The subject under discussion developed severe and persistent symptoms of TD suggesting moderate to high doses of both traditional and conventional antipsychotics, characterized by potent dopamine D_2 affinity, which in turn induce similar degrees of increases in D_2 receptor binding. Thus, it can be inferred that increased D_2 binding has clinical implications in humans and may be useful to identify patients at risk. dopamine D_2 receptor upregulation is a reliable biological parameter that can now be detected in vivo enabling us to research individual vulnerability factors as well as compensatory or preventive mechanisms (Silvestri et al., 2000).

2.2. MAGNETIC RESONANCE IMAGING

Magnetic resonance imaging (MRI) is particularly useful as it can distinguish gray matter from white matter with significantly more clarity than computed tomography (CT). More specifically, striatal structures can be measured volumetrically and differences in the size of caudate, putamen, or even the globus pallidus may be detected in patients with TD.

The results of 33 studies suggest that antipsychotics act regionally rather than globally on the brain. These volumetric changes indicate a more frequent effect on the basal ganglia and a less frequent effect on the thalamus and the cortex concerning typical rather than atypical antipsychotic use (Navari and Dazzan, 2009). In patients with chronic schizophrenia, morphological changes in brain structures have been noted. These usually are observed in lateral and third ventricle as enlargement as well as in temporal regions as reduction in their size (Lawrie et al., 1998).

Nopoulos et al. (2001) studied a sample of 45 drug-free male patients at their first episode of psychosis previously exposed to typical antipsychotics, while four of them were additionally exposed to atypicals. A reduction in the midbrain area was induced and was still present 3 weeks after withdrawal of typical antipsychotics. Shihabuddin et al. (1998) evaluated the volume of striatum in a small sample of naive and drug-free schizophrenia patients compared to healthy individuals. They compared three patient groups (drug-naïve versus drug-free versus controls) by level (ventral versus dorsal side) and by structure (putamen versus caudate). The main difference was a larger dorsal putamen volume in drug-free patients as opposed to controls and to an even greater degree compared to drug-naive patients. Findings on the caudate size were opposite, as drug-free patients showed a smaller caudate volume than both drug-naive patients and controls. It is suggested that the posttreatment enlargement might last longer after treatment discontinuation for the putamen than for the caudate, most likely due to a higher density of D_2 receptors in the putamen.

The potential reversibility of antipsychotic effects has to be considered, as observed, for example, for the caudate enlargement induced by typical antipsychotic and reversed by clozapine (Chakos et al., 1995; Frazier et al., 1996). The potentially different effects of typical and atypical antipsychotics on brain structures result most likely from different mechanisms of action (Lieberman et al., 1995; Scherk and Falkai, 2006). Atypical drugs, such as clozapine and olanzapine, could increase cellular resilience and therefore act on the pathophysiology of psychosis through an agonistic effect on *N*-methyl-D-aspartate receptors (Duncan et al., 1999; Millan, 2005), increasing the expression of trophic factors (Fumagalli et al., 2004; Angelucci et al., 2005) and stimulating neurogenesis (Halim et al., 2004; Wang et al., 2004).

In addition, typical antipsychotics, such as haloperidol, may be potentially toxic and induce oxidative stress and excitatory neurotoxicity (Post et al., 1998; Wright et al., 1998). The dose has also to be considered, since low doses of typical antipsychotics may produce effects similar to those of atypical ones (Oosthuizen et al., 2004). Alterations in neuronal and nonneuronal tissue compartments, in addition to physiological alterations in brain tissue (e.g. changes in tissue perfusion, fat, and water content) and in body weight, alcohol intake, steroid administration, and hormonal status may change the MRI volume, as well (Weinberger and McClure, 2002). Therefore, MRI, with its superior gray/white matter differentiation and its ability to detect subtle differences in tissue may discover abnormalities in TD.

2.3. COMPUTED TOMOGRAPHY

Several studies in the past had evaluated the neuropathology in TD, primarily in the striatum. Prior CT studies of middle-aged and older subjects using linear measurements have reported primarily negative results. Four CT studies of striatal and periventricular pathology in patients with TD had shown some positive findings in comparison of TD patients with different control groups, while five studies had negative findings (Gelenberg, 1976; Famuyiwa et al., 1976; Jeste et al., 1984, 1986; Bartels and Themelis, 1983; Brainin et al., 1983; Albus et al., 1985; Waddington and Youssef, 1985; Kaufmann et al., 1986). However, the positive studies were not in agreement as to which brain regions were pathogenic and only one study showed involvement of basal ganglia structures in TD patients. Swayze et al. (1988) attempted to identify neuropathological abnormalities in the periventricular region of young chronic schizophrenics. However, neuropathological abnormalities may be present even if not detected by CT imaging. Therefore, CT does not appear to contribute significantly at the time being in the assessment of TD.

3. CONCLUSIONS

Imaging of TD has moved from the anatomical imaging of the striatum with the use initially of CT and eventually of MRI to a more functional imaging of the cellular metabolism. There is now significant evidence that imaging of DA metabolism with the use of SPECT or PET shows promise in the evaluation of TD.

REFERENCES

Albus, M., Naber, D., Muller-Spahn, F., Douillet, P., Reinertshofer, T., Ackenheil, M., 1985. Tardive dyskinesia: relation to computer-tomographic, endocrine and psychopathological variables. Biol. Psychiatry Oct;20 (10), 1082–1089.

Angelucci, F., Aloe, L., Iannitelli, A., Gruber, S.H., Mathe, A.A., 2005. Effect of chronic olanzapine treatment on nerve growth factor and brain-derived neurotrophic factor in the rat brain. Eur. Neuropsychopharmacol. 15, 311–317.

Bakker, R.P., van Harten, P., van Os, J., 2006. Antipsychotic-induced tardive dyskinesia and the ser9Gly polymorphism in the DRD3 gene: a meta analysis. Schizophr. Res. 83, 185–192.

Bartels, M., Themelis, J., 1983. Computerized tomography in tardive dyskinesia. Arch. Psychiatr. Nervenkr. 233 (5), 371–379.

Booij, J., Hemelaar, J.T., Speelman, J.D., de Bruin, K., Janssen, A.G., Van Royen, E.A., 1999. One-day protocol for imaging of the nigrostriatal dopaminergic pathway in Parkinson's disease by [^{123}I]FPCIT SPECT. J. Nucl. Med. 40, 753–761.

Brainin, M., Reisner, Th., Zeitlhofer, J., 1983. Tardive dyskinesia: clinical correlation with computed tomography in patients aged less than 60 years. J. Neurol. Neurosurg. Psychiatr. 46, 1037.

Broussolle, E., Dentresangle, C., Landais, P., Garcia-Larrea, L., Pollak, P., Croisile, B., Hibert, O., Bonnefoi, F., Galy, G., Froment, J.C., Comar, D., 1999. The relation of putamen and caudate nucleus ^{18}F-Dopa uptake to motor and cognitive performances in Parkinson's disease. J. Neurol. Sci. 166, 141–151.

Chatziioannou, S., Rizos, E.N., Kallergi, M., Douzenis, A., Apostolopoulos, A., Bacalis, S., Lykouras, L., 2011. Tardive dyskinesia related to quetiapine and confirmed by a DAT scan. J. Clin. Psychopharmacol. 31, 535–538.

Chakos, M.H., Lieberman, J.A., Alvir, J., Bilder, R., Ashtari, M., 1995. Caudate nuclei volumes in schizophrenic patients treated with typical antipsychotics or clozapine. Lancet 345, 456–457.

Chen, S., Seeman, P., Liu, F., 2011. Antipsychotic drug binding in the substantia nigra: an examination of high metoclopramide binding in the brains of normal, Alzheimer disease, Huntington's disease, and multiple sclerosis patients, and its relation to tardive dyskinesia. Synapse 65, 119–124.

Correl, C.U., Leucht, S., Kane, J.M., 2004. Lower risk for tardive dyskinesia associated with second generation antipsychotics: a systematic review of 1-year studies. Am. J. Psychiatry 161, 414–425.

Duncan, G.E., Zorn, S., Lieberman, J.A., 1999. Mechanisms of typical and atypical antipsychotic drug action in relation to dopamine and NMDA receptor hypofunction hypotheses of schizophrenia. Mol. Psychiatry 4, 418–428.

Fachinetto, R., Villarinho, J.G., Wagner, C., Pereira, R., Avila, D.S., Burger, M., et al., 2007. *Valeriana officinalis* does not alter the orofacial dyskinesia induced by haloperidol in rats: role of dopamine transporter. Prog. Neuropsychopharmacol. Biol. Psychiatry 31, 1478–1486.

Famuyiwa, O., Eccleston, D., Donaldson, A., Garside, R., 1976. Tardive dyskinesia and dementia. Br. J. Psychiatry 135, 500.

Farde, L., Eriksson, L., Blomquist, G., Halldin, C., 1989. Kinetic analysis of central [^{11}C] raclopride binding to D_2-dopamine receptors studied by PET–a comparison to the equilibrium analysis. J. Cereb. Blood Flow Metab. 9 (5), 696–708.

Frazier, J.A., Giedd, J.N., Kaysen, D., Albus, K., Hamburger, S., Alaghband-Rad, J., Lenane, M.C., McKenna, K., Breier, A., Rapoport, J.L., 1996. Childhood-onset schizophrenia: brain MRI rescan after 2 years of clozapine maintenance treatment. Am. J. Psychiatry 153, 564–566.

Fumagalli, F., Molteni, R., Bedogni, F., Pennarelli, M., Perez, J., Racagni, G., Riva, M.A., 2004. Quetiapine regulates FGF-2 and BDNF expression in the hippocampus of animals treated with MK-801. Neuroreport 15, 2109–2112.

Ganzini, L., Casey, D.E., Hoffman, W.F., McCall, A.L., 1993. The prevalence of metoclopramide-induced tardive dyskinesia and acute extrapyramidal movement disorders. Arch. Intern. Med. 153, 1469–1475.

Garnett, E.S., Firnau, G., Nahmias, C., 1983. Dopamine visualized in the basal ganglia of living man. Nature 305, 137–138.

Gelenberg, A.J., 1976. Computerized tomography in patients with tardive dyskinesia. Am. J. Psychiatry 133 (5).

Halim, N.D., Weickert, C.S., McClintock, B.W., Weinberger, D.R., Lipska, B.K., 2004. Effects of chronic haloperidol and clozapine treatment on neurogenesis in the adult rat hippocampus. Neuropsychopharmacology 29, 1063–1069.

Hoergen, M., 2007. Psychiatry Online 13, 18–26.

Jeste, D.V., Wagner, R.L., Weinberger, D.R., Reith, K.G., Wyatt, R.J., 1984. Evaluation of CT scans in tardive dyskinesia. Am. J. Psychiatry 137, 247.

Jeste, D.V., Weinberger, D.R., Zaloman, S.J., Wyatt, R.J., 1986. Computed tomography in tardive dyskinesia. Br. J. Psychiatry 136, 606.

Jeste, D.V., Caligiuri, M.P., 1993. Tardive dyskinesia. Schizophr. Bull. 18, 303–315.

Jibson, M.D., Tandom, R., 1998. New atypical antipsychotic medications. J. Psychiatr. Res. 32, 215–228.

Kaufmann, C.A., Jeste, D.V., Shelton, R.C., Linnoila, M., Kafka, M.S., Wyatt, R.J., 1986. Noradrenergic and neuroradiological abnormalities in tardive dyskinesia. Biol. Psychiatry 21, 799.

Kahlig, K.M., Galli, A., 2003. Regulation of dopamine transporter function and plasma membrane expression by dopamine, amphetamine, and cocaine. Eur. J. Pharmacol. 479, 153–158.

Kane, J.M., 2004. Tardive dyskinesia rates with atypical antipsychotics in adults: prevalence and incidence. J. Clin. Psychiatry 65 (Suppl. 9), S16–S20.

Kapur, S., Seeman, P., 2001. Does fast dissociation from the dopamine D_2 receptor explain the action of atypical antipsychotics? A new hypothesis. Am. J. Psychiatry 158, 360–369.

Kim, J.H., Son, Y.D., Kim, H.K., Lee, S.Y., Cho, S.E., Kim, Y.B., Cho, Z.H., 2011. Antipsychotic-associated mental side effects and their relationship to dopamine D_2 receptor occupancy in striatal subdivisions: a high-resolution PET study with [^{11}C] raclopride. J. Clin. Psychopharmacol. 31 (4), 507–511.

Lavalaye, J., Sarlet, A., Booij, J., Linszen, D.H., Reneman, L., Gersons, B.P., van Royen, E.A., 2001a. Dopamine transporter density in patients with tardive dyskinesia: a single photon emission computed tomography study. Psychopharmacology 155, 107–109.

Lavalaye, J., Linszen, D.H., Booij, J., Dingemans, P.M., Reneman, L., Habraken, J.B., Gersons, B.P., Van Royen, E.A., 2001b. Dopamine transporter density in young patients with schizophrenia assessed with [^{123}I] FP-CIT SPECT. Schizophr. Res. 47, 59–67.

Lawrie, S.M., Abukmeil, S.S., Lieberman, J., Chakos, M., Wu, H., Alvir, J., Hoffman, E., 1998. Brain abnormality in schizophrenia. A systematic and quantitative review of volumetric magnetic resonance imaging studies. Br. J. Psychiatry 172, 110–120.

Lieberman, J.A., Kinon, B.L., Loebel, A.D., 1990. Dopaminergic mechanisms in idiopathic and drug-induced psychoses. Schizophr. Bull. 16, 97–110.

Lieberman, J.A., Tollefson, G.D., Charles, C., Zipursky, R., Sharma, T., Kahn, R.S., Keefe, R.S., Green, A.I., Gur, R.E., McEvoy, J., Perkins, D., Hamer, R.M., Gu, H., Tohen, M., 1995. Antipsychotic drug effects on brain morphology in first-episode psychosis. Arch. Gen. Psychiatry 62, 361–370.

Llorca, P.M., Cherau, I., Bayle, F.J., Lancon, C., 2002. Tardive dyskinesia and antipsychotics: a review. Eur. Psychiatry 17, 129–138.

Lykouras, L., Malliori, M., Christodoulou, G.N., 1999. Improvement of tardive dyskinesia following treatment with olanzapine. Eur. Neuropsychopharmacol. 9 (4), 367–368.

Millan, M.J., 2005. *N*-Methyl-D-aspartate receptors as a target for improved antipsychotic agents: novel insights and clinical perspectives. Psychopharmacology (Berlin) 179, 30–53.

Navari, S., Dazzan, P., 2009. Do antipsychotic drugs affect brain structure? A systematic and critical review of MRI findings. Psychol. Med. 39, 1763–1777.

Nemeroff, C.B., Kinkead, B., Goldstein, I., 2002. Quetiapine: preclinical studies, pharmacokinetics, drug interactions and dosing. J. Clin. Psychiatry 63 (Suppl. 13), 5–11.

Nopoulos, P.C., Ceilley, J.W., Gailis, E.A., Andreasen, N.C., 2001. An MRI study of midbrain morphology in patients with schizophrenia: relationship to psychosis, neuroleptics, and cerebellar neural circuitry. Biol. Psychiatry 49, 13–19.

Oosthuizen, P., Emsley, R., Jadri Turner, H., Keyter, N., 2004. A randomized, controlled comparison of the efficacy and tolerability of low and high doses of haloperidol in the treatment of first-episode psychosis. Int. J. Neuropsychopharmacol. 7, 125–131.

Post, A., Holsboer, F., Behl, C., 1998. Induction of NF-kappaB activity during haloperidol-induced oxidative toxicity in clonal hippocampal cells: suppression of NF-kappaB and neuroprotection by antioxidants. J. Neurosci. 18, 8236–8246.

Puñal-Riobóo, J., Serena-Puig, A., Varela-Lema, L., Alvarez-Páez, A.M., Ruano-Ravina, A., 2009. Clinical utility of ^{18}F-DOPA-PET in movement disorders. A systematic review. Rev. Esp. Med. Nucl. 28 (3), 106–113.

Rinne, J.O., Laihinen, A., Ruottinen, H., Ruotsalainen, U., Någren, K., Lehikoinen, P., Oikonen, V., Rinne, U.K., 1995. Increased density of dopamine D_2 receptors in the putamen, but not in the caudate nucleus in early Parkinson's disease: a PET study with [^{11}C] raclopride. J. Neurol. Sci. 132, 156–161.

Rizos, E.N., Chatziioannou, S., Douzenis, A., Siafakas, N., Katsantoni, E., Nikolaidou, P., Papathanasiou, M., Lykouras, L., February 2010a. Signs of striatal dopamine transporter density increase in association with improvement of tardive dyskinesia in a patient with schizophrenia, as demonstrated by a DAT scan. Eur. Neuropsychopharmacol. 20, 132–136.

Rizos, E.N., Chatziioannou, S., Siafakas, N., Douzenis, A., Katsantoni, E., Mandrapilia, A., Bacalis, S., Fotiadis, C., Lykouras, L., August 2010b. Increased striatal dopamine transporter levels, as indicated by a DAT scan, induced by ziprasidone in association to improvement of tardive dyskinesia— a case report. Prog. Neuropsychopharmacol. Biol. Psychiatry 34, 1135–1136.

Scherk, H., Falkai, P., 2006. Effects of antipsychotics on brain structure. Curr. Opin. Psychiatry 19, 145–150.

Seeman, P., Bzowej, N.H., Guan, H.C., Bergeron, C., Reynolds, G.P., Bird, E.D., Riederer, P., Jellinger, K., Tourtellotte, W.W., 1987. Human brain D_1 and D_2 dopamine receptors in schizophrenia, Alzheimer's, Parkinson's and Huntington's diseases. Neuropsychopharmacology 1 (1), 5–15.

Seeman, P., 1988. Tardive dyskinesia, dopamine receptors, and neuroleptic damage to cell membranes. J. Clin. Psychopharmacol. 8, 3S–9S.

Seeman, M.V., 2011. Antipsychotics and physical attractiveness. Clin. Schizophr. Relat. Psychoses 5 (3), 142–146 (Review).

Shihabuddin, L., Buchsbaum, M.S., Hazlett, E.A., Haznedar, M.M., Harvey, P.D., Newman, A., Schnur, D.B., Spiegel-Cohen, J., Wei, T., Machac, J., Knesaurek, K., Vallabhajosula, S., Biren, M.A., Ciaravolo, T.M., Luu-Hsia, C., 1998. Dorsal striatal size, shape, and metabolic rate in never-medicated and previously medicated schizophrenics performing a verbal learning task. Arch. Gen. Psychiatry 55, 235–243.
Silvestri, S., Seeman, M.V., Negrete, J.C., Houle, S., Shammi, C.M., Remington, G.J., Kapur, S., Zipursky, R.B., Wilson, A.A., Christensen, B.K., Seeman, P., 2000. Increased dopamine D_2 receptor binding after long-term treatment with antipsychotics in humans: a clinical PET study. Psychopharmacology 152, 174–180.
Swayze 2nd, V.W., Yates, W.R., Andreasen, N.C., Alliger, R.J., 1988. CT abnormalities in tardive dyskinesia. Psychiatry Res. 26 (1), 51–58.
Tatsch, K., 2001. Imaging of the dopaminergic system in parkinsonism with SPECT. Nucl. Med. Commun. 22, 819–827.
Vingerhoets, F.J., Schulzer, M., Calne, D.B., Snow, B.J., 1997. Which clinical sign of Parkinson's disease best reflects the nigrostriatal lesion? Ann. Neurol. 41, 58–64.
Wang, H.D., Dunnavant, F.D., Jarman, T., Deutch, A.Y., 2004. Effects of antipsychotic drugs on neurogenesis in the forebrain of the adult rat. Neuropsychopharmacology 29, 1230–1238.
Waddington, J.L., Youssef, H.A., 1985. Late-onset involuntary movements in chronic schizophrenia: age-related vulnerability to tardive dyskinesia independent of extent of neuroleptic medication. Ir. Med. J. 78, 143.
Weinberger, D.R., McClure, R.K., 2002. Neurotoxicity, neuroplasticity, and magnetic resonance imaging morphometry: what is happening in the schizophrenic brain? Arch. Gen. Psychiatry 59, 553–558.
Wright, A.M., Bempong, J., Kirby, M.L., Barlow, R.L., Bloomquist, J.R., 1998. Effects of haloperidol metabolites on neurotransmitter uptake and release: possible role in neurotoxicity and tardive dyskinesia. Brain. Res. 788, 215–222.
Yang, Y.Q., Sun, S., Yu, Y.Q., et al., 2011. Decreased serum brain-derived neurotrophic factor levels in schizophrenic patients with tardive dyskinesia. Neurosci. Lett. 502, 37–40.

CHAPTER FIFTEEN

Human Brain Imaging of Autism Spectrum Disorders

James Robert Brašić[1] and Mona Mohamed[2]

[1]Section of High Resolution Brain Positron Emission Tomography Imaging, Division of Nuclear Medicine, The Johns Hopkins University School of Medicine, Baltimore, Maryland, USA

[2]Division of Neuroradiology, The Russell H. Morgan Department of Radiology and Radiological Science, The Johns Hopkins University School of Medicine, Baltimore, Maryland, USA

1. INTRODUCTION

1.1. Autism Spectrum Disorders

Autism is a developmental disability afflicting as many as one in 166 children in the United States (Rossignol, 2007).

The diagnostic criteria for autism and related conditions (American Psychiatric Association, 1994, 2000) have just been revised (American Psychiatric Association, 2013), so the effects of the new nomenclature are currently in flux (Brašić, 2012a; Brašić et al., 2012b). Rather than focusing on narrow diagnostic entities, investigators may benefit from casting a broad net to include participants with the vast spectrum of conditions characterized by traits of autism. Thus, we shall widely include in our presentation the various conditions under the umbrella term "autism spectrum disorders". Although readers may be confused by the use of many terms to describe autism spectrum disorders, the state of the art is reflected by the use of nebulous terms that are being redefined.

The definitions of autism and related conditions have changed over the past century, so categorizing patients by different investigators at various times is arbitrary and uncertain (Brašić and Holland, 2006, 2007). Changes in diagnostic criteria for autism spectrum disorders likely contribute to the apparent increases in incidence over recent years (Newschaffer et al., 2007). Nevertheless, the incidence of autism spectrum disorders appears to be markedly increasing around the world for unknown reasons. The prevalence of autism spectrum disorders is approximately as high as 0.3–12.3% (Bryson et al., 1988; Delobel et al., 2013; Horder and Murphy, 2012; Hsu et al., 2012; Isaksen et al., 2012; Lai et al., 2012; Wickenhauser, 2007; Wong and Hui, 2008). While linkage and family studies suggest that the risk of developing autism is associated with the X chromosome (Gauthier et al., 2006; Jamain et al., 2003; Vincent et al., 2005), the mechanism has not been identified. While some studies associate the likelihood of developing autism with the age of the parents (Croen et al., 2002, 2007; Gillberg, 1980;

Imaging of the Human Brain in Health and Disease
http://dx.doi.org/10.1016/B978-0-12-418677-4.00015-4

Glasson et al., 2004; Lauritsen et al., 2005; Reichenberg et al., 2006, 2010), others do not confirm this finding (Cryan et al., 1996; Larsson et al., 2005).

Autism spectrum disorders include (1) autism or autistic disorder, a condition with a ratio of boys to girls of 4 to 1 (Bryson, 1996) with the onset in early childhood characterized by marked deficits in social interaction and communication and by a restricted and unusual range of activities and interests (Brašić, 2003b, 2013; Brašić et al., 1998b); (2) Asperger syndrome, a condition with impaired social interactions and markedly restricted interests without mental retardation (Brašić, 2012a); (3) Heller syndrome, a disintegration in the development of children between ages 2 and 10 years resulting in the autism syndrome; (4) Rett syndrome, a condition developing in early childhood leading to profound mental retardation occurring primarily in girls due to mutations in the X-linked methyl-CpG-binding protein 2 (MeCP2) located at the Xq28 region (Brašić et al., 2012b; Verhoeven et al., 2010); and (5) fragile X syndrome, a condition caused by a single genetic defect (Clapp and Tranfaglia, 2012; Vincent et al., 2005); (6) pervasive developmental disorders, conditions with some traits of autism without the full diagnostic criteria for the syndromes listed above. Symptoms of autism are also common in some specific neurological disorders, including tuberous sclerosis (American Psychiatric Association, 2000).

Thus, autism spectrum disorders include conditions characterized by marked abnormalities in social interactions (Eigsti and Shapiro, 2003) and communication and by markedly restricted and peculiar ranges of interests and activities (American Psychiatric Association, 2000). While the broad term of "autism spectrum disorders" may appear to include apples and oranges, the category encompasses individuals with the key traits of marked abnormalities in social interactions and markedly restricted ranges of interest and activities (American Psychiatric Association, 2000). Autism spectrum disorders constitute a heterogeneous group of disorders. While specific mutations characterize fragile X syndrome and Rett syndrome, biomarkers do not exist to identify most autism spectrum disorders. Thus, overlapping among subgroups of autism spectrum disorders exists. The lack of precise boundaries among most autism spectrum disorders explains the uncertainties in diagnosis. Advance in the field will probably result in many updates to this book's chapter.

1.2. COGNITIVE PSYCHOLOGY

Typically, children learn to hypothesize reasons as to why people take actions as a means to explain the behaviors of other people. Thus, normally, children learn to develop a theory of mind to explain the behavior of others. That is, children learn how to hypothesize how other people likely feel, think, and act in specific situations, including horror, anger, love, hate, and disgust. The ability to comprehend the likely interpretations of actions and behaviors is often impaired in people with autistic spectrum disorders. In particular, people with autistic spectrum disorders may not readily interpret the predictable emotional responses of other people in situations likely to produce a strong feeling. Thus, people with autistic spectrum disorders may appear to be callous to others. People with autistic spectrum disorders may not provide the comfort and empathy to others in times of distress due to the failure to recognize that the other person likely experiences emotional pain. When a friend experiences an unpleasant life event, including illness or loss, a person with autistic spectrum disorders may not immediately recognize that the friend is undoubtedly upset and needing consolation and reassurance.

The person with autistic spectrum disorders may appear indifferent to the joys and sorrows of others due to an impairment in the ability to guess the likely emotions of others in the pleasant and unpleasant situations of life. By the same token, children with autistic spectrum disorders may not seek comfort from their parents when they are injured or ill. Some parents may not know that their child with an autistic spectrum disorder is hurt until the parent physically examines the child to see the injury. Several tasks have been proposed to assess the ability of children, adolescents, and adults to develop a theory of mind for others. Procedures can be given by experienced examiners to assess the ability of a person with a possible autistic spectrum disorder to develop a theory of mind (Brašić et al., 2012c). In contrast to healthy normal adults, adults with autism or Asperger syndrome display deficits in performing mentalizing tasks (Castelli et al., 2002).

1.3. NEUROPATHOLOGY

Neuropathological examination of the brains of people with autism provides evidence of the anatomic abnormalities in the cerebrum. Postmortem examination of the brains of people with autism has demonstrated the thinning of the neurons in the anterior cingulate (Bauman and Kemper, 2005), abnormal microglial–neuronal spatial organization in the dorsolateral prefrontal cortex (Morgan et al., 2012), and increased neuronal cell counts in the prefrontal cortex (Courchesne et al., 2011). These histological studies of the brains of people with autism spectrum disorders provide clues to the regions of the brain to investigate in the living human brain.

1.4. BRAIN IMAGING

Brain imaging techniques can broadly be classified according to the source of energy for the procedure as follows: (1) ultrasound, including carotid Doppler and transcranial Doppler; (2) X-rays, including angiography, computed tomography (CT), and myelography; (3) magnetic fields, including magnetoencephalography (MEG), magnetic resonance imaging (MRI), magnetic resonance spectroscopy (MRS), and magnetic resonance spectroscopic imaging (MRSI); (4) radioactivity, including single-photon emission-computed tomography (SPECT) and positron-emission tomography (PET); and (5) electrical activity including quantitative electroencephalography (QEEG) (Wong and Brašić, 2001). Brain imaging techniques can further be categorized as structural and functional procedures. Structural procedures identify anatomic regions. Functional procedures demonstrate the physiology and metabolism of the regions of interest (Brašić, 2013). Structural imaging techniques include angiography, CT, Doppler, MRI, and myelography. Functional imaging techniques include functional MRI (fMRI), MEG, MRS, PET, QEEG, and SPECT (Brašić, 2013; Wong and Brašić, 2001; Wong et al., 2007).

While many studies on autism spectrum disorders have been accomplished utilizing a variety of imaging techniques, variability in the diagnoses of the study populations limits the comparability of reports. The focus of the current contribution will be the identification of the key recent developments in the use of brain imaging to evaluate people with autism spectrum disorders. Brain imaging studies have identified abnormalities in the frontal, temporal, and parietal cortices in autism spectrum disorders (Mana et al., 2010). We anticipate advances soon in the refinement of the use of imaging techniques with the various forms of autism spectrum disorders. Thus, updates to the use of brain imaging for autism spectrum disorders are likely in the near future.

2.1. STRUCTURAL HUMAN BRAIN IMAGING OF AUTISM SPECTRUM DISORDERS

A variety of anatomic anomalies have been identified by the application of structural brain imaging techniques to people with autism spectrum disorders. Corpus callosal thinning is an example of the reduced white matter observed in some people with autism (Courchesne et al., 1993). The reduced volume of the corpus callosum observed in several subgroups with autism spectrum disorders provides an anatomic basis for the reduced connectivity between the cerebral hemispheres proposed to explain some clinical features (Alexander et al., 2007; Boger-Megiddo et al., 2006; Travers et al., 2012; Vidal et al., 2006).

MRI of the brain demonstrated increased volumes of the right superior temporal gyrus in boys with high-functioning autism as represented in Figure 1 (Jou et al., 2010).

The superior temporal gyrus is associated with the processing of language and the perceptions of social interactions (Jou et al., 2010). Thus, this finding suggests an anatomic basis for some of the language and social deficits in autism.

Structural brain imaging has been investigated in people with autism spectrum disorders and other developmental disorders. Lobules VI and VII of the posterior–superior cerebellar vermis demonstrate reduced volumes in people with autism and fragile X syndrome and even greater reductions in volume in people with autism without the fragile X syndrome (Brambilla et al., 2003; Kaufmann et al., 2003). The volume of the white matter in the brain stem and the cerebellum is greater in people with autism and Down syndrome, a disorder associated with trisomy 21, than in people with the Down syndrome without autism (Carper et al., 2002; Carter et al., 2008).

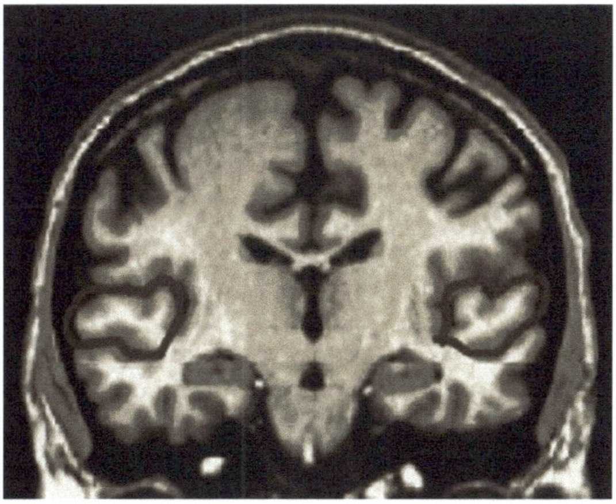

Figure 1 The boundaries of the superior temporal gyrus are manually outlined on this coronal section of the magnetic resonance imaging of a human brain. *Reproduced from Jou et al. (2010), Figure 3, page 211, with permission.*

Thus, structural brain imaging procedures have identified several abnormalities in the subgroups of people with autism spectrum disorders.

2.2. FUNCTIONAL HUMAN BRAIN IMAGING OF AUTISM SPECTRUM DISORDERS

Over the past decades, investigations have been conducted with a variety of functional imaging techniques on people with autism spectrum disorders. Chugani (2012) recently reviewed the contributions of functional brain imaging to the understanding of the neurochemistry of autism spectrum disorders. We shall concentrate on the salient recent developments in the applications of functional brain imaging to autism spectrum disorders.

2.2.1. MRS and MRSI

MRS noninvasively profiles brain metabolism. MRS dynamically represents the molecular and metabolic pathophysiology of cerebral disorders (Stanley, 2002). MRS quantitatively estimates the concentration of brain metabolites including *N*-acetylaspartate (NAA), phospholipids containing choline (Cho), creatine (Cr), myoinositol (mI), glutamate (Glu), glutamine (Gln), and lactate (Barker, 2001; Page et al., 2006).

2.2.1.1. N-*Acetylaspartate*

The metabolism of mitochondrial energy correlates with the synthesis of NAA in neurons and axons. Decrements of the concentration of NAA apparently represent the loss of neuronal integrity. The proton signal of NAA is a key indicator of healthy neurons. Thus, reductions of the concentration of NAA represent markers for neuronal dysfunction (Barker, 2001).

2.2.1.2. Creatine

Cellular oxidative metabolism is represented by the spectral resonance of Cr and phosphocreatine (PCr). The concentration of Cr is generally stable in a healthy brain. Thus, constant cellular metabolism in the healthy brain is reflected in the stability of the concentration of Cr. Stability in the concentration of Cr indicates constant cellular energy metabolism. Reductions in the concentration of Cr indicate deficient cellular energy metabolism (Ross and Bluml, 2001).

2.2.1.3. Choline

Phosphoryl choline and glycerophosphoryl choline, components of the manufacture and the breakdown of cellular membranes, make up the largest part of the total Cho resonance. Increments in Cho likely represent a greater cell growth and gliosis. Thus, Cho is a marker of cell membrane turnover and an increase in its level means gliosis or increased cellular proliferation (Govindaraju et al., 2000; Ross and Bluml, 2001).

2.2.1.4. Myoinositol

mI, a cerebral osmolyte, regulates the volume of cells. mI represents asotroglia. Additionally, mI takes part in the second-messenger system of phophoinositide (Ross and Bluml, 2001).

2.2.1.5. Concentrations and Ratios of Cerebral Metabolites

MRS provides quantitative estimates of the concentration of cerebral metabolites in the various regions of the brain. Furthermore, normal cerebral metabolism is characterized by the ratios of the concentrations of specific metabolites including NAA/Cho, NAA/Cr, Cho/Cr, and mI/Cr. The norms for individuals at specific ages and at particular developmental levels provide guidelines to assess deviations from health.

2.2.1.6. MRS Protocols

General anesthesia is usually required for children to obtain scans with minimal movement artifacts. Typically, patients lie supine with the head positioned in a head coil. 1.5-T units are common; 3.0-T scanners provide a higher resolution. Reconstructions of signals are typically provided with sagittal, oblique coronal, and oblique axial planes. Both localizing structural (T2) and fluid attenuation inversion recovery series are commonly obtained. Spectroscopic data are acquired with a single voxel technique (MRS) or multiple voxel imaging (MRSI) techniques. The signals are obtained in specific 10- to 40-mm^3 volumes of interests (VOIs) in anterior–posterior, right to left, and craniocaudal dimensions, including frontal white matter, amygdyla, hippocampus, and cerebellum. To minimize the artifacts of blood and cerebrospinal fluid (CSF), VOIs commonly exclude vasculature structures containing CSF. A point resolved spectroscopy sequence with long and short echo times of approximately 140 and 30 ms, respectively, is usually obtained after shimming and chemical shift selective water suppression. Sequences are run to estimate the concentrations of relevant metabolites and pertinent metabolic ratios, particularly NAA/Cho, NAA/Cr, and Cho/Cr using long or short echo time. The estimation of metabolites with a small brain concentration is facilitated with sequences with short echo times.

2.2.1.7. MRS in Autism Spectrum Disorders

Subgroups of people with autism spectrum disorders exhibit reduced NAA concentrations in the amygdala–hippocampal region (Hisaoka et al., 2001; Otsuka et al., 1999; Sokol et al., 2002), the cerebellum (Chugani et al., 1999; Otsuka et al., 1999; Sokol et al., 2002), the parietal cortex (Hashimoto et al., 1997), the temporal cortex (Sokol et al., 2002), and areas 41 and 42 of the temporal region (Hisaoka et al., 2001). Additionally, subgroups of children with autism demonstrate reduced ratios of NAA/Cho and NAA/Cr ratios in the brain (Filippi et al., 2002; Friedman et al., 2003; Hisaoka et al., 2001; Murphy et al., 2002; Volkmar and Pauls, 2003), particularly the amygdala–hippocampus (Gabis et al., 2008). A group of people with the Asperger syndrome demonstrated increased NAA/Cho in the anterior cingulate cortex (ACC; Oner et al., 2007). Another group of people with autism spectrum disorders demonstrated increased mI/Cr throughout the brain (Gabis et al., 2008).

A group of people with autism demonstrated a reduced signal intensity indicating a reduced concentration of NAA and corresponding neuronal loss or hypofunction in the left amygdala–hippocampal region and the left cerebellar hemisphere (Mori et al., 2001). Another group of people with autism spectrum disorders demonstrated reduced concentrations of NAA suggesting general neuronal damage in all brain regions and particularly in the left frontal cortex (Kleinhans et al., 2007). As some children with autism mature, the concentrations of NAA gradually increase to attain levels comparable to those of normal adults in adulthood (Aoki et al., 2012).

A group of people with autism exhibited reduced concentrations of PCr in the dorsolateral prefrontal cortex suggesting a hypermetabolic energy state and inadequate synthesis of brain membranes (Minshew et al., 1993).

A group of people with autism exhibited marked reductions of NAA and Glx, the combined concentrations of Glu, the primary excitatory neurotransmitter, and Gln, in the gray matter of the cerebral hemispheres and the cerebellum (DeVito et al., 2007). Another group of adults with autism demonstrated reduced Glx in the right anterior cingulate (Bernardi et al., 2011). A group of people with autism spectrum disorders exhibited increased concentrations of Cr, Glu, and Gln in the amygdala–hippocampal region, but not in the parietal region (Chugani et al., 1999). An adult with an autism spectrum disorder exhibited an increased concentration of Glx in the right amygdala–hippocampal region contrasted to a control brain region of the same person in the parietal lobe (Page et al., 2006). Other groups of people with autism spectrum disorders demonstrated elevated concentrations of Glu in the hippocampus (Aoki et al.,

2012; During and Spencer, 1993; Frith, 2004). The elevated concentrations of the excitatory neurotransmitter, Glu, provide a possible explanation for the seizure disorders of some people with autism spectrum disorders (During and Spencer, 1993; Maragakis and Rothstein, 2001; Niederhofer, 2007).

While some symptoms of autism spectrum disorders may result from excessive concentrations of Glu, the primary excitatory neurotransmitter, and other amino acids in the brain, some symptoms may result from abnormally low concentrations of gamma-amino butyric acid (GABA), the primary inhibitory neurotransmitter (Coghlan et al., 2012; Mendez et al., 2013). The low concentrations of GABA in the normal living human brain are challenging to estimate by MRS. A group of 12 people with autism demonstrated reductions of GABA, [GABA]/[NAA], and [GABA]/[Glu] ratios in the frontal lobe (Coghlan et al., 2012). Furthermore, another group of people with autism spectrum disorders also demonstrated reduced binding for the GABA type A ($GABA_A$) receptor throughout the brain with PET after the administration of the radioligand [^{11}C]Ro15-4513 (Mendez et al., 2013). Please refer to Section 2.2.4.2.3 on $GABA_A$ Neurotransmission.

Although dysfunction of the normal mechanisms of apoptosis and synaptic pruning to produce a greater neuronal density (Friedman et al., 2003) was hypothesized to account for the increased volumes of the frontal lobes (Carper et al., 2002; Courchesne, 2004; Courchesne et al., 2001; Hazlett et al., 2005; Redcay and Courchesne, 2005) of 10–15% of preschool children with autism spectrum disorders demonstrated by measurements of head circumference and MRI, the increased volumes may be better explained by a decreased synaptic neuronal density consistent with the lower concentrations of NAA, Cho, Cr, and mI demonstrated on MRS (Casanova et al., 2002) and gliosis and growth of glia and nonneuronal cells. Gliosis and glial proliferation also likely provide the basis for the brain expansion observed in some people with autism spectrum disorders in infancy (Vargas et al., 2005). In the first years of life, children with autism spectrum disorders often exhibit neuronal overgrowth, lack of pruning, and reduced connectivity (Carper and Courchesne, 2005; Courchesne, 2004; Courchesne and Pierce, 2005a,b; Courchesne et al., 2004, 2005; Frith, 2003, 2004).

The severity of symptoms of a group of people with autism was associated with the Cho/Cr ratios, a measure of membrane turnover or gliosis (Sokol et al., 2002). Another group of people with autism spectrum disorders exhibited reduced mI concentrations (Friedman et al., 2003).

On the other hand, other groups of people with autism spectrum disorders did not differ from healthy people in NAA/Cho, NAA/Cr, and Cho/Cr ratios (Hashimoto et al., 1997, 1998; Zeegers et al., 2007) and metabolite ratios of Cr and mI (Fayed and Modrego, 2005).

Overall, some people with autism spectrum disorders display impaired amino acid neurotransmission in autism and suppressed GABAergic inhibition. Please refer to Section 2.2.4.2.3 on $GABA_A$ Neurotransmission.

Additionally, dysfunction of the frontal white matter, the limbic system, the amygdala, the hippocampus, and the cerebellum provides an explanation for the cognitive and emotional symptoms of some people with autism spectrum disorders (Mori et al., 2001; Saitoh et al., 2001).

MRS and MRSI represent helpful tools to evaluate the severity of dysfunction of neurotransmitters and other key compounds of people with autism spectrum disorders. Additionally, MRS and MRSI can be used as follow-up tools for the assessment of the severity through measuring brain metabolites in children with autism spectrum disorders.

2.2.2. Diffusion Tensor Imaging

Diffusion tensor imaging (DTI) is an advanced magnetic resonance neuroimaging technique that provides promising biomarkers for autism spectrum disorders. DTI relies on the pattern of the diffusion of water molecules thus reflecting information about the architecture of white matter (Chenevert et al., 1990; Pierpaoli and Basser, 1996). The apparent diffusion coefficient, a measurement of the degree of restriction to water diffusion, and the fractional anisotropy (FA), a measurement of the directionality of water diffusivity, are key parameters to characterize fiber tract integrity generated using DTI techniques (Pierpaoli and Basser, 1996).

DTI can detect the changes in white matter integrity and connectivity of some people who later demonstrate symptoms of autism spectrum disorders. Some people with autism spectrum disorders during complex brain tasks exhibit frontoparietal underconnectivity on DTI (Kana et al., 2011). On DTI, some people with autism spectrum disorders demonstrate overconnectivity in the frontal and posterior regions due to compensatory adaptation for underconnectivity (Belmonte et al., 2004a,b; Courchesne and Pierce, 2005b; Monk et al., 2009).

A group of boys with autism exhibited bilateral reductions in many association tracts including the inferior frontooccipital fasiculus and the superior longitudinal fasiculus, as well as projection and commissural tracts particularly the forceps minor (Jou et al., 2011b). A group of infants who later developed autism spectrum disorders demonstrated high FA at 6 months with reductions in FA at 24 months (Wolff et al., 2012). A groups of boys with autism spectrum disorders demonstrated impaired neural connections in the corpus callosum, cingulum, and temporal lobes on DTI. FA was reduced in the inferior longitudinal fasciculus/inferior frontooccipital fasciculus, superior longitudinal fasciculus, and corpus callosum/cingulum. The largest reductions in the FA were observed in the right anterior radiation of the corpus callosum/cingulum area and the right fusiform face area (Jou et al., 2011a).

2.2.3. Functional Magnetic Resonance Imaging

fMRI studies of people with autism spectrum disorders provide evidence for network dysfunction, reduced connectivity, and impaired coordination between different parts of the brain needed for complex social and language problems (Allen and Courchesne,

2003; Baron-Cohen et al., 1999; Dalton et al., 2005; Just et al., 2004; Silk et al., 2006; Turner et al., 2006). Please refer to Section 2 on Cognitive Psychology. A mental rotation task demonstrated frontostriatal dysfunction in a group of people with autism spectrum disorders (Silk et al., 2006). A visuomotor task demonstrated dysfunctional connectivity about the orbitofrontal, oculomotor, and motor regions, the caudate nucleus (Turner et al., 2006), and thalamocortical regions (Mizuno et al., 2006) in other groups of people with autism spectrum disorders. A Tower of London test demonstrated underconnectivity between the frontal and parietal regions in another group of people with autism spectrum disorders (Just et al., 2004).

Functional imaging studies have also demonstrated dysfunctional prefrontal cortices during psychological tasks requiring recognition of facial displays of emotion in some people with autism spectrum disorders (Dalton et al., 2005). Another fMRI experiment found an association between the response to emotional faces in the amygdala and social anxiety in people with autism spectrum disorders (Kleinhans et al., 2010).

2.2.4. PET and SPECT

Since the application of PET to autism spectrum disorders has been reviewed recently (Brašić and Wong, 2011; Chugani, 2012), we shall focus on recent key developments. We shall also include updates utilizing SPECT, another functional imaging technique utilizing radiotracers.

2.2.4.1. Metabolism

PET provides a tool to quantitatively evaluate physiological processes in living organisms. PET provides the ability to estimate the rate at which the brain utilizes nutrients. Since dysfunction in the metabolism of endogenous compounds in specific regions of the brain may play a role in the pathophysiology of autism spectrum disorders, the measurement of metabolic activity in the brain is a valuable tool to foster knowledge of autism spectrum disorders.

2.2.4.1.1. Glucose Metabolism

The performance of PET after the administration of 2-deoxy-2-[^{18}F]fluoro-D-glucose ([^{18}F]FDG) is safe, effective tool to examine the utilization of sugar by the body. This procedure can be performed widely in community hospitals and in portable imaging units. Since glucose is metabolized by rapidly growing cells, the application of this procedure for whole-body tomography is suited for people with cancer to identify the location and size of primary malignancies and metastases.

Chugani et al. (2007) identified characteristic abnormalities in children with autism and facial port-wine stains without leptomeningeal angiomata. Four children with autism and facial port-wine stains lacked abnormalities on structural brain imaging with CT or MRI. They may represent a *forme fruste* of the Sturge–Weber syndrome. [^{18}F]FDG

PET demonstrated decreased metabolism in bilateral medial temporal regions, bilateral anterior cingulate gyri, and the right temporal cortex, and increased metabolism in the right parietal cortex in children with autism and facial port-wine stains (Chugani et al., 2007). Children with autism and facial port-wine stains demonstrated a lower metabolism in the bilateral anterior cingulate gyri and the left medial temporal region than did children with autism without stains (Chugani et al., 2007). On the other hand, children with autism without stains demonstrated reduced metabolism in the right temporal cortex in contrast to children with autism with facial port-wine stains (Chugani et al., 2007) (Figure 2).

While healthy normal control children demonstrate a greater metabolism in the right hemisphere in the frontal and temporal regions, children with autism without port-wine stains demonstrate increased metabolism in the left hemisphere in the frontal and temporal regions (Chugani et al., 2007). Children with autism, who also have facial port-wine stains, demonstrate an increased metabolism in the right parietal region that

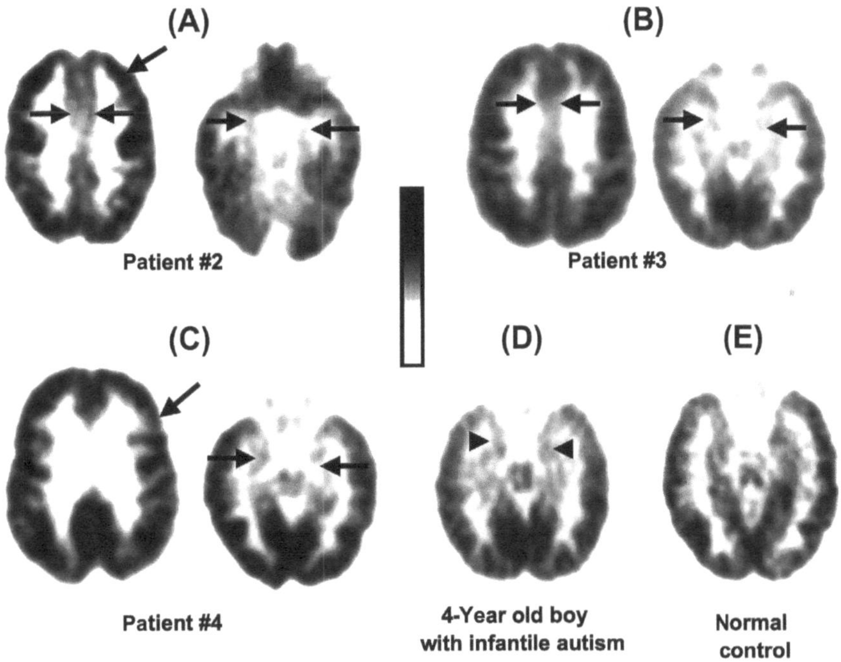

Figure 2 Representative transverse images of positron-emission tomography (PET) after the administration of 2-deoxy-2-[^{18}F]fluoro-D-glucose ([^{18}F]FDG) to children with autism and facial port-wine stains (A–C), a child with autism without facial port-wine stains (D), and a healthy child (E). Focal glucose hypometabolism is indicated by the arrows and arrowheads. While all the children with autism demonstrate bilateral temporal glucose hypometabolism, the hypometabolism is greater in the children with autism with facial port-wine stains. *Reproduced from Chugani et al. (2007), Figure 1, page 196, with permission.*

is absent in healthy control children, and children with autism without stains (Chugani et al., 2007) (Figure 2). This report provides converging evidence that dysfunction of glucose metabolism in the frontal and temporal regions characterizes a subgroup of children with autism.

2.2.4.1.2. Leucine Metabolism

Just as the metabolism of carbohydrates can be estimated by [^{18}F]FDG PET, protein synthesis can be estimated by PET after the administration of other radiotracers. The metabolism leucine, the essential amino acid, in the living human brain can be estimated by PET with L-[1-^{11}C]-leucine.

Protein synthesis in the brain of children with developmental delays was evaluated by means of PET after the administration of L-[1-^{11}C]-leucine (Shandal et al., 2011). Children with developmental delays and pervasive developmental disorder, an autism spectrum disorder, demonstrated a higher protein synthesis rate in the left middle temporal region than children with developmental delays without pervasive developmental disorder. For the combined group of children with developmental delays with and without pervasive developmental disorder, the protein synthesis rate of the left middle temporal region was proportional to the severity of the symptoms of autism. This small study suggests that the language dysfunction of children with developmental delays may be associated with increased protein synthesis in the left middle temporal region (Shandal et al., 2011).

2.2.4.2. Neurotransmission

2.2.4.2.1. Dopaminergic Neurotransmission

Chugani (2012) has reviewed the published literature about nuclear dopaminergic neuroimaging in autism to conclude that dopaminergic function is altered in the frontal cortex, but not in the striatum, in autism.

2.2.4.2.2. Serotonergic Neurotransmission

Chugani (2012) has reviewed the literature about nuclear serotonergic neuroimaging in autism to conclude that profound changes differentiate the serotonergic system in autism in contrast to health (Chugani et al., 1999). Normally, the capacity to synthesize serotonin is high in early childhood and then around age 5 it decreases to adult levels. By contrast, children with autism demonstrate reductions in the ability to synthesize serotonin early in life followed by a gradual increase in the capacity to synthesize serotonin from ages 2–15 years; adults with autism maintain an abnormally high capacity to synthesize serotonin. Asymmetries in the synthesis of serotonin in the frontal, thalamic, and cerebellar regions characterized subgroups of children with autism (Chugani, 2012). Additionally, decrements in the serotonin transporter have been reported in children with autism (Chugani, 2012).

Additional investigation provides evidence to contradict the hypothesis that dysfunction of the serotonergic system in the frontal and limbic regions plays a role in autism spectrum disorders. PET after the administration of [^{11}C]MDL 100907, a selective radioligand for the $5HT_{2A}$ receptor, was administered to adults with Asperger syndrome, an autism spectrum disorder characterized by impairments of social interactions and restricted interests and activities without intellectual impairments. No differences in the density and the distribution of $5HT_{2A}$ receptors were detected in a small sample of 17 adults with Asperger syndrome and 17 healthy control adults (Girgis et al., 2011). The conflicting evidence has several possible explanations. Different procedures in the administration of various radioligands by individual investigators using different equipment at different centers may contribute to conflicting reports. Additionally, the diagnostic procedures to identify patients with autism spectrum disorders may differ among other studies. In particular, adults with the Asperger syndrome may differ in many ways from people with autism. Additionally there may be subgroups of patients with findings at particular locations. The age, sex, and other characteristics of the participants may vary and contribute to conflicting results. Future studies with adequate sample sizes likely will confirm the increased serotonergic activity in adulthood of people with autism spectrum disorders demonstrated by prior studies (Chugani, 2012).

As we have discussed, many lines of enquiry suggest that dysfunction of the serotonergic system in the frontal and limbic regions plays a role in autism spectrum disorders. PET after the administration of [^{11}C]DASB, a selective radioligand for the serotonin transporter, was administered to adults with the Asperger syndrome, an autism spectrum disorder characterized by impairments of social interactions and restricted interests and activities without intellectual impairments. PET after the administration of [^{11}C]DASB detected no differences in the density and the distribution of serotonin transporters in a small sample of 17 adults with Asperger syndrome and 17 healthy control adults (Girgis et al., 2011). The conflicting evidence has several possible explanations. Different procedures in the administration of various radioligands by individual investigators using different equipment at different centers may contribute to conflicting reports. Furthermore, the diagnostic procedures to identify patients with autism spectrum disorders may differ in various studies. Additionally, there may be subgroups of patients with findings at particular locations. Adults with the Asperger syndrome may differ from other adults with autism spectrum disorders. The age, sex, and other characteristics of participants may vary to contribute to conflicting results. Future studies with adequate sample sizes likely will confirm the increased serotonergic activity demonstrated by prior studies (Chugani, 2012).

2.2.4.2.3. $GABA_A$ Neurotransmission

Chugani (2012) has reviewed the literature about nuclear $GABA_A$ neuroimaging in disorders with symptoms of autism to conclude that receptor binding for $GABA_A$ is reduced. Deficits in the neurotransmission of the inhibitory transmitter, $GABA_A$, have been proposed as neurochemical bases for the hyperexcitability observed in some people with autism spectrum disorders (Mendez et al., 2013). [^{123}I]iomazenil, an SPECT radiotracer for the $GABA_A$ receptor, demonstrated markedly reduced uptake in the superior and medial frontal cortices (Figure 3), regions hypothesized to be associated with the ability to infer the thoughts and feelings of others (Mori et al., 2012).

[^{11}C]Ro15-4513, a radiotracer to measure the density and the distribution of the α_1 and $\alpha_5 GABA_A$ subtype receptors, demonstrated much reduced uptake in three men with high-functioning autism spectrum disorder in contrast to healthy normal control men (Mendez et al., 2013). The distribution volume (V_T) for [^{11}C]Ro15-4513 was reduced in all brain regions of men with autism spectrum disorders in contrast to healthy normal control men (Mendez et al., 2013). Employing the Bonferroni criteria to correct for multiple comparisons, Mendez et al. (2013) demonstrated that men with autism spectrum disorders had a significantly reduced V_T for [^{11}C]Ro15-4513 in the right and left nucleus accumbens and the right and left subcallosal areas. Mendez et al. (2013) demonstrated that the $\alpha_5 GABA_A$ subtype was markedly reduced in autism spectrum disorders. The varying regions reported by different investigators suggest that differences between the participants and the protocols resulted in different outcomes. However, the studies suggest that dysfunction of the neurotransmission of $GABA_A$ likely plays a role in autism spectrum disorders.

2.2.4.2.4. Neurotransmission of Acetylcholine

Since deficits of facial recognition, an activity hypothesized to be associated with cholinergic activity in the fusiform gyrus, occur in some people with autism spectrum disorders, central cholingergic activity was evaluated in people with autism spectrum disorders. In contrast to healthy adults, adults with autism spectrum disorders demonstrated

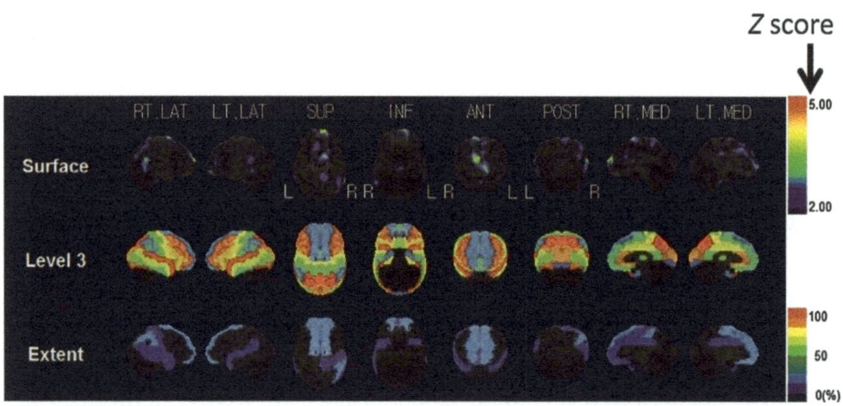

Figure 3 Representations of the differences in the single-photon emission-computed tomography (SPECT) of 24 children with autism spectrum disorders and 10 nonsymptomatic children with partial epilepsy after the intravenous administration of 167 MBq (4.51 mCi) [^{123}I]iomazenil a radiotracer for the $GABA_A$ receptor. Level 3 classifies the volumes of the brain at the level of the lobule. Extent indicates the regions with Z scores ≥2. *Reproduced from Mori et al. (2012), Figure 1, page 651, with permission.*

reductions in acetylcholinesterase activity in the fusiform gyrus on PET after the administration of *N*-[^{11}C]methylpiperidin-4-yl acetate ([^{11}C]MP4A), a radiotracer analog of acetylcholine that undergoes selective hydrolysis by acetylcholinesterase. The reductions in acetylcholinesterase in the fusiform gyrus correlate with the social deficits of the adults with autism spectrum disorders (Suzuki et al., 2011).

Acetylcholinergic dysfunction, specifically the density of acetylcholine vesicular transporters, has been examined in the living human brains of people with the Rett syndrome, an autism spectrum disorder, by means of SPECT after the administration of (2)-(2*R*,3*R*)-2-hydroxy-3-(4-phenylpiperidino)-5-[^{123}I]iodotetralin, (2)-5-[^{123}I]iodobenzovesamicol ([^{123}I]IBVM) (Brašić et al., 2012b). The density of acetylcholine vesicular transporters in the striata was reduced in people with the Rett syndrome in contrast to healthy adults; however, the reduction was not significantly different probably due to the small sample sizes of the study (Figure 4) (Brašić et al., 2012b).

The ability to perform some activities of daily living (ADLs), specifically walking without assistance and feeding self without assistance, was correlated with the density of acetylcholine vesicular transporters in the striata of the women with Rett syndrome (Brašić et al., 2012b). Please refer to Video Clip 2 for a movement assessment of a 21-year-old

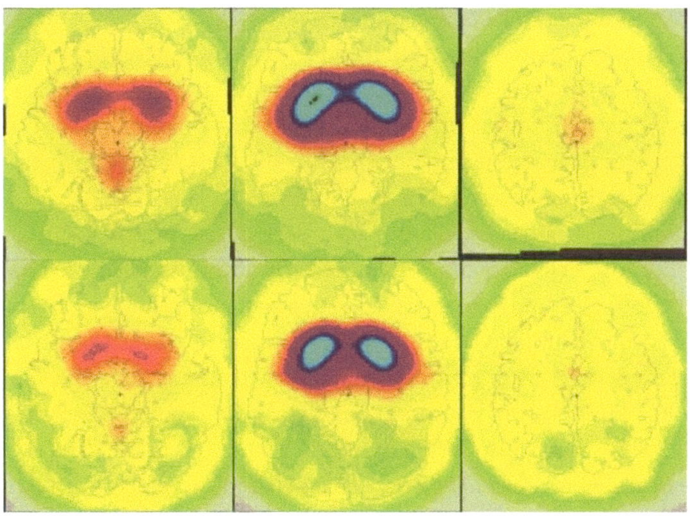

Figure 4 Visual representations of the average uptake on the photon emission-computed tomography (SPECT) 24 h following the intravenous injection of approximately 333 MBq (9 mCi) of (2)-5-[^{123}I] iodobenzovesamicol ([^{123}I]IBVM), a radiotracer for vesicular acetylcholine transporters to eight healthy adults (upper row) and four women with the Rett syndrome (lower row). The left side of each panel corresponds to the left of the brain. The panels represent transverse sections of the brain at the level of the cerebellum (left), striatum (center), and cingulate gyrus (right). The lower row illustrates the reduced uptake in the women with Rett syndrome in the vermis and the bilateral precentral cortices (lower left panel), the striatum (lower central panel), and the middle cingulate gyrus (right lower panel). *Reproduced from Brašić (2012a), Figure 6, page 478, with permission.*

woman whose computed tomography is represented in the composite images on the lower panels of Figure 4. The small sample sizes of the control and patient groups in this study likely hindered the identification of distinctions between groups. The utilization of SPECT with [^{123}I]IBVM provides a promising tool to evaluate acetylcholinergic function in autism spectrum disorders as well as in other neuropsychiatric disorders.

3. CONCLUSIONS

3.1. The Need for Precision and Accuracy in the Diagnosis of Autism Spectrum Disorders

The investigation of autism spectrum disorders has been hindered by the wide range of conditions included in published reports. The quantitative and qualitative review of studies over the past century is challenged by the use of imprecise and uncertain diagnostic procedures (Brašić and Holland, 2006). Currently, research on autism spectrum disorders is challenged by the development of new diagnostic criteria (Brašić, 2012a; Brašić et al., 2012c). Additionally, the identification of the criteria for the diagnosis of autism spectrum disorders requires great skill in the evaluation of children, adolescents, and adults with variable manifestations of characteristics, wide variations in ability levels, and other developmental disabilities (Belmonte et al., 2004a,b; Singhania, 2005). Thus, researchers will wonder if they are comparing apples and oranges with avocados and mangoes. The trend to lump all individuals with traits of impaired social relationships and restricted ranges of interests and activities in the class of autism spectrum disorders likely produces a group of heterogeneous diseases.

Precise diagnosis of autism spectrum disorders is crucial for optimal human brain imaging. While vague criteria may be justified for administrative purposes and for the development of policies and planning, investigators must themselves impose rigor in the characterization of subjects for research on autism spectrum disorders. Autism is a syndrome, a constellation of symptoms and signs with many possible causes (Goodwin and Guze, 1996). Thus, autism represents likely a heterogeneous group of underlying disorders with a vast spectrum of etiologies.

For this reason, researchers need to characterize participants with autism spectrum disorders as precisely as possible to glean possible correlations with demographic, psychological, social, educational, and biological characteristics.

A thorough identification of available characteristics is needed to categorize subgroups of autism spectrum disorders for identification of possible biomarkers for subclasses (Wong et al., 2002). The process of categorization of participants with autism spectrum disorders can be facilitated by the accurate recording of demographic (Brašić, 2003a), ethnic (Brašić, 2004), and educational information (Brašić et al., 2012d).

The various criteria for autism spectrum disorders likely identify different individuals in different study populations. The current diagnostic state is challenging due to the existence of multiple diagnostic tools and changing criteria (Hashimoto et al., 1997). A review of the many instruments to identify and characterize autism spectrum disorders is beyond the scope of this article. We suggest that researchers can benefit from the utilization of the Autism Diagnostic Interview-Revised (ADI-R) (Lord et al., 1994) and the Autism Diagnostic Observation Schedule (ADOS) (Lord et al., 2000) by trained examiners to accurately record with precision the key characteristics of autism spectrum disorders.

Additionally, investigators can benefit from readily administered tools to characterize the stereotypies and other movement disorders of people with autism spectrum disorders (Brašić, 1999). A Movement Assessment Battery (Barnes, 1989; Brašić, 2000, 2001, 2003b; Brašić and Barnett, 1997; Brašić and Gianutsos, 2000; Brašić et al., 1994, 1997, 1998a, 2000, 2001, 2004, 2012c; Campbell, 1985; Fleischhacker et al., 1989; Leckman et al., 1988, 1989; National Institute of Mental Health, Alcohol, Drug Abuse, and Mental Health Administration, Public Health Service, Department of Health, Education, and Welfare, 1988; Sachdev, 1994; Simpson et al., 1979) has been developed for administration in clinical and research settings on a routine basis to characterize the movements of children, adolescents, and adults in health as represented in Video Clip 1 and in those with neuropsychiatric disorders such as the Rett syndrome as represented in Video Clip 2.

Supplementary video related to this chapter can be found at http://dx.doi.org/10.1016/B978-0-12-418677-4.00015-4.

The examiner states the date of birth and the date of the examination.

The participant has already provided written informed consent. The examiner orally obtains verbal consent from the participant for the presentation and publication of the video clip. The examiner informs that the name of the participant will not be revealed.

The examiner asks the participant about the use of medications, caffeine, nicotine, and nonprescribed drugs.

The examiner asks the participant to perform the aspects of the Lateral Preferences Examination to determine the preferred eye, hand, and foot (Brašić et al., 2009; Brašić and Wong, 2011; Denckla, 1985).

The examiner asks the participant to perform the components of the Abnormal Involuntary Movement Scale (Brašić et al., 2012b; National Institute of Mental Health, Alcohol, Drug Abuse, and Mental Health Administration, Public Health Service, Department of Health, Education, and Welfare, 1988). The participant has already been asked to remove his shoes and socks. Two distracting procedures are employed. First with hands on knees, the participant is instructed to speak out loud the digits from 30 backwards. Then, with hands hanging between legs, the participant is asked to give the details of his route to the location. Assessment of muscle tone cannot be determined by viewing the videotape. Therefore, the examiner explicitly states in the videotape that the tone was normal on examination.

Data for several scales of akathisia are obtained (Barnes, 1989; Brašić, 2003b; Brašić et al., 2012b; Fleischhacker et al., 1989; Sachdev, 1994; Simpson et al., 1979). After the participant has been standing and sitting for 1 min, he is asked, "Do you feel restless inside? Do you have an urge to move? Are you able to keep your feet still?" If there is evidence of akathisia then additional questions are asked to determine the frequency, duration, severity, and limitations of the akathisia.

After the participant is asked to sit still for 10 min for the Timed Stereotypies Rating Scale (Brašić, 2003b; Brašić et al., 2012b; Campbell, 1985), he is told that he may relax and put on his shoes and socks. Nevertheless, the videocamera is then kept on for several minutes to record movements. People with tics and other adventitious movements may then exhibit a burst of movements after the extended period of suppression for the procedure. After the protocol is completed, there may be an outpouring of movements previously kept under control by the participant.

Correlation of movements and dynamic scan results will facilitate the identification of distinct subgroups of autism spectrum disorders.

Movement Assessment Battery

At the time of the assessment she has not consumed anything by mouth since midnight in preparation for general anesthesia in the afternoon. At the time of the assessment her medications are carbamazepine, lactulose, mineral oil, oral contraceptive, supersaturated potassium iodide, and valproic acid.

Narrative depiction of Movement Assessment Battery

The patient responds to her name. The examiner assists her to rise from the bed. He asks her to stand still for two minutes. She exhibits writhing movements of her fingers, hands, and wrists. She has hand twirling and waving of the hands. The examiner verifies the absence of items in the mouth. The examiner verifies that the teeth are in good condition. From 3:00 to 3:05 her eyelids flutter suggesting an absence seizure. The examiner asks her to open her mouth and to stick out her tongue. He asks her to tap her finger to her thumb. He observes increased tone in the muscles of the upper extremities. She is asked to stand with her hands outstretched. She is asked to walk back and forth.

Clinical Vignette

She is a young woman with a forme fruste, an atypical form, of Rett syndrome. She has mild hand-wringing. Although she has stereotyped behaviors, she has some functional hand use. She can feed herself sloppily with a spoon, hold a cup, and turn pages of a book. She has irritability and self-injury. She has severe hyperventilation and breath-holding spells. Breath holding spells lead to oral and peripheral cyanosis. She has higher cognitive function than the majority of women with Rett syndrome; she expresses her needs with a communication board. She has some expressive language including "Mom" and other single words. She has facial grimacing and bruxism. She has occasional toe-walking. She has seizures consisting of hand shaking and staring occurring at least weekly. She has severe vasomotor instability and tremulousness. Please refer to the descriptions included in Tables I and III and Figures 3 and 6 of the article by Brašić and colleagues (2012a). Her medications are carbamazepine, lactulose, mineral oil, oral contraceptive, supersaturated potassium iodide, and valproic acid.

Growth and development. She is the product of a normal pregnancy, labor, and delivery. Her mother was G2, P1.

Family history. A maternal aunt has a psychiatric disorder. A maternal cousin has a learning disability. Her sisters, aged 19 and 26 are both in good health.

Laboratory examination. Her karyotype is 46, XX. This woman exhibits the R306C MeCP2 mutation. Her electroencephalogram demonstrates moderate abnormalities with intermittent rhythmic slowing, generalized continuous slowing, and focal left central sharp transients.

Lateral preferences can also readily be assessed in clinical practices and research settings (Brašić et al., 2009; Brašić and Wong, 2011; Denckla, 1985) to identify the underlying neurological characteristics of people with autism spectrum disorders (Video Clip 2). Since obsessions and compulsions characterize a class of people with autism spectrum disorders, the administration of several tools for obsessions and compulsions is appropriate to accurately identify the extent of these phenomena (Dominguez et al., 1989; Goodman et al., 1989a,b; Marks, 1986; Pato and Pato, 1991). Additionally, the utilization of instruments to identify attention deficits (Ward et al., 1993; Wender, 1995; Wender et al., 2001), anxiety (First et al., 1997; Marks, 1986), nicotine dependence (Balfour and Fagerström, 1996; Kawakami et al., 1999), and other psychiatric disorders (Blacker, 2009; First et al., 1997; Folstein et al., 1975; Guy, 1976; Marder, 1995; McMahon et al., 2002; Overall and Gorham, 1962; Pato and Pato, 1991; Woerner et al., 1988) helps to characterize the neurological, psychological, and psychiatric status of people with autism spectrum disorders.

3.2. Human Brain Imaging of Autism Spectrum Disorders

A goal of research on the human brain imaging of autism spectrum disorders is to identify the general characteristics as well as the characteristics of the individual groups. Although some findings have not been confirmed, we shall provide an overview of salient findings. We present suggestions to aid investigators, clinicians, and administrators for the benefit of people with autism spectrum disorders.

Anomalies of the frontal and temporal regions have been identified in subgroups of people with autism spectrum disorders.

Abnormalities of the temporal lobe have been identified in some people with autism spectrum disorders. Subgroups of participants with autism demonstrated increased volumes of the right superior temporal gyrus (Figure 1) (Jou et al., 2010), dysfunction of glucose metabolism in the temporal regions (Figure 2) (Chugani et al., 2007), and an increased protein synthesis rate in the left middle temporal region (Shandal et al., 2011). Impaired temporal cortical development for language characterizes a subset of children with autism spectrum disorders (Eyler et al., 2012; Singhania, 2005).

Abnormalities of the frontal lobe have been identified in some people with autism spectrum disorders. Subgroups of people with autism spectrum disorders demonstrate dysfunction in the metabolism of glucose (Figure 2) (Chugani et al., 2007) and in the neurotransmission of dopamine (Chugani, 2012) and $GABA_A$ in the frontal regions (Chugani, 2012; Mendez et al., 2013; Mori et al., 2012).

Hypoactivation of the ACC in people with autism spectrum disorders provides a likely physiological basis for the social and nonsocial deficits of people with autism spectrum disorders (Di Martino et al., 2011).

A subgroup of children, adolescents, and adults with autism demonstrate altered serotonin metabolism (Chugani, 2012).

Dysfunction of cholinergic neurotransmission has been identified in subgroups of people with autism spectrum disorders.

A subgroup of adults with autism spectrum disorders demonstrate reductions in acetylcholinesterase in the fusiform gyrus, a region associated with social skills (Suzuki et al., 2011).

A group of women with the Rett syndrome demonstrating mutations in the MeCP2 demonstrate reductions of striatal acetylcholine in proportion to reductions in the ability to perform some ADLs (Figure 4) (Brašić et al., 2012b).

The findings of functional anomalies in multiple physiological systems throughout the brain, particularly in the frontal and temporal regions provide the bases for hypotheses about the pathogenesis and pathophysiology of autism spectrum disorders. The characteristics and the timing of the anomalies in brain development in autism spectrum disorders await clarification (Nicolson and Szatmari, 2003). It is likely that a subgroup of people with autism spectrum disorders experience altered or delayed migration of neurons in the usual developmental sequence similar to some people with schizophrenia (Lafargue and Brasic, 2000). An overgrowth of myelination in the first couple of years of life followed by an arrest in myelination may lead to anatomical and physiological anomalies taking place during the years in which the symptoms of the disorders become apparent (Hughes, 2007).

These abnormalities in neuronal development likely result in reductions in the connections between key portions of the brain responsible for facial recognition, identification of emotions, and communication. Reduced connectivity among the frontal and fusiform face regions and the corpus callosum in children with autism spectrum disorders (Hughes, 2007) may impair interactions between perceptual processes of higher and lower orders (Castelli et al., 2002). Disrupted connectivity of long-distance connections particularly in the frontal lobe may then lead to impaired social awareness and cognitive functions in people with autism spectrum disorders (Kana et al., 2011). Hypoactivation of the Autism Spectrum Disorder (ACC) in people with autism spectrum disorders provides confirmation of a physiological basis for the social and nonsocial deficits of people with autism spectrum disorders (Di Martino et al., 2011).

3.3. The Need for Human Brain Imaging Investigations of Autism Spectrum Disorders

Many studies are hindered by the small sample sizes. In particular, studies that did not detect a difference between people with autism spectrum disorders and healthy control participants likely lacked the adequate power to detect changes (Brašić et al., 2003). Multicenter investigations provide a means to develop a sample size large enough to detect subtle distinctions between experimental and control groups. Conducting the protocol in an identical manner at all sites is crucial to obtain optimal data.

Utilization of the various rating scales for autism spectrum disorders helps to identify a uniform population. For example, selection of participants with similar findings on the ADI-R (Lord et al., 1994) and the ADOS (Lord et al., 2000) by trained examiners will facilitate the identification of anomalies unique to the subgroup. In other words, by selecting participants with uniform clinical findings, investigators will improve the likelihood to identify the biological characteristics unique to the subgroup.

Reports of findings on brain imaging studies of subgroups of people with autism spectrum disorders merit replication by other investigators with different populations and in different locations. Conducting the studies in the same manner as in the original protocol will facilitate comparison and contrast of data from different groups. High-resolution research tomography of the brain (Sossi et al., 2005; Rahmim et al., 2005) optimizes the likelihood to detect subtle differences between people with autism spectrum disorders and healthy controls. Analysis of data utilizing comparable procedures (Ashburner and Friston, 1997; Ashburner et al., 1997) will facilitate comparison and contrast with other groups. Utilization of the current nomenclature (Innis et al., 2007) will facilitate an understanding of reported imaging studies.

Groups of participants with known biological bases provide particular promise for brain imaging studies. Research on people with subgroups with known genetic mutations, including the Rett syndrome (Brašić et al., 2012b) and fragile X syndrome (Clapp and Tranfaglia, 2012; Vincent et al., 2005), is facilitated by the similar biological characteristics of the participants. Participants with known genetic mutations likely represent relatively homogeneous biological subgroups of autism spectrum disorders. Each subgroup likely has distinct anomalies on brain imaging.

The ethics of conducting research on people with intellectual impairments merits discussion (Chun et al., 2002). Brain imaging studies of women with the Rett syndrome, a condition with profound mental retardation, requires general anesthesia to permit stabilization of the head to avoid motion artifact (Brašić et al., 2012b). Thus, more than minimal risk is required to perform brain imaging on people with the Rett syndrome and other impaired conditions. Brain imaging does not have a therapeutic effect on participants. Brain imaging with more than minimal risk without benefit can be justified as a means to gain crucial data about the structure and function of the brain in impaired populations. Although the research participants will not be directly helped by the protocol, the findings will likely provide the knowledge to develop interventions to treat others with the disorder. Legislation banning research on impaired populations without therapeutic component stifles investigation to identify the biological bases of the conditions (Brašić et al., 2012b).

3.4. Future Directions for Human Brain Imaging Investigations of Autism Spectrum Disorders

We have reviewed several promising procedures to elucidate the pathophysiology of autism spectrum disorders by means of human brain imaging. These findings merit replication by other investigators with different ethnic groups in different locations to confirm the results. Furthermore, the techniques of brain imaging developed to investigate schizophrenia (Brašić, 2013) and other neuropsychiatric disorders (Wong and Brašić, 2001, 2005; Wong et al., 2003, 2007, 2009, 2011) likely apply well to autism spectrum disorders. Some specific studies offer special promise to understand autism spectrum disorders.

Clarification of the hypothesized dysfunction of the transmission of Glu and GABA in people with autism spectrum disorders will provide the basis for an improved understanding of the abnormalities in the neurochemistry (Horder and Murphy, 2012). In particular, the investigation of the neurotransmission of Glu and GABA in the same participants will likely help to identify the joint modulation of these compounds. Please refer to Section 2.2.4.2.3 on $GABA_A$ Neurotransmission.

The identification of alterations in nicotinic cholinergic receptors in people with mild cognitive impairment and people with Alzheimer's disease (Kendziorra et al., 2011) and in people with Parkinson's disease (Meyer et al., 2009) suggests that there may be a dysfunction of these receptors in people with autism spectrum disorders. The existence of radiotracers for nicotinic cholinergic receptors (Horti and Wong, 2009; Horti et al., 2010) provides tools to identify the possible alterations in other neuropsychiatric disorders. SPECT after the administration of (*S*)-5-[^{123}I]iodo-3-(2-azetidinylmethoxy) pyridine (5-[^{123}I]IA) (Brašić et al., 2009; Zhou et al., 2001) can likely be accomplished in the nuclear medicine departments of community hospitals providing the possible development of a valuable diagnostic tool for some subgroups of people with autism spectrum disorders (Figure 5).

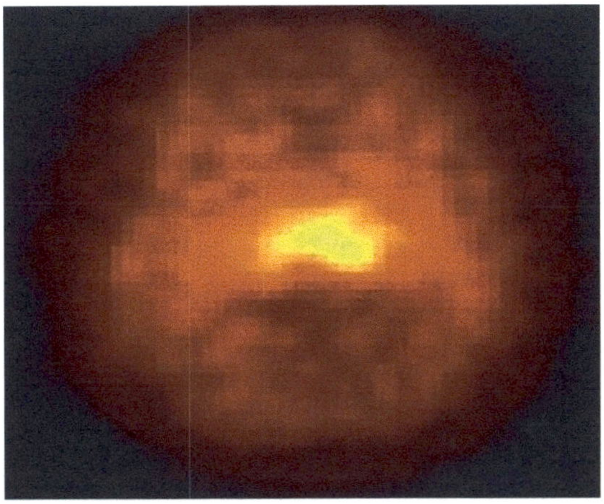

Figure 5 Single-photon emission-computed tomography (SPECT) representation of nicotinic acetylcholine receptors (nAChRs) on a transverse section at the level of the thalamus after the injection of 299.7 MBq (8.1 mCi) (*S*)-5-[^{123}I]iodo-3-(2-azetidinylmethoxy) pyridine (5-[^{123}I]IA) to a normal control 35-year-old male nonsmoker. The top of the image represents the anterior portion of the brain. The left side of the image represents the right side of the brain. *Reproduced from Brašić et al. (2009), Figure 4, page 350, with permission.*

PET after the administration of 2-[^{18}F]fluoro-3-(2(*S*)-azetidinylmethoxy)pyridine (2-[^{18}F]FA) (Brašić et al., 2012d; Kimes et al., 2003; Leung, 2006; Obrzut et al., 2005; Schildan et al., 2007; Shumway et al., 2007; Sorger et al., 2007) also can likely be accomplished in the nuclear medicine departments of community hospitals providing the possible development of a valuable diagnostic tool for some subgroups of people with autism spectrum disorders (Figure 6).

3.5. Therapeutic Implications of Human Brain Imaging in Autism Spectrum Disorders

A key reason for subjecting participants with autism spectrum disorders to the risks and adverse effects of imaging is the need for biological markers to identify the subgroups of the population of autism spectrum disorders. The dearth of effective interventions for autism spectrum disorders (Brašić, 2012a; Brašić et al., 2012b) is evidence for the need for the development of pharmacological and other therapies for autism spectrum disorders. Several novel procedures have been proposed to help people with autism spectrum disorders. The cerebral hypoperfusion particularly in the temporal lobes of people with autism demonstrated by imaging suggests several possible interventions including the administration of hyperbaric oxygen therapy (Rossignol, 2007). The imaging procedures reviewed in this chapter likely will provide the tools to identify people with autism spectrum disorders who may and may not benefit from specific targeted interventions. Additionally, the imaging tools likely will help to monitor the effects of interventions

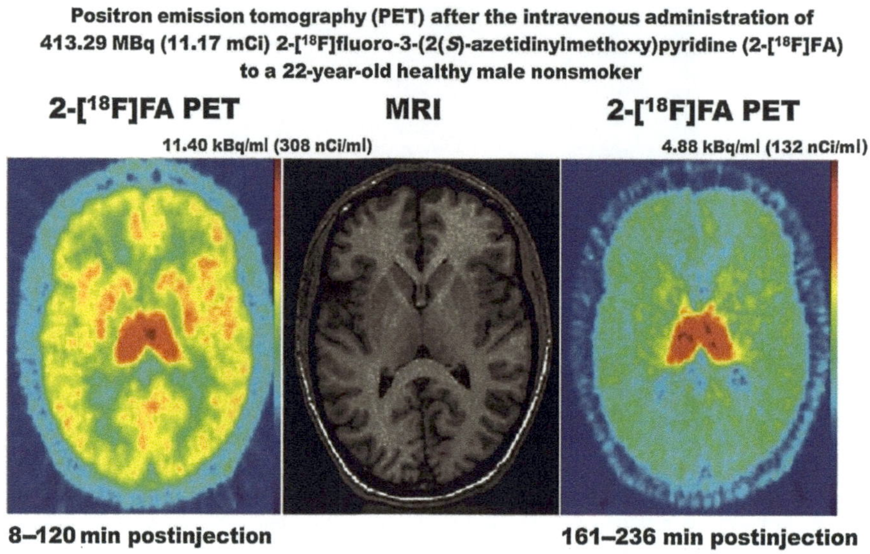

Figure 6 Representations of Transverse Sections of the Brain at the Level of the Thalamus of a Normal Control 22-Year-Old Male Nonsmoker. Magnetic resonance imaging (MRI) (middle) and mean representations of positron-emission tomography (PET) 8–120 min (left) and 161–236 min (right) following the injection as a bolus plus continuous infusion of 413.29 MBq (11.17 mCi) of 2-[^{18}F]fluoro-3-(2(*S*)-azetidinylmethoxy)pyridine (2-[^{18}F]FA) are presented. The intense density of the nicotinic receptors in the thalamus is evident in the left and right panels. *Reproduced from Brašić et al. (2012d), Figure 2, page 361, with permission.*

before, during, and after clinical trials. Thus, we aim to contribute to the treatment and cure of autism spectrum disorders.

ACKNOWLEDGMENTS

This research was sponsored by the Brain and Behavior Research Foundation (NARSAD) (JRB), the Essel Foundation (JRB), and the National Center for Advancing Translational Science (NCATS) Grant Number UL1RR025005 (Clinical Research Unit). Dr Brašic is a member of the Medical Advisory Board of the Tourette Syndrome Association of Greater Washington, Silver Spring, Maryland.

REFERENCES

Alexander, A.L., Lee, J.E., Lazar, M., Boudos, R., DuBray, M.B., Oakes, T.R., Miller, J.N., Lu, J., Jeong, E.K., McMahon, W.M., Bigler, E.D., Lainhart, J.E., 2007. Diffusion tensor imaging of the corpus callosum in autism. NeuroImage 34, 61–73.

Allen, G., Courchesne, E., 2003. Differential effects of developmental cerebellar abnormality on cognitive and motor functions in the cerebellum: an fMRI study of autism. Am. J. Psychiatry 160, 262–273.

American Psychiatric Association, 1994. Diagnostic and Statistical Manual of Mental Disorders, fourth ed. (DSM-IV). American Psychiatric Association, Washington, DC.

American Psychiatric Association, 2000. Diagnostic and Statistical Manual of Mental Disorders, fourth ed. Text Revision (DSM-IV-TR™). American Psychiatric Association, Washington, DC.

American Psychiatric Association, 2013. Diagnostic and Statistical Manual of Mental Disorders, fifth ed. (DSM-5). American Psychiatric Association, Washington, DC.

Aoki, Y., Kasai, K., Yamasue, H., January 17, 2012. Age-related change in brain metabolite abnormalities in autism: a meta-analysis of proton magnetic resonance spectroscopy studies. Transl. Psychiatry 2, e69. http://dx.doi.org/10.1038/tp.2011.65.

Ashburner, J., Friston, K., 1997. Multimodal image coregistration and partitioning—a unified framework. NeuroImage 6, 209–217.

Ashburner, J., Neelin, P., Collins, D.L., Evans, A., Friston, K., 1997. Incorporating prior knowledge into image registration. NeuroImage 6, 344–352.

Balfour, D.J.K., Fagerström, K.O., 1996. Pharmacology of nicotine and its therapeutic use in smoking cessation and neurodegenerative disorders. Pharmacol. Ther. 72, 51–81.

Barker, P.B., 2001. *N*-acetyl aspartate–a neuronal marker? Ann. Neurol. 49, 423–424.

Barnes, T.R.E., 1989. A rating scale for drug-induced akathisia. Br. J. Psychiatry 154, 672–676.

Baron-Cohen, S., Ring, H.A., Wheelwright, S., Bullmore, E.T., Brammer, M.J., Simmons, A., Williams, S.C., 1999. Social intelligence in the normal and autistic brain: an fMRI study. Eur. J. Neurosci. 11, 1891–1898.

Bauman, M.L., Kemper, T.L., 2005. Neuroanatomic observations of the brain in autism: a review and future directions. Int. J. Dev. Neurosci. 23, 183–187.

Belmonte, M.K., Allen, G., Beckel-Mitchener, A., Boulanger, L.M., Carper, R.A., Webb, S.J., 2004a. Autism and abnormal development of brain connectivity. J. Neurosci. 24, 9228–9231.

Belmonte, M.K., Cook Jr., E.H., Anderson, G.M., Rubenstein, J.L., Greenough, W.T., Beckel-Mitchener, A., Courchesne, E., Boulanger, L.M., Powell, S.B., Levitt, P.R., Perry, E.K., Jiang, Y.H., DeLorey, T.M., Tierney, E., 2004b. Autism as a disorder of neural information processing: directions for research and targets for therapy. Mol. Psychiatry 9, 646–663.

Bernardi, S., Anagnostou, E., Shen, J., Kolevzon, A., Buxbaum, J.D., Hollander, E., Hof, P.R., Fan, J., 2011. In vivo ^{1}H-magnetic resonance spectroscopy study of the attentional networks in autism. Brain Res. 1380, 198–205.

Blacker, D., 2009. In: Sadock, B.J., Sadock, V.A., Ruiz, P. (Eds.), Psychiatric Rating Scales. ninth ed. Kaplan and Sadock's Comprehensive Textbook of Psychiatry, vol. 1. Lippincott Williams & Wilkins, Philadelphia, PA, pp. 1032–1059.

Boger-Megiddo, I., Shaw, D.W., Friedman, S.D., Sparks, B.F., Artru, A.A., Giedd, J.N., Dawson, G., Dager, S.R., 2006. Corpus callosum morphometrics in young children with autism spectrum disorder. J. Autism Dev. Disord. 36, 733–739.

Brambilla, P., Hardan, A., di Nemi, S.U., Perez, J., Soares, J.C., Barale, F., 2003. Brain anatomy and development in autism: review of structural MRI studies. Brain Res. Bull. 61 (6), 557–569.

Brašić, J.R., 2012a. Asperger's syndrome. In: eMedicine from WebMD. http://emedicine.medscape.com/article/912296-overview. (accessed 28.08.12.).

Brašić, J.R., Bibat, G., Kumar, A., Zhou, Y., Hilton, J., Yablonski, M.E., Dogan, A.S., Guevara, M.R., Stephane, M., Johnston, M., Wong, D.F., Naidu, S., 2012b. Correlation of the vesicular acetylcholine transporter densities in the striata to the clinical abilities of women with Rett syndrome. Synapse 66, 471–482.

Brašić, J.R., Bronson, B., Chun, T.T., 2012c. Tardive Dyskinesia. . Medscape Reference. Updated February 9, 2012. Available at: http://emedicine.medscape.com/article/1151826-overview. (accessed 27.05.13.).

Brašić, J.R., Cascella, N., Kumar, A., Zhou, Y., Hilton, J., Raymont, V., Crabb, A., Guevara, M.R., Horti, A.G., Wong, D.F., 2012d. Positron emission tomography (PET) experience with 2-[^{18}F]fluoro-3-(2(S)-azetidinylmethoxy)pyridine (2-[^{18}F]FA) in the living human brain of smokers with paranoid schizophrenia. Synapse 66, 352–368.

Brašić, J.R., 1999. Movements in autistic disorder. Med. Hypotheses 53, 48–49.

Brašić, J.R., 2000. Differentiating myoclonus from tics. Psychol. Rep. 86 (1), 155–156.

Brašić, J.R., 2001. Clinical assessment of tics. Psychol. Rep. 89 (1), 48–50.

Brašić, J.R., 2003a. Documentation of demographic data. Psychol. Rep. 93, 151–152.

Brašić, J.R., 2003b. Treatment of movement disorders in autism spectrum disorders. In: Hollander, E. (Ed.), Autism Spectrum Disorders. The Medical Psychiatry Series, vol. 24. Marcel Dekker, Inc., New York, NY, pp. 273–346. ISBN 0-8247-0715-X.

Brašić, J.R., 2004. Documentation of ethnicity. Psychol. Rep. 95, 859–861.

Brašić, J.R., 2013. Autism. In: eMedicine from WebMD. http://emedicine.medscape.com/article/912781-overview. (accessed 28.08.12.).

Brašić, J.R., Barnett, J.Y., 1997. Hyperkinesias in a prepubertal boy with autistic disorder treated with haloperidol and valproic acid. Psychol. Rep. 80, 163–170.
Brašić, J.R., Barnett, J.Y., Aisemberg, P., Ahn, S.C., Nadrich, R.H., Kaplan, D., Ahmad, R., Mendonça, M.F., 1997. Dyskinesias subside off all medication in a boy with autistic disorder and severe mental retardation. Psychol. Rep. 81, 755–767.
Brašić, J.R., Barnett, J.Y., Kaplan, D., Sheitman, B.B., Aisemberg, P., Lafargue, R.T., Kowalik, S., Clark, A., Tsaltas, M.O., Young, J.G., 1994. Clomipramine ameliorates adventitious movements and compulsions in prepubertal boys with autistic disorder and severe mental retardation. Neurology 44, 1309–1312.
Brašić, J.R., Barnett, J.Y., Kowalik, S., Tsaltas, M.O., Ahmad, R., 2004. Neurobehavioral assessment of children and adolescents attending a developmental disabilities clinic. Psychol. Rep. 95, 1079–1086.
Brašić, J.R., Barnett, J.Y., Sheitman, B.B., Lafargue, R.T., Ahn, S.C., 1998a. Clinical assessment of adventitious movements. Psychol. Rep. 83, 739–750.
Brašić, J.R., Barnett, J.Y., Sheitman, B.B., Lafargue, R.T., Kowalik, S., Kaplan, D., Tsaltas, M.O., Ahmad, R., Nadrich, R.H., Mendonça, M.F., 1998b. Behavioral effects of clomipramine on prepubertal boys with autistic disorder and severe mental retardation. CNS Spectr. 3 (10), 39–46.
Brašić, J.R., Barnett, J.Y., Will, M.V., Nadrich, R.H., Sheitman, B.B., Ahmad, R., Mendonça, M.F., Kaplan, D., Brathwaite, C., 2000. Dyskinesias differentiate autistic disorder from catatonia. CNS Spectr. 5 (12), 19–22.
Brašić, J.R., Barnett, J.Y., Zelhof, R., Tarpley, H., 2001. Dopamine antagonists ameliorate the dyskinesias, aggression, and inattention of persons with mental retardation referred to psychiatric clinics. Ger. J. Psychiatry 4 (1), 9–16, http://www.gjpsy.uni-goettingen.de/gjp-article-brasic3.pdf (accessed 27.05.13.).
Brašić, J.R., Gianutsos, J.G., 2000. Neuromotor assessment and autistic disorder. Autism 4 (3), 287–298.
Brašić, J.R., Holland, J.A., 2006. Reliable classification of case–control studies of autistic disorder and obstetric complications. J. Dev. Phys. Disabil. 18, 355–381.
Brašić, J.R., Holland, J.A., 2007. A qualitative and quantitative review of obstetric complications and autistic disorder. J. Dev. Phys. Disabil. 19, 337–364.
Brašić, J.R., Rohde, C.A., Maris, M.A., Wong, D.F., 2003. Sample Size Determination for Studies of Rett Syndrome and Other Rare Conditions. 4th Annual Rett Syndrome Symposium, Rett Syndrome Research Foundation (www.rsrf.org), Baltimore, Maryland, June 23–25, 2003:46 [abstract].
Brašić, J.R., Wong, D.F., 2011. PET Scanning in Autism Spectrum Disorders. Medscape Reference. Updated June 7, 2011. Available at: http://emedicine.medscape.com/article/1155568-overview. (accessed 27.05.13.).
Brašić, J.R., Zhou, Y., Musachio, J.L., Hilton, J., Fan, H., Crabb, A., Endres, C.J., Reinhardt, M.J., Dogan, A.S., Alexander, M., Rousset, O., Maris, M.A., Galecki, J., Nandi, A., Wong, D.F., 2009. Single photon emission computed tomography (SPECT) experience with (*S*)-5-[^{123}I]iodo-3-(2-azetidinylmethoxy)pyridine (5-[^{123}I]IA) in the living human brain of smokers and nonsmokers. Synapse 63, 339–358.
Bryson, S.E., 1996. Brief report: epidemiology of autism. J. Autism Dev. Disord. 26, 165–167.
Bryson, S.E., Clark, B.S., Smith, I.M., 1988. First report of a Canadian epidemiological study of autistic syndromes. J. Child Psychol. Psychiatry 29, 433–445.
Campbell, M., 1985. The timed stereotypies rating scale. Psychopharmacol. Bull. 21, 1082.
Carper, R.A., Courchesne, E., 2005. Localized enlargement of the frontal cortex in early autism. Biol. Psychiatry 57, 126–133.
Carper, R.A., Moses, P., Tigue, Z.D., Courchesne, E., 2002. Cerebral lobes in autism: early hyperplasia and abnormal age effects. NeuroImage 16, 1038–1051.
Carter, J.C., Capone, G.T., Kaufmann, W.E., 2008. Neuroanatomic correlates of autism and stereotypy in children with Down syndrome. Neuroreport 19 (6), 653–656.
Casanova, M.F., Buxhoeveden, D.P., Switala, A.E., Roy, E., 2002. Minicolumnar pathology in autism. Neurology 58, 428–432.
Castelli, F., Frith, C., Happé, F., Frith, U., 2002. Autism, Asperger syndrome and brain mechanisms for the attribution of mental states to animated shapes. Brain 125, 1839–1849.
Chenevert, T.L., Brunberg, J.A., Pipe, J.G., 1990. Anisotropic diffusion in human white matter: demonstration with MR techniques in vivo. Radiology 177, 401–405.
Chugani, D.C., 2012. Neuroimaging and neurochemistry of autism. Pediatr. Clin. North Am. 59, 63–73.

Chugani, D.C., Muzik, O., Behen, M., Rothermel, R., Janisse, J.J., Lee, J., Chugani, H.T., 1999. Developmental changes in brain serotonin synthesis capacity in autistic and nonautistic children. Ann. Neurol. 45, 287–295.

Chugani, H.T., Juhász, C., Behen, M.E., Ondersma, R., Muzik, O., 2007. Autism with facial port-wine stain: a new syndrome? Pediatr. Neurol. 37, 192–199.

Chun, T.T., Brašic, J.R., Morgan, R.H., 2002. Is Research on Human Being with Rett Syndrome Ethical? 3rd Annual Rett Syndrome Symposium, Rett Syndrome Research Foundation (www.rsrf.org), Baltimore, Maryland, June 17–19, 2002:30 [abstract].

Clapp, K., Tranfaglia, M., 2012. What Is Fragile X? Available at: http://www.fraxa.org/fragilex.aspx. (accessed 27.05.13.).

Coghlan, S., Horder, J., Inkster, B., Mendez, M.A., Murphy, D.G., Nutt, D.J., 2012. GABA system dysfunction in autism and related disorders: from synapse to symptoms. Neurosci. Biobehav. Rev. 36 (9), 2044–2055.

Courchesne, E., 2004. Brain development in autism: early overgrowth followed by premature arrest of growth. Ment. Retard. Dev. Disabil. Res. Rev. 10 (2), 106–111.

Courchesne, E., Karns, C.M., Davis, H.R., Ziccardi, R., Carper, R.A., Tigue, Z.D., Chisum, H.J., Moses, P., Pierce, K., Lord, C., Lincoln, A.J., Pizzo, S., Schreibman, L., Haas, R.H., Akshoomoff, N.A., Courchesne, R.Y., 2001. Unusual brain growth patterns in early life in patients with autistic disorder: an MRI study. Neurology 57, 245–254.

Courchesne, E., Mouton, P.R., Calhoun, M.E., Semendeferi, K., Ahrens-Barbeau, C., Hallet, M.J., Barnes, C.C., Pierce, K., 2011. Neuron number and size in prefrontal cortex of children with autism. JAMA 306, 2001–2010.

Courchesne, E., Pierce, K., 2005a. Brain overgrowth in autism during a critical time in development: implications for frontal pyramidal neuron and interneuron development and connectivity. Int. J. Dev. Neurosci. 23, 153–170.

Courchesne, E., Pierce, K., 2005b. Why the frontal cortex in autism might be talking only to itself: local over-connectivity but long-distance disconnection. Curr. Opin. Neurobiol. 15, 225–230.

Courchesne, E., Press, G.A., Yeung-Courchesne, R., 1993. Parietal lobe abnormalities detected with MR in patients with infantile autism. AJR Am. J. Roentgenol. 160 (2), 387–393.

Courchesne, E., Redcay, E., Kennedy, D.P., 2004. The autistic brain: birth through adulthood. Curr. Opin. Neurol. 17, 489–496.

Courchesne, E., Redcay, E., Morgan, J.T., Kennedy, D.P., 2005. Autism at the beginning: microstructural and growth abnormalities underlying the cognitive and behavioral phenotype of autism. Dev. Psychopathol. 17, 577–597.

Croen, L.A., Grether, J.K., Selvin, S., 2002. Descriptive epidemiology of autism in a California population: who is at risk? J. Autism Dev. Disord. 32, 217–224.

Croen, L.A., Najjar, D.V., Fireman, B., Grether, J.K., 2007. Maternal and paternal age and risk of autism spectrum disorders. Arch. Pediatr. Adolesc. Med. 161, 334–340.

Cryan, E., Byrne, M., O'Donovan, A., O'Callaghan, E., 1996. A case–control study of obstetric complications and later autistic disorder. J. Autism Dev. Disord. 26, 453–460.

Dalton, K.M., Nacewicz, B.M., Johnstone, T., Schaefer, H.S., Gernsbacher, M.A., Goldsmith, H.H., Alexander, A.L., Davidson, R.J., 2005. Gaze fixation and the neural circuitry of face processing in autism. Nat. Neurosci. 8, 519–526.

Delobel, M., Van Bakel, M.-E., Klapouszczak, D., Vignes, C., Maffre, T., Raynaud, J.-P., Arnaud, C., Cans, C., 2013. Prévalence de l'autisme et autres troubles envahissants du développement: données des registres français de population. Générations 1995–2002. [Prevalence of autism and other pervasive developmental disorders: data from two French childhood disabilities registries for children born between 1995 and 2002] , Neuropsychiatr. Enfance Adolesc. 61, 23–30.

Denckla, M.B., 1985. Revised neurological examination for subtle signs (1985). Psychopharmacol. Bull. 21 (4), 773–800.

DeVito, T.J., Drost, D.J., Neufeld, R.W., Rajakumar, N., Pavlosky, W., Williamson, P., Nicolson, R., 2007. Evidence for cortical dysfunction in autism: a proton magnetic resonance spectroscopic imaging study. Biol. Psychiatry 61, 465–473.

Di Martino, A., Ross, K., Uddin, L.Q., Sklar, A.B., Castellanos, F.X., Milham, M.P., 2011. Functional brain correlates of social and nonsocial processes in autism spectrum disorders: an activation likelihood estimation meta-analysis. Biol. Psychiatry 65, 63–74.

Dominguez, R.A., Jacobson, A.F., de la Gandara, J., Goldstein, B.J., Steinbock, R.M., 1989. Drug response assessed by the Modified Maudsley Obsessive-Compulsive Inventory. Psychopharmacol. Bull. 25 (2), 215–218.

During, M.J., Spencer, D.D., 1993. Extracellular hippocampal glutamate and spontaneous seizure in the conscious human brain. Lancet 341, 1607–1610.

Eigsti, I.M., Shapiro, T., 2003. A systems neuroscience approach to autism: biological, cognitive, and clinical perspectives. Ment. Retard. Dev. Disabil. Res. Rev. 9, 205–215.

Eyler, L.T., Pierce, K., Courchesne, E., 2012. A failure of left temporal cortex to specialize for language is an early emerging and fundamental property of autism. Brain 135, 949–960.

Fayed, N., Modrego, P.J., 2005. Comparative study of cerebral white matter in autism and attention-deficit/hyperactivity disorder by means of magnetic resonance spectroscopy. Acad. Radiol. 12, 566–569.

Filippi, C.G., Uluğ, A.M., Deck, M.D., Zimmerman, R.D., Heier, L.A., 2002. Developmental delay in children: assessment with proton MR spectroscopy. AJNR Am. J. Neuroradiol. 23, 882–888.

First, M.B., Spitzer, R.L., Gibbon, M., Williams, J.B.W., 1997. Structured Clinical Interview for DSM-IV Axis I Disorders–Clinician Version (SCID-CV). American Psychiatric Press, Washington, DC.

Fleischhacker, W.W., Bergmann, K.J., Perovich, R., Pestreich, L.K., Borenstein, M., Lieberman, J., Kane, J., 1989. The Hillside Akathisia Scale: a new rating instrument for neuroleptic-induced akathisia. Psychopharmacol. Bull. 25, 222–226.

Folstein, M.F., Folstein, S.E., McHugh, P.R., 1975. "Mini-mental state". A practical method for grading the cognitive state of patients for the clinician. J. Psychiatr. Res. 12, 189–198.

Friedman, S.D., Shaw, D.W., Artru, A.A., Richards, T.L., Gardner, J., Dawson, G., Posse, S., Dager, S.R., 2003. Regional brain chemical alterations in young children with autism spectrum disorder. Neurology 60, 100–107.

Frith, C., 2004. Is autism a disconnection disorder? Lancet Neurol. 3, 577.

Frith, C., 2003. What do imaging studies tell us about the neural basis of autism? Novartis Found. Symp. 251, 149–176 281–297.

Gabis, L., Huang, W., Azizian, A., DeVincent, C., Tudorica, A., Kesner-Baruch, Y., Roche, P., Pomeroy, J., 2008. ^{1}H-magnetic resonance spectroscopy markers of cognitive and language ability in clinical subtypes of autism spectrum disorders. J. Child Neurol. 23, 766–774.

Gauthier, J., Joober, R., Dubé, M.P., St-Onge, J., Bonnel, A., Gariépy, D., Laurent, S., Najafee, R., Lacasse, H., St-Charles, L., Fombonne, E., Mottron, L., Rouleau, G.A., 2006. Autism spectrum disorders associated with X chromosome markers in French-Canadian males. Mol. Psychiatry 11, 206–213.

Gillberg, C., 1980. Maternal age and infantile autism. J. Autism Dev. Disord. 10, 293–297.

Girgis, R.R., Slifstein, M., Xu, X., Frankle, W.G., Anagnostou, E., Wasserman, S., Pepa, L., Kolevzon, A., Abi-Dargham, A., Laruelle, M., Hollander, E., 2011. The 5-HT_{2A} receptor and serotonin transporter in Asperger's Disorder: a PET study with [^{11}C]MDL 100907 and [^{11}C]DASB. Psychiatry Res. 194, 230–234.

Glasson, E.J., Bower, C., Petterson, B., de Klerk, N., Chaney, G., Hallmayer, J.F., 2004. Perinatal factors and the development of autism: a population study. Arch. Gen. Psychiatry 61, 618–627.

Goodman, W.K., Price, L.H., Rasmussen, S.A., Mazure, C., Delgado, P., Heninger, G.R., Charney, D.S., 1989a. The Yale-Brown Obsessive Compulsive Scale. II. Validity. Arch. Gen. Psychiatry 46, 1012–1016.

Goodman, W.K., Price, L.H., Rasmussen, S.A., Mazure, C., Fleischmann, R.L., Hill, C.L., Heninger, G.R., Charney, D.S., 1989b. The Yale-Brown Obsessive Compulsive Scale. I. Development, use, and reliability. Arch. Gen. Psychiatry 46, 1006–1011.

Goodwin, D.W., Guze, S.B., 1996. Psychiatric Diagnosis, fifth ed. Oxford University Press, New York, NY.

Govindaraju, V., Young, K., Maudsley, A.A., 2000. Proton NMR chemical shifts and coupling constants for brain metabolites. NMR Biomed. 13 (3), 129–153.

Guy, W., 1976. ECDEU Assessment Manual for Psychopharmacology, Revised, 1976. (DHEW Publ. No. (ADM) 76-338). United States Department of Health, Education, and Welfare, Public Health Service, Alcohol, Drug Abuse, and Mental Health Administration, Washington, DC, pp. 218–222, 534–537.

Hashimoto, T., Kawano, N., Fukuda, K., Endo, S., Mori, K., Yoneda, Y., Yamaue, T., Harada, M., Miyoshi, K., 1998. Proton magnetic resonance spectroscopy of the brain in three cases of Rett syndrome: comparison with autism and normal controls. Acta Neurol. Scand. 98, 8–14.

Hashimoto, T., Tayama, M., Miyazaki, M., Yoneda, Y., Yoshimoto, T., Harada, M., Miyoshi, H., Tanouchi, M., Kuroda, Y., 1997. Differences in brain metabolites between patients with autism and mental retardation as detected by in vivo localized proton magnetic resonance spectroscopy. J. Child Neurol. 12 (2), 91–96.

Hisaoka, S., Harada, M., Nishitani, H., Mori, K., 2001. Regional magnetic resonance spectroscopy of the brain in autistic individuals. Neuroradiology 43, 496–498.
Hazlett, H.C., Poe, M., Gerig, G., Smith, R.G., Provenzale, J., Ross, A., Gilmore, J., Piven, J., 2005. Magnetic resonance imaging and head circumference study of brain size in autism: birth through age 2 years. Arch. Gen. Psychiatry 62, 1366–1376.
Horder, J., Murphy, D.C., 2012. Recent advances in neuroimaging in autism. Neuropsychiatry 2, 221–229, http://www.futuremedicine.com/doi/abs/10.2217/npy.12.25 (accessed 27.05.13.).
Horti, A.G., Gao, Y., Kuwabara, H., Dannals, R.F., 2010. Development of radioligands with optimized imaging properties for quantification of nicotinic acetylcholine receptors by positron emission tomography. Life Sci. 86, 575–584.
Horti, A.G., Wong, D.F., 2009. Clinical perspective and recent development of PET radioligands for imaging cerebral nicotinic acetylcholine receptors. PET Clin. 4, 89–100.
Hsu, S.-W., Chiang, P.-H., Lin, L.-P., Lin, J.-D., 2012. Disparity in autism spectrum disorder prevalence among Taiwan National Health Insurance enrollees: age, gender and urbanization effects. Res. Autism. Spectr. Disord. 6, 836–841.
Hughes, J.R., 2007. Autism: the first firm finding = underconnectivity? Epilepsy Behav. 11, 20–24.
Innis, R.B., Cunningham, V.J., Delforge, J., Fujita, M., Gjedde, A., Gunn, R.N., Holden, J., Houle, S., Huang, S.C., Ichise, M., Iida, H., Ito, H., Kimura, Y., Koeppe, R.A., Knudsen, G.M., Knuuti, J., Lammertsma, A.A., Laruelle, M., Logan, J., Maguire, R.P., Mintun, M.A., Morris, E.D., Parsey, R., Price, J.C., Slifstein, M., Sossi, V., Suhara, T., Votaw, J.R., Wong, D.F., Carson, R.E., 2007. Consensus nomenclature for in vivo imaging of reversibly binding radioligands. J. Cereb. Blood Flow Metab. 27, 1533–1539.
Isaksen, J., Diseth, T.H., Schjølberg, S., Skjeldal, O.H., 2012. Observed prevalence of autism spectrum disorders in two Norwegian counties. Eur. J. Paediatr. Neurol. 16, 592–598.
Jamain, S., Quach, H., Betancur, C., Råstam, M., Colineaux, C., Gillberg, I.C., Soderstrom, H., Giros, B., Leboyer, M., Gillberg, C., Bourgeron, T. Paris Autism Research International Sibpair Study, 2003. Mutations of the X-linked genes encoding neuroligins $NLGN_3$ and $NLGN_4$ are associated with autism. Nat. Genet. 34, 27–29.
Jou, R.J., Jackowski, A.P., Papademetris, X., Rajeevan, N., Staib, L.H., Volkmar, F.R., 2011a. Diffusion tensor imaging in autism spectrum disorders: preliminary evidence of abnormal neural connectivity. Aust. N. Z. J. Psychiatry 45, 153–162.
Jou, R.J., Mateljevic, N., Kaiser, M.D., Sugrue, D.R., Volkmar, F.R., Pelphrey, K.A., 2011b. Structural neural phenotype of autism: preliminary evidence from a diffusion tensor imaging study using tract-based spatial statistics. AJNR Am. J. Neuroradiol. 32, 1607–1613.
Jou, R.J., Minshew, N.J., Keshavan, M.S., Vitale, M.P., Hardan, A.Y., 2010. Enlarged right superior temporal gyrus in children and adolescents with autism. Brain Res. 1360, 205–212.
Just, M.A., Cherkassky, V.L., Keller, T.A., Minshew, N.J., 2004. Cortical activation and synchronization during sentence comprehension in high-functioning autism: evidence of underconnectivity. Brain 127, 1811–1821.
Kana, R.K., Libero, L.E., Moore, M.S., 2011. Disrupted cortical connectivity theory as an explanatory model for autism spectrum disorders. Phys. Life Rev. 8, 410–437.
Kawakami, N., Takatsuka, N., Inaba, S., Shimizu, H., 1999. Developmental [sic] of screening questionnaire for tobacco/nicotine dependence according to ICD-10, DSM-III-R, and DSM-IV. Addict. Behav. 24, 155–166.
Kaufmann, W.E., Cooper, K.L., Mostofsky, S.H., Capone, G.T., Kates, W.R., Newschaffer, C.J., Bukelis, I., Stump, M.H., Jann, A.E., Lanham, D.C., 2003. Specificity of cerebellar vermian abnormalities in autism: a quantitative magnetic resonance imaging study. J. Child Neurol. 18, 463–470.
Kendziorra, K., Wolf, H., Meyer, P.M., Barthel, H., Hesse, S., Becker, G.A., Luthardt, J., Schildan, A., Patt, M., Sorger, D., Seese, A., Gertz, H.-J., Sabri, O., 2011. Decreased cerebral $\alpha_4\beta_2^*$ nicotinic acetylcholine receptor availability in patients with mild cognitive impairment and Alzheimer's disease assessed with positron emission tomography. Eur. J. Nucl. Med. Mol. Imaging 38, 515–525.
Kimes, A.S., Horti, A.G., London, E.D., Chefer, S.I., Contoreggi, C., Ernst, M., Friello, P., Koren, A.O., Kurian, V., Matochik, J.A., Pavlova, O., Vaupel, D.B., Mukhin, A.G., 2003. 2-[^{18}F]F-A-85380: PET imaging of brain nicotinic acetylcholine receptors and whole body distribution in humans. FASEB J. 17, 1331–1333.

Kleinhans, N.M., Richards, T., Weaver, K., Johnson, L.C., Greenson, J., Dawson, G., Aylward, E., 2010. Association between amygdala response to emotional faces and social anxiety in autism spectrum disorders. Neuropsychologia 48, 3665–3670.

Kleinhans, N.M., Schweinsburg, B.C., Cohen, D.N., Müller, R.A., Courchesne, E., 2007. *N*-acetyl aspartate in autism spectrum disorders: regional effects and relationship to fMRI activation. Brain Res. 1162, 85–97.

Lafargue, T., Brasic, J., 2000. Neurodevelopmental hypothesis of schizophrenia: a central sensory disturbance. Med. Hypotheses 55, 314–318.

Lai, D.-C., Tseng, Y.-C., Hou, Y.-M., Guo, H.-R., 2012. Gender and geographic differences in the prevalence of autism spectrum disorders in children: analysis of data from the national disability registry of Taiwan. Res. Dev. Disabil. 33, 909–915.

Larsson, H.J., Eaton, W.W., Madsen, K.M., Vestergaard, M., Olesen, A.V., Agerbo, E., Schendel, D., Thorsen, P., Mortensen, P.B., 2005. Risk factors for autism: perinatal factors, parental psychiatric history, and socioeconomic status. Am. J. Epidemiol. 161, 916–928.

Lauritsen, M.B., Pedersen, C.B., Mortensen, P.B., 2005. Effects of familial risk factors and place of birth on the risk of autism: a nationwide register-based study. J. Child Psychol. Psychiatry 46, 963–971.

Leckman, J.F., Riddle, M.A., Hardin, M.T., Ort, S.I., Swartz, K.L., Stevenson, J., Cohen, D.J., 1989. The Yale Global Tic Severity Scale: initial testing of a clinician-rated scale of tic severity. J. Am. Acad. Child Adolesc. Psychiatry 28, 566–573.

Leckman, J.F., Towbin, K.E., Ort, S.I., Cohen, D.J., 1988. Clinical assessment of tic disorder severity. In: Cohen, D.J., Bruun, R.D., Leckman, J.F. (Eds.), Tourette's Syndrome and Tic Disorders. Wiley, New York, NY, pp. 55–78.

Leung, K., 2006. 2-[^{18}F]Fluoro-3-[2(*S*)-2-azetidinylmethoxy]pyridine. Molecular Imaging and Contrast Agent Database (MICAD) [Internet]. National Center for Biotechnology Information (US), Bethesda, MD 2004–2010.

Lord, C., Risi, S., Lambrecht, L., Cook Jr., E.H., Leventhal, B.L., DiLavore, P.C., Pickles, A., Rutter, M., 2000. The autism diagnostic observation schedule-generic: a standard measure of social and communication deficits associated with the spectrum of autism. J. Autism Dev. Disord. 30, 205–223.

Lord, C., Rutter, M., Le Couteur, A., 1994. Autism Diagnostic Interview-Revised: a revised version of a diagnostic interview for caregivers of individuals with possible pervasive developmental disorders. J. Autism Dev. Disord. 24, 659–685.

Mana, S., Paillère Martinot, M.-L., Martinot, J.-L., 2010. Brain imaging findings in children and adolescents with mental disorders: a cross-sectional review. Eur. Psychiatry 25, 345–354.

Maragakis, N.J., Rothstein, J.D., 2001. Glutamate transporters in neurologic disease. Arch. Neurol. 58, 365–370.

Marder, S.R., 1995. Psychiatric rating scales. sixth ed. In: Kaplan, H.I., Sadock, B.J. (Eds.), Comprehensive Textbook of Psychiatry/VI, vol. 1. Williams & Wilkins, Baltimore, MD, pp. 619–635.

Marks, I.M., 1986. Behavioural Psychotherapy – Maudsley Pocket Book of Clinical Management. Wright, Bristol.

McMahon, R.P., Kelly, D.L., Kreyenbuhl, J., Kirkpatrick, B., Love, R.C., Conley, R.R., 2002. Novel factor-based symptom scores in treatment resistant schizophrenia: implications for clinical trials. Neuropsychopharmacology 26, 537–545.

Mendez, A.M., Horder, J., Myers, J., Coghlan, S., Stokes, P., Erritzoe, D., Howes, O., Lingford-Hughes, A., Murphy, D., Nutt, D., 2013. The brain GABA-benzodiazepine receptor alpha-5 subtype in autism spectrum disorder: a pilot [^{11}C]Ro15-4513 positron emission tomography study. Neuropharmacology 68, 195–201.

Meyer, P.M., Strecker, K., Kendziorra, K., Becker, G., Hesse, S., Woelpl, D., Hensel, A., Patt, M., Sorger, D., Wegner, F., Lobsien, D., Barthel, H., Brust, P., Gertz, H.J., Sabri, O., Schwarz, J., 2009. Reduced $\alpha_4\beta_2^*$-nicotinic acetylcholine receptor binding and its relationship to mild cognitive and depressive symptoms in Parkinson disease. Arch. Gen. Psychiatry 66, 866–877.

Minshew, N.J., Goldstein, G., Dombrowski, S.M., Panchalingam, K., Pettegrew, J.W., 1993. A preliminary ^{31}P MRS study of autism: evidence for undersynthesis and increased degradation of brain membranes. Biol. Psychiatry 33, 762–773.

Mizuno, A., Villalobos, M.E., Davies, M.M., Dahl, B.C., Müller, R.A., 2006. Partially enhanced thalamocortical functional connectivity in autism. Brain Res. 1104, 160–174.

Monk, C.S., Peltier, S.J., Wiggins, J.L., Weng, S.J., Carrasco, M., Risi, S., Lord, C., 2009. Abnormalities of intrinsic functional connectivity in autism spectrum disorders. NeuroImage 47, 764–772.

Morgan, J.T., Chana, G., Abramson, I., Semendeferi, K., Courchesne, E., Everall, I.P., 2012. Abnormal microglial-neuronal spatial organization in the dorsolateral prefrontal cortex in autism. Brain Res. 1456, 72–81.

Mori, K., Hashimoto, T., Harada, M., Yoneda, Y., Shimakawa, S., Fujii, E., Yamaue, T., Miyazaki, M., Saijo, T., Kuroda, Y., 2001. [Proton magnetic resonance spectroscopy of the autistic brain]. No To Hattatsu 33, 329–335.

Mori, T., Mori, K., Fujii, E., Toda, Y., Miyazaki, M., Harada, M., Hashimoto, T., Kagami, S., 2012. Evaluation of the GABAergic nervous system in autistic brain: ^{123}I-iomazenil SPECT study. Brain Dev. 34, 648–654.

Murphy, D.G., Critchley, H.D., Schmitz, N., McAlonan, G., Van Amelsvoort, T., Robertson, D., Daly, E., Rowe, A., Russell, A., Simmons, A., Murphy, K.C., Howlin, P., 2002. Asperger syndrome: a proton magnetic resonance spectroscopy study of brain. Arch. Gen. Psychiatry 59, 885–891.

National Institute of Mental Health, Alcohol, Drug Abuse, and Mental Health Administration, Public Health Service, Department of Health, Education, and Welfare, 1988. Abnormal Involuntary Movement Scale (AIMS). Psychopharmacol. Bull. 24 (4), 781–783.

Newschaffer, C.J., Croen, L.A., Daniels, J., Giarelli, E., Grether, J.K., Levy, S.E., Mandell, D.S., Miller, L.A., Pinto-Martin, J., Reaven, J., Reynolds, A.M., Rice, C.E., Schendel, D., Windham, G.C., 2007. The epidemiology of autism spectrum disorders. Annu. Rev. Public Health 28, 235–258.

Nicolson, R., Szatmari, P., 2003. Genetic and neurodevelopmental influences in autistic disorder. Can. J. Psychiatry 48, 526–537.

Niederhofer, H., 2007. Glutamate antagonists seem to be slightly effective in psychopharmacologic treatment of autism. J. Clin. Psychopharmacol. 27, 317–318.

Obrzut, S.L., Koren, A.O., Mandelkern, M.A., Brody, A.L., Hoh, C.K., London, E.D., 2005. Whole-body radiation dosimetry of 2-[^{18}F]Fluoro-A-85380 in human PET imaging studies. Nucl. Med. Biol. 32, 869–874.

Oner, O., Devrimci-Ozguven, H., Oktem, F., Yagmurlu, B., Baskak, B., Munir, K.M., 2007. Proton MR spectroscopy: higher right anterior cingulate *N*-acetylaspartate/choline ratio in Asperger syndrome compared with healthy controls. AJNR Am. J. Neuroradiol. 28, 1494–1498.

Otsuka, H., Harada, M., Mori, K., Hisaoka, S., Nishitani, H., 1999. Brain metabolites in the hippocampus-amygdala region and cerebellum in autism: an ^{1}H-MR spectroscopy study. Neuroradiology 41, 517–519.

Overall, J.E., Gorham, D.R., 1962. The brief psychiatric rating scale. Psychol. Rep. 10, 799–812.

Page, L.A., Daly, E., Schmitz, N., Simmons, A., Toal, F., Deeley, Q., Ambery, F., McAlonan, G.M., Murphy, K.C., Murphy, D.G., 2006. In vivo ^{1}H-magnetic resonance spectroscopy study of amygdala-hippocampal and parietal regions in autism. Am. J. Psychiatry 163, 2189–2192.

Pato, M.T., Pato, C.N., 1991. Psychometrics in obsessive-compulsive disorder. In: Zohar, J., Insel, T., Rasmussen, S. (Eds.), The Psychobiology of Obsessive-Compulsive Disorder. Springer, New York, NY, pp. 87–88.

Pierpaoli, C., Basser, P.J., 1996. Toward a quantitative assessment of diffusion anisotropy. Magn. Reson. Med. 36, 893–906.

Rahmim, A., Cheng, J.C., Blinder, S., Camborde, M.L., Sossi, V., 2005. Statistical dynamic image reconstruction in state-of-the-art high-resolution PET. Phys. Med. Biol. 50 (20), 4887–4912.

Redcay, E., Courchesne, E., 2005. When is the brain enlarged in autism? A meta-analysis of all brain size reports. Biol. Psychiatry 58, 1–9.

Reichenberg, A., Gross, R., Sandin, S., Susser, E.S., 2010. Advancing paternal and maternal age are both important for autism risk. Am. J. Public Health 100, 772–773.

Reichenberg, A., Gross, R., Weiser, M., Bresnahan, M., Silverman, J., Harlap, S., Rabinowitz, J., Shulman, C., Malaspina, D., Lubin, G., Knobler, H.Y., Davidson, M., Susser, E., 2006. Advancing paternal age and autism. Arch. Gen. Psychiatry 63, 1026–1032.

Ross, B., Bluml, S., 2001. Magnetic resonance spectroscopy of the human brain. Anat. Rec. 265, 54–84.

Rossignol, D.A., 2007. Hyperbaric oxygen therapy might improve certain pathophysiological findings in autism. Med. Hypotheses 68, 1208–1227.

Sachdev, P., 1994. A rating scale for acute drug-induced akathisia: development, reliability, and validity. Biol. Psychiatry 35, 263–271.

Saitoh, O., Karns, C.M., Courchesne, E., 2001. Development of the hippocampal formation from 2 to 42 years: MRI evidence of smaller area dentata in autism. Brain 124, 1317–1324.

Schildan, A., Patt, M., Sabri, O., 2007. Synthesis procedure for routine production of 2-[^{18}F]fluoro-3-(2(*S*)-azetidinylmethoxy)pyridine (2-[^{18}F]F-A-85380). Appl. Radiat. Isot. 65, 1244–1248.

Shandal, V., Sundaram, S.K., Chugani, D.C., Kumar, A., Behen, M.E., Chugani, H.T., 2011. Abnormal brain protein synthesis in language areas of children with pervasive developmental disorder: a L-[1-^{11}C]-leucine PET study. J. Child Neurol. 26, 1347–1354.

Shumway, D.A., Pavlova, O.A., Mukhin, A.G., 2007. A simplified method for the measurement of nonmetabolized 2-[^{18}F]F-A-85380 in blood plasma using solid-phase extraction. Nucl. Med. Biol. 34, 221–228.

Silk, T.J., Rinehart, N., Bradshaw, J.L., Tonge, B., Egan, G., O'Boyle, M.W., Cunnington, R., 2006. Visuospatial processing and the function of prefrontal-parietal networks in autism spectrum disorders: a functional MRI study. Am. J. Psychiatry 163, 1440–1443.

Simpson, G.M., Lee, J.H., Zoubok, B., Gardos, G., 1979. A rating scale for tardive dyskinesia. Psychopharmacology (Berlin) 64, 171–179.

Singhania, R., 2005. Autistic spectrum disorders. Indian J. Pediatr. 72, 343–351.

Sokol, D.K., Dunn, D.W., Edwards-Brown, M., Feinberg, J., 2002. Hydrogen proton magnetic resonance spectroscopy in autism: preliminary evidence of elevated choline/creatine ratio. J. Child Neurol. 17, 245–249.

Sorger, D., Becker, G.A., Patt, M., Schildan, A., Grossmann, U., Schliebs, R., Seese, A., Kendziorra, K., Kluge, M., Brust, P., Mukhin, A.G., Sabri, O., 2007. Measurement of the $\alpha_4\beta_2{}^*$ nicotinic acetylcholine receptor ligand 2-[^{18}F]fluoro-A-85380 and its metabolites in human blood during PET investigation: a methodological study. Nucl. Med. Biol. 34, 331–342.

Sossi, V., de Jong, H.W.A.M., Barker, W.C., Bloomfield, P., Burbar, Z., Camborde, M.L., Comtat, C., Eriksson, L.A., Houle, S., Keator, D., Knoss, C., Krais, R., Lammertsma, A.A., Rahmim, A., Sibomana, M., Teras, M., Thompson, C.J., Trebossen, R., Votaw, J., Walker, M., Wienhard, K., Wong, D.F., 2005. The second generation HRRT – a multi-centre scanner performance investigation. IEEE Nucl. Sci. Symp. Conf. Rec. 1–5, 2195–2199.

Stanley, J.A., 2002. In vivo magnetic resonance spectroscopy and its application to neuropsychiatric disorders. Can. J. Psychiatry 47, 315–326.

Suzuki, K., Sugihara, G., Ouchi, Y., Nakamura, K., Tsujii, M., Futatsubashi, M., Iwata, Y., Tsuchiya, K.J., Matsumoto, K., Takebayashi, K., Wakuda, T., Yoshihara, Y., Suda, S., Kikuchi, M., Takei, N., Sugiyama, T., Irie, T., Mori, N., 2011. Reduced acetylcholinesterase activity in the fusiform gyrus in adults with autism spectrum disorders. Arch. Gen. Psychiatry 68, 306–313.

Travers, B.G., Adluru, N., Ennis, C., Tromp, D.P., Destiche, D., Doran, S., Bigler, E.D., Lange, N., Lainhart, J.E., Alexander, A.L., July 11, 2012. Diffusion tensor imaging in autism spectrum disorder: a review. Autism Res.http://dx.doi.org/10.1002/aur. 1243 (Epub ahead of print).

Turner, K.C., Frost, L., Linsenbardt, D., McIlroy, J.R., Müller, R.A., 2006. Atypically diffuse functional connectivity between caudate nuclei and cerebral cortex in autism. Behav. Brain Funct. 2, 34. Available at: http://www.ncbi.nlm.nih.gov/pmc/articles/PMC1635430/ (accessed 27.05.13.).

Vargas, D.L., Nascimbene, C., Krishnan, C., Zimmerman, A.W., Pardo, C.A., 2005. Neuroglial activation and neuroinflammation in the brain of patients with autism. Ann. Neurol. 57, 67–81.

Verhoeven, J.S., De Cock, P., Lagae, L., Sunaert, S., 2010. Neuroimaging of autism. Neuroradiology 52, 3–14.

Vidal, C.N., Nicolson, R., DeVito, T.J., Hayashi, K.M., Geaga, J.A., Drost, D.J., Williamson, P.C., Rajakumar, N., Sui, Y., Dutton, R.A., Toga, A.W., Thompson, P.M., 2006. Mapping corpus callosum deficits in autism: an index of aberrant cortical connectivity. Biol. Psychiatry 60, 218–225.

Vincent, J.B., Melmer, G., Bolton, P.F., Hodgkinson, S., Holmes, D., Curtis, D., Gurling, H.M., 2005. Genetic linkage analysis of the X chromosome in autism, with emphasis on the fragile X region. Psychiatr. Genet. 15 (2), 83–90.

Volkmar, F.R., Pauls, D., 2003. Autism. Lancet 362 (9390), 1133–1141.

Ward, M.F., Wender, P.H., Reimherr, F.W., 1993. The Wender Utah Rating Scale: an aid in the retrospective diagnosis of childhood attention deficit hyperactivity disorder. Am. J. Psychiatry 150, 885–890.
Wender, P.H., 1995. Attention-Deficit Hyperactivity Disorder in Adults. Oxford University Press, New York, NY.
Wender, P.H., Wolf, L.E., Wasserstein, J., 2001. Adults with ADHD: an overview. Ann. N.Y. Acad. Sci. 931, 1–16.
Wickenhauser, A.J., 2007. More about autism. J. Am. Dent. Assoc. 138 (3) 288, 290, 292.
Woerner, M.G., Mannuzza, S., Kane, J.M., 1988. Anchoring the BPRS: an aid to improved reliability. Psychopharmacol. Bull. 24 (1), 112–117.
Wolff, J.J., Gu, H., Gerig, G., Elison, J.T., Styner, M., Gouttard, S., Botteron, K.N., Dager, S.R., Dawson, G., Estes, A.M., Evans, A.C., Hazlett, H.C., Kostopoulos, P., McKinstry, R.C., Paterson, S.J., Schultz, R.T., Zwaigenbaum, L., Piven, J., IBIS Network, 2012. Differences in white matter fiber tract development present from 6 to 24 months in infants with autism. Am. J. Psychiatry 169, 589–600.
Wong, D.F., Brašić, J.R., 2001. In vivo imaging of neurotransmitter systems in neuropsychiatry. Clin. Neurosci. Res. 1, 35–45.
Wong, D.F., Brašić, J.R., 2005. Progress in the neuropsychiatry of schizophrenia. Psychiatr. Times 22 (3) 57, 58, 60.
Wong, D.F., Brašić, J.R., Cascella, N., 2011. Neuroreceptor imaging of schizophrenia. In: Shenton, M.E., Turetsky, B.I. (Eds.), Understanding Neuropsychiatric Disorders: Insights from Neuroimaging. Cambridge University Press, Cambridge, UK, pp. 78–87.
Wong, D.F., Gründer, G., Brašić, J.R., 2007. Brain imaging research: does the science serve clinical practice? Int. Rev. Psychiatry 19, 541–558.
Wong, D.F., Gründer, G., Cascella, N.G., Brašić, J.R., 2009. Molecular brain imaging in schizophrenia. Chapter 12.9 in Volume 1, ninth ed. In: Sadock, B.J., Sadock, V.A., Ruiz, P. (Eds.), Kaplan & Sadock's Comprehensive Textbook of Psychiatry, Lippincott Williams & Wilkins, Philadelphia, PA, pp. 1519–1530.
Wong, D.F., Maini, A., Rousset, O.G., Brašić, J.R., 2003. Positron emission tomography–a tool for identifying the effects of alcohol dependence on the brain. Alcohol Res. Health 27, 161–173. Available at: http://pubs.niaaa.nih.gov/publications/arh27-2/161-173.pdf (accessed 27.05.13.).
Wong, D.F., Potter, W.Z., Brašić, J.R., 2002. Proof of concept: functional models for drug development in humans. In: Davis, K.L., Charney, D., Coyle, J.T., Nemeroff, C. (Eds.), Neuropsychopharmacology: The Fifth Generation of Progress. The American College of Neuropsychopharmacology and Lippincott Williams & Wilkins, Baltimore, MD, pp. 457–473.
Wong, V.C., Hui, S.L., 2008. Epidemiological study of autism spectrum disorder in China. J. Child Neurol. 23, 67–72.
Zeegers, M., van der Grond, J., van Daalen, E., Buitelaar, J., van Engeland, H., 2007. Proton magnetic resonance spectroscopy in developmentally delayed young boys with or without autism. J. Neural Transm. 114, 289–295.
Zhou, Y., Brašić, J.R., Musachio, J.L., Zukin, S.R., Kuwabara, H., Crabb, A.H., Endres, C.J., Hilton, J., Fan, H., Wong, D.F., 2001. Human [^{123}I]5-A-85380 dynamic SPECT studies in normals: kinetic analysis and parametric imaging. Nuclear Science Symposium Conference Record, vol. 3. Institute of Electrical and Electronics Engineers (IEEE), Incorporated. 1335–1340.

CHAPTER SIXTEEN

Radiotracers Used to Image the Brains of Patients with Alzheimer's Disease

Noble George[1], Emily Gean[1], Ayon Nandi[1], James Robert Brašić[1] and Dean F. Wong[1,2,3,4,5]

[1]The Russell H. Morgan Department of Radiology and Radiological Science, Johns Hopkins University School of Medicine, Baltimore, MD, USA
[2]Departments of Psychiatry, Johns Hopkins University School of Medicine, Baltimore, MD, USA
[3]Department of Neuroscience, Johns Hopkins University School of Medicine, Baltimore, MD, USA
[4]Department of Environmental Health Sciences, Johns Hopkins University School of Medicine, Baltimore, MD, USA
[5]Carey School of Business, Johns Hopkins University School of Medicine, Baltimore, MD, USA

1. INTRODUCTION

Alzheimer's disease (AD) is the most common form of dementia in individuals over the age of 65 years. More than five million people in the United States have been diagnosed with AD, and almost two-thirds of all Americans living with AD are women (Thies and Bleiler, 2011). AD is underdiagnosed, with primary care physicians failing to diagnose dementia in 24–72% of cases. In some cases, those with AD may delay initial physician consultation by 1–2 years after symptom onset. Typically, the lag between symptom onset and diagnosis is up to 4 years (Brodaty et al., 1993). In an aging global population, expected to double in the next 10 years, it is imperative that new technologies are developed to aid in early diagnosis. Although no curative treatment is available, early detection and diagnosis of dementia are helpful in improving potential benefits and alleviates unnecessary physical and emotional burdens on caregivers (Thies and Bleiler, 2011).

Based on the most recent recommendations from the National Institute on Aging-Alzheimer's Association work groups on diagnostic guidelines for AD, the stages of disease progression are defined as follows:

1. Preclinical AD—"Measurable changes in biomarkers (such as brain imaging and spinal fluid chemistry) that indicate the very earliest signs of disease, before outward symptoms are visible. Currently, there are no clinical diagnostic criteria for this phase, but the group provides a scientific framework to help researchers better define this stage of Alzheimer's."
2. Mild cognitive impairment (MCI) due to AD—"Mild changes in memory and thinking abilities, enough to be noticed and measured, but not impairment that compromises everyday activities and functioning."
3. Dementia due to AD—"Memory, thinking and behavioral symptoms that impair a person's ability to function in daily life" (Jack et al., 2011).

Imaging of the Human Brain in Health and Disease
http://dx.doi.org/10.1016/B978-0-12-418677-4.00016-6

Noninvasive imaging has been instrumental in confirming clinical diagnosis. The current American Academy of Neurology guidelines suggest conducting at least one structural brain scan with either unenhanced computed tomography or magnetic resonance imaging to identify structural changes that can be detected in the later stages of AD (Knopman et al., 2001).

These later-stage anatomic scans, however, do not provide adequate information early on in the disease process. Positron emission tomography (PET) imaging, which is used primarily in the oncologic field, however, has proven to be a promising imaging tool in the detection of AD in earlier stages (Bohnen et al., 2012).

Here, we discuss some of the commonly used radiotracers as well as the potential newer ones that may be seen soon in clinical use.

2.1. OVERVIEW

2.1.1. ^{18}F-Fluorodeoxyglucose

The most frequently used radiotracer in clinical settings is ^{18}F-fluorodeoxyglucose (^{18}F-FDG), a radiolabeled glucose analog readily taken up by the brain tissue that measures metabolic activity. In AD patients, there are distinct hypometabolic areas that are highly specific to the disease. In the early stages, there are areas of mild hypometabolism that involve the superior temporal regions more commonly seen in an asymmetric pattern but can occur symmetrically as well.

While in the later stages, there are moderate to severe hypometabolic regions in the parietal and temporal lobes with sparing of occipital, somatosensory, basal ganglia, thalamus, and cerebellum. These areas of hypometabolism are thought to represent the loss of synaptic activity and density.

It is therefore possible to differentiate between other forms dementia such as frontal temporal dementias (Picks disease, semantic dementia, primary progressive aphasia, and familial frontal temporal dementia), which demonstrate hypometabolic areas involving the frontal and anterior temporal lobes of the brain.

In a recent meta-analysis of case studies with clinical assessment as the reference standard, ^{18}F-FDG PET provided a diagnostic accuracy of 93% in its ability to discriminate AD subjects from healthy subjects, with a sensitivity of 96% and a specificity of 90% (Bohnen et al., 2012; Silverman et al., 2001).

When using pathologic confirmation as the reference standard, the sensitivity of ^{18}F-FDG PET was 94% (95% constant infusion (CI), 89–99%) and the specificity was 73% (95% CI, 60–87%) (Bohnen et al., 2012; Silverman et al., 2001).

FDG PET was accurate in differentiating AD patients from demented controls as well as MCI, with a 92% SN (95% CI, 84–96%) and 78% SP (95% CI, 69–85%) (Bloudek et al., 2011).

2.2. AMYLOID PLAQUES

2.2.1. Overview

The presence of amyloid plaques is a hallmark of AD diagnosis. Thus, various investigators have developed radioligands that bind to amyloid plaques. Additionally, there are various radionucleotides being investigated that may also be beneficial.

Below, we discuss the PET radioligands that have been developed to bind to amyloid plaques in neuronal brain tissue. Some are labeled with the radioisotope Carbon (^{11}C) (e.g., ^{11}C-Pittsburgh compound B (PIB), ^{11}C-BF-227, ^{11}C-AZD2184, ^{11}C-SB-13), which has a half-life of around 20 min, while others are labeled with fluorine (F-18), which has a half-life of around 110 min (e.g., ^{18}F-FDDNP (2-(1-{6-[(2-fluoroethyl(methyl) amino]-2-naphthyl}ethylidene)malononitrile), ^{18}F flumetamol, ^{18}F-AV-45, ^{18}F-BAY94-9172, and ^{18}F-AZD4694).

2.2.2. ^{11}C-Pittsburgh Compound B

The most investigated amyloid imaging agent is PIB [*N*-methyl-^{11}C]2-(4′-methylaminophenyl)-6-hydroxybenzothiazole, derived from thioflavin T. PIB like thioflavin T (pathologic stain) binds with high affinity to the beta-pleated sheet aggregates of the amyloid-beta (Aβ) peptide in vitro.

The first ^{11}C-PiB study on a human subject with a clinical diagnosis of AD was conducted in February 2002, at the Uppsala University. The initial human study of ^{11}C-PiB was expanded to include 16 AD subjects and nine cognitively normal controls, which demonstrated an increased cortical uptake corresponding to containing large amounts of known amyloid deposits (Klunk et al., 2004) (Figure 1).

The in vivo retention of ^{11}C-PIB in the brains of people with AD shows a regional distribution that is very similar to the distribution of Aβ deposits observed postmortem (Ikonomovic et al., 2008).

Brain PET imaging using ^{11}C-PIB has been instrumental in identifying patients with a higher deposition of amyloid plaques. In a large review study, ^{11}C-PIB-PET pooled estimates had a 93.5% sensitivity (95% CI, 71.3–99.9%), 56.2% specificity (95% CI, 47.2–64.8%), 2.01 LR+ (95% CI, 1.57–2.58%), and 0.17 LR− (95% CI, 0.08–0.36%) (Zhang et al., 2012).

^{11}C-PIB has proven to be instrumental in the quantification of Aβ deposition in MCI and AD, as well as in older adults without cognitive impairment (Sojkova and Resnick, 2011). ^{18}F-FDG PET and ^{11}C-PIB-PET have similar diagnostic accuracies in early cognitive impairment. However, PIB was able to discriminate significantly better between nonamnestic MCI and amnestic MCI subjects as compared to (^{18}F-FDG). This finding may suggest that early amyloid deposition occurs before cerebral metabolic disruption. Interpretation of ^{11}C-PIB images was more accurate than that of ^{18}F-FDG PET, both on the visual read

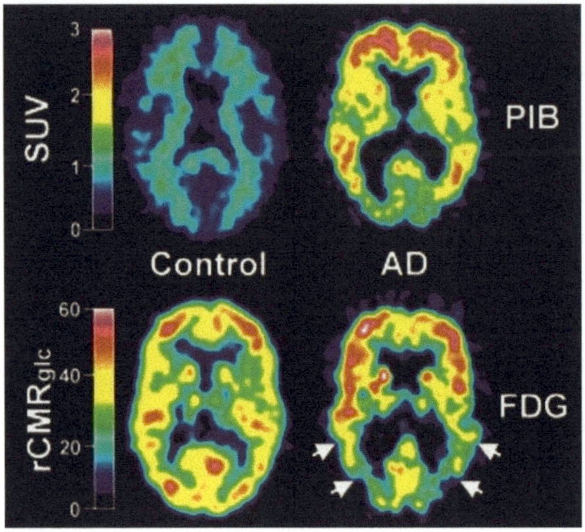

Figure 1 Comparison study of PIB and FDG PET in a healthy control (first column) vs an AD patient (second column). Note the increased uptake of PIB in the frontal and temporoparietal cortices of the AD patient (top right) along with the typical pattern of ^{18}FDG hypometabolism present in the temporoparietal cortex (white arrows) along with normal metabolic uptake in the frontal cortex. *Image Courtesy of Klunk et al. (2004).*

(accuracy, 90% vs 70%, $P=0.05$) and receiver operating characteristic (ROC) analysis (95% vs 83%, $P=0.02$) and accuracy declined more with ^{18}F-FDG than in older patients (Lowe et al., 2009).

Although ^{11}C-PIB possesses many characteristics desirable for a tracer for imaging amyloid, its very short half-life of 20 min limits its use to imaging centers that have access to a cyclotron.

Therefore, there was a significant need to develop newer amyloid tracers with a longer half-life, such as an F-18-labeled tracer, which would have a long-enough half-life that could be transported to imaging centers from remotely located cyclotrons. Below, we review ^{18}F-labeled radioligands for brain amyloid in advanced stages of clinical development. The objective in developing these tracers is to provide a reliable assessment of brain amyloid with a single scan.

2.2.3. ^{18}F-FDDNP

One of the earlier produced F-18 radiotracers is ^{18}F–FDDNP, which has a different binding site than ^{11}C-PIB and competes for binding with nonsteroidal antiphlogistics. It has significant affinity to tau deposits and possibly neurofibrillary tangles (NFTs).

In a study comparing the ^{18}F-FDDNP and ^{11}C-PIB uptake in patients with AD, MCI, and controls, ^{18}F-FDDNP uptake was higher in patients with AD than in controls, but MCI could not be distinguished from AD or from controls. ^{11}C-PIB demonstrated overall a better differentiation between the groups, which was most likely due to the nonspecific binding of ^{18}F-FDDNP to tau deposits and NFT as well as the lipophilic property, which increases background noise uptake (Tolboom et al., 2009).

2.2.4. ^{18}F-AV-45

The most recent agent that has been approved by the Food and Drug Administration for amyloid imaging, *Florbetapir* (^{18}F-AV-45), showed promise with high binding affinity and specificity to Aβ plaques (K_d, 3.72 +/− 0.30 nM). In vitro autoradiography of postmortem human brain sections showed substantial plaque labeling in AD brains and not in the control brains (Choi et al., 2009).

When administered to human subjects, ^{18}F-AV-45 demonstrated good binding to the cortical regions, which were expected to be high in amyloid deposition (e.g., precuneus and frontal and temporal cortices) in known AD patients, with minimal accumulation of tracer in the cortical regions of normal subjects (Wong et al., 2010). It was well tolerated with no adverse effects, and the PET images showed a significant discrimination between AD patients and health controls, using either a parametric reference region method or a simplified solar ultraviolet radiation calculated from 10 min of scanning (Wong et al., 2010) (Figure 2).

In a PET study of end-of-life patients, it was found that florbetapir–PET (^{18}F-AV-45) imaging was 93% sensitive and 100% specific for identifying Aβ-amyloid plaques in

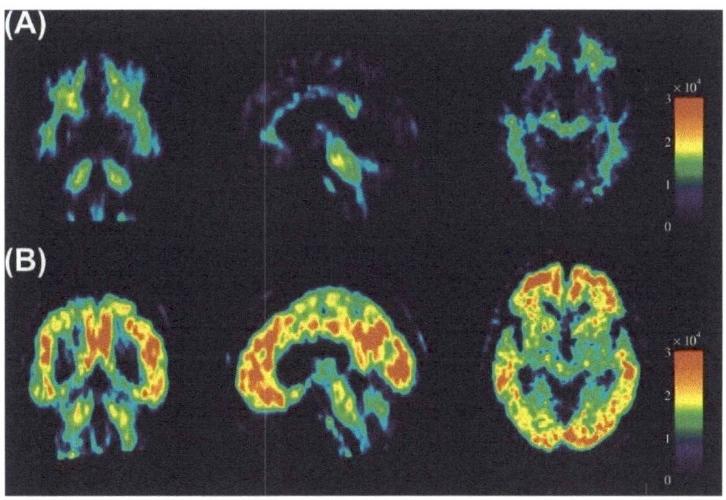

Top = Healthy Control 83 year old Male
Bottom = AD 78 year old Male

Figure 2 (A) Top images demonstrate an increased uptake of ^{18}F-AV-45 in an AD patient involving the frontal and temporal–parietal regions. (B) Bottom images demonstrate the physiologic distribution in a normal healthy control ^{18}F-AV-45. *Image Courtesy of Wong et al. (2010).*

the brain. Comparison studies between ^{18}F-AV-45/^{18}F-FDG are currently underway. A recent study shows that early-phase AV-45 images can provide perfusion information similar to the regional glucose metabolism of FDG, in addition to the amyloid deposition. In particular, the summed early-phase images of AV-45 within the time frame of 1–6 min can be considered as the optimal time frame for providing FDG-like information (Hsiao et al., 2012).

2.2.5. F-BAY94-9172

Another new F-18 Amyloid imaging agent is *F-BAY94-9172*, also known as *AV-1* or *^{18}F-Florbetaben*. This radioligand shares common structural features to ^{11}C PIB and demonstrated excellent specific binding to the Aβ plaques with a K_d of 6.7 nm. There was little binding of neurofibrillary tangles, Pick bodies, Lewy bodies, or glial cytoplasmic inclusions, which matched the reported postmortem distribution of Aβa plaques (Rowe et al., 2007, 2008; Villemagne et al., 2011).

All AD patients showed widespread neocortical binding, which was greater in the precuneus/posterior cingulate and frontal cortex than in the lateral temporal and parietal cortices. There was relative sparing of sensorimotor, occipital, and medial temporal cortices. In the majority of healthy controls and frontotemporal lobar degeneration (FTLD) patients, there was only white-matter binding.

^{18}F-BAY94-9172 (*^{18}F-florbetaben*) had a high sensitivity for AD, which clearly distinguished patients with FTLD from those with AD, and provided results comparable to those reported with ^{11}C-PIB in a variety of neurodegenerative diseases (Villemagne et al., 2011).

2.2.6. ^{18}F-Flutemetamol

^{18}F-Flutemetamol is currently in Phase 3 clinical trials and demonstrates excellent binding to amyloid plaques when compared to histopathologic tissue (Wolk et al., 2011).

A blinded visual assessment of ^{18}F-flutemetamol scans from AD subjects and from elderly healthy controls showed sensitivities of 93.1% and 93.3%, respectively, using clinical diagnosis as the standard (Vandenberghe et al., 2010).

^{18}F-Flutemetamol performs similarly to the ^{11}C-PIB parent molecule within the same subjects and provides high test–retest replicability and a potentially much wider accessibility for clinical and research use (Vandenberghe et al., 2010). More clinical data from additional trials are expected.

A recent study by Wong et al. (2012) assessed the association of [^{18}F]flutemetamol injection with the estimates of the quantitative uptake of the radioligand in the brain and amyloid levels using the immunohistochemical analysis of the parietal lobe biopsies from 12 subjects with normal pressure hydrocephalus. Blinded visual [^{18}F]flutemetamol image interpretations demonstrated a sensitivity of 100% and a specificity of 100% when

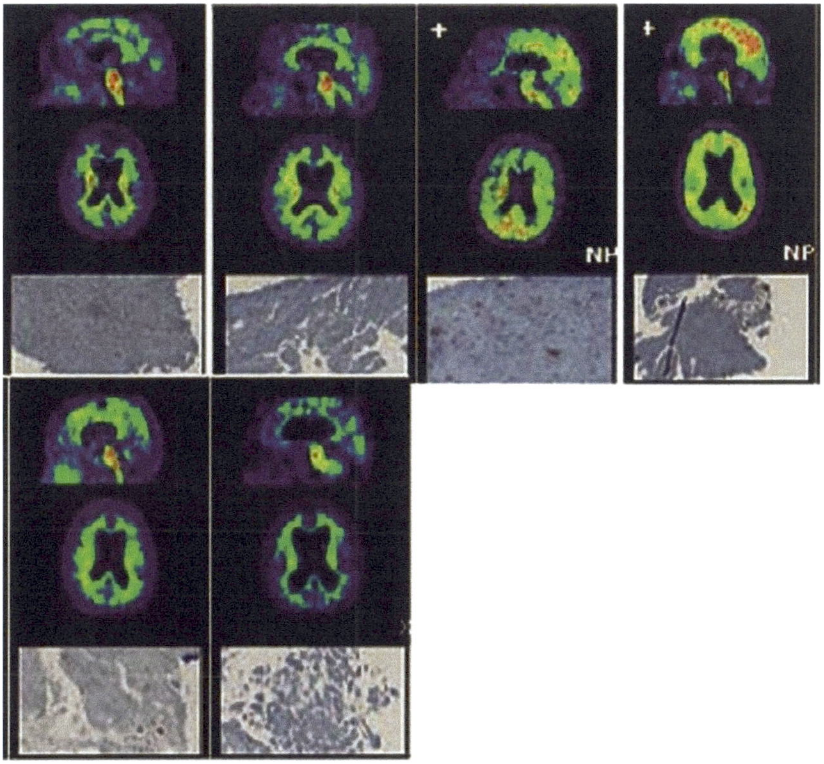

Figure 3 Individual PET images that correspond to the microscopic slides of biopsies (4G8 stain). PET images are shown in the sagittal and axial plane at the midlateral ventricular level. A plus (+) sign at the top left of the sagittal image indicates an abnormal scan. Amyloid β HC (4G8) is shown at 100× magnification for all biopsies. NP above the image indicates the presence of neuritic plaques based on a Bielschowsky silver stain (data not shown). A gray X indicates that adequate cortical gray matter was not available for Bielschowsky assessment. *Images courtesy of Wong et al. (2012).*

compared with pathology reads for amyloid plaque using Bielschowsky and Thioflavin S. staining (Figure 3).

2.2.7. ^{18}F-AZD4694

The most recent ^{18}F radiotracer to undergo clinical trials is ^{18}F-AZD4694. This radionuclide demonstrated an excellent binding of the novel radioligand ^{18}F-AZD4694, which shows high affinity for amyloid fibrils in vitro (K_d, 2.3 ± 0.3 nM). In cortical sections from human AD brain, [^{3}H]AZD4694 selectively labeled Aβ-amyloid deposits in gray matter, whereas there was a low level of nondisplaceable binding to plaque-devoid white matter.

In known AD patients, it demonstrated a regional distribution similar to other Aβ amyloid-binding radioligands (Cselényi et al., 2012). Further study of this compound is needed to access its potential.

2.2.8. [^{18}F]MK-3328

[^{18}F]MK-3328 was developed to exhibit a lower binding to white matter, compared to other tracers such as [^{18}F]AV-45, as binding to white matter can potentially hinder identification in early-onset AD (Hostetler et al., 2011). Low binding to the white matter may allow greater differentiation of amyloid deposition vs normal background uptake of white matter.

In initial PET studies in rhesus monkeys, [^{18}F]MK-3328 displayed high brain uptake with relatively low blood pressure in the white matter and cortical gray matter. The binding potential of [^{18}F]MK-3328 to the cortex and white matter is 0.13 and 0.15, respectively, compared to the binding potential of [^{18}F]AV-45 (0.11 to cortical and 0.33 to the white matter).

[^{18}F]MK-3328 may prove to be an ideal PET tracer for the in vivo quantification of amyloid plaques in humans. Future investigation of [^{18}F]MK-3328 in human healthy volunteers and AD patients is needed and is currently underway (Hostetler et al., 2011).

2.3. TAU IMAGING AGENTS

2.3.1. Overview

A recent addition to AD biomarkers is the evaluation of tau proteins, which are seen in AD as abnormal deposits of hyperphosphorylated tau which form NFTs, neuropil threads, and as dystrophic neuritis that cluster around the sheets of β-amyloid plaques (pathological hallmark of AD).

Although having a significant association with AD, tau deposits are also seen in other neurodegenerative diseases called tauopathies (sporadic corticobasal degeneration, progressive supranuclear palsy, Picks disease, as well as Frontotemporal Dementia and Parkinsonism linked to chromosome 17 (FTDP-17)).

Many studies have indicated that there is a significant correlation between the amount of neurodegeneration and cognitive impairment, which was assessed by quantifying NFT density (Duyckaerts et al., 1987; McLean et al., 1999).

Also, NFTs are not significantly demonstrated in individuals with none to minimal cognitive decline, while β-amyloid plaque deposition can be present in the normal population (Rowe et al., 2008).

As a result, neuroimaging with tau-specific radiotracers may provide a quantitative assessment of global and regional brain tau burden, which in turn can be used as a tool to monitor the progression or regression of disease.

2.3.2. ^{18}F-THK523

^{18}F-THK523 is a novel tau imaging radiotracer. In a recent study, the preclinical evaluation of ^{18}F-THK523 demonstrated adequate penetration of the blood–brain barrier due to its small size and lipophilic nature, along with a high affinity and selectivity for tau fibrils/pathology over β-amyloid fibrils/pathology in the brains of rTg4510 tau transgenic mice in vivo (Fodero-Tavoletti et al., 2011).

The potential clinical role of tau imaging will be to aid in the correlation of brain tau quantification with cognitive function, and future monitoring of the response to novel therapeutic agents.

3. CONCLUSIONS

In addition to the existing diagnostic tests, brain imaging with amyloid imaging agents will provide an excellent opportunity for an earlier and more specific diagnosis of AD. Amyloid imaging can aid in differentiating patients with dementia and can assign a probability of developing AD, from high to low, based on a visual or quantification method.

Patients with no clinical symptoms of dementia (preclinical AD) but with a higher deposition of amyloid plaques may benefit from early treatment (Pike et al., 2007). With the potential development of newer therapies and improvement in the clinical treatment of AD, amyloid imaging agents will prove to be even more essential in evaluating the prognosis of the treatment and in assisting as biomarkers for antidementia drug development. Further studies with head-to-head comparison of the amyloid imaging agents is needed in the future to properly identify the relative benefits and pitfalls.

REFERENCES

Brodaty, H., McGilchrist, C., Harris, L., et al., 1993. Time until institutionalization and death in patients with dementia: role of caregiver training and risk factors. Arch. Neurol. 50, 643–650.

Bohnen, N.I., Djang, D.S., Herholz, K., et al., 2012. Effectiveness and safety of ^{18}F-FDG PET in the evaluation of dementia: a review of the recent literature. J. Nucl. Med. 53 (1), 59–71.

Bloudek, L.M., Spackman, D.E., Blankenburg, M., et al., 2011. Review and meta-analysis of biomarkers and diagnostic imaging in Alzheimer's disease. J. Alzheimers Dis. 26 (4), 627–645.

Choi, S.R., Golding, G., Zhuang, Z., et al., 2009. Preclinical properties of ^{18}F-AV-45: a PET agent for A beta plaques in the brain. J. Nucl. Med. 50 (11), 1887–1894.

Cselényi, Z., Jönhagen, M.E., Forsberg, A., et al., 2012. Clinical validation of ^{18}F-AZD4694, an amyloid-β-specific PET radioligand. J. Nucl. Med. 53 (3), 415–424.

Duyckaerts, C., Brion, J.P., Hauw, J.J., Flament-Durand, J., 1987. Quantitative assessment of the density of neurofibrillary tangles and senile plaques in senile dementia of the Alzheimer type. Comparison of immunocytochemistry with a specific antibody and Bodian's protargol method. Acta Neuropathol. 73, 167–170.

Fodero-Tavoletti, M.T., Okamura, N., Villemagne, V.L., et al., 2011. ^{18}F-THK523: a novel *in vivo* tau imaging ligand for Alzheimer's disease. Oxford J. Med. Brain 134 (4), 1089–1100.

Hostetler, E.D., Sanabria-Bohórguez, S., Fan, H., et al., 2011. [^{18}F]Fluoroazabenzoxazoles as potential amyloid plaque PET tracers: synthesis and in vivo evaluation in rhesus monkey. Nucl. Med. Biol. 38 (8), 1193–1203.

Hsiao, I.T., Huang, C.C., Hsieh, C.J., et al., 2012. Correlation of early-phase ^{18}F-florbetapir (AV-45/Amyvid) PET images to FDG images: preliminary studies. Eur. J. Nucl. Med. Mol. Imaging 39 (4), 613–620.

Ikonomovic, M.D., Klunk, W.E., Abrahamson, E.E., et al., 2008. Post-mortem correlates of vivo PiB-PET amyloid imaging in a typical case of Alzheimer's disease. Brain 131 (Pt 6), 1630–1645.

Jack, C.R., Albert, M.S., Knopman, D.S., et al., 2011. Introduction to the recommendations from the National Institute on Aging-Alzheimer's Association workgroups on diagnostic guidelines for Alzheimer's disease. Alzheimers Dement. 7 (3), 257–262.
Klunk, W.E., Engler, H., Nordberg, A., et al., March 2004. Imaging brain amyloid in Alzheimer's disease with Pittsburgh Compound-B. Ann. Neurol. 55 (3), 306–319.
Knopman, D.S., DeKosky, S.T., et al., 2001. Practice parameter: diagnosis of dementia (an evidence-based review). Report of the Quality Standards Subcommittee of the American Academy of Neurology. Neurology 56 (9), 1143–1153.
Lowe, V.J., Kemp, B.J., Jack Jr., C.R., et al., 2009. Comparison of ^{18}F-FDG and PiB PET in cognitive impairment. J. Nucl. Med. 50 (6), 878–886.
McLean, C.A., Cherny, R.A., Fraser, F.W., Fuller, S.J., Smith, M.J., Beyreuther, K., et al., 1999. Soluble pool of Abeta amyloid as a determinant of severity of neurodegeneration in Alzheimer's disease. Ann. Neurol. 46, 860–866.
Pike, K.E., Savage, G., Villemagne, V.L., et al., 2007. Beta-amyloid imaging and memory in non-demented individuals: evidence for preclinical Alzheimer's disease. Brain 130 (Pt 11), 2837–2844.
Rowe, C.C., Ng, S., Ackermann, U., Gong, S.J., et al., 2007. Imaging beta-amyloid burden in aging and dementia. Neurology 68 (20), 1718–1725.
Rowe, C.C., Ackerman, U., Browne, W., et al., 2008. Imaging of amyloid beta in Alzheimer's disease with ^{18}F-BAY94-9172, a novel PET tracer: proof of mechanism. Lancet Neurol. 7 (2), 129–135.
Silverman, D.H., Small, G.W., Chang, C.Y., et al., 2001. Positron emission tomography in evaluation of dementia: regional brain metabolism and long-term outcome. JAMA 286, 2120–2127.
Sojkova, J., Resnick, S.M., 2011. In vivo human amyloid imaging. Curr. Alzheimer Res. 8 (4), 366–372.
Thies, W., Bleiler, L., 2011. Alzheimer's disease facts and figures overview of Alzheimer's disease. Alzheimers Dement. 7 (2), 208–244.
Tolboom, N., Yaqub, M., van der Flier, W.M., 2009. Detection of Alzheimer pathology in vivo using both ^{11}C-PIB and ^{18}F-FDDNP PET. J. Nucl. Med. 50, 191–197.
Vandenberghe, R., Van Laere, K., Ivanoiu, A., et al., 2010. ^{18}F-flutemetamol amyloid imaging in Alzheimer disease and mild cognitive impairment: a phase 2 trial. Ann. Neurol. 68 (3), 319–329.
Villemagne, V.L., Ong, K., Mulligan, R.S., et al., 2011. Amyloid imaging with ^{18}F-Florbetaben in Alzheimer disease and other dementias. J. Nucl. Med. 52 (8), 1210–1217.
Wolk, D.A., Grachev, I.D., Buckley, C., et al., 2011. Association between in vivo fluorine 18-labeled flutemetamol amyloid positron emission tomography imaging and in vivo cerebral cortical histopathology. Arch. Neurol. 68 (11), 1398–1403.
Wong, D.F., Rosenberg, P.B., Zhou, Y., et al., 2010. In vivo imaging of amyloid deposition in Alzheimer's disease using the novel radioligand [^{18}F]AV-45 (Florbetapir F 18). J. Nucl. Med. 51 (6), 913–920.
Wong, D.F., Moghekar, Abhay R., Brašić, James R., et al., 2012. An *in vivo* evaluation of cerebral cortical amyloid with [^{18}F]Flutemetamol using positron emission tomography compared with parietal biopsy samples in living normal pressure hydrocephalus patients. Mol. Imaging Bio. 15 (2), 230–237.
Zhang, S., Han, D., Tan, X., et al., 2012. Diagnostic accuracy of ^{18}F-FDG and ^{11}C-PIB-PET for prediction of short-term conversion to Alzheimer's disease in subjects with mild cognitive impairment. Int. J. Clin. Pract. 66 (2), 185–198.

CHAPTER SEVENTEEN

Human Brain Imaging of Anger

Sharmin Ghaznavi[1], Thilo Deckersbach[1] and Darin D. Dougherty[2]
[1]Department of Psychiatry, Massachusetts General Hospital, Boston, MA, USA
[2]Department of Psychiatry, Massachusetts General Hospital, Charlestown, MA, USA

1. INTRODUCTION

Anger is an emotion that we have all experienced and can relate to, whether that involves anger toward someone, something, or a situation. In fact, numerous emotion theorists include anger among the basic human emotions (Ekman, 1982; Oatley and Johnson-laird, 1987; Plutchik and Kellerman, 1980; Tomkins, 1984). Anger can range from mild annoyance to intense rage. It can be constructive, such as when we channel it to bring about an improvement in our circumstances or a change for the better, but it can also be destructive, as in instances where it leads to violence.

Despite how fundamental anger is to human experience, little is known about the neural correlates of anger. Until the last decade, most researchers have focused on the neural correlates of hostility and aggression, with an eye toward increasing our understanding of populations with a greater propensity for aggression and violence. In fact, until very recently, the terms *anger*, *hostility*, and *aggression* have been used interchangeably in the literature. Importantly, though, hostility and aggression are distinct from anger in that they are both behavioral sequelae of anger with particular motivations, often punitive and even destructive in nature, as opposed to the basic emotional state of anger (Spielberger et al., 1983). This chapter will focus on the neuroimaging studies related to the basic emotion of anger, in particular, the subjective experience of anger and perception of anger.

2.1. ANGERED: NEUROIMAGING OF THE SUBJECTIVE EXPERIENCE OF ANGER

In an effort to elucidate the neural correlates of the subjective experience of anger, there have been a handful of neuroimaging studies that involve inducing anger in participants in the laboratory setting. Overall, these studies suggest a significant role for prefrontal cortices, in particular, the orbitofrontal cortex.

One of the earliest studies to use this approach was a study by Kimbrell et al. (1995). They used $H_2\,^{15}O$ positron emission tomography (PET) to investigate the regional cerebral blood flow (rCBF) during an anger and anxiety induction paradigm in which healthy control participants were asked to recall either anger- or anxiety-inducing life

Imaging of the Human Brain in Health and Disease
http://dx.doi.org/10.1016/B978-0-12-418677-4.00017-8

events while viewing mood congruent faces; there was no neutral condition. They found that compared to the resting state, there was a greater rCBF in the left medial prefrontal cortex (PFC), bilateral primary visual cortex, and bilateral temporoparietal cortex, when participants underwent both the anger and anxiety induction. When they compared the two induction conditions, they found that anger was associated with relatively more midline septum, right anterior temporal, and right medial prefrontal activity, while anxiety corresponded to relatively more activity in the right posterior central gyrus.

In a follow-up study, also in healthy controls, using $H_2\ {}^{15}O$ PET, Kimbrell and colleagues compared the rCBF during the induction of transient anxiety, anger, as well as neutral emotions (Kimbrell et al., 1999). The induction paradigm was similar to that of their previous study: the participants were asked to recall particular emotion-inducing events while viewing mood congruent faces (i.e., an anger-inducing personal event while viewing angry faces); however, this time the trials included neutral personal events and neutral faces. They found that compared to the neutral emotion condition, the participants showed an increased rCBF in the left inferior frontal (orbitofrontal) and temporal pole regions and a decreased rCBF in the right posterior temporal/parietal and right superior frontal cortex during both the anger and anxiety inductions. Additionally, they found that the anger induction was uniquely associated with an increased rCBF in the right temporal pole and thalamus when compared to that in the neutral condition.

Dougherty and colleagues used ${}^{15}O$–CO_2 PET to investigate the rCBF in healthy men using a slightly different anger induction paradigm (Dougherty et al., 1999). In the study, the participants were asked to provide a written description of a personal event corresponding to when they were "the most angry you've been in your life," as well as two neutral events such as cooking a meal or walking the dog. These descriptions were then used to compose a script in the second person, present tense, and recorded in a neutral voice for playback during the study. During the PET scan, the participants were instructed to close their eyes and listen carefully to the script and imagine the event as vividly as possible as if they were participating in the event. Dougherty and colleagues found that the participants showed an increased rCBF in the left orbitofrontal cortex, the right anterior cingulate cortex, and the bilateral temporal poles during the anger-induction script compared to the neutral scripts.

In a series of ${}^{15}O$ PET studies investigating the neural basis of emotion, Damasio and colleagues examined the brain activity while healthy control participants were asked to recall and reexperience personal events that made them sad, happy, angry, and fearful, as well as neutral events (Damasio et al., 2000). They found that compared to the neutral event, recalling and reexperiencing an angering event resulted in a greater activity in the left anterior cingulate cortex and a decreased activity in the right orbitofrontal cortex.

More recently, Denson and colleagues used functional magnetic resonance imaging (fMRI) to investigate the neural activity during a novel anger provocation paradigm in healthy controls (Denson et al., 2009). The paradigm consists of presenting participants with four easy and eight difficult anagrams during which they are asked to state their answer out loud or say "no answer" if they do not know. The anger provocation consisted of the experimenter interrupting participants three times, each time asking them to speak louder and during the third interruption stating in a rude and condescending tone "Look this is the third time I have had to say this! Can't you follow directions?" Functional activity was obtained for the 2 min immediately following the third interruption and compared to the activity obtained while the participants were asked to stare at a fixation point. Denson and colleagues found that the anger provocation resulted in an increased activity in the dorsal, rostral, and posterior anterior cingulate, insula, medial frontal gyrus, lateral middle frontal gyrus, hippocampus, and thalamus. Additionally, they found that the activity in the left dorsal anterior cingulate cortex was positively correlated with self-reported feelings of anger.

While research on the neural correlates of the subjective experience of anger is far from being conclusive, it nonetheless suggests a significant involvement of prefrontal regions. Given that prefrontal regions are implicated in executive control, it is possible that the activity in the prefrontal regions in anger induction paradigms reflects efforts to exert control over the expression of anger. Of course, additional research is much needed to better characterize the neural underpinnings of our experience of anger.

2.2. ANGRY FACES, BODIES, AND VOICES: NEUROIMAGING OF THE PERCEPTION OF ANGER

Anger is an emotion that most of us have not only experienced but have also been subjected to as well. At some point, all of us have had the experience of having someone being angry with us. In those situations, it is critical that we are able to recognize that the individual is angry, so that we can take the best course of action. Aside from the verbal content relaying their anger and frustration, people let us know that they are angry in a number of other ways, through their facial expressions, their bodily motions, and even the prosody of their speech. Studies of the neural correlates of how we process or perceive anger have examined neural activity corresponding to processing all three of these dimensions of anger.

To date, most neuroimaging studies of anger perception have focused on processing facial expressions of anger. Sprengelmeyer et al. (1998) used fMRI to investigate the neural activity in healthy normal control participants as they performed a gender discrimination task with images of faces that expressed disgust, fear, and anger, as well as neutral faces. They found that compared to neutral faces, during the processing of angry faces, the participants showed a greater activation in the posterior part of the right cingulate cortex, left orbitofrontal cortex, and posterior part of the temporal lobe (Brodmann Area (BA) 21). Using a very similar paradigm, Blair et al. (1999) used $H_2\,^{15}O$ PET to examine the rCBF while healthy male control participants were asked to make a gender discrimination on faces with varying intensities of either sad or angry facial expressions, as well as neutral faces. They also found that compared to neutral faces, during processing of angry faces, the participants showed an increased activity in the right orbitofrontal cortex and the anterior cingulate cortex bilaterally. Additionally, the neural activity in these regions correlated with the intensity of the anger in the facial expression.

In the paradigm where participants are asked to make judgments about the gender of the face presented, the neural activity arguably corresponded to unconscious as well as conscious processing of the emotional dimension of the facial expressions. In an effort to study the neural correlates of conscious processing of emotional facial expressions, Kesler-West et al. (2001) used fMRI to examine the neural activity during a task in which normal healthy control participants were asked to explicitly "focus on each person's expression" while viewing a series of emotionally expressive faces and neutral faces. They also found that compared to the processing of neutral faces, when processing angry faces, the participants showed a greater activation in orbitofrontal cortex and right and left fusiform gyri.

Of note, in the studies described so far, none focused on activation in the amygdala, which is surprising given the role of the amygdala in the processing of emotional stimuli, and in particular, fear (Adolphs et al., 1995). After all, anger can suggest alarm and is arguably a close precursor of fear in situations where violence might ensue. Arguably, there are limitations to the ^{15}O PET studies and imaging the amygdala. To address this, Whalen et al. (2001) focused on examining the neural activity in the amygdala using fMRI in normal healthy controls while they viewed facial expressions of fear or anger as well as neutral faces. With this more focused approach on the amygdala, they found that compared to neutral faces, there was in fact greater activation within the right amygdala when processing angry faces, albeit in a circumscribed location.

Similarly, Beaver et al. (2008) have also shown increased amygdala activation during the processing of angry faces in normal healthy controls using fMRI, along with decreased activation in ventral anterior cingulate cortex. Additionally, they found that differences in the activity in both regions while processing angry faces were correlated with increased appetitive motivation or approach orientation as defined by the Behavioral Approach System Scale (BAS)—participants with higher BAS scores showed increased amygdala activation and decreased ventral anterior cingulate activation to angry faces compared to neutral and sad faces. In another fMRI study examining the relationship between BAS scores and neural activity during the processing of angry faces in normal healthy controls, Passamonti et al. (2008) also found that scores on the BAS were associated with a reduced negative connectivity between the ventral anterior cingulate and amygdala. The relationship between BAS scores and neural activation in these two studies suggests that an approach or aggressive stance may be reflected in greater amygdala activation and decreased prefrontal activation.

More recently, Carre et al. (2010) examined neural activity using fMRI in a large sample of normal healthy control participants ($n = 103$) while they performed a task in which they viewed a trio of faces and were asked to select one of two faces (presented at the bottom) which matched the target face (presented at the top). The faces that were presented were either all sad or angry faces. They contrasted the activity on the face processing trials to trials where participants viewed a trio of geometric faces and asked to select one of two shapes that matched the target shape. Carre and colleagues found that among male participants with elevated trait anxiety, trait anger as assessed by the Speilberger State Trait Anger Expression Inventory was positively correlated with bilateral dorsal amygdala activity to angry faces.

The involvement of PFC and amygdala in processing anger does not appear to be limited to processing facial expressions of anger. Pichon et al. (2008) examined the neural activity in normal healthy control participants using fMRI while they viewed static images as well as videos of bodies expressing fear or anger as well as neutral actions—faces were blurred so that facial expressions were not perceptible. They found that compared to images of bodies with neutral actions (static and dynamic), images of bodies expressing anger (static and dynamic) resulted in a greater activity in the right amygdala and left orbitofrontal cortex.

In two recent studies examining the neural correlates of anger prosody in speech, researchers also found that the amygdala and orbitofrontal cortex are involved. In one study, Sander et al. (2005) used fMRI to investigate the neural activity in normal healthy control participants while they underwent a dichotic listening task in which they were asked to identify the gender of the voice stating a pseudoword (e.g., "goster") in either a neutral or angry tone in either their right or left ear; one ear was the to-be-attended ear and the other the to-be ignored ear. They found that irrespective of whether the angry voice was to-be-attended or to-be-ignored, the participants showed a greater activation in the right amygdala and bilateral superior temporal sulcus when hearing the angry voice compared to the neutral voice. Sander and colleagues also found greater activation in the orbitofrontal cortex when the participants heard the angry voice, but only when it was presented to the to-be-attended ear.

Similarly, in another study, Quadflieg et al. (2008) also found greater activation in the orbitofrontal cortex during the processing of anger prosody. They used fMRI to examine the neural activity in healthy control participants and patients with social phobia while they performed two auditory listening tasks, one in which the participants were asked to identify the emotion expressed by the voice they were listening to and another in which they were asked to identify the gender of the voice they were listening to. In both tasks, the words, which were neutral, were stated in either an angry or neutral voice. Quadflieg and colleagues found that when listening to anger prosody, irrespective of the task, compared to the neutral voice, both the normal controls and patients with social phobia showed a greater activation in bilateral amygdala and orbitofrontal cortex, as well as the superior temporal region, anterior insula, caudate nucleus, and left putamen. Notably, when emotion was task relevant, all the participants showed a greater activation in the left orbitofrontal cortex only when listening to the angry voice compared to the neutral voice.

2.3. FROM ANGER TO AGGRESSION TO VIOLENCE

Given that the sequelae of anger can include aggression and violence, which take an enormous toll on the individuals affected as well as on society in general, it is only natural to ask what we might glean from the neuroscience of anger that might help us to better understand the neuroscience of aggression or violence. Toward this end, based on data from patients with brain lesions, and neuroimaging data in humans and animals, Davidson et al. (2000) proposed a model for how dysfunction in the neural circuitry underlying emotion regulation—specifically regulation of response to angering stimuli might lead to impulsive acts of aggression or violence. Their model highlights the importance of certain brain regions for emotion regulation, namely, the amygdala, orbital PFC, ventromedial PFC, dorsolateral PFC, and anterior cingulate cortex. They proposed that abnormalities in one or more of these regions and in the interconnections among them can lead to a higher propensity for impulsive aggression or violence. Specifically, abnormalities in the amygdala might manifest as too much or too little activation resulting in the inaccurate assessment of social situations, which in turn might result in unnecessary or excessive anger, while deficits in the circuits containing the orbitofrontal cortex, anterior cingulate, as well as ventromedial and dorsolateral PFC might result in failure to inhibit impulsive aggression in response to anger.

Consistent with this proposal, there are a few studies in patients with an increased propensity toward anger and aggression, which suggest that the role of prefrontal regions may indeed be to control the expression of anger. Dougherty et al. (2004) used ^{15}O–CO_2 PET to study the neural activity of patients with a major depressive disorder with anger attacks (MDD+A), patients with a major depressive disorder without anger attacks (MDD−A), and normal healthy controls during an anger induction. As in their previous study, the anger induction consisted of an autobiographical narrative corresponding to two events when they were "most" and "second most" angry. They found that during the anger induction, compared to patients with MDD+A, normal healthy controls showed a greater rCBF in the left ventromedial PFC. Additionally, during the anger induction, normal healthy controls showed an inverse relationship between the rCBF in the left ventromedial PFC and left amygdala, whereas patients with MDD+A showed a positive correlation between these brain regions. This pattern suggests that patients with MDD+A may not be able to recruit prefrontal regions as needed to control their anger and thus exhibit anger attacks.

In another study, Coccaro et al. (2007) investigated neural activity using fMRI in patients with intermittent explosive disorder (IED), which is characterized by impulsive aggressive behavior, and normal healthy controls, while they viewed faces with neutral and emotional expressions, including angry, disgusted, fearful, happy, and sad faces. They found that relative to normal healthy controls, individuals with IED showed greater activation in the amygdala and diminished activation in the orbitofrontal cortex when viewing angry faces. Additionally, the activation in these regions correlated with prior aggressive behavior in the patients with IED.

Finally, neuroimaging research in forensic populations (for a review see Hoskins, Roth, & Giancoloa as well as Pridmore, Chambers & McArthur) (Hoskins et al., 2010; Pridmore et al., 2005) also suggests a connection between problems in brain regions underlying emotion regulation and aggression. Much of that research points to the anatomic and metabolic differences in prefrontal regions associated with a history of aggression or violence, suggesting that an inability to regulate anger due to prefrontal dysfunction may be a precursor to violence.

3. CONCLUSIONS

In this chapter, we reviewed the neural correlates of the subjective experience of anger as well as neural correlates of the perception of anger. While additional research is much needed, the current body of research suggests that the perception of anger involves the amygdala and prefrontal cortical regions. Additionally, the subjective experience of anger involves prefrontal regions, although this may reflect efforts to exert control over expression of anger. Consistent with this, patient populations and forensic populations with a difficulty in controlling anger or a propensity toward aggression show differences in prefrontal functioning that suggest that an inability to control anger may underlie the inability to suppress anger and resultant aggression.

In line with this, research into how prefrontal cortical regions are involved in the control or "braking" of limbic activity holds significant promise for better understanding how we control anger and how we might help individuals who struggle with difficulty controlling anger or aggression. To date, there is one study by Pietrini et al. (2000), in which they used $H_2\,^{15}O$ PET to examine the rCBF in a group of healthy normal controls who were good at visual imagery while they performed a task where they had to imagine not acting and acting aggressively. The participants were asked to imagine four different scenarios with themselves riding with their mother on an elevator along with two men. In one scenario, they were asked to imagine that the ride was uneventful (emotionally neutral condition). In the remaining three scenarios, they were asked to imagine that the two men were attacking their mother. In one such scenario, they were asked to imagine to simply "observe" without taking any action (cognitive restraint condition). In the second scenario, they were asked to imagine attacking the men, but finding themselves restrained by one of the men but ending being restrained by one of the men (physical restraint). In the third scenario, the participants were asked to imagine that they were able to attack the men and hit them with all their strength " to seriously injure or kill them" (unrestrained aggression). Pietrini and colleagues found that during all three scenarios involving aggression, the participants showed a significant decrease in the rCBF in the medial orbitofrontal cortex and the largest decrease was associated with the scenario involving unrestrained aggression. Interestingly, relative to the unrestrained aggression scenario, during the cognitive restraint scenario, they found that there was a greater rCBF in the orbitofrontal cortex and right middle frontal cortex. This suggests that the role of orbitofrontal cortex may indeed be to control aggressive impulses, putting the "brakes" on aggression by way of top-down control over limbic regions. Additional neuroimaging studies of neural activity during restrained and unrestrained aggression, especially ones using psychophysical interactions analyses would be invaluable in determining the network of regions involved in controlling anger.

Finally, while it is clear that certain populations who might struggle with controlling their impulses to act on their anger or have a history of aggression improve with pharmacological treatment or cognitive behavioral therapy for anger management (e.g., patients with MDD and anger episodes), we do not know how these changes are realized in the brain. It may be that both interventions improve orbitofrontal cortical functioning to allow for a better control of their impulses to act on their anger or act aggressively. Research examining neural functioning before and after treatment in these populations would go a long way toward advancing our understanding of how our treatments for anger management and reducing aggression affect change.

REFERENCES

Adolphs, R., Tranel, D., Damasio, H., Damasio, A.R., 1995. Fear and the human amygdala. J. Neurosci. 15, 5879–5891.

Beaver, J.D., Lawrence, A.D., Passamonti, L., Calder, A.J., 2008. Appetitive motivation predicts the neural response to facial signals of aggression. J. Neurosci. 28, 2719–2725.

Blair, R.J., Morris, J.S., Frith, C.D., Perrett, D.I., Dolan, R.J., 1999. Dissociable neural responses to facial expressions of sadness and anger. Brain 122 (Pt 5), 883–893.

Carre, J.M., Fisher, P.M., Manuck, S.B., Hariri, A.R., 2010. Interaction between trait anxiety and trait anger predict amygdala reactivity to angry facial expressions in men but not women. Soc. Cogn. Affect. Neurosci.

Coccaro, E.F., McCloskey, M.S., Fitzgerald, D.A., Phan, K.L., 2007. Amygdala and orbitofrontal reactivity to social threat in individuals with impulsive aggression. Biol. Psychiatry 62, 168–178.

Damasio, A.R., Grabowski, T.J., Bechara, A., Damasio, H., Ponto, L.L., Parvizi, J., Hichwa, R.D., 2000. Subcortical and cortical brain activity during the feeling of self-generated emotions. Nat. Neurosci. 3, 1049–1056.

Davidson, R.J., Putnam, K.M., Larson, C.L., 2000. Dysfunction in the neural circuitry of emotion regulation—a possible prelude to violence. Science 289, 591–594.

Denson, T.F., Pedersen, W.C., Ronquillo, J., Nandy, A.S., 2009. The angry brain: neural correlates of anger, angry rumination, and aggressive personality. J. Cogn. Neurosci. 21, 734–744.

Dougherty, D.D., Rauch, S.L., Deckersbach, T., Marci, C., Loh, R., Shin, L.M., Alpert, N.M., Fischman, A.J., Fava, M., 2004. Ventromedial prefrontal cortex and amygdala dysfunction during an anger induction positron emission tomography study in patients with major depressive disorder with anger attacks. Arch. Gen. Psychiatry 61, 795–804.

Dougherty, D.D., Shin, L.M., Alpert, N.M., Pitman, R.K., Orr, S.P., Lasko, M., Macklin, M.L., Fischman, A.J., Rauch, S.L., 1999. Anger in healthy men: a PET study using script-driven imagery. Biol. Psychiatry 46, 466–472.

Ekman, P., 1982. Emotion in the Human Face. Cambridge University Press; Editions de la Maison des Sciences de l'Homme, Cambridge Cambridgeshire; New York, Paris, xvi, p. 439.

Hoskins, L.L., Roth, R.M., Giancola, P.R., 2010. Neuroimaging of Aggression: Empirical Findings and Implications.

Kesler-West, M.L., Andersen, A.H., Smith, C.D., Avison, M.J., Davis, C.E., Kryscio, R.J., Blonder, L.X., 2001. Neural substrates of facial emotion processing using fMRI. Brain Res. Cogn. Brain Res. 11, 213–226.

Kimbrell, T.A., George, M.S., Parekh, P.I., Ketter, T.A., Herscovitch, P., Post, R.M., 1995. Regional brain activity during self-induced anger and anxiety. Biol. Psychiatry 37, 617–618.

Kimbrell, T.A., George, M.S., Parekh, P.I., Ketter, T.A., Podell, D.M., Danielson, A.L., Repella, J.D., Benson, B.E., Willis, M.W., Herscovitch, P., Post, R.M., 1999. Regional brain activity during transient self-induced anxiety and anger in healthy adults. Biol. Psychiatry 46, 454–465.

Oatley, K., Johnson-laird, P.N., 1987. Towards a cognitive theory of emotions. Cogn. Emotion 1, 29–50.

Passamonti, L., Rowe, J.B., Ewbank, M., Hampshire, A., Keane, J., Calder, A.J., 2008. Connectivity from the ventral anterior cingulate to the amygdala is modulated by appetitive motivation in response to facial signals of aggression. NeuroImage 43, 562–570.

Pichon, S., de Gelder, B., Grezes, J., 2008. Emotional modulation of visual and motor areas by dynamic body expressions of anger. Soc. Neurosci. 3, 199–212.

Pietrini, P., Guazzelli, M., Basso, G., Jaffe, K., Grafman, J., 2000. Neural correlates of imaginal aggressive behavior assessed by positron emission tomography in healthy subjects. Am. J. Psychiatry 157, 1772–1781.

Plutchik, R., Kellerman, H., 1980. Emotion, Theory, Research, and Experience. Academic Press, New York.

Pridmore, S., Chambers, A., McArthur, M., 2005. Neuroimaging in psychopathy. Aust. N. Z. J. Psychiatry 39, 856–865.

Quadflieg, S., Mohr, A., Mentzel, H.J., Miltner, W.H., Straube, T., 2008. Modulation of the neural network involved in the processing of anger prosody: the role of task-relevance and social phobia. Biol. Psychol. 78, 129–137.

Sander, D., Grandjean, D., Pourtois, G., Schwartz, S., Seghier, M.L., Scherer, K.R., Vuilleumier, P., 2005. Emotion and attention interactions in social cognition: brain regions involved in processing anger prosody. NeuroImage 28, 848–858.

Spielberger, C.D., Johson, E.H., Russell, S.F., Crane, R.S., Jacobs, G., Worden, T.J., 1983. Assessment of anger: the state-trait scale. In: Butcher, J.N., Spielberger, C.D. (Eds.), Advances in Personality Assessment. Erlbaum, Hillside.

Sprengelmeyer, R., Rausch, M., Eysel, U.T., Przuntek, H., 1998. Neural structures associated with recognition of facial expressions of basic emotions. Proc. Biol. Sci. 265, 1927–1931.

Tomkins, S.S., 1984. Affect theory. In: Scherer, K.R., Ekman, P. (Eds.), Approaches to Emotion. Erlbaum, Hillsdale, pp. 163–195.

Whalen, P.J., Shin, L.M., McInerney, S.C., Fischer, H., Wright, C.I., Rauch, S.L., 2001. A functional MRI study of human amygdala responses to facial expressions of fear versus anger. Emotion 1, 70–83.

CHAPTER EIGHTEEN

Imaging Pain in the Human Brain

Marco L. Loggia[1,2] and Karin B. Jensen[2]
[1]Athinoula A. Martinos Center for Biomedical Imaging, Massachusetts General Hospital, Harvard Medical School, Boston, MA, USA
[2]Department of Psychiatry, Massachusetts General Hospital, Harvard Medical School, Boston, MA, USA

1. INTRODUCTION

The International Association for the Study of Pain defines pain as "an unpleasant sensory and emotional experience associated with actual or potential tissue damage, or described in terms of such damage" (Merksey et al., 1979). Although the primary role of the pain system is to warn us of the damaging potential of a noxious stimulus in order to protect our body from injury, the physical properties of the peripheral stimulation are not the only determinants of how intense or unpleasant the pain will be. If a relationship between the intensity of the stimulus and the intensity of the pain certainly exists (e.g. it is true that pinching one's finger is generally less bothersome than hitting it with a hammer), in some circumstances, the pain response can be surprisingly small or disproportionately large, in relation to the peripheral stimulation that has caused it (Melzack and Wall, 1988). It is largely due to the contribution of functional neuroimaging techniques that today we know that the brain, and especially the cerebral cortex, has a major role in determining the pain experience, above and beyond the intensity of peripheral stimulation itself.

Prior to the advent of brain imaging, the role of the cortex in the perception of pain was seriously doubted. In their 1911 seminal paper, Head and Holmes concluded that "a pure cortical lesion leads to no change in the threshold to measurable painful or uncomfortable stimuli" (Head and Holmes, 1911). The lack of cortical determinants in the genesis of pain was still postulated in the 1950s, when Penfield and Rasmussen reported that pain was rarely evoked by direct cortical stimulation (Penfield and Rasmussen, 1955). This notion, however, was challenged by the very first experimental investigation of the hemodynamic responses to pain stimulation (Lassen et al., 1978). This study, although characterized by very low spatial resolution, had revealed an increase in regional cerebral blood flow (rCBF) in frontal regions of the brain.

Over the past two decades, modern brain imaging techniques such as functional magnetic resonance imaging (fMRI), positron emission tomography (PET), magnetoencephalography (MEG), and electroencephalography (EEG) have revolutionized our understanding of pain. The first modern brain imaging studies in the 1990s (Apkarian

Imaging of the Human Brain in Health and Disease
http://dx.doi.org/10.1016/B978-0-12-418677-4.00018-X

et al., 1992; Jones et al., 1991; Talbot et al., 1991), as well as the exponentially increasing (Borsook et al., 2010) plethora of publications in more recent years, have unequivocally demonstrated not only that the cortex has a fundamental role in the processing of pain but also that a large proportion of the cortex appears to be dedicated to this activity. Thanks to these tools we now know that the complexity of the subjective experience of pain involves a multiplicity of different brain regions, including (but not limited to) primary and secondary somatosensory cortices, the anterior cingulate cortex (ACC), thalamus, insula, and prefrontal cortex (Apkarian et al., 2005) (Figure 1).

These regions have been proposed to be part of a distributed network often referred to as the "neuromatrix" (Melzack, 1989, 1990, 1999) or "pain matrix" (Albe-Fessard et al., 1985; Ingvar, 1999), which is generally, although not unanimously (Iannetti and Mouraux, 2010), considered to have at least some specificity for pain. Perhaps it is not that surprising that a very simple stimulus such as a brief heat pulse or pinprick on the skin is capable of eliciting activity changes within a very large number of cortical and subcortical regions. In fact, pain is considered to be a multidimensional experience (Melzack and Casey, 1968; Melzack and Wall, 1988), involving sensory, affective/motivational, cognitive, motor, and autonomic components. The use of imaging has started to disentangle these different components, by revealing, for instance, some regions that more likely support sensory processes such as spatiotemporal and intensive features of pain and others that appear more involved in coding the unpleasantness of the pain experience (Hofbauer et al., 2001; Rainville et al., 1999).

Brain imaging has also confirmed in humans the existence of a descending pain modulatory system, therefore indicating that the human brain not only is involved in the construction of the perceptual experience in response to noxious stimulation, but also

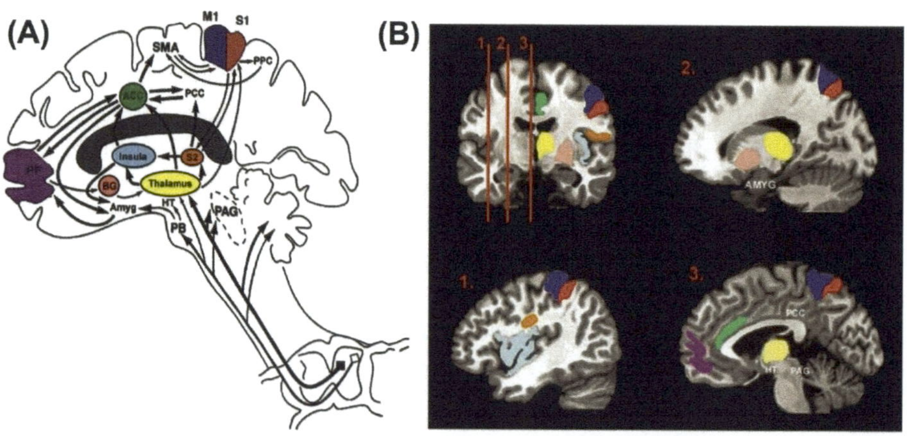

Figure 1 Pain induces activation of a multitude of cortical and subcortical brain regions. (A) Most commonly pain-activated regions, their interconnectivity, and afferent pathways. (B) Areas displayed on structural MRI image. *Reproduced with permission from Apkarian et al. (2005), © John Wiley and Sons.*

is capable of exerting a top-down control over pain (Schweinhardt and Bushnell, 2010). In more recent years, investigations of the neural correlates of pain have also dramatically changed our view of pain disorders, for instance, by demonstrating that chronic pain, far from being simply a disorder of the periphery, is associated with structural, functional, and neurochemical changes within the brain (Schweinhardt and Bushnell, 2010; Tracey and Bushnell, 2009). These observations have provided significant advances in our understanding of the pathophysiological mechanisms underlying pain conditions, and represent steps toward the identification of the much needed objective biomarkers for chronic pain.

2.1. PAIN PROCESSING IN THE BRAIN

The ascending pathways that relay nociceptive information from the periphery to the higher centers of the central nervous system originate in the nociceptive second-order projection neurons in the dorsal horn and project to the cortex mostly (although not exclusively) via the contralateral spinothalamic tract. These pathways are usually subdivided into two broad systems: the "lateral pain system", thought to be implicated in sensory-discriminative function (Hodge and Apkarian, 1990), which projects primarily to the lateral nuclei of the thalamus (e.g. ventroposterolateral and ventroposteromedial nuclei) and subsequently to S1 and S2 and insula; the "medial pain system", thought to primarily encode the emotional component of pain (Tracey and Mantyh, 2007), and which projects primarily to the medial and intralaminar thalamic nuclei (Hodge and Apkarian, 1990; Vogt et al., 1979) and then to medial regions such as the ACC (Sikes and Vogt, 1992; Vogt et al., 1993). Brain imaging studies have provided evidence in support of this functional dichotomy.

Recent MEG and EEG investigations suggest that the earliest brain responses to pain are located in the operculoinsular cortex (Frot and Mauguiere, 2003; Ploner et al., 1999b), as well as in S1 (Kanda et al., 2000). These observations suggest that such regions are involved in early sensory processing.

From a physiological point of view, both *S1 and S2*, as well as the *posterior insula*, exhibit several features expected to be present in structures that are involved in sensory-discriminative processing. One feature is that all these regions receive sensory input from all regions of the body and process and transmit the spatial localization of any incoming sensory stimulus according to a somatotopically organized map, i.e. a cortical representation of the body where adjacent body parts tend to be adjacent in the brain. Somatotopic representation in S1 has been found to be consistent with the traditional "sensory homunculus" identified by Penfield for tactile processing (Penfield and Rasmussen, 1955), i.e. a representation in which the foot is localized dorsomedially, the hand more ventrally (Ploner et al., 1999b; Tarkka and Treede, 1993), and the face closest to the lateral fissure (DaSilva et al., 2002). In the contralateral S2 and Insular Cortex (IC), the somatotopic organization is oriented rostrocaudally, with the face represented anterior to the foot/leg and forearm/hand (Brooks et al., 2005; Henderson et al., 2007; Vogel et al., 2003). More recently, studies have taken advantage of the relatively high spatial resolution of fMRI to investigate somatotopic representation of the pain responses within the operculoinsular cortex in even finer detail. For instance, a study revealed the existence of a within-limb somatotopy in the posterior insula, with the shoulder pain represented anterior to forearm pain and medial to hand pain (Henderson et al., 2010).

Another feature corroborating the role of S1, S2, and the posterior insula in sensory-discriminative processes is the encoding of pain intensity. These regions have been found to respond more strongly to stronger pain stimuli (Bornhovd et al., 2002; Coghill et al., 1999; Derbyshire et al., 1997; Kong et al., 2010; Loggia et al., 2012; Porro et al., 1998). This observation is compatible with electrophysiology studies, which have shown that a significant proportion of neurons in these cortical areas exhibit a progressive increase in discharge frequency as a function of stimulus intensity (Chudler et al., 1990; Kenshalo et al., 1988; Kenshalo and Isensee, 1983). While some primate studies have suggested that activity of S2 neurons reflects noxious stimulus intensity poorly (Dong et al., 1989), some neuroimaging studies have suggested that this region (Coghill et al., 1999; Loggia et al., 2012) does appear to encode stimulus intensity; other studies, however, have observed a sharp increase in amplitude only at stimulus intensities well above pain thresholds (Timmermann et al., 2001). Further support of the involvement of S1 and S2 in the sensory aspects of pain perception comes from a clinical study on a neuropsychological patient with selective ischemic lesions involving the right postcentral gyrus and the parietal operculum (Ploner et al., 1999a). When noxious laser stimuli were presented to the affected (left) hand, the patient did not report any pain sensation, but rather a "clearly unpleasant feeling" emerging from an ill-localized and extended area "somewhere between fingertips and shoulder". Thus, the lesion appeared to have impaired pain sensation but not pain affect, further corroborating the (at least partial) segregation of the processing of sensory and affective components of pain.

Although these regions are often activated in pain imaging studies, it is important to underline that none of them have been incontrovertibly shown to be "pain specific". An interesting observation, however, is that the posterior insula receives projections from the posterior portion of the recently identified ventral medial nucleus of the thalamus (VMpo) (Craig, 1995). The VMpo is a nucleus identified in both primates (Craig et al., 1994) and humans (Blomqvist et al., 2000) that receives dense, topographic projections from the lamina I spinothalamic fibers of the spinal cord and has been proposed to be specific to pain and temperature sensation (however, see Montes et al., 2005; Willis et al., 2002). Given the connectivity between the dorsal posterior insula and the VMpo, as well as imaging and other physiological observations, it has been proposed that "the dorsal posterior insula contains the primary cortical sensory representation of temperature and pain in humans" (Craig, 2011); however, this notion has been strongly debated.

While the posterior insula is thought to encode the primary sensory feature of pain stimuli, the *anterior insula* has been associated with the integration of the emotional and interoceptive states "in the representation of the subjective feeling of the moment" (Craig, 2009). Results from several imaging studies support this functional segregation of the insula along the rostrocaudal axis. For instance, studies have shown that while the posterior insula tends to be activated more strongly contralaterally (as expected from a region that encodes the sensory features of stimulation (Bingel et al., 2003; Brooks et al., 2002; Loggia et al., 2011; Peltz et al., 2011)), the mid and anterior insula tend to be bilaterally activated (Peltz et al., 2011). Furthermore, some authors have observed that the perception of a given heat pain stimulus induces the activation of both the posterior and anterior insula, while the subjective evaluation of that stimulus evokes activity only in the anterior insula (Kong et al., 2006b). Evidence from studies of anatomical connectivity also corroborate the theory that the posterior aspects of the insula are related mainly to sensory processing, including nociception, whereas the anterior insula is related predominantly to limbic, and visceroautonomic functions (Schnitzler and Ploner, 2000).

The vast number of conditions and behaviors with which the activation of the anterior insula has been associated (including awareness of sensations, attention, time perception, subjective expectations, maternal love, cigarette craving, etc.), together with the connectivity profile of the anterior insula, lead some to even propose that this region should be considered a potential neural correlate for awareness and consciousness (Craig, 2009).

Another region very commonly activated in pain studies, and very often coactivated with the anterior insula (Craig, 2009), is the ACC. The ACC receives projection from the midline and intralaminar thalamic nuclei (Vogt et al., 1979) and is therefore considered part of the "medial" pain system. The association of the ACC with emotional-motivational, rather than sensory-discriminative, aspects of pain is supported by lesional and imaging studies. For instance, patients who have undergone cingulotomy (i.e. surgical ablation of the cingulate gyrus) present attenuated emotional responses to pain (Foltz and White, 1968, 1962; Schnitzler and Ploner, 2000). Furthermore, hypnotically induced selective changes in pain unpleasantness have been shown to produce a selective modulation of ACC (but not S1) pain-induced activity, which correlated with subjective ratings of pain unpleasantness (Rainville et al., 1997). In order to examine whether the different cortical and subcortical areas activated during pain are involved in the processing of different components of the multidimensional nature of pain, Tölle and colleagues performed a regression analysis between pain-evoked regional blood flow increases and experimental pain parameters, including pain intensity and pain unpleasantness. The results of this analysis suggested that pain-evoked activation in the ACC is more closely related to affective components than to sensory components of pain (Tölle et al., 1999). As previously noted, no brain structures appear to be pain specific, a statement that is also valid for the ACC. In fact, the ACC is highly implicated in a number of different functions, including emotional processing and response selection (Vogt, 2005).

2.2. THE DESCENDING PAIN MODULATORY PATHWAY

The descending pain modulatory pathway is a highly organized network that enables top-down regulation of afferent nociceptive information (Figure 2).

It has been extensively described in animals and contributes to environmental analgesia (e.g. the pain relief observed during the fight-or-flight response) and opiate analgesia (Fields et al., 2006). Within this descending pathway, the hypothalamus and limbic forebrain structures, including the *amygdala*, ACC, and other frontal cortical areas, project to the *periaqueductal gray* (PAG), which in turn projects to the *rostral ventromedial medulla*

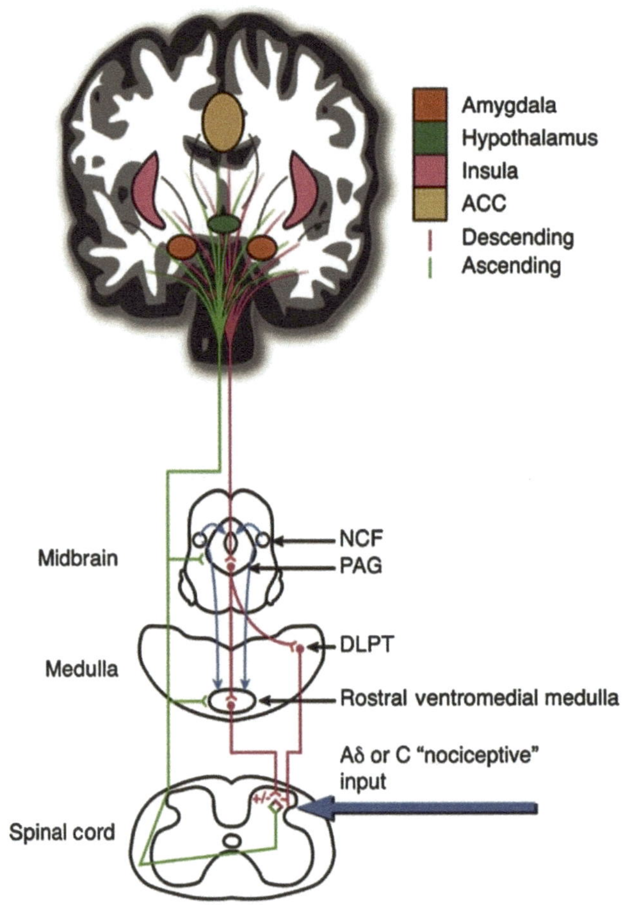

Figure 2 The descending pain modulatory system. NCF, nucleus cuneiformis; PAG, periaqueductal gray; DLPT, dorsolateral pontine tegmentum; ACC, anterior cingulate cortex. + indicates pronociceptive influences, whereas – indicates antinociceptive influences. *Reproduced with permission from Tracey and Mantyh (2007), © Cell Press.*

(RVM) and the *dorsolateral pontine tegmentum*. These two regions finally project to the dorsal horn laminae that house the nociceptive relay neurons, through the spinal cord dorsolateral funiculus. Stimulation studies have demonstrated that this pathway can significantly modulate the activity of nociceptive dorsal horn neurons. For example, stimulation of the PAG has been shown to reduce behavioral responses to noxious stimulation (Mayer and Liebeskind, 1974; Mayer et al., 1971; Reynolds, 1969) by inhibiting nociceptive dorsal horn neurons (Holden et al., 2002; Liebeskind et al., 1973). The involvement of this descending system in humans is supported by several imaging studies, particularly involving psychological modulation of pain (see Section 4). The existence of this network in humans has also been supported by structural connectivity experiments. A recent study (Hadjipavlou et al., 2006) used diffusion tractography to demonstrate that anatomical connections exist between the PAG, as well as the nucleus cuneiformis, and the prefrontal cortex, amygdala, thalamus, hypothalamus, and RVM, therefore providing support to the existence of an anatomical circuitry mediating the top-down influence on pain processing via brainstem structures in humans.

2.3. PSYCHOLOGICAL MODULATION OF PAIN

As mentioned in the introduction to this chapter, the physical properties of peripheral stimulation (e.g. its intensity) are not the only determinants of how intense or unpleasant the pain will be. In fact, animal and human experiments clearly indicate that pain can be significantly shaped by top-down factors, such as attentional focus, emotional state, psychosocial factors, etc. (Campbell et al., 2010; Loggia et al., 2008a–c; Rhudy et al., 2007; Villemure and Bushnell, 2002).

One psychological factor that has been shown to affect pain perception is the direction of attention (Villemure and Bushnell, 2002). Overall, studies suggest that when subjects direct their attention away from the noxious stimulation, they usually rate it as less painful than if they are attending to it. Although the neural mechanisms underlying the effects of attention on pain are not fully known, an increasing number of neuroimaging and neurophysiology studies suggest that various levels of the central nervous system might be involved. The reduction in pain ratings observed during distraction is associated with decreased activity in structures belonging to the thalamocortical ascending pain network (Apkarian et al., 2005), such as the thalamus, primary and secondary somatosensory cortices, insula, and ACC (Bantick et al., 2002; Brooks et al., 2002; Bushnell et al., 1999; Frankenstein et al., 2001; Hoffman et al., 2004; Longe et al., 2001; Seminowicz et al., 2004; Valet et al., 2004; Villemure and Bushnell, 2009; Wiech et al., 2005). For instance, Bushnell et al. (1999) used PET to measure the rCBF in subjects while they were simultaneously receiving pain and auditory stimuli. These authors observed that a significant pain-related increase in rCBF within S1 was detected when subjects directed their attention toward the thermal pain stimuli, but not when subjects directed their attention toward the auditory stimuli; this reduction in the rCBF was paralleled by a reduction in the pain ratings. Moreover, neurophysiological studies in monkeys have shown that nociceptive neurons in the medullary dorsal horn and the medial thalamus are less responsive to noxious stimuli when the monkey is attending to a distracter, compared to attending to the pain (Bushnell and Duncan, 1989; Bushnell et al., 1984). Distraction from pain is accompanied by not only reduced activity in some pain-processing structures but also increased activity in brain structures belonging to the descending pain modulatory pathway (Frankenstein et al., 2001; Tracey et al., 2002; Valet et al., 2004). For instance, using high-resolution echoplanar imaging at 4T, Tracey and colleagues observed that when subjects were instructed to distract from their pain, the activity of the PAG was significantly increased; furthermore, the level of PAG activity was predictive of the reductions in perceived pain intensity produced by distraction (Tracey et al., 2002).

Using connectivity analyses, Valet and colleagues showed that the cingulofrontal cortex (including the *orbitofrontal and perigenual* ACCs) exerts top-down influences on the PAG and posterior thalamus to gate pain modulation during distraction (Valet et al., 2004). Orbitofrontal and/or ACC activations during distraction from pain were also reported in two other studies (Bantick et al., 2002; Petrovic et al., 2000). Taken together, these studies suggest that distraction-related reductions in reported pain are associated with objective neurophysiological changes, some of which occur at early stages of sensory processing and therefore can not be explained in terms of report bias.

The neural bases of emotional modulation of pain in healthy individuals have been less extensively studied than those of attentional modulation. Studies aiming to understand how anticipation and anxiety induce an increased pain experience have implicated the *entorhinal cortex*, amygdala, anterior insula, and prefrontal cortices (Tracey and Mantyh, 2007). For instance, in an event-related fMRI study authors compared brain activity evoked by a mildly painful stimulus during a condition in which subjects were certain that they would receive only mild pain (low anxiety) and a condition in which they knew they might equally well receive an intense pain (high anxiety). Authors found that the same noxious stimulus was judged more painful, and evoked stronger activity in the entorhinal cortex, in the high- compared to low-anxiety condition. Furthermore, the time course of fMRI signal in the entorhinal cortex during conditions of low and high anxiety was significantly correlated with the time courses in the perigenual cingulate and midinsula, suggesting a possible source of pain modulation within the entorhinal cortex (Ploghaus et al., 2001). More direct evidence of the involvement of this descending pathway in emotional pain modulation comes from another fMRI experiment, which used pleasant and unpleasant odors to independently manipulate subjects' attention and mood (Villemure and Bushnell, 2009). In this study, the *lateral inferior frontal cortex* and the PAG were identified as possible emotion-related pain modulators, whereas the entorhinal and superior posterior parietal cortices were identified as possible attention-related pain modulators. The observation that in this study the PAG was implicated in emotional, but not attentional, modulation of pain suggests that the increased activity of this structure, which other authors have reported in association with distraction from pain (Tracey et al., 2002; Valet et al., 2004), might actually reflect changes in emotional state across conditions, rather than changes in the direction of attention per se. In fact, in one study, the distracting test used was the incongruent color-word Stroop task, which besides being distracting is also a stress task that can cause increased sympathetic autonomic arousal (Valet et al., 2004). The study by Villemure and Bushnell also shows that the reduction in pain unpleasantness, observed during experimentally induced positive mood, was associated with decreased pain-related activity within the ACC, medial thalamus, S1, S2, and midinsula, further confirming that the effect reflects a real modulation in pain perception and cannot be simply explained in terms of response bias. In a more recent study (Berna et al., 2010), authors have experimentally induced a negative mood in healthy subjects, a procedure known to elicit a preferential increase in the affective, rather than sensory, components of pain (Loggia et al., 2008b; Villemure and Bushnell, 2009; Villemure et al., 2003). By worsening the participants' mood, Berna and colleagues found increased activity in a broad network of brain areas in response to thermal pain, including the ACC, prefrontal areas, and hippocampus. Furthermore, participants who reported the largest increase in pain unpleasantness showed greater inferior frontal gyrus and amygdala activation. In conclusion, these studies suggest that regions such as the ACC, frontal cortex, and hippocampal formation are involved in emotional modulation of pain.

Another example of psychological modulation of pain comes from the placebo literature. It is known that psychosocial interventions, like placebo treatment (Kaptchuk et al., 2008) and expectancy manipulations (Colloca and Benedetti, 2006), can affect the subjective report of pain. For instance, when subjects are told that a "treatment" (in reality completely inert, such as a sugar pill) is a powerful analgesic, they often report a reduction in pain. As for other types of psychological pain modulation, however, it has long been debated whether such placebo-induced changes in reported pain truly reflect modulation of the pain experience or more simply a report bias. Brain imaging studies have been able to move beyond subjective reports and establish that placebo analgesia is associated with measurable brain changes (Faria et al., 2008; Kong et al., 2007). For example, Wager and colleagues demonstrated that placebo-induced pain relief is associated with a concomitant decrease in brain activity in pain-processing areas such as the thalamus and the insular cortex (Wager et al., 2004). In addition, many studies have found increased activation in the brain's descending pain inhibitory network during placebo analgesia, including prefrontal areas, rostral ACC, and brainstem structures involved in pain inhibition (Bingel et al., 2006; Eippert et al., 2009; Kong et al., 2006a, 2013; Petrovic et al., 2002; Wager et al., 2004). Common across these brain areas is their involvement in opioid-dependent pain inhibition (Fields, 2000), validating that placebo analgesia results in part from activation of endogenous opioids (Levine et al., 1978).

Recently, there has also been evidence pointing toward the involvement of the brain's reward circuitry during placebo analgesia (de la Fuente-Fernandez et al., 2001; Scott et al., 2009), suggesting that the release of dopamine in the ventral striatum may mediate the placebo response. Furthermore, more recent studies have suggested the involvement of additional neurotransmitter systems in the placebo effect, such as the endocannabinoid system (Benedetti et al., 2011).

In summary, placebo suggestions, as well as other types of cognitive or emotional manipulations, may induce the attenuation of the activity of nociceptive regions in the brain, possibly through the recruitment of brain regions implicated in opioid-, dopamine- and endocannabinoid-related modulation of pain. More in general, brain imaging studies on psychological modulation of pain have revealed that the human brain can exert a powerful top-down modulation on the processing of noxious stimuli. The investigation of the neural mechanisms underlying such modulation not only furthers our knowledge of the physiology of pain processing in the healthy brain but also allows us to achieve a better understanding of chronic pain, as an impairment of such top-down modulatory mechanisms has been proposed to be among the pathological mechanism underlying several pain conditions.

2.4. CHRONIC PAIN

2.4.1. Imaging the Brain in Chronic Pain

The ability to feel pain is essential for our survival as it functions as a warning system for threats to the organism (Treede et al., 1992). However, in chronic pain syndromes, pain persists even in the absence of acute tissue damage. The definition of chronic pain does not consider the different mechanisms that might lead to various forms of chronic pain. Instead, pain *duration* is considered when distinguishing acute pain from chronic pain (Merskey and Bogduk, 1994). According to definitions adopted internationally, pain associated with acute tissue damage, inflammation, or a disease process of short duration is referred to as acute pain (Turk and Melzack, 1992). This pain typically disappears when the tissue damage or inflammation has healed. When pain persists beyond the expected period of healing, usually taken to be 3–6 months (Merskey and Bogduk, 1994; Turk and Okifuji, 2010), it is referred to as chronic pain (Turk and Melzack, 1992). Chronic pain can accompany a long-term disease process (e.g. rheumatoid arthritis) or be associated with a localized peripheral pathology that does not resolve. Chronic pain can also be present despite the lack of any identifiable peripheral pathology, in which case it is referred to as idiopathic pain, in "functional pain syndromes" such as fibromyalgia.

In recent years, neuroimaging techniques have investigated the neural correlates of a number of different pain disorders, such as chronic back pain (Apkarian et al., 2004; Baliki et al., 2006, 2012; Derbyshire et al., 2002; Loggia et al., 2013; Wasan and Loggia et al., 2011), fibromyalgia (Gracely et al., 2002; Jensen et al., 2009, 2012a,b; Kuchinad et al., 2007; Schmidt-Wilcke et al., 2007), irritable bowel syndrome (Berman et al., 2008;

Davis et al., 2008; Naliboff et al., 2001), neuropathic pain (Cauda et al., 2009; Hsieh et al., 1995; Iadarola et al., 1995), vulvar vestibulitis (Pukall et al., 2005; Schweinhardt et al., 2008b), and facial pain (Ichesco et al., 2011; Niddam et al., 2007). Such studies have provided considerable evidence in support of the notion that chronic pain is associated with structural (Apkarian et al., 2004; Burgmer et al., 2009; Davis et al., 2008; Flor, 2003; Kuchinad et al., 2007; Ruscheweyh et al., 2011), functional (Giesecke et al., 2004; Gracely et al., 2002; Hsieh et al., 1995; Ichesco et al., 2011; Schweinhardt et al., 2008a), and neurochemical brain alterations (Harris et al., 2007, 2009). These observations have undoubtedly led to advances in our understanding of the pathophysiological mechanisms underlying pain conditions. Aside from revealing that chronic pain is associated with central alterations, imaging studies are also suggesting more specific hypotheses about the role of cortical and subcortical structures in the generation and maintenance of pathological pain. For instance, there is evidence for supraspinal mechanisms that may be responsible for the development of secondary hyperalgesia via pain facilitation and pronociceptive mechanisms (Zambreanu et al., 2005), thus indicating that at least some of the brain alterations observed in patients are not just a passive reflection of what happens in the periphery, but might have an active role in the pathology. It is now becoming more and more evident that the pathophysiology of chronic pain is probably due to a complex interaction between peripheral and central mechanisms. Thus, in the near future, neuroimaging methods might occupy a prominent role in the investigation of the biological substrates of chronic pain conditions.

2.4.2. Functional Imaging of Experimental Pain in Patients with Chronic Pain

Most functional neuroimaging studies in chronic pain patients have employed experimentally administered noxious stimuli to probe the pain system in disease (Apkarian, 2010). In many cases, researchers have chosen stimulation methods that are simply easy to administer, reliable, and minimally invasive or noninvasive stimuli, such as heat pain. In other cases, researchers have used a pain modality that attempts to approximate the clinical pain experienced by the population studied, in order to assess the underlying pathophysiology. For example, patients with Fibromyalgia (FM) can be investigated using pressure pain, due to their high level of tenderness to pressure (Wolfe et al., 1990); patients with Irritable Bowel Syndrome (IBS) can be subject to painful rectal distensions (Naliboff et al., 2001); and subjects with rheumatoid arthritis can undergo the squeezing of the affected joints (Schweinhardt et al., 2008a).

The evaluation of brain responses to experimental pain have suggested that the heightened sensitivity to noxious stimuli reported by chronic pain patients is effectively accompanied by enhanced brain responses (Giesecke et al., 2004; Gracely et al., 2002), and impaired functioning of the descending modulatory pain system (Berman et al., 2008; Jensen et al., 2009) (Figure 3). These studies therefore suggest that the augmentation of central processing of pain, disruption in the pain modulatory system, and/or an

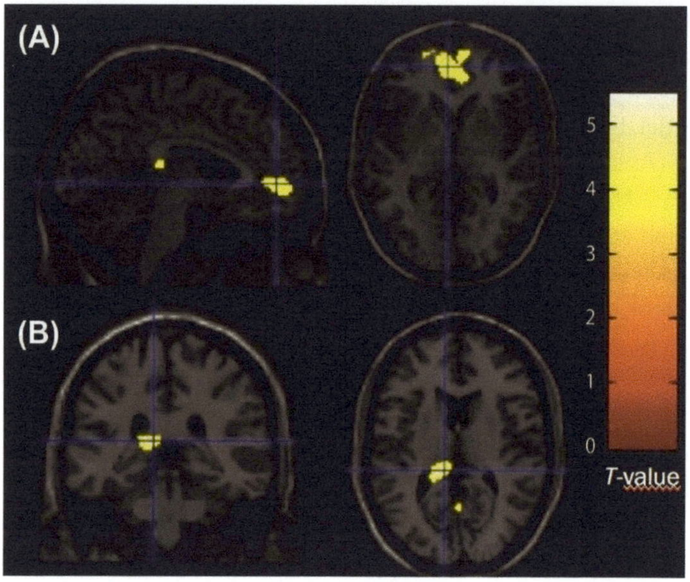

Figure 3 Regions of the brain where healthy controls have higher activity than FM patients during subjectively calibrated painful stimulation minus sensory stimulation. Clusters corresponding to (A) the rostral Anterior Cingulate Cortex (rACC) and (B) the pulvinar nucleus of thalamus. The exact anatomical locations (*x*, *y*, *z*) are given in Montreal Neurological Institute (MNI) coordinates. *Reproduced with permission from Jensen et al. (2009), © Elsevier.*

"overload" of the latter due to the ongoing chronic pain (Gracely et al., 2002) are all plausible mechanisms underlying the pathophysiology of chronic pain.

A study using real-time fMRI-based neurofeedback has shown that chronic pain patients trained to downregulate the activity of the ACC may derive clinical benefit (deCharms et al., 2005), which further suggests that hyperactivity of such region might contribute to the experience of chronic pain.

2.4.3. Functional Imaging of Clinical Pain in Patients with Chronic Pain

While the majority of pain imaging studies in patients focus on brain responses to externally administered pain stimuli (e.g. heat, electric shock, etc.), few studies have attempted to evaluate the neural correlates of clinical pain. In spite of the progress achieved thus far, the identification of neural measures underlying the perceptual experience of chronic pain itself presents methodological hurdles. Unlike experimental pain, chronic pain is difficult to elicit in a controlled manner (Apkarian et al., 2001), which makes it challenging to probe using the classical "two-state subtraction" designs (i.e. using block- and event-related states), particularly with Blood-oxygen-level dependent (BOLD) imaging. Recent developments in the field of neuroimaging, however, might allow us to bypass these methodological issues.

For instance, novel BOLD fMRI approaches have been developed in order to measure spontaneous pain in patients with chronic pain (Apkarian et al., 2001). Studies using these techniques have been performed in Chronic Back Pain (CBP) (Apkarian et al., 2009; Baliki et al., 2006) and in patients with knee osteoarthritis (Parks et al., 2011). The method of measuring spontaneous pain involves the continuous recording of ratings of ongoing pain during fMRI scanning. The pain ratings are convolved with the hemodynamic response and then incorporated in the analyses as regressors, such that the recorded pattern of increased and decreased pain observed over time serve as an individual "pain paradigm" for each patient. A study in CBP demonstrated that periods of sustained pain were associated with increased activation of the medial prefrontal cortex (mPFC), with less activity seen in the amygdala and the ventral striatum (Baliki et al., 2006). For periods when pain was increasing, however, a network of regions more commonly seen as active during acute pain, including the insula and ACC, became activated.

Arterial spin labeling (ASL) is a relatively novel perfusion-based fMRI technique (Aguirre et al., 2005) which presents several features that are particularly attractive for the study of chronic pain. First, while the BOLD signal exhibits increasing noise at low frequencies, the power spectrum of ASL is flat, which indicates that the detection of brain changes in ASL is not limited to relatively high frequency signals. This is important because the fluctuations in the symptoms of chronic pain patients are usually not of the order of seconds, but rather minutes or hours. Second, ASL allows the absolute quantification of rCBF in physiological units (ml/100 g/min), which allows for the direct comparisons of baselines, for example, without the need for a "contrast" with an experimental condition. ASL has only recently been applied to the study of pain in healthy volunteers and has been proved to detect brain responses to stimuli of a duration for which BOLD imaging would be suboptimal (e.g. 15-min hypertonic saline muscular infusions or 1-min heat pain stimuli (Owen et al., 2008, 2011, 2010)). Another study using ASL was also able to observe increases in rCBF in a model of postsurgical pain (Howard et al., 2011).

The first experiment applying ASL to a chronic pain condition was performed on chronic back pain patients by Wasan and Loggia et al. (2011). The authors scanned patients with cLBP and radicular pain as well as healthy pain-free subjects in two separate visits. During each visit, subjects received an ASL scan before and after a period of stimulation designed to either exacerbate the endogenous (clinical) low back pain in patients or induce exogenous heat pain in both patients and healthy controls. During the "clinical maneuvers" visit, the subjects underwent a series of clinical maneuvers such as straight leg raising or pelvic tilt, which, although painless in HCs, were painful for the patients. At the end of the stimulation period, and by the time of the second ASL scan, the patients reported a clinically relevant (>30%) increase in clinical pain. During the "heat pain" visit, subjects received calibrated heat pain stimuli on the leg, over the affected dermatome for cLBP and over an identical location in the matched controls. The heat pain stimuli were calibrated to approximate the pain intensity evoked by the clinical maneuvers in the patients. While the heat stimuli were themselves painful, they did not induce a clinically significant increase in clinical pain by the end of the stimulation period. The authors observed an increase in rCBF (of 6–10 ml/100 g of tissue/min) in a widespread network of regions including S1, S2, the anterior insula, and mPFC in patients after the clinical maneuvers; however, these results were neither observed in patients after heat pain nor in the controls undergoing similar clinical maneuvers (Figure 4).

The applicability of ASL in low-frequency paradigms as well the possibility of performing absolute quantification of the rCBF represent steps ahead toward the identification of the much needed biomarkers for chronic pain.

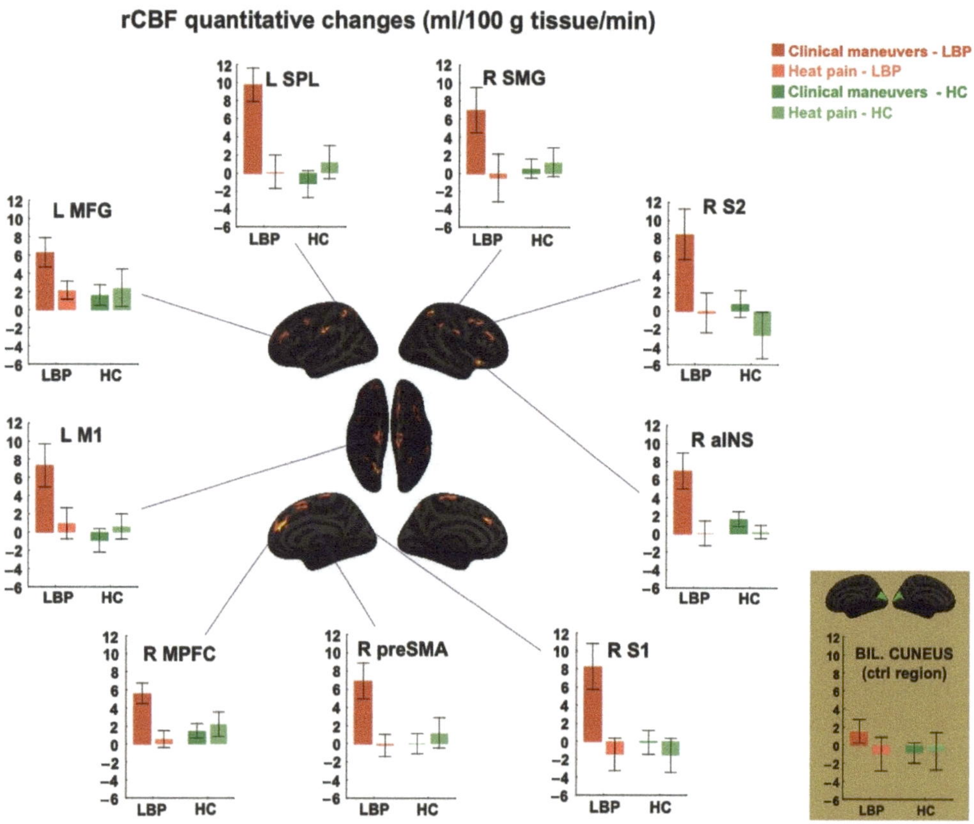

Figure 4 Clinical pain-induced increases in rCBF in cLBP patients, as assessed by Arterial Spin Labeling (ASL). Clinically significant worsening of ongoing chronic pain (≥30%) is associated with significant rCBF increases within a widespread network of brain regions. This effect was specific to changes in ongoing pain as it was observed during a (painful) resting period following acute exacerbation of endogenous cLBP with clinical maneuvers, but neither during the (painless) resting period following exogenous heat pain in the same patients nor in matched, pain-free healthy controls undergoing both procedures in identical fashion. The mean changes in rCBF (±Standard Error of the Mean) ±SEM across all experimental groups and conditions are shown for several regions. *Reproduced with permission from Wasan and Loggia, et al. (2011), © Lippincott, Williams and Wilkins.*

2.4.4. Imaging Brain Connectivity in Patients with Chronic Pain

A large number of studies have demonstrated that the presence of chronic pain also affects the functional connectivity among different brain regions. For instance, recent studies suggest that chronic pain affects the intrinsic brain activity of the Default Mode Network (DMN). The DMN is a network of brain areas, including the medial prefrontal, posterior cingulate, lateral temporal cortices, hippocampal formation, and inferior parietal lobules, which are frequently observed as deactivated during attention-demanding tasks and have extensive overlap with regions that are thought to be involved in self-referential cognitive processing (Buckner et al., 2008; Raichle et al., 2001; Shulman et al., 1997). Studies have shown alterations in DMN connectivity in patients with CBP (Baliki et al., 2008; Loggia et al., 2013), neuropathic pain (Cauda et al., 2009), and FM (Napadow et al., 2010). For instance, Napadow and colleagues demonstrated that FM patients exhibited greater connectivity between DMN and insular cortex and S2 and that the strength of connectivity between DMN and insula predicted spontaneous pain (Napadow et al., 2010). Interestingly, a very similar association between DMN–insula connectivity and clinical pain has also been found in patients with chronic low back pain, studied using ASL imaging (Loggia et al., 2013). The observation that a common neuroimaging metric appears to encode clinical pain in different patient populations raises the intriguing possibility that such measures may reflect a general feature of chronic pain.

2.4.5. Structural Changes in the Brain of Chronic Pain Patients

In addition to changes in the brain's activity patterns during pain and resting state, there is evidence for a linear relationship between structural brain changes and the duration of chronic pain (Kuchinad et al., 2007). Results from various studies indicate that long-term exposure to pain might cause decreases in gray matter in a number of brain regions (May, 2008), as demonstrated in patients with CBP (Apkarian et al., 2004), FM (Burgmer et al., 2009; Jensen et al. 2013; Kuchinad et al., 2007; Robinson et al., 2011; Schmidt-Wilcke et al., 2007), rheumatoid arthritis (Wartolowska et al., 2011), menstrual pain (Tu et al., 2010), vulvar pain (Schweinhardt et al., 2008b), and many other chronic pain conditions. While many studies suggest the presence of structural alterations, the specific brain regions associated with such changes, and even the direction of these changes, vary between studies. For instance, Apkarian and colleagues, who performed the first published study reporting structural alterations in the brain of a chronic pain condition, observed a reduction in gray matter density in the prefrontal cortex and thalamus in a cohort of patients with chronic back pain (Apkarian et al., 2004). Some studies, however, also report gray matter increases, for example, in the basal ganglia and the hippocampal formation (Schmidt-Wilcke et al., 2007; Schweinhardt et al., 2008b).

Interestingly, a very recent cortical thickness study (Seminowicz et al., 2011) demonstrated that successful treatment in patients with CBP may reverse not only abnormal brain function but also abnormal brain structure: after treatment, patients had increased cortical thickness in the left dorsolateral prefrontal cortex (DLPFC), which was thinner before treatment compared to controls. This observation suggests that brain plasticity in response to pain is bidirectional.

2.4.6. Imaging the Neurochemistry of Chronic Pain

Another promising line of research in the study of neural correlates of chronic pain is represented by the use of magnetic resonance spectroscopy (MRS). MRS allows noninvasive and in vivo exploration of the molecular composition of tissue, by identifying certain metabolites involved in physiological or pathological processes. By using techniques such as single voxel spectroscopy or chemical shift imaging, researchers were able to reveal the presence of biochemical alterations in the brain of chronic pain patients.

The most consistent observation across pain conditions appears to be the reduced concentration of *N*-acetylaspartate (NAA), which is usually interpreted as evidence of neuronal damage or loss. NAA reductions were observed in cLBP patients in the DLPFC, orbitofrontal cortex, insula, ACC, thalamus, and S1 (Grachev et al., 2000, 2002a; Gussew et al., 2011; Sharma et al., 2011; Siddall et al., 2006); in complex regional pain syndrome in the DLPFC (Grachev et al., 2002b); in FM patients in the hippocampal formation (Emad et al., 2008; Fayed et al., 2010; Wood et al., 2009); in neuropathic pain in the thalamus (Fukui et al., 2006; Pattany et al., 2002; Sorensen et al., 2008); and in migraineurs in the cerebellum (Dichgans et al., 2005).

Other studies have implicated changes in glutamate (Glu) or Glu/glutamine (glx) concentration. Glu or glx increases were observed in FM patients in the insula, posterior cingulate cortex, and amygdala (Fayed et al., 2010; Harris et al., 2009; Valdes et al., 2010), whereas decreases were observed in migraineurs in the cerebellum (Dichgans et al., 2005) and in cLBP patients in the ACC (Gussew et al., 2011).

Another metabolite often investigated is myoinositol, which is interpreted by some as a glial marker, although its specificity is strongly debated considering that it is involved in a large variety of functions including osmoregulation, glucose storage, and mobilization of cellular messaging systems. Myoinositol was found to be reduced in cLBP patients in the ACC and thalamus (Gussew et al., 2011); in FM patients, it was found to be decreased in the hippocampal formation (Fayed et al., 2010), but increased in the amygdala and thalamus, particularly in patients with more pain, fatigue, and depression (Valdes et al., 2010). Finally, recent advances in MRS pulse sequences and data preprocessing have opened the possibility of investigating metabolites previously difficult to isolate, such as γ-aminobutyric acid. This inhibitory neurotransmitter has been found to be reduced in the insula in fibromyalgia patients (Foerster et al., 2012), suggesting another possible mechanism underlying the pathophysiology of this condition.

In summary, a multitude of MRS studies have suggested that chronic pain is accompanied by neurochemical alterations, involving a variety of metabolites and across multiple brain regions. Future investigations will need to assess whether these changes represent a consequence of the long-term exposure to pain or might have a causal role in the development and maintenance of chronic pain.

3. CONCLUSIONS

Modern, noninvasive brain imaging techniques have revolutionized our understanding of pain, both in health and disease. Imaging studies are shedding new light into the brain correlates of the pain experience and their modulation by a variety of factors (e.g. emotional or cognitive state, psychosocial factors, genetic makeup, pharmacological or behavioral interventions, etc.). These studies are also increasing our knowledge of chronic pain, revealing that the brains of patients are characterized by structural, functional, and neurochemical alterations, some of which might have causal roles in the pathophysiology of pain conditions. While we are certainly still very far from having achieved a complete understanding of the neural substrates of pain, the progress observed in the past decades is undeniable. Furthermore, the rapid technical and methodological evolution we have glimpsed just in the past few years is very promising and suggests that in the years to come imaging of the human brain will play a role which might become more and more central in the quest for objective biomarkers of chronic pain. Further investigations of cerebral pain processing, in both patients and controls, are needed to increase our understanding of the complex dynamics that create the subjective experience of pain. Ultimately, an increased knowledge of the neural correlates of pain is likely to be translated into the clinic someday, leading to better treatments of chronic pain conditions.

ACKNOWLEDGMENTS

The authors would like to thank Rosa Spaeth for her helpful editorial suggestions.

REFERENCES

Aguirre, G.K., Detre, J.A., Wang, J., 2005. Perfusion fMRI for functional neuroimaging. Int. Rev. Neurobiol. 66, 213–236.

Albe-Fessard, D., Berkley, K.J., Kruger, L., Ralston 3rd, H.J., Willis Jr., W.D., 1985. Diencephalic mechanisms of pain sensation. Brain Res. 356, 217–296.

Apkarian, A.V., 2010. Human brain imaging studies of chronic pain. In: Kruger, L., Light, A. (Eds.), Translational Pain Research: From Mouse to Man. CRC Press, Boca Raton, FL.

Apkarian, A.V., Baliki, M.N., Geha, P.Y., 2009. Towards a theory of chronic pain. Prog. Neurobiol. 87, 81–97.

Apkarian, A.V., Bushnell, M.C., Treede, R.D., Zubieta, J.K., 2005. Human brain mechanisms of pain perception and regulation in health and disease. Eur. J. Pain 9, 463–484.

Apkarian, A.V., Krauss, B.R., Fredrickson, B.E., Szeverenyi, N.M., 2001. Imaging the pain of low back pain: functional magnetic resonance imaging in combination with monitoring subjective pain perception allows the study of clinical pain states. Neurosci. Lett. 299, 57–60.

Apkarian, A.V., Sosa, Y., Sonty, S., Levy, R.M., Harden, R.N., Parrish, T.B., Gitelman, D.R., 2004. Chronic back pain is associated with decreased prefrontal and thalamic gray matter density. J. Neurosci. 24, 10410–10415.

Apkarian, A.V., Stea, R.A., Manglos, S.H., Szeverenyi, N.M., King, R.B., Thomas, F.D., 1992. Persistent pain inhibits contralateral somatosensory cortical activity in humans. Neurosci. Lett. 140, 141–147.

Baliki, M.N., Chialvo, D.R., Geha, P.Y., Levy, R.M., Harden, R.N., Parrish, T.B., Apkarian, A.V., 2006. Chronic pain and the emotional brain: specific brain activity associated with spontaneous fluctuations of intensity of chronic back pain. J. Neurosci. 26, 12165–12173.

Baliki, M.N., Geha, P.Y., Apkarian, A.V., Chialvo, D.R., 2008. Beyond feeling: chronic pain hurts the brain, disrupting the default-mode network dynamics. J. Neurosci. 28, 1398–1403.

Baliki, M.N., Petre, B., Torbey, S., Herrmann, K.M., Huang, L., Schnitzer, T.J., Fields, H.L., Apkarian, A.V., 2012. Corticostriatal functional connectivity predicts transition to chronic back pain. Nat. Neurosci. 15, 1117–1119.

Bantick, S.J., Wise, R.G., Ploghaus, A., Clare, S., Smith, S.M., Tracey, I., 2002. Imaging how attention modulates pain in humans using functional MRI. Brain 125, 310–319.

Benedetti, F., Amanzio, M., Rosato, R., Blanchard, C., 2011. Nonopioid placebo analgesia is mediated by CB_1 cannabinoid receptors. Nat. Med. 17, 1228–1230.

Berman, S.M., Naliboff, B.D., Suyenobu, B., Labus, J.S., Stains, J., Ohning, G., Kilpatrick, L., Bueller, J.A., Ruby, K., Jarcho, J., Mayer, E.A., 2008. Reduced brainstem inhibition during anticipated pelvic visceral pain correlates with enhanced brain response to the visceral stimulus in women with irritable bowel syndrome. J. Neurosci. 28, 349–359.

Berna, C., Leknes, S., Holmes, E.A., Edwards, R.R., Goodwin, G.M., Tracey, I., 2010. Induction of depressed mood disrupts emotion regulation neurocircuitry and enhances pain unpleasantness. Biol. Psychiatry 67, 1083–1090.

Bingel, U., Lorenz, J., Schoell, E., Weiller, C., Büchel, C., 2006. Mechanisms of placebo analgesia: rACC recruitment of a subcortical antinociceptive network. Pain 120, 8–15.

Bingel, U., Quante, M., Knab, R., Bromm, B., Weiller, C., Buchel, C., 2003. Single trial fMRI reveals significant contralateral bias in responses to laser pain within thalamus and somatosensory cortices. NeuroImage 18, 740–748.

Blomqvist, A., Zhang, E.T., Craig, A.D., 2000. Cytoarchitectonic and immunohistochemical characterization of a specific pain and temperature relay, the posterior portion of the ventral medial nucleus, in the human thalamus. Brain 123 (Pt 3), 601–619.

Bornhovd, K., Quante, M., Glauche, V., Bromm, B., Weiller, C., Buchel, C., 2002. Painful stimuli evoke different stimulus-response functions in the amygdala, prefrontal, insula and somatosensory cortex: a single-trial fMRI study. Brain 125, 1326–1336.

Borsook, D., Sava, S., Becerra, L., 2010. The pain imaging revolution: advancing pain into the 21st century. Neuroscientist 16, 171–185.

Brooks, J.C., Nurmikko, T.J., Bimson, W.E., Singh, K.D., Roberts, N., 2002. fMRI of thermal pain: effects of stimulus laterality and attention. NeuroImage 15, 293–301.

Brooks, J.C., Zambreanu, L., Godinez, A., Craig, A.D., Tracey, I., 2005. Somatotopic organisation of the human insula to painful heat studied with high resolution functional imaging. NeuroImage 27, 201–209.

Buckner, R.L., Andrews-Hanna, J.R., Schacter, D.L., 2008. The brain's default network: anatomy, function, and relevance to disease. Ann. N.Y. Acad. Sci. 1124, 1–38.

Burgmer, M., Gaubitz, M., Konrad, C., Wrenger, M., Hilgart, S., Heuft, G., Pfleiderer, B., 2009. Decreased gray matter volumes in the cingulo-frontal cortex and the amygdala in patients with fibromyalgia. Psychosom. Med. 71, 566–573.

Bushnell, M.C., Duncan, G.H., 1989. Sensory and affective aspects of pain perception: is medial thalamus restricted to emotional issues? Exp. Brain Res. 78, 415–418.

Bushnell, M.C., Duncan, G.H., Dubner, R., He, L.F., 1984. Activity of trigeminothalamic neurons in medullary dorsal horn of awake monkeys trained in a thermal discrimination task. J. Neurophysiol. 52, 170–187.

Bushnell, M.C., Duncan, G.H., Hofbauer, R.K., Ha, B., Chen, J.I., Carrier, B., 1999. Pain perception: is there a role for primary somatosensory cortex? Proc. Natl. Acad. Sci. USA 96, 7705–7709.

Campbell, C.M., Witmer, K., Simango, M., Carteret, A., Loggia, M.L., Campbell, J.N., Haythornthwaite, J.A., Edwards, R.R., 2010. Catastrophizing delays the analgesic effect of distraction. Pain 149, 202–207.
Cauda, F., Sacco, K., Duca, S., Cocito, D., D'Agata, F., Geminiani, G., Canavero, S., 2009. Altered resting state in diabetic neuropathic pain. PLoS One 4, 4542.
Chudler, E.H., Anton, F., Dubner, R., Kenshalo Jr., D.R., 1990. Responses of nociceptive SI neurons in monkeys and pain sensation in humans elicited by noxious thermal stimulation: effect of interstimulus interval. J. Neurophysiol. 63, 559–569.
Coghill, R.C., Sang, C.N., Maisog, J.M., Iadarola, M.J., 1999. Pain intensity processing within the human brain: a bilateral, distributed mechanism. J. Neurophysiol. 82, 1934–1943.
Colloca, L., Benedetti, F., 2006. How prior experience shapes placebo analgesia. Pain 124, 126–133.
Craig, A.D., 1995. Supraspinal projections of lamina I neurons. In: Besson, J.M., Guilbaud, G., Ollat, H. (Eds.), Forebrain Areas Involved in Pain Processing. John Libbey Eurotext, Paris.
Craig, A.D., 2009. How do you feel–now? The anterior insula and human awareness. Nat. Rev. Neurosci. 10, 59–70.
Craig, A.D., 2011. Interoceptive cortex in the posterior insula: comment on Garcia-Larrea et al. 2010 Brain 133, 2528. Brain 134 e166; author reply e165.
Craig, A.D., Bushnell, M.C., Zhang, E.T., Blomqvist, A., 1994. A thalamic nucleus specific for pain and temperature sensation. Nature 372, 770–773.
DaSilva, A.F., Becerra, L., Makris, N., Strassman, A.M., Gonzalez, R.G., Geatrakis, N., Borsook, D., 2002. Somatotopic activation in the human trigeminal pain pathway. J. Neurosci. 22, 8183–8192.
Davis, K.D., Pope, G., Chen, J., Kwan, C.L., Crawley, A.P., Diamant, N.E., 2008. Cortical thinning in IBS: implications for homeostatic, attention, and pain processing. Neurology 70, 153–154.
de la Fuente-Fernandez, R., Ruth, T., Sossi, V., Schulzer, M., Calne, D., Stoessl, A., 2001. Expectation and dopamine release: mechanism of the placebo effect in Parkinson's disease. Science 293, 1164–1166.
deCharms, R.C., Maeda, F., Glover, G.H., Ludlow, D., Pauly, J.M., Soneji, D., Gabrieli, J.D., Mackey, S.C., 2005. Control over brain activation and pain learned by using real-time functional MRI. Proc. Natl. Acad. Sci. USA 102, 18626–18631.
Derbyshire, S.W., Jones, A.K., Creed, F., Starz, T., Meltzer, C.C., Townsend, D.W., Peterson, A.M., Firestone, L., 2002. Cerebral responses to noxious thermal stimulation in chronic low back pain patients and normal controls. NeuroImage 16, 158–168.
Derbyshire, S.W., Jones, A.K., Gyulai, F., Clark, S., Townsend, D., Firestone, L.L., 1997. Pain processing during three levels of noxious stimulation produces differential patterns of central activity. Pain 73, 431–445.
Dichgans, M., Herzog, J., Freilinger, T., Wilke, M., Auer, D.P., 2005. ^{1}H-MRS alterations in the cerebellum of patients with familial hemiplegic migraine type 1. Neurology 64, 608–613.
Dong, W.K., Salonen, L.D., Kawakami, Y., Shiwaku, T., Kaukoranta, E.M., Martin, R.F., 1989. Nociceptive responses of trigeminal neurons in SII-7b cortex of awake monkeys. Brain Res. 484, 314–324.
Eippert, F., Bingel, U., Schoell, E.D., Yacubian, J., Klinger, R., Lorenz, J., Büchel, C., 2009. Activation of the opioidergic descending pain control system underlies placebo analgesia. Neuron 63, 533–543.
Emad, Y., Ragab, Y., Zeinhom, F., El-Khouly, G., Abou-Zeid, A., Rasker, J.J., 2008. Hippocampus dysfunction may explain symptoms of fibromyalgia syndrome. A study with single-voxel magnetic resonance spectroscopy. J. Rheumatol. 35, 1371–1377.
Faria, V., Fredrikson, M., Furmark, T., 2008. Imaging the placebo response: A neurofunctional review. Eur. Neuropsychopharmacol. 18, 473–485.
Fayed, N., Garcia-Campayo, J., Magallon, R., Andres-Bergareche, H., Luciano, J.V., Andres, E., Beltran, J., 2010. Localized ^{1}H-NMR spectroscopy in patients with fibromyalgia: a controlled study of changes in cerebral glutamate/glutamine, inositol, choline, and *N*-acetylaspartate. Arthritis Res. Ther. 12, R134.
Fields, H.L., Basbaum, A.I., Heinricher, M.M., 2006. Central nervous system mechanisms of pain modulation. In: McMahon, S.B., Koltzenburg, M. (Eds.), Wall and Melzack's Textbook of Pain. Elsevier – Churchill Livingstone, London.
Flor, H., 2003. Cortical reorganisation and chronic pain: implications for rehabilitation. J. Rehabil. Med. 41, 66–72.

Foerster, B.R., Petrou, M., Edden, R.A., Sundgren, P.C., Schmidt-Wilcke, T., Lowe, S.E., Harte, S.E., Clauw, D.J., Harris, R.E., 2012. Reduced insular γ-aminobutyric acid in fibromyalgia. Arthritis Rheum. 64, 579–583.

Foltz, E.L., White, L.E., 1968. The role of rostral cingulumotomy in "pain" relief. Int. J. Neurol. 6, 353–373.

Foltz, E.L., White Jr., L.E., 1962. Pain "relief" by frontal cingulumotomy. J. Neurosurg. 19, 89–100.

Frankenstein, U.N., Richter, W., McIntyre, M.C., Remy, F., 2001. Distraction modulates anterior cingulate gyrus activations during the cold pressor test. NeuroImage 14, 827–836.

Frot, M., Mauguiere, F., 2003. Dual representation of pain in the operculo-insular cortex in humans. Brain 126, 438–450.

Fukui, S., Matsuno, M., Inubushi, T., Nosaka, S., 2006. *N*-acetylaspartate concentrations in the thalami of neuropathic pain patients and healthy comparison subjects measured with ^{1}H-MRS. Magn. Reson. Imaging 24, 75–79.

Giesecke, T., Gracely, R.H., Grant, M.A., Nachemson, A., Petzke, F., Williams, D.A., Clauw, D.J., 2004. Evidence of augmented central pain processing in idiopathic chronic low back pain. Arthritis Rheum. 50, 613–623.

Gracely, R.H., Petzke, F., Wolf, M., Clauw, D., 2002. Functional magnetic resonance imaging evidence of augmented pain processing in fibromyalgia. Arthritis Rheum. 46, 1333–1343.

Grachev, I.D., Fredrickson, B.E., Apkarian, A.V., 2000. Abnormal brain chemistry in chronic back pain: an in vivo proton magnetic resonance spectroscopy study. Pain 89, 7–18.

Grachev, I.D., Fredrickson, B.E., Apkarian, A.V., 2002a. Brain chemistry reflects dual states of pain and anxiety in chronic low back pain. J. Neural Transm. 109, 1309–1334.

Grachev, I.D., Thomas, P.S., Ramachandran, T.S., 2002b. Decreased levels of *N*-acetylaspartate in dorsolateral prefrontal cortex in a case of intractable severe sympathetically mediated chronic pain (complex regional pain syndrome, type I). Brain Cogn. 49, 102–113.

Gussew, A., Rzanny, R., Gullmar, D., Scholle, H.C., Reichenbach, J.R., 2011. ^{1}H-MR spectroscopic detection of metabolic changes in pain processing brain regions in the presence of non-specific chronic low back pain. NeuroImage 54, 1315–1323.

Hadjipavlou, G., Dunckley, P., Behrens, T.E., Tracey, I., 2006. Determining anatomical connectivities between cortical and brainstem pain processing regions in humans: a diffusion tensor imaging study in healthy controls. Pain 123, 169–178.

Harris, R.E., Clauw, D.J., Scott, D.J., McLean, S.A., Gracely, R.H., Zubieta, J.K., 2007. Decreased central μ-opioid receptor availability in fibromyalgia. J. Neurosci. 27, 10000–10006.

Harris, R.E., Sundgren, P.C., Craig, A.D., Kirshenbaum, E., Sen, A., Napadow, V., Clauw, D.J., 2009. Elevated insular glutamate in fibromyalgia is associated with experimental pain. Arthritis Rheum. 60, 3146–3152.

Head, H., Holmes, G., 1911. Sensory disturbances from cerebral lesions. Brain 34, 102–254.

Henderson, L.A., Gandevia, S.C., Macefield, V.G., 2007. Somatotopic organization of the processing of muscle and cutaneous pain in the left and right insula cortex: a single-trial fMRI study. Pain 128, 20–30.

Henderson, L.A., Rubin, T.K., Macefield, V.G., 2010. Within-limb somatotopic representation of acute muscle pain in the human contralateral dorsal posterior insula. Hum. Brain Mapp.

Hodge Jr., C.J., Apkarian, A.V., 1990. The spinothalamic tract. Crit. Rev. Neurobiol. 5, 363–397.

Hofbauer, R.K., Rainville, P., Duncan, G.H., Bushnell, M.C., 2001. Cortical representation of the sensory dimension of pain. J. Neurophysiol. 86, 402–411.

Hoffman, H.G., Richards, T.L., Coda, B., Bills, A.R., Blough, D., Richards, A.L., Sharar, S.R., 2004. Modulation of thermal pain-related brain activity with virtual reality: evidence from fMRI. Neuroreport 15, 1245–1248.

Holden, J.E., Van Poppel, A.Y., Thomas, S., 2002. Antinociception from lateral hypothalamic stimulation may be mediated by NK_1 receptors in the A7 catecholamine cell group in rat. Brain Res. 953, 195–204.

Howard, M.A., Krause, K., Khawaja, N., Massat, N., Zelaya, F., Schumann, G., Huggins, J.P., Vennart, W., Williams, S.C., Renton, T.F., 2011. Beyond patient reported pain: perfusion magnetic resonance imaging demonstrates reproducible cerebral representation of ongoing post-surgical pain. PLoS One 6, e17096.

Hsieh, J.-C., Belfrage, M., Stone-Elander, S., Hansson, P., Ingvar, M., 1995. Central representation of chronic ongoing neuropathic pain studied by positron emission tomography. Pain 63, 225–236.

Iadarola, M.J., Max, M.B., Berman, K.F., Byas-Smith, M.G., Coghill, R.C., Gracely, R.H., Bennett, G.J., 1995. Unilateral decrease in thalamic activity observed with positron emission tomography in patients with chronic neuropathic pain. Pain 63, 55–64.

Iannetti, G.D., Mouraux, A., 2010. From the neuromatrix to the pain matrix (and back). Exp. Brain Res. 205, 1–12.

Ichesco, E., Quintero, A., Clauw, D.J., Peltier, S., Sundgren, P.M., Gerstner, G.E., Schmidt-Wilcke, T., 2011. Altered functional connectivity between the insula and the cingulate cortex in patients with temporomandibular disorder: a pilot study. Headache.

Ingvar, M., 1999. Pain and functional imaging. Philos. Trans. R. Soc. Lond. B Biol. Sci. 354, 1347–1358.

Jensen, K.B., Kosek, E., Petzke, F., Carville, S., Fransson, P., Marcus, H., Williams, S.C.R., Choy, E., Giesecke, T., Mainguy, Y., Gracely, R., Ingvar, M., 2009. Evidence of dysfunctional pain inhibition in Fibromyalgia reflected in rACC during provoked pain. Pain 144, 95–100.

Jensen, K.B., Srinivasan, P., Spaeth, R., Tan, Y., Kosek, E., Petzke, F., Carville, S., Fransson, P., Marcus, H., Williams, S.C., Choy, E., Vitton, O., Gracely, R., Ingvar, M., Kong, J., 2013. Overlapping structural and functional brain changes in patients with long-term exposure to fibromyalgia. Arthritis Rheum. Aug 27 http://dx.doi.org/10.1002/art.38170 [Epub ahead of print].

Jensen, K.B., Kosek, E., Wicksell, R., Kemani, M., Olsson, G., Merle, J.V., Kadetoff, D., Ingvar, M., 2012a. Cognitive Behavioral Therapy increases pain-evoked activation of the prefrontal cortex in patients with fibromyalgia. Pain 153, 1495–1503.

Jensen, K.B., Loitoile, R., Kosek, E., Petzke, F., Carville, S., Fransson, P., Marcus, H., Williams, S.C., Choy, E., Mainguy, Y., Vitton, O., Gracely, R.H., Gollub, R., Ingvar, M., Kong, J., 2012b. Patients with fibromyalgia display less functional connectivity in the brain's pain inhibitory network. Mol. Pain 8, 32.

Jones, A.K., Brown, W.D., Friston, K.J., Qi, L.Y., Frackowiak, R.S., 1991. Cortical and subcortical localization of response to pain in man using positron emission tomography. Proc. Biol. Sci. 244, 39–44.

Kanda, M., Nagamine, T., Ikeda, A., Ohara, S., Kunieda, T., Fujiwara, N., Yazawa, S., Sawamoto, N., Matsumoto, R., Taki, W., Shibasaki, H., 2000. Primary somatosensory cortex is actively involved in pain processing in human. Brain Res. 853, 282–289.

Kaptchuk, T.J., Kelley, J.M., Conboy, L.A., Davis, R.B., Kerr, C.E., Jacobson, E.E., Kirsch, I., Schyner, R.N., Nam, B.H., Nguyen, L.T., Park, M., Rivers, A.L., McManus, C., Kokkotou, E., Drossman, D.A., Goldman, P., Lembo, A.J., 2008. Components of placebo effect: randomised controlled trial in patients with irritable bowel syndrome. BMJ 336, 999–1003.

Kenshalo Jr., D.R., Chudler, E.H., Anton, F., Dubner, R., 1988. SI nociceptive neurons participate in the encoding process by which monkeys perceive the intensity of noxious thermal stimulation. Brain Res. 454, 378–382.

Kenshalo Jr., D.R., Isensee, O., 1983. Responses of primate SI cortical neurons to noxious stimuli. J. Neurophysiol. 50, 1479–1496.

Kong, J., Gollub, R.L., Rosman, I.S., Webb, J.M., Vangel, M.G., Kirsch, I., Kaptchuk, T.J., 2006a. Brain activity associated with expectancy-enhanced placebo analgesia as measured by functional magnetic resonance imaging. J. Neurosci. 26, 381–388.

Kong, J., White, N.S., Kwong, K.K., Vangel, M.G., Rosman, I.S., Gracely, R.H., Gollub, R.L., 2006b. Using fMRI to dissociate sensory encoding from cognitive evaluation of heat pain intensity. Hum. Brain Mapp. 27, 715–721.

Kong, J., Kaptchuk, T.J., Polich, G., Kirsch, I., Gollub, R.L., 2007. Placebo analgesia: findings from brain imaging studies and emerging hypotheses. Rev. Neurosci. 18, 173–190.

Kong, J., Loggia, M.L., Zyloney, C., Tu, P., Laviolette, P., Gollub, R.L., 2010. Exploring the brain in pain: activations, deactivations and their relation. Pain 148, 257–267.

Kong, J., Jensen, K., Loiotile, R., Cheetham, A., Wey, H.Y., Tan, Y., Rosen, B., Smoller, J.W., Kaptchuk, T.J., Gollub, R.L., 2013. Functional connectivity of the frontoparietal network predicts cognitive modulation of pain. Pain 154, 459–467.

Kuchinad, A., Schweinhardt, P., Seminowicz, D.A., Wood, P.B., Chizh, B.A., Bushnell, M.C., 2007. Accelerated brain gray matter loss in fibromyalgia patients: premature aging of the brain? J. Neurosci. 27, 4004–4007.

Lassen, N.A., Ingvar, D.H., Skinhoj, E., 1978. Brain function and blood flow. Sci. Am. 239, 62–71.

Levine, J., Gordon, N., Fields, H., 1978. The mechanism of placebo analgesia. Lancet 23, 654–657.
Liebeskind, J.C., Guilbaud, G., Besson, J.M., Oliveras, J.L., 1973. Analgesia from electrical stimulation of the periaqueductal gray matter in the cat: behavioral observations and inhibitory effects on spinal cord interneurons. Brain Res. 50, 441–446.
Loggia, M.L., Edwards, R.R., Kim, J., Vangel, M.G., Wasan, A.D., Gollub, R.L., Harris, R.E., Park, K., Napadow, V., 2012. Disentangling linear and nonlinear brain responses to evoked deep tissue pain. Pain 153, 2140–2151.
Loggia, M.L., Jensen, K., Gollub, R.L., Wasan, A.D., Edwards, R.R., Kong, J., 2011. The catechol-O-methyltransferase (COMT) val158met polymorphism affects brain responses to repeated painful stimuli. PLoS One 6, e27764.
Loggia, M.L., Kim, J., Gollub, R.L., Vangel, M.G., Kirsch, I., Kong, J., Wasan, A.D., Napadow, V., 2013. Default mode network connectivity encodes clinical pain: an arterial spin labeling study. Pain 154, 24–33.
Loggia, M.L., Mogil, J.S., Bushnell, M.C., 2008a. Empathy hurts: compassion for another increases both sensory and affective components of pain perception. Pain 136, 168–176.
Loggia, M.L., Mogil, J.S., Bushnell, M.C., 2008b. Experimentally induced mood changes preferentially affect pain unpleasantness. J. Pain 9, 784–791.
Loggia, M.L., Schweinhardt, P., Villemure, C., Bushnell, M.C., 2008c. Effects of psychological state on pain perception in the dental environment. J. Can. Dent. Assoc. 74, 651–656.
Longe, S.E., Wise, R., Bantick, S., Lloyd, D., Johansen-Berg, H., McGlone, F., Tracey, I., 2001. Counter-stimulatory effects on pain perception and processing are significantly altered by attention: an fMRI study. Neuroreport 12, 2021–2025.
May, A., 2008. Chronic pain may change the structure of the brain. Pain 137, 7–15.
Mayer, D.J., Liebeskind, J.C., 1974. Pain reduction by focal electrical stimulation of the brain: an anatomical and behavioral analysis. Brain Res. 68, 73–93.
Mayer, D.J., Wolfle, T.L., Akil, H., Carder, B., Liebeskind, J.C., 1971. Analgesia from electrical stimulation in the brainstem of the rat. Science 174, 1351–1354.
Melzack, R., 1989. Labat lecture. Phantom limbs. Reg. Anesth. 14, 208–211.
Melzack, R., 1990. Phantom limbs and the concept of a neuromatrix. Trends Neurosci. 13, 88–92.
Melzack, R., 1999. From the gate to the neuromatrix. Pain (Suppl. 6), S121–S126.
Melzack, R., Casey, K.L., 1968. Sensory, motivational and central control determinants of pain: a new conceptual model. In: Kenshalo, D.R. (Ed.), The Skin Senses. Thomas, Springfield, IL.
Melzack, R., Wall, P.D., 1988. The Challenge of Pain. Basic Books, London.
Merksey, H., Albe-Fessard, D., Bonica, J.J., Carmon, A., Dubner, R., Kerr, F.W.L., Lindblon, U., Pagni, C.A., Renaer, M.J., Sternback, R.A., Sunderland, S., 1979. Pain terms: a list with definitions and notes on usage. Recommended by the IASP Subcommittee on Taxonomy. Pain 6, 249.
Merskey, H., Bogduk, N., 1994. Classification of Chronic Pain, Part III: Pain Terms, A Current List with Definitions and Notes on Usage. second ed. IASP Press, Seattle, WA.
Montes, C., Magnin, M., Maarrawi, J., Frot, M., Convers, P., Mauguiere, F., Garcia-Larrea, L., 2005. Thalamic thermo-algesic transmission: ventral posterior (VP) complex versus VMpo in the light of a thalamic infarct with central pain. Pain 113, 223–232.
Naliboff, B.D., Derbyshire, S.W.G., Munakata, J., Berman, S., Mandelkern, M., Chang, L., Mayer, E.A., 2001. Cerebral activation in patients with irritable bowel syndrome and control subjects during rectosigmoid stimulation. Psychosom. Med. 63, 365–375.
Napadow, V., LaCount, L., Park, K., AsSeine, S., Clauw, D., Harris, R.E., 2010. Intrinsic brain connectivity in fibromyalgia is associated with chronic pain intensity. Arthritis Rheum. 62, 2545–2555.
Niddam, D., Chan, R.-C., Lee, S.-H., Yeh, T.-C., Hsieh, J.-C., 2007. Central modulation of pain evoked from myofascial trigger point. Clin. J. Pain 23, 440–448.
Owen, D.G., Bureau, Y., Thomas, A.W., Prato, F.S., St Lawrence, K.S., 2008. Quantification of pain-induced changes in cerebral blood flow by perfusion MRI. Pain 136, 85–96.
Owen, D.G., Clarke, C.F., Bureau, Y., Ganapathy, S., Prato, F.S., St Lawrence, K.S., 2011. Measuring the neural response to continuous intramuscular infusion of hypertonic saline by perfusion MRI. J. Magn. Reson. Imaging.
Owen, D.G., Clarke, C.F., Ganapathy, S., Prato, F.S., St Lawrence, K.S., 2010. Using perfusion MRI to measure the dynamic changes in neural activation associated with tonic muscular pain. Pain 148, 375–386.

Parks, E.L., Geha, P.Y., Baliki, M.N., Katz, J., Schnitzer, T.J., Apkarian, A.V., 2011. Brain activity for chronic knee osteoarthritis: dissociating evoked pain from spontaneous pain. Eur. J. Pain 15, 843 e1–e14.
Pattany, P.M., Yezierski, R.P., Widerstrom-Noga, E.G., Bowen, B.C., Martinez-Arizala, A., Garcia, B.R., Quencer, R.M., 2002. Proton magnetic resonance spectroscopy of the thalamus in patients with chronic neuropathic pain after spinal cord injury. AJNR Am. J. Neuroradiol. 23, 901–905.
Peltz, E., Seifert, F., DeCol, R., Dorfler, A., Schwab, S., Maihofner, C., 2011. Functional connectivity of the human insular cortex during noxious and innocuous thermal stimulation. NeuroImage 54, 1324–1335.
Penfield, W., Rasmussen, T., 1955. The Cerebral Cortex of Man. The MacMillan Company, New York.
Petrovic, P., Kalso, E., Petersson, K.M., Ingvar, M., 2002. Placebo and opioid analgesia – imaging a shared neuronal network. Science 295, 1737–1740.
Petrovic, P., Petersson, K.M., Ghatan, P.H., Stone-Elander, S., Ingvar, M., 2000. Pain-related cerebral activation is altered by a distracting cognitive task. Pain 85, 19–30.
Ploghaus, A., Narain, C., Beckmann, C.F., Clare, S., Bantick, S., Wise, R., Matthews, P.M., Rawlins, J.N., Tracey, I., 2001. Exacerbation of pain by anxiety is associated with activity in a hippocampal network. J. Neurosci. 21, 9896–9903.
Ploner, M., Freund, H.J., Schnitzler, A., 1999a. Pain affect without pain sensation in a patient with a postcentral lesion. Pain 81, 211–214.
Ploner, M., Schmitz, F., Freund, H.J., Schnitzler, A., 1999b. Parallel activation of primary and secondary somatosensory cortices in human pain processing. J. Neurophysiol. 81, 3100–3104.
Porro, C.A., Cettolo, V., Francescato, M.P., Baraldi, P., 1998. Temporal and intensity coding of pain in human cortex. J. Neurophysiol. 80, 3312–3320.
Pukall, C.F., Strigo, I.A., Binik, Y.M., Amsel, R., Khalife, S., Bushnell, M.C., 2005. Neural correlates of painful genital touch in women with vulvar vestibulitis syndrome. Pain 115, 118–127.
Raichle, M.E., MacLeod, A.M., Snyder, A.Z., Powers, W.J., Gusnard, D.A., Shulman, G.L., 2001. A default mode of brain function. Proc. Natl. Acad. Sci. USA 98, 676–682.
Rainville, P., Carrier, B., Hofbauer, R.K., Bushnell, M.C., Duncan, G.H., 1999. Dissociation of sensory and affective dimensions of pain using hypnotic modulation. Pain 82, 159–171.
Rainville, P., Duncan, G.H., Price, D.D., Carrier, B., Bushnell, M.C., 1997. Pain affect encoded in human anterior cingulate but not somatosensory cortex. Science 277, 968–971.
Reynolds, D.V., 1969. Surgery in the rat during electrical analgesia induced by focal brain stimulation. Science 164, 444–445.
Rhudy, J.L., McCabe, K.M., Williams, A.E., 2007. Affective modulation of autonomic reactions to noxious stimulation. Int. J. Psychophysiol. 63, 105–109.
Robinson, M.E., Craggs, J.G., Price, D.D., Perlstein, W.M., Staud, R., 2011. Gray matter volumes of pain-related brain areas are decreased in fibromyalgia syndrome. J. Pain 12, 436–443.
Ruscheweyh, R., Deppe, M., Lohmann, H., Stehling, C., Floel, A., Ringelstein, E.B., Knecht, S., 2011. Pain is associated with regional grey matter reduction in the general population. Pain 152, 904–911.
Schmidt-Wilcke, T., Luerding, R., Weigand, T., Jurgens, T., Schuierer, G., Leinisch, E., Bogdahn, U., 2007. Striatal grey matter increase in patients suffering from fibromyalgia – a voxel-based morphometry study. Pain 132 (Suppl. 1), S109–S116.
Schnitzler, A., Ploner, M., 2000. Neurophysiology and functional neuroanatomy of pain perception. J. Clin. Neurophysiol. 17, 592–603.
Schweinhardt, P., Bushnell, M.C., 2010. Pain imaging in health and disease – how far have we come? J. Clin. Invest. 120, 3788–3797.
Schweinhardt, P., Kalk, N., Wartolowska, K., Chessell, I., Wordsworth, P., Tracey, I., 2008a. Investigation into the neural correlates of emotional augmentation of clinical pain. NeuroImage 40, 759–766.
Schweinhardt, P., Kuchinad, A., Pukall, C.F., Bushnell, M.C., 2008b. Increased gray matter density in young women with chronic vulvar pain. Pain 140, 411–419.
Scott, D.J., Stohler, C.S., Egnatuk, C.M., Wang, H., Koeppe, R.A., Zubieta, J.-K., 2009. Individual differences in reward responding explain placebo-induced expectations and effects. Neuron 55, 325–336.
Seminowicz, D.A., Mikulis, D.J., Davis, K.D., 2004. Cognitive modulation of pain-related brain responses depends on behavioral strategy. Pain 112, 48–58.

Seminowicz, D.A., Wideman, T.H., Naso, L., Hatami-Khoroushahi, Z., Fallatah, S., Ware, M.A., Jarzem, P., Bushnell, M.C., Shir, Y., Ouellet, J.A., Stone, L.S., 2011. Effective treatment of chronic low back pain in humans reverses abnormal brain anatomy and function. J. Neurosci. 31, 7540–7550.
Sharma, N.K., McCarson, K., Van Dillen, L., Lentz, A., Khan, T., Cirstea, C.M., 2011. Primary somatosensory cortex in chronic low back pain – a H-MRS study. J. Pain Res. 4, 143–150.
Shulman, G.L., Fiez, J.A., Corbetta, M., Buckner, R.L., Miezin, F.M., Raichle, M.E., Peterson, S.E., 1997. Common blood flow changes across Visual Tasks: II. Decreases in cerebral cortex. J. Cogn. Neurosci. 9, 648–663.
Siddall, P.J., Stanwell, P., Woodhouse, A., Somorjai, R.L., Dolenko, B., Nikulin, A., Bourne, R., Himmelreich, U., Lean, C., Cousins, M.J., Mountford, C.E., 2006. Magnetic resonance spectroscopy detects biochemical changes in the brain associated with chronic low back pain: a preliminary report. Anesth. Analg. 102, 1164–1168.
Sikes, R.W., Vogt, B.A., 1992. Nociceptive neurons in area 24 of rabbit cingulate cortex. J. Neurophysiol. 68, 1720–1732.
Sorensen, L., Siddall, P.J., Trenell, M.I., Yue, D.K., 2008. Differences in metabolites in pain-processing brain regions in patients with diabetes and painful neuropathy. Diabetes Care 31, 980–981.
Talbot, J.D., Marrett, S., Evans, A.C., Meyer, E., Bushnell, M.C., Duncan, G.H., 1991. Multiple representations of pain in human cerebral cortex. Science 251, 1355–1358.
Tarkka, I.M., Treede, R.D., 1993. Equivalent electrical source analysis of pain-related somatosensory evoked potentials elicited by a CO_2 laser. J. Clin. Neurophysiol. 10, 513–519.
Timmermann, L., Ploner, M., Haucke, K., Schmitz, F., Baltissen, R., Schnitzler, A., 2001. Differential coding of pain intensity in the human primary and secondary somatosensory cortex. J. Neurophysiol. 86, 1499–1503.
Tölle, T.R., Kaufmann, T., Siessmeier, T., Lautenbacher, S., Berthele, A., Munz, F., Zieglgansberger, W., Willoch, F., Schwaiger, M., Conrad, B., Bartenstein, P., 1999. Region-specific encoding of sensory and affective components of pain in the human brain: a positron emission tomography correlation analysis. Ann. Neurol. 45, 40–47.
Tracey, I., Bushnell, M.C., 2009. How neuroimaging studies have challenged us to rethink: is chronic pain a disease? J. Pain 10, 1113–1120.
Tracey, I., Mantyh, P.W., 2007. The cerebral signature for pain perception and its modulation. Neuron 55, 377–391.
Tracey, I., Ploghaus, A., Gati, J.S., Clare, S., Smith, S., Menon, R.S., Matthews, P.M., 2002. Imaging attentional modulation of pain in the periaqueductal gray in humans. J. Neurosci. 22, 2748–2752.
Treede, R.-D., Meyer, R.A., Raja, S.N., Campbell, J.N., 1992. Peripheral and central mechanisms of cutaneous hyperalgesia. Prog. Neurobiol. 38, 397–421.
Tu, C.H., Niddam, D.M., Chao, H.T., Chen, L.F., Chen, Y.S., Wu, Y.T., Yeh, T.C., Lirng, J.F., Hsieh, J.C., 2010. Brain morphological changes associated with cyclic menstrual pain. Pain 150, 462–468.
Turk, D.C., Melzack, R., 1992. The Measurement of Pain and the Assessment of People Experiencing Pain. The Guilford Press, New York.
Turk, D.C., Okifuji, A., 2010. Pain terms and taxonomies. In: Fishman, S.M., Ballantyne, J.C., Rathmell, J.P. (Eds.), Bonica's Management of Pain. third ed. Kluwer/Lippincott, Williams & Wilkins, Philadelphia, PA.
Valdes, M., Collado, A., Bargallo, N., Vazquez, M., Rami, L., Gomez, E., Salamero, M., 2010. Increased glutamate/glutamine compounds in the brains of patients with fibromyalgia: a magnetic resonance spectroscopy study. Arthritis Rheum. 62, 1829–1836.
Valet, M., Sprenger, T., Boecker, H., Willoch, F., Rummeny, E., Conrad, B., Erhard, P., Tolle, T.R., 2004. Distraction modulates connectivity of the cingulo-frontal cortex and the midbrain during pain – an fMRI analysis. Pain 109, 399–408.
Villemure, C., Bushnell, M.C., 2002. Cognitive modulation of pain: how do attention and emotion influence pain processing? Pain 95, 195–199.
Villemure, C., Bushnell, M.C., 2009. Mood influences supraspinal pain processing separately from attention. J. Neurosci. 29, 705–715.
Villemure, C., Slotnick, B.M., Bushnell, M.C., 2003. Effects of odors on pain perception: deciphering the roles of emotion and attention. Pain 106, 101–108.

Vogel, H., Port, J.D., Lenz, F.A., Solaiyappan, M., Krauss, G., Treede, R.D., 2003. Dipole source analysis of laser-evoked subdural potentials recorded from parasylvian cortex in humans. J. Neurophysiol. 89, 3051–3060.

Vogt, B.A., 2005. Pain and emotion interactions in subregions of the cingulate gyrus. Nat. Rev. Neurosci. 6, 533–544.

Vogt, B.A., Rosene, D.L., Pandya, D.N., 1979. Thalamic and cortical afferents differentiate anterior from posterior cingulate cortex in the monkey. Science 204, 205–207.

Vogt, B.A., Sikes, R.W., Vogt, L.J., 1993. Anterior cingulate cortex and the medial pain system. In: Vogt, B.A., Gabriel, M. (Eds.), Neurobiology of Cingulate Cortex and Limbic Thalamus. Birkhäuser, Boston.

Wager, T.D., Rilling, J.K., Smith, E.E., Sokolik, A., Casey, K.L., Davidson, R.J., Kosslyn, S.M., Rose, R.M., Cohen, J.D., 2004. Placebo-induced changes in fMRI in the anticipation and experience of pain. Science 303, 1162–1167.

Wartolowska, K., Hough, M.G., Jenkinson, M., Andersson, J., Wordsworth, B.P., Tracey, I., 2011. Structural brain changes in rheumatoid arthritis. Arthritis Rheum.

Wasan, A.D., Loggia, M.L., Chen, L.Q., Napadow, V., Kong, J., Gollub, R.L., 2011. Neural correlates of chronic low back pain measured by arterial spin labeling. Anesthesiology 115, 364–374.

Wiech, K., Seymour, B., Kalisch, R., Stephan, K.E., Koltzenburg, M., Driver, J., Dolan, R.J., 2005. Modulation of pain processing in hyperalgesia by cognitive demand. NeuroImage 27, 59–69.

Willis Jr., W.D., Zhang, X., Honda, C.N., Giesler Jr., G.J., 2002. A critical review of the role of the proposed VMpo nucleus in pain. J. Pain 3, 79–94.

Wolfe, F., Smythe, H., Yunus, M., Bennett, R., Bombardier, C., Goldenberg, D., Tugwell, P., Campbell, S., Abeles, M., Clark, P., 1990. The American College of Rheumatology 1990 Criteria for the Classification of Fibromyalgia. Report of the Multicenter Criteria Committee. Arthritis Rheum. 33, 160–172.

Wood, P.B., Ledbetter, C.R., Patterson 2nd, J.C., 2009. Changes in hippocampal metabolites after effective treatment for fibromyalgia: a case study. Clin. J. Pain 25, 810–814.

Zambreanu, L., Wise, R.G., Brooks, J.C., Iannetti, G.D., Tracey, I., 2005. A role for the brainstem in central sensitisation in humans. Evidence from functional magnetic resonance imaging. Pain 114, 397–407.

CHAPTER NINETEEN

Imaging of Neurochemical Transmission in the Central Nervous System

Lora Deuitch[1] and Rajesh Narendran[1,2]
[1]Departments of Radiology, University of Pittsburgh, Pittsburgh, PA, USA
[2]Department of Psychiatry, University of Pittsburgh, Pittsburgh, PA, USA

1. INTRODUCTION

Since the introduction of imaging techniques such as single photon emission tomography (SPECT) and positron emission tomography (PET), it has been established that changes in endogenous neurochemical transmission can be detected in the human brain. These fluctuations in neurotransmitter release have allowed researchers to explore receptor–ligand interactions in various neuropsychiatric disorders such as schizophrenia and drug addiction (Laruelle and Abi-Dargham, 1999; Volkow et al., 1999). By utilizing these techniques, changes in neurotransmitter release are measured in vivo, and thus, the parameters under which the subjects are being scanned can be manipulated in order to derive a more complete understanding of the neurochemical in question. This noninvasive measure of activity allows for the same subject to be scanned under different conditions, as well as using different radiotracers, which grants the researcher the ability to measure differences in activity within subject as well as between subjects. When performing PET (or SPECT) in combination with in vivo animal modalities, techniques such as microdialysis can be utilized to confirm the pharmacological relationships between radiotracers and corresponding challenges. Increases in extracellular dopamine (DA) levels can be correlated with changes in the binding potential (BP) of the radiotracer, in which the results can be used to validate the use of the ligand in humans. In addition, PET imaging of neurotransmitter release can be used to validate findings in previous postmortem in vitro studies. In this chapter, we review the SPECT and PET imaging techniques that are being used in clinical and translational research to measure neurotransmitter release in the central nervous system. Over the past two decades, SPECT and PET have been developed as tools to image endogenous ligands and their relation to cognitive or pharmacological challenges, which has enabled researchers to explore the neurological mechanisms behind various neuropsychiatric disorders. It is vital however that these novel-imaging paradigms be expanded upon in order to strengthen our understanding of in vivo ligand–receptor relationships as they apply to clinical disorders such as schizophrenia, addiction, Parkinson's, attention deficit hyperactivity disorder (ADHD), and Tourette's.

Imaging of the Human Brain in Health and Disease
http://dx.doi.org/10.1016/B978-0-12-418677-4.00019-1

2.1. IMAGING OF DA TRANSMISSION

2.1.1. $D_{2/3}$ Radiotracers

2.1.1.1. Stimulant-Induced DA Release

SPECT and PET imaging of DA release following the administration of a psychostimulant drug challenge such as D-amphetamine or methylphenidate was first demonstrated in animal studies with substituted benzamide radiotracers such as [^{123}I]IBZM and [^{11}C]raclopride that bind to DA $D_{2/3}$ receptors (Dewey et al., 1993; Innis et al., 1992; Seeman et al., 1989). As the binding of [^{123}]IBZM and [^{11}C]raclopride to $D_{2/3}$ receptors was less following a stimulant challenge compared to baseline, it was inferred to reflect the competition between DA that was released by the stimulant drug and the radiotracer that was previously bound to $D_{2/3}$ receptors. The replication of this observation in humans led to the widespread use of SPECT and PET endogenous competition binding techniques to characterize stimulant-induced DA release in both healthy and diseased conditions (Laruelle et al., 1995; Volkow et al., 1994). In contrast to healthy humans who showed a 10–15% reduction in the BP (BP_{ND}) of [^{123}I]IBZM and [^{11}C]raclopride in the striatum, subjects with various neuropsychiatric disorders (Table 1) were shown to have either a higher or lower change in BP_{ND} following the same dose intravenous D-amphetamine (0.3 mg/kg) or methylphenidate (0.5 mg/kg). The relationship between this decrease in $D_{2/3}$ receptor BP_{ND} and increase in extracellular fluid DA following an amphetamine challenge was then established as 1:40 (a 1% reduction in radiotracer BP_{ND} corresponds to a 40% increase in extracellular DA) using combined imaging and microdialysis in nonhuman primates (Breier et al., 1997; Laruelle et al., 1997b). Furthermore, the introduction of high-resolution PET scanners allowed for the characterization of stimulant-induced DA release not only at the level of the whole striatum but also in its substructures such as the caudate, putamen, and ventral striatum (VST) as well (Drevets et al., 2001; Martinez et al., 2003). Using this imaging methodology, recent studies have reported a greater stimulant-induced DA release in the limbic-related VST as opposed to the associative (caudate and precommissural putamen) and sensorimotor (postcommissural putamen) parts of the striatum in humans. This methodology has also allowed for numerous imaging studies to characterize abnormalities in DA transmission that are specific to these striatal subdivisions in numerous disorders such as addiction (Table 1).

More recent PET investigations have used oral as opposed to the intravenous D-amphetamine (and methylphenidate) to displace $D_{2/3}$ radiotracer BP_{ND} as the stimulant-induced alterations in vital signs and cardiac parameters (such as electrocardiography) are less pronounced with the oral route of administration. The results of these studies suggest that the mean decrease in oral amphetamine-induced (0.5 mg/kg) displacement of $D_{2/3}$ radiotracer BP_{ND} (such as [^{11}C]raclopride and [^{18}F]fallypride) are comparable to that induced by intravenous D-amphetamine (0.3 mg/kg) at the level of the caudate and the putamen. But, there are significant differences at the level of the VST in which oral amphetamine leads to less displacement compared to the intravenous route (Table 2). Slifstein and colleagues have

Table 1 Differences in Stimulant-Induced Displacement of [^{11}C]Raclopride in the Striatum in Clinical Conditions

Change	Clinical Condition	Challenge	n	Control ΔBP	Patient ΔBP	References
Decreased	Alcohol[§]	Amphetamine 0.3 mg/kg i.v.	15	13	5	Martinez et al. (2005)
	Alcohol	Methylphenidate 0.5 mg/kg i.v.	20	~14–28	~8–11%	Volkow et al. (2007)
	Cocaine	Amphetamine 0.3 mg/kg i.v.	24	10	3.0	Martinez et al. (2007)
	Cocaine	Methylphenidate 0.5 mg/kg i.v.	20	~22	~10	Volkow et al. (1997)
	Heroin	Methylphenidate 60 mg oral	16	12	4	Martinez et al. (2012)
	Methamphetamine[*]	Methylphenidate 60 mg oral	15	6–10	4–6	Wang et al. (2011)
Increased	Tourette's disorder[§]	Amphetamine 0.3 mg/kg i.v.	14	~8 to 10	14–18	Wong et al. (2007)
	Schizophrenia	Amphetamine 0.2 mg/kg i.v.	11	16	22	Breier et al. (1997)

ΔBP—Amphetamine-induced change in the BP (values shown are approximately when extracted from the graphs and also rounded to the nearest whole number).
*Values shown are from the left putamen, which was the only striatal subdivision that was significantly different.
§Values shown are from the VST, which was the only striatal subdivision that was significantly different.

Table 2 BP of Intravenous vs Oral Amphetamine in Healthy Controls Using [^{11}C]Raclopride and [^{18}F]Fallypride

Region	Intravenous 0.3 mg/kg [^{11}C]Raclopride (Martinez et al., 2003)	Intravenous 0.3 mg/kg [^{11}C]Raclopride (Drevets et al., 2001)	Intravenous 0.3 mg/kg [^{18}F]Fallypride (Slifstein et al., 2010)	Oral 0.5 mg/kg [^{11}C]Raclopride (Narendran et al., 2010a)	Oral 0.43 mg/kg [^{18}F]Fallypride (Riccardi et al., 2005)	Oral 0.5 mg/kg [^{18}F]Fallypride (Cropley et al., 2008)
Caudate	-5.5 ± 7.9[*,§]	-4.5 ± 8.1[§]	-6.0 ± 17.0[§]	-8.4 ± 4.2[*]	-5.6 ± 4.6[*]	-11.5 ± 2.7[*,§]
Putamen	-16.6 ± 9.9[*,¶]	-10.2 ± 10.6[*,¶]	-12.0 ± 16.0[*,¶]	-14.7 ± 4.8[*]	-11.2 ± 4.3[*]	-13.6 ± 3.0[*,¶]
VST	-17.8 ± 13.8[*]	-15.4 ± 10.6[*]	-13.0 ± 16.0[*]	-9.7 ± 4.4[*]	-7.2 ± 5.3[*]	-8.5 ± 2.8[*]
Striatum	-10.5 ± 7.5[*]	–	–	-12.3 ± 4.4[*]	–	–

VST, Ventral striatum.
*$p \leq 0.05$.
§Pre-DCA.
¶Post-PUT.

hypothesized that this difference in the displacement after oral and intravenous administration is related to the difference in the timing of the rise of amphetamine concentration in brain tissue, which in theory impacts the profile of DA release in striatal subdivisions (Slifstein et al., 2010). Despite this limitation, the use of oral amphetamine to stimulate DA release offers a relatively safe alternative to image amphetamine-induced DA release with PET in individuals who may be at a higher risk of developing cardiovascular events.

The exact mechanisms that contribute to the displacement of $D_{2/3}$ radiotracer binding following an acute amphetamine challenge is unknown and shown to involve DA-mediated competitive inhibition and cellular trafficking (internalization of the $D_{2/3}$ receptors from the cellular surface to the interior) at $D_{2/3}$ receptors (Ginovart et al., 2004; Guo et al., 2010; Skinbjerg et al., 2010; Sun et al., 2003). Nevertheless, the use of the stimulant-induced DA release-imaging paradigm has allowed for the characterization of DA transmission abnormalities in various neuropsychiatric disorders such as schizophrenia, Tourette's disorder, ADHD, and drug abuse.

2.1.1.2. Nonstimulant-Induced DA Release

The ability to measure stimulant-induced displacement of [^{123}I]IBZM and [^{11}C]raclopride in the striatum with SPECT and PET led to a series of human investigations to measure DA release following nonstimulant challenges such as caffeine, smoking (nicotine), alcohol, and Δ^9-tetrahydrocannabinol (THC). The results of these experiments with nonstimulant challenges that lead to DA release via indirect mechanisms have been somewhat mixed and inconsistent in reporting a decrease in $D_{2/3}$ radiotracer BP_{ND}.

Caffeine Challenge

No significant displacement of [^{11}C]raclopride binding in the striatum was reported in habitual coffee drinkers following an oral dose of 200 mg of caffeine. In this same study, a 11% decrease in [^{11}C]raclopride BP_{ND} was reported in the thalamus following the caffeine challenge. Nevertheless, as [^{11}C]raclopride BP_{ND} in the thalamus comprised a relatively high fraction of nonspecific binding and is associated with poor reproducibility, this observation needs to be interpreted with caution until replicated with a high affinity DA $D_{2/3}$ radioligand such as [^{18}F]fallypride or [^{11}C]FLB 457 (Mawlawi et al., 2001). Furthermore, a follow-up study by the same authors reported a 15% decrease in [^{11}C]raclopride BP_{ND} in the thalamus in habitual coffee drinkers who expected caffeine and received a placebo challenge, thereby questioning the robustness of their original observation. In summary, the available data suggest that caffeine-induced DA release is not sufficient to be measured with endogenous competition techniques that utilize [^{11}C]raclopride and PET.

Nicotine Challenge

Five out of seven studies (Table 3) that evaluated nicotine-induced DA release with [^{11}C]raclopride and PET report a significant decrease in the [^{11}C]raclopride binding in

Table 3 Nicotine Challenge in Smokers and Nonsmokers Using [^{11}C]Raclopride

References	Human	Challenge	Dose	n	Control Condition	Region	Effect on BP (% ± SD)
Barrett et al. (2004)	HC	Nicotine (smoking)	Abstinent for 12 h, followed by 3–6 (6.3–12.6 mg of nicotine) cigarettes (1 every 12 min) in scanner	10	Nonsmoking	CAD PUT posterior PUT anterior VST	−1.9 −1.6 2.9 3.12
Brody et al. (2004)	Smokers	Nicotine (smoking)	Abstinent for 3 h, followed by smoking 1 cigarette during a break	20	Nonsmoking	VPU R VST L VST R	−29.7* −36.6* −25.9*
Brody et al. (2006)	Smokers	Nicotine (smoking)	Abstinent for 3 h, followed by smoking 1 cigarette during a break	35	Nonsmoking	VST	−8.4 ± 13.8*
			Abstinent for 3 h, followed by no smoking during a break	10	Nonsmoking	VST	1.8 ± 12.5
Montgomery et al. (2007b)	Smokers	Nicotine intranasal	Abstinent for 12 h, followed by 2 mg of nicotine intranasal	10	Nonsmoking	CAD and PUT anterior PUT posterior VST	−1.0 ± 5.0 1.1 ± 4.5 −0.1 ± 7.2
Takahashi et al. (2008)	Smokers	Nicotine gum (oral)	Abstinent for 24 h, followed by 4 mg of nicotine oral	6	Placebo gum	STR	−7.3*
	HC	Nicotine gum (oral)	Abstinent for 24 h, followed by 4 mg of nicotine oral	6	Placebo gum	STR	1.6
Scott et al. (2007a)	Smokers	Nicotine (smoking)	Abstinent for 12 h, followed by a nicotine cigarette (nicotine 1.0 mg)	6	Nonnicotine cigarette (0.08 mg)	VST L	10.0*
Brody et al. (2009)	Smokers	Nicotine cigarette (smoking)	Abstinent for 3 h, followed by smoking their own cigarette during a break (nicotine 1.0 mg)	46	Nonsmoking	VST	−8.4 ± 1.5*
		Nonnicotine cigarette (smoking)	Abstinent for 3 h, followed by smoking a denicotinized cigarette (nicotine 0.05 mg) during a break	16	Nonsmoking	VST	−1.2 ± 2.5

HC—Healthy Control; BP—Binding Potential; NS—Not Significant; Regions of interest: CAD—Caudate; PUT—Putamen; VPU—Ventral PUT; STR—Striatum; VST—Ventral STR.
*$p \leq 0.05$.

the striatum and its subdivisions. These studies reported a nicotine effect on DA release despite differences in the nicotine-free period that was enforced in smokers prior to their PET scans (range from 3 to 24 h), route of administration of the nicotine challenge (oral, inhalation, and smoking), and image analysis methodology (bolus plus constant infusion vs bolus administration of radiotracer; equilibrium vs graphical vs simplified reference tissue image analysis). The mean displacement of [^{11}C]raclopride following the nicotine challenge ranged from 7 to 10% in the studies that reported an effect (Table 4), other than one study by Brody and colleagues that reported a much larger effect (30–36%) (Brody et al., 2004). While the reasons for the discrepancy between the two negative studies and the one study that reported a relatively large effect are not readily apparent from the review of these manuscripts, it might be related to the quality of equilibrium attained during the scan and image analysis methods. Nevertheless, a majority of the published studies support the ability to detect nicotine-induced DA release in humans with [^{11}C]raclopride and PET.

In contrast to combined PET and microdialysis studies that have established the relationship between a decrease in [^{11}C]raclopride BP_{ND} and an increase in extracellular fluid DA following an amphetamine challenge as 1:40 (Breier et al., 1997), data following a nicotine challenge indicate this ratio to be 1:1.4 (Kim and Han, 2009). One possible explanation for the increased sensitivity to detect DA release following an indirect challenge such as nicotine as opposed to a direct challenge such as amphetamine (which acts on the DA transporter to release DA into the synaptic and extrasynaptic spaces) may be related to the fact that PET/SPECT endogenous competition binding measures DA release in the synapse as opposed to microdialysis, which measures DA in the extrasynaptic space as discussed in previous reports (Kim and Han, 2009; Schiffer et al, 2006; Tsukada et al., 1999).

Alcohol Challenge

Two early studies that reported a lack of effect on striatal DA release for an acute alcohol challenge suffered from limitations such as relatively small sample sizes and failure to use modeling methodology to quantitate receptor BP_{ND} (Salonen et al., 1997; Wong et al., 1993). Three out of four studies (Table 4) that evaluated this issue more recently were successful in detecting an alcohol-induced effect on [^{11}C]raclopride binding in the striatum (Boileau et al., 2003; Urban et al., 2010; Yoder et al., 2009). It is possible that factors such as expectation of alcohol and a relatively lower dose of intravenous alcohol may have led to the recent negative report in the literature (Yoder et al., 2005). Finally, in the largest study that has reported on this issue, there was a strong trend to suggest that males release significantly more DA relative to females following an oral alcohol challenge (Urban et al., 2010). In summary, the available data support the ability of PET to detect DA release following an oral alcohol challenge in humans, whereas more conclusive data are necessary for the intravenous alcohol challenge.

Table 4 Alcohol Challenge in Healthy Controls Using [^{11}C]Raclopride

References	Challenge	Dose	n	Control Condition	Region	Effect on BP (% ± SD)
Boileau et al. (2003)	ETOH (oral)	1 ml/kg or 95% USP ETOH	7	Orange juice	CAD	−4.0 ± 16.4
					PUT	−5.2 ± 17.5
					VPU	−13.7 ± 16.4*
					VST	−15.0 ± 15.9*
Yoder et al. (2005)	ETOH (i.v)	Alcohol cue exposure, followed by infusion of i.v alcohol to target a BAC of 60 mg% ±5 mg%	8	No alcohol (baseline scan)	CAD L	−1.0 ± 10.0
					CAD R	−6.0 ± 10.0
					PUT L anterior	−1.0 ± 6.0
					PUT R anterior	−10.0 ± 18
					PUT L posterior	−6.0 ± 13.0
					PUT R posterior	−15.0 ± 28.0
Yoder et al. (2009)	ETOH (i.v)	Alcohol cue exposure, followed by infusion of i.v alcohol to target a BAC of 80 mg% ±5 mg%	8	Baseline scan with neutral cues Baseline scan with alcohol cues	VST L	−12.0 ± 8.0*
Urban et al. (2010)	ETOH (oral)	0.75 g/kg body water 100 proof Vodka	21	Cranberry juice/soda	CAD anterior§	−5.0 ± 8.0*
					CAD posterior	−6.0 ± 8.0*
					PUT anterior¶	−7.0 ± 8.0*
					PUT posterior	−5.0 ± 6.0*
					VST	−9.0 ± 8.0*
					STR	−6.0 ± 7.0*

BAC—Blood Alcohol Level; BP—Binding Potential; Regions of interest; CAD—Caudate; PUT—Putamen; VPU—Ventral PUT; STR—Striatum; VST—Ventral STR.
*$p \leq 0.05$.
§Pre-DCA.
¶Pre-DPU.

THC Challenge

Two studies that evaluated the DA release in the striatal subdivisions are consistent in reporting relatively small ($\Delta BP_{ND} < 4\%$, which only reached statistical significance in the VST and precommissural dorsal putamen) to no decrease in [^{11}C]raclopride BP_{ND} after a THC challenge (8-mg inhalation; 10-mg oral dose) (Bossong et al., 2009; Stokes et al., 2009). One of these studies performed an exploratory analysis of their published data set and reported a 30–50% reduction in [^{11}C]raclopride BP_{ND} in the frontal and temporal cortical regions after THC (Stokes et al., 2010). Future studies with a high affinity $D_{2/3}$ radioligand such as [^{11}C]FLB 457 are necessary to confirm these results prior to ascertaining much significance as the signal-tonoise ratio of [^{11}C]raclopride is too low to reliably quantify the DA release in the human cortex. Thus, the available data suggest that THC unlike other drugs of abuse mediates its psychotomimetic effects in humans via nondopaminergic mechanisms.

Overall, in contrast to the numerous reports that have emerged confirming the ability to measure stimulant-induced DA release in humans with PET, the ability to measure DA release following a nonstimulant challenge has proven to be more difficult and has been met only with partial success. The primary reason for failure to detect DA release following a nonstimulant challenge in studies appears to be the relatively small effect size associated with these challenges—that is, less DA release as well as a higher variability in an individual's response to the nonstimulant challenge. Thus, it is essential for imaging studies in clinical disorders that propose to use these nonstimulant challenges to measure DA release to be adequately powered to detect between-group differences.

2.1.1.3. Placebo (Expectation)-Induced DA Release

In recent years, there have been numerous imaging reports suggesting that the expectation to receive a drug such as caffeine or amphetamine, or a medication to alleviate symptoms such as pain can cause DA release in conditioned individuals. This phenomenon, which was first reported in individuals with Parkinson's' disease by Fuente-Fernandez and colleagues, has been replicated and extended to healthy individuals as well by other groups (de la Fuente-Fernandez et al., 2002; de la Fuente-Fernandez et al., 2001; Lidstone et al., 2010). In the first investigation, a placebo challenge led to a $23 \pm 11\%$ reduction of [^{11}C]raclopride binding in the striatum of individuals with Parkinson's disease, who were informed that they had a greater chance to receive apomorphine rather than placebo (de la Fuente-Fernandez et al., 2002; de la Fuente-Fernandez et al., 2001). Similar decreases in [^{11}C]raclopride BP_{ND} in the VST after placebo have been reported in subjects conditioned to receive amphetamine (Boileau et al., 2007) and in individuals who were informed that they would receive a medication for relief or pain prior to exposure to pain stimuli (Scott et al., 2008). These recent findings, besides advancing our understanding of the neurobiology of expectation and placebo response in clinical trials, also highlight the need for investigators to be attentive to study design when they use challenges to study stimulant-induced DA release in clinical populations that may be previously conditioned to these challenges.

2.1.1.4. Cognitive/Reward Task-Induced DA Release

More recently, numerous investigations have attempted to characterize physiological DA transmission with [^{11}C]raclopride and PET. The long list of studies that have reported the ability to detect striatal DA release following a physiological challenge paradigm include a motor sequence learning task (Badgaiyan et al., 2007), pain challenge (Scott et al., 2007b), psychosocial stress tasks (Pruessner et al., 2004), eating favorite food (Small et al., 2003), meditating (Kjaer et al., 2002), and performing a finger movement (Goerendt et al., 2003) or foot tapping task (Ouchi et al., 2002). These studies, which have generated enormous interest for the promise to study in vivo DA transmission during reward processing or cognitive learning, could provide a valuable insight into the pathophysiology of psychiatric and addictive disorders. On the other hand, the lack of methodological rigor in several of the published studies and inability to replicate these findings with more robust methodology have sparked a controversy in the field with respect to the feasibility of measuring physiological DA transmission (Lappin et al., 2009; Laruelle, 2000; Montgomery et al., 2006; Volkow et al., 2002). Although a comprehensive and critical review of all the task-induced DA release studies in the literature is beyond the scope of this paper, we will focus on the reward task-induced DA release studies to illustrate the methodological challenges and problems associated with the proposed use of these physiological challenge paradigms. In Table 5, we provide a summary of the previously published studies that have assessed DA release following monetary reward tasks. The methodological limitations that relate to several of these published studies are (1) the use of a single PET scan to measure both baseline and posttask BP_{ND} (Martin-Soelch et al., 2011; Pappata et al., 2002), (2) the use of bolus injection paradigms that are vulnerable to changes in the rate of cerebral blood flow (or peripheral clearance) that occur during the task (Hakyemez et al., 2008; Koepp et al., 1998; Pappata et al., 2002; Schott et al, 2008; Zald et al., 2004), (3) the use of reference tissue methods that fail to quantify the changes in nonspecific binding that may occur during the task (Hakyemez et al., 2008; Koepp et al., 1998; Pappata et al., 2002; Schott et al, 2008; Zald et al., 2004), (4) the lack of a control for motor movements during the reward task (Koepp et al., 1998; Martin-Soelch et al., 2011; Pappata et al., 2002; Schott et al, 2008), which have been shown to release DA in itself (Badgaiyan et al., 2007; Goerendt et al., 2003; Ouchi et al., 2002), and (5) relatively small sample sizes. Given these methodological limitations, it is no surprise that the mean change in [^{11}C]raclopride BP lacks consistency and ranges from −17 to +22% in the striatal subdivisions in the studies listed in Table 5. A recent study by Urban and colleagues used rigorous methodology and analysis to address several limitations that are evident in the literature (Urban et al., 2011). The authors of this study concluded that they were unable to

Table 5 Monetary Reward Tasks in Healthy Controls Using [¹¹C]Raclopride

References	Task Paradigm	N	Method	Control Condition	Region	Effect on BP Mean ± SD (%)
Koepp et al. (1998)	Video game-for reward	8	DB (SRTM)	Blank Screen	DST	−7.5*
					VST	−13*
Zald et al. (2004)	Variable ratio reward	9	DB (SRTM)	Motor	CAD L	−6.3 (16)*
					PUT L	+9.1 (20.3)*
					VST	No change
Pappata et al. (2002)	Number guessing task, with reward	12	SB (KIN)	None	VST	Decrease§
Hakyemez et al. (2008)	(A) Roulette wheel game, with passive reward	10	DB (SRTM)	Motor	CAD R	+18 (23)
	(B) Roulette wheel game, with novel stimuli	10	DB (SRTM)	Motor	CAD L	+11 (16)
					PUT L	+14 (10)*
					VST	No change
					PUT R	+11 (11)
Schott et al. (2008)	Delayed monetary incentive task, reward	11	DB (G)	Neutral trial	CAD	+0.5
					PUT	−3
					VST L	−17 (7)*
					VST R	−2 (1)
Martin-Soelch et al. (2011)	Slot machine, reward task	24	BCI (E)	None	CAD dorsal R	−1.0 (5.8)
					CAD dorsal L	−2.4 (6.8)
					CAD middle R	−1.5 (8.6)
					CAD middle L	−2.8 (7.2)
					PUT dorsal R	−1.5 (5.0)
					PUT dorsal L	−0.6 (4.2)
					VST R	−5.6 (9.5)*
					VST L	−2.3 (6.7)
Urban et al. (2011)	Delayed monetary incentive task, reward	18	BCI (E)	NA¶	CAD precommissural	−3.5 (8.9)
					CAD postcommissural	−6.4 (10.2)*
					PUT precommissural	−2.0 (6.9)
					PUT postcommissural	−1.9 (7.1)
					VST	−1.0 (9.7)

DB—Dual Bolus; SB—Single Bolus; SRTM—Simplified reference tissue method; KIN—Kinetic; G—Graphical Logan with reference input; BCI—(E) Bolus plus constant infusion with equilibrium analysis; VST—Ventral Striatum; DST—Dorsal Striatum; PUT—Putamen; CAD—Caudate; L—left; R—right.
*$p < 0.05$.
§No percentage provided.
¶As task was performed between the baseline and posttask PET scans.

replicate the previous studies that had reported DA release in the VST following reward tasks (Martin-Soelch et al., 2011; Schott et al, 2008) because their results showed a modest reduction in [^{11}C]raclopride BP_{ND} in the postcommissural caudate that failed to survive multiple corrections, and no change in the other subdivisions of the striatum. Despite the debates in the literature over whether [^{11}C]raclopride and PET can detect DA release during or following cognitive and reward processing, the data in Table 5 suggest that the relatively unreliable and small magnitude of DA release associated with these physiological challenges do not support their use for neurochemistry in clinical populations.

2.1.1.5. Alpha-Methyl-Para-Tyrosine–Induced DA Depletion

Previous investigations have scanned human subjects before and after the administration of the DA depleting drug Alpha-Methyl-Para-Tyrosine (AMPT) (1.0 g every 6 h for 48 h) and reported an increase in [^{123}I]IBZM and [^{11}C]raclopride BP_{ND} (Laruelle et al., 1997a). This paradigm, which is essentially the converse of the amphetamine-induced DA release paradigm, has been successfully used to estimate the level of DA at rest by measuring the fraction of $D_{2/3}$ receptors that are occupied by it at baseline. For example, studies using this paradigm have reported increased and decreased baseline $D_{2/3}$ receptor occupancy in patients with schizophrenia (Abi-Dargham et al., 2000; Kegeles et al., 2010) and cocaine addiction (Martinez et al., 2009) relative to matched controls. One study that evaluated both amphetamine-induced DA release and AMPT-induced DA depletion in schizophrenia and matched controls reported a tight correlation between both outcome measures in patients, but not in controls (Abi-Dargham et al., 2009). The lack of correlation between the two paradigms in controls suggests that while shared mechanisms may contribute to abnormalities in both the stimulated and nonstimulated DA transmission in schizophrenia, they likely represent two distinct modes of neurochemical transmission in the brain that need to be investigated in clinical disorders.

2.1.1.6. Agonist vs Antagonist Radiotracer Imaging

As discussed previously, [^{11}C]raclopride is an established striatal radiotracer used frequently in PET imaging; however, the tracer's antagonistic properties prevent it from effectively capturing all the dopaminergic activity in the striatum. Like most receptors, $D_{2/3}$ receptors (hereafter referred to as D_2) can be divided into two categories based on agonist affinity: D_{2high} (G-protein coupling) and D_{2low} (no G-protein coupling). Since [^{11}C]raclopride is an antagonist radiotracer, it binds with equal affinity to both receptor states; therefore, the binding fails to differentiate between the two states of the receptors at any given time. Thus, an agonist radiotracer would be expected to have increased sensitivity to endogenous DA since it is thought that DA competes more efficiently at D_{2high} receptor states. (−)-*N*-propyl-norapomorphine (NPA) is an agonist radiotracer first developed by Hwang et al. because of the tracer's different affinities for the D_{2high}

Table 6 Agonist vs Antagonist Tracer Displacement Following an Oral Amphetamine Challenge

	Amphetamine (0.5 mg/kg) Induced Δ in BP_{ND}			Amphetamine (0.3 mg/kg) Induced Δ in BP_{ND}		
Region	**[^{11}C]RAC (Narendran et al., 2010a)**	**[^{11}C]NPA**	**NPA/RAC**	**[^{11}C]RAC (Shotbolt et al., 2011)**	**[^{11}C]PHNO**	**PHNO/RAC**
Caudate	−8.4 ± 4.2	−16.1 ± 6.1	1.92	−7.0 ± 4.0	−12.0 ± 7.0	1.71
Putamen	−14.7 ± 4.8	−21.9 ± 4.9	1.49	−11.0 ± 6.0	−16.0 ± 4.0	1.45
Ventral striatum	−9.7 ± 4.4	−16.0 ± 7.0	1.65	−13.0 ± 7.0	−21.0 ± 6.0	1.62
Striatum	−12.3 ± 4.4	−19.6 ± 4.3	1.59	–	–	–

Antagonist radiotracer: [^{11}C]raclopride (RAC); Agonist radiotracers [^{11}C]NPA, [^{11}C]PHNO.

and D_{2low} receptor states (K_{Dhigh} = 0.4 and K_{Dlow} = 23 nM) (Hwang et al., 2000). When compared to the agonistic radiotracer [^{11}C]raclopride, [^{11}C]NPA displayed a higher change in BP_{ND} after an amphetamine administration in baboons (Narendran et al., 2004). It was concluded that the effect of amphetamine on [^{11}C]NPA-specific binding was greater by 42% when compared to that of [^{11}C]raclopride under the same conditions (Narendran et al., 2004). In a follow-up study, [^{11}C]NPA and [^{11}C]raclopride were used to image the human striatum before and after the administration of 0.5 mg/kg of amphetamine in the same subjects (Narendran et al., 2010a). As shown in Table 6, the results reflected similar findings in which [^{11}C]NPA specific binding showed a 50–90% greater decrease after the amphetamine challenge as compared to [^{11}C]raclopride in striatal subdivisions (Narendran et al., 2010a). Based on the results of these studies, it can be inferred that the displacement of $D_{2/3}$ agonist radiotracers is more sensitive than $D_{2/3}$ antagonist tracers after an amphetamine challenge.

In addition to [^{11}C]NPA, the agonist radiotracer [^{11}C]MNPA (a methoxy-tagged version) was compared to the antagonist radiotracer [^{11}C]raclopride in order to determine its vulnerability to endogenous competition. [^{11}C]MNPA is more easily synthesized and has a higher in vitro affinity than [^{11}C]NPA for D_{2high} receptors (Gao et al., 1990; Neumeyer et al., 1990; Seneca et al., 2006). Seneca et al. compared the two tracers in four cynomolgus monkeys at varying doses of intravenous amphetamine (0.1–1.0 mg/kg) and found that [^{11}C]MNPA has an increased sensitivity to an acute amphetamine challenge when compared to [^{11}C]raclopride (Seneca et al., 2006). Thus, this result is in agreement with previous agonist vs antagonist comparison studies (Cumming et al., 2002; Narendran et al., 2004).

Another agonist radiotracer [^{11}C](+)-PHNO was introduced by Wilson et al. as a viable D_{2high} tracer for PET imaging (Wilson et al., 2005). In cats, pretreatment with D_1, D_2, and D_3 specific antagonists found that [^{11}C](+)-PHNO binding is specific to D_2 receptors in the striatum (Ginovart et al., 2006). The same group also provided a within-subject comparison, which showed that [^{11}C](+)-PHNO binding in the striatum is 2.5 times greater than [^{11}C]NPA binding (Ginovart et al., 2006). In addition, the change in BP_{ND} for [^{11}C](+)-PHNO was 50% higher than that for [^{11}C]raclopride after 0.5 mg/kg of amphetamine, and 70% higher after a challenge of 2.0 mg/kg of amphetamine (Ginovart et al., 2006). It was concluded that [^{11}C](+)-PHNO is the best candidate for D_{2high} PET imaging in the striatum given the tracer's higher BP values and increased sensitivity to competition by endogenous DA when compared to not only the antagonist [^{11}C]raclopride but also to another agonist tracer [^{11}C]NPA as well (Ginovart et al., 2006). A later study however suggested that [^{11}C](+)-PHNO has a higher affinity for D_3 receptors than D_{2high} receptors based on the pattern of distribution when compared to [^{11}C]raclopride and [^{11}C]NPA (Narendran et al., 2006). Rabiner and colleagues subsequently confirmed these results in baboons, knockout mice, and humans by performing D_3 blocking studies with the use of highly selective D_3 and D_2 antagonists (Rabiner et al., 2009; Searle et al., 2010).

When compared to the antagonist radiotracer [^{11}C]raclopride in humans, Shotbolt et al. reported that the reduction in [^{11}C](+)-PHNO binding was approximately 50% greater than [^{11}C]raclopride binding after 0.3 mg/kg of oral amphetamine (Shotbolt et al., 2011). This result is consistent with that reported for the [^{11}C]NPA and [^{11}C]raclopride comparison after amphetamine was administered in humans (Table 6). Based on these head-to-head comparison studies, it can be concluded that $D_{2/3}$ agonist radiotracers are superior to $D_{2/3}$ antagonist radiotracers to image DA release in humans. The use of the $D_{2/3}$ agonist as opposed to antagonist radiotracers is likely to be beneficial when investigators are evaluating DA release in a patient population in which recruitment can be challenging (because the greater effect size to detect a DA effect will allow for the detection of significant differences with a smaller number of subjects) or relatively small increases in DA release are expected after a challenge drug or paradigm (such as a cognitive or reward task).

2.1.1.7. Imaging of Striatal vs Extrastriatal DA Receptors

Despite prominent results with the imaging of striatal DA in schizophrenia and addictive disorders, a need arose for a radiotracer with the ability to measure extrastriatal DA. The imaging of extrastriatal DA is made difficult due to the decreased levels of the $D_{2/3}$ receptor density in the cortex as compared to that in the striatum. Therefore, a radiotracer is needed with a sufficient signal-to-noise ratio, that is, a high affinity for DA $D_{2/3}$ receptors. The high signal-to-noise ratio would ensure that the radiotracer would be visible even under the conditions with low cortical DA $D_{2/3}$ density.

[^{11}C]FLB 457 was first developed by Halldin et al. in order to measure the $D_{2/3}$ receptors in extrastriatal brain regions (Halldin et al., 1995). Using two monkeys, Halldin et al. administered raclopride and haloperidol as challenge drugs to prove the specificity of [^{11}C]FLB 457 as a $D_{2/3}$ radiotracer (Halldin et al., 1995). Upon administration of 1.0 mg/kg of intravenous cold raclopride and 3.0 mg/kg of intravenous haloperidol, [^{11}C]FLB 457 binding decreased significantly in the cortex as well as in the striatum, which shows that the radiotracer is specific to $D_{2/3}$ receptors (Halldin et al., 1995). This study also resulted in the highest specific to nonspecific ratio of a $D_{2/3}$ PET receptor study with a 15 times higher accumulation of radioactivity in the striatum as compared to that in the cerebellum (4 and 2 times higher in the thalamus and cortex, respectively) (Halldin et al., 1995). Given that [^{11}C]FLB 457 demonstrates a high affinity ($1/K_D$, where K_D is the equilibrium disassociation constant) for DA $D_{2/3}$ receptors (in vitro $K_D = 0.018$ nM; raclopride $K_D = 1.30$ nM), a relatively high signal-to-noise ratio and reversible kinetics it is suitable for imaging extrastriatal $D_{2/3}$ receptors (Halldin et al., 1995).

Reproducibility Studies

In order to confirm the use of [^{11}C]FLB 457 as a reliable measure of $D_{2/3}$ receptor binding for PET imaging analysis, studies to test the reproducibility of the radiotracer were necessary. Vilkman et al. performed a test/retest of [^{11}C]FLB 457 on seven human subjects using graphical and reference tissue methods, which resulted in a test/retest variability of ≤10% in both the frontal and temporal cortexes (Vilkman et al., 2000). The test/retest variability is calculated as the difference in the BP_{ND} between the two scans and then normalized to the mean. In a similar study, Sudo et al. ($n = 8$) determined a test/retest variability of ≤9.4% in all regions of interest (temporal cortex, frontal cortex, occipital cortex, parietal cortex, and thalamus) except in the hippocampus (variability = 15.5%) (Sudo et al., 2001). This discrepancy in the results is largely based on the small size of the hippocampus when compared to the other brain regions of interest. Specifically for the frontal cortex, the variability was 7.7% (Sudo et al., 2001), which further validates [^{11}C]FLB 457 as a reliable measure of the $D_{2/3}$ receptor occupancy in cortical regions. Nevertheless, the results failed to inform in regard to the reproducibility of the radiotracer when applied to the stimulant-induced PET imaging paradigms typically used. Future studies performed by our group would evaluate the radiotracer using test retest in the same paradigm used in the measurement of amphetamine-induced DA release in humans.

[^{18}F]Fallypride vs [^{11}C]FLB 457

Mukherjee et al. introduced [^{18}F]fallypride in 1995 as an extrastriatal $D_{2/3}$ receptor radiotracer (Mukherjee et al., 1995). With a reported variability of ≤10% as well as a similar in vitro affinity to [^{11}C]FLB 457 (in vitro K_D of 0.03 nM) (Mukherjee et al., 1995) and the ability to measure both striatal and extrastriatal regions within the same scan, the question arose as to whether [^{11}C]FLB 457 or [^{18}F]fallypride was a more optimal $D_{2/3}$ radiotracer. Thus, it was crucial to determine which radiotracer would have a sufficient signal-to-noise ratio to measure DA release following a stimulant-induced challenge. Several primate studies demonstrated that the binding of [^{18}F]fallypride in vivo is vulnerable to endogenous DA following an amphetamine challenge in several extrastriatal regions (Mukherjee et al., 2005, 1997; Slifstein et al., 2004). However, three human studies concluded that while [^{18}F]fallypride can measure the DA release in some extrastriatal regions (such as thalamus, hippocampus, substantia nigra, and medial orbital frontal cortex) it cannot reliably measure the DA release in all four subdivisions of the prefrontal cortex (that include dorsolateral prefrontal cortex, medial prefrontal cortex, orbital frontal cortex, and anterior cingulate cortex) (Cropley et al., 2008; Riccardi et al., 2005; Slifstein et al., 2007, Table 7). The failure to measure the DA release in the prefrontal cortex is related to a lower signal-to-noise ratio, which is due to the higher in vivo K_D for [^{18}F]fallypride relative to the reported in vitro K_D (Slifstein et al., 2004). In addition, limitations exist with the radiotracer [^{18}F]fallypride however, which makes the interpretation of results across studies difficult. The initial issue when scanning with [^{18}F]fallypride is the inherently long scanning protocol required to obtain a reasonable washout of the radiotracer in the striatum. Therefore, the longer scan makes the resulting image more susceptible to head movement. In addition the [^{18}F] labeling does not allow for the baseline and postchallenge scan to occur on the same day. These variables lead to an increased within-subject variability, which in turn decreases the reliability of the radiotracer's ability to measure DA release.

On the other hand, multiple groups reported mixed conclusions as to the effectiveness of [^{11}C]FLB 457 measuring DA release postchallenge. In nonhuman primates, Chou et al. reported a decrease in the [^{11}C]FLB 457 binding after a 2.0-mg/kg intravenous amphetamine challenge (6.9–15.5% after 15 min; 5.9–16.3% after 3 h) (Chou et al., 2000). From these early results, it was concluded that [^{11}C]FLB 457 was a suitable extrastriatal radiotracer because the decreases observed were similar to that of [^{11}C]raclopride in the striatum (Chou et al., 2000). In contrast, [^{11}C]FLB 457 was reported as having no change in binding after a 1.0 mg/kg of methamphetamine challenge (Okauchi et al., 2001) or after an acute nicotine challenge in nonhuman primates (Tsukada et al., 2005). When using a cognitive challenge in humans, Aalto et al. reported that the in vivo binding of [^{11}C]FLB 457 decreased in the ventrolateral frontal cortex and anterior cingulate cortex as a result of working memory and attention tasks (Aalto et al., 2005). Montgomery et al. observed a decrease in the binding of [^{11}C]FLB 457 in the cortex after methylphenidate (Montgomery et al., 2007a). However the administration of [^{11}C]FLB 457 at nontracer doses makes the interpretation of these data difficult (Table 7). However, the aforementioned [^{11}C]FLB 457 studies had limitations; the use of nontracer doses greatly affects the $D_{2/3}$ receptor occupancy for it has been reported that doses should be restricted to 1.6 nM per 70 kg (for human subjects) in order to achieve a <5% occupancy at the $D_{2/3}$ receptors (Olsson et al., 1999; Sudo et al., 2001). In addition, a failure to quantify stimulant-induced changes in the cerebellum distribution volume (V_T) by using reference region analysis could result in an underestimation of

Table 7 Amphetamine-Induced Displacement of $[^{18}F]/[^{11}C]$Fallypride and $[^{11}C]$FLB 457 in Cortical Regions

Radiotracer	References	Challenge	Dose	n	Region	Effect on BP (% ± SD)
$[^{18}F]$fallypride	Cropley et al. (2008)	Amphetamine	0.5 mg/kg oral	14	Medial orbitofrontal cortex	−13.0 ± 4.8*
					Temporal cortex	−7.0 ± 3.1
					Medial temporal cortex	−7.8 ± 2.8*
	Riccardi et al. (2005)	Amphetamine	0.43 mg/kg oral	14	Amygdala	4.4 ± 3.4*
					Temporal cortex	3.7 ± 3.7*
	Narendran et al. (2009)	Amphetamine	0.5 mg/kg oral	12	Medial temporal lobe	−4.4 ± 11.8
					Anterior cingulate cortex	−8.7 ± 21.2
					Dorsolateral prefrontal cortex	−9.7 ± 25.3
					Orbital frontal cortex	−8.0 ± 27.4
					Medial prefrontal cortex	−6.9 ± 17.6
					Temporal cortex	−7.5 ± 10.9
					Parietal cortex	−10.0 ± 33.4
					Occipital cortex	−8.7 ± 30.0
$[^{11}C]$FLB 457	Narendran et al. (2009)	Amphetamine	0.5 mg/kg oral	12	Medial temporal lobe	−7.0 ± 6.1*
					Anterior cingulate cortex	−8.1 ± 8.2*
					Dorsolateral prefrontal cortex	−12.6 ± 15.2*
					Orbital frontal cortex	−5.1 ± 21.6
					Medial prefrontal cortex	−11.3 ± 14.3*
					Temporal cortex	−4.2 ± 9.1
					Parietal cortex	−12.4 ± 13.4*
					Occipital cortex	−4.7 ± 19.6
	Montgomery et al. (2007a)	Methylphenidate	40 or 60 mg oral	12	Amygdala/hippocampus	−3.0 ± 7.8
					Temporal cortex	−6.3 ± 9.2*
					Frontal cortex	−7.2 ± 10.6*
					Anterior cingulate	−5.3 ± 9.1

BP—Binding Potential.
*$p \leq 0.05$.

changes in cortical BP_{ND}. Lastly, many scans performed in previous studies were under 90 min, which may not be sufficient time to reliably estimate the [^{11}C]FLB 457 BP_{ND} in cortical regions.

University of Pittsburgh Studies

In 2009, our group performed a study in order to assess the signal-to-noise ratio of [^{11}C] FLB 457 and [^{11}C]fallypride in cortical regions of interest (temporal cortex, medial temporal lobe, anterior cingulate cortex, medial prefrontal cortex, orbitofrontal cortex, dorsolateral prefrontal cortex, parietal cortex, and occipital cortex) as well as the vulnerability of [^{11}C]FLB 457 and [^{11}C]fallypride binding to endogenous competition by DA following an acute (0.5 mg/kg) amphetamine challenge (Table 7) (Narendran et al., 2009). Twelve healthy volunteers were scanned under baseline and postamphetamine conditions (3 h postoral amphetamine administration) using both [^{11}C]FLB 457 and [^{11}C]fallypride (Narendran et al., 2009). The sequence of the radiotracers was counterbalanced for consecutive subjects in order to prevent any bias with subjects studied a minimum of 1 week and a maximum of 3 weeks apart (Narendran et al., 2009). Using a single bolus intravenous administration, the results were analyzed with the cerebellum as the reference region and using kinetic analysis with arterial input function (Narendran et al., 2009). In addition, the study used carbon-labeled fallypride in order to allow the baseline and postchallenge scan to occur on the same day (Narendran et al., 2009). The results of this study demonstrated that [^{11}C]FLB 457 is superior for imaging cortical DA postamphetamine challenge when compared to [^{11}C]fallypride (Narendran et al., 2009). The average BP_{ND} for [^{11}C]fallypride in most of the cortical regions is <0.5, which means the noise is twice as high as the signal (a $BP_{ND}=1$ indicates an equal ratio of specific to nonspecific signal), while the BP_{ND} for [^{11}C]FLB 457 is a reported 1.6 times greater than that for [^{11}C]fallypride (Narendran et al., 2009). When compared to one another, the amphetamine challenge produced a statistically significant displacement of [^{11}C]FLB 457 than [^{11}C]fallypride from baseline in five out of eight cortical regions of interest ([^{11}C]FLB 457 BP_{ND} displacement: medial temporal lobe, $-7 \pm 6.1\%$; anterior cingulate cortex, $-8.1 \pm 8.2\%$; dorsal lateral prefrontal cortex, $-12.6 \pm 15.2\%$; medial frontal cortex, $-11.3 \pm 14.3\%$; parietal cortex, $-4.7 \pm 19.6\%$) (Narendran et al., 2009). In a follow-up study, we measured the reproducibility of [^{11}C]FLB 457 in the absence of amphetamine to detect the impact of carryover mass, if any, from the baseline to postamphetamine scan on amphetamine-induced Δ BP_{ND}. Using the same methods as in the aforementioned study ($n=6$), our group reported a variability of ≤15% for BP_{ND}, which is consistent with previously published studies (Narendran et al., 2010b). In addition, no carryover mass-induced decrease in [^{11}C]FLB 457 BP_{ND} was observed. These results suggest that the impact of carryover mass, if any, on BP_{ND} after an acute amphetamine challenge is likely of a small magnitude in the paradigm that is used to measure the DA release in the cortex (Narendran et al., 2010b).

A similar PET study published in 2008 by Aalto et al. assessed the vulnerability of [^{11}C]FLB 457 at baseline and when faced with a challenge of 0.3 mg/kg of amphetamine or placebo (Aalto et al., 2008). They concluded that the cortical radiotracer is not a useful tool in measuring amphetamine-induced DA release in humans (Aalto et al., 2008). However, a greater variability in amphetamine plasma levels (range from 38 to 138 ng/ml) in the Aalto study as opposed to our study (range from 62 to 84 ng/ml) (Narendran et al., 2010b) explains the inability of amphetamine to displace [^{11}C]FLB 457 binding in this particular study. In addition, a higher injected FLB 457 mass, that is, nontracer doses, may have decreased the ability to detect an effect for amphetamine compared to the placebo challenge condition. These limitations among others (scan time, method of analysis) could be the reason why their study resulted in negative findings.

Microdialysis Study

In order to validate the use of [^{11}C]FLB 457 as a tool to measure amphetamine-induced DA release, our group performed a study in 2012 to demonstrate that the changes in DA binding following an amphetamine challenge were in fact related to the increases in DA concentration as measured by microdialysis. Five rhesus monkeys underwent a series of microdialysis and PET experiments with varying doses of amphetamine (0.3, 0.5, and 1.0 mg/kg) in order to derive a relationship between increases in dialysate DA in the frontal cortex and decreases in the [^{11}C]FLB 457 binding (Narendran et al., 2013). As a result, amphetamine was found to have a dose-dependent effect on increases in DA and decreases in [^{11}C]FLB 457 BP_{ND} (Narendran et al., 2013). This linear outcome suggests the use of this paradigm as a way to characterize changes in prefrontal cortical DA release not only in healthy controls but also in a wide variety of neuropsychiatric disorders.

Reference Regions

A reoccurring issue when imaging with [^{11}C]FLB 457 is whether or not to use the cerebellum as a reference region. Although some studies have not detected a decrease in cerebellar V_T following the administration of amphetamine in humans (Aalto et al., 2008; Narendran et al., 2009), others have suggested that up to 75% of binding of [^{11}C]FLB 457 in the cerebellum is specific to $D_{2/3}$ receptors (Ahmad et al., 2006; Asselin et al., 2007; Delforge et al., 2001; Montgomery et al., 2007a). In order to answer the question as to whether the cerebellum is a suitable reference region for analysis, we performed a blocking study using aripiprizole (a $D_{2/3}$ receptor agonist) as the challenge drug (Narendran et al., 2011). Six volunteers were scanned at baseline and again 3 h after a single oral dose of 15 mg of aripiprazole (Narendran et al., 2011). Not only was the cerebellum used as a reference region, but the pons and centrum semiovale were also assessed (Narendran et al., 2011). The percent displacement of [^{11}C]FLB 457 binding by aripiprazole was found to be significant in the cortex as well as in the cerebellum ($-33 \pm 42\%$; $-17 \pm 12\%$, respectively) but not in the pons or the centrum semiovale ($-10 \pm 10\%$; $-3 \pm 12\%$, respectively;

Narendran et al., 2011). Therefore, in order to further interpret these results, Lassen plots were used. Briefly, Lassen et al. introduced Lassen plots as a means to derive receptor occupancy by plotting regional changes in the V_T after drug administration, as well as providing an independent estimate of nonspecific binding (Lassen et al., 1995). This type of analysis allows for occupancy to be determined without a true reference region. The results of this study showed that cerebellar V_T is highly correlated with nonspecific binding as derived using centrum semiovale V_{ND} (an estimate of nonspecific binding) and the Lassen-derived V_{ND} (Narendran et al., 2011). Although the pons and centrum semiovale showed a significantly less displacement after aripiprazole administration, the contribution of specific binding to $D_{2/3}$ receptors in the cerebellum was much lower than in the cortical regions of interest (Narendran et al., 2011).

The question then arose as to whether the pons and/or the centrum semiovale could be used as reference regions for [^{11}C]FLB 457, and thus possibly improve the reproducibility of the tracer and allow for a greater detection of amphetamine-induced displacement. The pons and the centrum semiovale were then applied as reference regions in the previous test/retest study (Narendran et al., 2010b) as well as in the amphetamine challenge study (Narendran et al., 2009). However, when the test/retest results were reanalyzed using the pons and the centrum semiovale, the variability of the cortical regions of interest BP_{ND} increased as compared to that of the cerebellum (Narendran et al., 2011). In addition, when the same reference regions were applied to the previous amphetamine challenge study, an overall loss of amphetamine-induced Δ [^{11}C]FLB 457 binding was observed (pons V_T: 0/8 regions of interest; centrum semiovale V_T: 3/8 regions of interest; cerebellum V_T 5/8 regions of interest) (Narendran et al., 2011). The negative findings can be explained by the size of the pons and the centrum semiovale, in which both are smaller than the cerebellum, which can render the areas more vulnerable to head movement (Narendran et al., 2011). Furthermore, the centrum semiovale comprises exclusively white matter and thus has slower kinetics (Narendran et al., 2011). Based on these data sets as well as previous studies, it was concluded that the cerebellum despite its limitations is the best available option for a reference region for [^{11}C]FLB 457 (Narendran et al., 2011). Nevertheless, investigators should be cautious in relying solely on the cerebellum as an input to quantify [^{11}C]FLB 457 BP_{ND} and amphetamine-induced DA release in the human cortex (Narendran et al., 2011).

2.1.1.8. D_1 Radiotracers

In contrast to that observed with DA $D_{2/3}$ receptors radiotracers, studies in nonhuman primates and humans report that the in vivo binding of D_1 radiotracers such as [^{11}C]NNC 112, [^{11}C]NNC 756, and [^{11}C]SCH 23390 are not vulnerable to the endogenous competition by DA (Abi-Dargham et al., 1999; Chou et al., 1999; Verhoeff et al., 2002). While the exact reasons for the lack of vulnerability of D_1 radiotracer binding to endogenous competition by DA is not known, it does highlight the fact that other issues besides the simple occupancy model such as synaptic proximity of receptors, agonist-mediated endocytosis, and fraction of receptors configured in a high-affinity state for the agonist might be involved in the displacement of a PET radiotracer after a pharmacological challenge (Laruelle, 2000).

2.2. IMAGING OF SEROTONIN TRANSMISSION

2.2.1. $5HT_{1A}$ Receptor Radioligands

The successful use of DA $D_{2/3}$ antagonist radioligands to measure DA transmission led to investigations to evaluate the potential of $5HT_{1A}$ receptor antagonist radioligands such as [^{11}C] *N*-(2-(4-(2-Methoxyphenyl)-1-piperazinyl)ethyl)-*N*-(2- pyridyl)cyclohexanecarboxamide trihydrochloride (WAY) 100635 and [^{18}F] 2′-methoxyphenyl-(*N*-2′-pyridinyl)-p-fluoro-benzamidoethyipiperazine (MPPF) to measure the serotonin release in the brain. Four animal studies and one human study are consistent in reporting that relatively large alterations in endogenous serotonin concentration do not lead to significant changes in the binding of [^{11}C]WAY 100635 (Hume et al., 2001; Maeda et al., 2001; Parsey et al., 1999; Rabiner et al., 2002; Rice et al., 2001). In contrast, studies evaluating the in vivo binding of [^{18}F]MPPF before and after the administration of the serotonin releaser fenfluramine have reported mixed results. In rodents using beta-sensitive microprobe and microdialysis, Zimmer et al. observed a dose-dependent decrease in the binding of [^{18}F] MPPF that correlated with increases in serotonin concentration after a fenfluramine challenge (Zimmer et al., 2002). Udo de Haes and colleagues however were unable to replicate these results in primates (Udo de Haes et al., 2006) and humans (Udo de Haes et al., 2002), thereby casting doubt on the proposed use of this radiotracer as a tool to measure serotonin transmission. A more recent study reported a 44% decrease in the [^{18}F]MPPF binding in the dorsal raphe nucleus, but no change in the cortical regions of interest after an oral dose of fluoxetine (Sibon et al., 2008). While the exact reason for this discrepancy is unclear, the authors hypothesized that the presynaptic $5HT_{1A}$ receptors, which predominantly populate the dorsal raphe nucleus as opposed to the postsynaptic receptors in the cortex, may be more vulnerable to endogenous competition (or receptor internalization) by serotonin. In summary, the data with [^{11}C]WAY 100635 and [^{18}F]MPPF do not support their use as a reliable PET probe to measure serotonin release in humans.

Milak and colleagues investigated the potential of a $5HT_{1A}$ agonist radioligand, [^{11}C]CUMI-101 to measure serotonin release in nonhuman primates as $D_{2/3}$ agonist radioligands were shown to be superior to $D_{2/3}$ antagonist radioligands to measure DA transmission (Milak et al., 2011). The results of their study in which the in vivo binding of [^{11}C]CUMI was 24% lower after fenfluramine (2.5 mg/kg) and 15–30% lower after citalopram (2–4 mg/kg) suggest that $5HT_{1A}$ agonist radiotracers, unlike antagonists, might be vulnerable to endogenous competition by serotonin. Further validation and replication of this observation in humans are necessary prior to adaptation of this technique to measure serotonin release in clinical disorders.

2.2.2. $5HT_{2A}$ Receptor Radiotracers

Two $5HT_{2A}$ antagonist radioligands—[^{11}C]MDL 100907 and [^{18}F]altanserin have been assessed for their vulnerability to an acute serotonin challenge. For [^{11}C]MDL 100907, a rodent study with fenfluramine and a human study with tryptophan depletion (a serotonin depleter) concluded that its in vivo binding is insensitive to alterations in serotonin concentrations (Hirani et al., 2003; Talbot et al., 2012). For [^{18}F]altanserin and its deuterated analog, the published data are mixed with two out of three imaging studies reporting that its binding is insensitive to fluctuations in serotonin following fenfluramine (Staley et al., 2001) and citalopram (Pinborg et al., 2004). The third [^{18}F]altanserin study, which was performed in humans with dexfenfluramine 40 and 60 mg, reported a dose-dependent decrease of approximately 10 and 15% in V_T in both the cortical and subcortical regions (Quednow et al., 2011). Interestingly, in this study, the fenfluramine-induced change in the plasma BP (BP_P) was of a lesser magnitude than that observed for V_T and reached statistical significance only in a few cortical regions of interest. Thus, the relatively modest change in BP_P, the author's failure to report the fenfluramine-induced change for a more widely used PET outcome measure BP_{ND}, and discrepancies in the literature highlight the challenges in using [^{18}F]altanserin to measure endogenous serotonin levels in clinical studies.

2.2.3. $5HT_{1B}$ Receptor Radioligands

The advancement of the PET radiotracers [^{11}C]AZ10419369 and [^{11}C]P943 allowed for the in vivo measurement of $5HT_{1B}$ receptors in humans (Gallezot et al., 2010; Pierson et al., 2008). In a study in cynolomogus monkeys, Finnema and colleagues were able to show that the in vivo binding of [^{11}C]AZ10419369 was reduced in a dose-dependent fashion after an acute fenfluramine challenge (Finnema et al., 2010). This preliminary observation was recently confirmed in a second cohort of animals using more robust equilibrium analysis methods. In this study, the authors reported a mean reduction of −12 and −33% in [^{11}C]AZ10419369 BP_{ND} across the eight regions of interest, which included the dorsolateral prefrontal cortex, hippocampus, occipital cortex, caudate, putamen, thalamus, midbrain, and globus pallidus (Finnema et al., 2011). Interestingly, these exciting preliminary results have been confirmed in nonhuman primates with another $5HT_{1B}$ radiotracer ([^{11}C]P943) as well (Cosgrove et al., 2011; Ridler et al., 2011). Riddler et al., in an in depth characterization of the radiotracer [^{11}C]P943, were able to demonstrate $5HT_{1B}$ receptor occupancy by serotonin released following a fenfluramine (13–40%), citalopram (9–24%), and amphetamine (25%) challenges. The relatively large displacements in radiotracer binding that are observed after multiple serotonin-releasing drugs for both the available $5HT_{1B}$ radiotracers suggest that if this technique is replicated in humans, it is likely to represent a breakthrough in the imaging of serotonin release in clinical disorders.

2.3. IMAGING OF GAMMA-AMINOBUTYRIC ACID TRANSMISSION

Preclinical studies suggest that in the presence of elevated γ-aminobutyric acid (GABA) levels, the affinity of the GABA-A receptor for benzodiazepine substrates is enhanced through a conformational change in the receptor sites. Previous imaging groups utilized this concept to evaluate the in vivo vulnerability of the PET radiotracer [^{11}C]flumazenil binding to changes in GABA transmission. For example, two independent groups have reported an increase in [^{11}C]flumazenil BP_{ND} following isoflurane (+16–28%), sevoflurane (+20%), and propofol (+8%)—all volatile anesthetics that presumably work via the enhancement of GABA transmission at GABA-A receptors, thereby resulting in a dose-dependent increase in the binding of the benzodiazepine substrate, [^{11}C]flumazenil (Gyulai et al., 2001; Salmi et al., 2004). Interestingly, this increase in flumazenil binding was not observed following the administration of ketamine, an anesthetic that works via the *N*-Methyl-D-aspartate (NMDA) system, suggesting that the observed effect is specific to anesthetics that work the GABA-ergic system. In a more recent imaging study, we studied eight individuals with [^{11}C]flumazenil before and after the administration of tiagabine (oral, 16 mg), a selective GABA transporter reuptake blocker that leads to increased levels of GABA in the brain (Frankle et al., 2009). Consistent with previous imaging studies with anesthetics that have measured a GABA shift, we observed an approximately ~15% increase in [^{11}C]flumazenil binding (V_T and BP_{ND}) across the various cortical regions of interest. In a replication study, we were able to demonstrate that this increase in the [^{11}C]flumazenil binding (V_T) after tiagabine is dependent on the administered dose consistent with greater elevations in GABA levels (Frankle et al., in press). These results are also consistent with another [^{11}C]flumazenil imaging study in humans that reported trend level increases in BP_{ND} that ranged from 0.6 to 22% after the administration of the irreversible GABA-transaminase inhibitor, Gamma-vinyl-GABA that elevates GABA levels as well (Weber et al., 1999). Pending further validation of this technique, the ability to image a GABA shift with benzodiazepine site-specific PET radiotracers will provide an effective tool to assess endogenous GABA transmission in both health and disease.

2.4. IMAGING OF GLUTAMATE TRANSMISSION

The availability of metabotropic glutamate receptor 5 radiotracers such as [^{11}C]ABP 688 and [^{18}F]FPEB has led to investigations exploring their utility to measure endogenous glutamate levels (Miyake et al., 2011). A PET study with [^{11}C]ABP 688 used in a small number of baboons reported a significant decrease (10–20%) in the in vivo BP of this radiotracer after the administration of *N*-acetyl cysteine (50 mg/kg). In addition, no significant changes were observed in the cerebellum V_{ND} and free fraction in the plasma (f_P), thereby suggesting that the observed decrease in BP_{ND} is related to changes in either receptor density (B_{max}) or affinity of the radiotracer. As *N*-acetyl cysteine facilitates the activity of the cysteine glutamate antiporter and increases glutamate concentrations in the extrasynaptic space, the authors hypothesized that the decrease in BP_{ND} may have led to an affinity shift in the binding of the radiotracer for the allosteric site. Future replication studies in humans and an understanding of the molecular mechanisms that lead to this affinity shift may allow for the *N*-acetyl cysteine and [^{11}C]ABP 688 PET paradigm to characterize glutamate transmission in clinical disorders.

2.5. IMAGING OF OPIOID TRANSMISSION

Opioid receptors and their naturally occurring peptides (such as enkephalins and dinorphins) have long been implicated in chronic pain and addiction. However, until recently, there has been no PET imaging paradigms to measure endogenous opioid transmission to study these disorders. In a recent PET imaging study, Mitchell and colleagues used the μ-opioid receptor radioligand [^{11}C]carfentanil to investigate alcohol-induced opioid release in alcoholics and matched controls. In this study, an oral alcohol challenge led to a statistically significant decrease in [^{11}C]carfentanil BP_{ND} in the VST (mean reduction in BP_{ND} ~ 30%) in both healthy controls and alcoholics. In addition to the observation, a significant decrease in the [^{11}C]carfentanil BP_{ND} was also noted in the medial orbital frontal cortex in healthy controls, but not in alcoholic subjects. Using a somewhat related paradigm, Colasanti et al. used [^{11}C]carfentanil in another recent PET study to measure μ-opioid receptor BP_{ND} before and after an amphetamine administration (0.5 mg/kg) in healthy humans (Colasanti et al., 2012). Despite using amphetamine, a monoamine stimulant that has no appreciable affinity for μ-opioid receptors, as their challenge they reported a significant decrease in [^{11}C]carfentanil BP_{ND} in several cortical and striatal regions of interest as compared to baseline (mean ΔBP_{ND}: cortical regions, 6.5–7.0%; striatal subdivisions, 6.2–10.2%) (Colasanti et al., 2012). Based on this result and prior in vitro work, the authors (Colasanti et al., 2012) hypothesized that amphetamine-induced increases in DA are associated with endogenous opioid release, which in turn can be detected as a reduction in [^{11}C]carfentanil BP_{ND} (Colasanti et al., 2012). In response to these studies showing some success with the novel radiotracer, Gusterstam and colleagues examined the change [^{11}C]carfentanil binding after amphetamine administration in healthy controls (Guterstam et al., 2013). In a double-blind paradigm in which 10 healthy men were scanned with [^{11}C]carfentanil after receiving either 0.3 mg/kg i.v. amphetamine or a placebo, the authors found no statistically significant changes in tracer binding, thus concluding that the stimulant has no appreciable effect on the μ-opioid system (Guterstam et al., 2013). Despite the mixed findings in these studies, further studies combining microdialysis and PET imaging are needed to clarify that the relationship between [^{11}C]carfentanil displacement is in fact related to endogenous opioid release, as well as whether stimulant challenges such as amphetamine have any effect on this endorphin pathway.

3. CONCLUSIONS

In summary, recent advances in PET imaging paradigms with both well-established and novel PET radiotracers have an expanded ability of investigators to examine neurochemical abnormalities in various psychiatric and addictive disorders. The future use of these novel-imaging paradigms is likely to broaden our understanding of neurochemical transmission in vivo and allow for the advancement of better treatments for these disorders.

REFERENCES

Aalto, S., Bruck, A., Laine, M., Nagren, K., Rinne, J.O., 2005. Frontal and temporal dopamine release during working memory and attention tasks in healthy humans: a positron emission tomography study using the high-affinity dopamine D_2 receptor ligand [^{11}C]FLB 457. J. Neurosci. 25 (10), 2471–2477.

Aalto, S., Hirvonen, J., Kaasinen, V., Hagelberg, N., Kajander, J., Nagren, K., Seppala, T., Rinne, J.O., Scheinin, H., Hietala, J., 2008. The effects of d-amphetamine on extrastriatal dopamine D_2/D_3 receptors: a randomized, double-blind, placebo-controlled PET study with [^{11}C]FLB 457 in healthy subjects. Eur. J. Nucl. Med. Mol. Imaging.

Abi-Dargham, A., Rodenhiser, J., Printz, D., Zea-Ponce, Y., Gil, R., Kegeles, L., Weiss, R., Cooper, T., Mann, J.J., Van Heertum, R., Gorman, J., Laruelle, M., 2000. Increased baseline occupancy of D_2 receptors by dopamine in schizophrenia. Proc. Natl. Acad. Sci. USA 97 (14), 8104–8109.

Abi-Dargham, A., Simpson, N., Kegeles, L., Parsey, R., Hwang, D.R., Anjilvel, S., Zea-Ponce, Y., Lombardo, I., Van Heertum, R., Mann, J.J., Foged, C., Halldin, C., Laruelle, M., 1999. PET studies of binding competition between endogenous dopamine and the D_1 radiotracer [^{11}C]NNC 756. Synapse 32 (2), 93–109.

Abi-Dargham, A., van de Giessen, E., Slifstein, M., Kegeles, L.S., Laruelle, M., 2009. Baseline and amphetamine-stimulated dopamine activity are related in drug-naive schizophrenic subjects. Biol. Psychiatry 65 (12), 1091–1093.

Ahmad, R., Hirani, E., Grasby, P.M., Hume, S.P., 2006. Effect of reduction in endogenous dopamine on extrastriatal binding of [^{11}C]FLB 457 in rat brain—an ex vivo study. Synapse 59 (3), 162–172.

Asselin, M.C., Montgomery, A.J., Grasby, P.M., Hume, S.P., 2007. Quantification of PET studies with the very high-affinity dopamine D_2/D_3 receptor ligand [^{11}C]FLB 457: re-evaluation of the validity of using a cerebellar reference region. J. Cereb. Blood Flow Metab. 27 (2), 378–392.

Badgaiyan, R.D., Fischman, A.J., Alpert, N.M., 2007. Striatal dopamine release in sequential learning. NeuroImage 38 (3), 549–556.

Barrett, S.P., Boileau, I., Okker, J., Pihl, R.O., Dagher, A., 2004. The hedonic response to cigarette smoking is proportional to dopamine release in the human striatum as measured by positron emission tomography and [^{11}C]raclopride. Synapse 54 (2), 65–71.

Boileau, I., Assaad, J.M., Pihl, R.O., Benkelfat, C., Leyton, M., Diksic, M., Tremblay, R.E., Dagher, A., 2003. Alcohol promotes dopamine release in the human nucleus accumbens. Synapse 49 (4), 226–231.

Boileau, I., Dagher, A., Leyton, M., Welfeld, K., Booij, L., Diksic, M., Benkelfat, C., 2007. Conditioned dopamine release in humans: a positron emission tomography [^{11}C]raclopride study with amphetamine. J. Neurosci. 27 (15), 3998–4003.

Bossong, M.G., van Berckel, B.N., Boellaard, R., Zuurman, L., Schuit, R.C., Windhorst, A.D., van Gerven, J.M., Ramsey, N.F., Lammertsma, A.A., Kahn, R.S., 2009. Delta 9-tetrahydrocannabinol induces dopamine release in the human striatum. Neuropsychopharmacology 34 (3), 759–766.

Breier, A., Su, T.P., Saunders, R., Carson, R.E., Kolachana, B.S., deBartolomeis, A., Weinberger, D.R., Weisenfeld, N., Malhotra, A.K., Eckelman, W.C., Pickar, D., 1997. Schizophrenia is associated with elevated amphetamine-induced synaptic dopamine concentrations: evidence from a novel positron emission tomography method. Proc. Natl. Acad. Sci. USA 94 (6), 2569–2574.

Brody, A.L., Mandelkern, M.A., Olmstead, R.E., Allen-Martinez, Z., Scheibal, D., Abrams, A.L., Costello, M.R., Farahi, J., Saxena, S., Monterosso, J., London, E.D., 2009. Ventral striatal dopamine release in response to smoking a regular vs a denicotinized cigarette. Neuropsychopharmacology 34 (2), 282–289.
Brody, A.L., Mandelkern, M.A., Olmstead, R.E., Scheibal, D., Hahn, E., Shiraga, S., Zamora-Paja, E., Farahi, J., Saxena, S., London, E.D., McCracken, J.T., 2006. Gene variants of brain dopamine pathways and smoking-induced dopamine release in the ventral caudate/nucleus accumbens. Arch. Gen. Psychiatry 63 (7), 808–816.
Brody, A.L., Olmstead, R.E., London, E.D., Farahi, J., Meyer, J.H., Grossman, P., Lee, G.S., Huang, J., Hahn, E.L., Mandelkern, M.A., 2004. Smoking-induced ventral striatum dopamine release. Am. J. Psychiatry 161 (7), 1211–1218.
Chou, Y.H., Halldin, C., Farde, L., 2000. Effect of amphetamine on extrastriatal D_2 dopamine receptor binding in the primate brain: a PET study. Synapse 38 (2), 138–143.
Chou, Y.H., Karlsson, P., Halldin, C., Olsson, H., Farde, L., 1999. A PET study of D_1-like dopamine receptor ligand binding during altered endogenous dopamine levels in the primate brain. Psychopharmacology 146, 220–227.
Colasanti, A., Searle, G.E., Long, C.J., Hill, S.P., Reiley, R.R., Quelch, D., Erritzoe, D., Tziortzi, A.C., Reed, L.J., Lingford-Hughes, A.R., Waldman, A.D., Schruers, K.R., Matthews, P.M., Gunn, R.N., Nutt, D.J., Rabiner, E.A., 2012. Endogenous opioid release in the human brain reward system induced by acute amphetamine administration. Biol. Psychiatry.
Cosgrove, K.P., Kloczynski, T., Nabulsi, N., Weinzimmer, D., Lin, S.F., Staley, J.K., Bhagwagar, Z., Carson, R.E., 2011. Assessing the sensitivity of [^{11}C]P943, a novel 5-HT_{1B} radioligand, to endogenous serotonin release. Synapse 65 (10), 1113–1117.
Cropley, V.L., Innis, R.B., Nathan, P.J., Brown, A.K., Sangare, J.L., Lerner, A., Ryu, Y.H., Sprague, K.E., Pike, V.W., Fujita, M., 2008. Small effect of dopamine release and no effect of dopamine depletion on [^{18}F] fallypride binding in healthy humans. Synapse 62 (6), 399–408.
Cumming, P., Wong, D.F., Gillings, N., Hilton, J., Scheffel, U., Gjedde, A., 2002. Specific binding of [^{11}C] raclopride and N-[^{3}H]propyl-norapomorphine to dopamine receptors in living mouse striatum: occupancy by endogenous dopamine and guanosine triphosphate-free G protein. J. Cereb. Blood Flow Metab. 22 (5), 596–604.
de la Fuente-Fernandez, R., Phillips, A.G., Zamburlini, M., Sossi, V., Calne, D.B., Ruth, T.J., Stoessl, A.J., 2002. Dopamine release in human ventral striatum and expectation of reward. Behav. Brain. Res. 136 (2), 359–363.
de la Fuente-Fernandez, R., Ruth, T.J., Sossi, V., Schulzer, M., Calne, D.B., Stoessl, A.J., 2001. Expectation and dopamine release: mechanism of the placebo effect in Parkinson's disease. Science 293 (5532), 1164–1166.
Delforge, J., Bottlaender, M., Loc'h, C., Dolle, F., Syrota, A., 2001. Parametric images of the extrastriatal D_2 receptor density obtained using a high-affinity ligand (FLB 457) and a double-saturation method. J. Cereb. Blood Flow Metab. 21 (12), 1493–1503.
Dewey, S.L., Smith, G.S., Logan, J., Brodie, J.D., Fowler, J.S., Wolf, A.P., 1993. Striatal binding of the PET ligand ^{11}C-raclopride is altered by drugs that modify synaptic dopamine levels. Synapse 13 (4), 350–356.
Drevets, W.C., Gautier, C., Price, J.C., Kupfer, D.J., Kinahan, P.E., Grace, A.A., Price, J.L., Mathis, C.A., 2001. Amphetamine-induced dopamine release in human ventral striatum correlates with euphoria. Biol. Psychiatry 49 (2), 81–96.
Finnema, S.J., Varrone, A., Hwang, T.J., Gulyas, B., Pierson, M.E., Halldin, C., Farde, L., 2010. Fenfluramine-induced serotonin release decreases [^{11}C]AZ10419369 binding to 5-HT1B-receptors in the primate brain. Synapse 64 (7), 573–577.
Finnema, S.J., Varrone, A., Hwang, T.J., Halldin, C., Farde, L., 2011. Confirmation of fenfluramine effect on 5-HT(1B) receptor binding of [^{11}C]AZ10419369 using an equilibrium approach. J. Cereb. Blood Flow Metab.
Frankle, W.G., Cho, R.Y., Mason, N.S., Chen, C.M., Himes, M., Walker, C., Lewis, D.A., Mathis, C.A., Narendran, R., 2012. [^{11}C]flumazenil binding is increased in a dose-dependent manner with tiagabine-induced elevations in GABA levels. PloS one 7 (2), e32443.
Frankle, W.G., Cho, R.Y., Narendran, R., Mason, N.S., Vora, S., Litschge, M., Price, J.C., Lewis, D.A., Mathis, C.A., 2009. Tiagabine increases [^{11}C]flumazenil binding in cortical brain regions in healthy control subjects. Neuropsychopharmacology 34 (3), 624–633.

Gallezot, J.D., Nabulsi, N., Neumeister, A., Planeta-Wilson, B., Williams, W.A., Singhal, T., Kim, S., Maguire, R.P., McCarthy, T., Frost, J.J., Huang, Y., Ding, Y.S., Carson, R.E., 2010. Kinetic modeling of the serotonin 5-HT_{1B} receptor radioligand [^{11}C]P943 in humans. J. Cereb. Blood Flow Metab. 30 (1), 196–210.
Gao, Y.G., Baldessarini, R.J., Kula, N.S., Neumeyer, J.L., 1990. Synthesis and dopamine receptor affinities of enantiomers of 2-substituted apomorphines and their N-n-propyl analogues. J. Med. Chem. 33 (6), 1800–1805.
Ginovart, N., Galineau, L., Willeit, M., Mizrahi, R., Bloomfield, P.M., Seeman, P., Houle, S., Kapur, S., Wilson, A.A., 2006. Binding characteristics and sensitivity to endogenous dopamine of [^{11}C]-(+)-PHNO, a new agonist radiotracer for imaging the high-affinity state of D_2 receptors in vivo using positron emission tomography. J. Neurochem. 97 (4), 1089–1103.
Ginovart, N., Wilson, A.A., Houle, S., Kapur, S., 2004. Amphetamine pretreatment induces a change in both D_2-Receptor density and apparent affinity: a [^{11}C]raclopride positron emission tomography study in cats. Biol. Psychiatry 55 (12), 1188–1194.
Goerendt, I.K., Messa, C., Lawrence, A.D., Grasby, P.M., Piccini, P., Brooks, D.J., 2003. Dopamine release during sequential finger movements in health and Parkinson's disease: a PET study. Brain 126 (Pt 2), 312–325.
Guo, N., Guo, W., Kralikova, M., Jiang, M., Schieren, I., Narendran, R., Slifstein, M., Abi-Dargham, A., Laruelle, M., Javitch, J.A., Rayport, S., 2010. Impact of D_2 receptor internalization on binding affinity of neuroimaging radiotracers. Neuropsychopharmacology 35 (3), 806–817.
Guterstam, J., Jayaram-Lindstrom, N., Cervenka, S., Frost, J.J., Farde, L., Halldin, C., Franck, J., 2013. Effects of amphetamine on the human brain opioid system—a positron emission tomography study. Int. J. Neuropsychopharmacol. 16 (4), 763–769.
Gyulai, F.E., Mintun, M.A., Firestone, L.L., 2001. Dose-dependent enhancement of in vivo GABA(A)-benzodiazepine receptor binding by isoflurane. Anesthesiology 95 (3), 585–593.
Hakyemez, H.S., Dagher, A., Smith, S.D., Zald, D.H., 2008. Striatal dopamine transmission in healthy humans during a passive monetary reward task. NeuroImage 39 (4), 2058–2065.
Halldin, C., Farde, L., Hogberg, T., Mohell, N., Hall, H., Suhara, T., Karlsson, P., Nakashima, Y., Swahn, C.G., 1995. Carbon-11-FLB 457: a radioligand for extrastriatal D_2 dopamine receptors. J. Nucl. Med. 36 (7), 1275–1281.
Hirani, E., Sharp, T., Sprakes, M., Grasby, P., Hume, S., 2003. Fenfluramine evokes 5-HT2A receptor-mediated responses but does not displace [^{11}C]MDL 100907: small animal PET and gene expression studies. Synapse 50 (3), 251–260.
Hume, S., Hirani, E., Opacka-Juffry, J., Myers, R., Townsend, C., Pike, V., Grasby, P., 2001. Effect of 5-HT on binding of [^{11}C] WAY 100635 to 5-HT(IA) receptors in rat brain, assessed using in vivo microdialysis and PET after fenfluramine. Synapse 41 (2), 150–159.
Hwang, D.R., Kegeles, L.S., Laruelle, M., 2000. (-)-N-[^{11}C]propyl-norapomorphine: a positron-labeled dopamine agonist for PET imaging of D_2 receptors. Nucl. Med. Biol. 27 (6), 533–539.
Innis, R.B., Malison, R.T., Al-Tikriti, M., Hoffer, P.B., Sybirska, E.H., Seibyl, J.P., Zoghbi, S.S., Baldwin, R.M., Laruelle, M.A., Smith, E., Charney, D.S., Heninger, G., Elsworth, J.D., Roth, R.H., 1992. Amphetamine-stimulated dopamine release competes in vivo for [^{123}I]IBZM binding to the D_2 receptor in non-human primates. Synapse 10, 177–184.
Kegeles, L.S., Abi-Dargham, A., Frankle, W.G., Gil, R., Cooper, T.B., Slifstein, M., Hwang, D.R., Huang, Y., Haber, S.N., Laruelle, M., 2010. Increased synaptic dopamine function in associative regions of the striatum in schizophrenia. Arch. Gen. Psychiatry 67 (3), 231–239.
Kim, S.E., Han, S.M., 2009. Nicotine- and methamphetamine-induced dopamine release evaluated with in-vivo binding of radiolabelled raclopride to dopamine D_2 receptors: comparison with in-vivo microdialysis data. Int. J. Neuropsychopharmacol. 12 (6), 833–841.
Kjaer, T.W., Bertelsen, C., Piccini, P., Brooks, D., Alving, J., Lou, H.C., 2002. Increased dopamine tone during meditation-induced change of consciousness. Brain Res. Cogn. Brain Res. 13 (2), 255–259.
Koepp, M.J., Gunn, R.N., Lawrence, A.D., Cunningham, V.J., Dagher, A., Jones, T., Brooks, D.J., Bench, C.J., Grasby, P.M., 1998. Evidence for striatal dopamine release during a video game. Nature 393 (6682), 266–268.
Lappin, J.M., Reeves, S.J., Mehta, M.A., Egerton, A., Coulson, M., Grasby, P.M., 2009. Dopamine release in the human striatum: motor and cognitive tasks revisited. J. Cereb. Blood Flow Metab. 29 (3), 554–564.

Laruelle, M., 2000. Imaging synaptic neurotransmission with in vivo binding competition techniques: a critical review. J. Cereb. Blood Flow Metab. 20 (3), 423–451.
Laruelle, M., Abi-Dargham, A., 1999. Dopamine as the wind of the psychotic fire: new evidence from brain imaging studies. J. Psychopharmacol. 13 (4), 358–371.
Laruelle, M., Abi-Dargham, A., van Dyck, C.H., Rosenblatt, W., Zea-Ponce, Y., Zoghbi, S.S., Baldwin, R.M., Charney, D.S., Hoffer, P.B., Kung, H.F., Innis, R.B., 1995. SPECT imaging of striatal dopamine release after amphetamine challenge. J. Nucl. Med. 36, 1182–1190.
Laruelle, M., D'Souza, C.D., Baldwin, R.M., Abi-Dargham, A., Kanes, S.J., Fingado, C.L., Seibyl, J.P., Zoghbi, S.S., Bowers, M.B., Jatlow, P., Charney, D.S., Innis, R.B., 1997a. Imaging D_2 receptor occupancy by endogenous dopamine in humans. Neuropsychopharmacology 17 (3), 162–174.
Laruelle, M., Iyer, R.N., Al-Tikriti, M.S., Zea-Ponce, Y., Malison, R., Zoghbi, S.S., Baldwin, R.M., Kung, H.F., Charney, D.S., Hoffer, P.B., Innis, R.B., Bradberry, C.W., 1997b. Microdialysis and SPECT measurements of amphetamine-induced dopamine release in nonhuman primates. Synapse 25, 1–14.
Lassen, N.A., Bartenstein, P.A., Lammertsma, A.A., Prevett, M.C., Turton, D.R., Luthra, S.K., Osman, S., Bloomfield, P.M., Jones, T., Patsalos, P.N., et al., 1995. Benzodiazepine receptor quantification in vivo in humans using [^{11}C]flumazenil and PET: application of the steady-state principle. J. Cereb. Blood Flow Metab. 15 (1), 152–165.
Lidstone, S.C., Schulzer, M., Dinelle, K., Mak, E., Sossi, V., Ruth, T.J., de la Fuente-Fernandez, R., Phillips, A.G., Stoessl, A.J., 2010. Effects of expectation on placebo-induced dopamine release in Parkinson disease. Arch. Gen. Psychiatry 67 (8), 857–865.
Maeda, J., Suhara, T., Ogawa, M., Okauchi, T., Kawabe, K., Zhang, M.R., Semba, J., Suzuki, K., 2001. In vivo binding properties of [carbonyl-^{11}C]WAY-100635: effect of endogenous serotonin. Synapse 40 (2), 122–129.
Martin-Soelch, C., Szczepanik, J., Nugent, A., Barhaghi, K., Rallis, D., Herscovitch, P., Carson, R.E., Drevets, W.C., 2011. Lateralization and gender differences in the dopaminergic response to unpredictable reward in the human ventral striatum. Eur. J. Neurosci. 33 (9), 1706–1715.
Martinez, D., Gil, R., Slifstein, M., Hwang, D.R., Huang, Y., Perez, A., Kegeles, L., Talbot, P., Evans, S., Krystal, J., Laruelle, M., Abi-Dargham, A., 2005. Alcohol dependence is associated with blunted dopamine transmission in the ventral striatum. Biol. Psychiatry 58 (10), 779–786.
Martinez, D., Greene, K., Broft, A., Kumar, D., Liu, F., Narendran, R., Slifstein, M., Van Heertum, R., Kleber, H.D., 2009. Lower level of endogenous dopamine in patients with cocaine dependence: findings from PET imaging of D_2/D_3 receptors following acute dopamine depletion. Am. J. Psychiatry 166 (10), 1170–1177.
Martinez, D., Narendran, R., Foltin, R.W., Slifstein, M., Hwang, D.R., Broft, A., Huang, Y., Cooper, T.B., Fischman, M.W., Kleber, H.D., Laruelle, M., 2007. Amphetamine-induced dopamine release: markedly blunted in cocaine dependence and predictive of the choice to self-administer cocaine. Am. J. Psychiatry 164 (4), 622–629.
Martinez, D., Saccone, P.A., Liu, F., Slifstein, M., Orlowska, D., Grassetti, A., Cook, S., Broft, A., Van Heertum, R., Comer, S.D., 2012. Deficits in dopamine D_2 receptors and presynaptic dopamine in heroin dependence: commonalities and differences with other types of addiction. Biol. Psychiatry 71 (3), 192–198.
Martinez, D., Slifstein, M., Broft, A., Mawlawi, O., Hwang, D.R., Huang, Y., Cooper, T., Kegeles, L., Zarahn, E., Abi-Dargham, A., Haber, S.N., Laruelle, M., 2003. Imaging human mesolimbic dopamine transmission with positron emission tomography. Part II: amphetamine-induced Dopamine Release in the Functional Subdivisions of the Striatum. J. Cereb. Blood Flow Metab. 23 (3), 285–300.
Mawlawi, O., Martinez, D., Slifstein, M., Broft, A., Chatterjee, R., Hwang, D.R., Huang, Y., Simpson, N., Ngo, K., Van Heertum, R., Laruelle, M., 2001. Imaging human mesolimbic dopamine transmission with positron emission tomography: I. Accuracy and precision of D_2 receptor parameter measurements in ventral striatum. J. Cereb. Blood Flow Metab. 21 (9), 1034–1057.
Milak, M.S., Severance, A.J., Prabhakaran, J., Kumar, J.S., Majo, V.J., Ogden, R.T., Mann, J.J., Parsey, R.V., 2011. In vivo serotonin-sensitive binding of [^{11}C]CUMI-101: a serotonin 1A receptor agonist positron emission tomography radiotracer. J. Cereb. Blood Flow Metab. 31 (1), 243–249.
Miyake, N., Thompson, J., Skinbjerg, M., Abi-Dargham, A., 2011. Presynaptic dopamine in schizophrenia. CNS Neurosci. Ther. 17 (2), 104–109.

Montgomery, A.J., Asselin, M.C., Farde, L., Grasby, P.M., 2007a. Measurement of methylphenidate-induced change in extrastriatal dopamine concentration using [^{11}C]FLB 457 PET. J. Cereb. Blood Flow Metab. (2), 369–377.

Montgomery, A.J., Lingford-Hughes, A.R., Egerton, A., Nutt, D.J., Grasby, P.M., 2007b. The effect of nicotine on striatal dopamine release in man: a [^{11}C]raclopride PET study. Synapse 61 (8), 637–645.

Montgomery, A.J., Mehta, M.A., Grasby, P.M., 2006. Is psychological stress in man associated with increased striatal dopamine levels?: a [^{11}C]raclopride PET study. Synapse 60 (2), 124–131.

Mukherjee, J., Christian, B.T., Narayanan, T.K., Shi, B., Collins, D., 2005. Measurement of d-amphetamine-induced effects on the binding of dopamine D_2/D_3 receptor radioligand, 18F-fallypride in extrastriatal brain regions in non-human primates using PET. Brain Res. 1032 (1–2), 77–84.

Mukherjee, J., Yang, Z.Y., Das, M.K., Brown, T., 1995. Fluorinated benzamide neuroleptics—III. Development of (S)-N-[(1-allyl-2-pyrrolidinyl)methyl]-5-(3-[^{18}F]fluoropropyl)-2, 3-dimethoxybenzamide as an improved dopamine D_2 receptor tracer. Nucl. Med. Biol. 22 (3), 283–296.

Mukherjee, J., Yang, Z.Y., Lew, R., Brown, T., Kronmal, S., Cooper, M.D., Seiden, L.S., 1997. Evaluation of d-amphetamine effects on the binding of dopamine D_2 receptor radioligand, F-18-fallypride in nonhuman primates using positron emission tomography. Synapse 27 (1), 1–13.

Narendran, R., Frankle, W.G., Mason, N.S., Rabiner, E.A., Gunn, R.N., Searle, G.E., Vora, S., Litschge, M., Kendro, S., Cooper, T.B., Mathis, C.A., Laruelle, M., 2009. Positron emission tomography imaging of amphetamine-induced dopamine release in the human cortex: a comparative evaluation of the high affinity dopamine $D_{2/3}$ radiotracers [^{11}C]FLB 457 and [^{11}C]fallypride. Synapse 63 (6), 447–461.

Narendran, R., Hwang, D.R., Slifstein, M., Talbot, P.S., Erritzoe, D., Huang, Y., Cooper, T.B., Martinez, D., Kegeles, L.S., Abi-Dargham, A., Laruelle, M., 2004. In vivo vulnerability to competition by endogenous dopamine: comparison of the D_2 receptor agonist radiotracer (-)-N-[^{11}C]propyl-norapomorphine ([^{11}C]NPA) with the D_2 receptor antagonist radiotracer [^{11}C]-raclopride. Synapse 52 (3), 188–208.

Narendran, R., Jedema, H.P., Lopresti, B.J., Mason, N.S., Gurnsey, K., Ruszkiewicz, J., Chen, C.M., Deuitch, L., Frankle, W.G., Bradberry, C.W., 2013. Imaging dopamine transmission in the frontal cortex: a simultaneous microdialysis and [C]FLB 457 PET study. Mol. Psychiatry.

Narendran, R., Mason, N.S., Chen, C.M., Himes, M., Keating, P., May, M.A., Rabiner, E.A., Laruelle, M., Mathis, C.A., Frankle, W.G., 2011. Evaluation of dopamine $D_{2/3}$ specific binding in the cerebellum for the PET radiotracer [^{11}C]FLB 457: implications for measuring cortical dopamine release. Synapse.

Narendran, R., Mason, N.S., Laymon, C., Lopresti, B., Velasquez, N., May, M., Kendro, S., Martinez, D., Mathis, C., Frankle, G., 2010a. A comparative evaluation of the dopamine $D_{2/3}$ agonist radiotracer [^{11}C] NPA and antagonist [^{11}C]raclopride to measure amphetamine-induced dopamine release in the human striatum. J. Pharmacol. Exp. Ther. 63 (7), 574–584.

Narendran, R., Mason, N.S., May, M.A., Chen, C.M., Kendro, S., Ridler, K., Rabiner, E.A., Laruelle, M., Mathis, C.A., Frankle, W.G., 2010b. PET imaging of dopamine $D_{2/3}$ receptors in the human cortex with [^{11}C]FLB 457: reproducibility studies. Synapse 65 (1), 35–40.

Narendran, R., Slifstein, M., Guillin, O., Hwang, Y., Hwang, D.R., Scher, E., Reeder, S., Rabiner, E., Laruelle, M., 2006. Dopamine ($D_{2/3}$) receptor agonist positron emission tomography radiotracer [^{11}C]-(+)-PHNO is a D_3 receptor preferring agonist in vivo. Synapse 60 (7), 485–495.

Neumeyer, J.L., Gao, Y.G., Kula, N.S., Baldessarini, R.J., 1990. Synthesis and dopamine receptor affinity of (R)-(-)-2-fluoro-N-n-propylnorapomorphine: a highly potent and selective dopamine D_2 agonist. J. Med. Chem. 33 (12), 3122–3124.

Okauchi, T., Suhara, T., Maeda, J., Kawabe, K., Obayashi, S., Suzuki, K., 2001. Effect of endogenous dopamine on endogenous dopamine on extrastriated [^{11}C]FLB 457 binding measured by PET. Synapse 41 (2), 87–95.

Olsson, H., Halldin, C., Swahn, C.G., Farde, L., 1999. Quantification of [^{11}C]FLB 457 binding to extrastriatal dopamine receptors in the human brain. J. Cereb. Blood Flow Metab. 19 (10), 1164–1173.

Ouchi, Y., Yoshikawa, E., Futatsubashi, M., Okada, H., Torizuka, T., Sakamoto, M., 2002. Effect of simple motor performance on regional dopamine release in the striatum in Parkinson disease patients and healthy subjects: a positron emission tomography study. J. Cereb. Blood Flow Metab. 22 (6), 746–752.

Pappata, S., Dehaene, S., Poline, J.B., Gregoire, M.C., Jobert, A., Delforge, J., Frouin, V., Bottlaender, M., Dolle, F., Di Giamberardino, L., Syrota, A., 2002. In vivo detection of striatal dopamine release during reward: a PET study with [^{11}C]raclopride and a single dynamic scan approach. NeuroImage 16 (4), 1015–1027.

Parsey, R.V., Hwang, D.R., Simpson, S., Guo, N., Mawlawi, O., Kegeles, L.S., Bergert, B., Montoya, J., Popilskis, S., Mann, J.J., Laruelle, M., 1999. PET studies of competition between [^{11}C]carbonyl-WAY 100635 and endogenous serotonin. J. Nucl. Med. 40 (5) 29.

Pierson, M.E., Andersson, J., Nyberg, S., McCarthy, D.J., Finnema, S.J., Varnas, K., Takano, A., Karlsson, P., Gulyas, B., Medd, A.M., Lee, C.M., Powell, M.E., Heys, J.R., Potts, W., Seneca, N., Mrzljak, L., Farde, L., Halldin, C., 2008. [^{11}C]AZ10419369: a selective 5-HT1B receptor radioligand suitable for positron emission tomography (PET). Characterization in the primate brain. NeuroImage 41 (3), 1075–1085.

Pinborg, L.H., Adams, K.H., Yndgaard, S., Hasselbalch, S.G., Holm, S., Kristiansen, H., Paulson, O.B., Knudsen, G.M., 2004. [^{18}F]altanserin binding to human 5HT2A receptors is unaltered after citalopram and pindolol challenge. J. Cereb. Blood Flow Metab. 24 (9), 1037–1045.

Pruessner, J.C., Champagne, F., Meaney, M.J., Dagher, A., 2004. Dopamine release in response to a psychological stress in humans and its relationship to early life maternal care: a positron emission tomography study using [^{11}C]raclopride. J. Neurosci. 24 (11), 2825–2831.

Quednow, B.B., Treyer, V., Hasler, F., Dorig, N., Wyss, M.T., Burger, C., Rentsch, K.M., Westera, G., Schubiger, P.A., Buck, A., Vollenweider, F.X., 2011. Assessment of serotonin release capacity in the human brain using dexfenfluramine challenge and [^{18}F]altanserin positron emission tomography. NeuroImage.

Rabiner, E.A., Messa, C., Sargent, P.A., Husted-Kjaer, K., Montgomery, A., Lawrence, A.D., Bench, C.J., Gunn, R.N., Cowen, P., Grasby, P.M., 2002. A database of [^{11}C]WAY-100635 binding to 5-HT(1A) receptors in normal male volunteers: normative data and relationship to methodological, demographic, physiological, and behavioral variables. NeuroImage 15 (3), 620–632.

Rabiner, E.A., Slifstein, M., Nobrega, J., Plisson, C., Huiban, M., Raymond, R., Diwan, M., Wilson, A.A., McCormick, P., Gentile, G., Gunn, R.N., Laruelle, M.A., 2009. In vivo quantification of regional dopamine-D_3 receptor binding potential of (+)-PHNO: studies in non-human primates and transgenic mice. Synapse 63 (9), 782–793.

Riccardi, P., Li, R., Ansari, M.S., Zald, D., Park, S., Dawant, B., Anderson, S., Doop, M., Woodward, N., Schoenberg, E., Schmidt, D., Baldwin, R., Kessler, R., 2005. Amphetamine-induced displacement of [^{18}F] fallypride in striatum and extrastriatal regions in humans. Neuropsychopharmacology 31 (5), 1016–1026.

Rice, O.V., Gatley, S.J., Shen, J., Huemmer, C.L., Rogoz, R., DeJesus, O.T., Volkow, N.D., Gifford, A.N., 2001. Effects of endogenous neurotransmitters on the in vivo binding of dopamine and 5-HT radiotracers in mice. Neuropsychopharmacology 25 (5), 679–689.

Ridler, K., Plisson, C., Rabiner, E.A., Gunn, R.N., Easwaramoorthy, B., Abi-Dargham, A., Laruelle, M., Slifstein, M., 2011. Characterization of in vivo pharmacological properties and sensitivity to endogenous serotonin of [^{11}C] P943: a positron emission tomography study in *Papio anubis*. Synapse 65 (11), 1119–1127.

Salmi, E., Kaisti, K.K., Metsahonkala, L., Oikonen, V., Aalto, S., Nagren, K., Hinkka, S., Hietala, J., Korpi, E.R., Scheinin, H., 2004. Sevoflurane and propofol increase ^{11}C-flumazenil binding to gamma-aminobutyric acid A receptors in humans. Anesth. Analg. 99 (5), 1420–1426 (table of contents).

Salonen, I., Hietala, J., Laihinen, A., Lehikoinen, P., Leino, L., Nagren, K., Ruotsalainen, U., Oikonen, V., Tuokkola, T., Nanto, V., 1997. A PET study on the acute effect of ethanol on striatal D_2 dopamine receptors with [C-11]raclopride in healthy males. Hum. Psychopharmacol. Clin. Exp. 12 (2), 145–152.

Schiffer, W.K., Volkow, N.D., Fowler, J.S., Alexoff, D.L., Logan, J., Dewey, S.L., 2006. Therapeutic doses of amphetamine or methylphenidate differentially increase synaptic and extracellular dopamine. Synapse 59 (4), 243–251.

Schott, B.H., Minuzzi, L., Krebs, R.M., Elmenhorst, D., Lang, M., Winz, O.H., Seidenbecher, C.I., Coenen, H.H., Heinze, H.J., Zilles, K., Duzel, E., Bauer, A., 2008. Mesolimbic functional magnetic resonance imaging activations during reward anticipation correlate with reward-related ventral striatal dopamine release. J. Neurosci. 28 (52), 14311–14319.

Scott, D.J., Domino, E.F., Heitzeg, M.M., Koeppe, R.A., Ni, L., Guthrie, S., Zubieta, J.K., 2007a. Smoking modulation of mu-opioid and dopamine D_2 receptor-mediated neurotransmission in humans. Neuropsychopharmacology 32 (2), 450–457.

Scott, D.J., Stohler, C.S., Egnatuk, C.M., Wang, H., Koeppe, R.A., Zubieta, J.K., 2008. Placebo and nocebo effects are defined by opposite opioid and dopaminergic responses. Arch. Gen. Psychiatry 65 (2), 220–231.

Scott, D.J., Stohler, C.S., Koeppe, R.A., Zubieta, J.K., 2007b. Time-course of change in [^{11}C]carfentanil and [^{11}C]raclopride binding potential after a nonpharmacological challenge. Synapse 61 (9), 707–714.

Searle, G., Beaver, J.D., Comley, R.A., Bani, M., Tziortzi, A., Slifstein, M., Mugnaini, M., Griffante, C., Wilson, A.A., Merlo-Pich, E., Houle, S., Gunn, R., Rabiner, E.A., Laruelle, M., 2010. Imaging dopamine D_3 receptors in the human brain with positron emission tomography, [^{11}C]PHNO, and a selective D_3 receptor antagonist. Biol. Psychiatry 68 (4), 392–399.

Seeman, P., Guan, H.-C., Niznik, H.B., 1989. Endogenous dopamine lowers the dopamine D_2 receptor density as measured by [^{3}H]raclopride: implications for positron emission tomography of the human brain. Synapse 3, 96–97.

Seneca, N., Finnema, S.J., Farde, L., Gulyas, B., Wikstrom, H.V., Halldin, C., Innis, R.B., 2006. Effect of amphetamine on dopamine D_2 receptor binding in nonhuman primate brain: a comparison of the agonist radioligand [^{11}C]MNPA and antagonist [^{11}C]raclopride. Synapse 59 (5), 260–269.

Shotbolt, P., Tziortzi, A.C., Searle, G.E., Colasanti, A., van der Aart, J., Abanades, S., Plisson, C., Miller, S.R., Huiban, M., Beaver, J.D., Gunn, R.N., Laruelle, M., Rabiner, E.A., 2011. Within-subject comparison of [^{11}C]-(+)-PHNO and [^{11}C]raclopride sensitivity to acute amphetamine challenge in healthy humans. J. Cereb. Blood Flow Metab.

Sibon, I., Benkelfat, C., Gravel, P., Aznavour, N., Costes, N., Mzengeza, S., Booij, L., Baker, G., Soucy, J.P., Zimmer, L., Descarries, L., 2008. Decreased [^{18}F]MPPF binding potential in the dorsal raphe nucleus after a single oral dose of fluoxetine: a positron-emission tomography study in healthy volunteers. Biol. Psychiatry 63 (12), 1135–1140.

Skinbjerg, M., Liow, J.S., Seneca, N., Hong, J., Lu, S., Thorsell, A., Heilig, M., Pike, V.W., Halldin, C., Sibley, D.R., Innis, R.B., 2010. D_2 dopamine receptor internalization prolongs the decrease of radioligand binding after amphetamine: a PET study in a receptor internalization-deficient mouse model. NeuroImage 50 (4), 1402–1407.

Slifstein, M., Hwang, D.R., Huang, Y., Guo, N., Sudo, Y., Narendran, R., Talbot, P., Laruelle, M., 2004. In vivo affinity of [^{18}F]fallypride for striatal and extrastriatal dopamine D_2 receptors in nonhuman primates. Psychopharmacology 175 (3), 274–286.

Slifstein, M., Kegeles, L., Hackett, E., Castrillon, J., Gonzales, R., Bae, J., Laruelle, M., Abi-Dargham, A., 2007. Effect of amphetamine challenge on the binding of [^{18}F]fallypride in the striatum and extrastriatal brain regions: a study in healthy human volunteers. In: 46th Annual Meeting of the American College of Neuropsychopharmacology. Boca Raton, FL.

Slifstein, M., Kegeles, L.S., Xu, X., Thompson, J.L., Urban, N., Castrillon, J., Hackett, E., Bae, S.A., Laruelle, M., Abi-Dargham, A., 2010. Striatal and extrastriatal dopamine release measured with PET and [^{18}F] fallypride. Synapse 64 (5), 350–362.

Small, D.M., Jones-Gotman, M., Dagher, A., 2003. Feeding-induced dopamine release in dorsal striatum correlates with meal pleasantness ratings in healthy human volunteers. NeuroImage 19 (4), 1709–1715.

Staley, J.K., Van Dyck, C.H., Tan, P.Z., Al Tikriti, M., Ramsby, Q., Klump, H., Ng, C., Garg, P., Soufer, R., Baldwin, R.M., Innis, R.B., 2001. Comparison of [^{18}F]altanserin and [^{18}F]deuteroaltanserin for PET imaging of serotonin(2A) receptors in baboon brain: pharmacological studies. Nucl. Med. Biol. 28 (3), 271–279.

Stokes, P.R., Egerton, A., Watson, B., Reid, A., Breen, G., Lingford-Hughes, A., Nutt, D.J., Mehta, M.A., 2010. Significant decreases in frontal and temporal [^{11}C]-raclopride binding after THC challenge. NeuroImage 52 (4), 1521–1527.

Stokes, P.R., Mehta, M.A., Curran, H.V., Breen, G., Grasby, P.M., 2009. Can recreational doses of THC produce significant dopamine release in the human striatum? NeuroImage 48 (1), 186–190.

Sudo, Y., Suhara, T., Inoue, M., Ito, H., Suzuki, K., Saijo, T., Halldin, C., Farde, L., 2001. Reproducibility of [^{11}C]FLB 457 binding in extrastriatal regions. Nucl. Med. Commun. 22 (11), 1215–1221.

Sun, W., Ginovart, N., Ko, F., Seeman, P., Kapur, S., 2003. In vivo evidence for dopamine-mediated internalization of D_2-receptors after amphetamine: differential findings with [^{3}H]raclopride vs [^{3}H]spiperone. Mol. Pharmacol. 63 (2), 456–462.

Takahashi, H., Fujimura, Y., Hayashi, M., Takano, H., Kato, M., Okubo, Y., Kanno, I., Ito, H., Suhara, T., 2008. Enhanced dopamine release by nicotine in cigarette smokers: a double-blind, randomized, placebo-controlled pilot study. Int. J. Neuropsychopharmacol. 11 (3), 413–417.

Talbot, P.S., Slifstein, M., Hwang, D.R., Huang, Y., Scher, E., Abi-Dargham, A., Laruelle, M., 2012. Extended characterisation of the serotonin 2A (5-HT2A) receptor-selective PET radiotracer 11C-MDL100907 in humans: quantitative analysis, test-retest reproducibility, and vulnerability to endogenous 5-HT tone. NeuroImage 59 (1), 271–285.

Tsukada, H., Miyasato, K., Nishiyama, S., Fukumoto, D., Kakiuchi, T., Domino, E.F., 2005. Nicotine normalizes increased prefrontal cortical dopamine D_1 receptor binding and decreased working memory performance produced by repeated pretreatment with MK-801: a PET study in conscious monkeys. Neuropsychopharmacology 30 (12), 2144–2153.

Tsukada, H., Nishiyama, S., Kakiuchi, T., Ohba, H., Sato, K., Harada, N., 1999. Is synaptic dopamine concentration the exclusive factor which alters the in vivo binding of [^{11}C]raclopride?: PET studies combined with microdialysis in conscious monkeys (In Process Citation). Brain Res. 841 (1–2), 160–169.

Udo de Haes, J.I., Bosker, F.J., Van Waarde, A., Pruim, J., Willemsen, A.T., Vaalburg, W., Den Boer, J.A., 2002. 5-HT(1A) receptor imaging in the human brain: effect of tryptophan depletion and infusion on [^{18}F] MPPF binding. Synapse 46 (2), 108–115.

Udo de Haes, J.I., Harada, N., Elsinga, P.H., Maguire, R.P., Tsukada, H., 2006. Effect of fenfluramine-induced increases in serotonin release on [^{18}F]MPPF binding: a continuous infusion PET study in conscious monkeys. Synapse 59 (1), 18–26.

Urban, N.B., Kegeles, L.S., Slifstein, M., Xu, X., Martinez, D., Sakr, E., Castillo, F., Moadel, T., O'Malley, S.S., Krystal, J.H., Abi-Dargham, A., 2010. Sex differences in striatal dopamine release in young adults after oral alcohol challenge: a positron emission tomography imaging study with [^{11}C]raclopride. Biol. Psychiatry 68 (8), 689–696.

Urban, N.B., Slifstein, M., Meda, S., Xu, X., Ayoub, R., Medina, O., Pearlson, G.D., Krystal, J.H., Abi-Dargham, A., 2011. Imaging human reward processing with positron emission tomography and functional magnetic resonance imaging. Psychopharmacology.

Verhoeff, N.P., Hussey, D., Lee, M., Tauscher, J., Papatheodorou, G., Wilson, A.A., Houle, S., Kapur, S., 2002. Dopamine depletion results in increased neostriatal D_2, but not D_1, receptor binding in humans. Mol. Psychiatry 7 (3) 233, 322–328.

Vilkman, H., Kajander, J., Nagren, K., Oikonen, V., Syvalahti, E., Hietala, J., 2000. Measurement of extrastriatal D_2-like receptor binding with [^{11}C]FLB 457—a test-retest analysis. Eur. J. Nucl. Med. 27 (11), 1666–1673.

Volkow, N.D., Fowler, J.S., Wang, G.J., 1999. Imaging studies on the role of dopamine in cocaine reinforcement and addiction in humans. J. Psychopharmacol. 13 (4), 337–345.

Volkow, N.D., Wang, G.-J., Fowler, J.S., Logan, J., Schlyer, D., Hitzemann, R., Lieberman, J., Angrist, B., Pappas, N., MacGregor, R., Burr, G., Cooper, T., Wolf, A.P., 1994. Imaging endogenous dopamine competition with [^{11}C]raclopride in the human brain. Synapse 16, 255–262.

Volkow, N.D., Wang, G.J., Fowler, J.S., Logan, J., Gatley, S.J., Hitzemann, R., Chen, A.D., Dewey, S.L., Pappas, N., 1997. Decreased striatal dopaminergic responsiveness in detoxified cocaine-dependent subjects. Nature 386, 830–833.

Volkow, N.D., Wang, G.J., Fowler, J.S., Logan, J., Jayne, M., Franceschi, D., Wong, C., Gatley, S.J., Gifford, A.N., Ding, Y.S., Pappas, N., 2002. "Nonhedonic" food motivation in humans involves dopamine in the dorsal striatum and methylphenidate amplifies this effect. Synapse 44 (3), 175–180.

Volkow, N.D., Wang, G.J., Telang, F., Fowler, J.S., Logan, J., Jayne, M., Ma, Y., Pradhan, K., Wong, C., 2007. Profound decreases in dopamine release in striatum in detoxified alcoholics: possible orbitofrontal involvement. J. Neurosci. 27 (46), 12700–12706.

Wang, G.J., Smith, L., Volkow, N.D., Telang, F., Logan, J., Tomasi, D., Wong, C.T., Hoffman, W., Jayne, M., Alia-Klein, N., Thanos, P., Fowler, J.S., 2011. Decreased dopamine activity predicts relapse in methamphetamine abusers. Mol. Psychiatry.

Weber, O.M., Verhagen, A., Duc, C.O., Meier, D., Leenders, K.L., Boesiger, P., 1999. Effects of vigabatrin intake on brain GABA activity as monitored by spectrally edited magnetic resonance spectroscopy and positron emission tomography. Magn. Reson. Imaging 17 (3), 417–425.

Wilson, A.A., McCormick, P., Kapur, S., Willeit, M., Garcia, A., Hussey, D., Houle, S., Seeman, P., Ginovart, N., 2005. Radiosynthesis and evaluation of [^{11}C]-(+)-4-propyl-3,4,4a,5,6,10b-hexahydro-2H-naphtho[1,2-b][1,4]oxazin-9-ol as a potential radiotracer for in vivo imaging of the dopamine D_2 high-affinity state with positron emission tomography. J. Med. Chem. 48 (12), 4153–4160.

Wong, D.F., Brasic, J.R., Singer, H.S., Schretlen, D.J., Kuwabara, H., Zhou, Y., Nandi, A., Maris, M.A., Alexander, M., Ye, W., Rousset, O., Kumar, A., Szabo, Z., Gjedde, A., Grace, A.A., 2007. Mechanisms of dopaminergic and serotonergic neurotransmission in Tourette syndrome: clues from an in vivo neurochemistry study with PET. Neuropsychopharmacology.

Wong, D.F., Wand, G.S., Yung, B., Dannals, R.F., HT R, Chan B., Chen, C., 1993. The effects of intravenous ethanol on intrasynaptic dopamine measures in human basal ganglia. J. Nucl. Med. 34, 132.

Yoder, K.K., Kareken, D.A., Seyoum, R.A., O'Connor, S.J., Wang, C., Zheng, Q.H., Mock, B., Morris, E.D., 2005. Dopamine D_2 receptor availability is associated with subjective responses to alcohol. Alcohol. Clin. Exp. Res. 29 (6), 965–970.

Yoder, K.K., Morris, E.D., Constantinescu, C.C., Cheng, T.E., Normandin, M.D., O'Connor, S.J., Kareken, D.A., 2009. When what you see isn't what you get: alcohol cues, alcohol administration, prediction error, and human striatal dopamine. Alcohol. Clin. Exp. Res. 33 (1), 139–149.

Zald, D.H., Boileau, I., El-Dearedy, W., Gunn, R., McGlone, F., Dichter, G.S., Dagher, A., 2004. Dopamine transmission in the human striatum during monetary reward tasks. J. Neurosci. 24 (17), 4105–4112.

Zimmer, L., Mauger, G., Le Bars, D., Bonmarchand, G., Luxen, A., Pujol, J.F., 2002. Effect of endogenous serotonin on the binding of the 5-hT1A PET ligand 18F-MPPF in the rat hippocampus: kinetic beta measurements combined with microdialysis. J. Neurochem. 80 (2), 278–286.

CHAPTER TWENTY

Characterizing Recovery of the Human Brain following Stroke: Evidence from fMRI Studies

Christian La[1,2,3], Brittany M. Young[1,3,6], Camille Garcia-Ramos[2], Veena A. Nair[1] and Vivek Prabhakaran[1,3,4,5,6]

[1]Department of Radiology, University of Wisconsin–Madison, WI, USA
[2]Department of Medical Physics, University of Wisconsin–Madison, WI, USA
[3]Neuroscience Training Program, University of Wisconsin–Madison, WI, USA
[4]Department of Neurology, University of Wisconsin–Madison, WI, USA
[5]Department of Psychiatry, University of Wisconsin–Madison, WI, USA
[6]Medical Scientist Training Program, University of Wisconsin–Madison, WI, USA

1. INTRODUCTION

Stroke remains a perennial source of human morbidity and mortality in the United States and ranks as the No. 1 cause of disability and the fourth leading cause of mortality in the United States. Nearly 800,000 people suffer from stroke each year with approximately 4.7 million stroke survivors in the United States (Roger et al., 2012). Stroke is usually caused by a compromise of the blood circulation (common etiologies include thrombosis, embolism, and systemic hypoperfusion), inflicting a focal area of damage. Although the lesions remain fairly focal in stroke, remote areas of the brain are affected involving a widespread network of brain regions, inflicting impairment in multiple functional domains.

Stroke survival has improved with mortality dropping from No. 3 to No. 4 in the United States, leading to more survivors. However, this has led to more survivors with persistent and pervasive disabilities, such as motor deficits, aphasia, spatial neglect, and other functional impairments that reduce patients' activity, participation, and productivity in daily life. Spontaneous recovery does occur, but this is highly variable and is rarely complete in most patients (6 months poststroke, 50% of survivors experience residual hemiparesis, only 5% show full recovery, and 26% are dependent in activities of daily living) (Kelly-Hayes et al., 2003). The estimated costs of stroke, based on health care expenses and lost productivity, were estimated to be $57 billion in 2003 to $73.7 billion in 2010, and from 2005 to 2050 this is projected to be $2.2 trillion (Heidenreich et al., 2011).

Imaging of the Human Brain in Health and Disease
http://dx.doi.org/10.1016/B978-0-12-418677-4.00020-8

2.1. fMRI AND STROKE

The use of invasive investigational studies in humans at the microlevel is rather difficult and potentially not feasible. However, at the macrolevel, neuroimaging techniques such as positron emission tomography (PET) and functional magnetic resonance imaging (fMRI) have allowed us to study these brain changes in vivo and with minimal invasiveness. While PET imaging makes use of the tracking of exogenous biologically active tracers after injection, fMRI utilizes the inherent magnetic properties of oxygenated and deoxygenated hemoglobin circulating in the blood vessels of the brain. This ratio between the two states of hemoglobin, which is modulated by the increase in blood flow to areas of the brain that are in use, gives rise to a signal: the blood-oxygenated-level-dependent (BOLD) signal. With the use of fMRI, we can attain indirect measurement of cortical activation in response to a task, but also an assessment of the intrinsic functional networks within the brain; two measures that are exceptionally valuable in the understanding and investigation of stroke-related injury.

Following an ischemic injury such as stroke, cells in the region of the lesion are deprived of adequate blood supply, causing them to undergo cell death (Figures 1 and 2). The normal functioning of the brain system is disrupted, forcing a change in its organization away from normal mechanisms of brain functioning. Although there are many variables (e.g. stroke topography, time post stroke, age, hemispheric dominance, infarct volume, medical comorbidities, etc.) influencing the mechanism and rate of recovery, this chapter will focus on a conceptual overview of our current understanding of post-stroke re-organization. For this purpose, we will review the application of functional magnetic resonance imaging (fMRI) in three primary domains of motor, language, and attention as a guide to system-level brain plasticity following ischemic stroke injury.

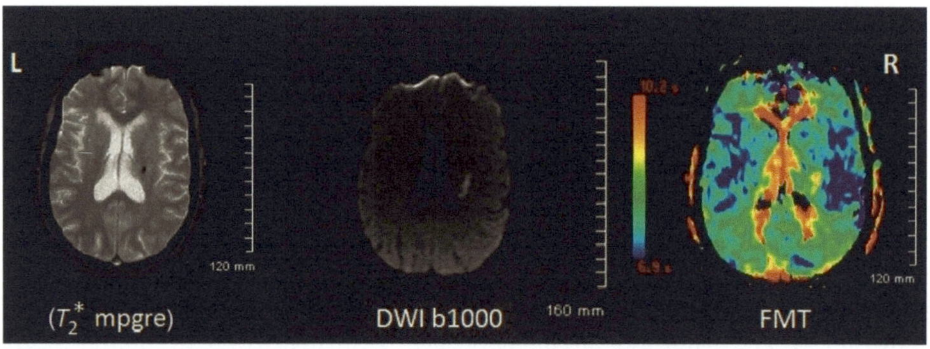

Figure 1 Acute stroke patient (76-year-old female) with minimal hemorrhagic transformation in the right retrolenticular region. Left to right: minimal hemorrhagic transformation (T_2^* multiplanar gradient-recalled echo (MPGRE)), restricted diffusion (Diffusion Weighted Imaging (DWI)), and increase transit time (first moment transit time (FMT)) in the right retrolenticular region (this and all other figures follow neurological convention).

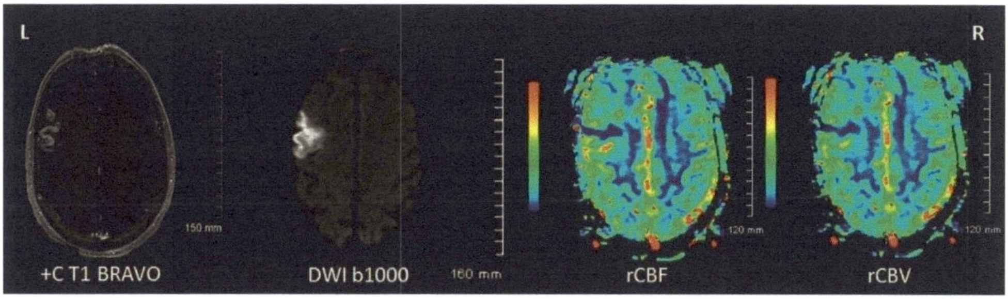

Figure 2 Acute to subacute stroke patient (79-year-old male) with left frontal infarct. Left to right: postcontrast enhancement (T1 BRAin VOlume imaging (BRAVO)), restricted diffusion (DWI), decreased blood flow (regional Cerebral Blood Flow (rCBF)), and blood volume (regional Cerebral Blood Volume (rCBV)) in the left frontal region.

2.1.1. Motor Deficit

Chollet et al. (1991) were among the first to report the motor activation patterns in poststroke conditions. Activation, as recorded by regional cerebral blood flow, was demonstrated primarily in the contralateral primary sensorimotor cortex and the ipsilateral cerebellum with respect to movement of the normal, unaffected limb. In contrast, movement of the recovered, but previously impaired, hand elicited activation in both the contra- and ipsilesional primary sensorimotor cortex and both cerebellar hemispheres. Additional secondary regions of the insula, inferior parietal cortex, and premotor cortex (PMC) were also bilaterally activated in the process. This study, using PET, shed initial light on the potential mechanism of brain recovery after injury. With movement of the paretic limb, numerous other PET studies have consistently reported similar activity pattern in areas of primary motor cortex (M1), PMC, supplementary motor area (SMA), parietal and prefrontal cortex, thalamus, and the cerebellum in both the affected and unaffected hemispheres (Weiller et al., 1995; Buckner et al., 1996). These findings were also replicated in fMRI studies, as described in later sections (Figure 3).

Impaired motor function is one of the primary ways in which stroke patients are persistently affected. Following injury, neural activity in the lesioned necrotic tissues is reduced to nearly nonexistent levels during movement of the paretic hand; yet, neural activity in the perilesional areas has been recorded. Both animal and human studies have conveyed the importance of the perilesional areas in adaptive reorganization after stroke (Nudo and Milliken, 1996; Xerri et al., 1998; Dong et al., 2007). For instance, Dong et al. (2007) have suggested that this activity in the perilesional M1 might reflect an assimilation of the impaired functions by the perilesional areas. After 2 weeks of arm-focused training, their stroke subjects exhibited higher activation in those areas of the brain than their healthy control participants in both pre- and posttreatment conditions.

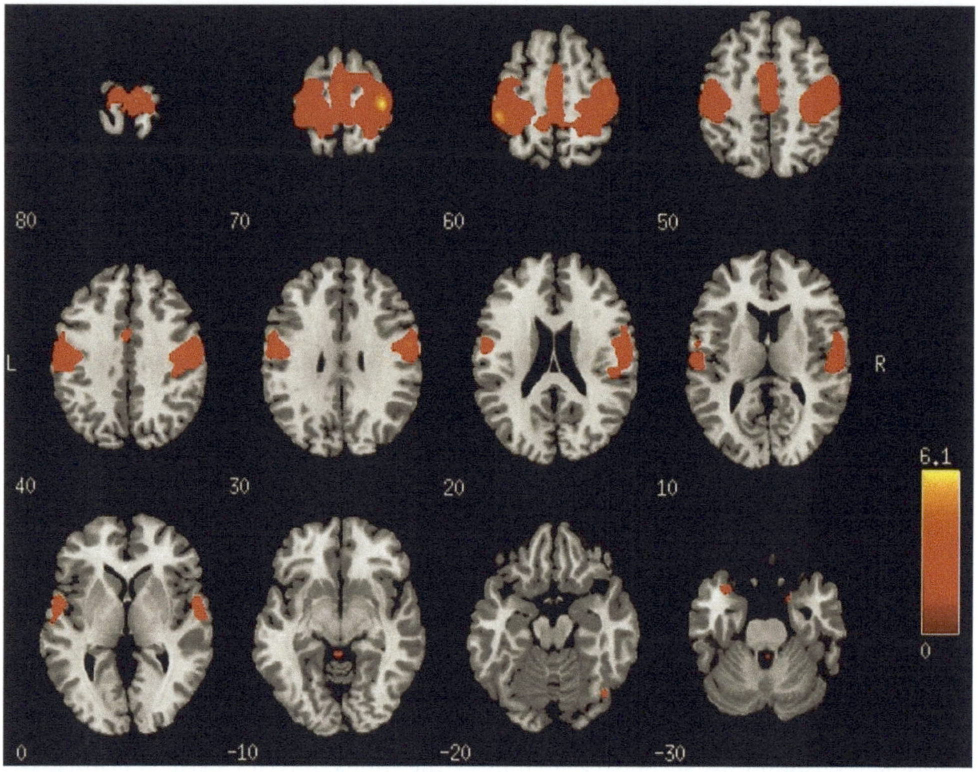

Figure 3 Motor network extracted by independent component analysis of resting fMRI data in a group of healthy control subjects ($n = 15$), featuring bilateral primary motor cortices (M1) and supplementary motor area (SMA).

Moreover, areas that were anatomically remote from the region of infarcted brain tissue became relevant and were recruited in the process of activation of the paretic hand (Cramer et al., 2000; Ward et al., 2004). Additionally, Johansen-Berg and colleagues have shown that in chronic stroke patients (i.e. patients at least 6 months after onset of infarct) training-induced improvements in motor performance are associated with increases in neural activity of the ipsilesional dorsal PMC (Johansen-Berg et al., 2002; Pineiro et al., 2002). Collectively, these findings suggest an initial compensatory mechanism in the perilesional regions responding to the lack of functionality in the damaged area.

About 3 years after the original PET study, Cao et al. (1994) transitioned to implementing the application of fMRI for mapping motor activation during motor function with poststroke hemiparesis. Numerous studies have followed (Cramer et al., 1997; Cao et al., 1998; Marshall et al., 2000; Takahashi et al., 2005), all of which confirmed the PET findings of increased activation in the motor network in the contralesional hemisphere centered around the primary sensorimotor cortex, SMAs, and the area along the rim

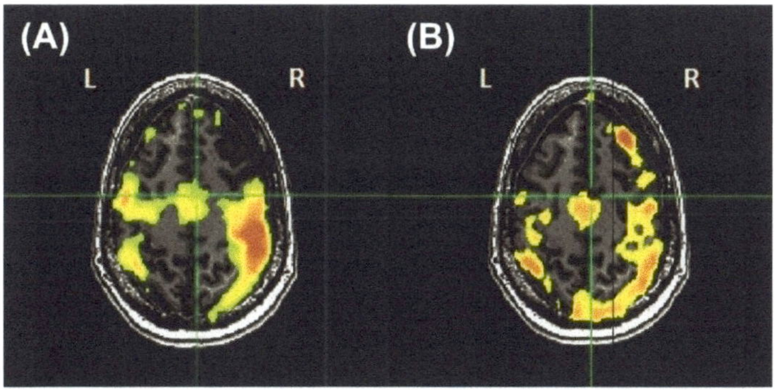

Figure 4 Task-derived activation from single stroke subject with motor deficit in the right upper extremity: activation from movement of unimpaired left hand (A) and from movement of the impaired right hand (B) both show right-lateralized motor cortex activity.

of the cortical infarct. However, the mechanism and purpose of this overactivity of the undamaged hemisphere is still unclear.

Such activity in the undamaged hemisphere contributes to a reduction of observed lateralization during motor activation. This finding is consistent with the proposed homologous area adaptation model suggested by Grafman (2000). In this model, the homologous area in the undamaged hemisphere is recruited to assist and assume the function of the damaged hemisphere when the latter is incapable of independently fulfilling its function. Although this model of rearrangement in cortical activity pattern is logical, the mechanism by which recovery happens may not stop here (Figure 4).

Longitudinal studies have revealed that although bilateral activity of these areas appears as early as within the first few days after stroke, concomitant with the early phase of recovery (Rehme et al., 2011b), this overactivity of the contralesional hemisphere typically returns to a normal physiological level after 6 months to 1 year in well-recovered patients (Loubinoux et al., 2003; Ward et al., 2003; Rehme and Grefkes, 2013). Furthermore, this decrease in task-related brain activation of the undamaged hemisphere correlates with both functional recovery of the primary and secondary motor brain areas and with improvement in behavioral outcomes (Carey et al., 2002; Ward et al., 2003; Zemke et al., 2003). This correlation was also shown to persist, independent of the rate of recovery or initial severity of the stroke (Ward et al., 2003). Together, these findings suggest that after an initial compensatory increase in activity, a return to a "normal-like" preinjury pattern of lateralized brain signal associated with the movement or intended movement of the impaired side is optimal for better functional outcomes.

Nonetheless, some authors have made a suggestion to the contrary, proposing that the enhanced contralesional activity observed in chronic stroke patients may develop in order to support or facilitate motor functions of the paretic hand (Lotze et al., 2006). Such interpretations have been heavily challenged, particularly with the development of noninvasive brain stimulation protocols in the study of stroke rehabilitation (Fregni et al., 2005; Takeuchi et al., 2005; Nowak Da et al., 2008). In these studies, suppression of activity in the unaffected hemisphere by means of repetitive transcranial magnetic stimulation or transcranial direct current stimulation has been shown on multiple occasions to improve the motor performance of the paretic hand, suggesting that such activity of the unaffected hemisphere might be detrimental to the stroke recovery process. This evidence has further stimulated the debate over whether increased activation in the unaffected hemisphere is disruptive or facilitative to motor recovery after stroke, a question which has yet to be conclusively addressed.

There is also evidence that other factors such as time poststroke and lesion location (cortical vs subcortical) may also influence the role of the contralesional hemisphere in motor recovery (Shelton and Reding, 2001; Kwon et al., 2007; Buma et al., 2010). Another determining factor influencing recovery is the extent of corticospinal tract damage after injury. The more damage to fibers originating from the M1, the less likely a full motor recovery will be achieved. Recovery in such a situation would generate the need for greater recruitment of higher level motor areas such as the SMA or the PMC to compensate for M1 deficiency (Newton et al., 2006; Ward et al., 2006; Stinear et al., 2007; Jayaram and Stinear, 2008). A more comprehensive approach examining the role of multiple regions likely involved in motor function is needed to understand the mechanisms underlying enhanced contralesional activity and its effect on recovery.

One approach to assessing the interactions among functionally relevant regions is the examination of brain connectivity. This approach conceptualizes the brain as a collection of discrete functional networks with spatial elements or nodes that interact with each other in a temporally and spatially specific fashion. One study examining such brain connectivity employed a method of structural equation modeling (SEM) and identified reduced effective connectivity between the ipsilesional SMA and the dorsal M1 that persisted even in well-recovered stroke patients (Sharma et al., 2009). This study also found significantly enhanced positive influences from the prefrontal cortex onto bilateral premotor areas, which correlated with increases in motor function (Sharma et al., 2009). A later study similarly found a strong initial reduction of premotor to ipsilateral M1 excitatory influences that later recovered as a function of behavioral recovery over the course of 3 to 6 months following the ischemic event (Rehme et al., 2011a). Together, these findings of increased effective connectivity between premotor areas and the ipsilesional M1 in association with better motor performance outcomes suggest that the reestablishment of greater top-down control over the ipsilesional primary motor areas may facilitate motor recovery (Rehme and Grefkes, 2013).

A mathematical method of analyzing pairwise relations between objects is graph theory. When such theoretical analyses are applied to functional connectivity measures, it allows for an examination of clusterwise reorganization, characterizing the "efficiency" of information exchange. This *efficiency* of information exchange converges on the idea of an optimal balance between local specialization and global integration of brain functional activity (Tononi et al., 1998). A study conducted by Wang et al. (2010) using such a method found decreased regional *betweenness centrality*, a measure of the functional importance of a node within a cluster, in the ipsilesional M1 and contralesional cerebellar regions during the first few days following stroke. This decrease was then followed by an increase in centrality during motor recovery over the next 12 months, implying that the reestablishment of these regions was significant in the brain organization serving the recovered limb. Simultaneously, local *clustering*, or the extent of local cliquishness, decreased with better recovery, which suggests that neural reorganization after stroke promotes motor recovery that progresses through the establishment of nonoptimal network structures (Wang et al., 2010). The authors hypothesized that this pattern may be due to unspecific axonal sprouting in the unaffected regions, increasing the spatial randomness of this organization.

Globally, measures of functional connectivity also aid in the investigation of interhemispheric influences. This approach has been used in a number of studies, with a particular emphasis on functional connectivity between homologous areas of the two hemispheres (Carter et al., 2010, 2012; Wang et al., 2010; Park et al., 2011). It has been shown that stroke can induce disturbances in the interhemispheric functional interactions between ipsilesional M1 and contralesional nodes of the motor network in the PMC and posterior parietal cortex (Wang et al., 2010). Carter et al. (2010) used a seed-based examination of the sensorimotor network to reveal that interhemispheric M1 connectivity positively correlated to motor performance during the subacute stage of stroke. Similarly, increases in interhemispheric connectivity within the motor network between the ipsilesional M1 and nodes in the contralesional thalamus, SMA, and middle prefrontal cortex as measured within the first few days after stroke have been shown to be predictive of motor recovery over the next 6 months (Park et al., 2011). An increase in functional connectivity, however, is not necessarily synonymous with positive influences, but rather just a measure of the integrity of the functional network.

2.1.2. Language Deficit: Aphasia

Following an ischemic stroke, patients also tend to develop disturbances in the comprehension and formulation of language, a deficit known as aphasia. With the language network comprising multiple regions of the language-dominant hemisphere, it is not unlikely that an ischemic stroke in the dominant hemisphere will result in the development of such dysfunction. It is commonly accepted that the language network is left lateralized in the majority of individuals, although it appears in a continuous spectrum of laterality, few having bilateral or even right-dominant (rare) representation of language (Knecht et al., 2002). Approximately 96% of right-handed individuals have language lateralized in the left hemisphere, whereas about 70% of left-handed individuals have language lateralized in the left hemisphere (Knecht et al., 2000) (Figure 5).

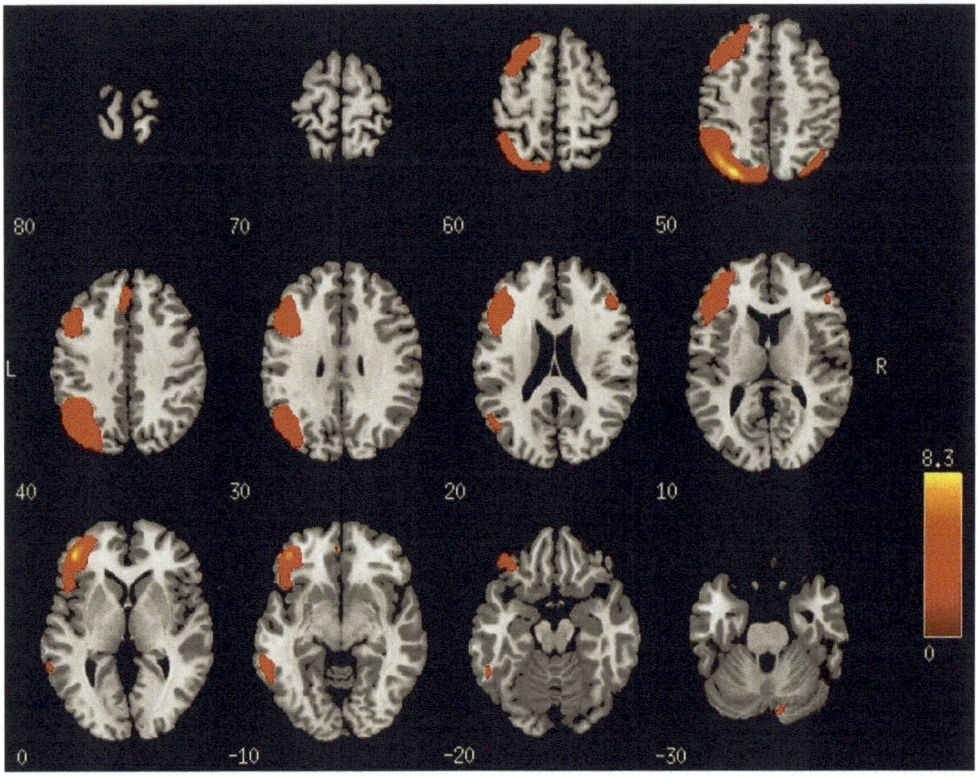

Figure 5 Language network extracted by independent component analysis of resting fMRI data in a group of healthy control subjects ($n=15$) shows predominantly left-lateralized frontotemporal involvement of brain regions.

Regions that are primarily associated with the comprehension and production of language include areas of the SMA and areas in the perisylvian network of the language-dominant hemisphere: inferior and middle frontal gyrus, angular gyrus, supramarginal gyrus, as well as superior, middle, and inferior temporal gyrus. Therefore, damage to these regions could potentially lead to the development of aphasic symptoms.

Although language is traditionally considered to be left lateralized, mounting evidence in the past few decades has pointed to a more active role of the right hemisphere than was previously believed. The right hemisphere is actively involved in certain aspects of language processing, such as contextual processing and the appreciation of metaphors (Vigneau et al., 2011; Yang, 2012). Right hemisphere activity could also occur as a result of overlearned categories. This was shown in a study by Birn et al. (2010), where repeated retrieval of the same highly overlearned sequence of words, such as continuous naming of the months of the year, led to enhanced right hemisphere activity. Regions of the right hemisphere are therefore thought to be associated with automatic speech. Several groups have also hypothesized that contralateral activity in cognitive tasks (e.g. language) is a product of compensatory mechanisms in normal aging, i.e. the provision that tasks can be successfully completed with necessary help from the non-dominant hemisphere to the dominant one (Reuter-Lorenz et al., 2000; Cabeza, 2002; Rossi et al., 2004; Philipose et al., 2007; Park and Reuter-Lorenz, 2009; Meier et al., 2011). Others have suggested that contralateral engagement in executive function tends to be related to impaired language function (Wierenga et al., 2008; Meinzer et al., 2009). Additionally, a shift of language function to homologous areas has been demonstrated in a study where regions of the language networks remained structurally intact but were exposed to hypoperfusion from a neighboring subcortical infarct (Prabhakaran et al., 2007).

Cao et al. (1999) were also among the first to study aphasia using fMRI. In their study, language recovery in patients was associated with mainly bilateral activation, and they concluded that both the restoration of left hemisphere networks and the recruitment of right hemisphere networks were important to language recovery. However, it was not until 2006 that a group reviewed the reorganization pattern of the language network at different phases of recovery (Saur et al., 2006). Until then, language recovery was assumed to occur by upregulation of the temporofrontal network of the undamaged hemisphere, which includes recruitment of perilesional tissue (Warburton et al., 1999) and involvement of homologous areas (Weiller et al., 1995; Musso et al., 1999; Rosen et al., 2000).

After a stroke, the region surrounding the necrotic tissues exhibits lower perfusion than areas farther away from the lesion. However, this perilesional region tends to play an important role in the recovery of function. In a study by Fridriksson et al. (2012), BOLD activation in these areas of hypoperfusion, particularly in the left frontal lobe, predicted increases in correct naming following an anomic treatment in individuals with chronic aphasia. Activity in the perilesional frontal lobe was the most robust predictor of correct naming improvement, seen most strongly in the residual naming areas and also in perilesional areas that were not previously recruited. Additionally, increased frontal activity after the anomic treatment was associated with an increase in correctly named items, whereas reduction of activity in the posterior regions was associated with reduced errors (Fridriksson et al., 2012).

In 2006, Saur et al. (2006) proposed the investigation of the reorganizational process instead of the final arrangement of the reorganized brain. They proposed a model of brain reorganization involving multiple stages of spontaneous language recovery processes. Initially, little early activation was seen, and the activation which was observed centered mainly around the noninfarcted regions of left hemispheric language structures. This initial phase was then followed by a large increase in bilateral language network activations with a peak in the right hemisphere region homologous to Broca's area. Then, in the chronic phase, activation normalized with a reshift of peak activation to left hemisphere language areas. These observed changes in cortical activation patterns were all accompanied by a gradual increase in language recovery scores, synonymous with better recovery of language function (Saur et al., 2006).

In the early phase, the widespread dysfunction as recorded by the lack of activity in either hemisphere causes a severe language deficit often resulting in global aphasia. In the second, subacute phase, the diaschisis resolves through upregulation of activity in the left hemisphere language network along with its contrahemispheric homolog. It remains unclear whether this strong right hemispheric activation represents functionally relevant activation compensating for lesioned left hemisphere areas or if it is the result of a breakdown in interhemispheric control, leading to disinhibition of the right hemisphere (Rosen et al., 2000). In the chronic phase, activation normalizes with a reshift of activation toward the left hemisphere, a pattern resembling that in healthy subjects. Furthermore, other studies have suggested a similar return to left lateralization of the language network activity to correspond to successful recovery, while recruitment of homologous regions was deemed an unsuccessful reorganization (Naeser et al., 2005; Meinzer et al., 2008; Szaflarski et al., 2013).

Although activations in the inferior frontal gyrus and superior temporal gyrus in the dominant hemisphere are the main areas associated with the recovery of language functions, other areas within the left hemisphere not previously engaged in language tasks before stroke (e.g. left superior parietal cortex and precuneus) can be recruited in the process of recovery. Increased activation in such areas has also been linked to successful recovery (Musso et al., 1999; Fridriksson et al., 2007; Fridriksson, 2010). However, despite the seemly compensatory nature of the involvement of these new areas in language tasks, more evidence is needed to support this mechanism as a possible route of reorganization (Saur and Hartwigsen, 2012).

Functional connectivity analysis might yield further insights into the mechanisms of language function and recovery in the context of aphasia. As discussed above, a number of regions in the left hemisphere have been implicated as integral to the language network, and this network has been verified using functional connectivity analyses applied to task fMRI (Lee et al., 2012) and resting fMRI (Pravatà et al., 2011; Shirer et al., 2012; Tie et al., 2012). These functional connectivity analyses in healthy individuals showed an extensive network of functionally connected areas within the left hemisphere, which were also associated with homologous cortical areas in the right hemisphere (Turken and Dronkers, 2011; Shirer et al., 2012), and were consistent with studies characterizing the functional anatomy of the verbal semantic system (Binder et al., 2009). The arcuate fasciculus appears to be of particular importance in the language network, with a high number of functional and structural connections identified consistently across a number of studies (Powell et al., 2006; Glasser and Rilling, 2008; Rilling et al., 2008), possibly suggesting that this fasciculus and the regions subserved by this fiber tract are especially vulnerable to impairment-inducing damage with the occurrence of a brain injury such as stroke. This is consistent with the severe language comprehension deficits observed clinically in patients suffering from lesions of this area (Turken and Dronkers, 2011).

To date, there have been a very limited number of studies reviewing functional connectivity disruption in the language network of patients suffering from stroke-induced

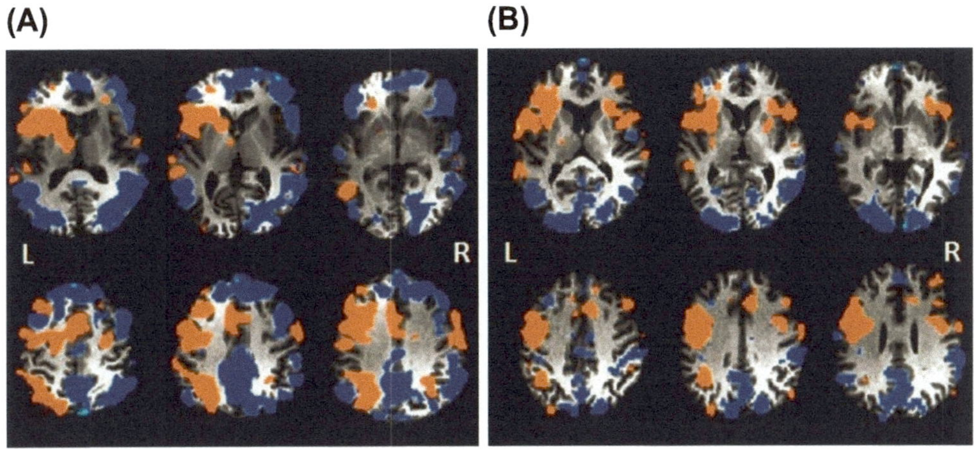

Figure 6 Task-derived (phonemic fluency) activation in normal individual (A) and aphasic stroke patient (B). Red represents task-positive activation and includes the language network. Blue represents task-negative activation and represents the default mode network.

aphasia. However, structural connectivity measures from diffusion tensor imaging have revealed greater connectivity between the temporal lobe and frontal lobe after therapy, suggesting the importance of the integrity of the arcuate fibers—findings later verified by structural connectivity measures (Schlaug et al., 2009). In that study, the fiber number and volume of the arcuate fasciculus correlated to the treatment outcome. Incidentally, an association between an increase in connection strengths and behavioral improvement among the left hemisphere language areas was revealed using dynamic causal modeling and SEM. Patients with larger lesions had more connections strengthened in the right hemisphere, whereas patients with smaller lesions had more connections strengthened in the left hemisphere (Vitali et al., 2010) (Figure 6).

2.1.3. Attention Deficit: Spatial Neglect

Attention deficits are yet another common set of impairments that seriously hinder the daily living of a patient after an ischemic stroke. Arising primarily from a right middle cerebral artery occlusion, neglect is highly lateralized with 90% of the patients suffering hemispatial neglect on the left side of the visual field or body (Pedersen et al., 1997; Appelros et al., 2002; Gabrieli and Whitfield-Gabrieli, 2007; He et al., 2007). This complex syndrome imposes the inability to attend and respond to stimuli located on the side of space opposite to the side of the brain affected by the brain lesion. Such a syndrome is not due to a lack of incoming sensory input but rather due to a deficit in the attention directed toward and salience ascribed to input from a specific area of space (Figure 7).

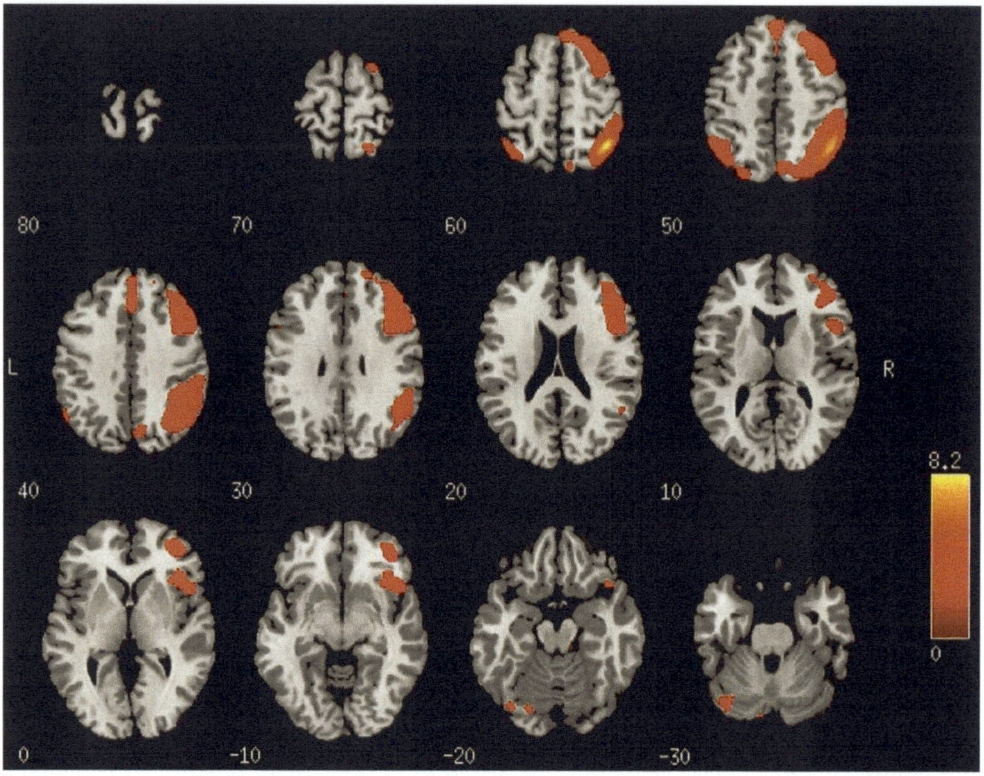

Figure 7 Attention network extracted by independent component analysis of resting fMRI data in a group of healthy control subjects (*n* = 15) shows a predominantly a right-lateralized frontoparietal involvement of brain regions.

Neuroimaging from various task-evoked activity studies have isolated areas of the dorsal frontal and posterior parietal cortex to be involved in the directing of attention to spatial locations (Corbetta et al., 1993; Nobre et al., 1997). These regions form the dorsal attention system (Knecht et al., 2000) and are believed to maintain endogenous signals regarding goals to exert top-down modulatory signals biasing the processing of appropriate stimulus (Ruff et al., 2006; Bressler et al., 2008). However, these regions are not the areas typically damaged in patients who suffer from spatial neglect.

In contrast to the dorsal frontal and posterior parietal cortical regions identified in the task-evoked attention neuroimaging studies of normal healthy individual as mentioned above, areas of damage identified in patients whose stroke symptoms include neglect have tended to be more ventrally located (Friedrich et al., 1998; He et al., 2007; Shulman et al., 2010), particularly in the temporoparietal junction and the ventral frontal cortex (Corbetta et al., 2000; Downar et al., 2000). These areas constitute what is known as the ventral attention network, which is a secondary attention system necessary for directing attention to salient sensory events that is often disrupted in stroke patients with neglect (Corbetta and Shulman, 2002; Fox et al., 2006). This ventral attention network is not activated by expectations or task preparations but instead responds along with the dorsal attention network when behaviorally relevant objects are detected (Corbetta et al., 2000). This ventral attention network then redirects the dorsal attention network toward behaviorally relevant stimuli, especially when these stimuli are initially unattended (Corbetta and Shulman, 2002). Corbetta and colleagues have proposed a theory in which, during the reorienting of attention, outputs from the ventral attention network act as "circuit-breaker" signals interrupting ongoing selective attention in the dorsal system and shifting attention toward the novel object of interest (Corbetta and Shulman, 2002). Additionally, the ventral network is well activated by stimuli that are important, even when they are not very distinctive (Indovina and Macaluso, 2007), and may also be activated by irrelevant objects that are similar to the target or by false positives (Serences et al., 2005).

Following a stroke, it has been proposed that the spatial attention deficit arises from a dysfunction between the two previously described dorsal and ventral attention networks (Corbetta and Shulman, 2002; Corbetta et al., 2005). As described above, the ventral network works by disengaging the dorsal attention network in order to allow it to redirect its attention to the novel stimulus of interest. However, with a lesion affecting regions of this ventral network, no outputs are generated from the ventral attention network onto the dorsal attention network. This results in a lack of redirection of top-down modulation from the dorsal attention network onto the visual or somatic areas, which in turn cannot attend to the stimuli located in the opposite hemifield or on the opposite side of the body (Figure 8).

Almost 90% of spatial neglect patients suffer from a left hemifield deficit (Corbetta et al., 2005). This right hemispheric dominance of the attention network, translating to left hemifield neglect when the hemisphere is damaged, might reflect an asymmetry of the attention system in the two hemispheres. Similar to the left lateralization of the language network, spatial attention observes a predominantly right lateralization, with a smaller system and spatial representation in the left hemisphere, particularly for the

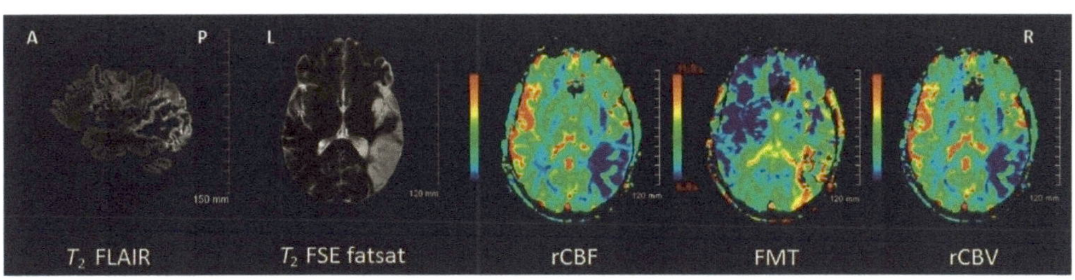

Figure 8 Chronic hemispatial neglect stroke patient (33-year-old female) with lesion affecting right Middle Cerebral Artery (MCA) territory.

ventral attention network (Arrington et al., 2000; Corbetta et al., 2000; Downar et al., 2000; Fox et al., 2006). The standard theory of neglect (Mesulam, 1981) postulates that while the right hemisphere controls shifts of attention to both sides of space, the left hemisphere only controls attention to the right side. This theory of neglect would help explain why there is this persistence of hemispatial neglect impacting the left hemifield of space.

Furthermore, a right hemisphere stroke that structurally damages the ventral network also has remote effects (Corbetta et al., 2005; He et al., 2007). In their 2005 study, Corbetta et al. (2005) explained the diaschisis effect by which brain damage in a particular region may have far-reaching effects in remote areas that are structurally intact but belong to the same functional network system. This study showed that patients with hemispatial neglect 4 weeks after stroke demonstrated weak or absent task-related activity in structurally intact areas of the attention network, including the occipital visual cortex, posterior parietal cortex, and dorsolateral prefrontal cortex, despite robust activity in neighboring nonnetwork regions such as the parietal and sensorimotor cortex. Furthermore, many of these areas in the attention network showed a resurgence of activity by 39 weeks poststroke, which correlated with improved functional outcomes. The authors concluded that this correlation between functional recovery of attention deficits and the restoration of regional network activity support a model of recovery in stroke-induced neglect in which the reestablishment of weighted activity within the distributed attention network mediates functional gains (Corbetta et al., 2005).

From a brain connectivity perspective, a breakdown of interhemispheric functional connectivity was found in patients expressing spatial neglect. He et al. (2007) demonstrated that interhemispheric connectivity in these patients was disrupted particularly between the right posterior parietal cortex and its left hemisphere homolog, despite these regions being structurally intact. In this study, the authors also showed that in the acute stage, the degree of breakdown in interhemispheric communication between the posterior parietal cortices correlated with the severity of impairment, but such connectivity dysfunction was reduced later in the chronic phase. Regions of the ventral attention network, some of which were damaged by the infarct, were diffusely disrupted in the acute phase and never recovered. Results presented in this study stress the relevance of the functional connectivity disruption between structurally intact regions of the dorsal attention network with impaired performance in a spatial attention task in stroke patients with spatial neglect.

2.1.4. Hypotheses of Post-Stroke Recovery Mechanism

Gains in interhemispheric connectivity following stroke are not only associated with better outcomes in the context of the motor system. Similar patterns have been documented in the areas of attention (He et al., 2007; Carter et al., 2010) and aphasia (Warren et al., 2009), where increased interhemispheric connectivity between parietal cortices and increased interhemispheric connectivity in the inferior frontal cortices were correlated with increased functional performance after stroke. However, increased functional connectivity does not necessarily imply increased magnitude in activation, as influences between the hemispheres could be either excitatory or inhibitory in nature.

Different hypotheses have been suggested regarding the mechanisms by which functional network organization facilitates cognitive processing. These hypotheses include the ideas that (1) interhemispheric interactions allow for the coordination of parallel processing between the hemispheres (Banich and Brown, 2000), (2) that lateralizing functions allows for competing processes to be insulated from one another (Kosslyn et al., 1992), and (3) that the spatial separation of brain areas within or between functional networks may allow for the prevention of internetwork interference and competition (Gee et al., 2011). Models of brain organization have also been proposed to help address these questions. These models include the theory of interhemispheric rivalry and competitive feedback inhibition (Kinsbourne, 1977, 2006), as well as the concept of oscillatory patterns as a mechanism for the support, propagation, and coordination of interactions at the neuronal level (Llinas et al., 1999; Buzsaki and Draguhn, 2004; Logothetis et al., 2007; Hoerzer et al., 2010). Despite differences in these formulated hypotheses, ongoing therapeutic strategies, especially in the motor domain, including those with constraint-induced therapy (Schaechter et al., 2002; Dromerick et al., 2009; Wolf et al., 2010) and noninvasive brain stimulation therapy (Fregni et al., 2005; Takeuchi et al., 2005; Nowak Da et al., 2008), have targeted the suppression of the contralesional hemisphere, facilitating the ipsilesional, and have shown positive effect in the reduction of motor and language impairment after treatment. Whether or not this latter approach of noninvasive brain stimulation is equally productive in other domains of language and neglect remains to be studied.

2.1.5. Intervention Studies and Therapeutic Prospects

Following the derived model from the previously described three functional domains of motor deficit, aphasia, and spatial neglect, a recurring theme of a return to preinjury patterns of activation emerges. This includes a promotion of perilesional and other ipsilesional hemisphere activation, as well as a suppression of the overactivation of the contralesional hemisphere, limiting the compensatory mechanism that could otherwise be harmful in long-term recovery.

Various interventional methods, such as constraint-induced therapy and noninvasive brain stimulation therapy, have targeted the brain with this approach. Constraint-induced methods (constraint-induced movement therapy (CIMT), for motor deficit treatment, and constraint-induced language therapy for aphasia treatment) involve: (1) the constraining of the impaired modality, (2) the restraint of the unimpaired modality, and (3) massed practice occurring in an enriched environment using behavioral shaping (Schaechter et al., 2002; Wolf et al., 2010). Early data suggesting clinical effectiveness of this approach has led to the transition to single-blinded, randomized controlled trials, one of which, the EXtremity Constraint Induced Therapy Evaluation (EXCITE) trial, is a multisite clinical trial involving seven US academic institutions and the recruitment of 222 patients with a first ischemic or hemorrhagic stroke (Wolf et al., 2005, 2007, 2010; Cramer, 2007) and others such as the Very Early Constraint Induced Movement during Stroke Rehabilitation (VECTORS) trial (Dromerick et al., 2009) show optimal dosing for CIMT.

Alternatively, noninvasive brain stimulation techniques work by two different, potentially complementary approaches: (1) the facilitation of the activity in ipsilesional hemisphere (Fregni et al., 2005; Hummel et al., 2005; Khedr et al., 2005; Boggio et al., 2007) and (2) the suppression of the "overactivity" in the contralesional hemisphere (Fregni et al., 2005, 2006; Mansur et al., 2005; Naeser et al., 2005; Boggio et al., 2007; Nowak Da et al., 2008). One example of applying such methods to stroke rehabilitation therapy has shown that suppression of the right homologous pars triangularis by this method of repetitive TMS in chronic, nonfluent aphasia patients leads to a temporary improvement in the picture naming task (Naeser et al., 2011).

However, the use of neuroimaging for the monitoring of these treatment strategies has yet to become a standard in intervention studies. To this day, the use of more direct behavioral outcomes, such as clinical assessment, remains the preferred convention for documenting functional recovery. Only a few studies have combined noninvasive brain stimulation and functional brain imaging (Grefkes et al., 2010; Stagg et al., 2012). However, this combination holds huge potential in the study of stroke recovery, offering a means of tracking reorganizational dynamics.

Nevertheless, although not a widespread practice or standard of care, speculations regarding the future therapeutic and prognostic potential of fMRI for stroke patients have surfaced (Dobkin, 2003; Cramer et al., 2011). These therapeutic applications focus largely on the use of fMRI to provide real-time feedback on regional brain activation (deCharms, 2008), which may then be used to teach the subject volitional modulation of brain function during a targeted task activity (Cramer et al., 2011). Such modulation of neural activity using real-time feedback has been performed with other populations, such as healthy subjects or patients with chronic pain (deCharms et al., 2005; Caria et al., 2007). Given that there appears to exist favorable patterns of brain activation associated with better functional improvement after stroke, it is possible to conceive of fMRI as a means of administering neuromodular therapy for stroke patients hoping to regain function.

Similarly, while established fMRI-based biomarkers for prognostic purposes in the context of functional rehabilitation following stroke are currently lacking, recent research has begun utilizing fMRI to characterize patterns of brain activity associated with motor function improvements both spontaneously and in the context of rehabilitative therapy as described above. A second benefit to these studies is the potential to use findings from fMRI data to identify subpopulations of stroke patients who may benefit more from some therapies than others. It is then feasible to consider that fMRI holds potential for future prognostic applications in guiding rehabilitative therapy.

3. CONCLUSIONS

Although there are many variables that can influence the mechanism and rate of recovery following stroke injury, task-based and resting fMRI have taught us a great deal about neural reorganization poststroke. In the initial stage, consistent observations in the literature point toward enhanced activation in the contralesional hemisphere. This widespread activation in the recruitment of perilesional areas and regions of the contralesional hemisphere serves as a compensatory mechanism in an initial attempt to minimize the functional deficit. Cross-sectional and longitudinal studies have revealed that, although important in the initial phase of recovery, persistent overactivation of the undamaged hemisphere could act as a nonoptimal mechanism of recovery and potentially have a negative effect on the trajectory of functional recovery. This model is not only the most agreed upon model for motor function recovery but also a valid model in language and spatial attention deficits.

Therapeutic strategies in all three of these domains are in development to make full use of our current understanding of poststroke reorganization. Whether it is achieved using constraint-induced therapy, noninvasive brain stimulation, or even real-time neuroimaging feedback, the theory underlying these therapies remains the same. The model of a return to normal-like preinjury lateralization and specialization of regional brain activation prevails as the best and optimal mechanism for poststroke recovery.

The use of brain imaging and fMRI has helped us gained consistent ground toward an understanding of stroke injury and brain recovery mechanisms. This gained understanding has, to some extent, and may continue to translate into better rehabilitation strategies, reducing the burden of stroke-related impairment in the stroke patient's life.

ACKNOWLEDGMENTS

This work is supported by NCATS UL1TR00047 TL1 Trainee Award, UW ICTR NIH/UL1RR025011 KL2 Scholar Award, NCATS 9U54TR000021, NIGMS R25GM083252, NIH/NIMH RC1MH090912-01 Challenge Grant, American Heart Association Midwest Affiliate Postdoctoral fellowship, NLM training grant NLM 2T115LM007359-11, and UW Medical Scientist Training Program T32GM008692. We would also like to acknowledge Lou Tetlan, PhD, for her assistance in reviewing and editing this chapter.

REFERENCES

Appelros, P., Karlsson, G.M., et al., 2002. Neglect and anosognosia after first-ever stroke: incidence and relationship to disability. J. Rehabil. Med. 34 (5), 215–220.

Arrington, C.M., Carr, T.H., et al., 2000. Neural mechanisms of visual attention: object-based selection of a region in space. J. Cogn. Neurosci. 12 (Suppl. 2), 106–117.

Banich, M.T., Brown, W.S., 2000. A life-span perspective on interaction between the cerebral hemispheres. Dev. Neuropsychol. 18 (1), 1–10.

Binder, J.R., Desai, R.H., et al., 2009. Where is the semantic system? A critical review and meta-analysis of 120 functional neuroimaging studies. Cereb. Cortex 19 (12), 2767–2796.

Birn, R.M., Kenworthy, L., et al., 2010. Neural systems supporting lexical search guided by letter and semantic category cues: a self-paced overt response fMRI study of verbal fluency. NeuroImage 49 (1), 1099–1107.

Boggio, P.S., Nunes, A., et al., 2007. Repeated sessions of noninvasive brain DC stimulation is associated with motor function improvement in stroke patients. Restor. Neurol. Neurosci. 25 (2), 123–129.

Bressler, S.L., Tang, W., et al., 2008. Top-down control of human visual cortex by frontal and parietal cortex in anticipatory visual spatial attention. J. Neurosci. 28 (40), 10056–10061.

Buckner, R.L., Corbetta, M., et al., 1996. Preserved speech abilities and compensation following prefrontal damage. Proc. Natl. Acad. Sci. USA 93 (3), 1249–1253.

Buma, F.E., Lindeman, E., et al., 2010. Functional neuroimaging studies of early upper limb recovery after stroke: a systematic review of the literature. Neurorehabil. Neural Repair 24 (7), 589–608.

Buzsaki, G., Draguhn, A., 2004. Neuronal oscillations in cortical networks. Science 304 (5679), 1926–1929.

Cabeza, R., 2002. Hemispheric asymmetry reduction in older adults: the HAROLD model. Psychol. Aging 17 (1), 85–100.

Cao, Y., D'Olhaberriague, L., et al., 1998. Pilot study of functional MRI to assess cerebral activation of motor function after poststroke hemiparesis. Stroke 29 (1), 112–122.

Cao, Y., Vikingstad, E.M., et al., 1999. Cortical language activation in stroke patients recovering from aphasia with functional MRI. Stroke 30 (11), 2331–2340.

Cao, Y., Vikingstad, E.M., et al., 1994. Functional magnetic resonance studies of the reorganization of the human hand sensorimotor area after unilateral brain injury in the perinatal period. Proc. Natl. Acad. Sci. USA 91 (20), 9612–9616.

Carey, J.R., Kimberley, T.J., et al., 2002. Analysis of fMRI and finger tracking training in subjects with chronic stroke. Brain 125 (Pt 4), 773–788.

Caria, A., Veit, R., et al., 2007. Regulation of anterior insular cortex activity using real-time fMRI. NeuroImage 35 (3), 1238–1246.

Carter, A.R., Astafiev, S.V., et al., 2010. Resting interhemispheric functional magnetic resonance imaging connectivity predicts performance after stroke. Ann. Neurol. 67 (3), 365–375.

Carter, A.R., Patel, K.R., et al., 2012. Upstream dysfunction of somatomotor functional connectivity after corticospinal damage in stroke. Neurorehabil. Neural Repair 26 (1), 7–19.

Chollet, F., DiPiero, V., et al., 1991. The functional anatomy of motor recovery after stroke in humans: a study with positron emission tomography. Ann. Neurol. 29 (1), 63–71.

Corbetta, M., Kincade, J.M., et al., 2000. Voluntary orienting is dissociated from target detection in human posterior parietal cortex. Nat. Neurosci. 3 (3), 292–297.

Corbetta, M., Kincade, M.J., et al., 2005. Neural basis and recovery of spatial attention deficits in spatial neglect. Nat. Neurosci. 8 (11), 1603–1610.

Corbetta, M., Miezin, F.M., et al., 1993. A PET study of visuospatial attention. J. Neurosci. 13 (3), 1202–1226.

Corbetta, M., Shulman, G.L., 2002. Control of goal-directed and stimulus-driven attention in the brain. Nat. Rev. Neurosci. 3 (3), 201–215.

Cramer, S.C., 2007. The EXCITE trial: a major step forward for restorative therapies in stroke. Stroke 38 (7), 2204–2205.

Cramer, S.C., Moore, C.I., et al., 2000. A pilot study of somatotopic mapping after cortical infarct. Stroke 31 (3), 668–671.

Cramer, S.C., Nelles, G., et al., 1997. A functional MRI study of subjects recovered from hemiparetic stroke. Stroke 28 (12), 2518–2527.

Cramer, S.C., Sur, M., et al., 2011. Harnessing neuroplasticity for clinical applications. Brain 134 (Pt 6), 1591–1609.

deCharms, R.C., 2008. Applications of real-time fMRI. Nat. Rev. Neurosci. 9 (9), 720–729.

deCharms, R.C., Maeda, F., et al., 2005. Control over brain activation and pain learned by using real-time functional MRI. Proc. Natl. Acad. Sci. USA 102 (51), 18626–18631.

Dobkin, B.H., 2003. The Clinical Science of Neurologic Rehabilitation. Oxford University Press, New York.

Dong, Y., Winstein, C.J., et al., 2007. Evolution of FMRI activation in the perilesional primary motor cortex and cerebellum with rehabilitation training-related motor gains after stroke: a pilot study. Neurorehabil. Neural Repair 21 (5), 412–428.
Downar, J., Crawley, A.P., et al., 2000. A multimodal cortical network for the detection of changes in the sensory environment. Nat. Neurosci. 3 (3), 277–283.
Dromerick, A.W., Lang, C.E., et al., 2009. Very early constraint-induced movement during stroke rehabilitation (VECTORS): a single-center RCT. Neurology 73 (3), 195–201.
Fox, M.D., Corbetta, M., et al., 2006. Spontaneous neuronal activity distinguishes human dorsal and ventral attention systems. Proc. Natl. Acad. Sci. USA 103 (26), 10046–10051.
Fregni, F., Boggio, P.S., et al., 2005. Transcranial direct current stimulation of the unaffected hemisphere in stroke patients. Neuroreport 16 (14), 1551–1555.
Fregni, F., Boggio, P.S., et al., 2006. A sham-controlled trial of a 5-day course of repetitive transcranial magnetic stimulation of the unaffected hemisphere in stroke patients. Stroke 37 (8), 2115–2122.
Fridriksson, J., 2010. Preservation and modulation of specific left hemisphere regions is vital for treated recovery from anomia in stroke. J. Neurosci. 30 (35), 11558–11564.
Fridriksson, J., Moser, D., et al., 2007. Neural correlates of phonological and semantic-based anomia treatment in aphasia. Neuropsychologia 45 (8), 1812–1822.
Fridriksson, J., Richardson, J.D., et al., 2012. Left hemisphere plasticity and aphasia recovery. NeuroImage 60 (2), 854–863.
Friedrich, F.J., Egly, R., et al., 1998. Spatial attention deficits in humans: a comparison of superior parietal and temporal-parietal junction lesions. Neuropsychology 12 (2), 193–207.
Gabrieli, J.D.E., Whitfield-Gabrieli, S., 2007. Attention to neglect. Neuron 53 (6), 776–777.
Gee, D.G., Biswal, B.B., et al., 2011. Low frequency fluctuations reveal integrated and segregated processing among the cerebral hemispheres. NeuroImage 54 (1), 517–527.
Glasser, M.F., Rilling, J.K., 2008. DTI tractography of the human brain's language pathways. Cereb. Cortex 18 (11), 2471–2482.
Grafman, J., 2000. Conceptualizing functional neuroplasticity. J. Commun. Disord. 33 (4), 345–356.
Grefkes, C., Nowak, D.A., et al., 2010. Modulating cortical connectivity in stroke patients by rTMS assessed with fMRI and dynamic causal modeling. NeuroImage 50 (1), 233–242.
He, B.J., Snyder, A.Z., et al., 2007. Breakdown of functional connectivity in frontoparietal networks underlies behavioral deficits in spatial neglect. Neuron 53 (6), 905–918.
Heidenreich, P.A., Trogdon, J.G., et al., 2011. Forecasting the future of cardiovascular disease in the United States: a policy statement from the American Heart Association. Circulation 123 (8), 933–944.
Hoerzer, G.M., Liebe, S., et al., 2010. Directed coupling in local field potentials of macaque v4 during visual short-term memory revealed by multivariate autoregressive models. Front Comput. Neurosci. 4, 14.
Hummel, F., Celnik, P., et al., 2005. Effects of non-invasive cortical stimulation on skilled motor function in chronic stroke. Brain 128 (3), 490–499.
Indovina, I., Macaluso, E., 2007. Dissociation of stimulus relevance and saliency factors during shifts of visuospatial attention. Cereb. Cortex 17 (7), 1701–1711.
Jayaram, G., Stinear, J.W., 2008. Contralesional paired associative stimulation increases paretic lower limb motor excitability post-stroke. Exp. Brain Res. 185 (4), 563–570.
Johansen-Berg, H., Rushworth, M.F., et al., 2002. The role of ipsilateral premotor cortex in hand movement after stroke. Proc. Natl. Acad. Sci. USA 99 (22), 14518–14523.
Kelly-Hayes, M., Beiser, A., et al., 2003. The influence of gender and age on disability following ischemic stroke: the Framingham study. J. Stroke Cerebrovasc. Dis. 12 (3), 119–126.
Khedr, E.M., Ahmed, M.A., et al., 2005. Therapeutic trial of repetitive transcranial magnetic stimulation after acute ischemic stroke. Neurology 65 (3), 466–468.
Kinsbourne, M., 1977. Hemi-neglect and hemisphere rivalry. Adv. Neurol. 18, 41–49.
Kinsbourne, M., 2006. From unilateral neglect to the brain basis of consciousness. Cortex 42 (6), 869–874.
Knecht, S., Dräger, B., et al., 2000. Handedness and hemispheric language dominance in healthy humans. Brain 123 (12), 2512–2518.
Knecht, S., Floel, A., et al., 2002. Degree of language lateralization determines susceptibility to unilateral brain lesions. Nat. Neurosci. 5 (7), 695–699.

Kosslyn, S.M., Chabris, C.F., et al., 1992. Categorical versus coordinate spatial relations: computational analyses and computer simulations. J. Exp. Psychol. Hum. Percept. Perform. 18 (2), 562–577.
Kwon, Y.H., Lee, M.Y., et al., 2007. Differences of cortical activation pattern between cortical and corona radiata infarct. Neurosci. Lett. 417 (2), 138–142.
Lee, M.H., Smyser, C.D., et al., 2012. Resting-State fMRI: a review of methods and clinical applications. ANJR Am. J. Neuroradiol.
Llinas, R.R., Ribary, U., et al., 1999. Thalamocortical dysrhythmia: a neurological and neuropsychiatric syndrome characterized by magnetoencephalography. Proc. Natl. Acad. Sci. USA 96 (26), 15222–15227.
Logothetis, N.K., Kayser, C., et al., 2007. In vivo measurement of cortical impedance spectrum in monkeys: implications for signal propagation. Neuron 55 (5), 809–823.
Lotze, M., Markert, J., et al., 2006. The role of multiple contralesional motor areas for complex hand movements after internal capsular lesion. J. Neurosci. 26 (22), 6096–6102.
Loubinoux, I., Carel, C., et al., 2003. Correlation between cerebral reorganization and motor recovery after subcortical infarcts. NeuroImage 20 (4), 2166–2180.
Mansur, C.G., Fregni, F., et al., 2005. A sham stimulation-controlled trial of rTMS of the unaffected hemisphere in stroke patients. Neurology 64 (10), 1802–1804.
Marshall, R.S., Perera, G.M., et al., 2000. Evolution of cortical activation during recovery from corticospinal tract infarction. Stroke 31 (3), 656–661.
Meier, T.B., Naing, L., et al., 2011. Validating age-related functional imaging changes in verbal working memory with acute stroke. Behav. Neurol. 24 (3), 187–199.
Meinzer, M., Flaisch, T., et al., 2008. Functional re-recruitment of dysfunctional brain areas predicts language recovery in chronic aphasia. NeuroImage 39 (4), 2038–2046.
Meinzer, M., Flaisch, T., et al., 2009. Neural signatures of semantic and phonemic fluency in young and old adults. J. Cogn. Neurosci. 21 (10), 2007–2018.
Mesulam, M.M., 1981. A cortical network for directed attention and unilateral neglect. Ann. Neurol. 10 (4), 309–325.
Musso, M., Weiller, C., et al., 1999. Training-induced brain plasticity in aphasia. Brain 122 (9), 1781–1790.
Naeser, M.A., Martin, P.I., et al., 2005. Improved picture naming in chronic aphasia after TMS to part of right Broca's area: an open-protocol study. Brain Lang. 93 (1), 95–105.
Naeser, M.A., Martin, P.I., et al., 2011. TMS suppression of right pars triangularis, but not pars opercularis, improves naming in aphasia. Brain Lang. 119 (3), 206–213.
Newton, J.M., Ward, N.S., et al., 2006. Non-invasive mapping of corticofugal fibres from multiple motor areas – relevance to stroke recovery. Brain 129 (Pt 7), 1844–1858.
Nobre, A.C., Sebestyen, G.N., et al., 1997. Functional localization of the system for visuospatial attention using positron emission tomography. Brain 120 (Pt 3), 515–533.
Nowak Da, G.C.D.M., et al., 2008. Effects of low-frequency repetitive transcranial magnetic stimulation of the contralesional primary motor cortex on movement kinematics and neural activity in subcortical stroke. Arch. Neurol. 65 (6), 741–747.
Nudo, R.J., Milliken, G.W., 1996. Reorganization of movement representations in primary motor cortex following focal ischemic infarcts in adult squirrel monkeys. J. Neurophysiol. 75 (5), 2144–2149.
Park, C.H., Chang, W.H., et al., 2011. Longitudinal changes of resting-state functional connectivity during motor recovery after stroke. Stroke 42 (5), 1357–1362.
Park, D.C., Reuter-Lorenz, P., 2009. The adaptive brain: aging and neurocognitive scaffolding. Annu. Rev. Psychol. 60, 173–196.
Pedersen, P.M., Jørgensen, H.S., et al., 1997. Hemineglect in acute stroke-incidence and prognostic implications: the Copenhagen Stroke Study1. Am. J. Phys. Med. Rehabil. 76 (2), 122–127.
Philipose, L.E., Alphs, H., et al., 2007. Testing conclusions from functional imaging of working memory with data from acute stroke. Behav. Neurol. 18 (1), 37–43.
Pineiro, R., Pendlebury, S., et al., 2002. Altered hemodynamic responses in patients after subcortical stroke measured by functional MRI. Stroke 33 (1), 103–109.
Powell, H.W.R., Parker, G.J.M., et al., 2006. Hemispheric asymmetries in language-related pathways: a combined functional MRI and tractography study. NeuroImage 32 (1), 388–399.

Prabhakaran, V., Raman, S.P., et al., 2007. Neural substrates of word generation during stroke recovery: the influence of cortical hypoperfusion. Behav. Neurol. 18 (1), 45–52.

Pravatà, E., Sestieri, C., et al., 2011. Functional connectivity MR imaging of the language network in patients with drug-resistant epilepsy. ANJR Am. J. Neuroradiol. 32 (3), 532–540.

Rehme, A.K., Eickhoff, S.B., et al., 2011a. Dynamic causal modeling of cortical activity from the acute to the chronic stage after stroke. NeuroImage 55 (3), 1147–1158.

Rehme, A.K., Fink, G.R., et al., 2011b. The role of the contralesional motor cortex for motor recovery in the early days after stroke assessed with longitudinal fMRI. Cereb. Cortex 21 (4), 756–768.

Rehme, A.K., Grefkes, C., 2013. Cerebral network disorders after stroke: evidence from imaging-based connectivity analyses of active and resting brain states in humans. J. Physiol. 591 (1), 17–31.

Reuter-Lorenz, P.A., Jonides, J., et al., 2000. Age differences in the frontal lateralization of verbal and spatial working memory revealed by PET. J. Cogn. Neurosci. 12 (1), 174–187.

Rilling, J.K., Glasser, M.F., et al., 2008. The evolution of the arcuate fasciculus revealed with comparative DTI. Nat. Neurosci. 11 (4), 426–428.

Roger, V.L., Go, A.S., et al., 2012. Heart disease and stroke statistics—2012 update a report from the American Heart association. Circulation 125 (1), e2–e220.

Rosen, H.J., Petersen, S.E., et al., 2000. Neural correlates of recovery from aphasia after damage to left inferior frontal cortex. Neurology 55 (12), 1883–1894.

Rossi, S., Miniussi, C., et al., 2004. Age-related functional changes of prefrontal cortex in long-term memory: a repetitive transcranial magnetic stimulation study. J. Neurosci. 24 (36), 7939–7944.

Ruff, C.C., Blankenburg, F., et al., 2006. Concurrent TMS-fMRI and psychophysics reveal frontal influences on human retinotopic visual cortex. Curr. Biol. 16 (15), 1479–1488.

Saur, D., Hartwigsen, G., 2012. Neurobiology of language recovery after stroke: lessons from neuroimaging studies. Arch. Phys. Med. Rehabil. 93 (Suppl. 1), S15–S25.

Saur, D., Lange, R., et al., 2006. Dynamics of language reorganization after stroke. Brain 129 (6), 1371–1384.

Schaechter, J.D., Kraft, E., et al., 2002. Motor recovery and cortical reorganization after constraint-induced movement therapy in stroke patients: a preliminary study. Neurorehabil. Neural Repair 16 (4), 326–338.

Schlaug, G., Marchina, S., et al., 2009. Evidence for plasticity in white-matter tracts of patients with chronic Broca's aphasia undergoing intense intonation-based speech therapy. Ann. N. Y. Acad. Sci. 1169 (1), 385–394.

Serences, J.T., Shomstein, S., et al., 2005. Coordination of voluntary and stimulus-driven attentional control in human cortex. Psychol. Sci. 16 (2), 114–122.

Sharma, N., Baron, J.-C., et al., 2009. Motor imagery after stroke: relating outcome to motor network connectivity. Ann. Neurol. 66 (5), 604–616.

Shelton, F.N., Reding, M.J., 2001. Effect of lesion location on upper limb motor recovery after stroke. Stroke 32 (1), 107–112.

Shirer, W.R., Ryali, S., et al., 2012. Decoding subject-driven cognitive states with whole-brain connectivity patterns. Cereb. Cortex 22 (1), 158–165.

Shulman, G.L., Pope, D.L., et al., 2010. Right hemisphere dominance during spatial selective attention and target detection occurs outside the dorsal frontoparietal network. J. Neurosci. 30 (10), 3640–3651.

Stagg, C.J., Bachtiar, V., et al., 2012. Cortical activation changes underlying stimulation-induced behavioural gains in chronic stroke. Brain 135 (1), 276–284.

Stinear, C.M., Barber, P.A., et al., 2007. Functional potential in chronic stroke patients depends on corticospinal tract integrity. Brain 130 (Pt 1), 170–180.

Szaflarski, J.P., Allendorfer, J.B., et al., 2013. Recovered vs. not-recovered from post-stroke aphasia: the contributions from the dominant and non-dominant hemispheres. Restor. Neurol. Neurosci.

Takahashi, C.D., Der Yeghiaian, L., et al., 2005. Stroke recovery and its imaging. Neuroimaging Clin. N. Am. 15 (3), 681–695 xii.

Takeuchi, N., Chuma, T., et al., 2005. Repetitive transcranial magnetic stimulation of contralesional primary motor cortex improves hand function after stroke. Stroke 36 (12), 2681–2686.

Tie, Y., Rigolo, L., et al., 2012. Defining language networks from resting-state fMRI for surgical planning—a feasibility study. Hum. Brain Mapp.

Tononi, G., Edelman, G.M., et al., 1998. Complexity and coherency: integrating information in the brain. Trends Cogn. Sci. 2 (12), 474–484.

Turken, A.U., Dronkers, N.F., 2011. The neural architecture of the language comprehension network: converging evidence from lesion and connectivity analyses. Front Syst. Neurosci. 5.

Vigneau, M., Beaucousin, V., et al., 2011. What is right-hemisphere contribution to phonological, lexico-semantic, and sentence processing? Insights from a meta-analysis. NeuroImage 54 (1), 577–593.

Vitali, P., Tettamanti, M., et al., 2010. Generalization of the effects of phonological training for anomia using structural equation modelling: a multiple single-case study. Neurocase 16 (2), 93–105.

Wang, L., Yu, C., et al., 2010. Dynamic functional reorganization of the motor execution network after stroke. Brain 133 (4), 1224–1238.

Warburton, E., Price, C.J., et al., 1999. Mechanisms of recovery from aphasia: evidence from positron emission tomography studies. J. Neurol. Neurosurg. Psychiatry 66 (2), 155–161.

Ward, N.S., Brown, M.M., et al., 2003. Neural correlates of motor recovery after stroke: a longitudinal fMRI study. Brain 126 (Pt 11), 2476–2496.

Ward, N.S., Brown, M.M., et al., 2004. The influence of time after stroke on brain activations during a motor task. Ann. Neurol. 55 (6), 829–834.

Ward, N.S., Newton, J.M., et al., 2006. Motor system activation after subcortical stroke depends on corticospinal system integrity. Brain 129 (Pt 3), 809–819.

Warren, J.E., Crinion, J.T., et al., 2009. Anterior temporal lobe connectivity correlates with functional outcome after aphasic stroke. Brain 132 (12), 3428–3442.

Weiller, C., Isensee, C., et al., 1995. Recovery from Wernicke's aphasia: a positron emission tomographic study. Ann. Neurol. 37 (6), 723–732.

Wierenga, C.E., Benjamin, M., et al., 2008. Age-related changes in word retrieval: role of bilateral frontal and subcortical networks. Neurobiol. Aging 29 (3), 436–451.

Wolf, S.L., Newton, H., et al., 2007. The Excite Trial: relationship of intensity of constraint induced movement therapy to improvement in the wolf motor function test. Restor. Neurol. Neurosci. 25 (5–6), 549–562.

Wolf, S.L., Thompson, P.A., et al., 2005. The EXCITE trial: attributes of the wolf motor function test in patients with subacute stroke. Neurorehabil. Neural Repair 19 (3), 194–205.

Wolf, S.L., Thompson, P.A., et al., 2010. The EXCITE stroke trial: comparing early and delayed constraint-induced movement therapy. Stroke 41 (10), 2309–2315.

Xerri, C., Merzenich, M.M., et al., 1998. Plasticity of primary somatosensory cortex paralleling sensorimotor skill recovery from stroke in adult monkeys. J. Neurophysiol. 79 (4), 2119–2148.

Yang, J., 2012. The role of the right hemisphere in metaphor comprehension: a meta-analysis of functional magnetic resonance imaging studies. Hum. Brain Mapp.

Zemke, A.C., Heagerty, P.J., et al., 2003. Motor cortex organization after stroke is related to side of stroke and level of recovery. Stroke 34 (5), e23–e26.

INDEX

Note: Page numbers followed by "f" denote figures; "t" tables.

B

C

D

Q

R

S

T

V

MIX

Papier | Fördert gute Waldnutzung

FSC® C083411

Kolibri 360 GmbH
Ferdinand-Jühlke-Straße 7
99095 Erfurt, Deutschland
produktsicherheit@kolibri360.de